PTC Creo Parametric 3.0 for Designers

(3rd Edition)

CADCIM Technologies
525 St. Andrews Drive
Schererville, IN 46375, USA
(www.cadcim.com)

Contributing Author
Sham Tickoo
Professor
Department of Mechanical Engineering Technology
Purdue University Calumet
Hammond, Indiana
USA

CADCIM Technologies

PTC Creo Parametric 3.0 for Designers
Sham Tickoo

CADCIM Technologies
525 St Andrews Drive
Schererville, Indiana 46375, USA
www.cadcim.com

ISBN 978-1-936646-92-0

www.cadcim.com

DEDICATION

To teachers, who make it possible to disseminate knowledge
to enlighten the young and curious minds
of our future generations

To students, who are dedicated to learning new technologies
and making the world a better place to live in

THANKS

To the faculty and students of the MET department of
Purdue University Calumet for their cooperation

To employees of CADCIM Technologies for their valuable help

Online Training Program Offered by CADCIM Technologies

CADCIM Technologies provides effective and affordable virtual online training on various software packages including Computer Aided Design and Manufacturing (CAD/CAM), computer programming languages, animation, architecture, and GIS. The training is delivered 'live' via Internet at any time, any place, and at any pace to individuals as well as the students of colleges, universities, and CAD/CAM training centers. The main features of this program are:

Training for Students and Companies in a Classroom Setting

Highly experienced instructors and qualified engineers at CADCIM Technologies conduct the classes under the guidance of Prof. Sham Tickoo of Purdue University Calumet, USA. This team has authored several textbooks that are rated "one of the best" in their categories and are used in various colleges, universities, and training centers in North America, Europe, and in other parts of the world.

Training for Individuals

CADCIM Technologies with its cost effective and time saving initiative strives to deliver the training in the comfort of your home or work place, thereby relieving you from the hassles of traveling to training centers.

Training Offered on Software Packages

CADCIM provides basic and advanced training on the following software packages:

CAD/CAM/CAE: *CATIA, Pro/ENGINEER Wildfire, Creo Parametric, Creo Direct, SolidWorks, Autodesk Inventor, Solid Edge, NX, AutoCAD, AutoCAD LT, AutoCAD Plant 3D, Customizing AutoCAD, EdgeCAM, and ANSYS*

Architecture and GIS: *Autodesk Revit Architecture, AutoCAD Civil 3D, Autodesk Revit Structure, AutoCAD Map 3D, Revit MEP, Navisworks, Primavera Project Planner, and Bentley STAAD Pro*

Animation and Styling: *Autodesk 3ds Max, Autodesk 3ds Max Design, Autodesk Maya, Autodesk Alias, Foundry NukeX, and MAXON CINEMA 4D*

Computer Programming: *C++, VB.NET, Oracle, AJAX, and Java*

For more information, please visit the following link: ***http://www.cadcim.com***

Note
If you are a faculty member, you can register by clicking on the following link to access the teaching resources: ***http://www.cadcim.com/Registration.aspx***. The student resources are available at ***http://www.cadcim.com***. We also provide **Live Virtual Online Training** on various software packages. For more information, write us at ***sales@cadcim.com***.

Table of Contents

Chapter 3: Creating Sketches in the Sketch Mode-II

Chapter 4: Creating Base Features

Chapter 5: Datums

Chapter 6: Options Aiding Construction of Parts-I

Chapter 7: Options Aiding Construction of Parts-II

Chapter 8: Advanced Modeling Tools-I

Chapter 9: Advanced Modeling Tools-II

Chapter 10: Advanced Modeling Tools-III

Chapter 11: Assembly Modeling

Chapter 12: Generating, Editing, and Modifying the Drawing Views

Chapter 13: Dimensioning the Drawing Views

Chapter 14: Other Drawing Options

Chapter 15: Surface Modeling

Chapter 16: Working with Sheetmetal Components

Preface

PTC Creo Parametric 3.0

PTC Creo Parametric, developed by Parametric Technology Corporation, is a new technology in the series of Pro/ENGINEER. It provides a broad range of powerful and flexible CAD capabilities that can address even the most tedious design challenges. Being a parametric feature-based solid modeling tool, it not only integrates the 3D parametric features with 2D tools, but also assists in every design-through-manufacturing process. This software is remarkably user-friendly and it contributes to the enhanced of the entire design process.

This solid modeling software allows you to easily import the standard format files with an amazing compatibility. The 2D drawing views of the components are automatically generated in the **Drawing** mode. Using this software, you can generate detailed, orthographic, isometric, auxiliary, and section views. Additionally, you can use any predefined drawing standard files for generating the drawing views. You can display the model dimensions in the drawing views or add reference dimensions whenever you want. The bidirectionally associative nature of this software ensures that any modification made in the model is automatically reflected in the drawing views. Similarly, any modification made in the dimensions of the drawing views is automatically updated in the model.

The **PTC Creo Parametric 3.0 for Designers** textbook has been written to enable the readers to use the modeling power of PTC Creo Parametric 3.0 effectively. The latest surfacing techniques like Freestyle and Style are explained in detail in this book. The textbook also covers the Sheetmetal module with the help of relevant examples and illustrations. The mechanical engineering industry examples and tutorials are used in this textbook to ensure that the users can relate the knowledge of this book with the actual mechanical industry designs. The salient features of this textbook are as follows:

- **Tutorial Approach**

  The author has adopted the tutorial point-of-view and the learn-by-doing approach throughout the textbook. This approach guides the users through the process of creating the models in the tutorials.

- **Real-World Projects as Tutorials**
  The author has used the real-world mechanical engineering projects as tutorials in this textbook so that the readers can correlate them with the real-time models in the mechanical engineering industry.

- **Tips and Notes**
  Additional information related to various topics is provided in the form of tips and notes.

- **Learning Objectives**
  The first page of every chapter summarizes the topics that will be covered in that chapter. This helps the users to easily refer to a topic.

- **Self-Evaluation Test, Review Questions, and Exercises**
  Every chapter ends with a Self-Evaluation test so that the users can assess their knowledge of the chapter. The answers to the Self-Evaluation test are given at the end of the chapter. Also, the Review Questions and Exercises are given at the end of each chapter and they can be used by the Instructors as test questions and exercises.

- **Heavily Illustrated Text**
  The text in this book is heavily illustrated with the help of around 1400 line diagrams and screen capture images that support the tools section and tutorials.

Symbols Used in the Text

Note
The author has provided additional information to the users about the topic being discussed in the form of notes.

Tip
Special information and techniques are provided in the form of tips that helps in increasing the efficiency of the users.

New
This symbol indicates that the command or tool being discussed is new.

Enhanced
This symbol indicates that the command or tool being discussed has been enhanced in PTC Creo Parametric 3.0.

Formatting Conventions Used in the Text

Please refer to the following list for the formatting conventions used in this textbook.

- Names of tools, buttons, options, groups, tabs, slide-down panels, and Ribbon are written in boldface.

  Example: The **Extrude** tool, the **OK** button, the **Editing** group, the **Sketch** tab, and so on.

- Names of dialog boxes, drop-downs, drop-down lists, dashboards, areas, edit boxes, check boxes, and radio buttons are written in boldface.

  Example: The **Revolve** dashboard, the **Chamfer** drop-down of **Engineering** group in the **Model** tab, the **Thickness** drop-down of the **Shell** dashboard, the **Extended intersect surfaces** check box in the **Options** slide-down panel of the **Draft** dashboard, and so on.

- Values entered in edit boxes are written in boldface.

  Example: Enter **5** in the **Radius** edit box.

- Names and paths of the files are written in italics.

  Example: *C:\Creo-3.0\c03*, *c03tut03.prt*, and so on.

- Different options available for invoking a tool are given in a shaded command box.

  Ribbon: Get Started > Launch > New

Naming Conventions Used in the Text

Tool

If you click on an item in a toolbar or a group of the **Ribbon** and a dashboard or dialog box is invoked to create/edit an object or perform some action, then that item is termed as **tool**.

For example:
Line tool, **Normal** tool, **Extrude** tool
Fillet tool, **Draft** tool, **Delete Segment** tool
If you click on an item in a toolbar or a group of the **Ribbon** and a dialog box is invoked wherein you can set the properties to create/edit an object, then that item is also termed as **tool**, refer to Figure 1.

For example:
To Create: **Extrude** tool, **Sweep** tool, **Round** tool
To Edit: **Extend** tool, **Trim** tool

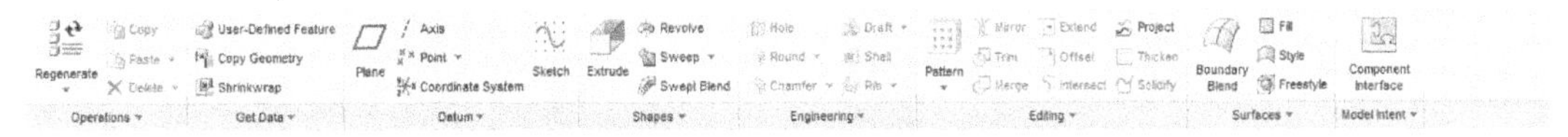

Figure 1 Various tools in the Ribbon

Button

The item in a dialog box that has a 3D shape like a button is termed as **Button**. For example, **OK** button, **Cancel** button, **Apply** button, and so on.

Dialog Box

The naming conventions for the components in a dialog box are shown in Figure 2.

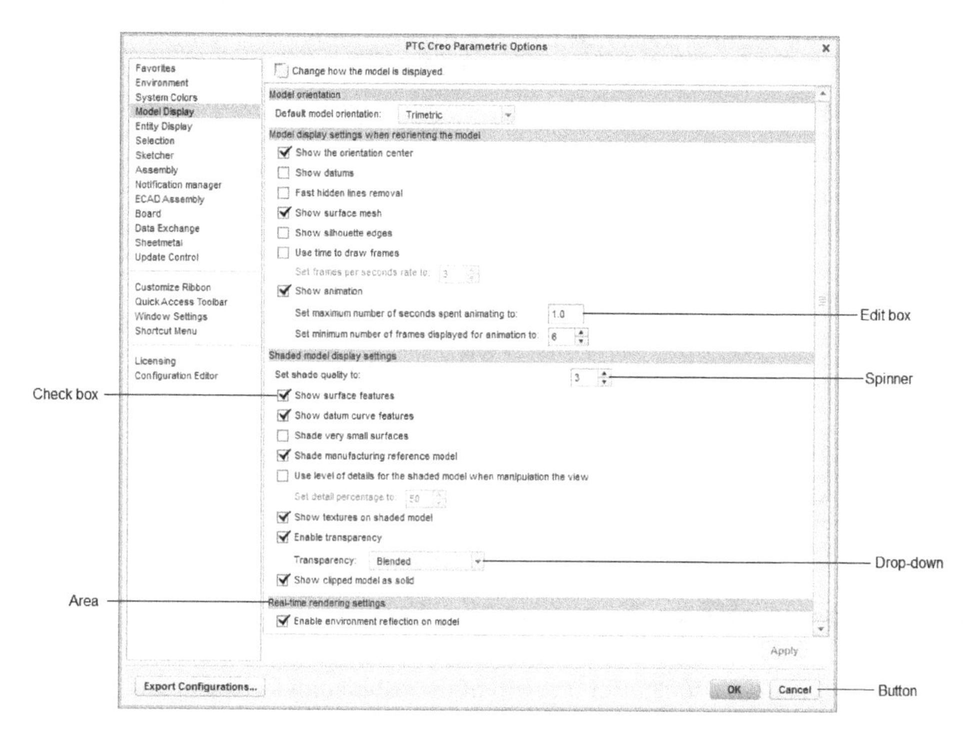

Figure 2 The components in a dialog box

Drop-down

A drop-down is one in which a set of common tools are grouped together for creating an object. You can identify a drop-down with a down arrow on it. These drop-downs are given a name based on the tools grouped in them. For example, **Arc** drop-down (refer to Figure 3), **Chamfer** drop-down (refer to Figure 4), **Draft** drop-down (refer to Figure 5), and so on.

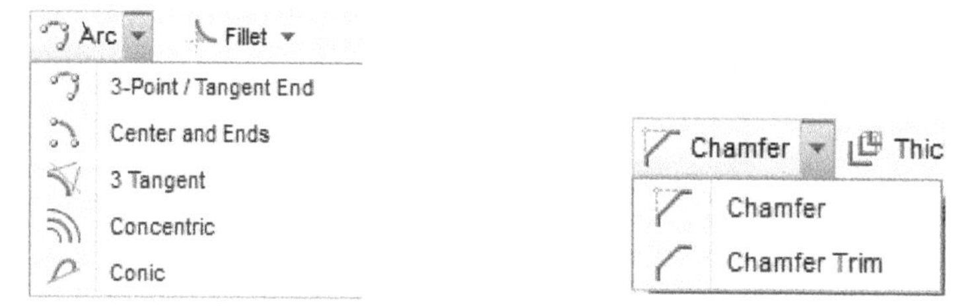

*Figure 3 The **Arc** drop-down*

*Figure 4 The **Chamfer** drop-down*

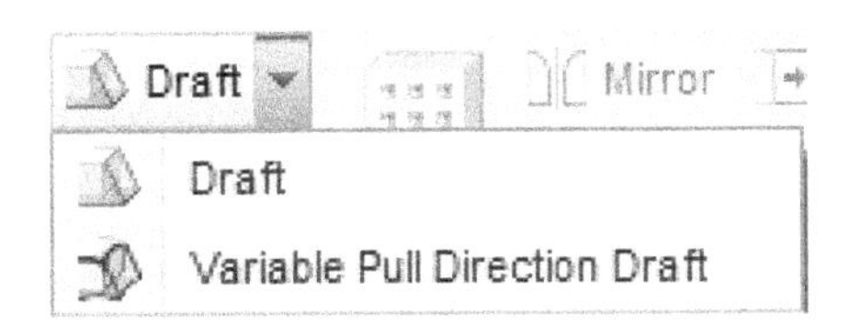

Figure 5 *The* ***Draft*** *drop-down*

Drop-down List

A drop-down list is the one in which a set of options are grouped together. You can set various parameters using these options. You can identify a drop-down list with a down arrow on it. For example, **Dimension** drop-down list, **Clear Appearance** drop-down list, and so on, refer to Figure 6.

Figure 6 *The* ***Dimension*** *and* ***Clear Appearance*** *drop-down lists*

Options

Options are the items that are available in shortcut menu, drop-down list, dialog boxes, and so on. For example, choose the **Front** option from the **View Manager** dialog box, refer to Figure 7; choose the **New** option from the **File** menu, refer to Figure 8.

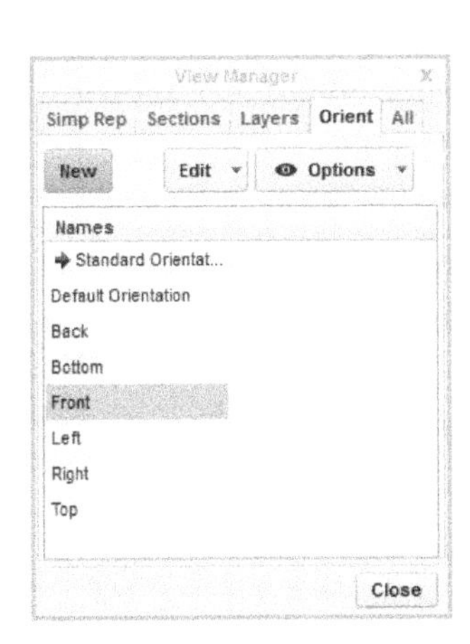

Figure 7 *The* ***Front*** *option in the* ***View Manager*** *dialog box*

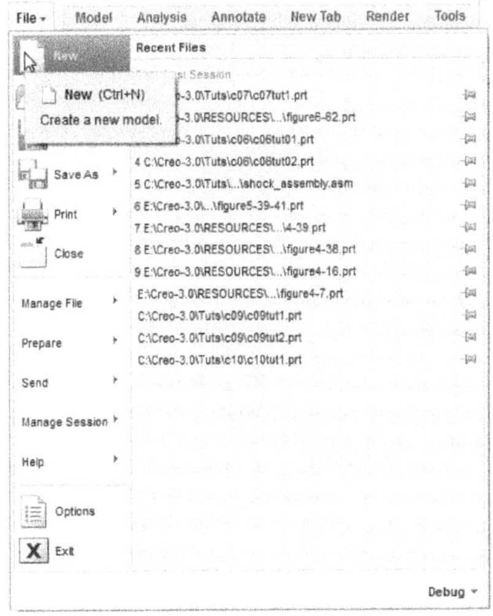

Figure 8 *The* ***New*** *option in the* ***File*** *menu*

Free Companion Website

It has been our constant endeavor to provide you the best textbooks and services at affordable price. In this endeavor, we have come out with a Free Companion website that will facilitate the process of teaching and learning of PTC Creo Parametric 3.0. If you purchase this textbook from our website (***www.cadcim.com***), you will get access to the files on the Companion website.

The following resources are available for the faculty and students in this website:

Faculty Resources

- **Technical Support**
  You can get online technical support by contacting ***techsupport@cadcim.com***.

- **Instructor Guide**
  Solutions to all review questions and exercises in the textbook are provided in this link to help the faculty members test the skills of the students.

- **PowerPoint Presentations**
  The contents of the book are arranged in PowerPoint slides that can be used by the faculty for their lectures.

- **Part Files**
  The part files used in illustration, tutorials, and exercises are available for free download.

Student Resources

- **Technical Support**
  You can get online technical support by contacting ***techsupport@cadcim.com***.

- **Part Files**
  The part files used in illustrations and tutorials are available for free download.

- **Additional Students Projects**
  Various projects are provided for the students to practice.

If you face any problem in accessing these files, please contact the publisher at ***sales@cadcim.com*** or the author at ***stickoo@purduecal.edu*** or ***tickoo525@gmail.com***.

Stay Connected

You can now stay connected with us through Facebook and Twitter to get the latest information about our textbooks, videos, and teaching/learning resources. To stay informed of such updates, follow us on Facebook (***www.facebook.com/cadcim***) and Twitter (@cadcimtech). You can also subscribe to our YouTube channel (***www.youtube.com/cadcimtech***) to get the information about our latest video tutorials.

Chapter 1

Introduction to PTC Creo Parametric 3.0

Learning Objectives

After completing this chapter, you will be able to:

- *Understand the advantages of using PTC Creo Parametric.*
- *Know the system requirements of PTC Creo Parametric.*
- *Get familiar with important terms and definitions in PTC Creo Parametric.*
- *Understand important options in the File menu.*
- *Understand the importance of Model Tree.*
- *Understand the functions of mouse buttons.*
- *Use the options of default toolbars.*
- *Customize the Ribbon.*
- *Understand the functions of browser.*
- *Understand the use of Appearance Gallery.*
- *Render stages in PTC Creo Parametric.*
- *Change the color scheme of the background in PTC Creo Parametric.*

INTRODUCTION TO PTC Creo Parametric 3.0

Welcome to PTC Creo Parametric. If you are a new user of PTC Creo Parametric software, you are going to join hands with thousands of users of this high-end CAD/CAM/CAE tool worldwide. If you are a user of the previous releases of this software, you are going to upgrade your designing skills because of the tremendous improvement in this latest release such as flexible modeling, freestyle modeling, and so on. Also, the interface of PTC Creo Parametric is very user friendly. You will find a tremendous reduction in the time taken to complete a design using this solid modeling tool.

PTC Creo Parametric is a powerful software used to create complex designs with great precision. The design intent of a three-dimensional (3D) model or an assembly is defined by its specification and its use. You can use the powerful tools of PTC Creo Parametric to capture the design intent of a complex model by incorporating intelligence into the design. Once you understand the feature-based, associative, and parametric nature of PTC Creo Parametric, you can appreciate its power as a solid modeling tool.

To make the designing process simple and quick, the designing process have been divided into different modules in this software package. This means each step of the designing is completed in a different module. For example, generally a design process consists of the following steps:

- Sketching using the basic sketch entities
- Converting the sketch into features and parts
- Assembling different parts and analyzing them
- Documenting parts and the assembly in terms of drawing views
- Manufacturing the final part and assembly

All these steps are divided into different modes of PTC Creo Parametric namely, the **Sketch** mode, **Part** mode, **Assembly** mode, **Drawing** mode, and **Manufacturing** mode.

Despite making various modifications in a design, the parametric nature of this software helps preserve the design intent of a model with tremendous ease. PTC Creo Parametric allows you to work in a 3D environment and calculates the mass properties directly from the created geometry. You can also switch to various display modes like wireframe, shaded, hidden, and no hidden at any time with ease as it does not affect the model but only changes its appearance.

FEATURES OF PTC CREO PARAMETRIC

Different features of the software are discussed next.

Feature-Based Nature

PTC Creo Parametric is a feature-based solid modeling tool. A feature is defined as the smallest building block and a solid model created in PTC Creo Parametric is an integration of a number of these building blocks. Each feature can be edited individually to bring in the desired change in the solid model. The use of feature-based property provides greater flexibility to the parts created. For example, consider the part shown in Figure 1-1. It consists of one counterbore hole at the center and six counterbore holes around the Bolt Circle Diameter (BCD).

Now, consider a case where you need to change all the outer counterbore holes to drill holes keeping the central counterbore hole and the BCD for the outer holes same. Also, you need to change the number of holes from six to eight. In a non feature-based software package, you will have to delete the entire part and then create a new part based on the new specifications. Whereas, PTC Creo Parametric allows you to make this modification by just modifying some values in the same part, see Figure 1-2. This shows that the solid parts created in PTC Creo Parametric are a combination of various features that can be modified individually at any time.

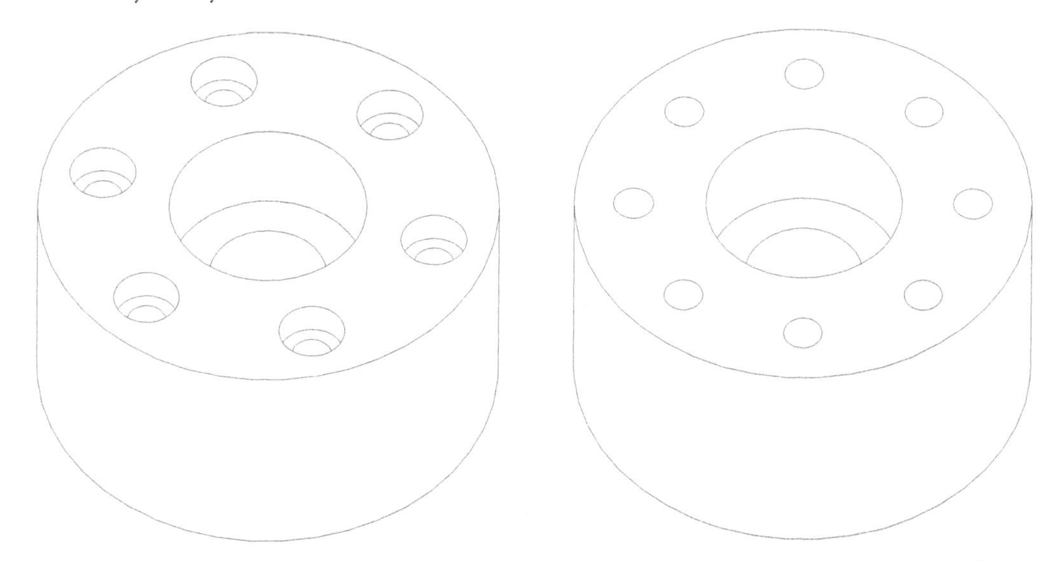

Figure 1-1 *Model displaying the counterbore holes* ***Figure 1-2*** *Model after making the modifications*

Bidirectional Associative Property

There is a bidirectional associativity between all modes of PTC Creo Parametric. The bidirectional associative nature of a software package is defined as its ability to ensure that if any modifications are made in a particular model in one mode, then those modifications are also reflected in the same model in other modes. For example, if you make any change in a model in the **Part** mode and regenerate it, the changes will also be highlighted in the **Assembly** mode. Similarly, if you make a change in a part in the **Assembly** mode, after regeneration, the change will also be highlighted in the **Part** mode. This bidirectional associativity also correlates the two-dimensional (2D) drawing views generated in the **Drawing** mode and the solid model created in the **Part** mode of PTC Creo Parametric. This means that if you modify the dimensions of the 2D drawing views in the **Drawing** mode, the change will be automatically reflected in the solid model and also in the assembly after regeneration. Likewise, if you modify the solid model in the **Part** mode, the changes will also be seen in the 2D drawing views of that model in the **Drawing** mode. Thus, bidirectional associativity means that if a modification is made to one mode, it changes the output of all the other modes related to the model. This bidirectional associative nature relates various modes in PTC Creo Parametric.

Figure 1-3 shows the drawing views of the part shown in Figure 1-1 generated in the **Drawing** mode. The views show that the part consists of a counterbore hole at the center and six counterbore holes around it.

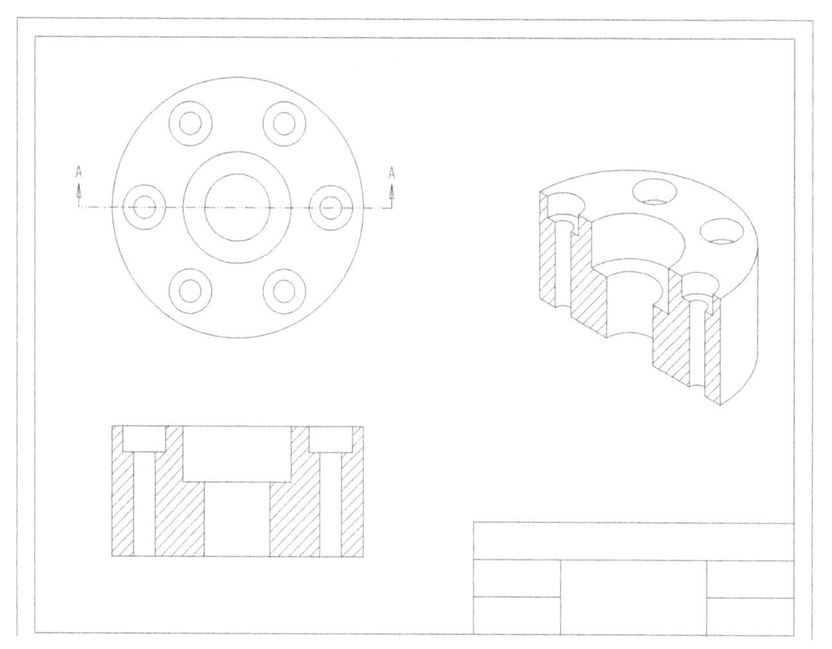

Figure 1-3 *Drawing views of the model before modifications*

Now, when the part is modified in the **Part** mode, the modifications are automatically reflected in the **Drawing** mode, as shown in Figure 1-4. The views in this figure show that all outer counterbore holes are converted into drilled holes and the number of holes is increased from six to eight.

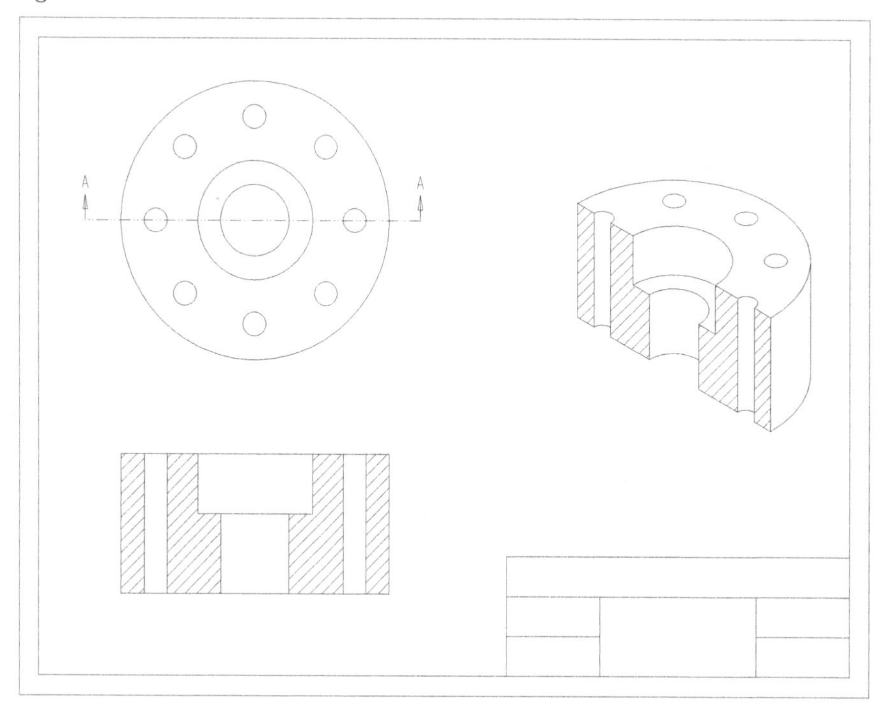

Figure 1-4 *Drawing views of the model after modifications*

Figure 1-5 shows the Crosshead assembly. It is clear from the assembly that the diameter of the hole is more than what is required (shown using dotted lines). In an ideal case, the diameter of the hole should be equal to the diameter of the bolt.

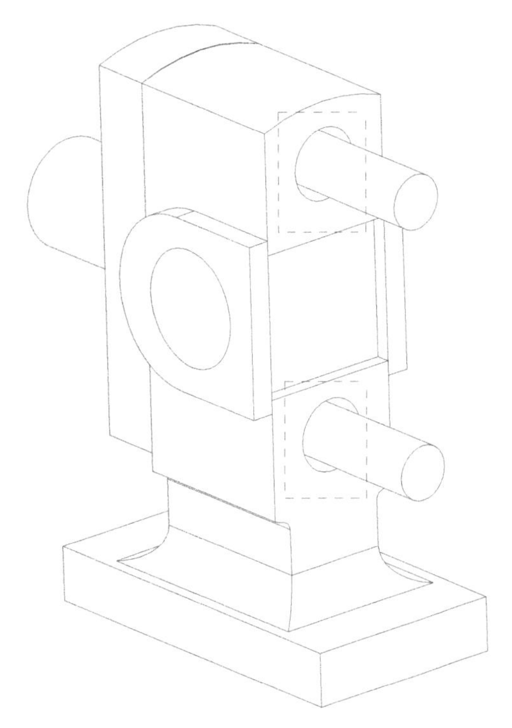

***Figure 1-5** Crosshead assembly illustrating difference in diameter of the hole and the bolt*

The diameter of the hole can be changed easily by opening the file in the **Part** mode and making the necessary modifications in the part. This modification is reflected in the assembly, as shown in Figure 1-6. This is due to the bidirectional associative nature of PTC Creo Parametric.

Since all modes of PTC Creo Parametric are interrelated, it becomes very easy to modify your model at any time.

Parametric Nature

PTC Creo Parametric is parametric in nature, which means that the features of a part become interrelated if they are drawn by taking the reference of each other. You can redefine the dimensions or the attributes of a feature at any time. The changes will propagate automatically throughout the model. Thus, they develop a relationship among themselves. This relationship is known as the parent-child relationship. So if you want to change the placement of the child feature, you can make alterations in the dimensions of the references and hence change the design as per your requirement. The parent-child relationship will be discussed in detail while discussing the datums in later chapters.

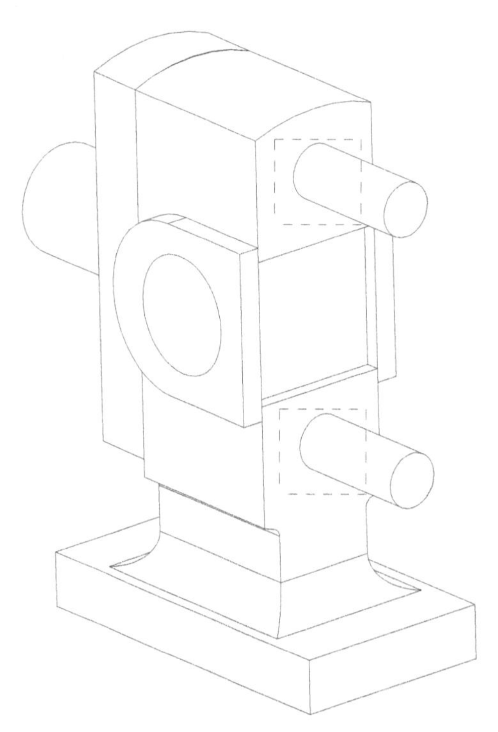

Figure 1-6 *Assembly after modifying the diameter of the hole*

SYSTEM REQUIREMENTS

The system requirements for PTC Creo Parametric are given below.

1. Operating System: Windows XP Professional Edition, XP Professional x64 Edition, Windows 7 (Professional, Ultimate, and Enterprise editions) or later.
2. Monitor: 1280 x 1024 (or higher) resolution support with 32-bit color.
3. Processor: 3.0 GHz minimum (Core 2 Duo or higher).
4. Memory: 2GB RAM minimum (3GB RAM or higher).
5. Hard disk space: 3.0GB minimum (4.0GB or higher).
6. An ethernet adapter interface card or network card.
7. Microsoft approved 3-button mouse.
8. Microsoft Internet Explorer 6.0 or later.
9. A certified and supported graphics card.

GETTING STARTED WITH PTC Creo Parametric

Install the PTC Creo Parametric on your system; the shortcut icon of PTC Creo Parametric will automatically be created on the desktop. You can start it by double-clicking on its shortcut icon on the desktop.

You can also start it by choosing the **Start** button at the lower left corner of the screen and then choose **All Programs > PTC Creo > PTC Creo Parametric 3.0**, if you are using Windows 7.

Figure 1-7 shows the initial screen that appears when you start PTC Creo Parametric.

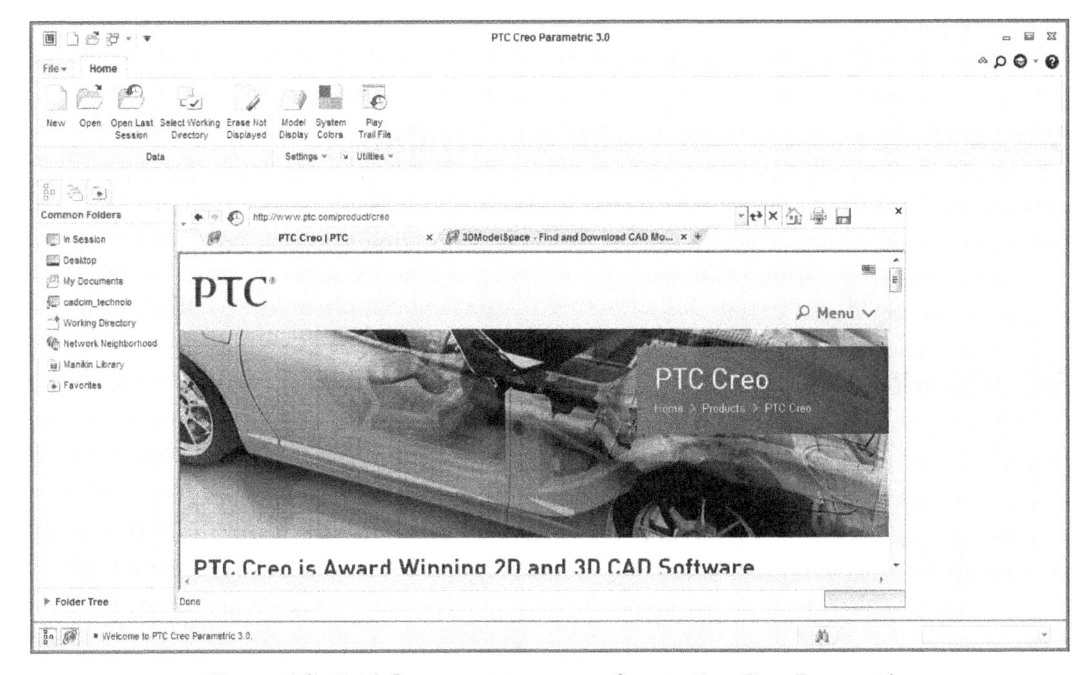

Figure 1-7 Initial screen appearance after starting Creo Parametric

IMPORTANT TERMS AND DEFINITIONS

Some important terms that will be used in this book while working with PTC Creo Parametric are discussed next.

Entity

An element of the section geometry is called an entity. The entity can be an arc, line, circle, point, conic, coordinate system, and so on. When one entity is divided at a point, then the total number of entities are said to be two.

Dimension

It is the measurement of one or more entities.

Constraint

Constraints are logical operations that are performed on the selected geometry to make it more accurate in defining its position and size with respect to the other geometry.

Parameter

It is defined as a numeric value or a word that defines a feature. For example, all dimensions in a sketch are parameters. The parameters can be modified at any time.

Relation

A relation is an equation that relates two entities.

Weak Dimensions and Weak Constraints

Weak dimensions and weak constraints are temporary dimensions or constraints that appear in light blue color. These are automatically applied to the sketch. They are removed from the sketch without any confirmation from the user. The weak dimensions or the weak constraints should be changed to strong dimensions or constraints if they seem to be useful for the sketch. This only saves an extra step of dimensioning the sketch or applying constraints to it.

Strong Dimensions and Strong Constraints

Strong dimensions and strong constraints appear in dark blue color. These dimensions and constraints are not removed automatically. All dimensions added manually to a sketch are strong dimensions.

Tip: *When several strong dimensions or constraints conflict, PTC Creo Parametric makes the constraints and dimensions appear in blue box, and prompts you to delete one or more of them.*

File MENU OPTIONS

The options that are displayed when you choose **File** from the menu bar are discussed next.

Select Working Directory

A working directory is a directory on your system where you can save the work done in the current session of PTC Creo Parametric. You can set any directory existing on your system as the working directory. Before starting the work in PTC Creo Parametric, it is important to specify the working directory. If the working directory is not selected before saving an object file, then the object file will be saved in a default directory. This default directory is set at the time of installing PTC Creo Parametric. If the working directory is selected before saving the object files that you create, it becomes easy to organize them. In PTC Creo Parametric, the working directory can be set in the following two ways:

Using the Navigator

When you start a PTC Creo Parametric session, the navigator is displayed on the left of the drawing area. This navigator can be used to select a folder and set it as the working directory.

To do so, click on the **Folder Tree** node displayed at the bottom of the navigator; the expanded **Folder Tree** area will be displayed. Browse to the required location using the nodes available next to the folders and then select the desired folder. The selected folder will become the working directory for the current session. Alternatively, right-click on the folder that you need to set as the working directory; a shortcut menu will be displayed, as shown in Figure 1-8. Choose the **Set Working Directory** option from this shortcut menu to set the selected folder as the working directory. To make a new folder, choose the **New Folder** option from the shortcut menu.

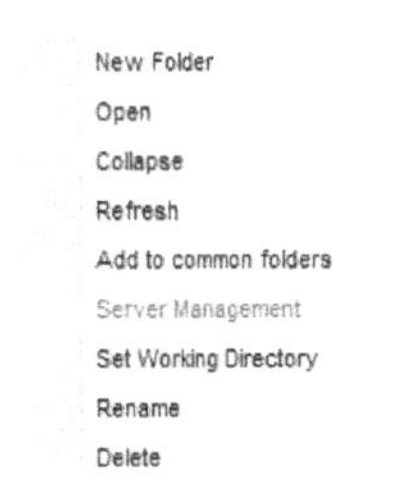

Figure 1-8 *Shortcut menu*

Using the Select Working Directory Dialog Box

To specify a working directory, choose **File > Manage Session > Select Working Directory** from the menu bar; the **Select Working Directory** dialog box will be displayed, as shown in Figure 1-9. Using this dialog box, you can set any directory as the working directory.

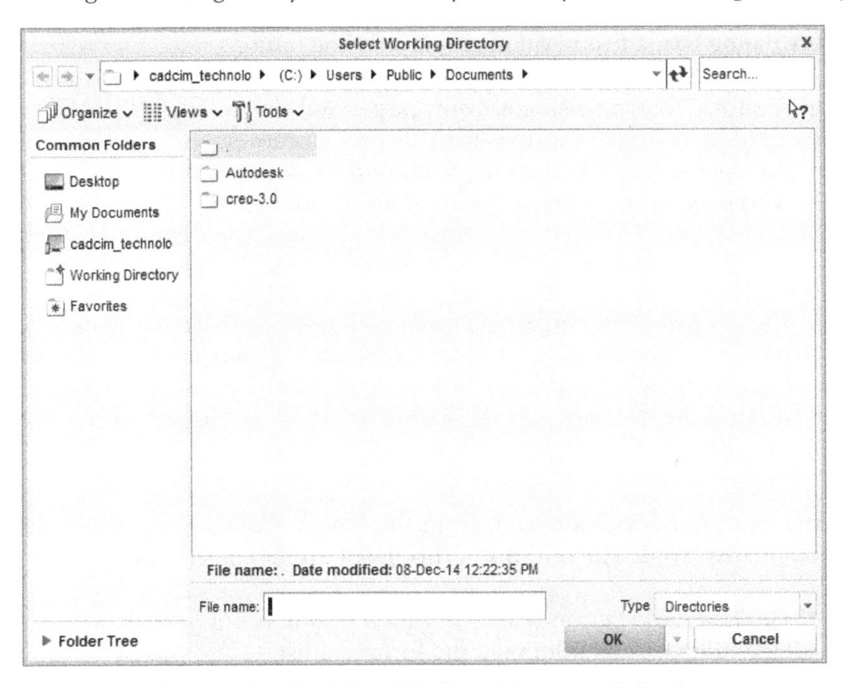

Figure 1-9 *The **Select Working Directory** dialog box*

Choose the arrow at the upper left corner of the **Select Working Directory** dialog box; a flyout will be displayed, as shown in Figure 1-10. This flyout displays some of the drives present on your computer along with the **Favorites** folder. The **Favorites** folder contains all directories that you saved as favorites. The procedure to save the favorite

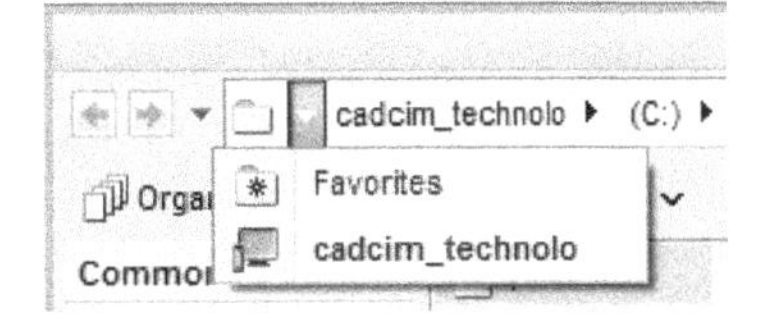

Figure 1-10 *The flyout displayed*

directories will be discussed later. When the **Select Working Directory** dialog box is invoked by default, it displays the contents of the default directory. However, you can change the default directory that appears every time you open this dialog box. Various options in the **Select Working Directory** dialog box are discussed next.

File name

The **File name** edit box displays the name of the directory selected in the **Select Working Directory** dialog box. You can select a directory using the flyout, as discussed earlier or by entering the path of any existing directory in this edit box.

Type

The **Type** drop-down list has two options, **Directories** and **All Files (*)**. If you select the **Directories** option, all directories present get listed, and if you select the **All Files (*)** option, then all files along with the directories are listed in the dialog box.

Organize

When you choose the **Organize** button from the **Select Working Directory** dialog box, a flyout will be displayed. The options in this flyout are used to create a new directory or rename an existing directory. You can also cut, copy, paste, and delete the existing folders using the options in the flyout. Moreover, you can add any existing folder in the **Common Folders** by using the **Add to common folders** option in this flyout, refer to Figure 1-11.

Figure 1-11 *The **Organize** flyout*

Tip: *The **Select Working Directory** dialog box has some of the properties of the Microsoft Windows operating system. You can set the working directory using this dialog box by browsing through directories and folders. You can also rename a file, directory, or a folder in this dialog box. Also, you can create a new directory using this dialog box.*

Views

When you choose the **Views** button from the **Select Working Directory** dialog box, a flyout will be displayed. The options in this flyout are discussed next.

List: The **List** radio button is used to view the contents of the current folder or drive. These include files and folders in the form of a list.

Details: The **Details** radio button is used to view the contents of the current folder or drive in the form of a table, which displays the name, size, and date on which it was last modified.

Tools

When you choose the **Tools** button from the **Select Working Directory** dialog box, a flyout will be displayed, as shown in Figure 1-12. The options in this flyout are discussed next.

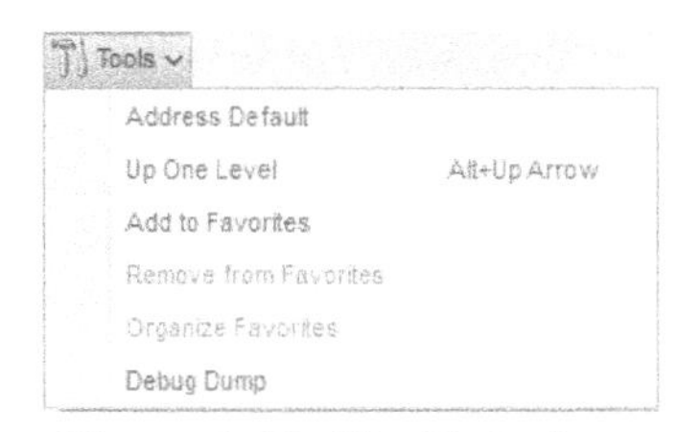

*Figure 1-12 The **Tools** flyout*

Address Default: When you choose this option, the **'Look In' Default** dialog box will be displayed. Figure 1-13 shows this dialog box with the options in the drop-down list. If you select the **Default** option from the drop-down list and then invoke the **File Open** dialog box, it will display the directory that is set as default. If you select the **Working Directory** option from the drop-down list and then invoke the **File Open** dialog box, it will display the working directory that is set. If you select the **In Session** option and then invoke the **File Open** dialog box, the **File Open** dialog box will open with the **In Session** folder selected by default. Similarly, you can set the **Pro/Library** as the working directory.

Up One Level: The **Up One Level** option allows you to move one level up in the directory. Choose this option; a directory that is one level above the current directory will be displayed. Alternatively, press ALT+UP arrow keys or **BACKSPACE** key to move one level up. You can also choose the arrow button on the left of the required directory in the address bar to display all folders in it.

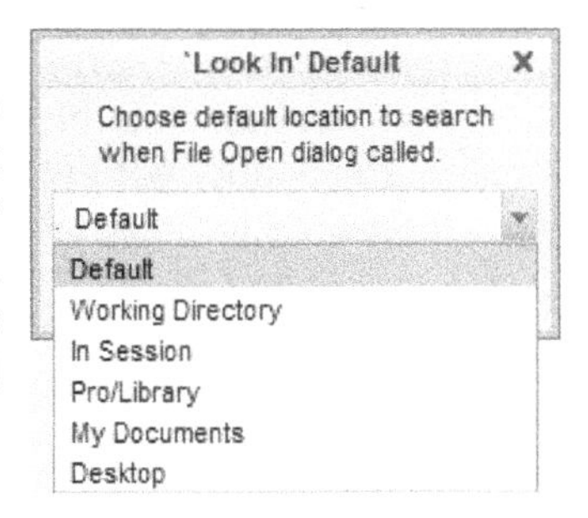

*Figure 1-13 The **'Look In' Default** dialog box with options in the drop-down list*

Add to Favorites: The **Add to Favorites** option allows you to add the folders in the **Favorites** folder.

Remove from Favorites: The **Remove from Favorites** option allows you to remove the folders from the **Favorites** folder. This option is not enabled by default. To enable this option, you first need to add the folder in the **Favorites** folder using the **Add to Favorites** option.

Sort By: In the **Select Working Directory** dialog box, the **Directories** option in the **Type** drop-down list is displayed by default. From this drop-down list, if you select the **All Files** option and then choose the **Tools** button, a flyout will be displayed with the **Sort By** option.

The **Sort By** option is used to list all files in the directory in an order to facilitate the process of searching a file. When you choose the **Sort By** option, a cascading menu is displayed. In the cascading menu, there are two radio buttons, **Model Name** and **Markup/Instance Name**. If you select the **Model Name** radio button, the file list will be sorted out alphabetically by the model name in the **Select Working Directory** dialog box. The **Markup/Instance Name** radio button sorts out the file list by specific markups or instance names in the **Select Working Directory** dialog box.

Common Folders and Folder Tree

The **Common Folders** and **Folder Tree** tabs are available on the left of the **Select Working Directory** dialog box. The Common Folders contains folders such as Desktop, My Documents, Computer, Working Directory, and Favorites. You can add more folders in the **Common Folders** by using the **Add to common folders** option available in the **Organize** flyout.

The **Folder Tree** contains all the drives available on your computer along with their contents. You can also set the working directory by using the **Folder Tree**. By default, the **Folder Tree** is in the collapsed state. To expand it, you need to click on the node that is available on the left of the **Folder Tree**. The **Working Directory** and **Favorites** folders available in the **Common Folders** are discussed next.

Working Directory: This folder is used when you have already set the working directory. You may browse through the directories in the **Select Working Directory** dialog box, but when you choose this folder, the directory selected previously as the working directory is displayed in the list box.

Favorites: This folder is used to save the location of the directories that are to be used frequently. You just need to specify the working directory to be used frequently and save its location by selecting the **Favorites** folder.

If you want to select one of the favorite working directories, then select the **Favorites** folder from the **Common Folders**; the list of all directories that were saved as favorites will be displayed in the list box. Select the required favorite directory and choose **OK**; the selected favorite directory will be set as the current working directory.

Note

*In PTC Creo Parametric, an object can be created in different modes such as **Part**, **Drawing**, **Sketch**, and then saved as a file.*

New

To create a new object, choose the **New** tool from the **Data** group of the **Home** tab in the **Ribbon** or choose the **New** tool from the **Quick Access** toolbar; the **New** dialog box will be displayed, as shown in Figure 1-14. This dialog box displays various modes available in PTC Creo Parametric. In this dialog box, by default, the **Part** mode radio button is selected and the default name of the object file is displayed in the **Name** edit box. You can also enter a new name for the object file. Note that the name must not contain a special character.

When you select the **Part**, **Assembly**, or **Manufacturing** radio button in this dialog box, the subtypes of the respective modes will be displayed under the **Sub-type** area of this dialog box.

Accept the default settings in the **New** dialog box by choosing the **OK** button; the default template will be loaded. To load a template other than the default one, clear the **Use default template** check box and then choose the **OK** button; the **New File Options** dialog box will be displayed, as shown in Figure 1-15. Using this dialog box, you can select the predefined templates or create a user-defined template. You can also open an empty template provided in the **New File Options** dialog box. In this case, you need to create the datum planes and the coordinate systems manually.

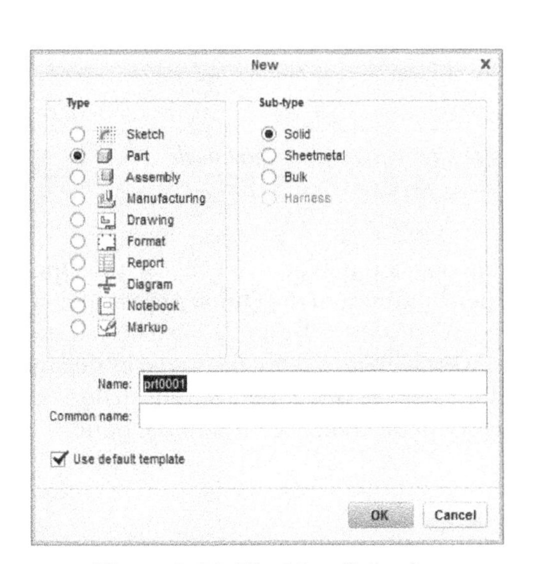

Figure 1-14 *The* ***New*** *dialog box*

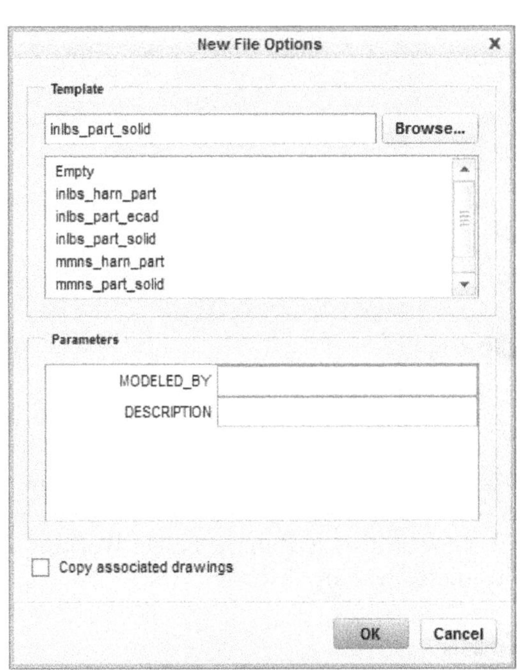

Figure 1-15 *The* ***New File Options*** *dialog box*

If the measuring units for creating models is inches, select **inlbs_part_solid** from the list in the **Template** area and then choose **OK** from the **New File Options** dialog box. On doing so, the three default datum planes and a coordinate system will be displayed in the drawing area. Also, the **Model Tree** will appear on the left of the screen, as shown in Figure 1-16.

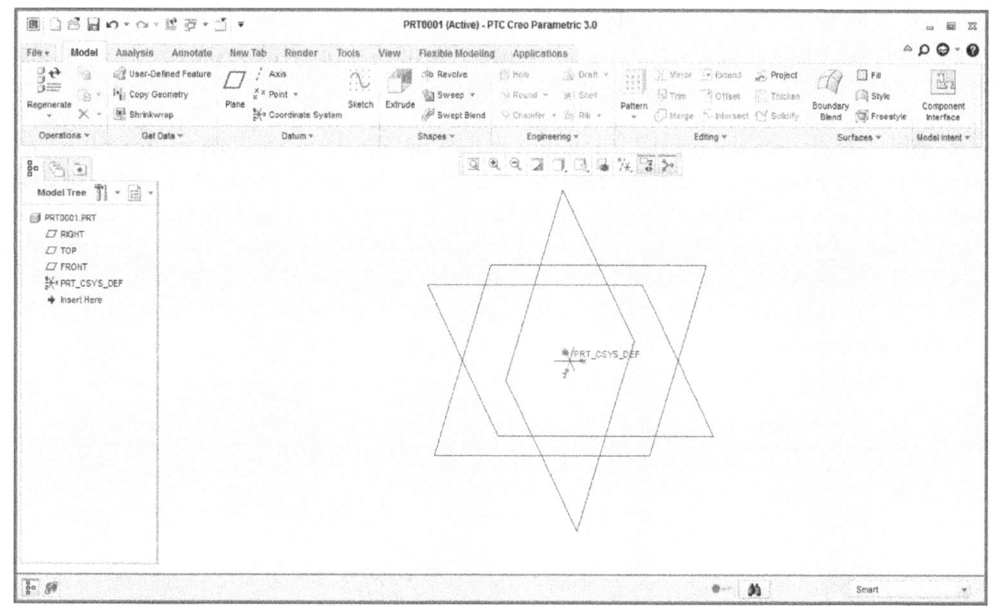

*Figure 1-16 The initial screen appearance after entering the **Part** mode*

Open

The **Open** button is used to open an existing object file. When you choose the **Open** option from the **File** menu or choose the **Open** button from the **Quick Access** toolbar, the **File Open** dialog box will be displayed, as shown in Figure 1-17. The selected working directory will be displayed in it. Note that the **Preview** area is not displayed by default. To view the **Preview** area, choose the **Preview** button. Most of the options in this dialog box are same as discussed in the **Select Working Directory** dialog box. The rest of the options in this dialog box are discussed next.

Tools

On choosing this button, a flyout will be displayed. The options available in this flyout are same as the options discussed in the **Tools** button of the **Select Working Directory** dialog box, except the **All Versions** and **Show Instances** options. These two options are discussed next.

All Versions

This option, when chosen, displays all versions of an object file. In PTC Creo Parametric, once a file is saved, its new version is generated with an entension that is incremented by 1. An object file is not copied on another object file but a new version of it is created. Therefore, every time you save an object using the **Save** option, a new version of it is created on the disk in the current working directory.

Show Instances

The **Show Instances** option, when chosen, displays all instances of the object file. Select the required file and then choose the **Show Instances** option from the **Tools** flyout; all the instances of the selected file will be displayed.

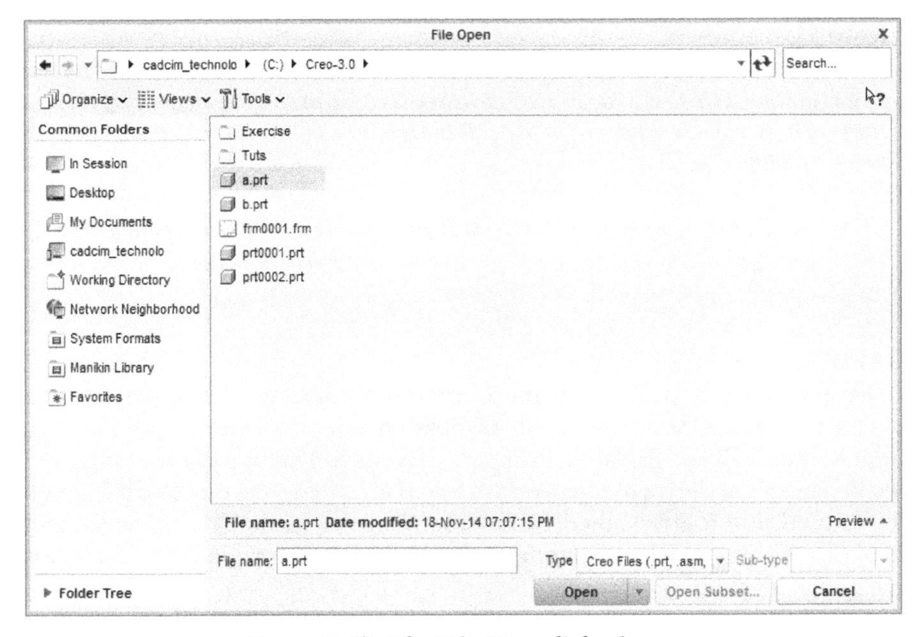

*Figure 1-17 The **File Open** dialog box*

File name

In the **File name** edit box, you can enter the name of the existing object file that you want to open.

Type

The **Type** drop-down list contains the file formats of various modes available in PTC Creo Parametric. It also contains many other file formats that can be imported in PTC Creo Parametric. These file formats include IGES, SET, STEP, DWG/DWF, Medusa, Inventor, Parasolid, Rhino, and so on.

By default, the **Creo Files** option is selected in this drop-down list. As a result, you can open the files created in any mode of PTC Creo Parametric. However, if you select a specific mode from this drop-down list, only the files of the corresponding mode will be displayed. For example, if you select **Part** from the drop-down list, then only the *.prt* files will be displayed. This makes the selection and opening of the files easy.

Preview

The **Preview** button is used to preview the model before opening it. On choosing this button, you can preview the selected model in the **File Open** dialog box. You can zoom, pan, and rotate the model in the preview. Also, you can change the appearance (shaded, wireframe, no hidden, and hidden line) of the model, switch the model between the orthographic and perspective views, change the orientation type (dynamic, anchored, delayed, velocity, fly through, and standard) of the model, set the number of frames per second, and refit the preview in the preview screen.

Note

Assembly files with the file extension .asm can also be previewed by using the ***Preview*** *button. If you are not able to see preview of the assembly files in the* ***Preview*** *area, then choose the* ***Refresh*** *button on the upper right of the* ***File Open*** *dialog box to resize the assembly according to the* ***Preview*** *area.*

There is no Command prompt in PTC Creo Parametric. However, you are provided with prompts in the message area. Whenever you have to enter a numerical value or text, a message input window will be displayed in the message area.

In Session

The **In Session** folder is available in the **Common Folders** on the upper left of the **File Open** dialog box. When you choose the **In Session** folder, all the object files that are in the current session will be displayed in the display area. The object files that you open in PTC Creo Parametric in the current session are stored in its temporary memory. This temporary memory is stored in a folder named **In Session**. Once you exit PTC Creo Parametric, the contents of this folder are deleted automatically. However, the original files are not removed from their actual location.

Erase

As discussed earlier, all files opened in a session of PTC Creo Parametric are saved in the temporary memory. There are three options to erase objects in the temporary memory: **Erase Current**, **Erase Not Displayed**, and **Erase Unused Model Reps**. These options are available in the **Manage Session** flyout in the **File** menu. The options that are displayed in this flyout are discussed next.

Tip: *Suppose you open an assembly that has a component named Nut. Close the assembly and now open another assembly that also has a component named Nut. Now, there are chances that the second assembly you choose to open may open with the Nut that was present in the previous assembly. This is because the component with the file named Nut was already present in the memory of PTC Creo Parametric (in session).*

To avoid such type of errors, you should erase the files in the current session of PTC Creo Parametric before opening the next assembly.

Erase Current

The **Erase Current** option is used to erase the file opened and displayed in the drawing area. On choosing this option, the **Erase Confirm** message box will be displayed, prompting you to confirm the erasing of the file, as shown in Figure 1-18.

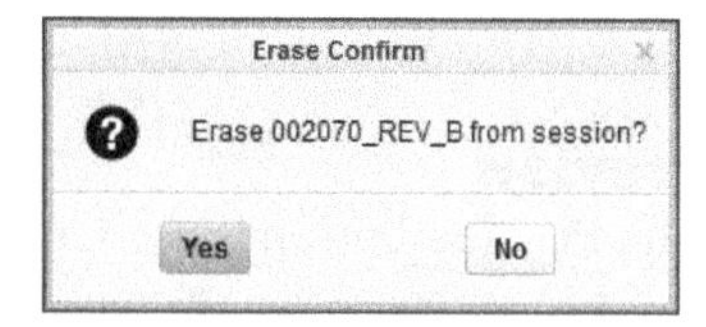

Figure 1-18 *The* ***Erase Confirm*** *message box*

Erase Not Displayed

The **Erase** option is used to delete the files stored in the temporary memory. To do so, choose **File > Manage Session > Erase Not Displayed** from the menu bar; the **Erase Not Displayed** dialog box will be displayed, as shown in Figure 1-19. The files that are not open in the current session will be displayed in this dialog box. Choose the **OK** button from this dialog box to remove these files.

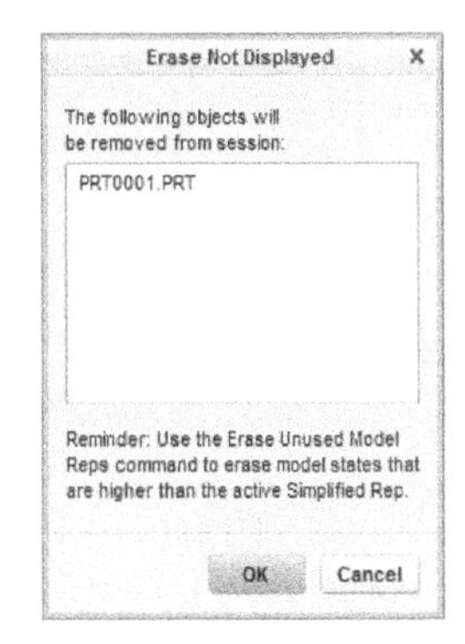

*Figure 1-19 The **Erase Not Displayed** dialog box*

Erase Unused Model Reps

This option is used to remove the unused simplified representations from the **In Session** folder. When you choose this option from the **File** menu, a message box will be displayed with the message that all the objects which were not displayed have been erased.

Delete

There are two options to delete files permanently from the hard disk. These options are available in the **Manage File** flyout in the **File** menu. The options in this flyout are discussed next.

Delete Old Versions

This option is used to delete all old versions of the current file. When you choose the **Delete Old Versions** option, the **Delete Old Versions** message box is displayed confirming to delete all old versions files. Choose **Yes** from the message box. All versions of that file will be deleted from the hard disk except the latest version.

Delete All Versions

This option is used to delete all versions including the current file from the hard disk. When you choose the **All Versions** option, a warning is displayed stating that performing this function can result in the loss of data. This option is chosen when the file is opened and is displayed in the drawing area.

Note

*If there are simplified representations in the memory, then **Object Erase** menu will be displayed. Choose the **Current Obj** option if you want to delete object files only or you can choose the **Simpfld Reps** option if you want to delete the representations only.*

Play Trail File

The **Play Trail File** option is used to recover data in case of a program crash or if you exit the session without saving the data. While working in PTC Creo Parametric, a file called *trail.txt* is created in its default working directory (for example, *C:\Users\Public\Documents*). The file *trail.txt* contains all the working steps used in the last session.

To retrieve the data, open the file *trail.txt* in Notepad. If there is more than one such file in the directory, open the latest one. Scroll to the end of this file and change the **Activate 'transient_balloon'** from **'no'** to **'yes'**, as shown in Figure 1-20. Save this file by choosing **File > Save As** with another name like *crash.txt* and exit.

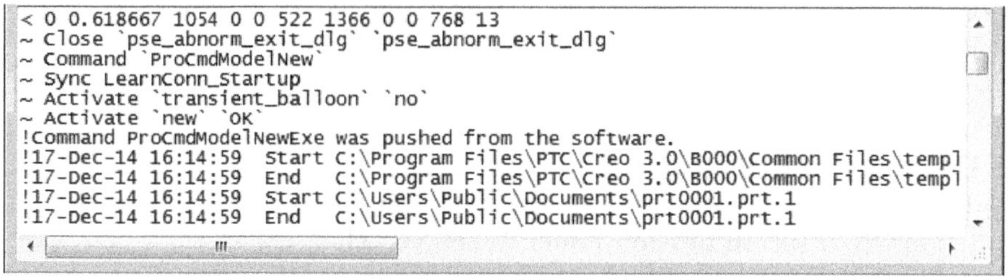

```
< 0 0.618667 1054 0 0 522 1366 0 0 768 13
~ Close `pse_abnorm_exit_dlg` `pse_abnorm_exit_dlg`
~ Command `ProCmdModelNew`
~ Sync LearnConn_Startup
~ Activate `transient_balloon` `no`
~ Activate `new` `OK`
!Command ProCmdModelNewExe was pushed from the software.
!17-Dec-14 16:14:59  Start C:\Program Files\PTC\Creo 3.0\B000\Common Files\templ
!17-Dec-14 16:14:59  End   C:\Program Files\PTC\Creo 3.0\B000\Common Files\templ
!17-Dec-14 16:14:59  Start C:\Users\Public\Documents\prt0001.prt.1
!17-Dec-14 16:14:59  End   C:\Users\Public\Documents\prt0001.prt.1
```

***Figure 1-20** The **trail.txt** file modified in Notepad*

Now, restart PTC Creo Parametric and choose the **Play Trail File** option from **Manage Session**. Select the file *crash.txt* from the **Open** dialog box and then choose the **Open** button; PTC Creo Parametric will repeat every step you made in the last session and restore the data.

Save

The **Save** option is used to save the objects present in the **In Session** folder or an object in the drawing area. When you choose the **Save** option from the **File** menu or the **Save** button from the **Quick Access** toolbar, the **Save Object** dialog box will be displayed. Also, the name of the current object will be displayed in the **Model Name** edit box. Choose the **OK** button from the **Save Object** dialog box to save the object.

Save a Copy

The **Save a Copy** option is used to save a copy of the current object in the same working directory or in some other directory. When you choose this option from **File > Save As**, the **Save a Copy** dialog box will be displayed. Now, you need to specify the new name of the object file to be saved as a copy and the name of the target directory in the **Save a Copy** dialog box. You can browse through the directories and select the target directory. The file will be saved in the selected directory.

Using this option, you can also export a file in other file formats such as Inventor file, pdf, ACIS, Wavefront file, and so on. After specifying the name of the new file and the target directory, choose the file format in which you want to export the file from the **Type** drop-down in the **Save a Copy** dialog box.

Save a Backup

The **Save a Backup** option in the **Save As** flyout is used to create a backup copy of an object file in the memory. When you choose this option, the **Backup** dialog box will be displayed, as shown in Figure 1-21.

In the **Model Name** edit box of the **Backup** dialog box, the name of the file for which you want to create a backup is displayed. In the **Backup To** edit box, the name of the directory is specified where the object will be saved as a backup. If you create the backup of an assembly or a drawing object, PTC Creo Parametric will save all its dependent files in the specified directory. If you create the backup of a drawing file in a different directory, then the part file of the drawing will also be created automatically in the same directory where backup of the drawing file is created.

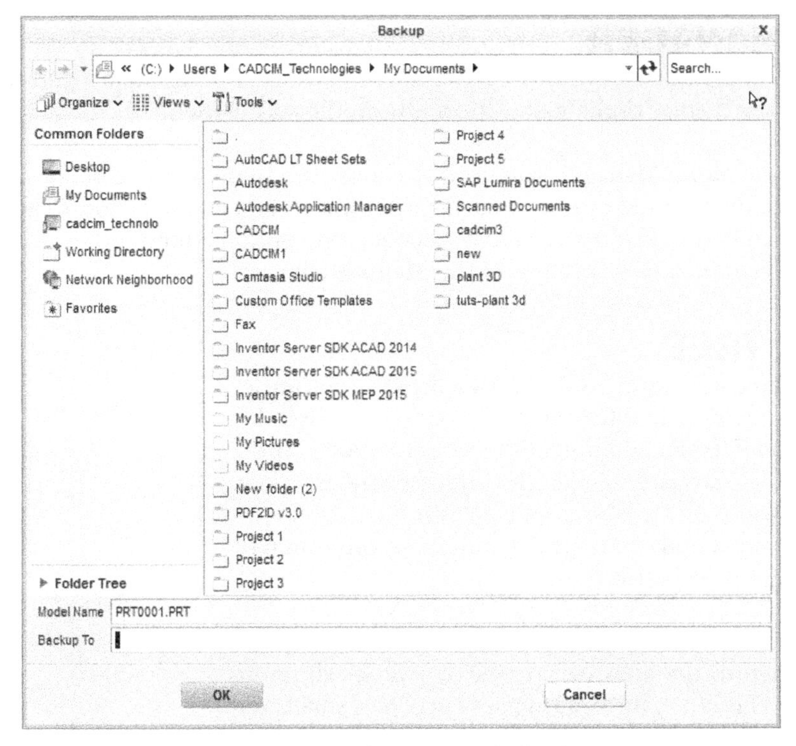

*Figure 1-21 The **Backup** dialog box*

Rename

The **Rename** option in the **Manage File** flyout of the **File** menu is used to rename the currently active object on the screen. To rename an object, choose this option; the **Rename** dialog box will be displayed, as shown in Figure 1-22. Specify the new name of the object in the **New Name** edit box and choose **OK**.

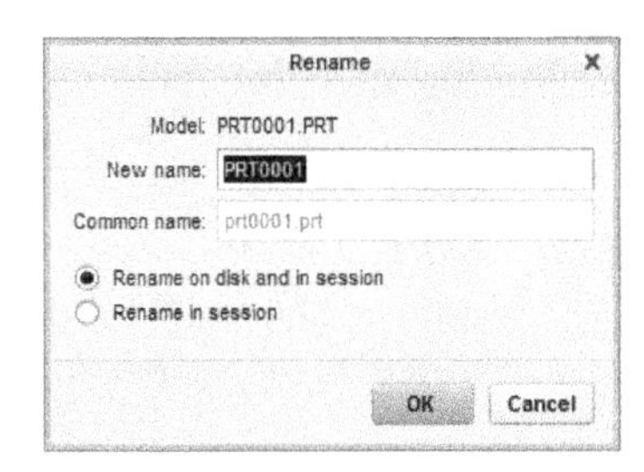

*Figure 1-22 The **Rename** dialog box*

MANAGING FILES

As discussed earlier, a new file is generated whenever you save an object. The number of files generated are directly proportional to the number of times you save that object. So, these files occupy a lot of disk space. The latest version of the file which is currently being used should be stored. Latest version refers to the highest number that is suffixed with the file name of that object. The rest of the files are old versions and should be deleted from the hard disk, if they are not required.

Note

*To save disk space, you should keep deleting the old versions of a file. This can be done by using the **File > Manage File > Delete Old Versions** option from the menu bar.*

MENU MANAGER

The **Menu Manager** consist of a cascading menu in which a set of menus and submenus are embedded. The display of the menus depends on the task chosen.

While using the **Menu Manager**, you need to choose the **Done** or **Done Sel** option to execute and complete the selected option. This is important when you are in the **Drawing** mode of PTC Creo Parametric. If you are directly selecting one option after another, then it is easy to loose track of commands or options in the **Menu Manager**.

MODEL TREE

The **Model Tree** stores and displays all features in a chronicle. You can select any desired feature of a model or an assembly from the **Model Tree** and apply different operations on the selected feature. You can also select the feature by right-clicking on it; a shortcut menu will be displayed. Move the cursor on the shortcut menu and choose the required option from it by using the left mouse button.

When you create a new object file, the **Model Tree** appears and is attached to the drawing area by default, as shown in Figure 1-23. Therefore, the drawing area becomes small. You can hide the **Model Tree** by choosing the **Show Navigator** button available below the **Model Tree**. The **Model Tree** can slide in or slide out, thus increasing or decreasing the drawing area. It can also be stretched horizontally to cover the drawing area.

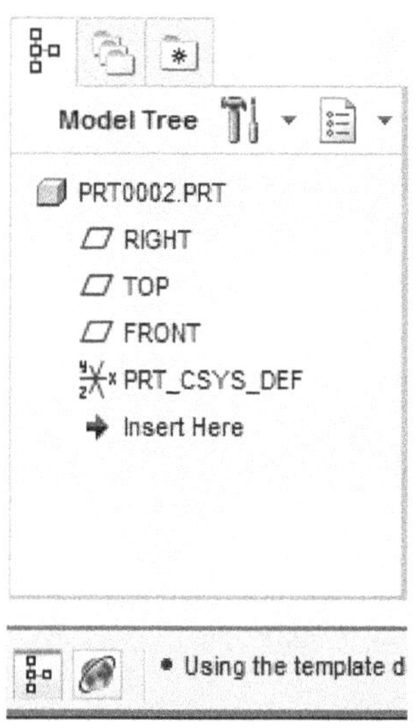

***Figure 1-23** Partial view of the Model Tree*

Note

*In PTC Creo Parametric, most of the features are created using the **Dashboard**. The **Dashboard** is displayed below the **Quick Access** toolbar and it contains all the options to complete the related operations on the model.*

Tip: *Looking at the **Model Tree**, you can understand the method and approach used to create the model. Using the **Model Tree**, you can modify the features of a model. Generally, when you import a model in a different file format in PTC Creo Parametric, the features of the model are not displayed in the **Model Tree** and therefore, you will not be able to modify it.*

UNDERSTANDING THE FUNCTIONS OF THE MOUSE BUTTONS

While working with PTC Creo Parametric, it is important to understand the function of the three buttons of the mouse to make an efficient use of this device. The various combinations of the keys and three buttons of the mouse are listed below:

1. Figure 1-24 shows the functions of the left mouse button. The left mouse button is used to make a selection. Using CTRL+left mouse button, you can add or remove items from the selection set.

2. Figure 1-25 shows the functions of the right mouse button. The right mouse button is used to invoke the shortcut menus and to query select the items. When you bring the cursor on an item, it is highlighted in green color. Now, if you hold the right mouse button, a shortcut menu is displayed. Choose the **Pick From List** option from the shortcut menu; the **Pick From List** dialog box will be displayed. You can make selections from this dialog box.

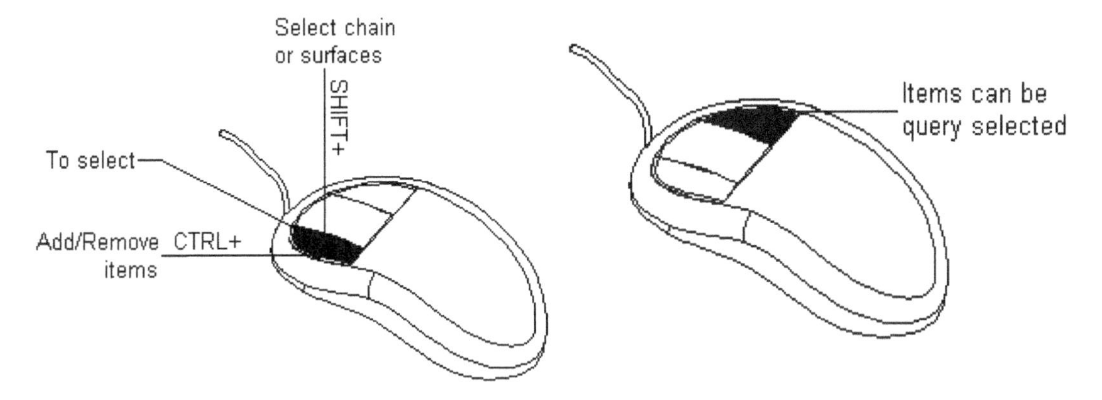

Figure 1-24 *Functions of the left mouse button* ***Figure 1-25*** *Functions of the right mouse button*

3. Figure 1-26 shows the functions of the middle mouse button in the 3D mode. The middle mouse button is used to spin the model in the drawing area and view it from different directions.

   The CTRL+middle mouse button is used to dynamically zoom in and out the view. When you press and hold the CTRL+middle mouse button and move the cursor up, the view is reduced and you zoom out. When the mouse is moved down, the view is enlarged and you zoom in.

   When you use CTRL+middle mouse button and move the mouse horizontally, the model is rotated about a point that is specified as center.

   The SHIFT+middle mouse button is used to pan the object on the screen.

4. Figure 1-27 shows the functions of the middle mouse button in the 2D mode (sketcher environment and **Drawing** mode). It is used to place dimensions in the drawing area. It is also used to confirm an option or to abort the creation of an entity.

   The middle mouse button is used to pan view in the **Sketch** mode and the **Drawing** mode.

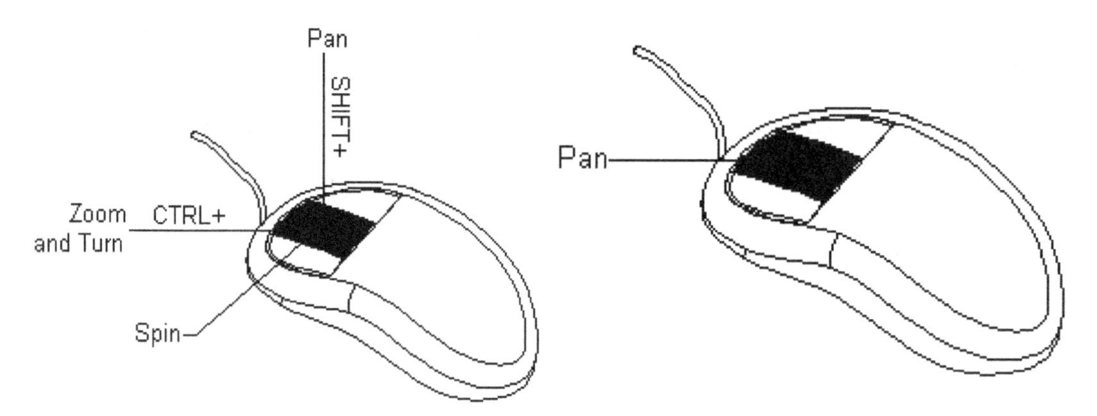

Figure 1-26 Functions of the middle mouse button in the 3D mode

*Figure 1-27 Functions of the middle mouse button in the **Sketch** mode*

Note

*When you spin the model with the **Spin Center** ON, the model is rotated about the spin center. If the spin center is turned off, then the model is rotated about the cursor.*

RIBBON

Before you start working on PTC Creo Parametric, it is very important for you to understand the default **Ribbon** and tools in the main window. Figure 1-28 shows various default interface components in PTC Creo Parametric. The **Ribbon** is composed of a series of groups, which are organized into tabs depending on their functionality. The groups in the **Ribbon** that initially appear on the screen are shown in Figure 1-28. You will notice that all the tools in the groups are not enabled. These tools will be enabled only after you create a part or open an existing file. However, the tools that are required for the current session are already enabled. As you proceed to enter one of the modes provided by PTC Creo Parametric, you will notice that the tools required by that mode are enabled. Additionally, to make the designing easy and user-friendly, this software package provides you with a number of groups. Different modes of PTC Creo Parametric display different groups. Some of the frequently used groups are shown in Figure 1-28.

TOOLBARS

In PTC Creo Parametric, there are two toolbars, **Quick Access** and **Graphics**. The toolbar on the top of the window is called the **Quick Access** toolbar and the toolbar on the top of the drawing area is called the **Graphics** toolbar.

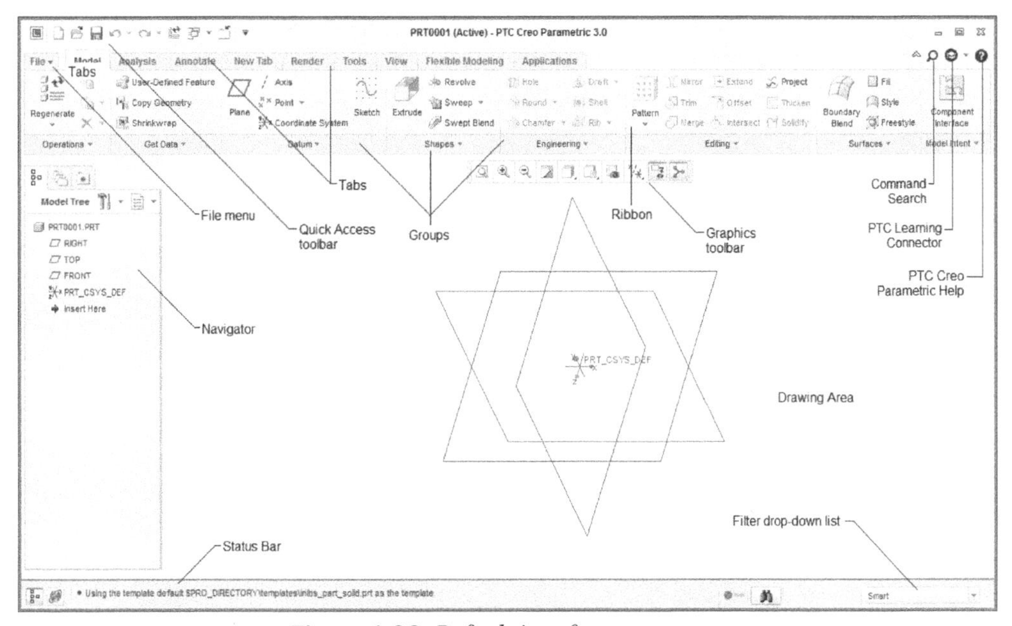

Figure 1-28 Default interface components

Graphics Toolbar

The **Graphics** toolbar, shown in Figure 1-29, has ten buttons by default. The first button is **Refit**, which is used to fit a model on the screen. The other buttons are **Zoom In** and **Zoom Out**, which are used to enlarge or diminish the model view, respectively. The **Repaint** button is used for repainting the screen, which helps in removing any temporary information from the drawing area. You can change the display style by using the **Display Style** drop-down. Using the **Named Views** drop-down, you can change the view.

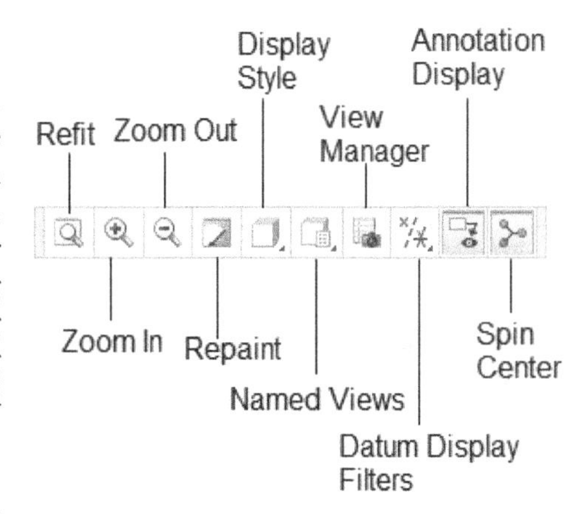

*Figure 1-29 The **Graphics** toolbar*

The next drop-down is **Datum Display Filters**. Using the options in this drop-down, you can toggle the display of datums to on/off. You can toggle the display of annotations to on/off using the **Annotation Display** button. The last button in this toolbar is **Spin Center**, which is used to toggle the visibility of the spin center to on/off.

File Menu

The **File** menu contains options shown in Figure 1-30. These options are used to create a new file, save a file, print the current file, or open an existing file. Various other options are also available in the menu to manage the current session and files. The **Options** tool is also available in the **File** menu. When you choose this tool, the **PTC Creo Parametric Options** dialog box is displayed. Using this dialog box, you can configure various parameters of PTC Creo Parametric. You can also customize the **Ribbon** interface.

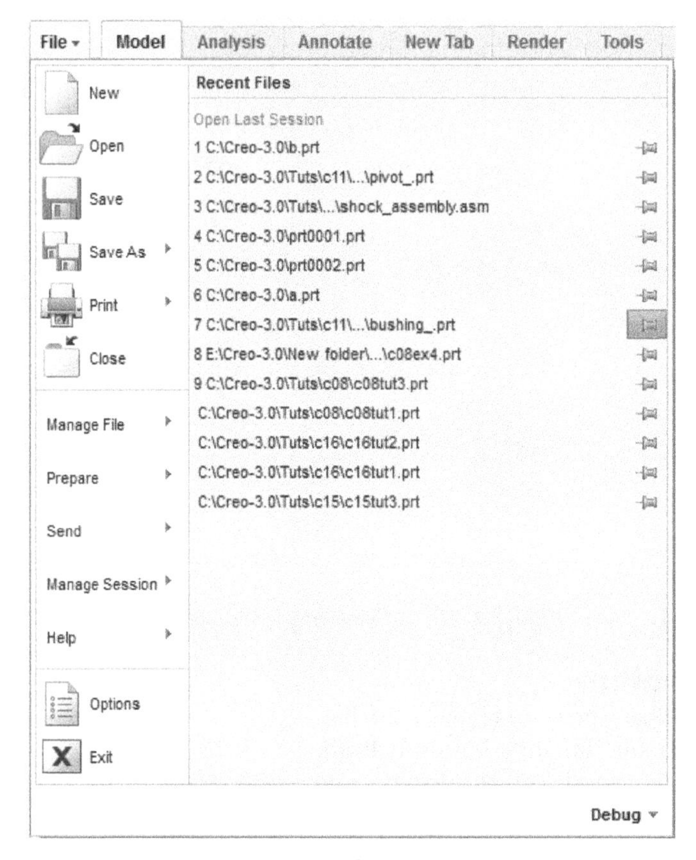

*Figure 1-30 The **File** menu*

Customizing the Ribbon

To customize the **Ribbon**, choose **Options** from the **File** menu; the **PTC Creo Parametric Options** dialog box will be displayed. Select the **Customize Ribbon** option from the left area of the dialog box; the dialog box will be modified. Now, choose **Commands Not in the Ribbon** option from the **Choose commands from** drop-down in the right area of the dialog box. Next, choose the **New Group** or **New Tab** button from the dialog box depending on the requirement; the new group or tab will be added in the list box below the **Customize the Ribbon** drop-down in the dialog box. Now, choose the button to be added in the new group from the list box available below the **Choose commands from** drop-down and then choose the **Add>>** button from the dialog box; the selected button will be added to the group, refer to Figure 1-31.

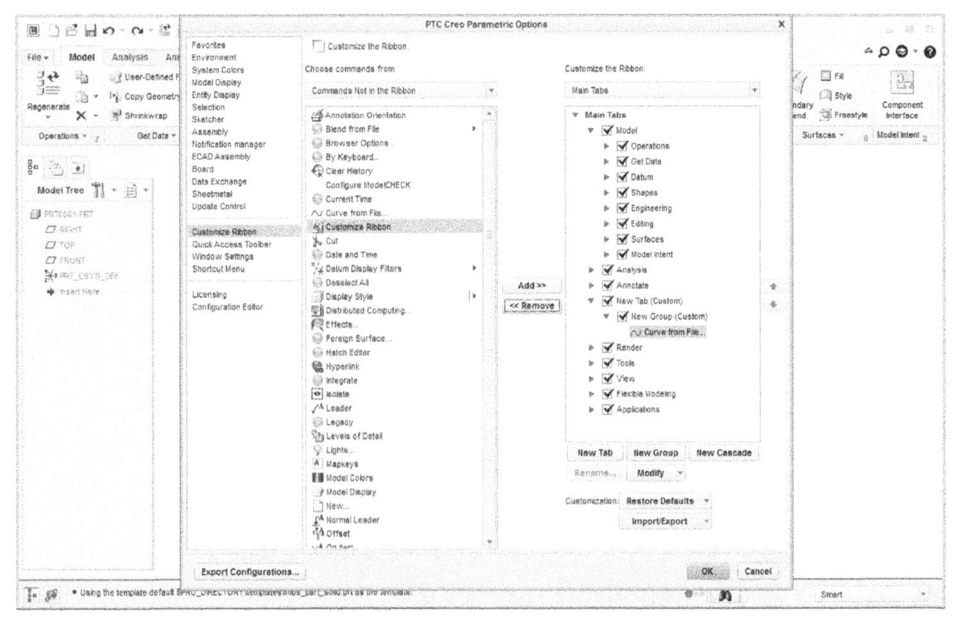

*Figure 1-31 Customizing the **Ribbon***

PTC Creo Parametric Help

When you choose the **PTC Creo Parametric Help** button, the **Creo Parametric Help** dialog box will be displayed and you will be prompted to enter the username and password. The help database is accessible only to the users who have customer account with PTC.

Command Search

The **Command Search** button is available at the top right corner of the program window. When you choose this button, the search box is displayed, as shown in Figure 1-32. You need to enter the first few letters of the command to be searched in the box. When you enter the first letter in the search box, a list of commands starting from that letter will be displayed. You can invoke any command by selecting it from the list, or you can enter the next letter to refine your search. On moving the cursor on any of the commands in the list, the command will get highlighted in the application window, as shown in Figure 1-33. Also, the path of the command will be displayed below the cursor.

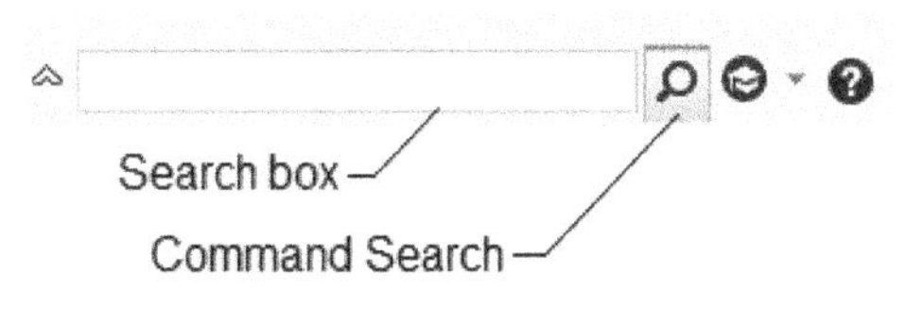

Figure 1-32 The command search box

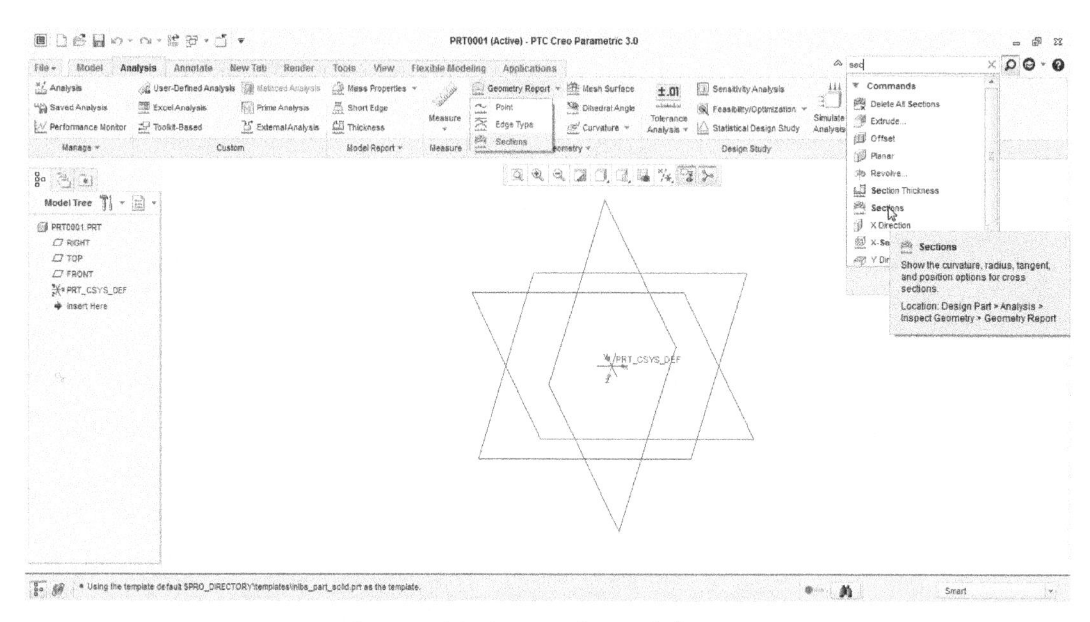

Figure 1-33 Command Search function

PTC Learning Connector Button

The **PTC Learning Connector** button is used to access video tutorials provided by PTC. When you choose this button, the **PTC Learning Connector** window will be displayed, as shown in Figure 1-34. When you invoke any tool from the **Ribbon**, the corresponding tutorial will be displayed in the **PTC Learning Connector** window.

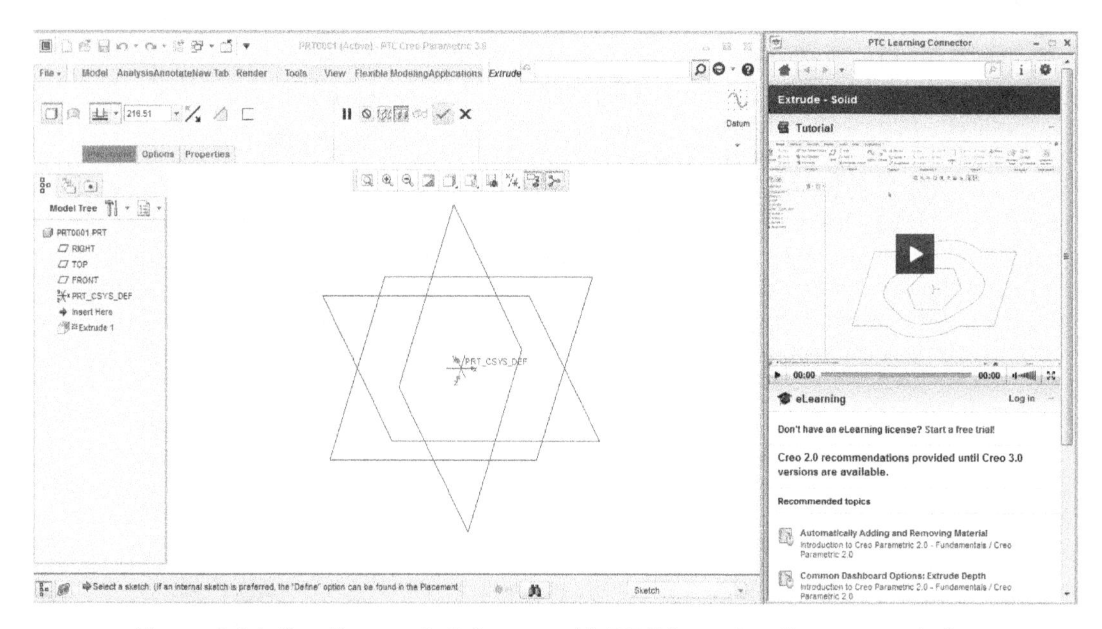

*Figure 1-34 Creo Parametric 3.0 screen with **PTC Learning Connector** window*

Note

If you want to change the predefined unit system, choose ***File > Prepare > Model Properties*** *from the menu bar to display the* ***Model Properties*** *dialog box. In this dialog box, click on the* ***change*** *on the right of the* ***Units*** *option in the* ***Materials*** *head; the* ***Units Manager*** *dialog box will be displayed with the* ***System of Units*** *tab chosen. Next, select the desired unit system from the list box and then choose the* ***Set*** *button; the* ***Changing Model Units*** *message box will be displayed. Choose the* ***OK*** *button from the message box; the new unit system will be set and displayed with the red arrow on the left.*

NAVIGATOR

The navigator is present on the left of the drawing area and can slide in or out of the drawing area. To make the navigator slide in or out, you need to select the **Show Navigator** button on the bottom left corner of the program window. A partial view of the navigator is shown in Figure 1-35. It has the following functions:

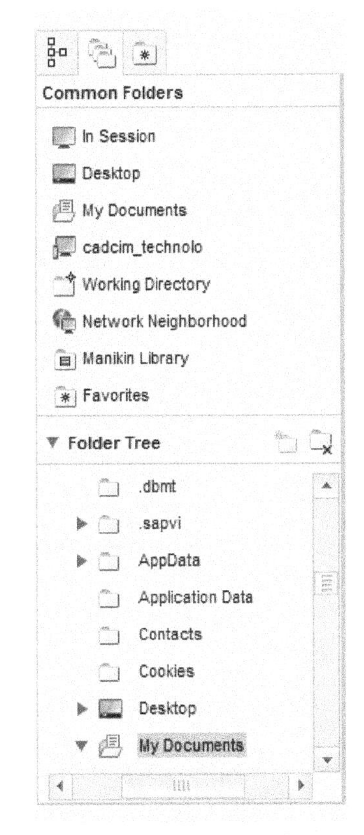

Figure 1-35 *Partial view of the navigator*

1. When you browse files using the navigator, the browser expands and the files in the selected folder are displayed in the browser.

2. When you open a model, the **Model Tree** is displayed in the navigation area.

3. The buttons on the top of the navigator are used to display different items in the navigation area. The **Model Tree** button is used to display the **Model Tree** in the navigation area. This button is available only when a model is opened. The **Folder Browser** button is used to display the folders that are in the local system. The **Favorites** button is used to display the contents of the **Personal Favorites** folder.

4. Any other location, if available on your system, can also be accessed by using the navigator.

PTC Creo Parametric BROWSER

The PTC Creo Parametric browser is present on the right of the navigator. You can slide in and out of the navigator by using the **Show Navigator** button at the bottom left corner of the application window. You can also stretch the browser window horizontally. When the PTC Creo Parametric session starts, the browser is displayed on the screen. The browser has two tabs. The homepage is linked to *ptc.com*, by default, and the other tab is linked to the online resources of PTC. You can add more tabs in it, switch between tabs, and view the browsing history similar to any other internet explorer. The main advantage of using this browser is that you can work on the PTC Creo Parametric files and navigate through the browser, simultaneously. Figure 1-36 shows the browser with some part files displayed in it. You can also browse to the required folder, change the display of files to thumbnails for previewing the file, dynamically view the models in a pop-up window, or drag and drop the selected file into the PTC Creo Parametric window.

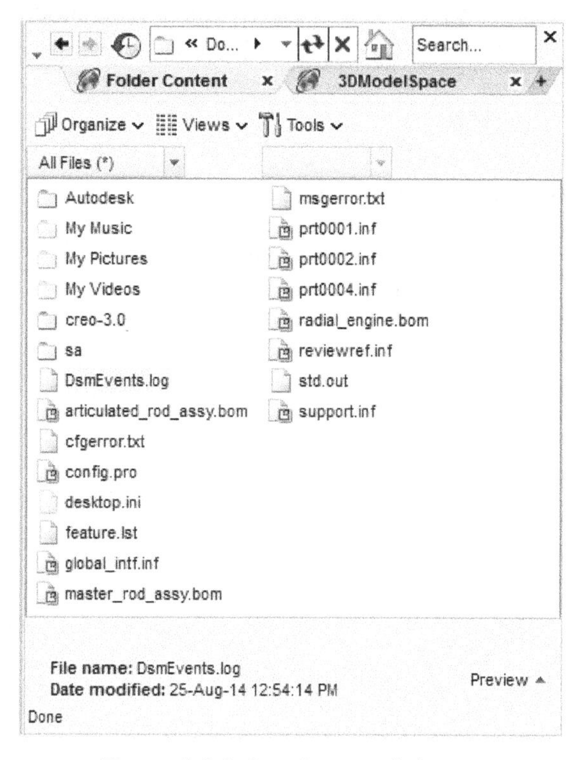

Figure 1-36 *Creo Parametric browser*

The functions of the browser are listed next.

1. It is used to preview the PTC Creo Parametric files and browse the file system.

2. A PTC Creo Parametric file can be opened by using the browser on double-clicking on it or by dragging it to the drawing area. When you open a file using the browser, the model is displayed on the screen and the browser is closed automatically.

3. You can access the PTC Creo Parametric user community site using the browser.

4. You can connect to your client's computer and jointly work with them using the browser.

APPEARANCE GALLERY

A new library of appearances and real world materials has been added in PTC Creo Parametric. You can also add shadows and reflections to the background of the model by using the appearance gallery. This is useful while rendering with a background image. This gives a realistic view to the model.

The **Appearance Gallery** button is available in the **Appearance** group of the **Render** tab in the **Ribbon**. It is used to assign different colors to the model and change its appearances. To change the appearance of the model, click on the down arrow beside the **Appearance Gallery** button; a palette will be displayed, as shown in Figure 1-37. Select the required appearance from the palette; the **Select** box will be displayed. Select the entities from the drawing area whose appearance you want to change; the appearance of the entities will change. You can assign a desired appearance to the entire model or to the individual face of the model, or entire assembly or individual component of an assembly. Various areas and options in the palette are discussed next.

Search

This option is used for searching a specific appearance you need. To search an appearance, type a keyword related to the appearance to be searched in the **Search** text box and press ENTER; all the appearances matching with the keyword will be searched from all the system libraries and they will be displayed in the palette. For example, if you type **glass** in the text box, then **ptc-glass** will be displayed in the **My Appearances** area and the **Glass** library is displayed in the **Library** area.

View Options

You can set the display of the appearance icons in the palette by using the **View Options** flyout. These options will be available when you click on the button at the top-right corner of the palette. The icons can be displayed as small thumbnails, large thumbnails, names and thumbnails, or only names. To do so, invoke the **View Options** flyout and then choose the required option from it. Choose **Rendered Samples** from the flyout to view the rendered icons of appearances. In this flyout, the **Show Tooltips** check box is selected by default. As a result, tooltips are always displayed for appearances.

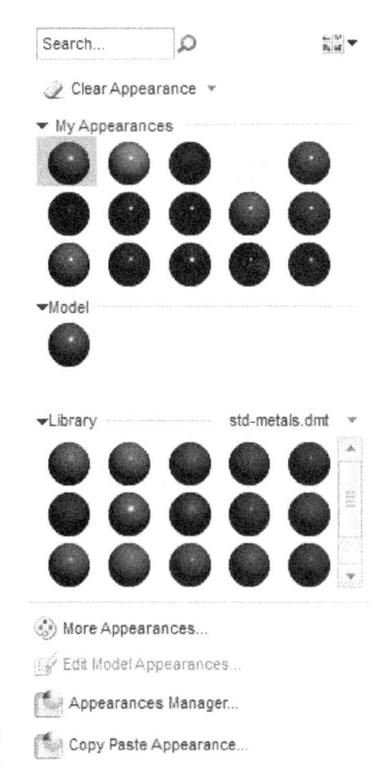

Figure 1-37 *Appearance Gallery*

Clear Appearance

This option is used to remove the selected appearance from the model. To do so, select the appearance and then choose this option. Alternatively, first choose this option and then select the appearance to be removed from the model.

To remove all appearances assigned to a model, click on the black arrow displayed besides the **Clear Appearance** button; a flyout will be displayed. Next, choose the **Clear All Appearances** option from this flyout.

My Appearance

The **My Appearance** area displays all appearances stored in the startup directory. You can also add new appearances to the palette, edit the existing appearance, or delete the selected appearance in the **Appearance** area. To do so, right-click on the appearance icon and then choose the required option from the shortcut menu displayed. Note that you cannot delete the active appearance.

Model

The **Model** area consists of the appearances stored and used in an active model. You can edit the selected appearance, add new appearances, or select the objects to apply the selected appearance by using the shortcut menu as discussed in the **My Appearance** area.

Library

This area displays the predefined appearances in the current library. The library contains appearances from the System library and the Photolux library. Appearances in the library are grouped in various classes such as metals, glass, wood, and so on, based on their similar characteristics. The name of the current library is displayed on the right of the **Library** head. To assign appearances from other libraries, click on the library name; a flyout will be displayed with a tree list of classes and libraries. Browse through the tree and select the required library; all appearances in the selected library will be listed. From the list displayed, you can assign an appearance to the model as well as create a new appearance, as discussed earlier.

More Appearances

Choose the **More Appearances** button from the **Appearance Gallery**; the **Appearance Editor** will be displayed, as shown in Figure 1-38. The **Appearance Editor** provides options for editing the appearance as well as for advanced rendering, thereby providing much more realistic images.

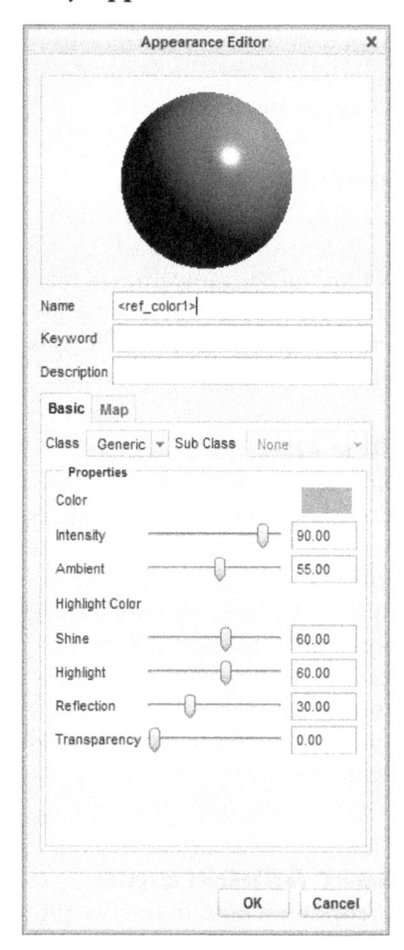

Figure 1-38 The ***Appearance Editor***

Using the **Appearance Editor**, you can edit the name, description, and keywords of the selected appearance, except the default one. The advanced options are given under two tabs, **Basic** and **Map**. The **Basic** tab provides various classes and sub-classes of appearances. By default, **Generic** is selected in the **Class** drop-down list. However, you can select the required class and their sub-classes. Also, the properties related to selected class such as color, illumination, reflection, transparency, glossiness, and so on are displayed in the **Properties** area. These properties can be modified by adjusting the sliders given next to them.

The **Map** tab allows you to define and apply texture such as wood, fabrics, metals, and so on to the model. You can preview the changes made to the model on changing the settings in the **Appearance Editor**.

Edit Model Appearances

On choosing the **Edit Model Appearances** button; the **Model Appearance Editor** will be displayed. In this editor, all appearances used in the model will be displayed along with their properties. The **Model Appearance Editor** is same as the **Appearances Editor**. Using this editor, you can modify the properties such as color, texture, shine, and gloss of the model.

Appearances Manager

Choose this button to display the **Appearances Manager** window. The **Appearances Manager** window displays two panels. The left panel displays the palette with appearances and libraries and the right panel displays the editor with the properties of the active appearance. You can set the required values in the **Appearances** palette and the **Appearances Editor**. Also, you can open an existing library file and add and save your own textured file.

RENDERING IN PTC Creo Parametric

Rendering is a process of generating two-dimensional image of a three-dimensional scene or an object to make it more realistic. A rendered image makes it easier to visualize the shape and size of 3D object as compared to a wireframe or a shaded image. Rendering also helps you express your design intent to other people. You need to alter environments, lights, textures, and so on to get a high quality rendered image. Generally, the rendering process includes the following stages:

Loading a File

The first step is to load the solid model or the assembly to be rendered.

Applying Appearances

Next step is to apply the appearances to the model. You can do so by using the **Appearance Gallery**, which has already been discussed.

Applying Scenes

After applying the appearances, you need to apply scenes to the model. Scenes are predefined render settings that include lights, rooms, and environmental effects. To set scenes, choose the **Scene** tool from the **Scene** group of the **Render** tab in the **Ribbon**; the **Scenes** dialog box will be displayed, as shown in Figure 1-39. Using this window, you can apply the predefined settings, edit the current settings, create new scenes, copy a scene, and save the scenes for future use. To apply a scene to the model, double-click on the required scene from the **Scene Gallery** area of the **Scenes** window.

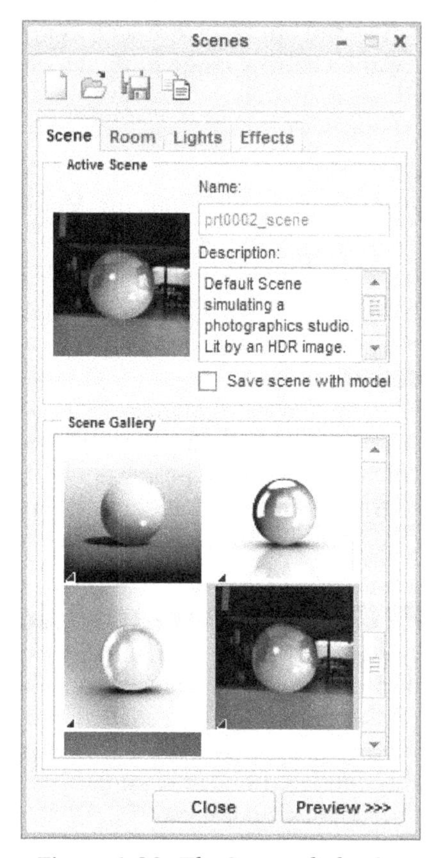

Figure 1-39 The ***Scenes*** *dialog box*

Creating the Rendering Room

A room sets the stage for rendering the model. The room can be cylindrical or rectangular. To create a rendering room, choose the **Room** tab from the **Scenes** window. Using the options in the **Room** tab, you can apply appearances to the wall, ceiling, and floor of the room, set the room orientation with respect to the model, adjust the size of the room, and create a background of the room, if required.

Applying Lights

The next step is to apply the lighting effects on the model. By adding lights to the model, you can illuminate the scene from various angles. To add lights, choose the **Lights** tab from the **Scenes** window. You can create a light source as well as position it by using the options in the **Lights** tab.

Shadows are enabled as soon as you add lights to the model. The direction of shadows cast by a light bulb, spot light, skylight, or environment type of lights is towards the center of the room.

Adding Effects

Finally, you can add the surrounding effects such as reflection, tone mapping, background, and depth of field. The tone of the image or scene can be set as per the presets such as **Studio Settings**, **Indoor Settings**, and **Outdoor Settings**.

Setting the Perspective View

You can set the perspective view of the model that can be used for rendering the model. To do so, choose the **Perspective view** button from the **Perspective** group in the **Render** tab of the **Ribbon**. To set the desired angle, invoke the **Perspective** dialog box by choosing the **Perspective Settings** button from the **Perspective** group in the **Render** tab of the **Ribbon**. Using this dialog box, you can adjust the perspective view settings such as **Walk Through**, **Fly Through**, **From To**, and **Follow Path**. You can also adjust the amount of perspective for viewing the model by adjusting the eye distance and the focal length.

Rendering the Model

The last step is to render the model. To render the model, choose the **Render Setup** button from the **Setup** group in the **Render** tab of the **Ribbon**; the **Render Setup** window will be displayed. Set the required rendering options in this window. Next, you need to render the model with specified settings. To do so, choose the **Render Window** button from the **Render** group in the **Render** tab of the **Ribbon**. If required, modify the render settings and render the model again.

Note

*You can toggle between the display of the rendering effects on the model by choosing the **Shading** button from the **Display Style** drop-down in the **Graphics** toolbar.*

COLOR SCHEME USED IN THIS BOOK

PTC Creo Parametric allows you to use various color schemes for the background screen color and for displaying the entities on the screen. This textbook will follow the white background of PTC Creo Parametric environment for the purpose of printing. However, for better understanding and clear visualization at various places, this book will follow other color schemes too. To change the color scheme of the background screen, choose **File > Options** from the menu bar; the **PTC Creo Parametric Options** window will be displayed. Select the **System Colors** category from the left pane of this window. Next, click on the node adjacent to the **Graphics** category of the **Colors** area in the right pane of this window; different object names will be displayed with color drop-downs adjacent to them, refer to Figure 1-40. Choose white color from the **Background** drop-down in this category and choose the **OK** button to apply the color.

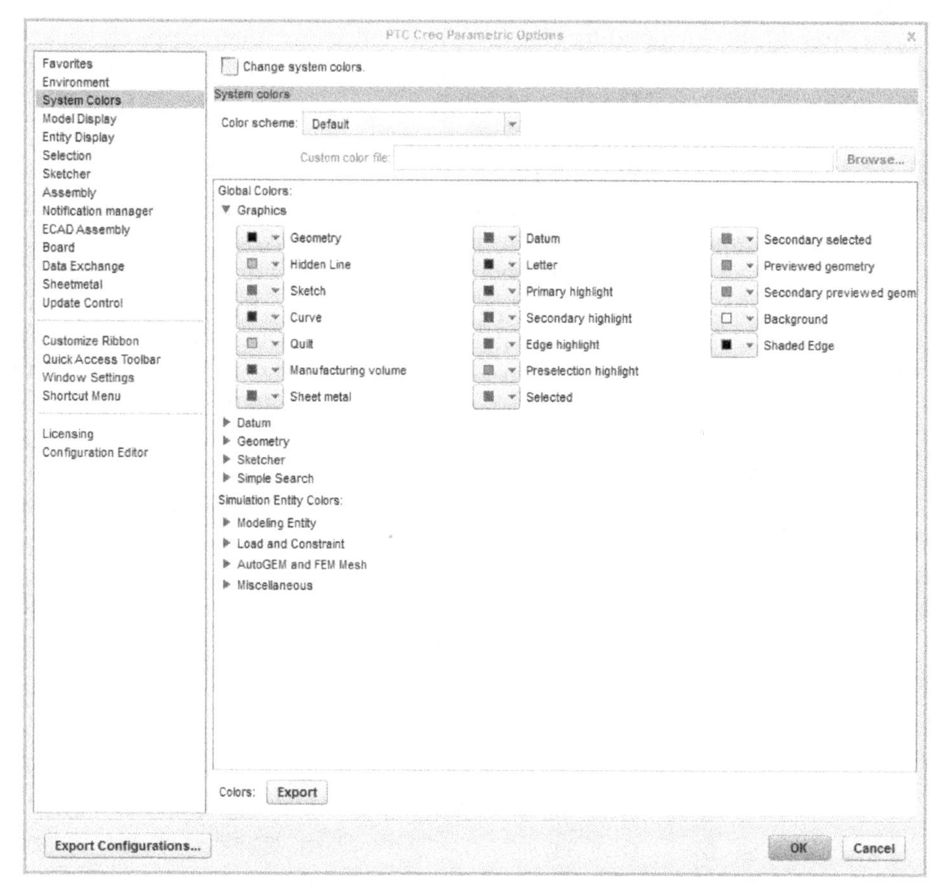

Figure 1-40 *The* ***System colors*** *category in the* ***PTC Creo Parametric Options*** *dialog box*

Chapter 2

Creating Sketches in the Sketch Mode-I

Learning Objectives

After completing this chapter, you will be able to:

- *Use various tools to create geometry.*
- *Dimension a sketch.*
- *Apply constraints to a sketch.*
- *Modify a sketch.*
- *Use the Modify Dimensions dialog box.*
- *Edit the geometry of a sketch by trimming.*
- *Mirror a sketch.*
- *Insert standard/user-defined sketches.*
- *Use the drawing display options.*

THE SKETCH MODE

Almost all models designed in PTC Creo Parametric consist of datums, sketched features, and placed features. For creating datums and placed features, you do not need to draw sketches. However, to create a three-dimensional (3D) feature, it is necessary to draw or import its two-dimensional (2D) sketch. When you enter the **Part** mode and select the options to create any sketched feature, the system automatically takes you to the sketcher environment. In the sketcher environment, the sketch of the feature is created, dimensioned, and constrained. The sketches created in the **Sketch** mode are stored in the *.sec* format. After creating the sketch, you need to return to the **Part** mode to create the required feature.

Note

You will learn about datums and placed features in later chapters.

In PTC Creo Parametric, a sketch can be drawn in the **Sketch** mode, in the sketcher environment or can be imported from other software. A designer can draw a 2D sketch of the product and assign the required dimensions and constraints to it. By assigning the dimensions, the designer can make sure that the 2D sketch of the product or model is satisfying the necessary conditions; later on the 3D model of the product can be created in the **Part** mode.

Working with the Sketch Mode

To create any section in the **Sketch** mode of PTC Creo Parametric, certain basic steps have to be followed. The following points outline the steps to draw a sketch in the **Sketch** mode:

1. Sketch the required section geometry

The different sketcher tools available in this mode can be used to sketch the required section geometry.

2. Add the constraints and dimensions to the sketched section

While sketching the section geometry, weak constraints and dimensions are automatically added to the section. The sketch can also be dimensioned and constrained manually. The dimensions that are applied manually are called strong dimensions. After adding the dimensions, you can modify them as required.

3. Add relations to the sketch if needed

The geometry of the sketch can be controlled by adding relations.

4. Regenerate the section

If the sketch is fully dimensioned and constrained, the sketch is automatically regenerated. Throughout this book, it is assumed that you are sketching in the **Sketch** mode with the **Intent Manager** turned on. PTC Creo Parametric has the capability to analyze the section, and if the section is not complete for any reason, the section will not be regenerated. You will learn about these reasons as you go through this chapter.

Tip: *Throughout this book, the sketcher environment is referred to the environment in PTC Creo Parametric where you can draw 2D geometries. Apart from the* ***Sketch*** *mode, the sketcher environment can be accessed in other modes of PTC Creo Parametric.*

Invoking the Sketch Mode

To invoke the **Sketch** mode, choose **New** from the **File** menu or choose the **New** button from the **Data** group in the **Home** tab of the **Ribbon**; the **New** dialog box will be displayed with different PTC Creo Parametric modes in the **Type** area. Select the **Sketch** radio button to start a new file in the **Sketch** mode, refer to Figure 2-1; a default name of the sketch file appears in the **Name** edit box. You can change the sketch name as required and then choose the **OK** button to enter the **Sketch** mode.

Figure 2-1 The New dialog box

THE SKETCHER ENVIRONMENT

When you invoke the **Sketch** mode, the initial screen displayed is similar to the one shown in Figure 2-2. This figure also shows the **Sketch** tab that will be displayed in the **Ribbon**. The drawing tools are available in the **Sketching** group of the **Sketch** tab. When you enter the sketcher environment, the **Intent Manager** is turned ON by default. Also, when you are in the selection mode, shortcut menus can be invoked by holding down the right mouse button in the drawing area. The options in these shortcut menus vary depending on the item selected. These shortcut menus also contain the tools to draw the sketches.

Note

*Datum planes are not displayed in the **Sketch** mode.*

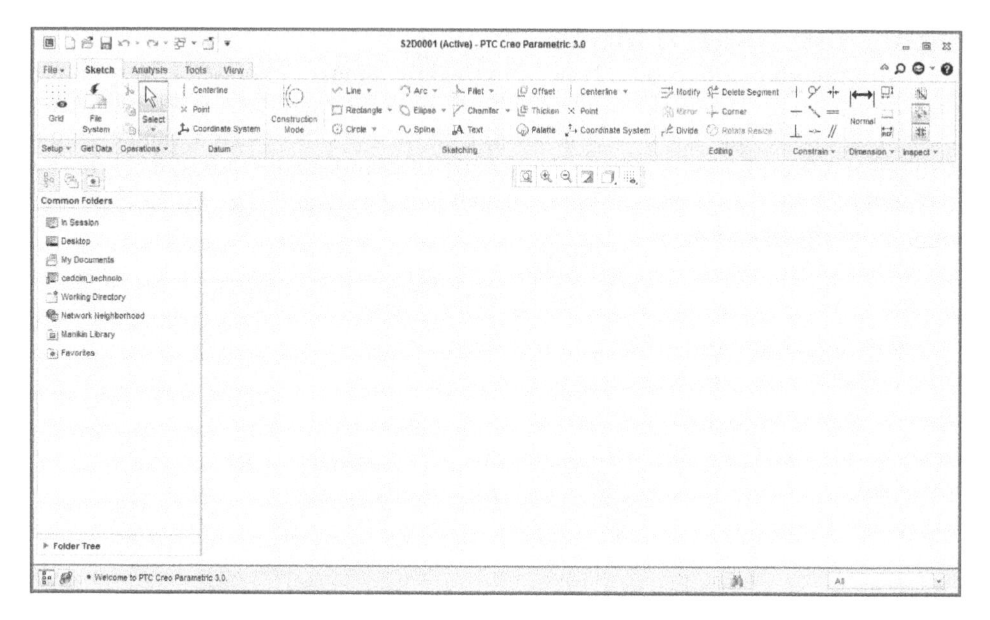

Figure 2-2 Initial screen appearance in the ***Sketch*** *mode*

The navigator is displayed on the left of the drawing area. In navigator, the **Folder Browser** tab is activated by default. It covers a part of the drawing area and therefore, the drawing area is decreased. You can increase the drawing area by clicking on the **Show Navigator** button, which is available at the bottom left corner of the window.

Note

The ***Folder Browser*** *tab is divided into two areas,* ***Common Folders*** *and* ***Folder Tree****. The functions of the* ***Common Folders*** *and* ***Folder Tree*** *areas have already been discussed in Chapter 1.*

WORKING WITH A SKETCH IN THE SKETCH MODE

When you invoke the sketcher environment, the **One-by-One** selection filter is chosen by default, in the **Select** drop-down. You can select other selection filters from the **Operations** group in the **Select** drop-down. Other selection filters available in this drop-down are the **Chain**, **All Geometry**, and **All**. By using the **One-by-One** selection filter, you can select each individual entity from the drawing area. By using the **Chain** selection filter, you can select a complete chain of entities linked with the selected entity. If the **All Geometry** selection filter is chosen then all the geometric entities available in the drawing area are selected automatically. If you choose the **All** selection filter then all entities available in the drawing area are selected automatically.

Note

1. The ***One-by-One*** *selection filter is activated by default. If you activate another selection filter, then that selection filter will get deactivated once it has been used and again the* ***One-by-One*** *selection filter will be activated.*

2. The sketch is saved with .sec file extension.

3. You can create a simple sketch by using the options available in the shortcut menu. To invoke the shortcut menu, press and hold the right mouse button in the drawing area. Note that, once the shortcut menu is displayed, the right mouse button can be released.

DRAWING A SKETCH USING THE TOOLS AVAILABLE IN THE SKETCH TAB

In the sketcher environment, the **Sketch** tab is active in the **Ribbon**, by default. There are various panels in this tab which contain the tools to draw a sketch, dimension it, and modify the dimensions. In this section, you will learn how to draw sketched entities using the tools available in the **Sketch** tab.

Creating a Point

Points are used to specify locations. You can use a point as reference for creating other geometric elements. In the sketcher environment of Creo, you can create sketch points and geometric points. The procedure of creating these is discussed next.

Creating a Sketching Point

Ribbon: Sketch > Sketching > Point

The sketching points can be used only in sketcher environment. The following steps explain the procedure to create a sketching point:

1. Choose the **Point** tool from the **Sketching** group; you will be prompted to select a location for the point.

2. Click in the drawing area; the point is placed at the specified location in the drawing area.

Creating a Geometric Point

Ribbon: Sketch > Datum > Point

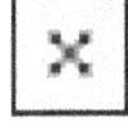

The **Point** tool available in the **Datum** group is used to create geometric point. The points created by using this tool can also be used as reference point in the Part modeling environment . The procedure to create a geometric point by using the **Point** tool from the **Datum** group is similar to the procedure of creating point from the **Sketching** group.

Note

1. To increase the number of visible command prompt lines in the message area, select the upper boundary line of the message area using the left mouse button and drag it upward, towards the screen.

2. When you place a single point, no dimensions appear. But, when you place two points, they are dimensioned with respect to each other.

Drawing a Line

To draw lines, there are two tools in the **Sketching** group. To view these tools, choose the down arrow on the right of the **Line** tools; a drop-down will appear with two tools. The first tool is the **Line Chain** tool. This tool is used to create a line or chain of lines by selecting two points in the drawing area. The second tool in the flyout is the **Line Tangent** tool. This tool is used to create a tangent line between two entities.

The procedure to create lines by using these tools is discussed next.

Drawing a Line Using the Line Chain Tool

Ribbon: Sketch > Sketching > Line > Line Chain

The following steps explain the procedure to create a line using the **Line Chain** tool available in the **Line** drop-down:

1. Choose the **Line Chain** tool. Click in the drawing area to start the line; a rubber-band line appears starting from the selected point with the other end attached to the cursor. The symbols **V** and **H** that appear while drawing the vertical and horizontal lines are called constraints. Constraints are discussed later in this chapter.

2. After specifying the start point for the line, move the cursor in the drawing area to the desired location and click to specify the endpoint of the line. The rubber-band line continues and you can draw the second line.

3. Repeat step 2 until all lines are drawn. You can end the line creation by pressing the middle mouse button. To abort the line creation, press the middle mouse button again. The start point and end point of the line or line chain are displayed in red colored dots.

Note

1. If you draw a single line, the color of the line drawn will be green. If you draw multiple lines, the color will be orange.

2. After drawing a line, when you press the middle mouse button twice to end the line creation, the line drawn is highlighted in green color. In the sketcher environment, the green color of an entity indicates that it is selected. If you press the DELETE key, the line will be erased from the drawing area. After drawing a line, weak dimensions are applied to the sketch and they appear in light blue color. These weak dimensions are applied automatically to the sketched entities as you draw them. The concept of weak dimensions is discussed later in this chapter.

Drawing a Line Using the Line Tangent Tool

Ribbon: Sketch > Sketching > Line > Line Tangent

The **Line Tangent** tool is used to draw a tangent line between two entities such as arcs, circles, or a combination of these. The following steps explain the procedure to draw a tangent using this tool:

1. Choose the **Line Tangent** tool from the **Line** drop-down in the **Sketching** group; you will be prompted to select the start point on an arc, a circle, or an ellipse.

2. Select the first entity from where the tangent line will be drawn; a rubber-band line appears with the cursor. Also, you will be prompted to select the end point on an arc, a circle, or an ellipse. On selecting the second entity, a line tangent to both the selected entities will be drawn.

Drawing a Centerline

You can draw a centerline by using the tools available in the **Centerline** drop-down. Click on the down arrow on the right of the **Centerline** tool; a menu will appear with two tools. The first tool is the **Centerline** tool, which is used to create a centerline by selecting two points in the drawing area. Centerline is used for creating revolved features, mirroring, and so on. The second tool is the **Centerline Tangent** tool that enables you to create a tangent centerline that can be referenced in the sketcher environment. The procedures to draw centerlines using these two tools are discussed next.

Drawing a Centerline Using the Centerline Tool

Ribbon: Sketch > Sketching > Centerline > Centerline

You can draw horizontal, vertical, or inclined centerlines using the **Centerline** tool. This tool is available in the drop-down that will be displayed when you choose the black arrow on the right of the **Centerline** tool in the **Sketching** group. The centerline in a sketch is used as an axis of rotation, for mirroring, aligning, and dimensioning entities. The steps discussed next explain the procedure to draw a centerline:

1. In the **Sketching** group, choose the **Centerline** tool from the **Centerline** drop-down; you will be prompted to select the start point.

2. Click in the drawing area to specify the start point; you will be prompted to select the end point.

3. Click in the drawing area to specify the endpoint; a centerline of infinite length is drawn.

Drawing a Centerline Using the Centerline Tangent Tool

Ribbon: Sketch > Sketching > Centerline > Centerline Tangent

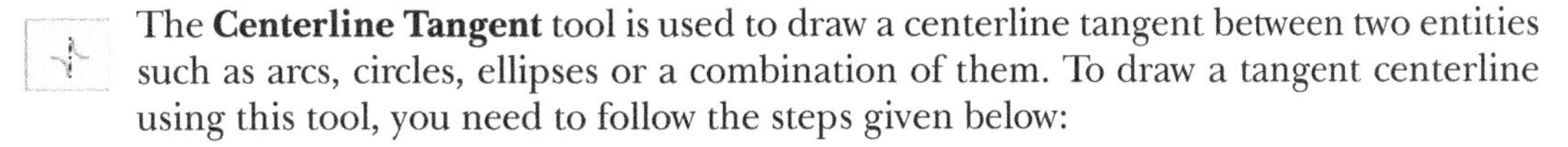

The **Centerline Tangent** tool is used to draw a centerline tangent between two entities such as arcs, circles, ellipses or a combination of them. To draw a tangent centerline using this tool, you need to follow the steps given below:

1. Choose the **Centerline Tangent** tool from the **Centerline** drop-down in the **Sketching** group; you will be prompted to select the start location on arc or circle.

2. Select the first entity from where the tangent line has to be drawn; a rubber-band line with the cursor will appear. Also, you will be prompted to select the end point on an arc

or circle. As soon as you select the second entity, a centerline with infinite length that is tangent to both the selected entities is drawn.

Drawing a Geometry Centerline

Ribbon: Sketch > Datum > Centerline

The **Centerline** tool available in the **Datum** group is used to create centerlines as a part of the geometry. The centerline created by using this tool can be referenced outside the sketcher environment. The procedure to create a centerline by using the **Centerline** tool from the **Datum** group or by using the **Centerline** tool from **Sketching** group is same. The only difference between them is that the centerline created by using the **Geometry Centerline** tool is a geometric entity and not a sketch. To draw a geometric centerline, choose the **Centerline** tool from the **Datum** group. Next, specify the start and end points of the geometry centerline; a centerline of infinite length will be created in the drawing area.

Drawing a Rectangle

Ribbon: Sketch > Sketching > Rectangle

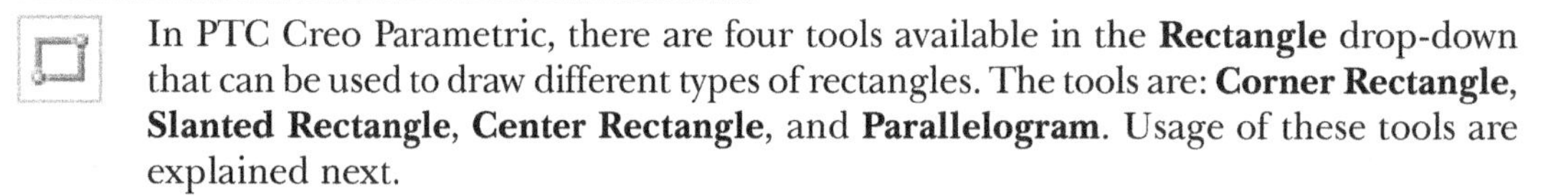

In PTC Creo Parametric, there are four tools available in the **Rectangle** drop-down that can be used to draw different types of rectangles. The tools are: **Corner Rectangle**, **Slanted Rectangle**, **Center Rectangle**, and **Parallelogram**. Usage of these tools are explained next.

Note

*When a non-overlapping closed sketch is drawn then it appears to be filled with light orange color. It indicates that the sketch is closed. This happens because, by default, the **Shade Closed Loops** button in the **Inspect** group of the **Sketch** tab is chosen. You can deactivate this button by choosing it.*

Creating a Corner Rectangle

Ribbon: Sketch > Sketching > Rectangle > Corner Rectangle

You can create a rectangle by using the two corner points. To do so, invoke the **Corner Rectangle** tool from the **Rectangle** drop-down in the **Sketching** group; you will be prompted to specify the first corner point of the rectangle. First corner point is the starting point of the rectangle and the second corner point is the end point of the rectangle. To create a rectangle by using the **Corner Rectangle** tool, you need to follow the steps given below:

1. Invoke the **Corner Rectangle** tool from the **Rectangle** drop-down; you will be prompted to select two points as corners of the rectangle. Click to specify the first point; a rubber-band box appears with the cursor attached to the opposite corner of the box.

2. Move the cursor to any non-collinear desired location in the drawing area and then click to specify the second point for the diagonal of the rectangle.

Creating a Slanted Rectangle

Ribbon: Sketch > Sketching > Rectangle > Slanted Rectangle

You can create an inclined rectangle by using the **Slanted Rectangle** tool. To create an inclined rectangle, you need to follow the steps given next.

1. Invoke the **Slanted Rectangle** tool from the **Rectangle** drop-down; you will be prompted to select two points to indicate the first side of the rectangle. Click to specify the first point; an orange rubber-band line will be displayed with the cursor attached to the end point of the line.

2. Click at any point to create the first side of the box; you will be prompted to specify the end point of the second line.

3. Move the cursor perpendicular to the first line in the drawing area and then click to specify the second side for the rectangle.

Creating a Center Rectangle

Ribbon: Sketch > Sketching > Rectangle > Center Rectangle

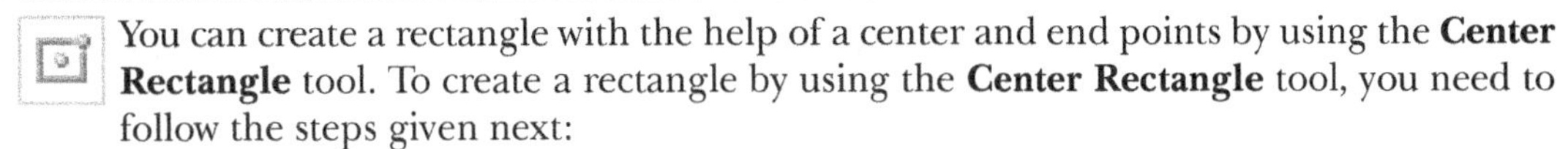

You can create a rectangle with the help of a center and end points by using the **Center Rectangle** tool. To create a rectangle by using the **Center Rectangle** tool, you need to follow the steps given next:

1. Invoke the **Center Rectangle** tool from the **Rectangle** drop-down available in the **Sketching** group; you will be prompted to specify the center point of the rectangle. Click to specify the center point; an orange rubber-band box appears with the cursor attached to the end point of the diagonal of the box.

2. Click at any desired point to create the rectangle.

Creating a Parallelogram

Ribbon: Sketch > Sketching > Rectangle > Parallelogram

You can create a parallelogram by using the **Parallelogram** tool. To create a parallelogram by using the **Parallelogram** tool, you need to follow the steps given next:

1. Invoke the **Parallelogram** tool from the **Rectangle** drop-down available in the **Sketching** group; you will be prompted to define the start point of the first side of parallelogram. Click to specify the start point; an orange rubber-band line will be displayed with the cursor attached to the end point of the line.

2. Click at any desired point to draw the first side of parallelogram; an orange rubber-band box will be displayed with the cursor attached to the end point of the second side of the parallelogram. Also, you will be prompted to specify the end point of the second line.

3. Click at any desired point to create the parallelogram.

Drawing a Circle

You can draw circles by using the **Center and Point**, **Concentric**, **3 Point**, and **3 Tangent** tools. These tools are available in the **Circle** drop-down in the **Sketching** group. The steps to create a circle by using these tools are given next.

Drawing a Circle Using the Center and Point Tool

Ribbon: Sketch > Sketching > Circle > Center and Point

The **Center and Point** tool is used to draw a circle by specifying the center of the circle and a point on its circumference. The following steps explain the procedure to draw a circle using this tool.

1. Choose the **Center and Point** tool; you will be prompted to select the center of the circle.

2. Click in the drawing area to specify the center point of the circle; you will be prompted to select a point on the circumference of the circle. Also, an orange rubber-band circle will be displayed with the center at the specified point and the cursor attached to its circumference.

3. Move the cursor to specify the size of circle. Click at a desired point to complete the creation of the circle; you will be prompted again to select the center of the circle.

4. Repeat steps 2 and 3 if you want to draw more circles, else press the middle mouse button to abort the process.

Drawing a Circle Using the Concentric Tool

Ribbon: Sketch > Sketching > Circle > Concentric

The following steps explain the procedure to draw a concentric circle using the **Concentric** tool:

1. Choose the **Concentric** tool from the **Circle** drop-down in the **Sketching** group; you will be prompted to select an arc to determine the center. You can select an arc or a circle to specify the center point.

2. Click on an arc or a circle to determine the concentricity of the circle to be drawn. Move the mouse and click at the required location to specify the size of circle.

3. To finish the creation of the circle, press the middle mouse button.

Drawing a Circle Using the 3 Point Tool

Ribbon: Sketch > Sketching > Circle > 3 Point

The following steps explain the procedure to draw a circle using the **3 Point** tool:

1. Choose the **3 Point** tool from the **Circle** drop-down; you will be prompted to specify the first point on the circle.

2. Click to specify the first point at the desired location in the drawing area; you will be prompted to select the second point on the circle. Move the cursor and click to specify the second point in the drawing area.

3. As you select the second point, an orange rubber-band circle appears with the cursor attached to it and you are prompted to select the third point. Move the mouse to size the circle and click to specify the third point; a circle is drawn and you are prompted again to select the first point on the circle to draw the next circle.

4. To abort the process of circle creation you can press the middle mouse button at any stage.

Drawing a Circle Using the 3 Tangent Tool

Ribbon: Sketch > Sketching > Circle > 3 Tangent

The **3 Tangent** tool is used to draw a circle tangent to three existing entities. This tool references other entities to draw a circle. The circle created using this tool is drawn irrespective of the points selected on the entities. The following steps explain the procedure to draw a circle using the **3 Tangent** tool:

1. Choose the **3 Tangent** tool from the **Circle** drop-down; you will be prompted to select the start location on an arc, circle, or line.

2. Select the first entity; the color of the entity changes to green and you will be prompted to select the end location on an arc, circle, or line. Select the second tangent entity; you will be prompted to select the third location on an arc, circle, or line. Select the third tangent entity; a circle tangent to these three entities is drawn.

3. To end the process of circle creation, press the middle mouse button.

Drawing a Construction Circle

Ribbon: Sketch > Sketching > Construction Mode

A construction circle is a circle that is used to align entities, create diametrical or radial dimensions, and to reference the entities. Figure 2-3 shows an application of a construction circle. In the sketch of a flange, centers of the circles lie on a particular bolt circle diameter (BCD) that is defined using a construction circle.

To create a construction circle, choose the **Construction Mode** toggle button and then draw the circle using tool available in the **Circle** drop-down. Alternatively, select a previously drawn circle. Then, hold the right mouse button to invoke the shortcut menu, as shown in Figure 2-4. Choose the **Construction** option from the shortcut menu; the circle will appear dotted in green color, indicating that it is a construction circle.

Note

*You can also create construction geometry by using the **Construction Mode** toggle button in the **Sketching** group. If you activate this button and then create a sketch entity, the entity thus formed will be a construction geometry.*

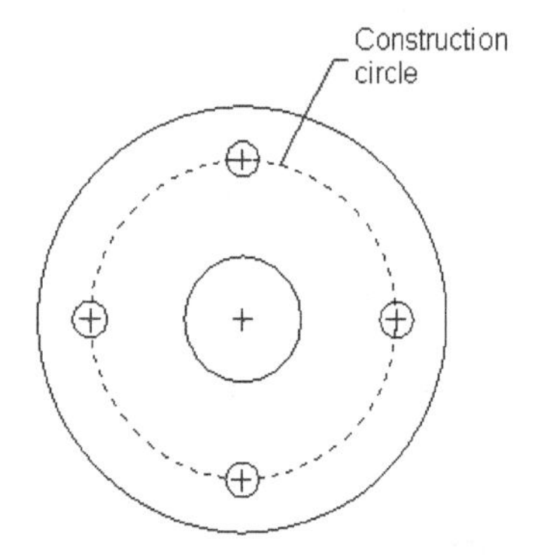

Figure 2-3 *Sketch of a flange*

Figure 2-4 *Partial view of **Construction** option in the shortcut menu displayed*

Tip: *To convert a construction circle back to a geometric entity, select the construction circle and hold the right mouse button to invoke a shortcut menu. From the shortcut menu, choose the **Geometry** option.*

Drawing an Ellipse

You can draw an ellipse by using the tools available in the **Ellipse** drop-down. The **Ellipse** drop-down is available in the **Sketching** group. The tools available in this drop-down are **Axis Ends Ellipse** and **Center and Axis Ellipse**. The steps to draw ellipse using these tools are discussed next.

Drawing an Ellipse Using the Axis Ends Ellipse Tool

Ribbon: Sketch > Sketching > Ellipse > Axis Ends Ellipse

The following steps explain the procedure to draw an ellipse by using the **Axis Ends Ellipse** tool:

1. Choose the **Axis Ends Ellipse** tool from the **Ellipse** drop-down; you will be prompted to select the start point of the major axis of the ellipse.

2. Click in the drawing area to specify the start point of the major axis; you will be prompted to select the endpoint of the major axis.

3. Click in the drawing area to specify the endpoint; an orange rubber-band ellipse will appear with the cursor attached to it. Also, you will be prompted to select a point on the minor axis to define the ellipse.

4. Click to specify the point; an ellipse will be created. You can press the middle mouse button to end the creation of the ellipse.

Drawing an Ellipse Using the Center and Axis Ellipse Tool

Ribbon: Sketch > Sketching > Ellipse > Center and Axis Ellipse

The following steps explain the procedure to draw an ellipse by using the **Center and Axis Ellipse** tool:

1. Choose the **Center and Axis Ellipse** tool from the **Ellipse** drop-down; you will be prompted to specify the center of the ellipse.

2. Click at the desired location in the drawing area to specify the center point; you will be prompted to select the endpoint of the major axis of the ellipse. Click in the drawing area to specify the point; an orange rubber-band ellipse will appear with the cursor attached to the ellipse. Move the cursor in the drawing area to size the ellipse.

3. Specify the endpoint on the minor axis of the ellipse; the ellipse is drawn. After exiting the command the dimensions for the major radius and the minor radius will be displayed in light blue color. The light blue color indicates that the dimensions are weak.

Drawing an Arc

There are five tools to draw an arc. These tools can be invoked from the **Arc** drop-down in the **Sketching** group. The procedures to draw arcs using these tools are discussed next.

Drawing an Arc Using the 3-Point / Tangent End Tool

Ribbon: Sketch > Sketching > Arc > 3-Point / Tangent End

The **3-Point / Tangent End** tool is used to draw arcs that are tangent from the endpoint of an existing entity, or by defining three points in the drawing area.

When you choose this tool to draw an arc from an endpoint, the **Target** symbol will be displayed as soon as you select an endpoint. This **Target** symbol is a green colored circle that is divided into four quadrants. The following steps explain the procedure to draw an arc from the endpoint of an existing entity by using this tool:

1. Choose the **3-Point / Tangent End** tool from the **Arc** drop-down; you will be prompted to select the start point of the arc.

2. Specify three points in the drawing area to draw an arc. If you want to draw an arc from the endpoint of an existing entity, select the endpoint of that entity. As soon as you select the endpoint, the **Target** symbol appears at the endpoint of the entity. Move the cursor along the tangent direction through a small distance, a rubber-band arc appears with one end attached to the endpoint of the entity and the other end attached to the cursor. Note that when you move the cursor out of the **Target** symbol perpendicular to the endpoint, an arc is drawn by specifying three points. In this case, the rubber-band arc does not appear, refer to Figure 2-5.

On the other hand, if you move the cursor out horizontally from one of the quadrants of the **Target** symbol, an arc is drawn tangent to the endpoint, as shown in Figure 2-6.

3. Move the cursor to the desired position in the drawing area to size the arc. Use the left mouse button to complete the arc.

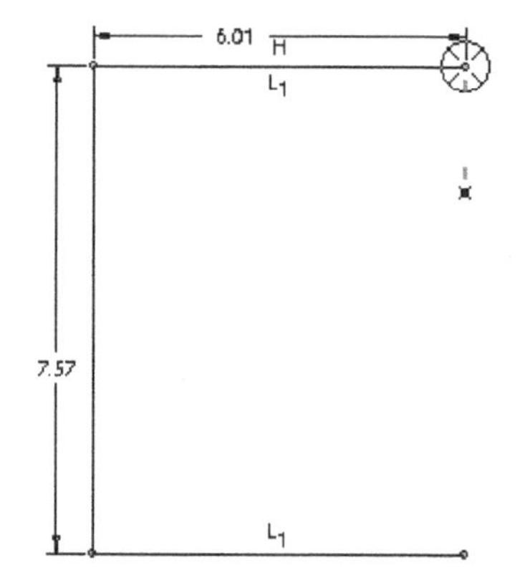

Figure 2-5 Cursor moved out of the ***Target*** *symbol perpendicular to the endpoint*

Figure 2-6 Cursor moved out of the ***Target*** *symbol along the tangent direction*

Tip: *If you do not want to draw a tangent arc, move the cursor out of the* ***Target*** *symbol perpendicular to the endpoint.*

Drawing an Arc Using the Center and Ends Tool

Ribbon: Sketch > Sketching > Arc > Center and Ends

The following steps explain the procedure to draw an arc using the **Center and Ends** tool:

1. Choose the **Center and Ends** tool from the **Arc** drop-down; you will be prompted to select the center of the arc.

2. Click to specify a center point for the arc in the drawing area; an orange colored center mark will appear at that point and you will be prompted to select the start point of the arc. As you move the cursor, a dotted circle appears attached to the cursor.

3. Specify the start point of the arc on the circumference of the dotted circle; an orange rubber-band arc will appear from the start point. The length of this arc will change dynamically as you move the cursor and you will be prompted to select the endpoint of the arc.

4. Move the cursor to specify the arclength and then click to select the endpoint of the arc; an arc will be drawn between the two specified points.

Note

You can draw only one arc with one center. If you want to draw another arc, you will have to select the center again.

Drawing an Arc Using the 3 Tangent Tool

Ribbon: Sketch > Sketching > Arc > 3 Tangent

The **3 Tangent** tool is used to draw an arc that is tangent to three selected entities. The following steps explain the procedure to draw an arc using this tool:

1. Choose the **3 Tangent** tool from the **Arc** drop-down; you will be prompted to select the start location on an arc, circle, or line.

2. As soon as you move the cursor on the first entity or select the first entity, the color of the entity will change to green and you will be prompted to select the end location on an arc, circle, or line.

3. On selecting the end location, you will be prompted to select a third location on an arc, circle, or line. Select the third entity; an arc is drawn tangent to the three selected entities.

You can continue drawing arcs or press the middle mouse button to abort arc creation.

Drawing an Arc Using the Concentric Tool

Ribbon: Sketch > Sketching > Arc > Concentric

The **Concentric** tool is used to draw an arc concentric to an existing arc. The entity selected must be an arc or a circle. The following steps explain the procedure to draw an arc using this tool:

1. Choose the **Concentric** tool from the **Arc** drop-down; you will be prompted to select an arc to determine the center of the arc to be created.

2. On moving the cursor on the first entity or selecting the entity; a dotted circle will appear on the screen and you will be prompted to select the start point of the arc. Click to specify the start point; an orange rubber-band arc will appear with one end attached to the start point. As you move the cursor, the length of the arc will change and you will be prompted to select the endpoint of the arc.

3. Click to specify the endpoint; the arc will be created.

You can continue drawing another arc or end the arc creation by pressing the middle mouse button.

Drawing an Arc Using the Conic Tool

Ribbon: Sketch > Sketching > Arc > Conic

The **Conic** tool is used to draw a conic arc. The following steps explain the procedure to draw a conic arc using this tool:

1. Choose the **Conic** tool from the **Arc** drop-down; you will be prompted to specify the start point of the conic entity.

2. Click to specify the start point in the drawing area; you will be prompted to specify the endpoint of the conic entity.

3. Click to specify the endpoint; a centerline will be drawn between the two points and you will be prompted to specify the shoulder point of the conic. Specify a point on the screen; the conic arc will be drawn.

Note

1. If you delete the centerline of the conic arc, the arc will not be deleted.

2. Note that if the conic arc is the only entity in the drawing area, then you cannot delete its centerline.

DIMENSIONING THE SKETCH

After you draw a sketch, the next step involves the dimensioning of the sketch. The basic purpose of dimensioning in PTC Creo Parametric is to control the size of the sketch and to locate it with some reference. In PTC Creo Parametric, a sketch cannot be regenerated unless it is fully dimensioned and constrained. The phrase "the sketch cannot be regenerated" means that the sketch is not accepted by PTC Creo Parametric.

By default, sketched entities are dimensioned and constrained automatically while sketching. However, you need to add additional dimensions to the sketch. Tools for dimensioning are available in the **Dimension** group of the **Sketch** tab, refer to Figure 2-7. The **Normal** tool in this group is used to manually dimension the entities.

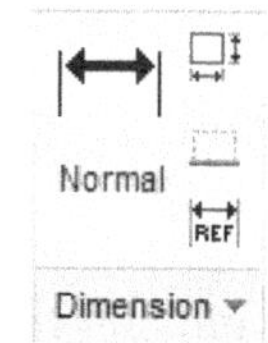

Figure 2-7 The ***Dimension*** *group*

Note

If you do not want the weak dimensions to be applied automatically then, clear the ***Show weak dimensions*** *check box from* ***File > Options > Sketcher > Object display settings****.*

Converting a Weak Dimension into a Strong Dimension

As discussed earlier, when you draw a sketch, some weak dimensions are automatically applied to the sketch. As you proceed to complete the sketch, these dimensions are automatically deleted from the sketch without any confirmation.

Select a weak dimension from the drawing area; the selected dimension will be highlighted in green. Press and hold the right mouse button to invoke the shortcut menu, as shown in Figure 2-8. Choose the **Strong** option from the shortcut menu and then click outside using the middle mouse button. Alternatively, press CTRL+T to convert the selected dimension to a strong dimension. Strong dimension is displayed in blue color.

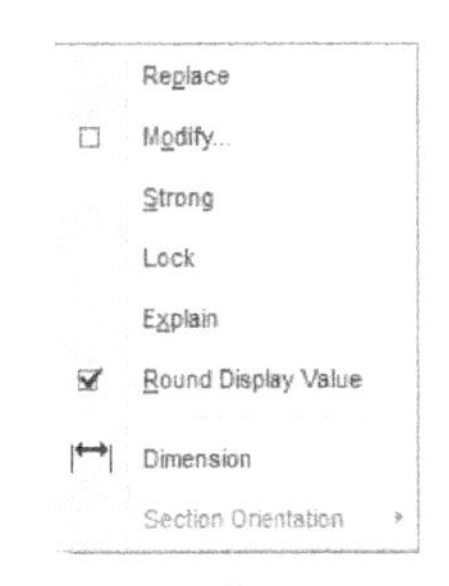

Figure 2-8 Shortcut menu to convert the weak dimensions to strong

Dimensioning a Sketch Using the Normal Tool

Ribbon: Sketch > Dimension > Normal

The **Normal** tool is used for normal dimensioning of the sketch. The following steps explain the procedure to dimension a sketch using this option:

1. Choose the **Normal** tool from the **Dimension** group. Click on the entity you want to dimension; the color of the entity changes from orange to green.

2. Move the cursor and place the dimension at the desired place by using the middle mouse button. You can modify the dimension values using the modifying options that will be discussed later in this chapter.

DIMENSIONING THE BASIC SKETCHED ENTITIES

Choose the **Normal** tool and follow the procedures given below to dimension the sketched entities.

Linear Dimensioning of a Line

You can dimension a line by selecting its endpoints or by selecting the line. After selecting the two endpoints or the line, press the middle mouse button to place the dimension. If the line is inclined and you select the two endpoints to dimension, then the location where you press the middle mouse button is important, because it defines the orientation of the dimension that will be displayed on the screen.

Figure 2-9 shows the three possible orientations of dimension that can be displayed when you dimension a line.

Note

It is not possible to dimension a line in three orientations simultaneously in the sketcher environment. The dimensions in Figure 2-9 are shown for explanation purpose only.

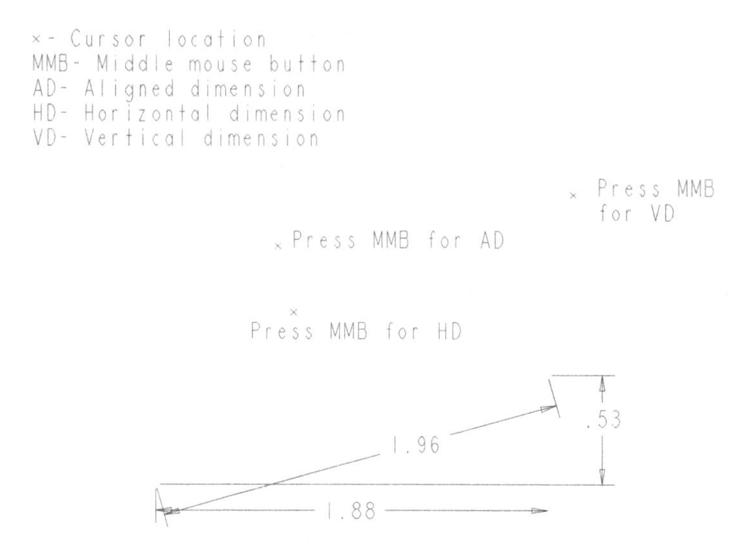

Figure 2-9 Approximate locations of the cursor to achieve different dimensions

Angular Dimensioning of an Arc

To add angular dimension to an arc, select both ends of the arc by using the left mouse button and then select a point on the arc. Next, place the dimension at the desired point by pressing the middle mouse button. The dimension placed can either be arclength or angular. If the dimension placed is arclength and you want the angular dimension then select the dimension, right-click and then choose the **Convert to Angle** option from the shortcut menu displayed. Figure 2-10 shows the angular dimension of an arc. You can modify the dimension by using other tools as well. These methods are discussed later on.

Diameter Dimensioning

For diameter dimensioning, click on a circle twice. Then place the dimension at the desired location by pressing the middle mouse button. The diameter dimension will be displayed, as shown in Figure 2-11. This method can also be used for dimensioning arcs.

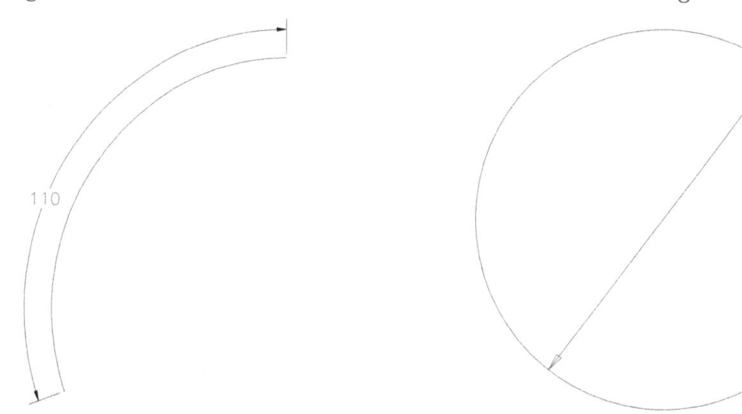

1.67

Figure 2-10 Angular dimensioning

Figure 2-11 Diameter dimensioning

Radial Dimensioning

For radial dimensioning, click on the entity once. Then, place the dimension at the desired location by pressing the middle mouse button. The radial dimension will be displayed, as shown in Figure 2-12.

Dimensioning Revolved Sections

Revolved sections are used to create revolved features such as flanges, couplings, and so on. To dimension a revolved section, click on the entity to be dimensioned. Next, select the centerline about which you want the section to be revolved. Once again, select the original entity that you want to dimension. Now, place the dimension at the desired location by pressing the middle mouse button. Figure 2-13 shows the dimension placed in a revolved section. This dimension represents the diameter of a revolved section.

Tip: *To add dimension to a revolved section, you can also first select the centerline. Next, the entity to dimension, and then again the centerline.*

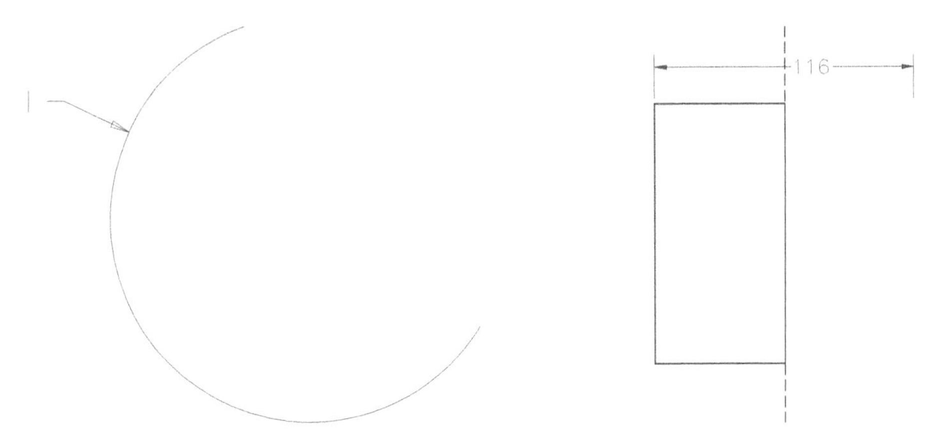

Figure 2-12 *Radial dimensioning*

Figure 2-13 *Dimensioning the revolved sections*

WORKING WITH CONSTRAINTS

In PTC Creo Parametric, the entities in a sketch have to be fully specified in terms of size, shape, orientation, and location. This is achieved by setting constraints. Using constraints in a sketch reduces the number of dimensions in that sketch.

Constraints are the logical operations that are performed on the selected geometry to make it more accurate in defining its position with respect to the other geometry. For example, if a line is nearly parallel to another line, PTC Creo Parametric snaps the parallel line and displays the parallel constraint symbol. Now, if you confirm the line creation, the line is drawn parallel to the other line. You can also apply constraints manually.

Types of Constraints

There are two types of constraints in PTC Creo Parametric: **Geometry** and **Assembly**. In this chapter, you will learn about the **Geometry** constraints only and the **Assembly** constraints will be discussed in later chapters.

The geometric constraints are available in the **Constrain** group of the **Sketch** tab. The tools available in this group are shown in Figure 2-14.

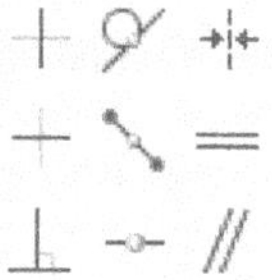

The constraints in this group are used to apply constraints manually. Although the constraints are applied automatically as you draw the sketch, you can use the tools in this group if you want to manually apply additional constraints to the sketch. The constraints in the group are discussed next.

Figure 2-14 *The constraints in the* ***Constrain*** *group*

Vertical

This constraint forces the selected line segment to become a vertical line. This constraint also forces the two vertices to be placed along a vertical line.

Horizontal

This constraint forces the selected line segment or two vertices that are apart by some distance to become horizontal or to lie in a horizontal line.

Perpendicular

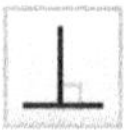

This constraint forces the selected entity to become normal to another selected entity.

Tangent

This constraint forces the two selected entities to become tangent to each other.

Mid-point

This constraint forces a selected point or vertex to lie on the middle of a line.

Coincident

This constraint can be used to force the two selected points to become coincident to constrain a point on the selected entity, and to make two selected entities collinear, so that they lie on the same line.

Symmetric

This constraint makes a section symmetrical about the centerline. When you select this constraint, you will be prompted to select a centerline and two vertices to make them symmetrical.

Equal

This constraint forces any two selected entities to become equal in dimension. When you select this constraint, you will be prompted to select two lines to make their lengths equal, or you will be prompted to select two arcs, circles, or ellipses to make their radii equal.

Parallel

This constraint is used to force two lines to become parallel. When selected, this constraint prompts you to select two entities that you want to make parallel.

Disabling the Constraints

The need to disable a constraint arises while drawing an entity. For example, consider a case where you want to draw a circle at some distance apart from another circle. While drawing it, the system tends to apply the equal radius constraint when the sizes of the two circles become equal. At this moment, if you do not want to apply the equal radius constraint, right-click twice to disable the equal radius constraint; a green line / will appear across the symbol. However, if you right-click once, the constraint will get locked and a circle will appear as its symbol.

MODIFYING THE DIMENSIONS OF A SKETCH

There are three ways to modify the dimensions of a sketch. These methods are discussed next.

Using the Modify Tool

Ribbon: Sketch > Editing > Modify

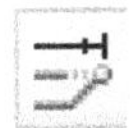

You can select one or more dimensions from the sketch to modify them. When you select dimension(s) from a sketch, they are highlighted in green. If you want to select more than one dimension, hold the CTRL key and select the dimensions by clicking on them. You can also use the CTRL+ALT+A keys or define a window to select the dimensions in the sketch. Next, choose the **Modify** tool from the **Editing** group to modify the dimensions; the **Modify Dimensions** dialog box will be displayed, as shown in Figure 2-15.

To modify dimensions by using this dialog box, you can either enter a value in the edit box or use the thumbwheel available on the right of the edit box. The **Sensitivity** slider is used to set the sensitivity of the thumbwheel.

By default, the **Regenerate** check box is selected. As a result, any modifications in the dimensions are automatically updated in the sketch. If you want to modify the dimensions of the sketch without regenerating the sketch you need to clear this check box. If this check box is cleared, the dimensions will not be

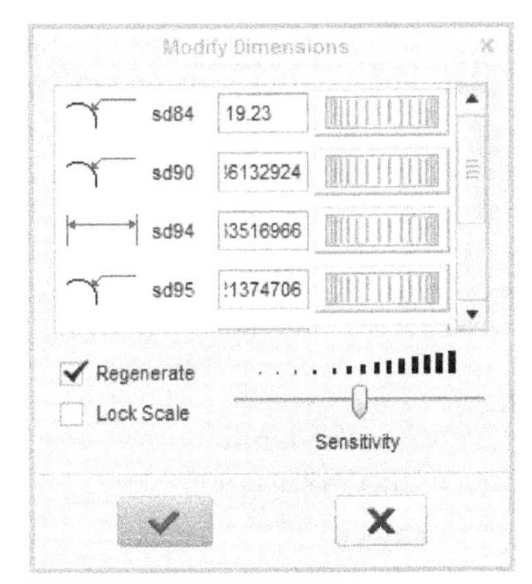

*Figure 2-15 The **Modify Dimensions** dialog box*

modified until you exit this dialog box. This means that PTC Creo Parametric allows you to make multiple modifications before updating the sketch.

Tip: *It is recommended that you clear the* ***Regenerate*** *check box and then modify the dimensions if you need to modify more than one dimension. Also, when the dimensions you specify vary drastically with the original ones.*

The **Lock Scale** check box is used to lock the scale of the selected dimensions. After locking the scale, if you modify any dimension, all other dimensions will also be modified by the same scale.

Modifying a Dimension by Double-Clicking on it

You can also modify a dimension by double-clicking on it. When you double-click on a dimension, the pop-up text field appears. Enter a new dimension value in this field and press ENTER or use the middle mouse button. Remember that you can select a dimension only when you choose **One-by-One** selection filter.

Modifying Dimensions Dynamically

In the sketcher environment, PTC Creo Parametric is always in the selection mode, unless you have invoked some other tool. When you bring the cursor to an entity, the color of the entity changes to green. Now, if you hold down the left mouse button, you can modify the entity by dragging the mouse. You will notice that as the entity is modified, the dimensions referenced to the selected entity are also modified.

RESOLVE SKETCH DIALOG BOX

While applying constraints or dimensions, sometimes the system may prompt you to delete one or more highlighted dimensions or constraints. This is because while adding dimensions or constraints, some strong dimensions or constraints conflict with the existing dimensions or constraints. As soon as a conflict occurs, the **Resolve Sketch** dialog box is displayed, as shown in Figure 2-16. When you select a dimension or a constraint from the **Resolve Sketch** dialog box, the corresponding dimension or constraint in the drawing area is enclosed in a blue box. The buttons in the **Resolve Sketch** dialog box are discussed next.

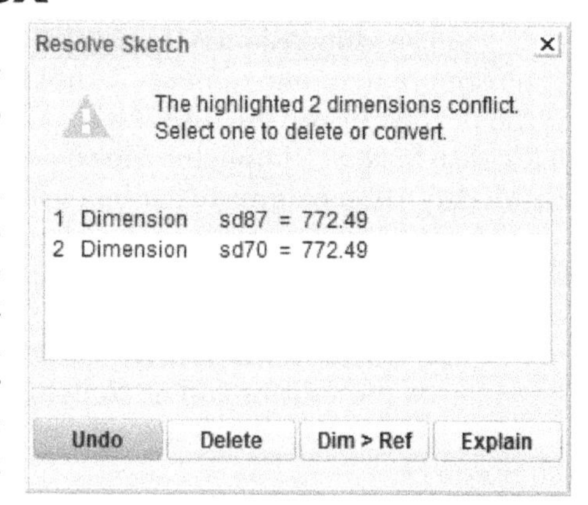

Figure 2-16 The ***Resolve Sketch*** *dialog box*

Undo

When you choose the **Undo** button, the section is brought back to the state that it was in just before the conflict occurred.

Delete

The **Delete** button is used to delete a selected dimension or constraint that is enclosed within the blue box. To delete dimension or the constraint select them from the blue box and choose the **Delete** button from the **Resolve Sketch** dialog box.

Dim > Ref

On choosing the **Dim > Ref** button, the selected dimension is converted to a reference dimension.

Note

The reference dimensions are used only for reference and are not considered in feature creation.

Explain

When you choose the **Explain** button, the system provides you with the information about the selected constraint or dimension. The information will be displayed in the message area.

DELETING THE SKETCHED ENTITIES

To delete a sketched entity, select it by defining a window. You can specify a window by picking two points so that the entity or entities are enclosed in the window. After specifying the window, the color of the selected entity changes to green. Hold down the right mouse button in the drawing area to invoke a shortcut menu. Now, choose the **Delete** option from this menu to delete the selected item.

You can also delete an item by selecting it and pressing the DELETE key.

To delete more than one item from the drawing area, press the CTRL key and click to select the entities to be deleted. Press the DELETE key to delete the selected entities. You can also specify a window to select the entities.

Note

It is necessary to be in the selection mode while selecting the items. The term "items" used in this chapter refers to dimensions and entities. The ***Geometry****,* ***Dimension****, and* ***Constraint*** *filters are available in the drop-down list located in the* ***Status Bar****. These filters narrow down your search and help you select the exact item. This means, if you want to select all constraints in the sketch, choose the* ***Constraint*** *filter and specify a window to select. You will notice that only the constraints are selected.*

To restore the last deleted item, choose the **Undo Delete** button. This button is available in the **Quick Access** toolbar.

TRIMMING THE SKETCHED ENTITIES

While creating a design, there are a number of places where you need to remove the unwanted and extended entities. You can do this by using the trimming tools that are available in the **Editing** group. These tools are discussed next.

Delete Segment

Ribbon: Sketch > Editing > Delete Segment

This tool is used to trim the entities that extend beyond the point of intersection. It is also used to delete the selected entities. After choosing the **Delete Segment** tool, when you move the cursor over an entity, the entity will be highlighted in green color. Press the left mouse button to trim the entity.

Corner

Ribbon: Sketch > Editing > Corner

The **Corner** tool is used to trim two entities at their corners. Note that when you trim entities using this option, the portion from where you select the entities is retained and the other portion is trimmed. The following steps explain the procedure to trim entities using this button:

1. Choose the **Corner** tool from the **Editing** group; you will be prompted to select two entities to be trimmed.

2. Click to select the two entities on the sides that you want to keep after the trimming action, see Figure 2-17. These two entities must be intersecting entities. The entities are trimmed from the point of intersection.

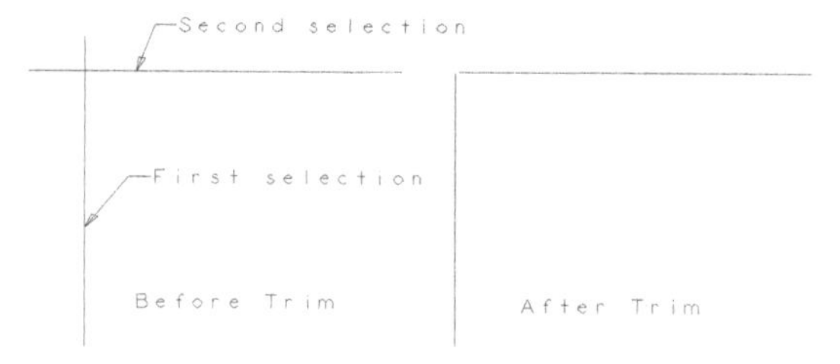

Figure 2-17 *Trimming the lines*

Divide

Ribbon: Sketch > Editing > Divide

The **Divide** tool is used to divide an entity into a number of parts or entities by specifying points on the entity.

The following steps explain the procedure to divide an entity:

1. Choose the **Divide** tool from the **Editing** group; you will be prompted to specify a point on an entity.

2. Click to select the entity at the point where you want to divide it; the entity is divided into two different entities. They can now be treated as two separate entities.

3. Similarly, you can break other entities like circles or arcs into several small entities.

MIRRORING THE SKETCHED ENTITIES

Ribbon: Sketch > Editing > Mirror

The **Mirror** tool is used to mirror sketched geometries about a centerline. This tool helps to reduce the time used for creation of symmetrical geometries and dimensioning them.

The following steps explain the procedure to mirror the sketched geometry:

1. Sketch a geometry and then sketch a centerline about which you need to mirror the geometry.
2. Select the entities that you need to mirror; the selected entities turn green in color.
3. Choose the **Mirror** tool from the **Editing** group; you will be prompted to select the centerline about which you need to mirror. Select the centerline; the selected entities will be mirrored about the centerline.

Tip: *In case of symmetrical parts, you can save time involved in dimensioning a sketch by dimensioning half of the section and then mirroring it. PTC Creo Parametric will assume that the mirrored half has the same dimensions as the sketched half.*

INSERTING STANDARD/USER-DEFINED SKETCHES

Ribbon: Sketch > Sketching > Palette

This tool helps you insert certain standard or user defined features such as polygons, profiles, shapes, stars, and other previously created sketches in the **Sketch** mode, thus minimizing the time for repetitive sketching.

The steps that explain the procedure to insert a foreign entity in the **Sketch** mode are given next.

1. Choose the **Palette** tool from the **Sketching** group; the **Sketcher Palette** dialog box will be displayed. The options in this dialog box are used to insert a previously created sketch.

Note

*If you have selected a working directory which contains only .sec files, a tab with the name of the working directory will be available in the **Sketcher Palette** dialog box. Also, this tab is chosen by default. On the other hand, if you have selected a working directory which contains other types of files, the same tab will be displayed in the end. In such cases, the **Polygon** tab will be chosen by default, as shown in Figure 2-18.*

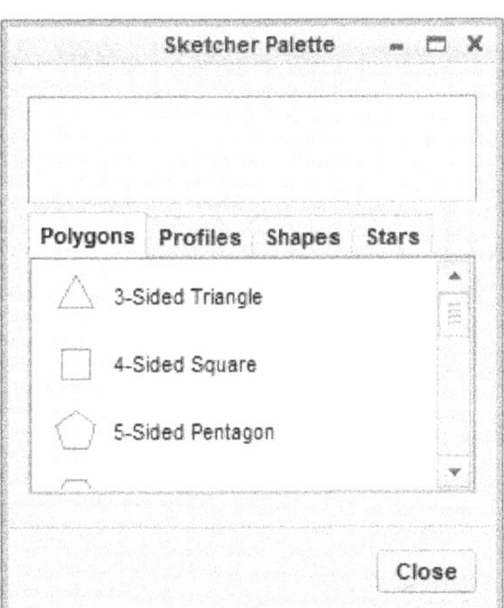

***Figure 2-18** The **Sketcher Palette** dialog box*

2. For inserting a sketch from the working directory tab into the sketch, double-click on the required sketch from the list in the dialog box; a + sign will be attached to the cursor. Click anywhere in the drawing area; the **Import Section** tab will be displayed in the **Ribbon**. Also, the move ⊗, rotate, and scale handles will be displayed on the imported sketch automatically.

3. In the **Import Section** tab, enter the scale value in the **Scale** edit box and the rotational angle value in the **Rotate** edit box. Alternatively, left-click on the required handle on the sketch and drag; the corresponding action will take place dynamically. The move handle also acts as the pivot point for rotation and scale. However, you can relocate the pivot point. To move the pivot point, right-click on the pivot point and drag it to the required location and then release the right-click; the pivot point will be relocated.

4. Next, choose the **Done** button to exit.

5. Choose the **Close** button from the **Sketcher Palette** dialog box to accept the sketch inserted. Else, repeat the steps 2-4 to continue inserting more sketches in the drawing area.

   Similarly, you can add the sketches from the **Polygon**, **Profiles**, **Shapes**, and **Stars** tabs.

DRAWING DISPLAY OPTIONS

While working with complex sketches, sometimes you need to increase the display of a particular portion of a sketch so that you can work on the minute details of the sketch. For example, you are drawing a sketch of a piston and you have to work on the minute details of the grooves for the piston rings. To work on these minute details, you have to enlarge the display of these grooves. You can enlarge or reduce the drawing display using various drawing display tools provided in PTC Creo Parametric. These tools are available in the **View** tab. Some of these drawing display options are discussed next. The remaining drawing display options will be discussed in the later chapters.

Zoom In

Ribbon: View > Orientation > Zoom In

This tool is used to enlarge the view of the drawing on the screen. After choosing the **Zoom In** tool from the **Orientation** group, you will be prompted to define a box. The area that you will enclose inside the box will be enlarged and displayed in the drawing area. Note that when you enlarge the view of the drawing, the original size of the entities is not changed. To exit the **Zoom** tool, right-click in the drawing area.

Zoom Out

Ribbon: View > Orientation > Zoom Out

This tool is used to reduce the view of the drawing on the screen, thus increasing the drawing display area. Each time you choose this button, the display of the sketch in the drawing area is reduced. This button is available in the **Orientation** group.

Refit

Ribbon: View > Orientation > Refit

This tool is available in the **Orientation** group. This tool is used to reduce or enlarge the display such that all entities that comprise the sketch are fitted inside the current display. Note that the dimensions may not necessarily be included in the current display.

Repaint

Ribbon: View > Display > Repaint

While working with complex sketches, some unwanted temporary information is retained on the screen. The unwanted information may include the shadows of the deleted sketched entities, dimensions, and so on. This unwanted information can be removed from the drawing area by using the **Repaint** button available in the **Display** group or from the **Graphics** toolbar in model display.

Sketcher Display Filters

While working with the sketches, sometimes you need to disable the display of some of the sketcher components such as dimensions, constraints, vertices, and so on. To disable or enable their display, you can use the toggle buttons available in the **Display** group of the **View** tab. These buttons are discussed next.

Disp Dims

Ribbon: View > Display > Disp Dims

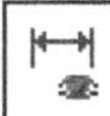

You can use this button to enable or disable the display of dimension on the screen.

Disp Constr

Ribbon: View > Display > Disp Constr

You can use this button to enable or disable the display of geometric constraints.

Disp Verts

Ribbon: View > Display > Disp Verts

You can use this button to enable or disable the display of vertices on a sketch or a model.

Disp Grid

Ribbon: View > Display > Disp Grid

You can use this button to enable or disable the display of sketching grid. Sometimes you may not be able to see the grid as it becomes very dense. To see the grid in such cases, you need to zoom in the screen.

Note

*1. To remove the temporary information, you can repaint the screen by choosing the **Repaint** tool from the **In-graphics** toolbar or pressing the CTRL+R keys.*

2. If you have a mouse that has a scroll wheel, then scrolling the wheel will zoom in and out the view. One more way to zoom in and out is to use the middle mouse button and the CTRL key. When you press CTRL+middle mouse button and drag the mouse upward, the sketch is zoomed out and when you drag the mouse downward, the sketch is zoomed in.

*3. In the **Sketch** mode, you can pan the sketch using the middle mouse button but in the **Part** mode, use SHIFT+middle mouse button to pan the model.*

TUTORIALS

Tutorial 1

In this tutorial, you will draw the model shown in Figure 2-19. The sketch of the model is shown in Figure 2-20. **(Expected time: 30 min)**

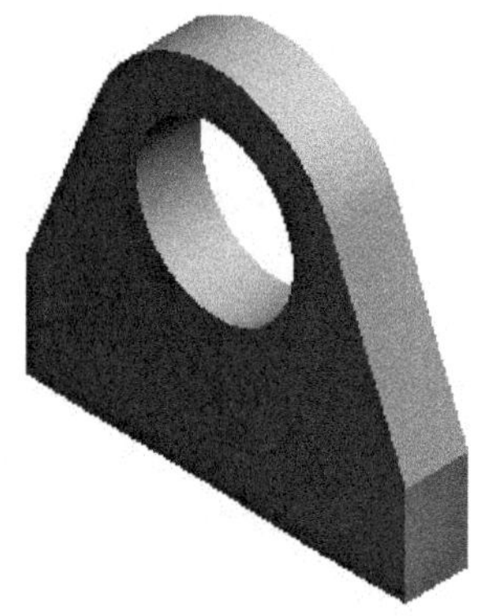

Figure 2-19 *Model for Tutorial 1*

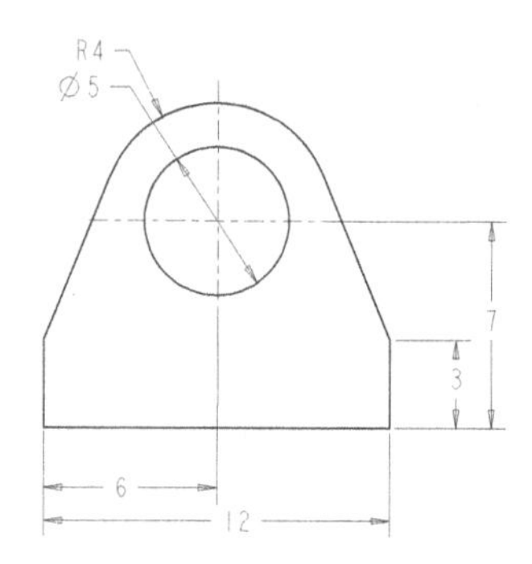

Figure 2-20 *Sketch of the model*

The following steps are required to complete this tutorial:

a. Start PTC Creo Parametric session.
b. Set the working directory and create a new sketch file.
c. Draw lines by using the **Line Chain** tool, refer to Figures 2-22 and 2-23.
d. Draw an arc and a circle, refer to Figures 2-24 through 2-26.
e. Dimension the sketch and then modify the dimensions of the sketch, refer to Figure 2-26
f. Save the sketch and close the file.

Starting PTC Creo Parametric

1. Start PTC Creo Parametric by double-clicking on the **PTC Creo Parametric** icon on the desktop of your computer.

Setting the Working Directory

After the PTC Creo Parametric session is started, the first task is to set the working directory. A working directory is a directory on your system where you can save the work done in the current session of PTC Creo Parametric. You can set any existing directory on your system as the working directory. Since it is the first tutorial of this chapter, you need to create a folder with the name *c02*, if it does not exist.

1. Choose the **Select Working Directory** option from the **Manage Session** flyout of the **File** menu; the **Select Working Directory** dialog box is displayed.

2. In dialog box, browse to *C:\Creo-3.0* folder. If this folder does not exist, create this folder before setting the working directory.

For selecting *C:\Creo-3.0,* click on the arrow at the right of the **C:** option in the top address box of the **Select Working Directory** dialog box; the folders in the **C** drive are displayed in a flyout. Now, choose **Creo-3.0** from the flyout, refer to Figure 2-21. Alternatively, you can use the **Folder Tree** node available at the bottom left corner of the screen to set the working directory. To do so, click on the **Folder Tree** node; the **Folder Tree** expands. In the **Folder Tree**, browse to the desired location using the nodes corresponding to the folders. Select the required folder; the **Folder Content** window is displayed. Close this window. The selected folder will become the current working directory.

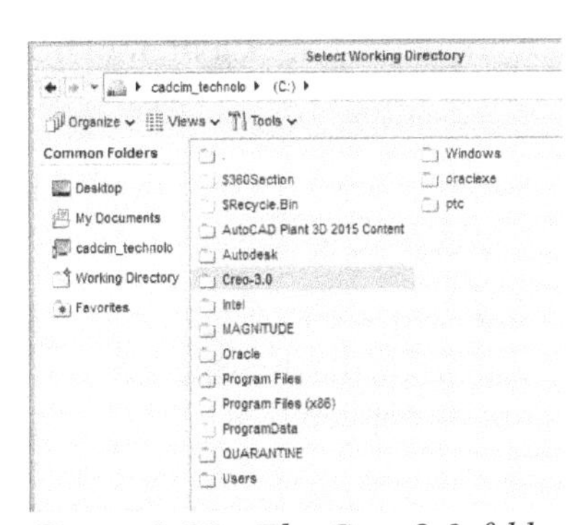

*Figure 2-21 The **Creo-3.0** folder chosen from the flyout*

3. Choose the **Organize** button from the **TOP** pane of the **Select Working Directory** dialog box to display the flyout. From the flyout, choose the **New Folder** option; the **New Folder** dialog box is displayed.

4. Enter **c02** in the **New Directory** edit box and choose **OK** from the **New Folder** dialog box; a new folder named *c02* is created in ***C:\Creo-3.0*** location.

5. Choose **OK** from the **Select Working Directory** dialog box to set the working directory to *C:\Creo-3.0\c02*; the message **Successfully changed to C:\Creo-3.0\c02 directory** is displayed in the message area.

Starting a New Object File

Any sketch drawn in the **Sketch** mode is saved with the *.sec* file extension. This file format is one of the file formats available in PTC Creo Parametric.

1. Choose the **New** button from the **Data** group in the **Ribbon** or **Quick Access** toolbar or press CTRL+N; the **New** dialog box is displayed. In this dialog box, select the **Sketch** radio button from the **Type** area; the default name of the sketch appears in the **Name** edit box.

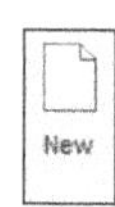

2. Enter **c02tut01** in the **Name** edit box and choose the **OK** button.

   You are in the sketcher environment of the **Sketch** mode. When the **Sketch** mode is invoked, the **Show Navigator** is displayed on the left in the drawing area.

3. Choose the **Show Navigator** button available at the bottom left corner of the screen to close the **Show Navigator**. On closing the tree, the drawing area is increased.

Drawing the Lines of the Sketch

You need to start drawing the sketch with the right vertical line.

1. Choose the **Line Chain** tool from the **Line** drop-down available in the **Sketching** group.

2. Specify the start point by clicking on the right in the drawing area. One end of the line is attached to the cursor. Move the cursor down to get an approximate size of the line.

   Notice that when the cursor moves vertically downward, a green colored constraint **V** appears in the drawing area, next to the line. Now, if you draw a line, the vertical constraint will be applied to it.

3. Click to specify the endpoint of the line. The vertical constraint **V** is applied to the line, but it is not visible in the drawing area until the line creation is active.

   Also, another rubber-band line is attached to the cursor with its start point at the endpoint of the last line.

4. Move the cursor horizontally towards the left; a horizontal rubber-band line extends to the left as you move the mouse.

Notice that when the cursor moves horizontally towards the left, a green colored constraint, **H** appears in the drawing area next to the line. Now, if you draw a line, a horizontal constraint will be applied to it.

5. After getting the desired size of the line created, click to end the line. The horizontal constraint **H** is applied to the line, but it is not visible in the drawing area until the line creation is active.

6. Move the cursor upward in the drawing area; a vertical rubber-band line extends as you move the mouse. As you move the cursor upward, notice that at a particular point where the length of the left vertical line is equal to the length of the right vertical line, an $\mathbf{L_1}$ symbol is displayed on both the vertical lines. This symbol suggests that the equal length constraint is applied to the two vertical lines.

7. When the $\mathbf{L_1}$ constraint appears on the vertical line, click to specify the endpoint of the vertical line. The rubber-band line is still attached to the cursor.

   You can also apply constraints later. However, to save an extra step of adding the constraints, you will use the constraints that are applied automatically while drawing.

8. Move the cursor to size the line and specify the endpoint of the left inclined line, as shown in Figure 2-22.

9. Press the middle mouse button to end the line creation.

10. The line option is still active. Move the cursor close to the top end of the right vertical line; the cursor snaps to the point that is at equal length of the right vertical line. Select the point by clicking.

11. Size the inclined line and specify the endpoint of the right inclined line. Press the middle mouse button twice; lines are created and all the constraints that you have applied become visible, refer to Figure 2-23. Now, you need to draw the arc and the circle.

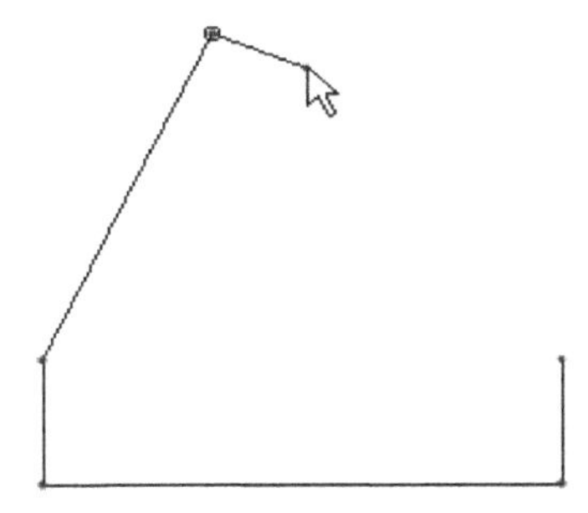

Figure 2-22 Partial sketch with left inclined line

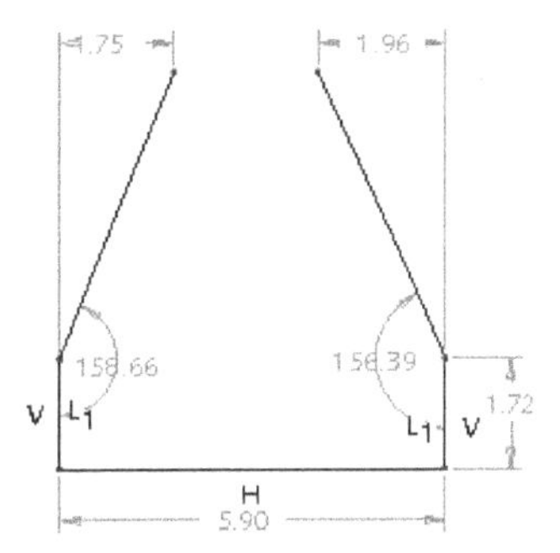

Figure 2-23 Partial sketch of the line drawn with weak constraints

Note

1. The labels such as L_2 or L_3 vary from sketch to sketch.

*2. The horizontal constraint **H** and the vertical constraint **V** appear in blue color. The blue color of the constraint indicates that the constraint is strong. This means you cannot change the orientation of the line until you delete the constraint applied to the line. The constraints displayed in light blue color indicates that they are week constraints, refer to Figure 2-23.*

Drawing the Arc

1. Choose the **3-Point/Tangent End** tool from the **Arc** drop-down in the **Sketching** group; you are prompted to select the start point of the arc.

2. Select the endpoint of the left inclined line; the **Target** symbol appears in green color.

3. Move the cursor along the tangent direction through a small distance; a rubber-band arc that is tangent to the inclined line appears. As you move the cursor to the endpoint of the right inclined line, at a particular point, the tangent constraint is applied to both ends of the arc. This is indicated by the symbol **T**, which appears at the endpoints of the inclined lines.

4. As the tangent constraint appears, click to end the arc creation. You will notice that the tangent constraint with the symbol **T** appears at the endpoints of the arc, as shown in Figure 2-24. Press the middle mouse button to end the arc creation.

   The tangent constraint **T** appears in blue color, which indicates that it is a strong constraint and the tangency of the inclined line with the arc cannot be modified until you delete the tangent constraint.

Note

1. You will see that in Figure 2-23, there are some weak dimensions that are not displayed in Figure 2-24. This is because weak dimensions get deleted without confirming their deletion. Hence, after drawing the arc, some weak dimensions get deleted automatically.

*2. If the tangent constraint symbol is not displayed on any of the inclined lines, apply the constraint manually by choosing the **Tangent** button from the **Constrain** group.*

Drawing the Circle

1. Choose the **Concentric** tool from the **Circle** drop-down; you are prompted to select an arc.

2. Select the arc by clicking on it. Move the mouse; a circle appears.

3. Click to select a point inside the sketch to draw the circle.

4. Press the middle mouse button to end the circle creation. The sketch is complete.

Dimensioning the Sketch

The right vertical line, the bottom horizontal line, the arc, and the circle are dimensioned automatically and the weak dimensions are applied to them. You will use these dimensions. Hence, there is no need to dimension these entities again.

1. Choose the **Normal** tool from the **Dimension** group.

2. Select the center of the arc and then the bottom horizontal line; the center turns red and the line turns green in color.

3. Place the dimension on the right of the sketch by pressing the middle mouse button.

4. Select the center of the arc and then the left vertical line; both the center and the vertical line turns green in color.

5. Press the middle mouse button to place the dimension below the sketch, refer to Figure 2-25.

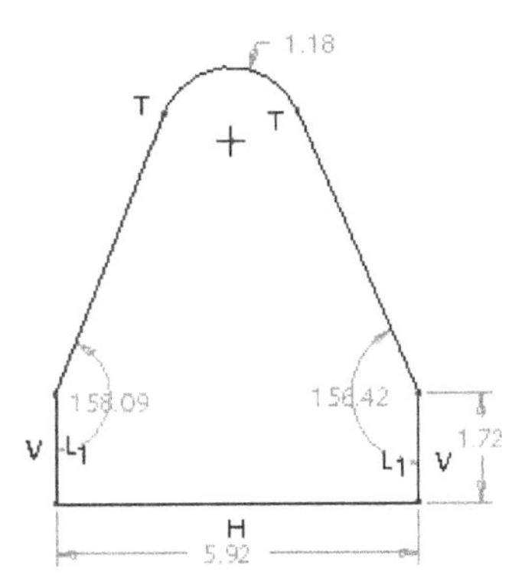

Figure 2-24 *Sketch with arc*

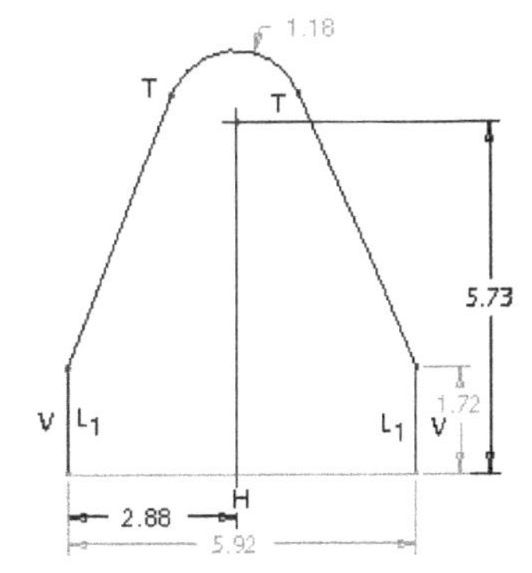

Figure 2-25 *Sketch with all the entities, weak dimensions, and weak constraints*

Modifying the Dimensions

The sketch is dimensioned with default values. You need to modify these values to the given values.

1. Choose the **One-by-One** selection filter from the **Select** drop-down available in the **Operations** group.

2. Select all dimensions by specifying a window around them.

Note

You can also use CTRL+ALT+A to select the entire sketch with dimensions.

3. When all dimensions turn green in color, choose the **Modify** tool from **Editing** group; the **Modify Dimensions** dialog box is displayed.

All dimensions in the sketch are displayed in this dialog box. Each dimension has a separate thumbwheel and an edit box. You can use the thumbwheel or the edit box to modify the dimensions. It is recommended that you use the edit boxes to modify the dimensions, if the change in the dimension value is large.

4. Clear the **Regenerate** check box and then modify the values of the dimensions.

   Once you clear this check box, any modification in a dimension value is not update in the sketch. It is recommended that you clear the **Regenerate** check box when more than one dimension has to be modified.

   Notice that the dimensions that you select in the **Modify Dimensions** dialog box gets enclosed in a blue box in the drawing area.

5. Modify all dimensions according to the dimensions, refer to Figure 2-26. After modifying the dimensions, choose the **Done** button from the **Modify Dimensions** dialog box; the message **Dimension modifications successfully completed** is displayed in the message area.

Tip: *You can modify the location of the dimensions as they appear on the screen by selecting and dragging them to a new location.*

The completed sketch is shown in Figure 2-26.

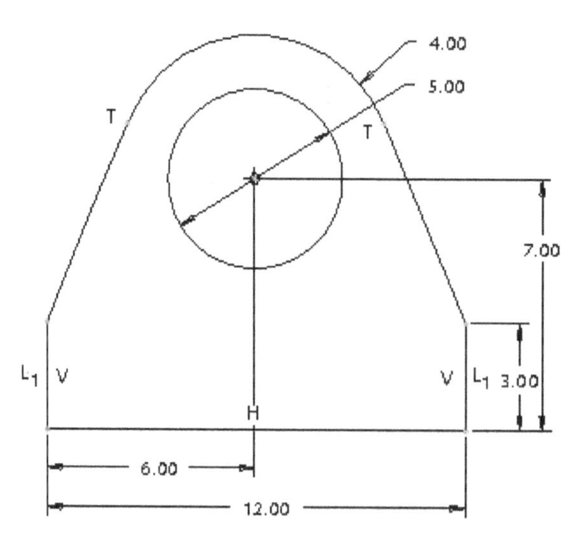

Figure 2-26 The complete sketch with dimensions and constraints

Saving the Sketch

Now, the sketch needs to be saved because you may need the sketch later in the **Part** mode to create a 3D model.

1. Choose the **Save** button from the **Quick Access** toolbar; the **Save Object** dialog box is displayed with the name of the sketch that you had entered earlier.

2. Choose the **OK** button; the sketch is saved.

3. After saving the sketch, choose the **Close** button from the **Quick Access** toolbar.

Tutorial 2

In this tutorial, you will draw the sketch for the model shown in Figure 2-27. The sketch of the model is shown in Figure 2-28. For your reference, all entities in the sketch are labeled alphabetically. **(Expected time: 30 min)**

Figure 2-27 *Model for Tutorial 2*

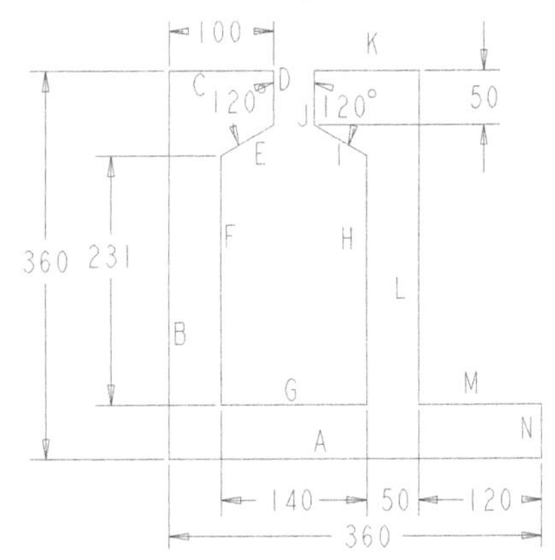

Figure 2-28 *Sketch of the model*

The following steps are required to complete this tutorial:

a. Set the working directory and create a new sketch file.
b. Draw the sketch by using the **Line Chain** tool, refer to Figure 2-29.
c. Dimension the required entities and then modify the dimensions of the sketch, refer to Figures 2-30 and 2-31.
d. Save the sketch and close the file.

Setting the Working Directory

The working directory was selected in Tutorial 1, therefore you need not to select the directory again. But, if a new session of PTC Creo Parametric is started, you need to set the working directory again by following the steps given next.

1. Open the Navigator (if it is in collapsed state) by clicking on the Show Navigator button on the bottom left corner of the PTC Creo Parametric main window; the Navigator slides out. At the bottom of the Navigator, the **Folder Tree** is displayed in the **Folder Browser** tab. Click on the **Folder Tree**; the **Folder Tree** expands.

2. Click on the right arrow adjacent to the *Creo-3.0* folder in the Navigator; the contents of the *Creo-3.0* folder are displayed.

3. Now, right-click on the *c02* folder to display a shortcut menu. From the shortcut menu, choose the **Set Working Directory** option; the working directory is set to *c02*.

4. Close the Navigator by clicking on the **Navigator** button located at the bottom left corner of the main window; the Navigator slides in.

Starting a New Object File

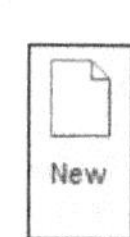

1. Choose the **New** button from the **Data** group; the **New** dialog box is displayed. Select the **Sketch** radio button from the **Type** area of the **New** dialog box; the default name of the sketch appears in the **Name** edit box.

2. Enter **c02tut02** in the **Name** edit box and choose **OK**; you are in the sketcher environment of the **Sketch** mode.

Drawing the Sketch

The sketch in Figure 2-28 consists of only lines. For ease of understanding, all lines in the sketch are labeled alphabetically.

1. Choose the **Line Chain** tool from the **Line** drop-down of the **Sketching** group. Select a point close to the lower right corner of the drawing area by clicking and start drawing the horizontal line A. You will notice that as you draw line A, the **H** symbol is displayed on the line. This indicates that the line is horizontally constrained. Move the cursor towards the left and specify the endpoint of the line.

2. Move the cursor vertically upward so that the **V** constraint appears on the line. When you get the appropriate size of the line, click to specify the endpoint of line B; line B is completed.

3. Move the cursor to the right in the drawing area and click to specify the endpoint of line C.

4. Now, to draw line D, move the cursor down and click to specify the endpoint of line D.

5. Move the cursor to size the line and click to specify the endpoint of the inclined line E.

6. The next line you need to draw is line F. Move the cursor vertically downward and click to specify the endpoint of line F.

7. Now, to draw line G, move the cursor horizontally toward the right and click to specify the endpoint of line G.

8. Move the cursor vertically upward and click to specify the endpoint of line H.

9. Now, continue drawing the remaining lines that are shown in Figure 2-28. When the sketch

is complete, end the line creation by pressing the middle mouse button twice. Notice that the sketched entities are dimensioned automatically as you draw them. These dimensions are weak dimensions and appear in light blue color.

Applying Constraints to the Sketch

Constraints are applied to the sketch to maintain the design intent of the feature and this might sometimes result in less dimensions in the sketch.

1. Choose the **Equal** tool from the **Constrain** group and select lines F and H. The equal length constraint $\mathbf{L_2}$ is applied to both the lines. The constraint labels such as L_2 or L_3 vary from sketch to sketch.
2. Now, select lines J and N; the equal length constraint is applied to both the lines. Press the middle button to make other selections.
3. Select lines C and K; the equal length constraint is applied to both the lines. Press the middle button to make other selections.
4. Select lines A and B; the equal length constraint is applied to both the lines. Press the middle button twice to exit.

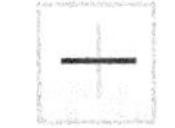

5. Choose the **Horizontal** tool from the **Constrain** group; you are prompted to select a line or two vertices.
6. Select the vertex that is joining the lines L and M. Now, select the vertex that is joining the lines G and H. For the placement of lines, refer to Figure 2-28. Both the vertices are aligned horizontally, as shown in Figure 2-29.
7. Select the vertex that is joining lines C and D and the vertex that is joining lines J and K, refer to Figure 2-28. Both the vertices are aligned horizontally, as shown in Figure 2-30.

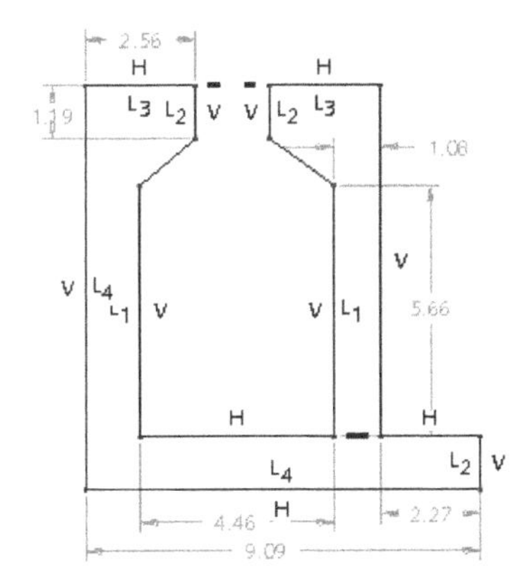

Figure 2-29 Vertices aligned horizontally

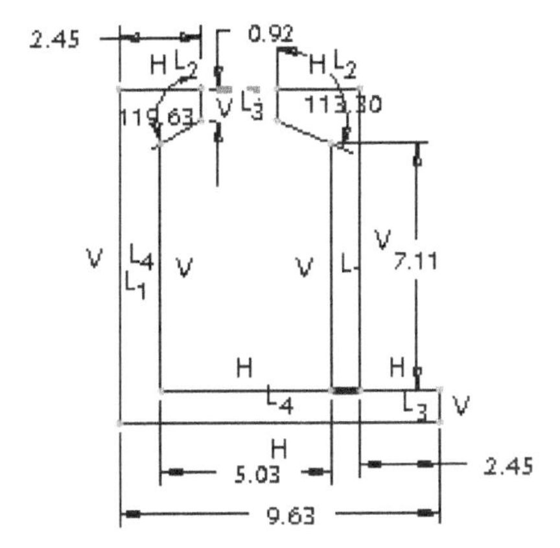

Figure 2-30 Vertices aligned horizontally

Dimensioning the Sketch

Weak dimensions have already been applied to the sketch while drawing. You need to dimension only the angle between lines D and E and lines J and I.

1. Choose the **Normal** tool from the **Dimension** group.

2. Select lines D and E by using the left mouse button; the selected lines turn green in color. Now, press the middle mouse button to place the dimension close to the vertex where lines D and E join.

3. Similarly, dimension the angle between lines J and I.

   Figure 2-30 shows the sketch after applying dimensions. If your sketch does not have all dimensions shown in this figure, apply them by using the **Normal** tool.

Modifying the Dimensions

The dimensions that are applied to the sketch need modification in dimension values.

1. Choose the **One-by-One** tool from the **Select** drop-down available in the **Operations** group and then select all dimensions by specifying a window around them.

2. When dimensions turn green in color, choose the **Modify** tool; the **Modify Dimensions** dialog box is displayed.

3. Clear the **Regenerate** check box and then modify the values of the dimensions. On clearing this check box, the sketch is not regenerated while you modify the dimensions.

   Notice that the dimension that you select in the **Modify Dimensions** dialog box is enclosed in a blue box in the drawing area.

4. When all dimensions are modified, choose the **Done** button from the **Modify Dimensions** dialog box; the message **Dimension modifications successfully completed** is displayed in the message area. The completed sketch is shown in Figure 2-31.

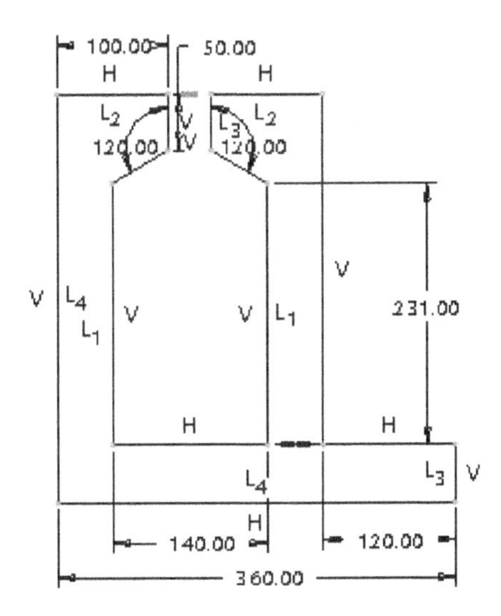

Figure 2-31 Complete sketch with dimensions and constraints

5. Save the sketch as discussed earlier. Next, choose the **Close** button from the **File** menu to exit the **Sketch** mode.

Note

You can also modify dimensions individually. However, individual modification of dimensions is recommended only when there is a minor change in the dimension value or when only one dimension is required to be modified.

Tutorial 3

In this tutorial, you will draw the sketch of the model shown in Figure 2-32. The sketch of the model is shown in Figure 2-33. For your reference, all entities in the sketch are labeled alphabetically. Also, you have to print the sketch. **(Expected time: 30 min)**

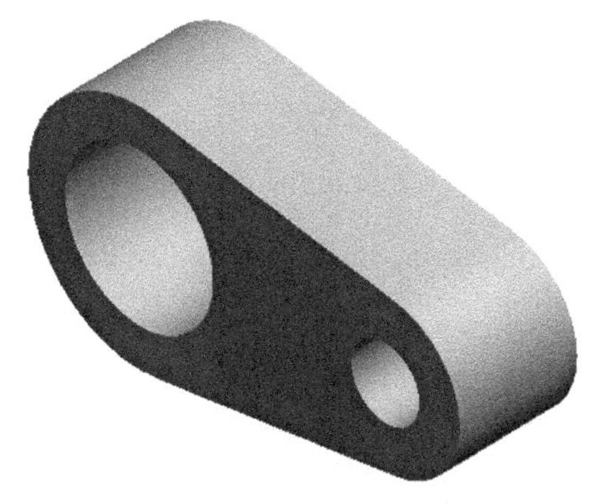

Figure 2-32 Model for Tutorial 3

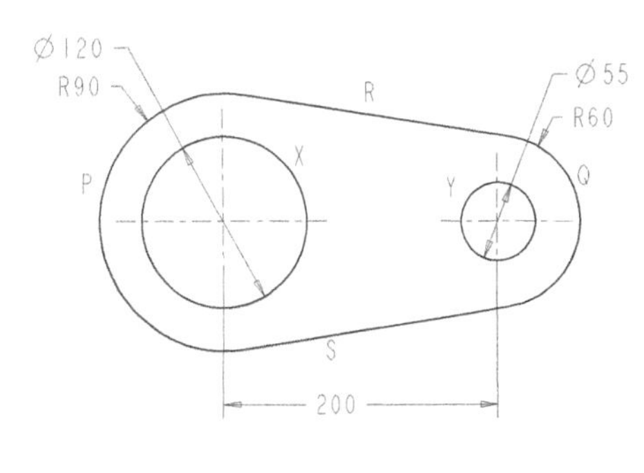

Figure 2-33 Sketch of the model

The following steps are required to complete this tutorial:

a. Set the working directory and create a new sketch file.
b. Draw the sketch by using the sketcher tools, refer to Figures 2-34 through 2-37.
c. Dimension the sketch and then modify the dimensions of the sketch, refer to Figure 2-38.
d. Save the sketch and print it.

Setting the Working Directory

The working directory was selected in Tutorial 1, therefore, you do not need to select the working directory again. But if a new session of PTC Creo Parametric is started, then you have to set the working directory again by following the steps given next.

1. Open the Navigator by sliding it out. In the Navigator, the **Folder Tree** is displayed at the bottom. Click on the black arrow available on the right of the **Folder Tree**; the **Folder Tree** expands. Click on the (+) sign adjacent to the *Creo-3.0* folder in the Navigator; the contents of the *Creo-3.0* folder are displayed.

2. Now, right-click on the *c02* folder to display a shortcut menu. From the shortcut menu, choose the **Set Working Directory** option; the working directory is set to *c02*. Close the Navigator.

Starting New Object File

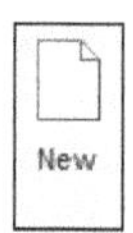

1. Choose the **New** button from the **Data** group; the **New** dialog box is displayed. Select the **Sketch** radio button from the **Type** area of the **New** dialog box; the default name of the sketch appears in the **Name** edit box.

2. Enter **c02tut03** in the **Name** edit box. Choose the **OK** button to enter the sketcher environment of the **Sketch** mode.

Drawing the Circles

1. Choose the **Center and Point** tool from the **Circle** drop-down in the **Sketching** group and specify the center of the circle.

2. Move the cursor to size the circle and then click to complete the circle.

3. Draw another circle whose center is collinear with the center of the previous circle.

Figure 2-34 shows the two collinear circles drawn by using the **Center and Point** tool.

Drawing the Tangent Lines

1. Choose the **Line Tangent** tool from the **Line** drop-down in the **Sketching** group; you are prompted to select the start location on the arc or the circle.

2. Select the left circle at the top; a rubber-band line appears whose one end is attached to the circle and the other end is attached to the cursor.

3. Click on the top of the right circle; a tangent connecting the two circles is drawn.

4. Similarly, draw a tangent by selecting the two circles at the bottom.

   Figure 2-35 shows the sketch after drawing the tangent lines.

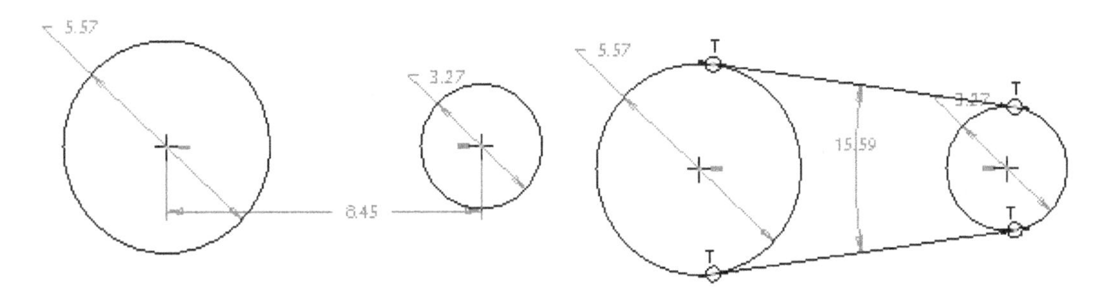

Figure 2-34 *Two collinear circles drawn using the **Centre and Point** tool*

Figure 2-35 *Circles joined by tangent lines*

Trimming the Circles

As evident from Figure 2-35, the tangents that are drawn intersect the circles at the point where they meet the circle. Therefore, the part of the circle that is not required can be dynamically trimmed.

1. Choose the **Delete Segment** tool from the **Editing** group.

2. Select the two circles individually to trim them at the locations shown in Figure 2-36. Figure 2-37 shows the two circles after deleting the unwanted portions of the circle.

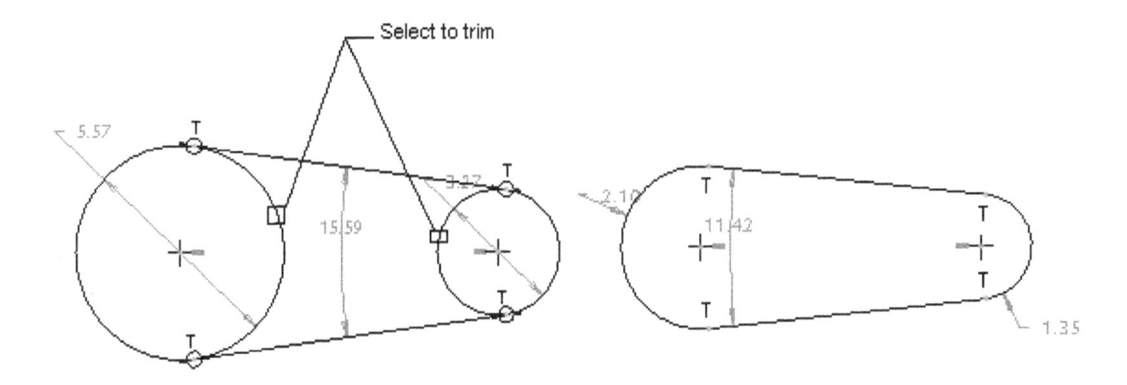

Figure 2-36 *Locations to trim*

Figure 2-37 *Sketch after trimming*

Drawing the Circles

1. Choose the down arrow on the right of the **Center and Point** tool to display the flyout. Choose the **Concentric** tool from the flyout; you are prompted to select an arc.

2. Select arc **P** and create circle X concentric to the arc. Similarly, select arc **Q** to create a concentric circle Y (refer to Figure 2-33).

   Notice that the radius dimension is applied to the two arcs, whereas the diameter dimension is applied to the circles. It is so because, the arcs are applied with radius dimension and circles are applied with diameter dimension, by default.

Dimensioning the Sketch

In order to fully define a sketch, you need to dimension it.

1. Choose the **Normal** tool.

2. Select the centers of the two circles and place the dimension at the bottom of the sketch.

Modifying the Dimensions

1. Choose the **One-by-One** button.

2. Select all dimensions by defining a window.

Note

You can also use CTRL+ALT+A from the keyboard to select all entities and items in the sketch.

3. When all dimensions turn green in color, choose the **Modify** tool; the **Modify Dimensions** dialog box is displayed.

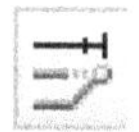

4. Clear the **Regenerate** check box and then modify the values of the dimensions. You will notice that the dimension that you edit in the **Modify Dimensions** dialog box is enclosed by a blue box in the drawing area.

5. When all dimensions are modified, choose the **Done** button from the **Modify Dimensions** dialog box; the message **Dimension modifications successfully completed** is displayed in the message area.

   The completed sketch is shown in Figure 2-38.

6. Save the sketch as discussed earlier. Next, you need to print the sketch.

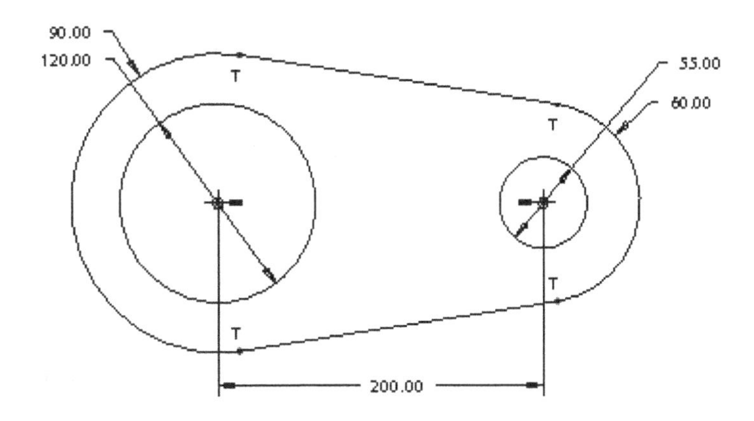

Figure 2-38 Complete sketch with dimensions and constraints

Printing the Sketch Using the Plot Option

1. Choose the **Print** button from the **File>Print** menu or press CTRL+P; the **Print** dialog box is displayed, as shown in Figure 2-39.

2. Choose the **Commands and Settings** button from this dialog box; a shortcut menu is displayed, as shown in Figure 2-40.
3. Choose the **Add Printer Type** option from the shortcut menu; the **Add Printer Type** dialog box is displayed.
4. From the printers listed in the **Add Printer Type** dialog box, select the printer that is installed on your system and choose the **OK** button.
5. From the **Print** dialog box, choose the **Configure** button; the **Shaded Image Configuration** dialog box is displayed. This dialog box allows you to set the paper size.

*Figure 2-39 The **Print** dialog box*

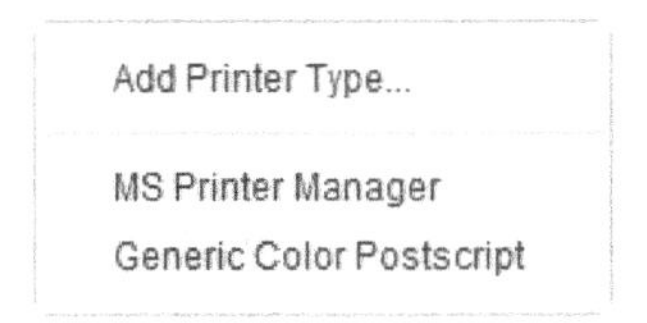

*Figure 2-40 Shortcut menu displayed on choosing the **Commands and Settings** button*

6. Select the **A** option from the **Size** drop-down list, if not already selected; the dimensions of the sheet are set, by default. Also, select the desired image resolution and the image depth from the dialog box.

7. Choose the **OK** button from the **Shaded Image Configuration** dialog box.

8. Now, choose the **OK** button from the **Print** dialog box to complete the printing.

Self-Evaluation Test

Answer the following questions and then compare them to those given at the end of this chapter:

1. When you draw a sketch, the dimensions and constraints are automatically applied to it. (T/F)

2. In PTC Creo Parametric, you can create lines that are tangent to two circles. (T/F)

3. If the **Intent Manager** is turned ON and you draw a line, the cursor snaps to the endpoint of the previous line. (T/F)

4. You can convert a weak constraint into strong constraint by using the shortcut menu that is displayed when you right-click on the weak constraint. (T/F)

5. For drawing a circle, first you need to specify its diameter. (T/F)

6. The **Modify** tool is located in the__________ group of the **Ribbon**.

7. A sketch can be modified by changing its __________.

8. The **Intent Manager** is __________ by default when you enter the **Sketch** mode. (ON/OFF)

9. In the **Sketch** mode, the tangent constraint is represented by a __________ symbol.

10. The file created in the **Sketch** mode is saved with a __________ file extension.

Review Questions

Answer the following questions:

1. You can dynamically modify the geometry of a sketch. (T/F)

2. You can use the **Rectangle** tool from the **Sketching** group to draw a square. (T/F)

3. The tools available in the __________ group are used to apply constraints manually.

4. You cannot undo a previous operation in the sketcher environment. (T/F)

5. You can also use the options in the right click shortcut menu to draw a sketch, when no other tool is chosen. (T/F)

6. The **Sketch** mode is important in PTC Creo Parametric. (T/F)

7. Points are used to specify locations. (T/F)

8. There are two types of lines can be sketched by using the buttons available in the **Sketching** group. (T/F)

9. The **Normal** tool is used for normal dimensioning of the sketch. (T/F)

10. You can select one or more dimensions from the sketch to modify them. (T/F)

EXERCISES

Exercise 1

In this exercise, you will draw the model shown in Figure 2-41. The sketch of the model is shown in Figure 2-42. **(Expected time: 30 min)**

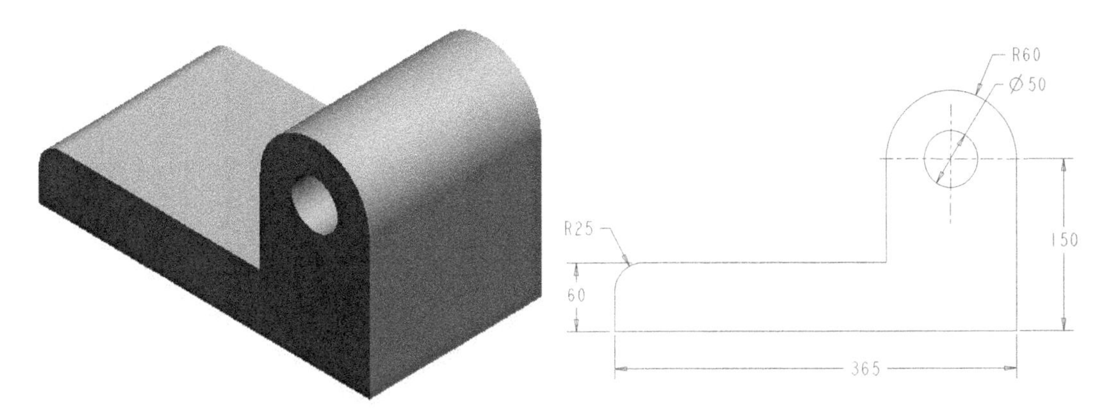

Figure 2-41 Solid model for Exercise 1

Figure 2-42 Sketch of the model

Exercise 2

In this exercise, you will draw the model shown in Figure 2-43. The sketch of the model is shown in Figure 2-44. **(Expected time: 30 min)**

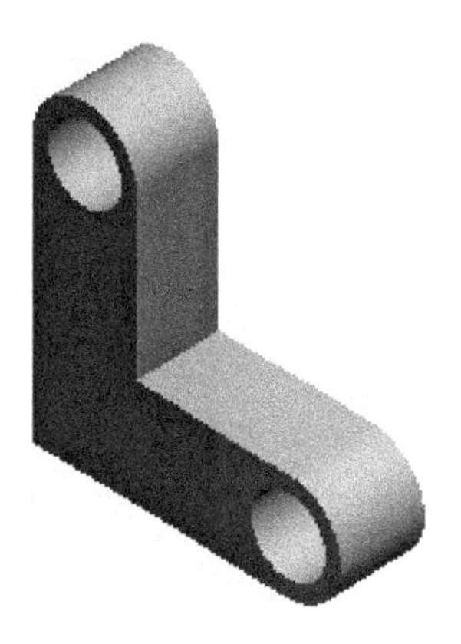

Figure 2-43 Solid model for Exercise 2

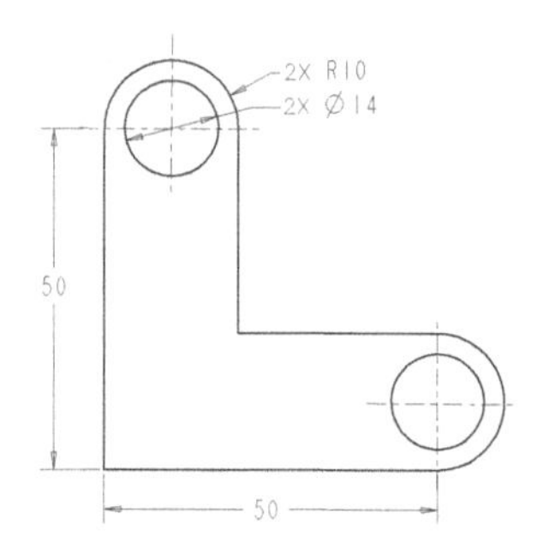

Figure 2-44 Sketch of the model

Exercise 3

In this exercise, you will draw the model shown in Figure 2-45. The sketch of the model is shown in Figure 2-46. **(Expected time: 30 min)**

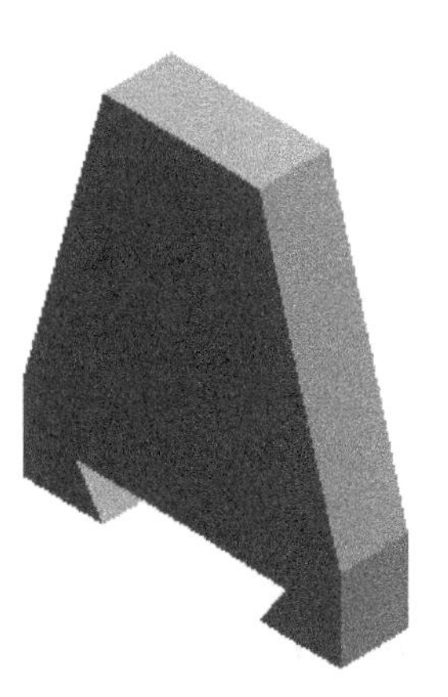

Figure 2-45 Solid model for Exercise 3

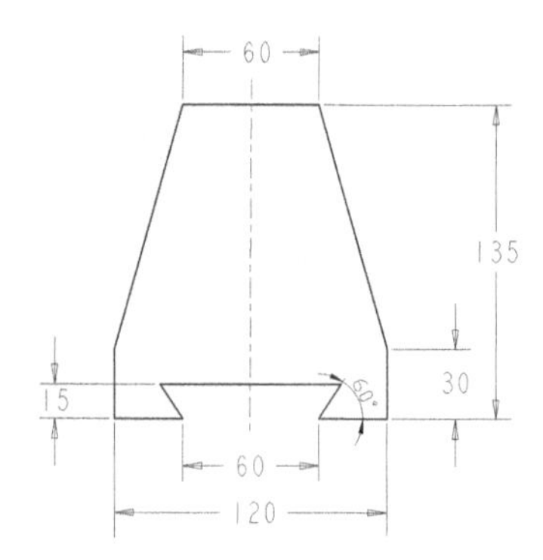

Figure 2-46 Sketch of the model

Exercise 4

In this exercise, you will draw the model shown in Figure 2-47. The sketch of the model is shown in Figure 2-48. **(Expected time: 30 min)**

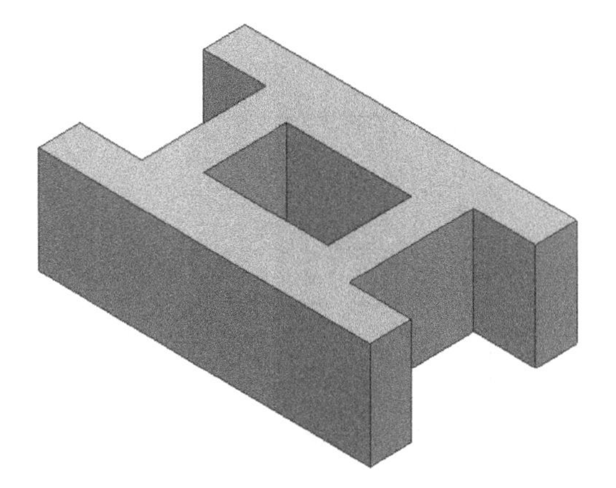

Figure 2-47 Solid model for Exercise 4

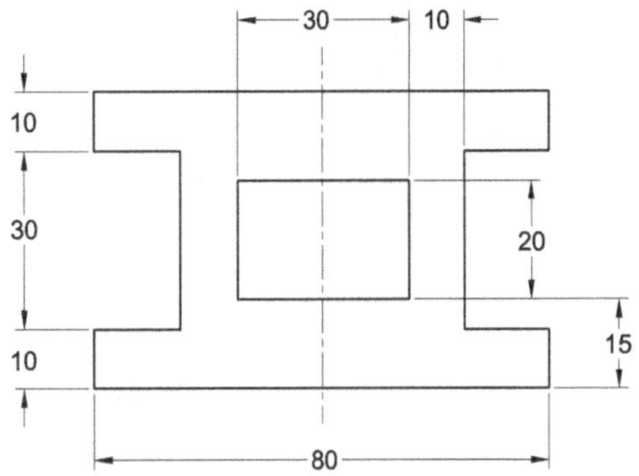

Figure 2-48 Sketch of the model

Exercise 5

In this exercise, you will draw the model shown in Figure 2-49. The sketch of the model is shown in Figure 2-50. **(Expected time: 30 min)**

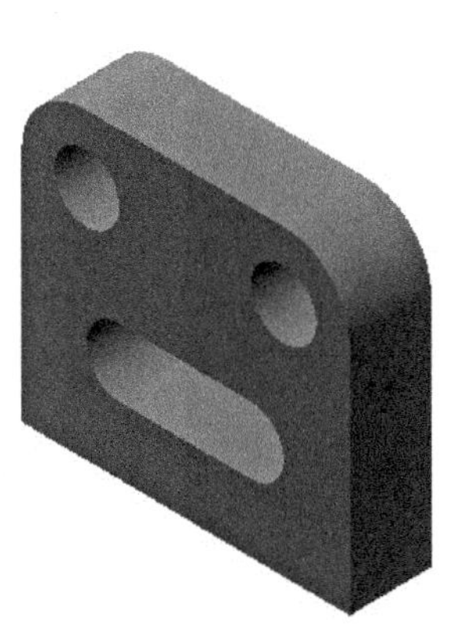

Figure 2-49 Solid model for Exercise 5

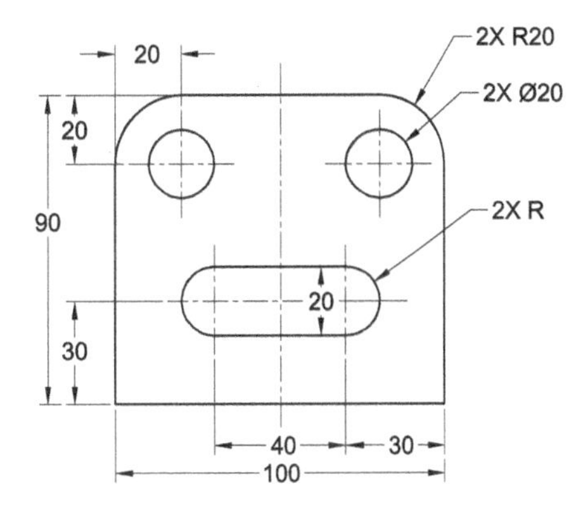

Figure 2-50 Sketch of the model

Exercise 6

In this exercise, you will draw the model shown in Figure 2-51. The sketch of the model is shown in Figure 2-52. **(Expected time: 30 min)**

Figure 2-51 *Solid model for Exercise 6*

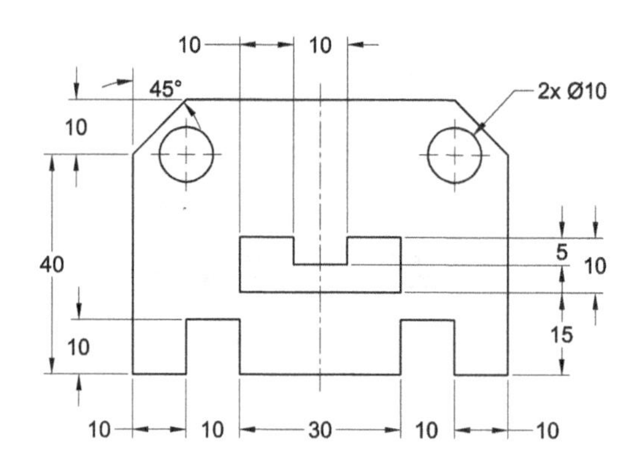

Figure 2-52 *Sketch of the model*

Answers to Self-Evaluation Test

1. T, **2.** T, **3.** T, **4.** T, **5.** F, **6. Editing**, **7.** dimensions, **8.** ON, **9.** T, **10.** *.sec*

Chapter 3

Creating Sketches in the Sketch Mode-II

Learning Objectives

After completing this chapter, you will be able to:

- *Use various options to dimension a sketch.*
- *Create fillets.*
- *Place a user-defined coordinate system.*
- *Create, dimension, and modify splines.*
- *Create text.*
- *Move and resize entities.*
- *Copy a sketch.*
- *Import 2D drawings.*

DIMENSIONING THE SKETCH

In Chapter 2, you learned dimensioning a sketch using the **Normal** tool from the **Dimension** group. In this chapter, you will learn the use of the **Baseline** tool for dimensioning a sketch.

Dimensioning a Sketch Using the Baseline Tool

Ribbon: Sketch > Dimension > Baseline

In PTC Creo Parametric, the **Baseline** tool is used to create dimensions in terms of horizontal and vertical distance values of an entity with respect to a specified baseline. This type of dimensioning in a drawing makes writing a CNC program for manufacturing a component easy.

The **Baseline** tool is available in the **Dimension** group. Using this tool, you can dimension a line, arc, conic, and so on. The following steps explain the procedure to create dimensions using the **Baseline** tool:

1. Choose the **Baseline** tool from the **Dimension** group, refer to Figure 3-1.

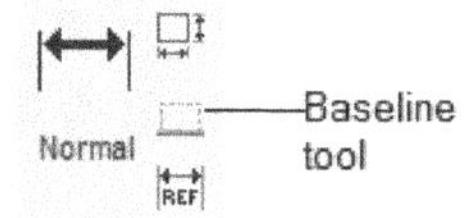

*Figure 3-1 The **Dimension** group*

2. Select the entity that will act as the baseline (origin or reference). Press the middle mouse button to place the dimension; the dimension **0.00** will be displayed where you place the dimension. Note that since the location value of the baseline is taken as the origin, the dimension value of the baseline entity will become 0.00. The dimension values of all other entities dimensioned with reference to the baseline will be measured from this origin.

   Depending upon the entity selected as the baseline, the horizontal or vertical location values of the entity will be placed. For example, if you select a vertical line, the value of its location will be placed vertically. Similarly, if you select a horizontal line, the value of its location will be placed horizontally.

   You can dimension arcs, circles, and splines by using the two options that are displayed on invoking the **Baseline** tool. If you select the center of a circle or an arc for baseline dimensioning and press the middle mouse button, the **Dim Orientation** dialog box will be displayed, as shown in Figure 3-2. Also, you will be prompted to select the orientation. Select the required radio button from this dialog box and choose the **Accept** button; the dimension will be placed based on the orientation selected.

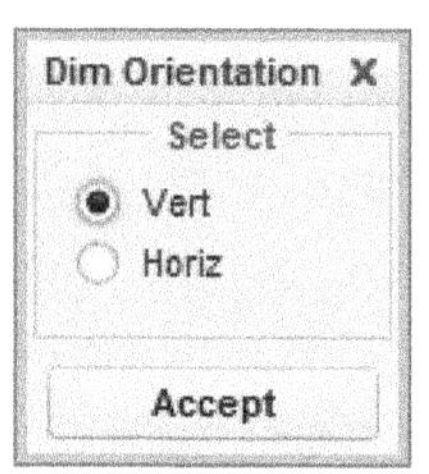

*Figure 3-2 The **Dim Orientation** dialog box*

3. Next, choose the **Normal** tool from the **Dimension** group. Select the baseline dimension that was placed earlier and then select the entity to dimension. Now, press the middle mouse button to place the dimension.

The orientation of the dimension will depend upon the baseline dimension and the entity selected. Figure 3-3 shows a sketch dimensioned using the above-mentioned method. In this figure, the two baselines are dimensioned using the **Baseline** tool. Therefore, the dimensions of these lines are displayed as 0.00. The remaining lines are dimensioned by selecting the baseline dimension and then the required entity by using the **Normal** tool.

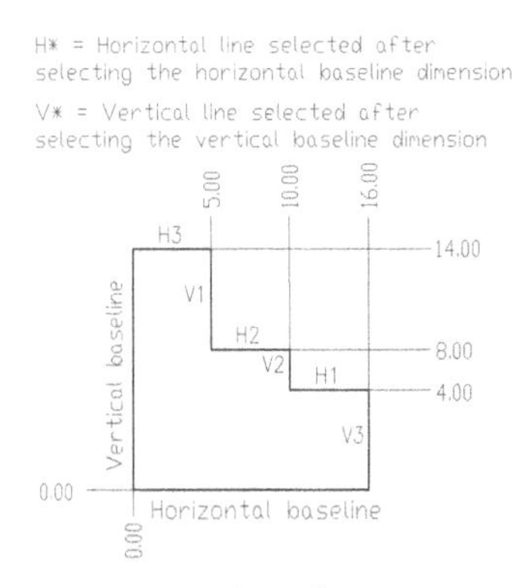

Figure 3-3 Baseline dimensioning of a sketch

Replacing the Dimensions of a Sketch Using the Replace Tool

Ribbon: Sketch > Operations > Replace

The **Replace** tool is used to replace a dimension with a new dimension in a sketch. To use this option, you must have a dimensioned sketch. The following steps explain the procedure to dimension a sketch using the **Replace** tool:

1. Choose the **Replace** tool from the menu displayed on clicking the down arrow adjacent to the **Operations** group; you will be prompted to select the dimension to be replaced.

2. Select the dimension to be replaced; the selected dimension will be deleted and you will be prompted to create a replacement dimension. Select the entities between which you want to create a new dimension; the previous dimension will be replaced by a new dimension.

CREATING FILLETS

In the sketcher environment, you can create the following two types of fillets:

1. Circular fillets
2. Elliptical fillets

Creating Circular Fillets

Ribbon: Sketch > Sketching > Fillet > Circular

A circular fillet is the arc formed at the intersection of two lines, a line and an arc, or two arcs. This type of fillet is controlled by the radius or diameter dimension of the fillet. The resulting fillet will depend on the location where the elements are selected.

Figure 3-4 shows two non-parallel lines and Figure 3-5 shows the circular fillet created between them. The circular fillet thus created is an arc with its endpoints tangent to the two lines. This is evident from the **T** symbol that is automatically applied to the endpoints of the fillet arc.

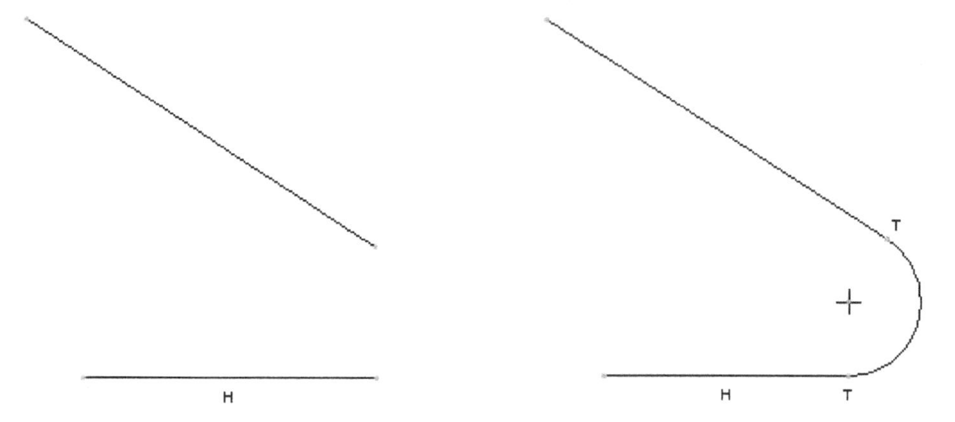

Figure 3-4 *Two lines that do not join*

Figure 3-5 *Fillet created between the two lines*

Figure 3-6 shows two lines that join at a point and Figure 3-7 shows the circular fillet created at the joint.

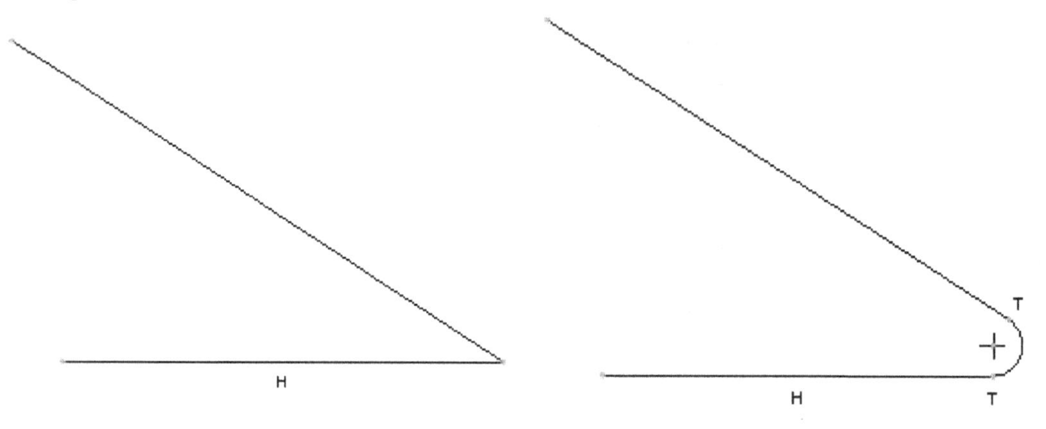

Figure 3-6 *Two lines joining at a point*

Figure 3-7 *Filleted corner*

Figure 3-8 shows two sets of arcs and Figure 3-9 shows the circular fillet created between the two arcs. The location where you select the arcs to create the fillet is important. The fillet is created tangent to the selection points on the arcs. Here, the endpoints of the arcs are selected to create the fillet.

If the points of selection on the two arcs are away from the endpoints of the arcs, then the fillet is created at the selection points. The portion of the arc that extends beyond the fillet should be manually deleted or trimmed. Figure 3-10 shows the points at which the two arcs are selected and Figure 3-11 shows the fillet created at the selection points.

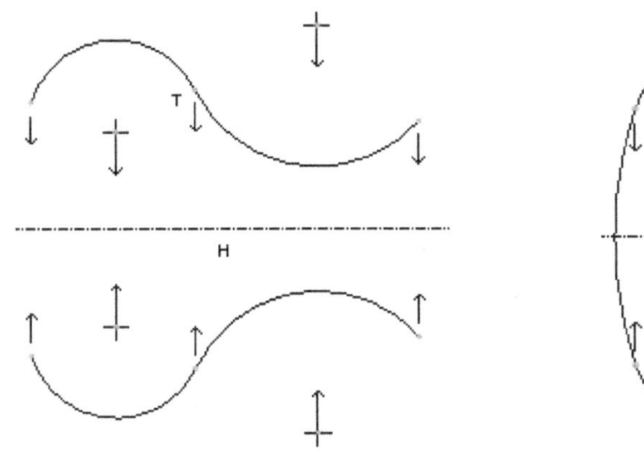

Figure 3-8 *Two sets of arcs*

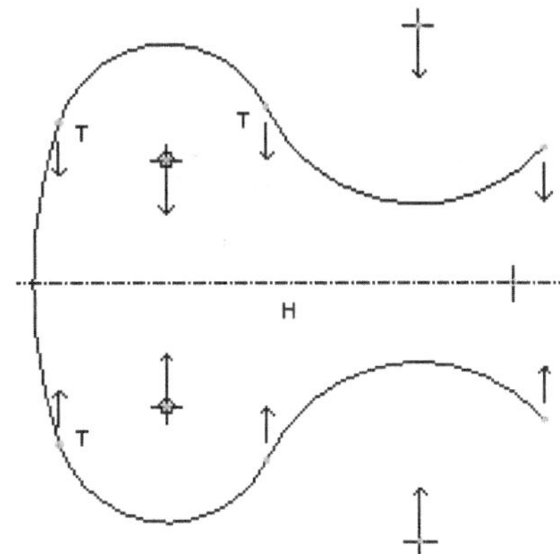

Figure 3-9 *Fillet created by selecting the endpoints*

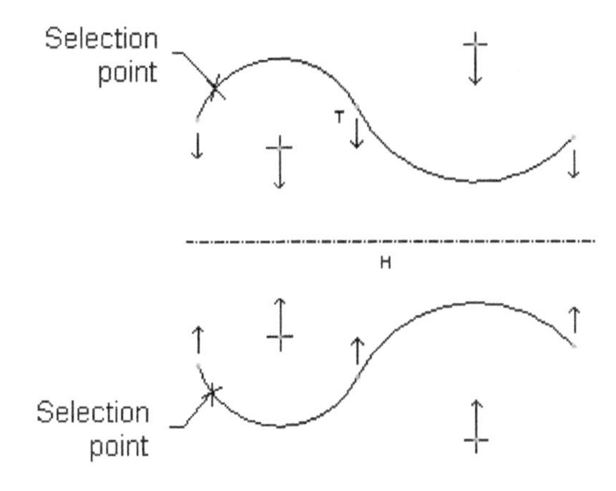

Figure 3-10 *Points selected on arcs*

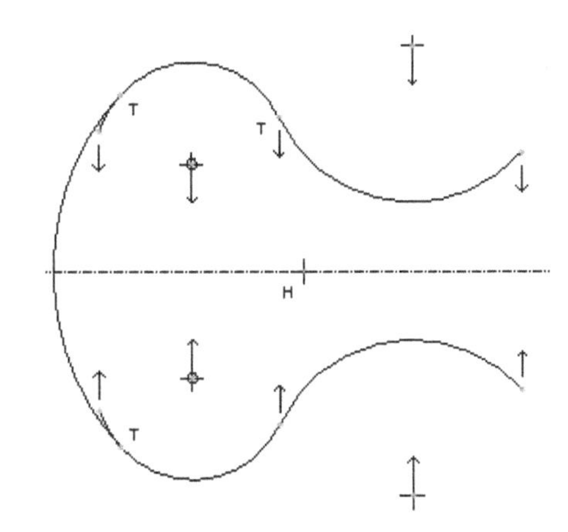

Figure 3-11 *Fillet created*

To create circular fillets, there are two options named **Circular** and **Circular Trim**. These options are available in the **Fillet** drop-down. These options are discussed next.

As mentioned earlier, the **Circular** option is used to create circular fillets. When you create fillet using this option, the filleting entities do not get trimmed but the part of entity after the fillet converts into construction geometry.

The **Circular Trim** option is also used to create circular fillets, but in this case the portion extending beyond the entities get trimmed automatically.

Procedure to create circular fillets using both the options is same and it is discussed next.

1. Choose the **Circular** or **Circular Trim** option from the **Fillet** drop-down in the **Sketching** group; you will be prompted to select two entities.

2. Select the first entity for filleting by using the left mouse button; the red color of the first entity changes to green. Now, select the second entity. It will be drawn between the two selected entities as soon as you select the second entity.

3. Repeat step 2 until you have created all fillets.

Creating Elliptical Fillets

Ribbon: Sketch > Sketching > Fillet > Elliptical

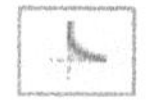

An elliptical fillet is the arc in the form of an ellipse that joins two lines, two arcs, or a line and an arc. The geometry of the elliptical fillet depends on the location where you select the entities to create a fillet.

The advantage of elliptical fillets over circular fillets is that the geometry of elliptical fillets can be controlled by dimensions in two directions. Therefore, when an elliptical fillet is dynamically modified, its geometry can be controlled in either the x-direction or the y-direction resulting in more curved geometric shape than a circular fillet.

Figures 3-12 and 3-13 illustrate the elliptical fillet. Notice that a strong tangent constraint **T** is automatically applied when you create a fillet.

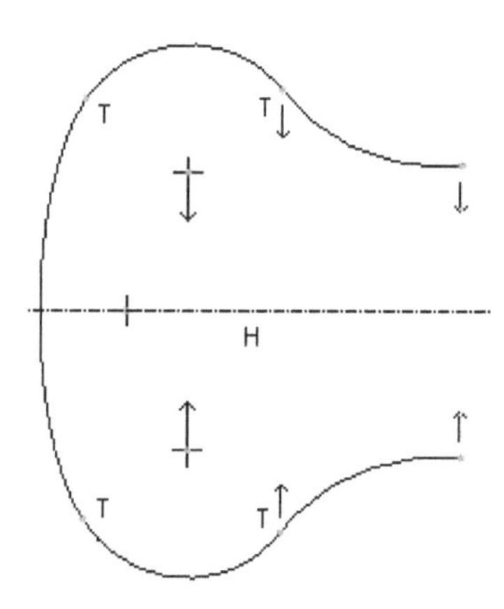

Figure 3-12 *Arcs to be filleted*

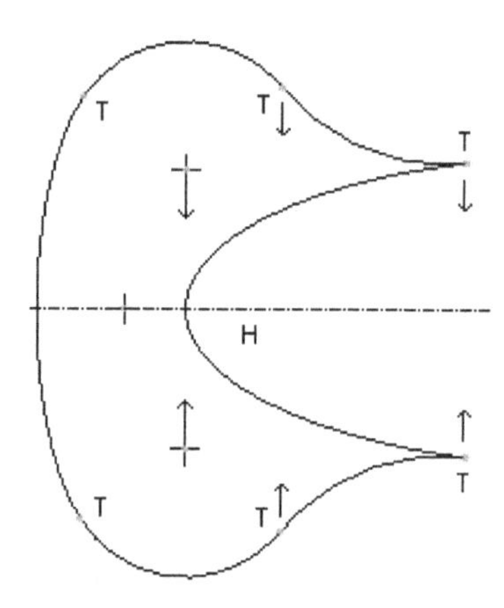

Figure 3-13 *Elliptical fillet created*

Similar to circular fillet, there are two options to create elliptical fillet, **Elliptical** and **Elliptical Trim**. The procedure to create elliptical fillets using any of these options is same. The procedure to create elliptical fillets is discussed next.

1. Choose the **Elliptical** or **Elliptical Trim** tool from the **Fillet** drop-down; you will be prompted to select two entities.

2. Select the first entity by clicking; the color of the entity changes to green.

3. Select the second entity. As soon as you select the second entity, the elliptical fillet is created. The shape of the elliptical fillet depends upon the specified points. After the fillet is created, you will be again prompted to select two entities for elliptical fillet.

4. Repeat steps 2 and 3 until you have created all fillets.

When you select an elliptical fillet to dimension, the **Ellipse Rad** dialog box will be displayed, as shown in Figure 3-14. There are two radio buttons in this dialog box. When the **Major Axis** radio button is selected, the elliptical fillet will be dimensioned radially along the X-direction. The **Minor Axis** radio button, when selected, dimensions the elliptical fillet radially in the Y-direction.

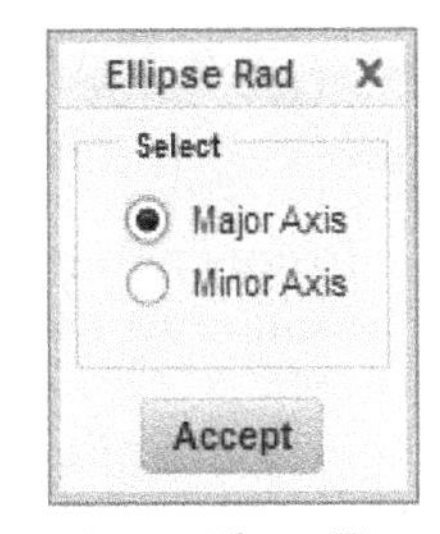

*Figure 3-14 The **Ellipse Rad** dialog box*

CREATING A REFERENCE COORDINATE SYSTEM

There are two types of coordinate systems, construction and geometric. As discussed earlier, the construction entities cannot be referenced outside the Sketcher environment whereas, the geometric entities can be. The **Coordinate System** tool in the **Datum** group is used to create a coordinate system that will act as a reference for dimensioning. You can dimension the splines using the coordinate system. Thus, it provides you the flexibility to modify the spline points by specifying different coordinates with respect to the coordinate system.

The user-defined coordinate system is used in blend features to align different sections in a blend. It is also used in the **Assembly** and **Manufacturing** modes of PTC Creo Parametric.

The following steps explain the procedure to create a coordinate system:

1. Choose the **Coordinate System** tool from the **Datum** group; you will be prompted to select the location for the coordinate system. The coordinate system symbol is attached to the cursor.

2. Place the coordinate system at the desired points on the screen by clicking the left mouse button. The coordinate system will be placed at as many places as you click in the graphics window. You can end the coordinate system creation by using the middle mouse button.

Note

If you add a coordinate system to a sketch, it must be dimensioned. But if the coordinate system is placed at the endpoints of a line, an arc, a spline, or at the center of an arc or a circle, it need not be dimensioned. In other words, a coordinate system must be referenced to an entity in a sketch.

WORKING WITH SPLINES

Splines are curved entities that pass through a number of intermediate points. Generally, splines are used to define the outer surface of a model. This is because the splines can provide different shape to curves and the flexibility to modify the surfaces that result from the splines. Splines find application in automobile and aeroplane body designing.

Creating a Spline

Ribbon: Sketch > Sketching > Spline

To draw a spline, choose the **Spline** tool from the **Sketching** group. The steps to create a spline are discussed next.

1. Choose the **Spline** tool from the **Sketching** group; you will be prompted to select the location for spline.

2. Use the left mouse button to select the start point for the spline. Similarly, select additional points in the graphics window; a spline will be drawn passing through all specified points. Press the middle mouse button to end the creation of spline. All points through which the spline passes are called interpolation points.

Dimensioning of Splines

When a spline is drawn, the weak dimensions are automatically applied to the spline. A spline can be dimensioned manually by:

1. Dimensioning the endpoints
2. Radius of curvature dimensioning
3. Tangency dimensioning
4. Coordinate dimensioning
5. Dimensioning the interpolation points

Dimensioning the Endpoints

To dimension a spline by selecting the endpoints, you need to follow the steps given below:

1. Choose the **Normal** tool from the **Dimension** group.

2. Select the two endpoints of the spline and place the horizontal or vertical dimension by pressing the middle mouse button. Figure 3-15 shows a spline that is dimensioned by selecting the endpoints.

Radius of Curvature Dimensioning

The radius of curvature of a spline can be dimensioned only if its tangency is defined. In other words, radius of curvature of a spline can be dimensioned only if the spline is tangent to an entity. For dimensioning the radius of curvature of a spline, you need to follow the steps given below:

1. Choose the **Normal** tool from the **Dimension** group.

2. Select the endpoint of the spline where the tangency is defined.

3. Press the middle mouse button to place the dimension. Figure 3-16 shows the radius of curvature dimensioning of a spline.

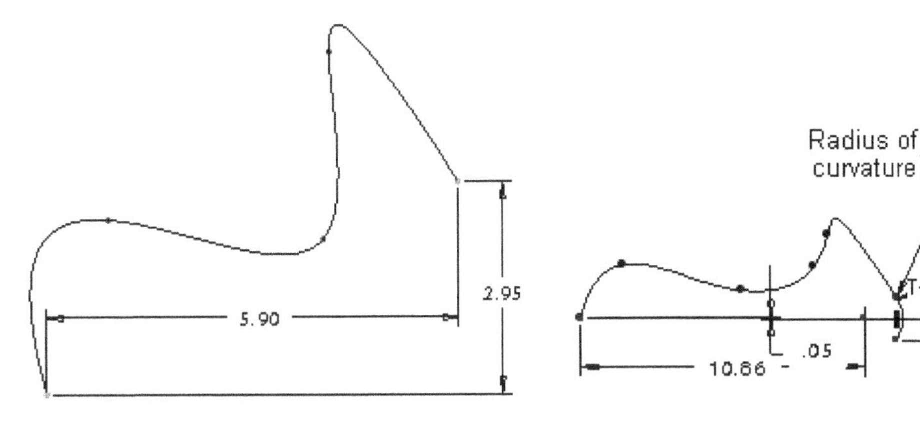

Figure 3-15 Endpoint dimensioning *Figure 3-16 Radius of curvature dimensioning*

Tangency or Angular Dimensioning

A spline can be dimensioned angularly with respect to a line tangent to it. This type of dimensioning is also called angular dimensioning. To angular dimension a spline and a line tangent to it, you need to follow the steps given below:

1. Choose the **Normal** tool from the **Dimension** group.
2. Select the spline by clicking the left mouse button.
3. Select the entity tangent to the spline by clicking the left mouse button.
4. Select the interpolation point of the spline that is to be dimensioned tangentially.
5. Press the middle mouse button to place the dimension.

Coordinate Dimensioning

The spline can be dimensioned with respect to a user-defined coordinate system. Choose the **Coordinate System** tool from the **Sketching** group. The coordinate system is attached to the cursor. Place the coordinate system in the graphics window. Now, the spline can be dimensioned with respect to the coordinate system.

Dimensioning the Interpolation Points

A spline can be dimensioned by dimensioning its interpolation points or vertices. This type of dimensioning is used when the designer wants the spline to be standard for all designs. This is because the exact curve can be duplicated if the interpolation points or the vertices of a spline are dimensioned.

Tip: *A dimension can be moved by choosing the* ***One-by-One*** *button from the* ***Operations*** *group and then pressing and holding the left mouse button on the dimension and moving it. The dimension text is replaced by a green colored box. You can drag the dimension to the desired location in the graphics window and release the left mouse button to place the dimension at that point.*

Modifying a Spline

In PTC Creo Parametric, a spline can be modified by:

1. Moving the interpolation points of the spline.
2. Adding points to a spline.
3. Deleting points of a spline.
4. Creating a control polygon and moving its control points.
5. Modifying the dimensions of the spline.

Moving the Points of a Spline

The position of the interpolation points can be dynamically modified. To modify a spline, select an interpolation point on the spline and then drag it to modify the shape of the spline.

Alternatively, select the spline and hold the right mouse button to invoke the shortcut menu. Choose the **Modify** option from the shortcut menu or double-click on the spline; the **Spline** tab will be displayed in the **Ribbon** with the options and buttons to modify a spline.

The interpolation points of the spline appear in black crossmarks in the graphics window. Drag the interpolation points to modify the shape of the spline.

Adding Interpolation Points to a Spline

To add interpolation points to a spline, choose the **Spline** tab. Now, right-click on the spline to invoke a shortcut menu. Next, choose the **Add Point** option; a point is added to the spline where the spline was selected. The new point appears in black crossmark. You cannot increase the length of the spline by adding points before the start point and after the endpoint of the spline.

Deleting Interpolation Points of a Spline

To delete a point or a vertex, choose the **Spline** tab. Next, right-click on the vertex to be removed to invoke the shortcut menu. Choose the **Delete Point** option from the shortcut menu; the selected point is deleted. You can continue deleting vertices or points from a spline until only two end points are left in the spline.

Creating a Control Polygon and Moving its Control Points

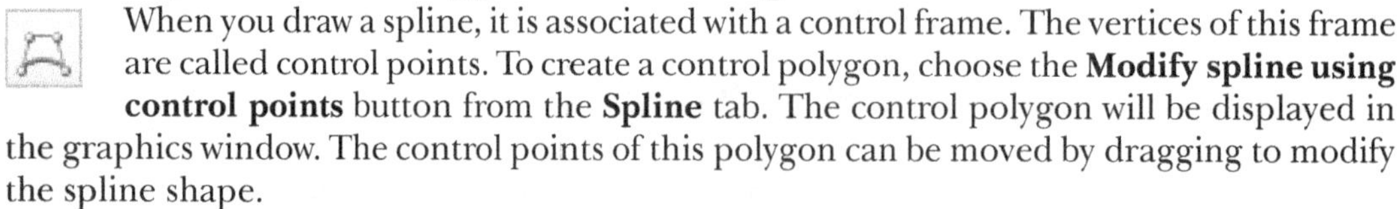

When you draw a spline, it is associated with a control frame. The vertices of this frame are called control points. To create a control polygon, choose the **Modify spline using control points** button from the **Spline** tab. The control polygon will be displayed in the graphics window. The control points of this polygon can be moved by dragging to modify the spline shape.

Tip: *To dynamically modify the shape of the sketch, you need to select an entity of the sketch and drag the mouse to modify the sketch. Remember that if the selected entity is constrained, then you cannot modify it. You can modify it only after disabling the constraints.*

Modifying the Dimensions of the Spline

The shape of the spline is controlled by the position of its interpolation points. Hence by modifying the dimensions, the position of the interpolation points are changed, which results in modification of the shape of the spline.

WRITING TEXT IN THE SKETCHER ENVIRONMENT

Ribbon: Sketch > Sketching > Text

There are various instances when a designer needs to write text on the model. For example, for creating a label, model number, company name, and so on. In PTC Creo Parametric, you can write this text in the sketcher environment.

In the sketcher environment, the text is written using the **Text** tool from the **Sketching** group. The following steps explain the procedure to write text in the sketcher environment:

1. Choose the **Text** tool from the **Sketching** group; you will be prompted to select the start point of line to determine the text height and orientation.

2. Specify the start point on the screen by clicking the left mouse button; you will be prompted to select the second point of line to determine the text height and orientation.

3. Note that to write the text upright, the second point should be above the start point and in a straight line. If the second point is below the start point, the text will be written down from right to left. Specify the second point on the screen by clicking the left mouse button; the **Text** dialog box will be displayed, as shown in Figure 3-17.

   After specifying the second point, a construction line is drawn having height equal to the distance between the two points. The height and orientation of the text depends on the height and angle of the construction line. If the construction line is drawn at an angle, then the text will be written at that angle.

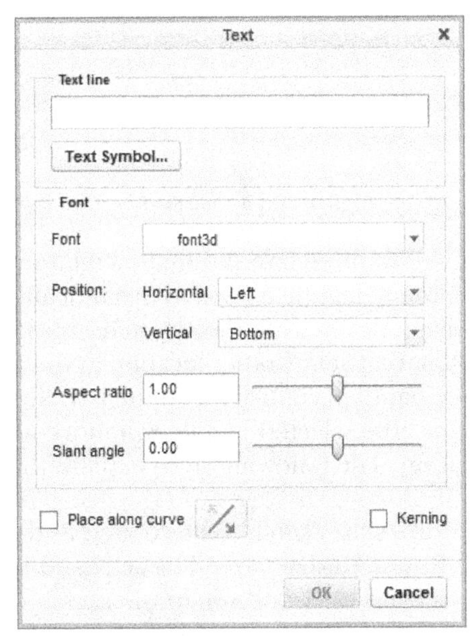

*Figure 3-17 The **Text** dialog box*

4. Enter the text in the **Text line** edit box, which can be up to 79 characters. As you enter the text, the text will be displayed dynamically in the graphics window. You can choose the desired font of the text from the **Font** drop-down list. The aspect ratio and the slant angle of the text can be controlled by using the slider bars.

5. Choose the **OK** button in the **Text** dialog box to exit it.

Note

In later chapters of this book, you will learn that there are other methods also to enter in to sketcher environment, besides entering through the sketch mode.

ROTATING AND RESIZING ENTITIES

Ribbon: Sketch > Editing > Rotate Resize

The sketches can be scaled or rotated by using the **Rotate Resize** tool available in the **Editing** group. Select a sketch and then choose the **Rotate Resize** tool from the **Editing** group. On choosing this tool, the sketch, which consists of various entities, will act as a single entity. Also, the sketch appears green in color and is enclosed within a boundary box, as shown in Figure 3-18.

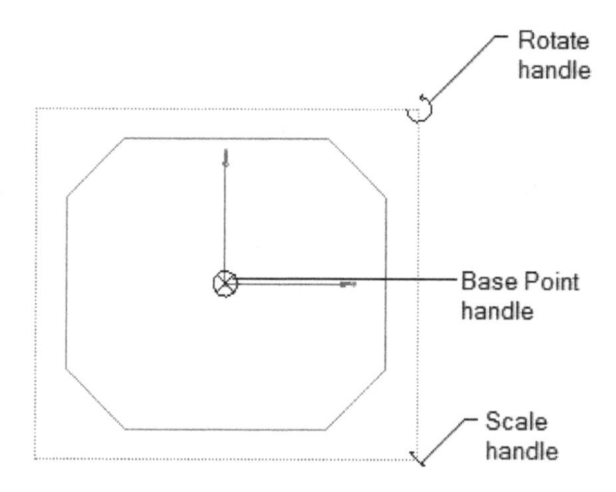

Figure 3-18 *Selected entities enclosed within a boundary box with three handles*

There are three handles that facilitate in scaling, rotating, and moving the selected sketch. The rotate handle is used to dynamically rotate the selected entities. The scale handle is used to dynamically scale the selected entities. The base point handle is used to pick the sketch and place it at any other location in the graphics window. To change the location of any of the three handles, right-click on the handle and drag it by pressing and holding the right mouse button; the selected handle will move along with the cursor. Place the symbol at the desired location. The following steps explain the procedure to move and resize a sketch:

1. Select the sketch to be rotated and scaled, and then choose the **Rotate Resize** tool; the **Rotate Resize** tab will be activated in the **Ribbon**. The options in these areas are used to move, scale, and rotate the sketch dynamically or by entering a value in the respective edit boxes. You can also select a reference about which you want to translate, rotate, or scale the sketch.

2. To move the sketch dynamically, select the move handle and then drag the handle to the required location; the sketch will be repositioned at the new location. You can also select a reference from the drawing area about which you want to move the sketch and the reference will be displayed in move collector of the dashboard.

3. To dynamically rotate the sketch, select the rotate handle and then move the cursor; the sketch will rotate as you move the cursor. You can also enter the rotation angle in the **Rotate** edit box.

4. To scale the sketch, select the scale handle and then move the cursor. As you move the cursor, the sketch is scaled dynamically in the graphics window. You can also enter the scale value in the **Scale** edit box.

5. After the sketch has been moved and resized, choose the **Done** button from the **Rotate Resize** tab.

IMPORTING 2D DRAWINGS IN THE SKETCH MODE

Ribbon: Sketch > Get Data > File System

The two-dimensional (2D) drawings when opened in the sketcher environment can be saved in the *.sec* format. The *.sec* file can be used to create a solid model. You can use a prestored sketch by importing it in the modeling environment. The **File System** button in the **Get Data** group is used to import 2D sketches. Using this button, you can save time in drawing the same or similar section again. The file formats from which the data can be imported are shown in Figure 3-19.

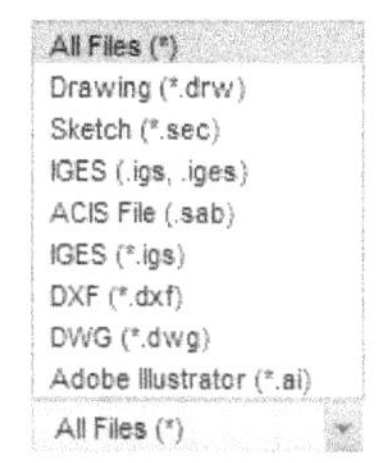

Figure 3-19 *File formats*

When you choose the **File System** button from the **Get Data** group; the **Open** dialog box will be displayed, as shown in Figure 3-20. You can use this dialog box to select and open the file.

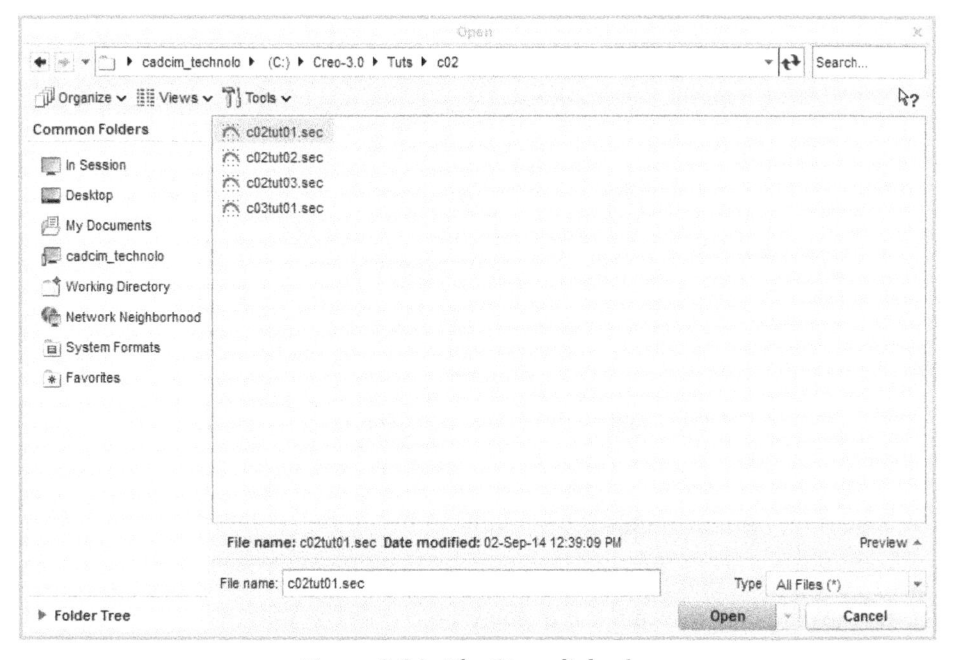

Figure 3-20 *The* ***Open*** *dialog box*

In the **Open** dialog box, if you select a drawing file created in the **Drawing** mode of PTC Creo Parametric, the draft entities of that file will be imported and selected drawing will be opened in a window. Also, you will be prompted to select the entities to copy from the window. Select the draft entities and then press the middle mouse button; the window disappears and a plus sign gets attached to the cursor indicating that you need to select a point on the PTC Creo Parametric screen to insert the file. Select a point on the screen; the selected entities get inserted and are displayed within an enclosed boundary. Also, the **Import Section** tab will be displayed. Use this dialog box to set the position, scale, and orientation of the sketch. Note that if the *.drw* file does not consist of draft entities, no data will be imported.

In the **Open** dialog box, if you select a *.sec* file that will be created in the sketcher environment, the sketch will be displayed in the graphics window enclosed within a boundary.

The section imported using the **File System** button in the current sketch is an independent copy. The imported section will be no longer associated with the source section. The units, dimensions, grid parameters, and accuracy are acquired from the current sketch.

Tip: *You can turn on or off the display of the vertices of the section, the dimensions, and the constraints from the* ***Sketching*** *group.*

TUTORIALS

Tutorial 1

In this tutorial, you will import an existing sketch that you had drawn in Tutorial 3 of Chapter 2. After placing the sketch, draw the keyway, as shown in Figure 3-21.

(Expected time: 15 min)

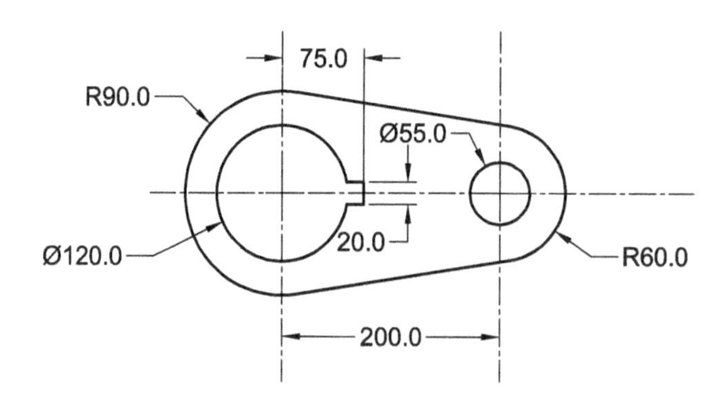

Figure 3-21 *Sketch for Tutorial 1*

The following steps are required to complete this tutorial:

a. Start PTC Creo Parametric.
b. Set the working directory and create a new object file.

c. Import the section by using the **File System** button, refer to Figure 3-22.
d. Draw the keyway and dimension it, refer to Figures 3-23 and 3-24.
e. Modify the dimensions, refer to Figure 3-25.
f. Save the sketch and exit the sketcher environment.

Starting PTC Creo Parametric

1. Start PTC Creo Parametric by double-clicking on the PTC Creo Parametric icon on the desktop of your computer.

Setting the Working Directory

When the PTC Creo Parametric session starts, the first task is to set the working directory. As mentioned earlier, working directory is a directory on your system where you can save the work done in the current session of PTC Creo Parametric. You can set any existing directory on your system as the working directory. Since this is the first tutorial of this chapter, you need to create a folder named *c03* in the *C:\Creo-3.0* folder.

1. Choose **Manage Session > Select Working Directory** option from the **File** menu; the **Select Working Directory** dialog box is displayed.

2. Select *C:>Creo-3.0*. If this folder does not exist, then first create it and then set the working directory. Alternatively, you can use the Folder Tree available on the bottom left corner of the screen to set the working directory. To do so, click on the **Folder Tree** node; the Folder Tree will expand. In the Folder Tree, browse to the desired location using the nodes corresponding to the folders and select the required folder. After selecting the folder, the **Folder Content** window will be displayed. Close this window and the selected folder will become your current working directory.

3. Choose the **Organize** button from the **Select Working Directory** dialog box or right-click in this dialog box to display a shortcut menu. From the shortcut menu, choose the **New Folder** option; the **New Folder** dialog box is displayed.

4. Enter **c03** in the **New Directory** edit box of the **New Folder** dialog box and then choose **OK**; a folder with the name *c03* is created at *C:\Creo-3.0*.

5. Choose **OK** from the **Select Working Directory** dialog box; *C:\Creo-3.0\c03* is set as the working directory.

Starting a New Object File

1. Choose the **New** button from the **Data** group; the **New** dialog box is displayed. Select the **Sketch** radio button from the **Type** area of the **New** dialog box; the default name of the sketch appears in the **Name** edit box.

2. Enter *c03tut1* in the **Name** edit box and choose the **OK** button.

   You are in the sketcher environment of the **Sketch** mode. When you enter the sketcher environment, the Navigator is displayed on the left of the graphics window. Slide-in the

Navigator by clicking on the sash present on its right edge. Now, the drawing area is increased.

Importing the Section

1. Choose **Get Data > File System** from the **Ribbon**; the **Open** dialog box is displayed with the working directory as the current directory.

2. Click on the black arrow beside the **Creo-3.0** option in the address bar and choose **c02** from the flyout displayed. Make sure the **Sketch (*.sec)** option is selected in the **Type** drop-down list. Select *c02tut3.sec* and choose the **Open** button from the **Open** dialog box.

3. Move the cursor in the drawing area. Notice that the cursor is attached with a plus mark. Now, click anywhere in the drawing area to place the sketch. The sketch is displayed in the drawing area and the **Import Section** tab is displayed in the **Ribbon**.

4. Enter **1** as a scale factor in the respective edit box and choose the **Done** button to complete importing the sketch.

5. Choose the **Refit** button from the **Graphics** toolbar. The sketch, similar to the one shown in Figure 3-22, is displayed in the drawing area.

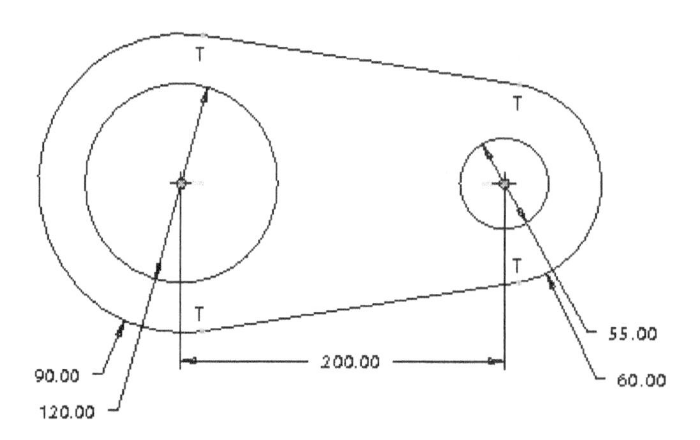

Figure 3-22 *Sketch imported and placed in the current file*

Drawing the Keyway

To create the keyway, you need to sketch a small rectangle and then remove the portion of the circle that lies between the rectangle.

1. Choose the **Line Chain** tool from **Line** drop-down in the **Sketching** group. 

2. Draw the keyway, as shown in Figure 3-23; the weak dimensions and constraints are automatically applied to the sketch of the keyway.

The horizontal lines of the keyway and the circle intersect at the points where the lines meet the circle. The portion of the circle that lies between the two horizontal lines of the keyway needs to be deleted from the circle.

3. Choose the **Zoom In** button from the **Orientation** group of the **View** tab; the cursor is converted into a magnifying glass symbol.

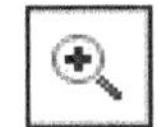

4. Draw a window around the keyway to zoom in it. Now, the display of the keyway is enlarged.

5. Choose the **Delete Segment** tool from the **Editing** group.

6. Click to select the part of the circle that lies between the two horizontal lines; the selected part is deleted.

7. Choose the **Refit** button from the **Graphics** toolbar to view the full sketch.

Dimensioning the Keyway

Now, you need to apply dimensions to the keyway.

1. Choose the **Normal** tool from the **Dimension** group.

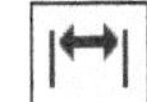

2. Dimension the keyway, as shown in Figure 3-24.

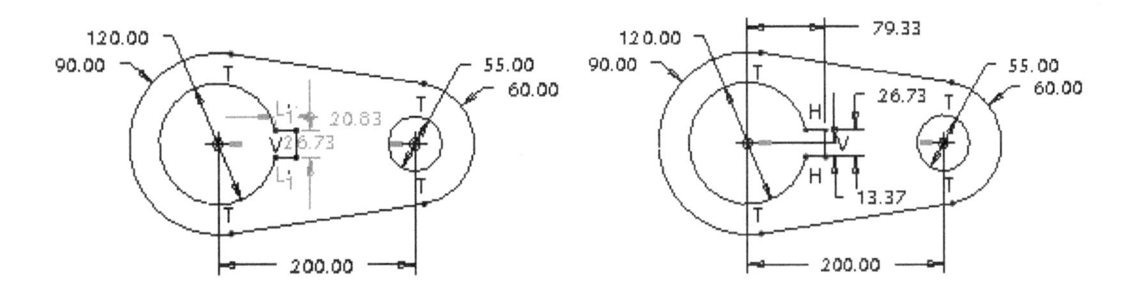

Figure 3-23 Sketch of the keyway with weak dimensions and constraints

Figure 3-24 Sketch after dimensioning the keyway

Modifying the Dimensions

The dimensions of the keyway need to be modified as per the given dimension values.

1. Select the three dimensions of the keyway by pressing CTRL+left mouse button.

2. Choose the **Modify** tool from the **Editing** group; the **Modify Dimensions** dialog box is displayed.

3. Clear the **Regenerate** check box and then modify the dimensions of the keyway. When you clear the check box, the sketch does not regenerate as you modify the dimensions.

   The dimension that you edit in the **Modify Dimensions** dialog box gets enclosed in a blue box in the sketch.

4. Modify all dimensions. Refer to Figure 3-21 for dimension values.

5. After the dimensions are modified, choose the **Regenerate the section and close the dialog** button from the **Modify Dimensions** dialog box; the message **Dimension modifications successfully completed** is displayed in the message area.

   The sketch after modifying the dimension values of the sketch is shown in Figure 3-25.

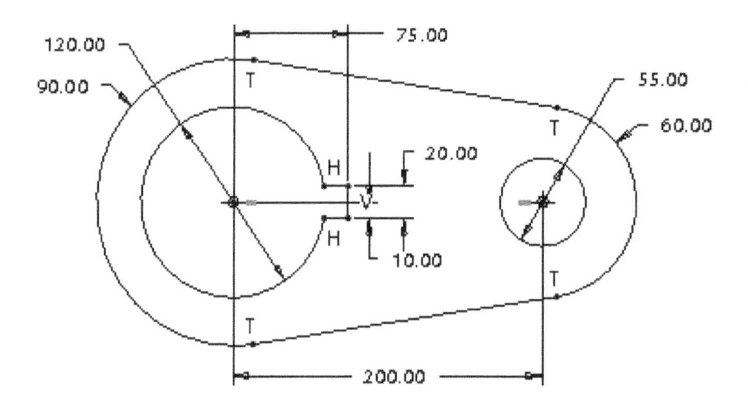

***Figure 3-25** Sketch after modifying the dimensions*

Saving the Sketch

As you may need the sketch later, you must save it.

1. Choose the **Save** button from the **File** menu; the **Save Object** dialog box is displayed with the name of the sketch entered earlier.

2. Choose the **OK** button; the sketch is saved.

3. After saving the sketch, choose the **Close** button from the **Quick Access** toolbar to exit the **Sketch** mode.

Tutorial 2

In this tutorial, you will draw the sketch for the model shown in Figure 3-26. The sketch is shown in Figure 3-27. **(Expected time: 30 min)**

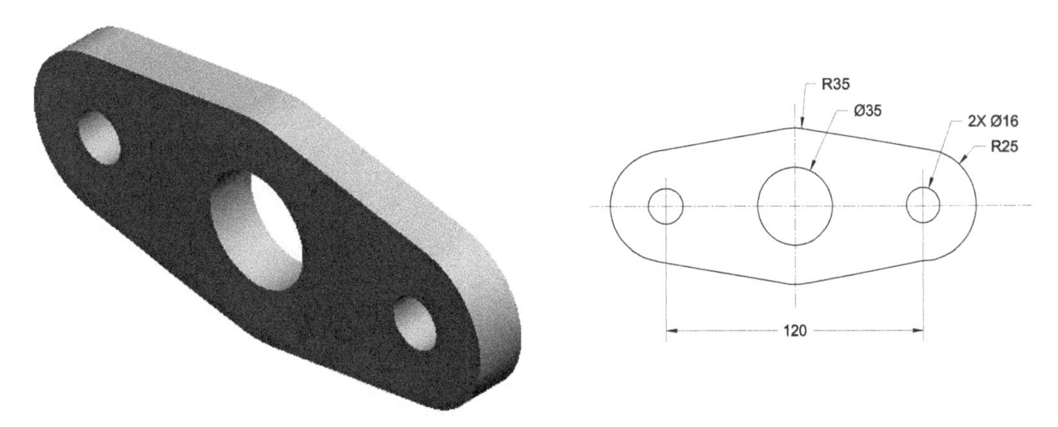

Figure 3-26 Model for Tutorial 2 *Figure 3-27 Sketch of the model*

The following steps are required to complete this tutorial:

a. Set the working directory and create a new object file.
b. Draw the sketch using sketcher tools, refer to Figures 3-28 and 3-29.
c. Apply the required constraints and dimensions to the sketched entities, refer to Figure 3-32.
d. Modify the dimensions of the sketch, refer to Figure 3-33.
e. Save the sketch and exit the **Sketch** mode.

Setting the Working Directory

The working directory was selected in Tutorial 1, and therefore there is no need to select the working directory again. But if a new session of PTC Creo Parametric is started, then you need to set the working directory again by following the steps given next.

1. Open the Navigator by choosing the **Show Navigator** button on the bottom left corner of the PTC Creo Parametric screen; the Navigator slides out. In the Navigator, the Folder Tree is displayed at the bottom. Click on the black arrow, which is available on the right of the Folder Tree; the Folder Tree expands.

2. Click on the node adjacent to the *Creo-3.0* folder in the Navigator; the contents of the *Creo-3.0* folder are displayed.

3. Now, right-click on the *c03* folder to display a shortcut menu. From this shortcut menu, choose the **Set Working Directory** option; *c03* is set as the working directory.

4. Close the Navigator by clicking the sash on the right edge of the Navigator; the Navigator slides in.

Starting a New Object File

1. Start a new object file in the **Sketch** mode. Name the file as *c03tut2*.

Drawing the Sketch

To draw the outer loop, you need to draw three circles and then draw lines tangent to them.

1. Choose the **Center and Point** tool from the **Circle** drop-down. Draw the circles in a horizontal line, as shown in Figure 3-28.

2. Choose the **Line Tangent** tool from the **Line** drop-down in the **Sketching** group.
3. Select the left and middle circles on their respective top points; a tangent is drawn from the top of the left circle to the top of the middle circle.
4. Next, select the right and middle circles on their respective top points; a tangent is drawn from the top of the right circle to the top of the middle circle.
5. Similarly, using the **Line Tangent** tool, draw the other tangents through the bottom-most points of the left, middle, and right circles, as shown in Figure 3-29.

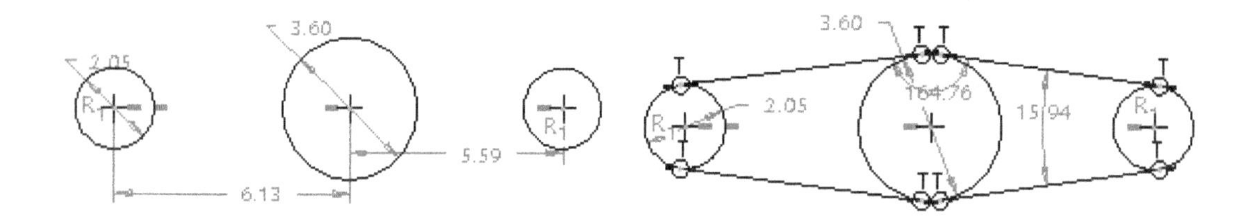

Figure 3-28 *Three circles with weak dimensions and constraints*

Figure 3-29 *Tangent lines drawn on the circles*

Trimming the Circles

The inner portions of the circles are not required. Therefore, you need to trim them.

1. Choose the **Delete Segment** tool from the **Editing** group.
2. Move the cursor close to the right portion of the left circle; the right part of the circle turns green in color. Next, click on it to delete it.
3. Similarly, trim the parts of the middle and right circles that are not required. The sketch after trimming the circles is shown in Figure 3-30.

Drawing the Circles

1. Choose the **Concentric** tool from the **Circle** drop-down in the **Sketching** group; you are prompted to select an arc.

2. Click on the left arc and move the mouse; a circle appears. Select a point inside the sketch to complete the circle. Press the middle mouse button.

Tip: *While drawing a concentric circle, sometimes the circle snaps to the other circle or arc and it becomes difficult to draw a circle of the size you need. In such a case, you can disable the snapping of the circle to the other circle or arc. To do so, hold the SHIFT key to disable the snapping or to disable the equal radii constraint that the system tends to apply while drawing the circle.*

Note

You may need to zoom in drawing to select the top arc in the next step.

3. Click on the top arc and move the mouse; a circle appears. Select a point inside the sketch to complete the circle. Press the middle mouse button to end the creation of circle.

4. Click on the right arc and move the mouse; a circle appears. Select a point inside the sketch to complete the circle. Press the middle mouse button to end the creation of circle.

   The sketch after drawing all three circles is shown in Figure 3-31.

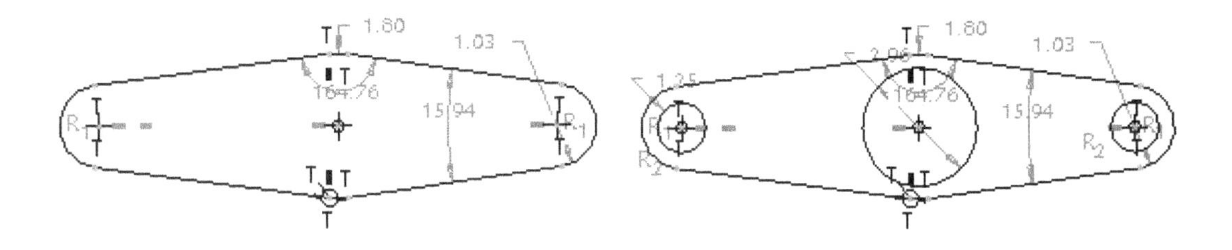

Figure 3-30 *Sketch after trimming the circles* ***Figure 3-31*** *Sketch after drawing all three circles*

Applying the Constraints

1. Choose the **Equal** tool from the **Constrain** group.

2. Select the left arc and then select the right arc to apply the equal radius constraint.

3. Select the left circle and the right circle to apply the equal radius constraint.

4. Choose the **Parallel** tool from the **Constrain** group.

5. Click to select the tangent line that connects the left arc and the middle arc at the top and then click to select the tangent line that connects the right arc and the middle arc at the bottom. It is evident from the parallel constraint symbol that the parallel constraint has been applied to the two tangent lines in the sketch.

Dimensioning the Sketch

PTC Creo Parametric applies weak dimensions to the sketch automatically. These dimensions are not the needed dimensions because these dimensions do not help in machining the model. Therefore, you need to dimension the sketch with the dimensions that will be used to machine the model. To do so, follow the steps given next.

1. Choose the **Normal** tool from the **Dimension** group.

2. Select the center of the right and left circles; the centers of the circles turn black in color. Now, using the middle mouse button, place the dimension below the sketch. The sketch after applying constraints is shown in Figure 3-32.

   The rest of the weak dimensions are the needed dimensions.

Modifying the Dimensions

All constraints and dimensions have been applied to the sketch. Now, you need to modify the dimensions.

1. Select all dimensions using the CTRL+ALT+A keys.

2. Choose the **Modify** tool from the **Editing** group; the **Modify Dimensions** dialog box is displayed.

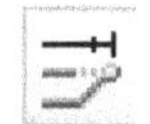

3. Clear the **Regenerate** check box and then modify the values of the dimensions, refer to Figure 3-27 for dimension values.

   When you clear the check box, the sketch does not regenerate while modifying the dimensions. The dimension that you edit in the **Modify Dimensions** dialog box is enclosed in a blue box in the sketch.

4. After the dimensions are modified, choose the **OK (Regenerate the section and close the dialog)** button from the **Modify Dimensions** dialog box; the message **Dimension modifications successfully completed** is displayed in the message area.

   The sketch after modifying the dimension values is shown in Figure 3-33.

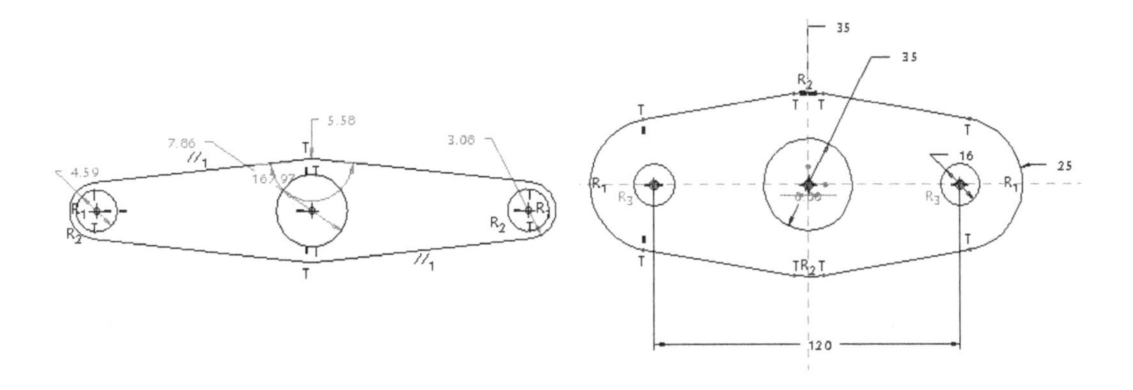

Figure 3-32 Sketch with constraints *Figure 3-33* Sketch after modifying the dimensions

Saving the Sketch

1. Choose the **Save** button from the **File** menu and save the sketch.

Exiting the Sketch Mode

1. Choose the **Close** button from the **Quick Access** toolbar to exit the Sketch.

Tutorial 3

In this tutorial, you will draw the sketch of the model shown in Figure 3-34. The sketch is shown in Figure 3-35. **(Expected time: 30 min)**

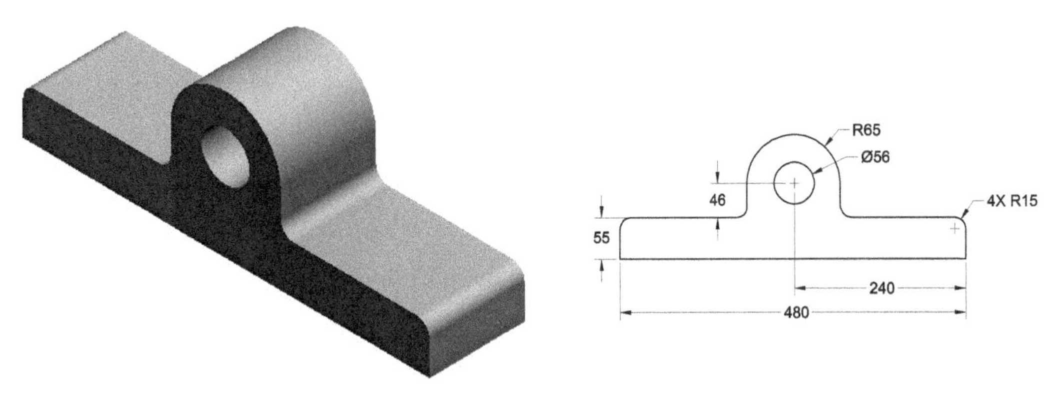

Figure 3-34 Model for Tutorial 3 *Figure 3-35* Sketch of the model

The following steps are required to complete this tutorial:

a. Set the working directory and create a new object file.
b. Draw the sketch using sketcher tools, refer to Figures 3-36 and 3-37.
c. Apply fillets at two corners of the sketch, refer to Figures 3-38 and 3-39.

d. Dimension the sketch, refer to Figure 3-40.
e. Modify dimensions of the sketch, refer to Figure 3-41.
f. Save the sketch and exit the **Sketch** mode.

Setting the Working Directory

The working directory was selected in Tutorial 1, and therefore there is no need to select the working directory again. But if a new session of PTC Creo Parametric start, then you have to set the working directory again by following the steps given next.

1. Open the Navigator by clicking on the sash in the left edge of the PTC Creo Parametric window; the Navigator slides out. In the Navigator, the Folder Tree is displayed at the bottom. Click on the black arrow that is available at the right-side of the Folder Tree; to expand it.

2. Click on the node adjacent to the *Creo-3.0* folder in the Navigator to display the content of this folder.

3. Now, right-click on the *c03* folder to display a shortcut menu. From this shortcut menu, choose the **Set Working Directory** option; *c03* is set as the working directory.

4. Close the Navigator by clicking on the sash on the right edge of the Navigator; the Navigator slides in.

Starting a New Object File

1. Start a new object file in the **Sketch** mode. Name the file as *c03tut3*.

Drawing the Sketch

1. Choose the **Line Chain** tool from the **Line** drop-down of the **Sketching** group.

2. Draw the lines with constraints, as shown in Figure 3-36.

3. Choose the **3-Point / Tangent End** tool from the **Arc** drop-down available in the **Sketching** group.

4. Select the endpoint of the left vertical line as the start point of the arc. Complete the arc at the endpoint of the right vertical line.

5. Choose the **Concentric** tool from the **Circle** drop-down available in the **Sketching** group; you are prompted to select an arc.

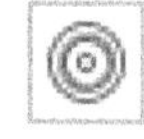

6. Click on the arc; a rubber-band circle appears. Size the circle by moving the cursor and click to complete it. The sketch after drawing the circle is shown in Figure 3-37.

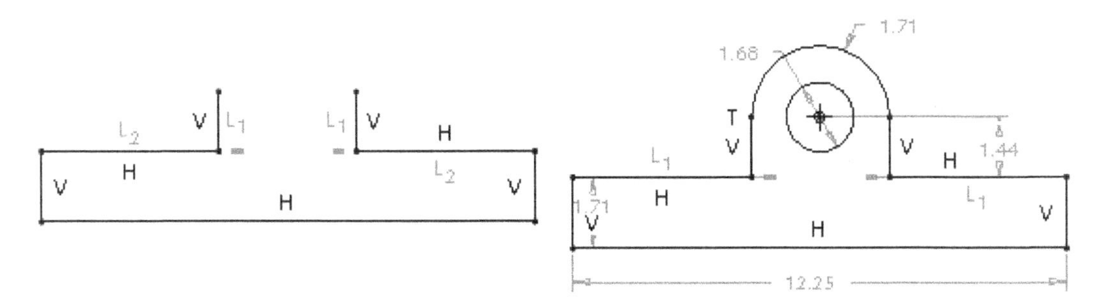

Figure 3-36 Lines in the sketch with the dimensions turned off for clarity

Figure 3-37 Sketch after drawing the arc and the circle

Note

*1. Choose the **Disp Dims** button from the **Display** group in the **View** tab to turn the dimensions on or off.*

2. PTC Creo Parametric does not have the options like midpoint, endpoint, or center of an arc or a circle. However, while drawing a sketch, these options are applied in the form of weak constraints. For example, while drawing an entity, the endpoint of the entity snaps to the cursor. The middle point constraint appears when you bring the cursor near to the middle point of the line to draw another line.

Filleting the Corners

1. Choose the **Circular Trim** option from the **Fillet** drop-down in the **Sketching** group; you are prompted to select the two entities to be filleted. The corners that you need to fillet are shown in Figure 3-38.

2. Select the two entities one by one using the left mouse button to fillet the corners of these entities.

   The sketch after creating the fillets is shown in Figure 3-39.

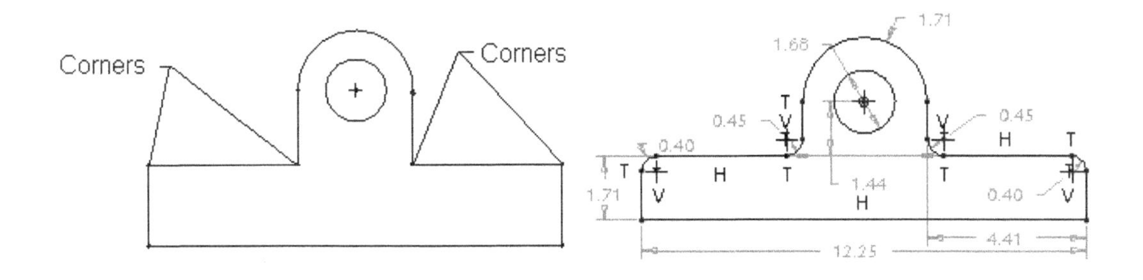

Figure 3-38 Corners to be filleted

Figure 3-39 Sketch after creating fillets

Applying the Constraints

1. Choose the **Equal** tool from the **Constrain** group.

2. Click to select the fillets and apply the equal constraint to all fillets.

Dimensioning the Sketch

The weak dimensions are applied to the sketch automatically. These are not the required dimensions and therefore, you need to dimension the sketch manually.

1. Choose the **Normal** tool from the **Dimension** group.

2. Dimension the sketch, as shown in Figure 3-40.

Modifying the Dimensions

You need to modify the dimension values that are assigned to the sketch.

1. Select all dimensions using CTRL+ALT+A.

2. Choose the **Modify** tool from the **Editing** group; the **Modify Dimensions** dialog box is displayed.

3. Clear the **Regenerate** check box and then modify the values of the dimensions. If this check box is cleared, the sketch does not regenerate while modifying the dimensions.

   The dimension that you edit in the **Modify Dimensions** dialog box is enclosed in a blue box in the sketch.

4. Modify all dimensions. Refer to Figure 3-35 for dimension values.

5. Choose the **OK (Regenerate the section and close the dialog)** button from the **Modify Dimensions)** dialog box.

   The sketch after modifying the dimension values is shown in Figure 3-41.

Saving the Sketch

1. Choose the **Save** button from the **File** menu and save the sketch.

Exiting the Sketch Mode

1. Choose the **Close** button from the **Quick Access** toolbar to exit the **Sketch** mode.

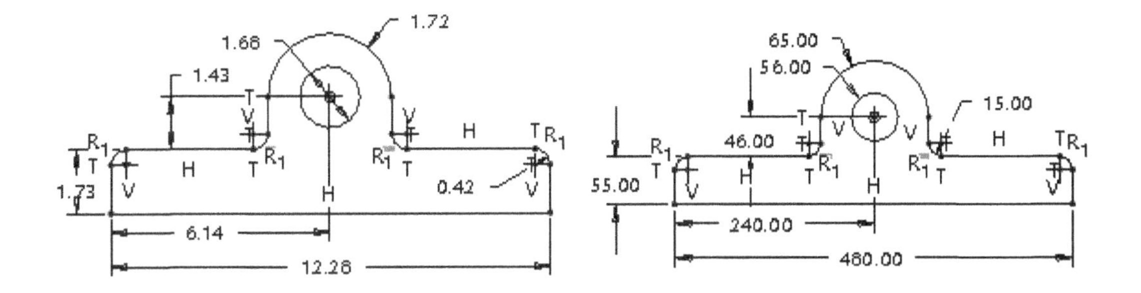

Figure 3-40 *Sketch after dimensioning*

Figure 3-41 *Sketch after modifying the dimensions*

Self-Evaluation Test

Answer the following questions and then compare them to those given at the end of this chapter:

1. The _________ dialog box is used to modify dimensions.

2. The display of dimensions and constraints can be turned on/off by choosing the _________ and _________ buttons respectively from the **Display** group.

3. The _________ tool is used to rotate selected entities.

4. You can delete entities by selecting them and then using the _________ key on the keyboard.

5. The shape of a spline can be modified dynamically. (T/F)

6. You can increase the length of the spline by adding points before the start point and after the endpoint of the spline. (T/F)

7. While copying the sketched entities, the **Paste** tab is also displayed. (T/F)

8. When you modify a weak dimension, it becomes strong. (T/F)

9. You can dimension the length of a centerline. (T/F)

10. The font of the text written in the sketcher environment cannot be modified. (T/F)

Review Questions

Answer the following questions:

1. How many handles are displayed in the graphics window while rotating and resizing entities?

   (a) one (b) two
   (c) three (d) four

2. Which of the following mouse buttons is used to place the dimension?

   (a) left (b) middle
   (c) right (d) mouse is not used for dimensioning

3. Which of the following is the default font for the text in the sketcher environment?

   (a) font (b) filled
   (c) font3d (d) isofont

4. Which of the following groups is used to toggle the display of dimensions and constraints in the sketcher environment?

   (a) **Sketching** (b) **Constrain**
   (c) **Display** (d) **Dimension**

5. In which type of dimensioning, the **Dim Orientation** dialog box is displayed while dimensioning the arcs and circles?

   (a) **Normal** (b) **Perimeter**
   (c) **Baseline** (d) None of the above

6. For placing a section in a new sketch, you can use the right mouse button. (T/F)

7. You can create elliptical fillets in PTC Creo Parametric. (T/F)

8. While creating text in the sketcher environment, you need to draw a construction line that will define the height of the text. (T/F)

9. You can modify the dimensions dynamically. (T/F)

10. You can modify a spline by moving its interpolation points. (T/F)

Exercises

Exercise 1

In this exercise, you will draw the sketch of the model shown in Figure 3-42. The sketch to be drawn is shown in Figure 3-43. **(Expected time: 30 min)**

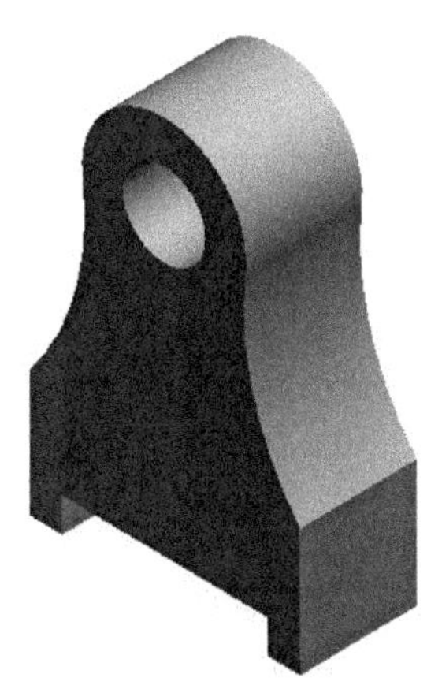

Figure 3-42 Solid model for Exercise 1

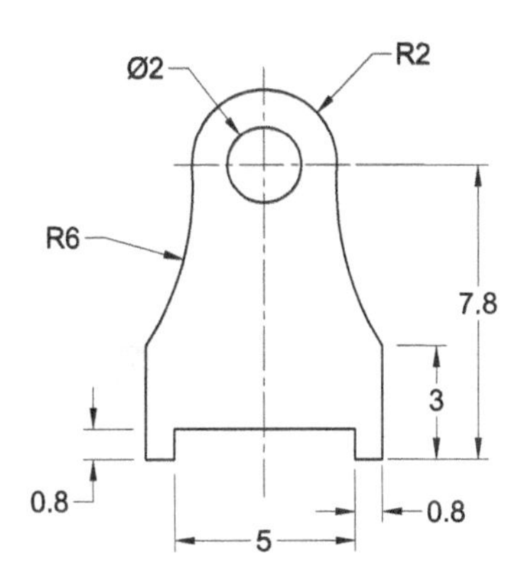

Figure 3-43 Sketch of the model

Exercise 2

In this exercise, you will draw the sketch of the model shown in Figure 3-44. The sketch to be drawn is shown in Figure 3-45. **(Expected time: 15 min)**

Figure 3-44 Solid model for Exercise 2

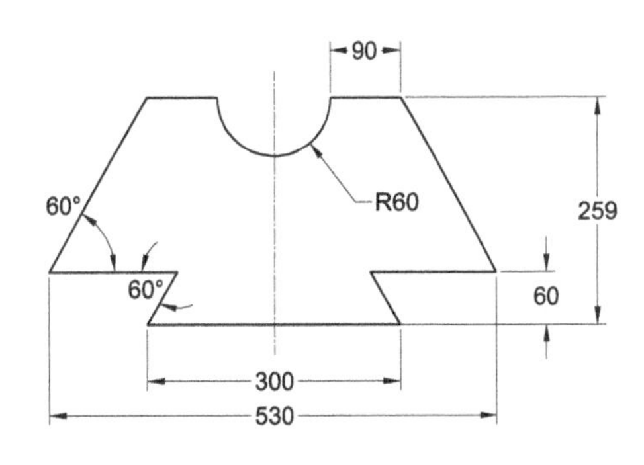

Figure 3-45 Sketch of the model

Exercise 3

In this exercise, you will draw the sketch of the model shown in Figure 3-46. The sketch to be drawn is shown in Figure 3-47. **(Expected time: 30 min)**

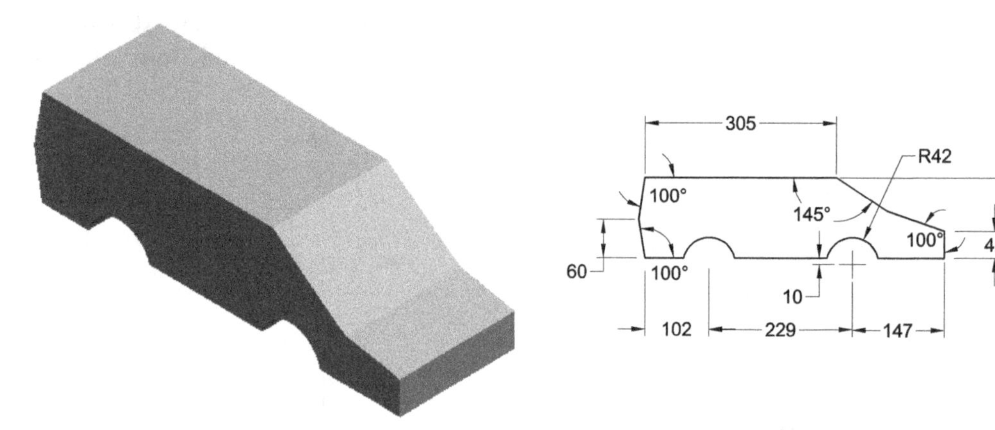

Figure 3-46 Solid model for Exercise 3

Figure 3-47 Sketch of the model

Answers to Self-Evaluation Test

1. **Modify Dimensions**, **2.** **Disp Dims**, **Disp Constr**, **3.** **Rotate Resize**, **4.** DELETE, **5.** T, **6.** F, **7.** T, **8.** T, **9.** F, **10.** F

Chapter 4

Creating Base Features

Learning Objectives

After completing this chapter, you will be able to:

- *Use default datums for creating base feature.*
- *Create a solid feature by using the Extrude tool.*
- *Create a thin feature by using the Extrude tool.*
- *Create a solid feature by using the Revolve tool.*
- *Create a thin feature by using the Revolve tool.*
- *Specify the depth of extrusion of a solid feature.*
- *Specify the angle of revolution of a revolved feature.*
- *Orient the datum planes.*
- *Understand the Parent/Child relationship.*
- *Understand the nesting of sketches.*
- *Create extrude and revolve cuts.*

CREATING BASE FEATURES

The base feature is the first solid feature created while creating a model in the **Part** mode. The base feature is created using the datum planes. Although you can create a base feature without using the datum planes, but in that case, you will not have proper control over the orientation of feature and direction of feature creation.

While creating the base feature of a model, the designer should be extra careful in specifying its attributes. This is because if the base feature is not created correctly, then the features created on it will also be inappropriate. This results in waste of time and effort. Although PTC Creo Parametric provides you with the options to redefine a feature, doing that also consumes additional time and effort. To create the base feature, you need to enter the **Part** mode.

Tip: *It is recommended that you set the working directory before starting a new file.*

Invoking the Part mode

To invoke the **Part** mode, choose the **New** option from the **File** menu or choose the **New** button from the **Data** group; the **New** dialog box will be displayed with various modes of PTC Creo Parametric. The **Part** radio button in the **Type** area and the **Solid** radio button in the **Sub-type** area of the **New** dialog box are selected by default. The default name of the part file also appears in the **Name** edit box. You can change the part name as desired and then choose the **OK** button to enter the **Part** mode.

In the **New** dialog box, the **Use default template** check box is selected by default. As a result, the default template provided in PTC Creo Parametric will be used for the part file. This template has certain parameters that will be assigned to the part file that you will create. For example, the units of this model will be Inch lbm Second; the length will be in inch; mass in lb; time in seconds, and temperature in Fahrenheit. If you clear the **Use default template** check box and choose the **OK** button from the **New** dialog box, the **New File Options** dialog box will be displayed, as shown in Figure 4-1.

From the **New File Options** dialog box, you can select the required template file. If you want the default system of units to be mmNs (millimeter Newton sec), then select the **mmns_part_solid** template and choose the **OK** button from this dialog box.

The **Part** mode is the most commonly used mode of PTC Creo Parametric. This is because solid modeling is done in this mode. It should be noted that a solid model is the start of a product development cycle. Product development cycle refers to the development of a product from scratch to its prototype. If you have created a solid model, then it can further be used to generate its drawing views, numerically controlled (NC) machining codes, core and cavity, analyze of the solid model, and so on.

The two-dimensional (2D) sketch drawn in the **Sketch** mode can be converted into a three-dimensional (3D) model in the **Part** mode. The **Part** mode contains the same sketcher environment with similar options to sketch as those available in the **Sketch** mode. There are some sketcher options in the **Part** mode that are not available in the **Sketch** mode because they do not have any use in the **Sketch** mode.

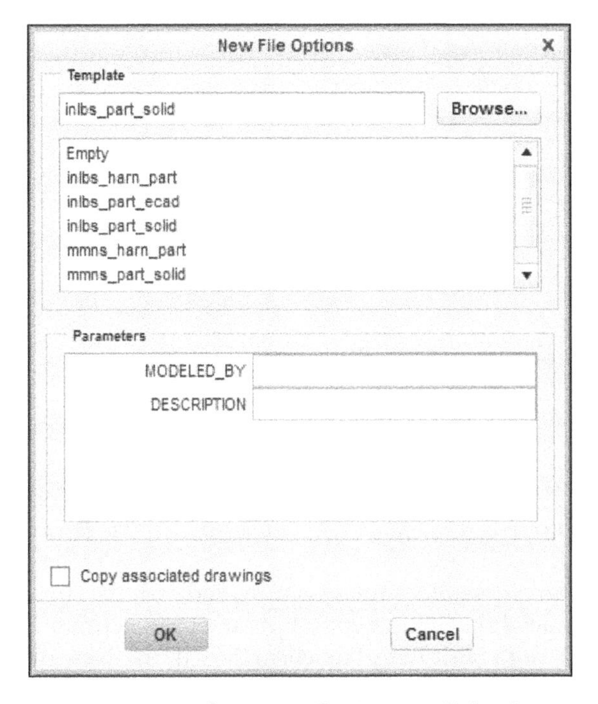

Figure 4-1 *The* ***New File Options*** *dialog box*

Figure 4-2 shows the initial screen appearance in the **Part** mode. It displays **Model Tree**, three default datum planes, **Ribbon**, and toolbars.

Tip: *It is recommended that you always use the datum planes to create a base feature. This is because the model created using the datum planes can be easily oriented. The uses of datum planes are discussed in Chapter 5.*

The Default Datum Planes

Generally, the three default datum planes act as the first feature in the **Part** mode. These datum planes are used to create the base feature. Also, they are used to draw a 2D sketch and then convert it into a 3D model by using the feature creation tools. The base feature that you create is referenced with default datum planes.

Note

Although, it is said that the three default datum planes are the first feature in the ***Part*** *mode, in the* ***Model Tree****,* ***RIGHT****,* ***TOP****, and* ***FRONT*** *datum planes appear as three separate features. If you delete any one of them, only that datum plane is deleted.*

These three default datum planes are mutually perpendicular to each other. They are not referenced to each other and are individual features. When a solid model is created, these datum planes adjust their size to the size of the model. You can create any number of datum planes as per your requirement. The creation of additional datum planes is discussed in Chapter 5.

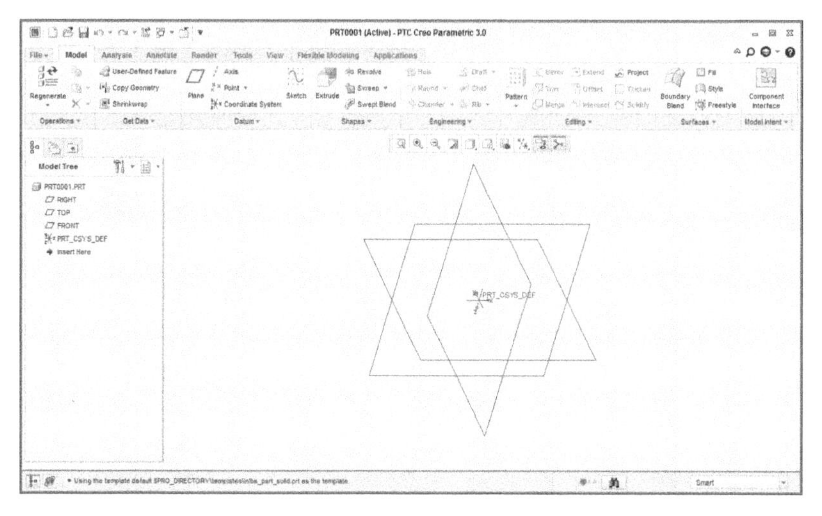

Figure 4-2 *The initial screen appearance in the* ***Part*** *mode*

In PTC Creo Parametric, a default template is provided with the three default datum planes. However, if you do not need the default datum planes, then at the time of creating a new file, you need to clear the **Use default template** check box in the **New** dialog box and select the **Empty** template from the **New File Options** dialog box.

Note

1. In this chapter, you will learn to use three default datum planes to create the base features. It is important to remember the feature-based nature of PTC Creo Parametric while working on this chapter. The feature-based nature of PTC Creo Parametric has been discussed in Chapter 1.

2. It is important to create a model in the correct orientation. The correct orientation is the one in which the model will be used later either in assembly or other modes. For example, the Casting component of the Plummer block is modeled in such a way that it always stands vertical.

CREATING A PROTRUSION

Ribbon: Model > Shapes

Protrusion is defined as a process of adding material defined by a sketched section. In PTC Creo Parametric, there are various options available for adding material to a sketch such as **Extrude**, **Revolve**, **Sweep**, and so on. These options can be chosen from the **Shapes** group in the **Model** tab. Figure 4-3 shows the options in the **Shapes** group.

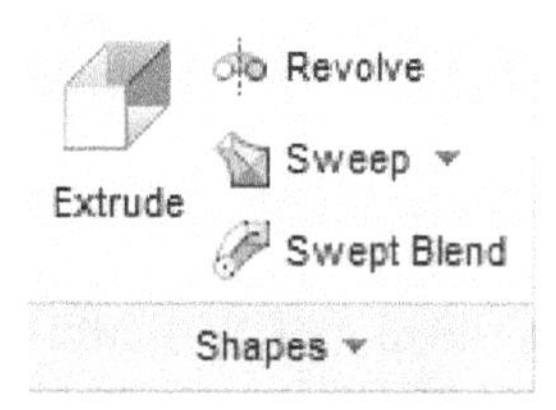

Figure 4-3 *The* ***Shapes*** *group*

Tip: *Remember that if you are creating a protrusion (a material addition process), then no matter what the direction of viewing you choose, the protrusion takes place in the direction toward the user, that is, out of the screen.*

Extruding a Sketch

Ribbon: Model > Shapes > Extrude

The **Extrude** tool in the **Shapes** group adds material to the area defined by a closed sketch drawn. The procedure of creating a base feature by using the **Extrude** tool is explained next.

1. Choose the **Extrude** tool from the **Shapes** group; a dashboard, named **Extrude** will be displayed on the top of the drawing area. This dashboard has all options to define the extrude feature, refer to Figure 4-4.

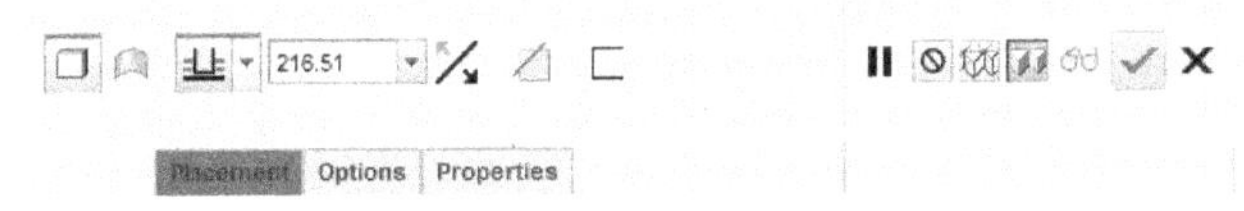

Figure 4-4 *Partial view of the* ***Extrude*** *dashboard*

2. Choose the **Placement** tab from the dashboard to display the slide-down panel. Choose the **Define** button from the slide-down panel to display the **Sketch** dialog box, as shown in Figure 4-5. Alternatively, right-click in the drawing area to display the shortcut menu. Choose the **Define Internal Sketch** option from the shortcut menu; the **Sketch** dialog box will be displayed. You will be prompted to select a sketching plane. Select the datum plane named **FRONT** from the drawing area as the sketching plane.

   In the **Sketch** dialog box, the name of the plane that you have selected appears in the **Plane** collector. At the same time, the reference plane is selected by default. The name of the reference datum plane appears in the **Reference** collector of the **Sketch** dialog box. PTC Creo Parametric sets the orientation of the reference plane by default. You can also set the orientation of the reference plane as required by selecting an appropriate option from the **Orientation** drop-down list.

3. After selecting the sketching plane and the reference plane, choose the **Sketch** button in the **Sketch** dialog box.

   Now, you have entered in the sketcher environment. The selected datum plane is the sketching plane and is not oriented parallel to the screen. To make it parallel to the screen, choose the **Sketch View** button from the **Graphic** toolbar. Alternatively, choose the **Sketch View** button from the **Setup** group of the **Sketch** tab. In PTC Creo Parametric, the references for the sketch are applied by default. So, you need not select references for the sketches.

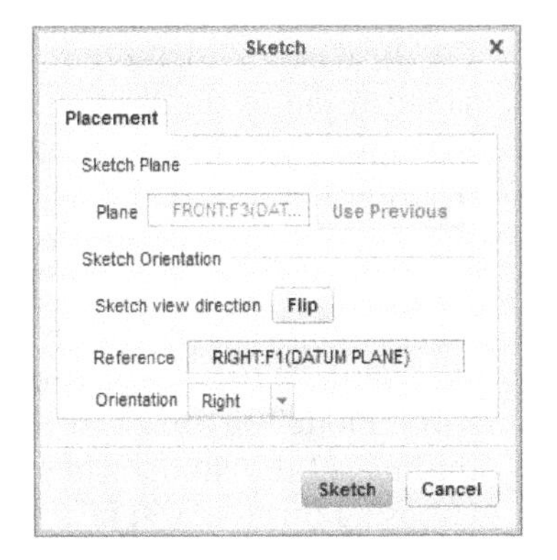

Figure 4-5 *The* ***Sketch*** *dialog box*

If you want to increase the drawing space in the drawing area, close the **Model Tree** by choosing the **Show Navigator** button at the bottom left corner of the screen, refer to Figure 4-6.

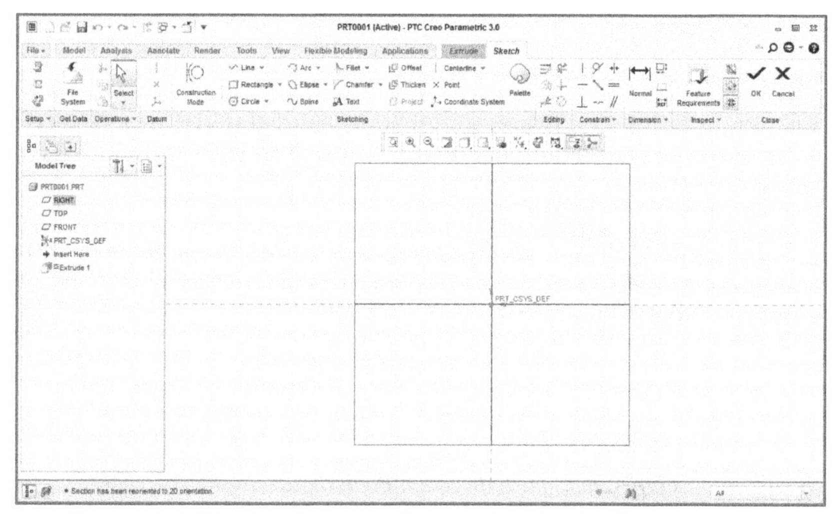

Figure 4-6 The drawing area after entering the sketcher environment

4. Now, choose the **Line Chain** tool from the **Line** drop-down in the **Sketching** group to draw the right-half of the I-section and then draw a vertical centerline aligned with the **RIGHT** datum plane. Select all lines on the right-half of the I-section and choose the **Mirror** button from the **Editing** group; you will be prompted to select a centerline. Select the centerline to mirror the right-half and to create the I-section, as shown in Figure 4-7. After the I-section is sketched, choose the **OK** button from the **Sketch** dashboard.

After doing so, the dashboard will be enabled and displayed above the drawing area. The model is created by assuming some default attributes and its preview is displayed in orange color. The attributes of a model are available on the dashboard and are discussed later in this chapter.

On the dashboard, there is an edit box containing a default value. This value is the depth of extrusion of the sketch that you have created. Enter the desired value in the edit box and press ENTER. You can also hold and drag the handle displayed on the model to dynamically specify the depth of extrusion. This drag handle will be visible if the model is rotated slightly using the middle mouse button.

Three buttons are also available in the dashboard, **No Preview**, **Un-attached**, and **Attached**. When you invoke the **No Preview** tool preview of the model is not displayed while modeling. You can invoke the **Un-attached** tool when you want to see only the outlines of the model while working with the **Extrude** tool. The preview displayed after choosing this tool will be a transparent. You can invoke the **Attached** tool when you want to see the final visualization of the model to be created.

5. Choose the **Build Feature** button; the required 3D model will be displayed in the drawing area. However, it will appear as a 2D entity. You need to change the display of the model such that the depth of the model can also be viewed. To do so, choose the **Named Views** button from the **Graphics** toolbar and choose the **Default Orientation** option from the flyout displayed. Alternatively, press CTRL+D to change the view to default orientation. Figure 4-7 shows the I-section after extruding it to a certain depth.

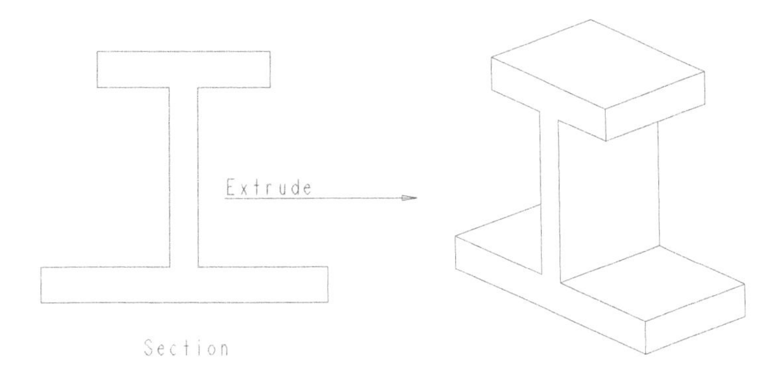

Figure 4-7 *The I-section extruded to a certain depth*

Figures 4-8 and 4-9 show some models created using the **Extrude** tool.

Figure 4-8 *Model created using the **Extrude** button*

Figure 4-9 *Model created using the **Extrude** button*

The above steps explain how to construct a 3D model using the **Extrude** tool. The **Extrude** dashboard and the **Sketch** dialog box that were used while creating the 3D model from the I-section are discussed next.

The Extrude Dashboard

The options and buttons in the **Extrude** dashboard of the **Ribbon**, shown in Figure 4-10, are used to extrude a sketch and to specify certain attributes related to the model. These attributes can be assigned to the model before or after drawing the sketch of the model. The tabs and the buttons available in the dashboard are discussed next.

Figure 4-10 Partial view of the ***Extrude*** *dashboard*

Placement Tab

Choose the **Placement** tab; a slide-down panel will be displayed, as shown in Figure 4-11. The collector in this panel displays **Select 1 item** because the section has not been defined yet. You can select a sketch that is drawn using a datum curve or you can choose the **Define** button to draw a sketch. Choose the **Define** button from this slide-down panel; the **Sketch** dialog box will be displayed.

Figure 4-11 The slide-down panel displayed on choosing the ***Placement*** *tab*

Options Tab

Choose the **Options** tab; a slide-down panel will be displayed, as shown in Figure 4-12. The slide-down panel is used to specify whether you want the sketch to extrude to one side or to both sides of the sketching plane. It has **Side 1** and **Side 2** drop-downs. The options in the **Side 1** drop-down are discussed next.

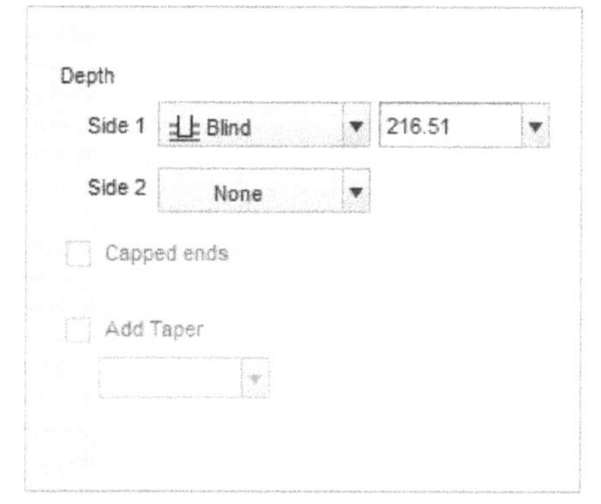

Figure 4-12 The slide-down panel displayed on choosing the ***Options*** *tab*

Blind: The **Blind** option is the most commonly used option to define the extrusion depth of the sketch by specifying a depth value. When you select this option from the **Side 1** drop-down, a default value appears in the edit box that is adjacent to this drop-down. Enter a value in this edit box and press ENTER to modify the depth of extrusion.

Symmetric: If you select the **Symmetric** option from the **Side 1** drop-down, the material will be equally added in both the directions of the sketching plane. When you select this option, the **Side 2** drop-down becomes inactive that means you cannot select any option from the this drop-down.

To Next: The **To Next** option is used to set the extrusion depth of the sketch up to the next surface of the model in the direction of extrusion.

Through All: The **Through All** option is used to set the extrusion depth of the sketch through all the surfaces of the model in the direction of extrusion.

Through Until: The **Through Until** option is used to select the surface up to which the sketch is to be extruded. While using this option, you cannot select the point or plane for reference.

To Selected: The **To Selected** option is used to select a point, surface, plane, or a curve up to which the section is to be extruded.

Note

*The **To Next**, **Through All**, and **Through Until** options are available in the drop-down list only after creating the base feature.*

The options in the **Side 2** drop-down are similar to the options in the **Side 1** drop-down. The **None** option in this drop-down allows you to extrude the sketch in the first direction only. The **Blind** and **To Selected** options provide the depth of extrusion to the sketch in the second direction.

Tip: *The features created by using the **To Selected** option do not have a dimension associated with them. Therefore, they cannot be modified by changing the dimension value. However, changing the terminating surface changes the depth of the feature. You will understand it better when you learn the modification of an existing feature.*

*The extrusion depth given using the **Symmetric** option does not appear when you generate dimensions in the drawing views of a model. The drawing views of the model are generated in the drawing mode of PTC Creo Parametric and are discussed in Chapter 12.*

Capped ends

The **Capped ends** check box is used only in case of surfaces. When a surface is created using the **Extrude** tool, then its ends are open. If you select this check box, then the open ends of the surface created will be closed by an auto generated surface. You will learn more about this option in later chapters.

Add Taper

If you select the **Add Taper** check box, an edit box below this check box will get activated. This edit box is used to add draft in the solid or surface created by using the **Extrude** tool. The value entered in this edit box is referenced to the datum on which the sketch was created for extrusion. The taper value ranges from -89.9 to 89.9.

Properties Tab

Choose the **Properties** tab; a slide-down panel will be displayed. This slide-down panel displays the feature name in the **Name** collector. When the **i** button is chosen from the slide-down panel, the browser will be opened and all the information about the protrusion feature gets displayed. The browser and its properties have been discussed in Chapter 1.

Extrude as solid

The **Extrude as solid** button is chosen by default and is used to create a solid by adding material to the section. When you select a sketch for extrusion using this button, the sketch should be a single closed loop.

Extrude as surface

The **Extrude as surface** button is used to create a surface by adding material to the section. Using this button, you can create surface models. The sketch that is drawn for the surface model need not be a closed loop. The surface modeling is discussed later in Chapter 15.

Note

*The tabs and the buttons available in the **Extrude** dashboard can either be used before drawing the sketch or after drawing it.*

Change depth direction of extrude to other side of sketch

The **Change depth direction of extrude to other side of sketch** button is used once the sketch is completed. When you choose this button, it toggles the direction of extrusion with reference to the sketch plane. The direction of extrusion is defined as the direction in which the feature is created with respect to the sketching plane. This direction is displayed by an arrow in the drawing area. When you choose this button, the arrow points in the reverse direction suggesting that the direction of feature creation has been reversed. You can also click on the arrow to flip the direction of extrusion.

Note

The direction of the arrow depends on the type of extrude feature being created. If material has to be added to a feature, then the arrow, by default, points in the direction toward the user, that is out of the screen. But, if a cut feature is being created in which material has to be removed from a feature, then by default, the arrow points into the screen.

If you spin the model using the middle mouse button, then the direction of the arrow can be easily recognized.

Remove Material

The **Remove Material** button is available in the **Extrude** dashboard only after the base feature is created. This is because this button is used to remove the material from an existing feature. Therefore, when you create a base feature, this button is not available. The use of this button is explained in later chapters.

Thicken Sketch

The **Thicken Sketch** button is used to create thin parts with specified thickness. When you choose this button, the **Change direction of extrude between one side, other side, or both sides of sketch** button and a dimension edit box appear on the right of this button.

The **Change direction of extrude between one side, other side, or both sides of sketch** button is used to set the direction with respect to the sketch where the thickness should be added. This button appears on the dashboard only when the sketch is completed and you exit the sketcher environment. This button serves three functions. It is used to set the thickness direction to either side of the sketch and even to both sides of the sketch symmetrically. When the sketch is completed, the preview of the model will be displayed in the drawing area. By default, the thickness to the sketch is applied on one side. Now, when you choose the **Change direction of extrude between one side, other side, or both sides of sketch** button, the thickness will be applied to the other side of the sketch. When you choose this button for the second time, the thickness is applied symmetrically to both sides of the sketch.

The dimension edit box that appears when you choose the **Thicken Sketch** button, is used to enter the thickness value of the feature to be created. This edit box is available only when you exit the sketcher environment.

Figure 4-13 shows the arrow pointing away from the side of the section wall where the material will be added. If you want to change the direction of the arrow, then choose the **Change direction of extrude between one side, other side, or both sides of sketch** button. Figure 4-14 shows the thickness of the material added to one side of the wall pointed by the arrow.

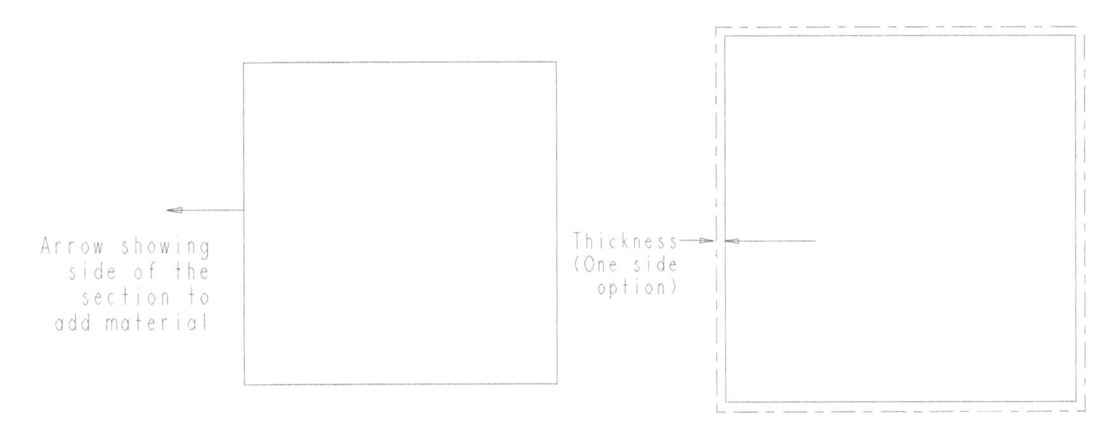

Figure 4-13 *The sketch drawn to thicken*

Figure 4-14 *Thickness of the material added to one side of the sketch*

You can choose the **Change direction of extrude between one side, other side, or both sides of sketch** button so that the thickness of the material added is symmetric to both sides of the section wall, as shown in Figure 4-15. Figure 4-16 shows the model created by using the **Thicken Sketch** button.

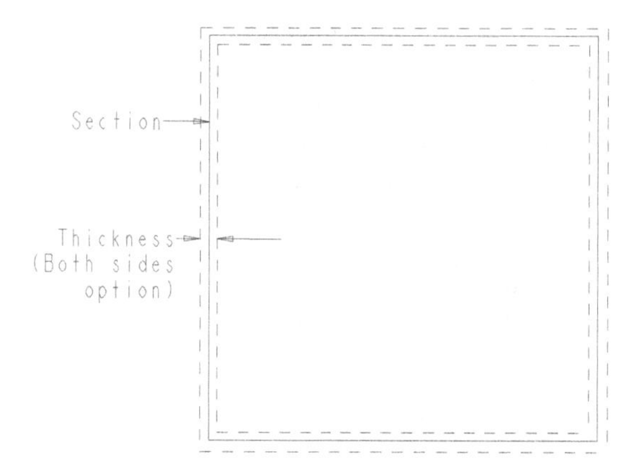

Figure 4-15 Thickness of the material added symmetrically to both sides of the sketch

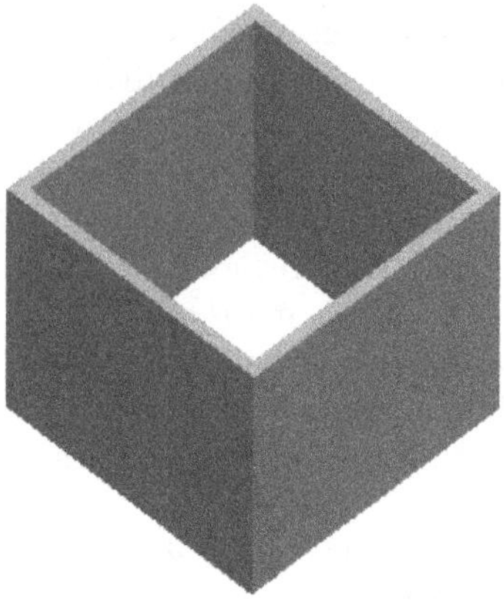

Figure 4-16 Model created using the ***Thicken Sketch*** *button*

Pause tool

This button is used to pause the current feature creation. After you choose this button, you can access other available tools such as the **Plane** tool. To resume the current feature creation, you can choose the **Resume** button that appears in place of the **Pause** tool.

Geometry preview/Feature preview

There are four tools to control the preview of a model while using the protusion commands. These tools are discussed next.

No Preview

On invoking this tool, the preview of the model will not be displayed while modeling.

Un-attached

Invoke this tool if you want to see only the outlines of the model while working with the **Extrude** tool. The preview displayed after choosing this tool will be transparent.

Attached

Invoke this tool if you want to see the final visualization of the model to be created.

Feature Preview

If you choose this button, the system allows you to preview the feature and its relationship with other features after finalizing the protusion. Note that the preview of the feature appears to be a solid model.

Build Feature

This button is used to confirm the feature creation and exit the current feature creation tool.

Close

This button is used to abort the creation of feature. When you choose this button, the dashboard is closed.

Note that you can modify the depth of extrusion of the model dynamically after the model is created. To do so, you need to drag the small, white colored square symbol displayed on the model. This method is called dynamic editing and is discussed later. You can also modify the direction of extrusion dynamically. To do so, click on the arrow displayed on the model.

The Sketch Dialog Box

The **Sketch** dialog box will be displayed when you choose the **Define** button from the **Placement** slide-down panel. Alternatively, right-click in the drawing area to display the shortcut menu. Choose the **Define Internal Sketch** option from it; the **Sketch** dialog box will be displayed, as shown in Figure 4-17.

*Figure 4-17 The **Sketch** dialog box*

This dialog box is used to select the sketching plane and to set its orientation. As soon as you select the sketching plane, the reference plane and its orientation are selected by default. If you want to change the sketching plane that is already selected, then hold the right mouse button in the drawing area to display a shortcut menu, as shown in Figure 4-18. There are three options in this shortcut menu. By default, the **View Orientation** option is selected. If you choose the **Clear** option, the reference plane and its orientation, which were selected by default, will be cleared and now you can select them manually. Similarly, to change the sketching plane, choose the **Placement** option in the shortcut menu. Again, invoke the shortcut menu and choose the **Clear** option.

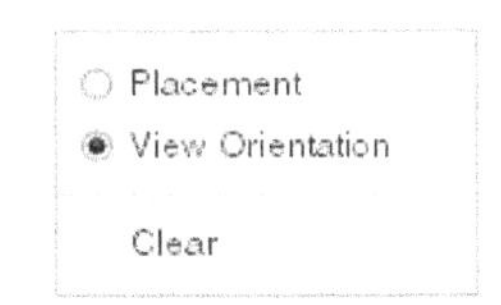

*Figure 4-18 Shortcut menu displayed when the **Sketch** dialog box is invoked*

However, after you select the sketching plane, no matter which reference plane is selected by default, you can select the reference plane manually. The reference you select manually replaces the old reference. If the background of the **Plane** collector or any other collector is displayed in green color, then it indicates that you can select its reference from the model. This point should be remembered because this is true with other dialog boxes that are available in PTC Creo Parametric. You will learn more about this later in this book.

The **Flip** button in the **Sketch** dialog box can be used after you select the sketching plane and its orientation. This button is used to set the direction of viewing the sketching plane. When you choose this button, the direction of the arrow, which is displayed on the sketching plane, will change.

In the **Sketch** dialog box, the options in the **Orientation** drop-down list are used to specify the orientation of the horizontal or vertical reference for the sketching plane. Generally, the view is normal to the sketching plane you have selected. In order to orient the sketching plane normal to the viewing direction, you have to specify a plane or a planar surface that is perpendicular to the sketching plane. For example, if you select the **RIGHT** datum plane as the reference plane and then select the **Top** option from the drop-down list, then the **RIGHT** datum plane will be on the top while sketching. The options in this drop-down list are common to other feature creation tools and are available whenever you need to draw a sketch. Before you proceed further, you need to understand the three default datum planes that are displayed when you open a new part file. You can view the feature that you have created from different directions. To do so, choose the **Named Views** button from the **Orientation** group in the **View** tab; a flyout will be displayed, as shown in Figure 4-19. In this flyout, there are some standard preset views provided by PTC Creo Parametric. If you see the feature from the right, your viewing direction will be normal to the **RIGHT** datum plane. If you see the feature from the top, your viewing direction will be normal to the **TOP** datum plane. Similarly, if you see the feature from the front, your viewing direction will be normal to the **FRONT** datum plane.

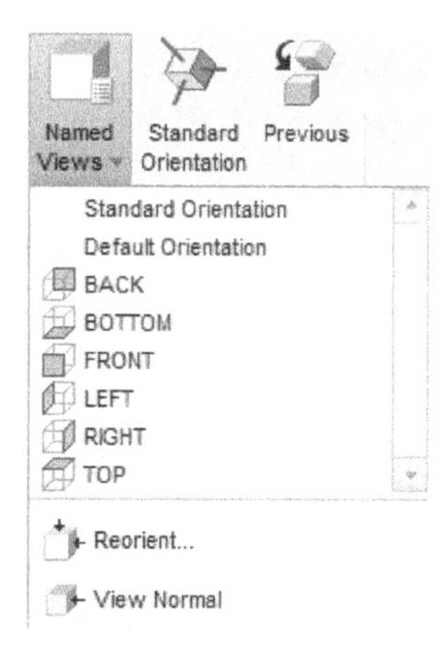

***Figure 4-19** The **Named Views** flyout*

The options in this flyout are used to set the orientation of the model in the drawing area.

The options in the **Orientation** drop-down list in the **Sketch** dialog box are very important for the orientation of the base feature. So, you need to select the reference plane very carefully, especially for the base feature, refer to Figure 4-20. This figure shows two cases, where the sketching plane for both the solid models is same, the same sketch is extruded, and the same option **Top** is selected from the **Orientation** drop-down list in the **Sketch** dialog box, but the reference planes selected are different. The same sketch is used to extrude both the models and the top curve in the sketch is on the top while sketching. Note the difference in orientation of the resulting models in their default trimetric orientations. The model can be oriented in its default orientation using CTRL+D or by selecting the **Default Orientation** option from the **Named Views** flyout.

Tip: *The **Orientation** dialog box, which is displayed when you choose the **Reorient** button from the **Orientation** group in the **View** tab, is used to orient the model in the drawing area. This dialog box provides some advanced options to orient the model and also allows you to save the orientation of the model. Using the options in this dialog box, you can dynamically orient a model, by selecting the **Dynamic orient** option from the **Type** drop-down list.*

Once a plane has been selected for sketching and the reference plane is oriented, you can enter the sketcher environment. To enter the sketcher environment, choose the **Sketch** button in the **Sketch** dialog box.

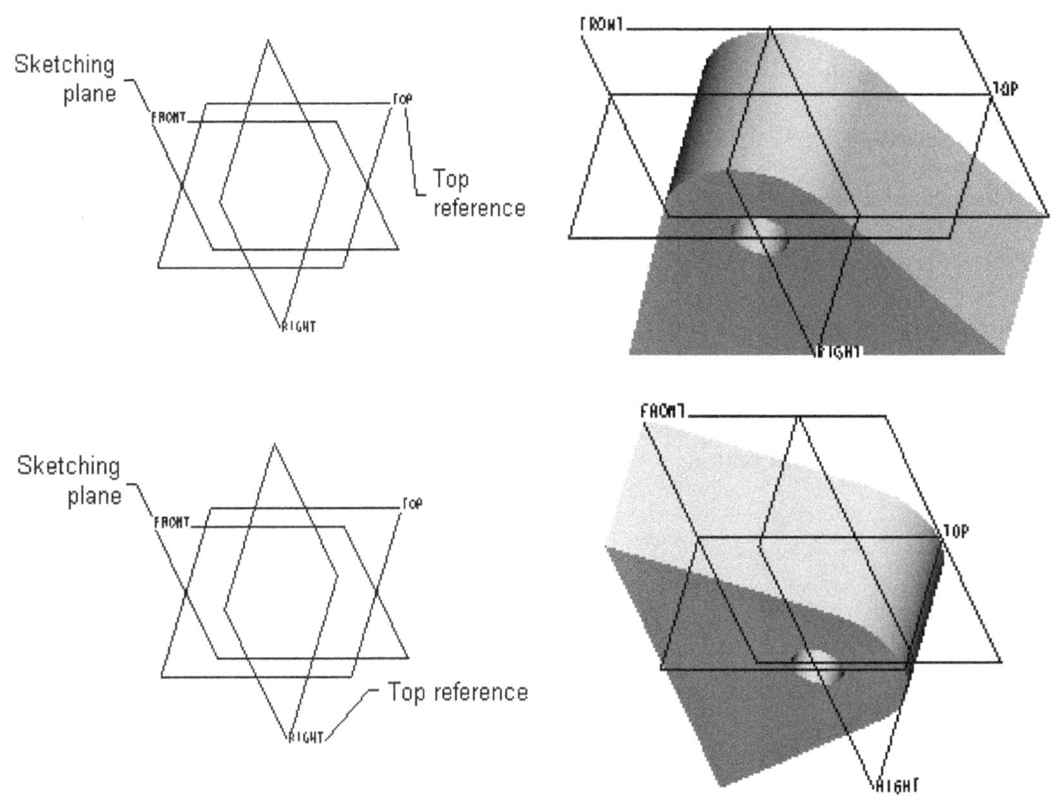

Figure 4-20 *Figure displaying importance of selecting reference planes*

Revolving a Sketch

Ribbon: Model > Shapes > Revolve

The **Revolve** tool allows you to revolve the sketched section through the specified angle about an axis of the coordinate system or of a geometric centerline. You can also select an existing straight curve or the edge of an existing part. By revolving a sketch, you can add material (protrusion) or remove material (cut). Here, you will learn to use the **Revolve** tool to add material. The revolved feature can be revolved on one side of the sketching plane or on both sides of the sketching plane. Besides this, other attributes can be specified to the sketch by using the **Revolve** dashboard. The **Revolve** dashboard is shown in Figure 4-21 and it will be displayed when you choose the **Revolve** tool from the **Shapes** group in the **Model** tab.

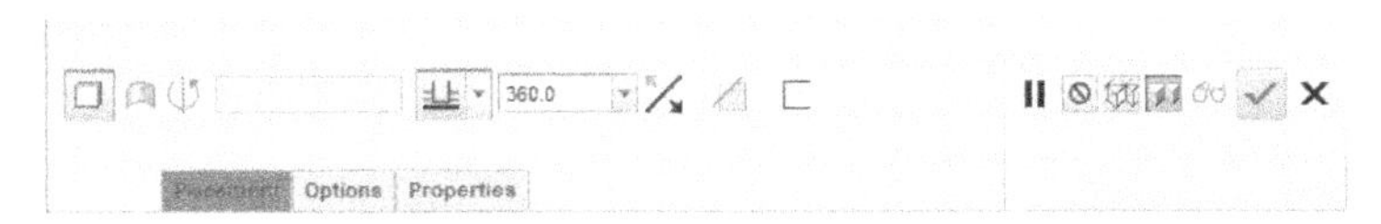

Figure 4-21 *Partial view of the **Revolve** dashboard*

Some of the points to be kept in mind while creating a revolve feature are given next.

1. If the revolve feature is a base feature, then the section drawn should be a closed section for revolving the sketch as a solid.

2. Draw a centerline to complete the sketch of the revolved feature or select an axis of the coordinate system or an existing straight curve or an edge of an existing feature as the axis of revolution.

3. The section should be drawn on one side of the centerline.

Revolve as Solid

The **Revolve as solid** button in the **Revolve** tab is used to revolve a sketch to create a solid feature. To revolve a sketch using this button, the sketch should be a closed loop. Figures 4-22 and 4-23 show two examples of solid features created by using the **Revolve as solid** button.

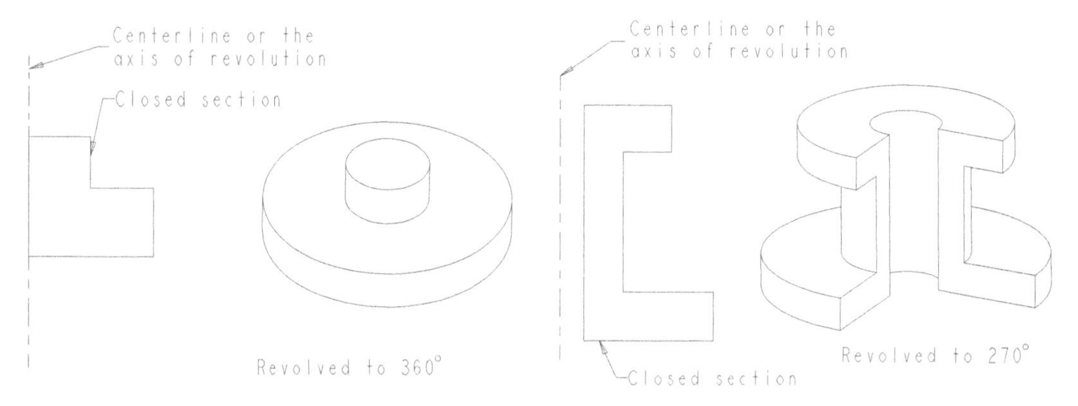

Figure 4-22 Model revolved 360 degree

Figure 4-23 Model revolved 270 degree

Revolving a Sketch with Thickness

The **Thicken Sketch** button is used in combination with the **Revolve as solid** button to create revolved features having certain thickness. Unlike revolving the sketch to create a solid feature, when you revolve a sketch with thickness, the section need not be a closed loop. When you exit the sketcher environment after the section is created, you can specify the side of the section where the material will be added. After specifying the side for the material addition, you can enter the thickness value in the dimension edit box that appears on the dashboard.

Figures 4-24 and 4-25 show the examples of the revolved sketch having some thickness.

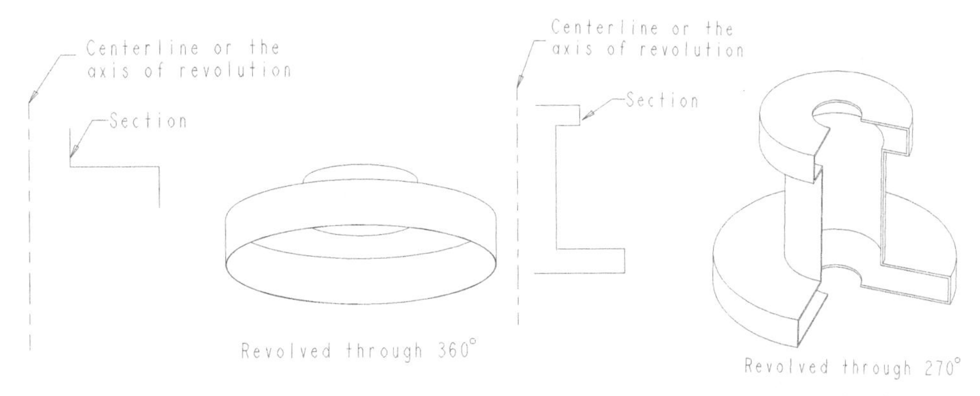

Figure 4-24 Model revolved 360 degree *Figure 4-25 Model revolved 270 degree*

UNDERSTANDING THE ORIENTATION OF DATUM PLANES

Consider a case where you need to revolve a sketch through an angle of 270-degree. The trimetric view of the solid model is shown in Figure 4-26. Now, when you start to create the solid model, the first step is to decide the datum plane on which you need to create the sketch.

Figure 4-26 The trimetric view of the solid model

As evident from the model, the axis of revolution of the model is perpendicular to the **TOP** datum plane. Therefore, the **TOP** datum plane will be selected to be at the top while drawing the sketch. Now, only two options are left for selecting the sketching plane, the **FRONT** and the **RIGHT** datum planes. You can draw the section of the revolved model on any of the two planes. This is because when you view the model, you will notice that the cross-section of the model is parallel to the **RIGHT** datum plane as well as the **FRONT** datum plane.

The procedure to draw the cross-section sketch on the two datum planes to achieve the desired orientation of the model is discussed next.

Case-1

Drawing a Sketch on the RIGHT Datum Plane

1. When the **Sketch** dialog box is displayed, select the **RIGHT** datum plane as the sketching plane.

   On doing so, the selected datum plane will get highlighted and the **TOP** datum plane will be automatically selected as the reference plane. Also, an arrow will appear on the sketching plane. This arrow indicates the direction of the feature creation. You can change the direction of arrow, but at this stage, you have to accept its default direction.

2. From the **Orientation** drop-down list, select the **Top** option, if it is not already selected.

   Now, when you draw the sketch, the **RIGHT** datum plane will be parallel to the screen and the **TOP** datum plane will be at the top.

3. Choose the **Sketch** button from the **Sketch** dialog box to enter the sketcher environment.

4. Next, draw the sketch of the cross-section, dimension it, and then modify the dimensions, as shown in Figure 4-27.

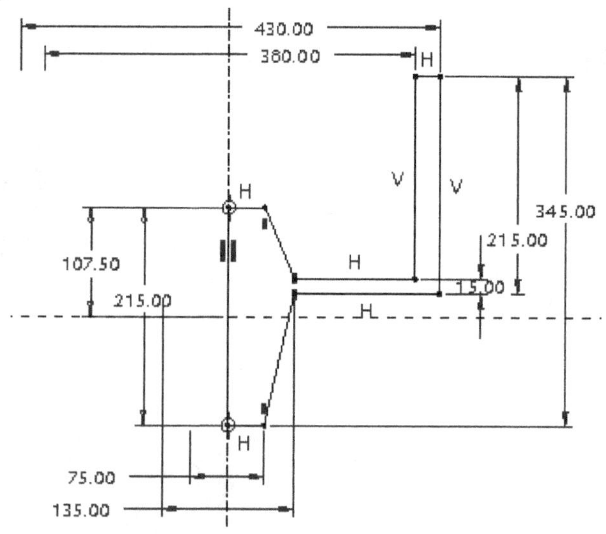

Figure 4-27 *Sketch with dimensions and constraints*

5. Exit the sketcher environment by choosing the **OK** button from the **Sketch** dashboard.

6. If you have drawn the geometric centerline while sketching, then the preview will be displayed. However, if you have not drawn the centerline, then you will be prompted to select a straight edge or an axis of the coordinate system. In this figure, the geometric centerline is drawn along the Y axis.

7. Select the angle value **270** in the drop-down list available on the dashboard.

8. From the **Named Views** flyout, choose the **Default Orientation** option or press CTRL+D. The model will orient to its default orientation, that is, trimetric view, as shown in Figure 4-28.

   However, the view that is shown in Figure 4-28 is not the required one. The model is not oriented correctly. This means the direction of feature creation that was selected is not correct in order to get the desired orientation.

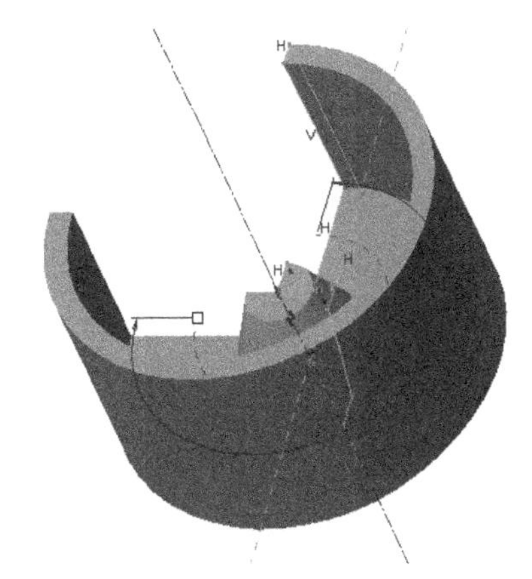

Figure 4-28 *Default trimetric view of the model*

Figure 4-29 shows the sketch drawn on the **RIGHT** datum plane. In such a case the sketch always rotates in the clockwise direction. Note that the arrow in Figure 4-29, shows the direction in which you will be viewing the sketching plane while in the sketcher environment. Considering the mentioned facts, to orient the sketching plane correctly, the arrow direction should have been flipped by using the **Flip** button in the **Sketch** dialog box. Since you have not achieved the desired orientation of the model, exit this feature creation tool. Then, again invoke the **Revolve** tool, and this time after you select the sketching plane, choose the **Flip** button to reverse the direction of viewing the sketch. Remember that the case discussed here is only applicable when you select the **RIGHT** datum plane as the sketching plane.

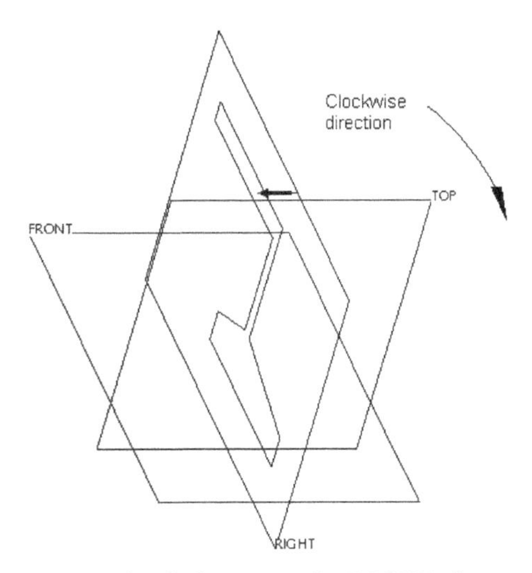

Figure 4-29 *Sketch drawn on the* ***RIGHT*** *datum plane*

Tip: *1. You might have difficulty in visualizing the cross-section of a revolved model that has an angle of revolution of 360-degree. This is because the cross-section of the model is not visible. Therefore, you cannot decide the sketch to be drawn for the cross-section of the model.*

2. Whenever you come across a revolved model, just imagine that if you cut the model along its axis of revolution and remove one quarter of the revolved feature, then the section obtained in one quarter of the revolved feature is the section of the revolved model. Therefore, you need to draw this section in the sketcher environment to create the desired model.

Figure 4-30 shows the sketch on the sketching plane with the direction of the arrow reversed. In this case, the sketch when rotated in the clockwise direction results in the desired orientation of the model, as shown in Figure 4-31.

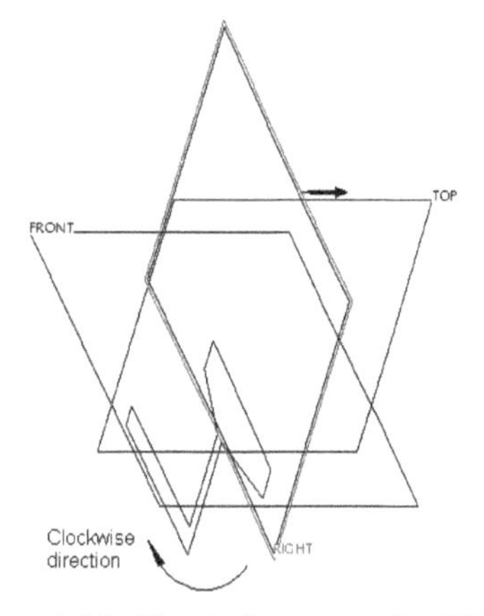

Figure 4-30 *Sketch drawn on the* ***RIGHT*** *datum plane wth reverse direction of arrow*

Figure 4-31 *The desired trimetric view of the model*

Tip: *In PTC Creo Parametric, when a revolved model is created, the material is added in the clockwise direction. You can revolve the model in the counterclockwise direction by using the* ***Change angle direction of revolve to other side of sketch*** *button from the* ***Revolve*** *tab. Also, to revolve the model in the counterclockwise direction, you can enter a negative value for the angle of revolution.*

Case-2

Drawing a Sketch on the FRONT Datum Plane

In this case, you will select the **FRONT** datum plane as the sketching plane and the **RIGHT** datum plane as the reference plane. It is recommended that you decide the direction of the arrow at this stage so that you can create the model in the required orientation, refer to Figure 4-26. Figure 4-32 shows the sketch drawn on the **FRONT** datum plane and the default direction of viewing the sketching plane. Remember that the arrow points in the direction along which you view the plane in the sketcher environment. Figure 4-33 shows the default trimetric view of the solid model created in this case.

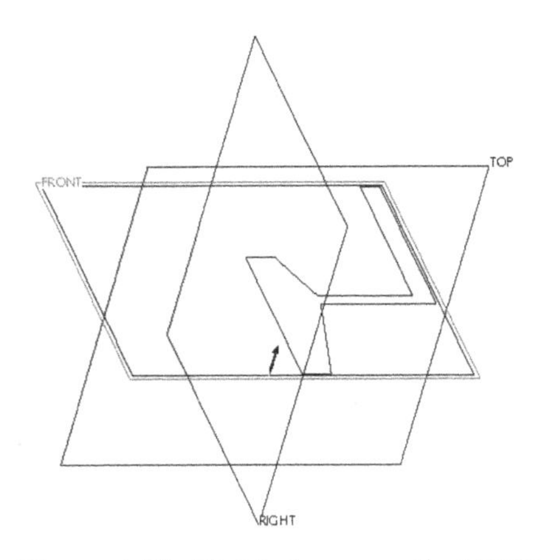

Figure 4-32 *Sketch drawn on the* ***FRONT*** *datum plane*

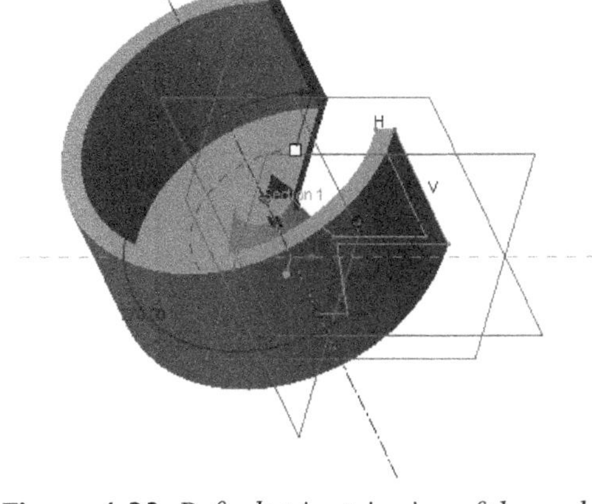

Figure 4-33 *Default trimetric view of the model*

This is not the desired default orientation of the model. You can still get the desired orientation of the model by choosing the **Change angle direction of revolve to other side of the sketch** button from the **Revolve** dashboard, but you should understand the default direction of the arrow on the sketching plane. The desired orientation of the model is shown in Figure 4-34.

Figure 4-35 shows the sketch drawn on the **FRONT** datum plane and the arrow is showing the direction of viewing the sketching plane. The direction of the arrow is reversed by choosing the **Flip** button in the **Sketch** dialog box. Figure 4-36 shows the default trimetric view of the model in this case.

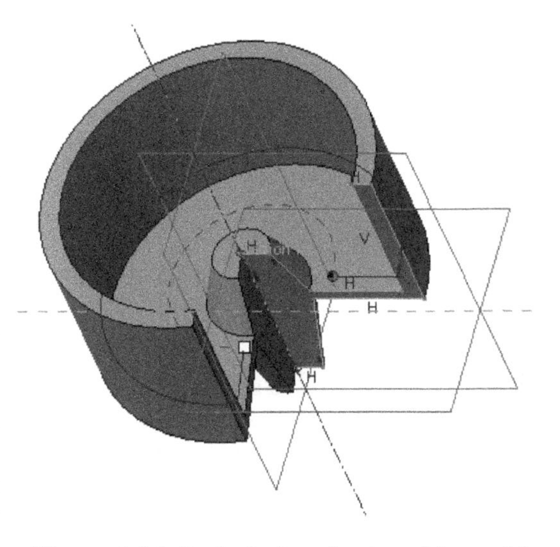

Figure 4-34 *Desired trimetric view of the model*

In Figure 4-35, the sketch is revolved in the clockwise direction to create the model shown in Figure 4-36.

From the two cases discussed, the conclusion is that the arrow that will be displayed when you select the sketching plane is of great importance in orienting the model. Also, the direction of feature creation can be changed once it is created using the **Change angle direction of revolve to other side of the sketch** button from the **Revolve** dashboard.

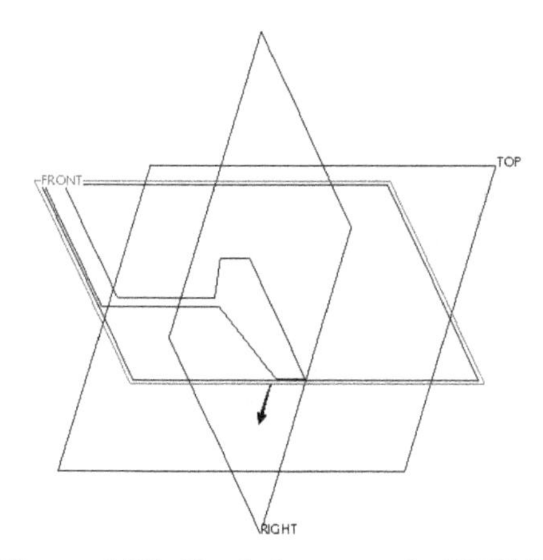

*Figure 4-35 Sketch drawn on the **FRONT** datum plane*

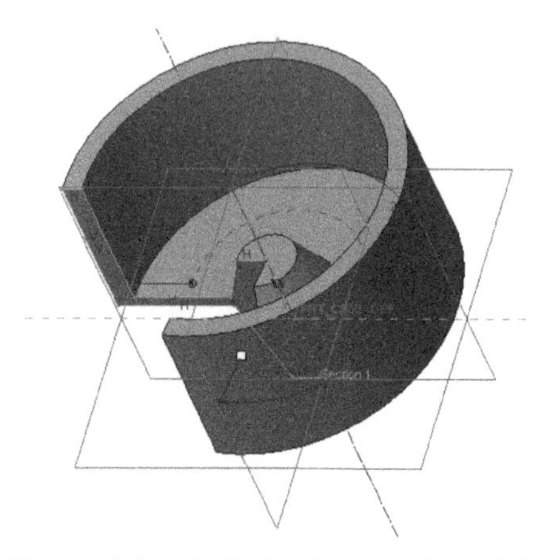

Figure 4-36 Default trimetric view of the model

PARENT-CHILD RELATIONSHIP

Every model created in PTC Creo Parametric is composed of features that in some way or the other are related to other features in the model. The feature that occurs first in the **Model Tree** is called the parent feature and any feature(s) that is related to this feature is called the child feature(s).

Tip: *If you want to check the parent-child relationship of features in a model, choose the **Feature Information** option from the **Investigate** group of the **Tools** tab in the **Ribbon**; you will be prompted to select a feature. After you select the feature, the **Feature Info** window will be displayed. The child features of the selected feature will be displayed in the **Children** area and all parent features will be displayed in the **Parents** area.*

There are two types of relationships that can exist between the features, Implicit relationship and Explicit relationship.

Implicit Relationship

This type of relationship exists when the two features are related through equations. These equations are formed using relations.

Explicit Relationship

The explicit relationship is developed when a feature is used as a reference to create another feature. For example, one of the planar surfaces of a feature is used to create another feature, or an edge of a feature is used to dimension the other feature. In this case, the first feature is called the parent feature and the second feature is called the child feature of the first feature. Another example of this type of relationship is, when a hole is referenced to the edges of the surface it is placed on, or to the edges of some other surfaces. In this case, the hole is the child feature of the feature it is referenced to.

Remember that if the parent feature is modified, then the child feature is also modified. Similarly, if the parent feature is deleted, then the child feature is also deleted. For example, if you have used the three default datum planes to create the base feature, and you delete any one of the datum planes, the base feature will also be deleted. In Chapter 6, you will learn how to break the parent-child relationship.

NESTING OF SKETCHES

If more than one closed loop is drawn, one inside the other in the sketch of a single feature, it is called nesting of sketches. These sketches are drawn in the sketcher environment. Figure 4-37 shows a sketch in which two circles are created on the base profile.

Advantages of Nesting the Sketches

1. One of the advantages of nesting the sketches is that the number of features used to create a model is reduced. In Figure 4-38, the model has two features. The base feature is the base plate and the second feature is the holes in the model. But, when you nest the two sketches to create the model, you are using only one feature to create the model.

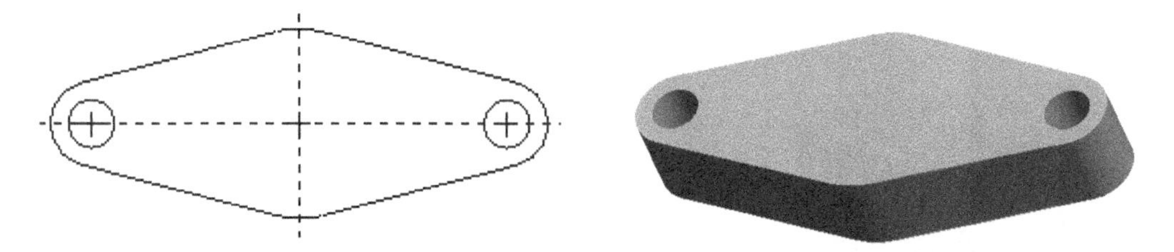

Figure 4-37 *Nested sketch* ***Figure 4-38*** *Solid model of the sketch*

2. There is no parent-child relationship that exists in a nested sketch.

3. The depth of the hole is equal to the depth of extrusion of the base feature. This is because the two circles and the base profile are part of the same sketch. This depends on the designer whether he wants to use the nested sketch to create a part.

Disadvantages of Nesting the Sketches

1. In nesting, since the two features on the model are combined into one feature, therefore there is no flexibility in editing the features of a model.

2. If at a later stage, the designer needs to convert the circular holes into elliptical holes with depth as half the depth of extrusion of the base feature, it consumes a lot of time to edit the model.

After understanding the advantages and disadvantages of nesting the sketches, it is recommended to divide a model into separate features. Draw all features as individual features so that the created model is flexible. However, it depends on the need of the designer and the need of the model how the model is approached for creating.

CREATING CUTS

In PTC Creo Parametric, you can create cuts in a model. Cutting is a material removal process and the option is available only when at least one base feature exists in the drawing area. You can create cuts in a feature by using the **Remove Material** button that is available on the dashboard of the feature creation tools. In this section, you will learn to create extrude and revolve cuts. The procedure to create a cut on a feature is similar to that of adding material or protrusion.

Removing Material by Using the Extrude Tool

You can create a cut by removing material from an existing feature using the **Extrude** tool. The material removed from the feature is defined by a sketch.

To create an extrude cut in a model, choose the **Extrude** tool from the **Shapes** group; a dashboard will be displayed. Choose the **Remove Material** button and enter the sketcher environment as discussed earlier in this chapter. After drawing the sketch for the cut feature, you can specify the direction of material removal with respect to the sketch using the arrow displayed in drawing area. For example, the arrow in Figure 4-39 shows the direction of material removal. After the required settings have been specified, choose the **Build Feature** button to create the cut feature, as shown in Figure 4-40.

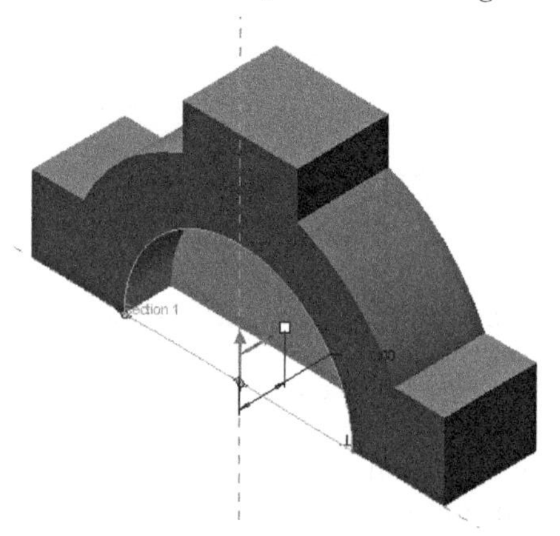

Figure 4-39 *Sketch for the extrude cut and arrow showing the direction of material removal*

Figure 4-40 *Cut feature created on the model*

Tip: *In the model shown in Figure 4-39, the sketching plane selected for the creation of extruded cut is the front face of the base feature and not a datum plane. You will learn to create a datum plane on the surface of an existing feature and select it as the sketching plane in Chapter 5. But it is not recommended to create a datum plane if the planar surface of the feature can be used as the sketching plane.*

You can also change the direction of material removal by choosing the **Change material direction of extrude to other side of sketch** button from the **Extrude** dashboard. On doing so, the arrow will point in the direction shown in Figure 4-41. In this case, the material on the plane selected for sketching will be removed, leaving the extruded cut feature, as shown in Figure 4-42.

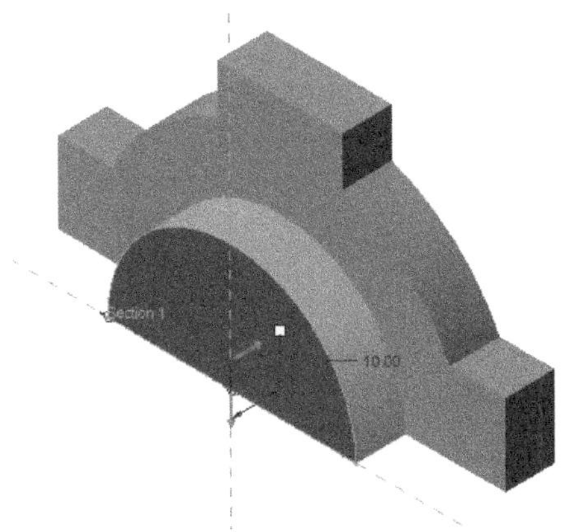

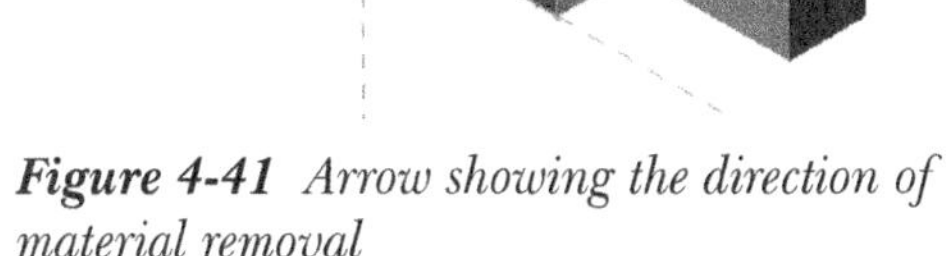

Figure 4-41 *Arrow showing the direction of material removal*

Figure 4-42 *Cut feature created in the direction of arrow*

Note

*A straight hole can also be created by drawing its cross-section, that is a circle, and then creating an extrude cut. But, PTC Creo Parametric provides predefined placement for a hole feature, which is more desirable than dimensioning the cross-section of a cut feature. Straight holes do not require a sketch if you use the **Hole** dashboard. The **Hole** dashboard is discussed in Chapter 6.*

Removing Material by Using the Revolve Tool

You can create a cut by removing material from an existing feature using the **Revolve** tool. The material to be removed will be defined by the sketch you draw.

To remove the material from an existing feature, choose the **Revolve** tool from the **Shapes** group, a dashboard will be displayed. Select the **Remove Material** button to enter the sketcher environment as discuss earlier in this chapter. Remember that the geometry centerline is necessary in a revolve feature. Figure 4-43 shows the section drawn to be revolved. The front face of the second extruded feature is selected as the sketching plane. Figure 4-44 shows the revolve cut created on the selected surface.

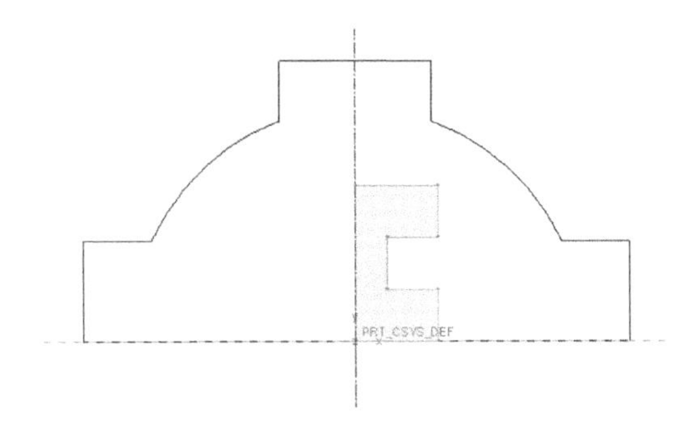

Figure 4-43 *The section for revolve cut*

Figure 4-44 *Revolve cut created*

TUTORIALS

Tutorial 1

In this tutorial, you will create the model shown in Figure 4-45. The dimensions of the model are shown in Figure 4-46. **(Expected time: 30 min)**

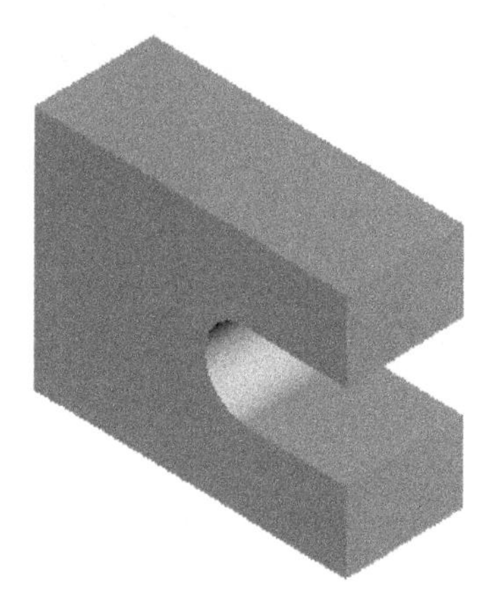

Figure 4-45 The isometric view of the model

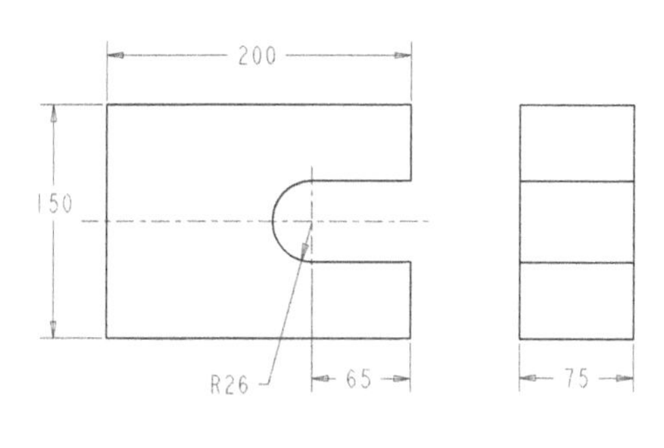

Figure 4-46 The front and right views of the model with dimensions

The following steps are required to complete this tutorial:

a. Set the working directory and create a new object file in the **Part** mode.
b. First examine the model and then determine the type of protrusion for the model. Select the sketching plane for the model and orient it parallel to the screen.
c. Draw the sketch by using the sketching tools and apply constraints and dimensions, refer to Figures 4-47 through 4-51.
d. Exit the sketcher environment and define the model attributes, refer to Figures 4-52 and 4-53.

Setting the Working Directory

After starting the PTC Creo Parametric session, the first task is to set the working directory. A working directory is a directory on your system where you can save the work done in the current session of PTC Creo Parametric. You can set any existing directory on your system as the working directory.

1. Choose the **Manage Session > Select Working Directory** from the **File** menu; the **Select Working Directory** dialog box is displayed. Browse to the *C:\Creo-3.0* folder.

2. Choose the **Organize** button from the **Select Working Directory** dialog box to display the flyout. Next, choose the **New Folder** option from the flyout; the **New Folder** dialog box is displayed.

3. Enter **c04** in the **New Directory** edit box and choose **OK** from the **New Folder** dialog box. Now, you have created a folder named *c04* in *C:\Creo-3.0.*

4. Next, choose **OK** from the **Select Working Directory** dialog box; the working directory is set to *C:\Creo-3.0\c04* and a message **Successfully changed to C:\Creo-3.0\c04 directory** is displayed in the message area.

Starting a New Object File

Solid models are created in the **Part** mode of PTC Creo Parametric. The file extension for the files created in this mode is *.prt*.

1. Choose the **New** button from the **File** menu; the **New** dialog box is displayed. The **Part** radio button is selected by default in the **Type** area and the **Solid** radio button is selected by default in the **Sub-type** area of the **New** dialog box.

2. Enter the file name as *c04tut1* in the **Name** edit box and choose the **OK** button. The three default datum planes are displayed in the drawing area. Also, the **Model Tree** appears on the left of the drawing area in the Navigator.

Tip: *By default, a Creo file opens in* ***inlbs_part_solid*** *unit. However, you can change it by choosing* ***Prepare > Model Properties > Units*** *option from the* ***File*** *menu.*

3. Close the **Model Tree** by double-clicking on the sash present at the right edge of the Navigator. Now, the drawing area is increased.

Selecting the Extrude Option

The given solid model is created by extruding the sketch to a distance of 75. Therefore, the sketch will be extruded as a solid to create the model.

1. Choose **Extrude** from the **Shapes** group in the **Model** tab; the **Extrude** dashboard is displayed on the top of the drawing area. All the attributes needed to create the model will be defined after the sketch is drawn.

Selecting the Sketching Plane

To create a sketch for the extruded feature, first you need to select the sketching plane for the model. The **FRONT** datum plane will be selected as the sketching plane. The sketching plane is selected such that the direction of extrusion of the solid model is perpendicular to it. From the isometric view of the model shown in Figure 4-45, it is evident that the direction of extrusion of the model is perpendicular to the **FRONT** datum plane.

1. Choose the **Placement** tab from the **Extrude** dashboard to display the slide-down panel. Next, choose the **Define** button from the slide-down panel; the **Sketch** dialog box is displayed.

2. Select the **FRONT** datum plane from the drawing area as the sketching plane. As you select the sketching plane, the reference plane and its orientation are set automatically. The reference plane is selected to orient the sketching plane.

The arrow appearing on the sketching plane indicates the direction of viewing the sketch.

In the **Sketch** dialog box, the **Reference** collector displays **RIGHT:F1(DATUM PLANE)**. This indicates that the **RIGHT** datum plane is selected as the reference plane. In the **Orientation** drop-down list, the **Right** option is selected by default. This means while drawing the sketch, the **RIGHT** datum plane will be on the right. The **RIGHT** datum plane will be perpendicular to the sketching plane and the sketching plane will be parallel to the screen.

3. Choose the **Sketch** button in the **Sketch** dialog box to enter the Sketcher environment.

Drawing the Sketch

You need to draw the sketch of the solid model that will be extruded later to create the 3D model.

1. Choose the **Rectangle** tool from the **Sketching** group.

2. Draw a rectangle by defining its lower left corner and upper right corner. The rectangle is created and weak dimensions are applied to it.

You will notice that strong vertical and horizontal constraints are applied to the lines composing the rectangle. This is because drawing a rectangle is itself a constraint to the lines composing the rectangle.

3. Choose the **One-By-One** button from the **Selection** Filter in the **Operations** group.

4. Select the right vertical line; the line turns green. Press the DELETE key to delete it. The sketch after deleting the vertical line is shown in Figure 4-47.

5. Now, draw the lines and arc. Some weak dimensions and constraints are applied to the sketch, as shown in Figure 4-48.

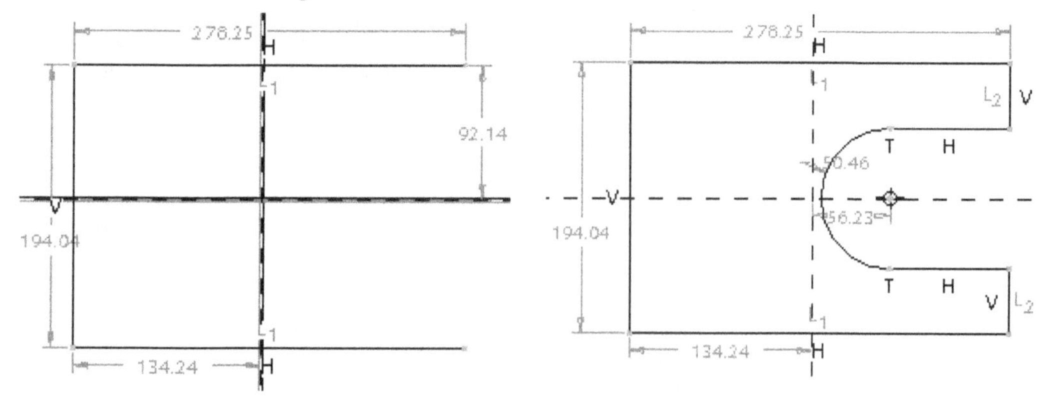

Figure 4-47 The sketch after deleting the vertical line

Figure 4-48 The sketch with weak dimensions and constraints

Note

1. As evident from Figure 4-48, the strong constraints are also applied to the sketch while drawing it. However, if these strong constraints are not applied when you draw the sketch, you need to apply these constraints manually.

*2. The center of the arc and the **TOP** datum plane are aligned by default. In case, they are not aligned, you need to align them. To do so, choose the **Coincident** button from the **Constrain** group. Select the center of the arc and then select the **TOP** datum plane. Now, the center and the datum plane are aligned.*

Applying Constraints to the Sketch

You need to apply equal length constraints to the sketch in order to maintain the design intent of the model.

1. Choose the **Equal** tool from the **Constrain** group in the **Sketch** tab and select the two vertical lines on the right of the sketch. The weak constraint L_2 was applied when the two vertical lines were drawn. Now, the equal length constraint L_2 is applied to both the lines. The constraint labels like L_2 or L_3 vary from sketch to sketch.

2. Select the two horizontal lines that are connected to the arc to apply the equal length constraint. The equal length constraint L_2 is applied to both the lines and the constraint symbol L_2 on the two right vertical lines is changed to L_1. The sketch after applying the equal length constraints is shown in Figure 4-49.

Dimensioning the Sketch

Although some weak dimensions are applied to the sketch, you need to add a dimension to the sketch.

1. Choose the **Normal** tool from the **Dimension** group.

2. Select the center of the arc and the upper right vertical line and then place the dimension by using the middle mouse button, as shown in Figure 4-50.

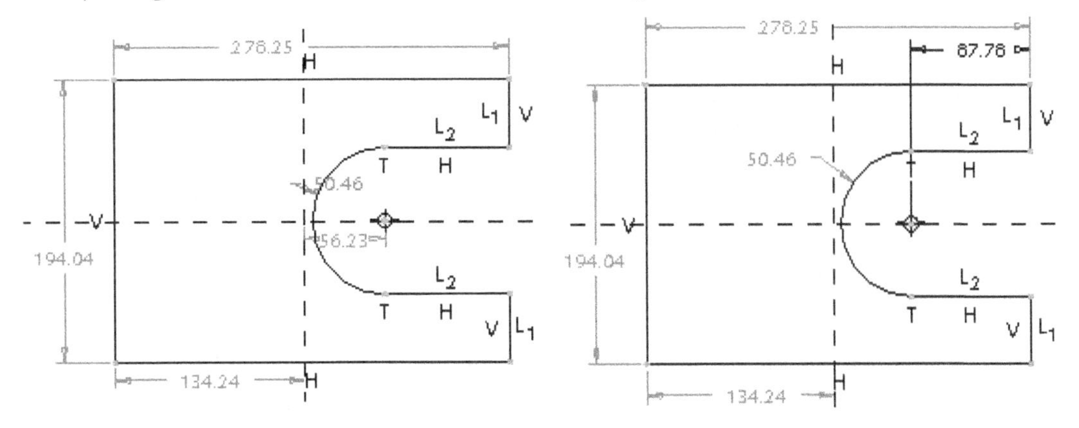

Figure 4-49 *Equal length constraints applied to the sketch*

Figure 4-50 *Dimension added to the sketch*

You need to dimension only these entities because the rest of the weak dimensions are useful dimensions and can be modified directly.

Note

1. The line numbers displayed in your sketch may be different from the one given in Figure 4-48.

2. If the dimensions in your sketch are different from those shown in Figure 4-50, add the missing dimensions and delete the dimensions that are not required.

Modifying the Dimensions

You need to modify the dimension values of the sketch. The default dimensions shown in Figure 4-50 also include the length and width of the rectangle and the distance of the sketched section from the selected references. It is recommended that you draw the base feature symmetrical with the other two datum planes. As a result, the distance from the **RIGHT** datum plane is 100 (200 divided by 2 is equal to 100).

1. Select the sketch and dimensions using the CTRL+ALT+A keys.

2. Choose the **Modify** tool from the **Editing** group; the **Modify Dimensions** dialog box is displayed. All dimensions in the sketch are displayed in this dialog box, and each dimension has a separate thumbwheel and an edit box. You can use the thumbwheel or the edit box to modify dimensions. It is recommended to use the edit boxes to modify dimensions, if the desired value is not obtained using the thumbwheel.

3. Clear the **Regenerate** check box; any modification done in a dimension value does not update the sketch during modification. The dimensions get modified after you exit the **Modify Dimensions** dialog box. It is recommended to clear the **Regenerate** check box when more than one dimension has to be modified.

4. Modify all dimensions one by one, as shown in Figure 4-51. You will notice that the dimension you select in the **Modify Dimensions** dialog box is enclosed in a blue box in the drawing area.

5. After modifying the dimensions, choose the **Regenerate the section and close the dialog** button from the **Modify Dimensions** dialog box; the message **Dimension modifications successfully completed** is displayed in the message area.

6. Choose the **OK** button from the **Close** group to exit the sketcher environment.

Specifying the Model Attributes

Next, you need to specify the attributes to create the model.

1. Choose the **Named Views** button from the **Orientation** group in the **View** tab; a flyout is displayed. Choose the **Default Orientation** option from the flyout or press CTRL+D; the default trimetric view of the model is displayed in the drawing area.

This display gives you a better view of the sketch in the 3D space. The model is displayed in orange. Also, the pink colored arrow is displayed on the model indicating the direction of extrusion. The model may not fit fully in the drawing area.

2. Press CTRL+middle mouse button and drag the mouse downward in the drawing area. Notice that a red rubber band line is attached to the cursor. The length of the rubber band line gives an idea about the extent you need to zoom the model. After the model is visible in the drawing area, release the middle mouse button and the CTRL key.

Note

*You can also fit the model into the drawing area by choosing the **Refit** button from the **Graphic** toolbar.*

The model appears, as shown in Figure 4-52. All attributes selected by default in the **Extrude** tab will be used to create the model. You need to change only the depth of extrusion.

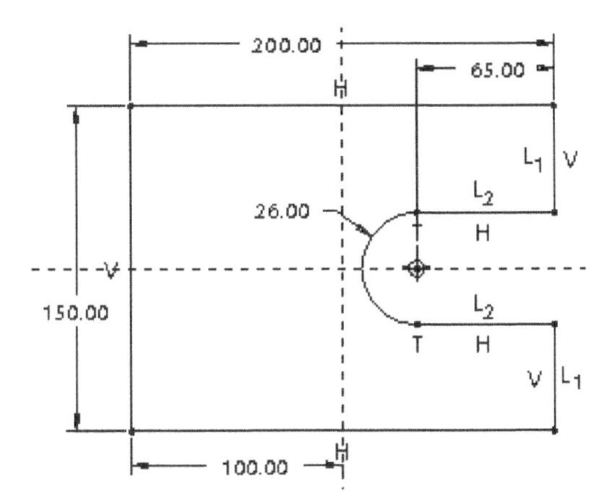

Figure 4-51 *Sketch after modifying the dimensions*

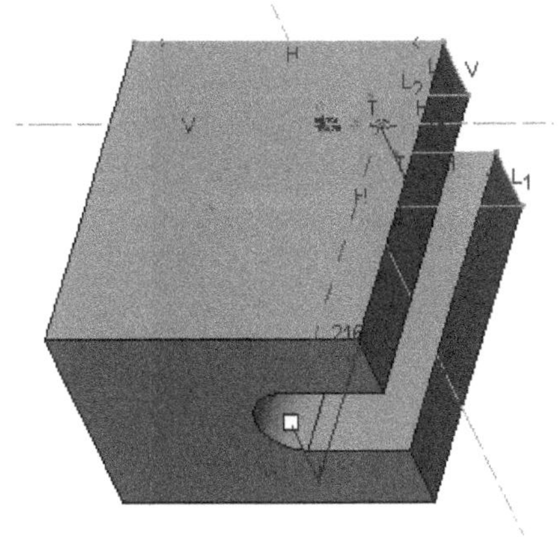

Figure 4-52 *Arrow showing the direction of feature creation*

3. Enter **75** in the edit box present on the **Extrude** tab, and press ENTER; the model in the drawing area is displayed with the specified depth of extrusion.

4. Choose the **Build feature** button from the **Extrude** dashboard to confirm the feature creation and exit the current feature creation tool. The trimetric view of the model is shown in Figure 4-53.

Note

*In Figure 4-53, the display of datum planes and the coordinate system are turned off by choosing the **Plane Display** and **Csys Display** buttons, respectively.*

Saving the Model and Closing the File

1. Choose the **Save** option from the **File** menu or choose the **Save** button from the **Quick Access** toolbar; the **Save Object** dialog box is displayed with the name of the object file that you have specified earlier.

2. Choose the **OK** button from the **Save Object** dialog box to save the file.

Note

*You can also change the current name of the file before saving it by choosing the **Rename** option from the **Manage File** flyout in the **File** menu. If you do so, the **Rename** dialog box is displayed. Enter the required name of the file in the **New Name** edit box and then choose the **OK** button to accept the new name.*

Figure 4-53 *Default trimetric view of the model*

3. Choose **File > Close** from the menu bar or choose the **Close** button from the **Window** group in the **View** tab to close the current window.

Tutorial 2

In this tutorial, you will create the model shown in Figure 4-54. The dimensions of the model are shown in Figure 4-55. **(Expected time: 30 min)**

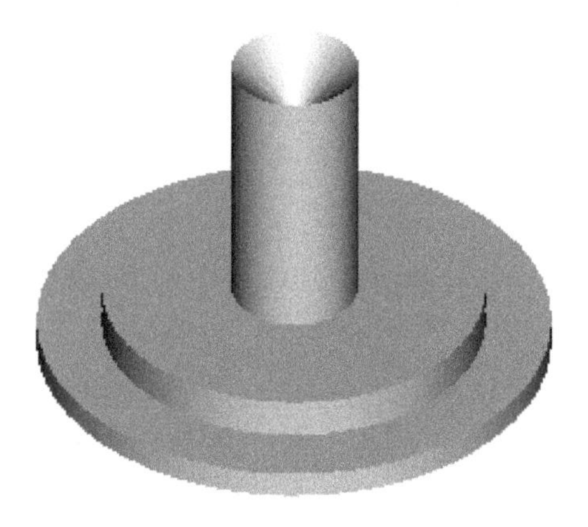

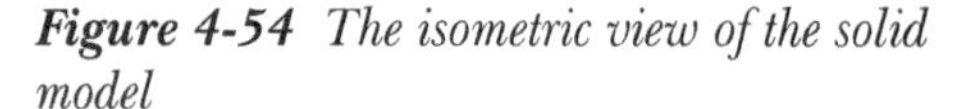

Figure 4-54 *The isometric view of the solid model*

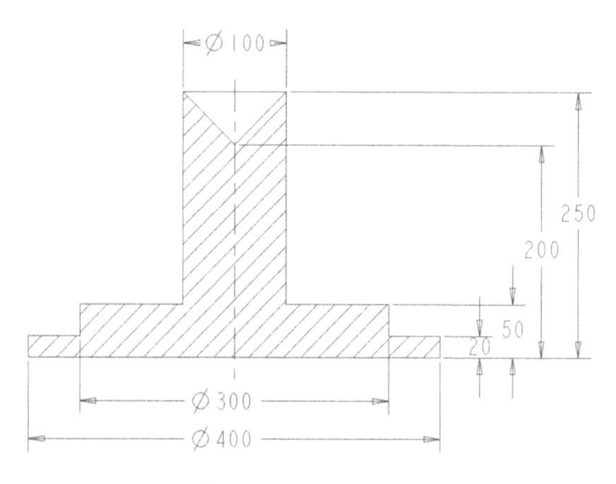

Figure 4-55 *The front section view of the solid model*

The following steps are required to complete this tutorial:

a. Create a new object file in the **Part** mode.
b. First examine the model and then determine the type of protrusion required for the model. Next, select the sketching plane for the model.

c. Draw the sketch for the revolved feature and a centerline to revolve it using the sketching tools. Next, apply dimensions to the model, refer to Figures 4-56 through 4-58.
d. Exit the sketcher environment and define the model attributes, refer to Figures 4-59 and 4-60.

Setting the Working Directory

The working directory was selected in Tutorial 1; therefore, there is no need to select it again. But if you are starting a new session of PTC Creo Parametric, you need to set the working directory again by following the steps given below.

1. Open the Navigator by clicking on the **Show Navigator** button at the bottom left corner of the main window of PTC Creo Parametric; the Navigator slides out. In the Navigator, choose the **Folder Browser** tab; the **Folder Tree** is displayed at the bottom. Click on the black arrow on the right of the **Folder Tree**; the **Folder Tree** expands.

2. Click on the node adjacent to the *Creo-3.0* folder in the Navigator; the contents of the *Creo-3.0* folder are displayed.

3. Now, right-click on the *c04* folder to display a shortcut menu. Choose the **Set Working Directory** option from the shortcut menu; the working directory is set to *c04*.

4. Close the Navigator by choosing the **Show Navigator** button.

Starting a New Object File

1. Open a new object file in the **Part** mode and then name the file as *c04tut2*.

   The three default datum planes are displayed in the drawing area. However, if the default datum planes were turned off in the previous tutorial, then they will not appear in the drawing area.

2. Turn on the display of datum planes by selecting the **Plane Display** option available in the **Datum Display Filter** drop-down list in the **Graphics** toolbar.

Selecting the Revolve Tool

The given solid model is a revolved feature that will be created by revolving the sketch through an angle of 360-degree about an axis. Therefore, the **Revolve** tool will be used to create the model. The procedure to use the **Revolve** tool from the **Shapes** group in the **Model** tab is given next.

1. Choose the **Revolve** tool from the **Shapes** group in the **Model** tab; the **Revolve** dashboard is displayed in the **Ribbon**.

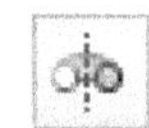

   Some of the attributes like rotation angle and direction of rotation of the sketch will be defined after the sketch has been created.

Selecting the Sketching Plane

To create the sketch of the model, first you need to select a sketching plane for the model. Note that the axis of revolution of the revolved feature is normal to the **TOP** datum plane in the model. Therefore, any of the other two datum planes, other than the **TOP** datum plane, can be selected as the sketching plane. Here, you will select the **FRONT** datum plane as the sketching plane. To do so, you need to follow the steps given next.

1. Choose the **Placement** tab from the dashboard to display the slide-down panel. Choose the **Define** button from the side-down group; the **Sketch** dialog box is displayed.

2. Select the **FRONT** datum plane as the sketching plane. As you select the sketching plane, the reference plane and its orientation are set automatically. The reference plane is selected in order to orient the sketching plane.

   In the **Sketch** dialog box, the **Reference** collector displays **RIGHT:F1(DATUM PLANE)**. This indicates that the **RIGHT** datum plane is selected as the reference plane. In the **Orientation** drop-down list, the **Right** option is selected by default. As a result, while drawing the sketch, the **RIGHT** datum plane will be on the right.

   The **RIGHT** datum plane will be perpendicular to the sketching plane and the sketching plane will be parallel to the screen.

3. Choose the **Sketch** button from the **Sketch** dialog box.

Drawing the Sketch

Now, you need to draw the sketch of the revolved feature. The sketch to be drawn is the cross-section of the revolved feature, which will be revolved about the centerline.

1. Choose the **Line Chain** tool from the **Sketching** group and then draw the sketch, refer to Figure 4-56. The sketch should be a closed loop and the bottom horizontal line should be aligned to the **TOP** datum plane.

   As you draw the sketch, weak dimensions and strong constraints are applied to the sketch.

2. Choose the **Centerline** tool from the **Datum** group. Draw a centerline for the axis of revolution. The centerline should be drawn such that it is aligned with the **RIGHT** datum plane, refer to Figure 4-56.

Tip: *If you have drawn centerline on both axes, then right click on the desired centerline to display a shortcut menu. Choose the* ***Designate Axis of Revolution*** *option from the shortcut menu to revolve the sketch around that particular centerline.*

Dimensioning the Sketch

Weak dimensions are automatically applied to the sketch. Since the model is a revolved feature, you need to manually apply the linear diameter dimensions to the sketch. The linear diameter dimensions are applied by using the centerline that was drawn in the sketch.

Tip: *The linear diameter dimensioning is necessary for all revolved features because mostly all revolved models are machined on a lathe. Therefore, while machining a revolved model, it is necessary that the operator of the machine has a drawing of the model that is diametrically dimensioned.*

1. Choose the **Normal** tool from the **Dimension** group.

2. Select the centerline, the first right vertical line, and then again the centerline.
3. Now, use the middle mouse button to place the dimension on the top of the sketch; the diameter dimension is placed.

   The method to dimension a sketch diametrically has been discussed in Chapter 2.

4. Select the centerline, the second right vertical line, and then again the centerline.
5. Now, press the middle mouse button to place the dimension below the sketch.
6. Select the centerline, the third right vertical line, and then again the centerline. Now, press the middle mouse button to place the dimension below the previous dimension.

   Dimension the remaining entities in the sketch, as shown in Figure 4-57.

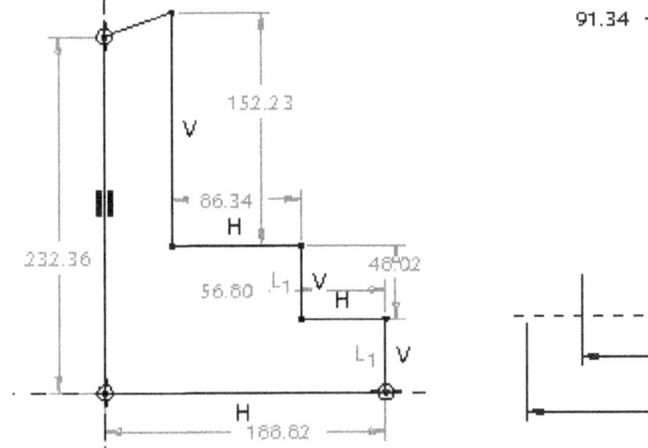

Figure 4-56 Sketch with weak dimensions

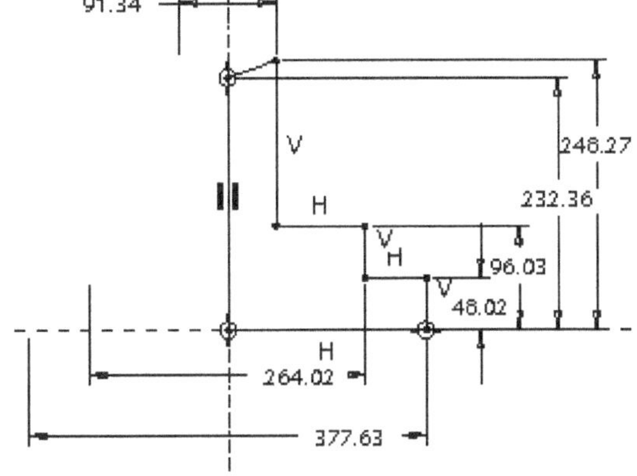

Figure 4-57 Sketch after dimensioning

Modifying the Dimensions

When you dimension a sketch, default dimension values are applied to the sketch. You need to modify the dimension values of the sketch.

1. Select the sketch and dimensions by using the CTRL+ALT+A keys.
2. Choose the **Modify** tool from the **Editing** group; the **Modify Dimensions** dialog box is displayed.

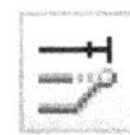

3. In this dialog box, clear the **Regenerate** check box and then modify the values of the dimensions, as shown in Figure 4-58. If you clear this check box, any modification of the dimension value will not update the sketch. It is recommended that you clear the **Regenerate** check box if more than one dimension has to be modified.

   You will notice that the dimension that you have selected in the **Modify Dimensions** dialog box is enclosed in a blue box in the drawing area.

4. After modifying all dimensions, choose the **Regenerate the section and close the dialog** button from the **Modify Dimensions** dialog box; the message **Dimension modifications successfully completed** is displayed in the message area.

5. Choose the **OK** button to exit the sketcher environment.

Specifying the Model Attributes

When you exit the sketcher environment, the **Revolve** dashboard above the drawing area is enabled again. Using this dashboard, you can specify the angle of revolution for the revolved feature.

1. Choose the **Named View** button from the **Orientation** group in the **View** tab to display a flyout. Choose the **Default Orientation** option from the flyout or press CTRL+D; the model is oriented in its default orientation, that is, trimetric view, as shown in Figure 4-59.

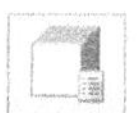

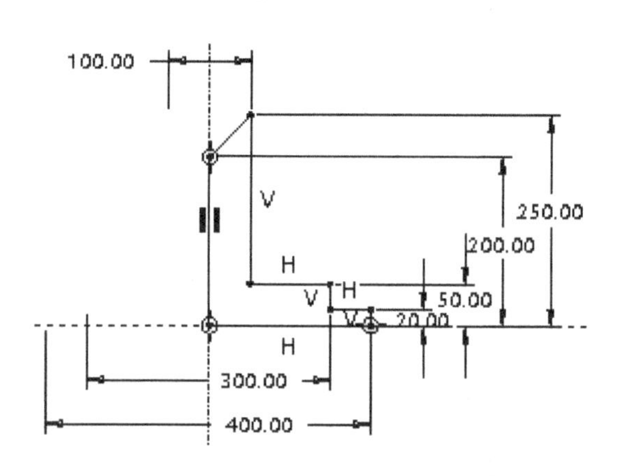

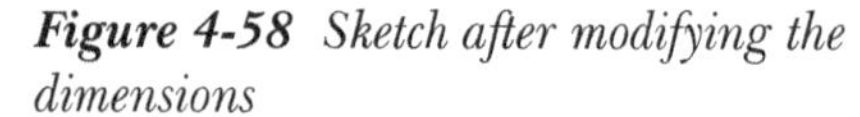
Figure 4-58 *Sketch after modifying the dimensions*

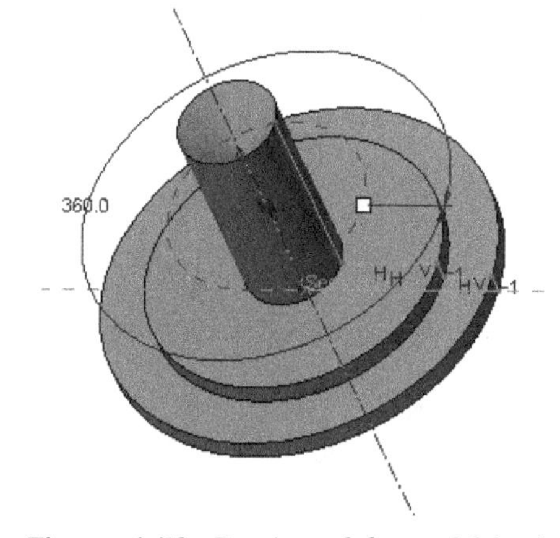

Figure 4-59 *Preview of the model in the default trimetric view*

This display gives you a better view of the sketch in the 3D space and the model appears in orange color. The drag handle is also available on the model that can be used to modify the angle of revolution dynamically.

All attributes needed to create a solid model are selected by default in the **Revolve** dashboard.

Note

If the model is not fully displayed in the drawing area, you may need to zoom it out.

2. Choose the **Build feature** button in the **Revolve** dashboard. The model appears as the one shown in Figure 4-60.

Figure 4-60 *The default trimetric view of the model*

Saving the Model

1. Choose the **Save** option from the **File** menu or choose the **Save** button from the **Quick Access** toolbar; the **Save Object** dialog box is displayed with the name of the object file specified earlier.

2. Choose the **OK** button to save the file.

Closing the Current Window

The given model is completed and is also saved. Now, you can close the current window.

1. Choose **File > Close** from the menu bar.

Note

*If you need to view the **Model Tree** of the model that you have created, then open it. Click on the **Show Navigator** button on the bottom left corner of the screen; the **Model Tree** slides out. To close the **Model Tree**, click on the button again; the **Model Tree** slides in.*

Tutorial 3

In this tutorial, you will create the model shown in Figure 4-61. The dimensions of the sketch are shown in Figure 4-62. The thickness of the model is 1 unit and depth of extrusion is 25 units. **(Expected time: 30 min)**

The following steps are required to complete this tutorial:

a. Create a new object file in the **Part** mode.
b. First examine the model and then determine the type of protrusion for the model. Next, select the sketching plane for the model.
c. Draw the sketch by using the sketching tools and then apply dimensions to it, refer to Figures 4-63 through 4-65.
d. Specify the attributes to create the model and then save it, refer to Figures 4-66 through 4-68.

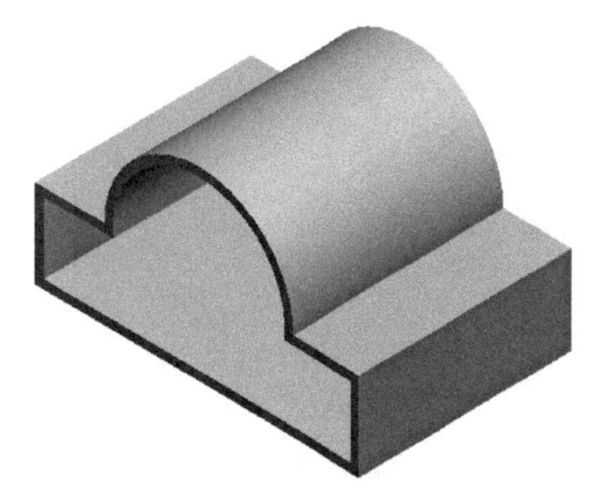

Figure 4-61 The isometric view of the model

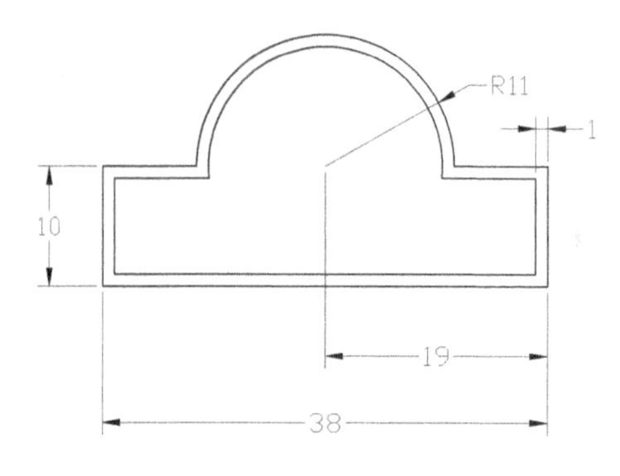

Figure 4-62 The front and the right views

The working directory was selected in Tutorial 1, and therefore, you do not need to select the working directory again. But if a new session of PTC Creo Parametric is started, then set the working directory using the Navigator.

Starting a New Object File

1. Open a new object file in the **Part** mode and then name it as *c04tut3*.

   If the default datum planes were not turned off in the previous tutorial, they will appear in the drawing area. If the datum planes are not displayed, turn them on by using the **Plane Display** button.

Selecting the Extrude Option

The given thin solid model is created by extruding the sketch to a distance of 25 units. So, you need to use the **Thicken Sketch** button on the **Extrude** dashboard to create the model.

1. Choose the **Extrude** tool from the **Shapes** group; the **Extrude** tab is displayed in the **Ribbon**.

   Most of the attributes needed to create the model will be defined after the sketch is drawn.

Selecting the Sketching Plane

From the isometric view of the model shown in Figure 4-61, it is clear that the direction of extrusion of the solid model is perpendicular to the **FRONT** datum plane. Therefore, you need to select the **FRONT** datum plane as the sketching plane.

1. Choose the **Placement** tab from the **Extrude** dashboard to display the slide-down panel. Choose the **Define** button from the slide-down panel; the **Sketch** dialog box is displayed.

2. Select the **FRONT** datum plane as the sketching plane; the reference plane and its orientation are set automatically.

   In the **Sketch** dialog box, the **Reference** collector displays **RIGHT:F1(DATUM PLANE)**. This indicates that the **RIGHT** datum plane is selected as the reference plane. In the **Orientation** drop-down list, the **Right** option is selected by default. As a result, while drawing the sketch, the **RIGHT** datum plane will be on the right.

   The **RIGHT** datum plane will be perpendicular to the sketching plane and the sketching plane will be parallel to the screen.

3. Choose the **Sketch** button from the **Sketch** dialog box to enter the sketching environment.

Drawing the Sketch

Using the sketcher tools, you need to draw the sketch for the thin extruded model.

1. Draw the sketch of the model, as shown in Figure 4-63, by using the **Line Chain** and **3-Point / Tangent End** tools from the **Sketching** group.

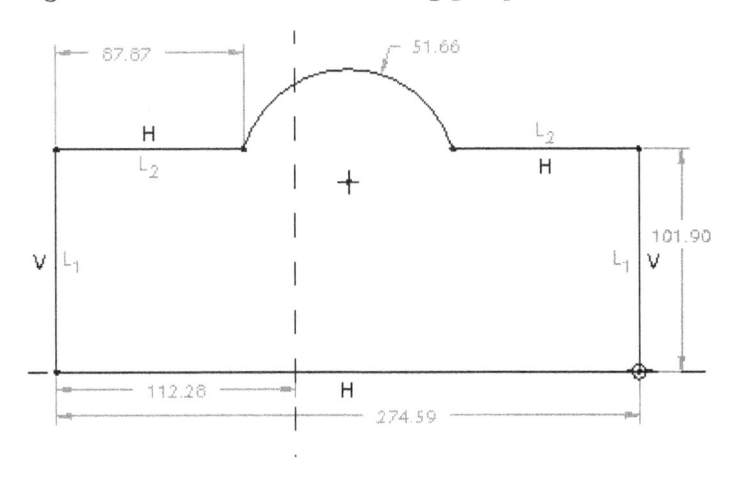

Figure 4-63 *Sketch of the thin extruded model*

Applying Constraints to the Sketch

While drawing the sketch, some weak dimensions are applied to the sketch. These dimensions appear in gray color. Although the weak dimensions are applied, you need to apply other constraints to the sketch.

1. Choose the **Coincident** tool from the **Constrain** group.

2. Select the center of the arc and then the **RIGHT** datum plane; the center of the arc is aligned with the **RIGHT** datum plane.

3. Select the bottom horizontal line and then the **TOP** datum plane. The bottom horizontal line is aligned with the **TOP** datum plane. Skip this point, if the two horizontal lines on the top are aligned with the **TOP** datum plane.

4. Choose the **Equal** tool from the **Constrain** group and then select the two horizontal lines on the top to apply the equal length constraint.

5. Choose the **Perpendicular** tool and select the left horizontal line and the arc; the perpendicular constraint symbol is applied. Similarly, select the right horizontal line and the arc; the perpendicular constraint symbol is applied. If these constraints have already been applied, the **Resolve Sketch** dialog box is displayed. Choose **Undo** from this dialog box.

   The sketch after applying constraints is shown in Figure 4-64.

Note

When you apply constraints to the sketch, some of the weak dimensions are automatically deleted. But, if you need certain constraints that were applied while drawing the sketch, you can use them and need not apply those constraints again.

Modifying Dimensions

You need to modify the dimension values of the weak dimensions.

1. Select the sketch and dimensions by using the CTRL+ALT+A keys.

2. Choose the **Modify** tool from the **Editing** group; the **Modify Dimensions** dialog box is displayed.

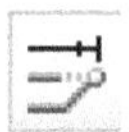

3. In this dialog box, clear the **Regenerate** check box and then modify the values of the dimensions, as shown in Figure 4-65. As mentioned earlier, it is recommended that you clear the **Regenerate** check box when more than one dimension has to be modified.

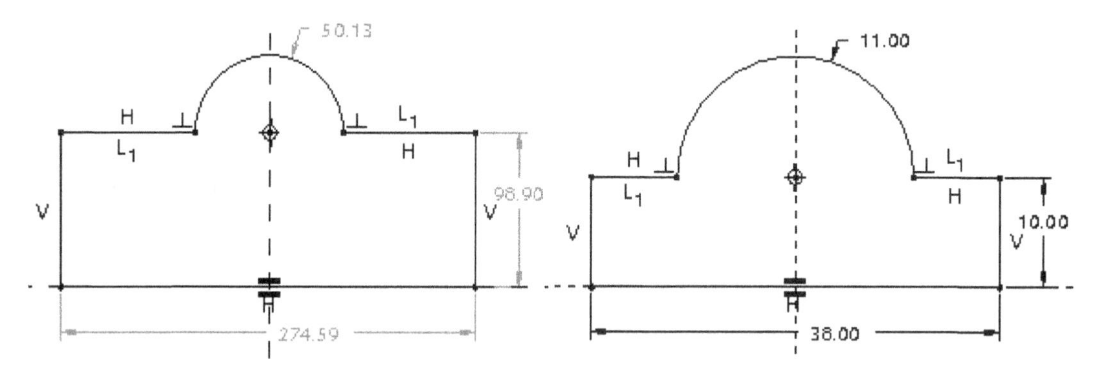

Figure 4-64 *Sketch after applying constraints*

Figure 4-65 *Sketch after modifying the dimensions*

You will notice that the dimension selected in the **Modify Dimensions** dialog box is enclosed in a blue box in the drawing area.

4. After modifying all dimensions, choose the **Regenerate the section and close the dialog** button from the **Modify Dimensions** dialog box; the message **Dimension modifications successfully completed** is displayed in the message area.

5. Choose the **OK** button to exit the sketcher environment.

Specifying the Model Attributes

The attributes required to create the model need to be selected from the **Extrude** dashboard.

1. Choose the **Thicken Sketch** button from the **Extrude** dashboard; the model changes to a thin model of default thickness, as shown in Figure 4-66.

2. Choose the **Named View** button from the **View** tab; a flyout is displayed. Choose the **Default Orientation** option from the flyout or you can press CTRL+D.

   The default trimetric view of the model is displayed. The model appears orange in color. If the model does not fit completely on the screen, then you need to zoom out. This display gives you a better view of the model in the 3D space.

   There are two drop-down lists in the **Extrude** dashboard. The drop-down list on the left is used to enter the depth of extrusion and the drop-down list on the right appears only when the **Thicken Sketch** button is chosen. This drop-down list is used to specify the thickness value of the thin model.

3. Enter **25** in the left edit box and press ENTER.

4. Enter **1** in the right edit box and press ENTER. The model appears, as shown in Figure 4-67. The model is completed and now you need to exit the **Extrude** dashboard.

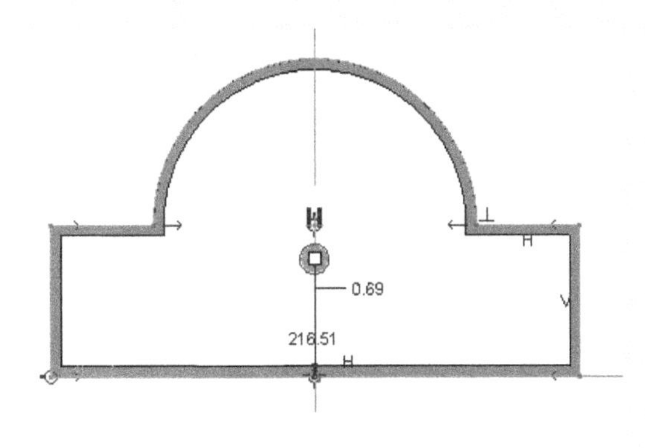

Figure 4-66 Arrow showing the direction of material addition with respect to the boundary of the section

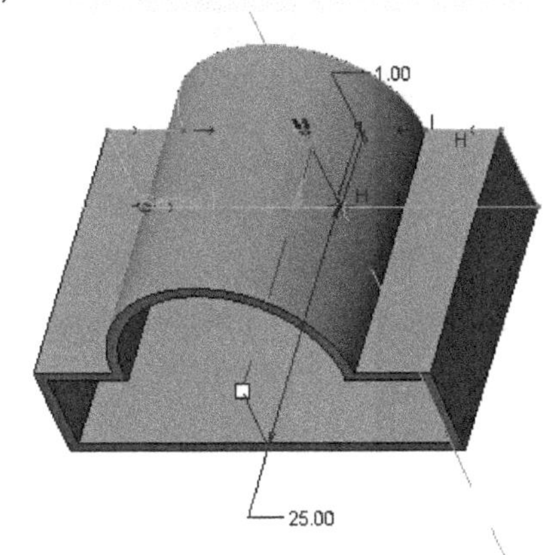

Figure 4-67 The default trimetric view of the model

5. Choose the **Build feature** button from the **Extrude** dashboard; the default trimetric view of the model is displayed, as shown in Figure 4-68.

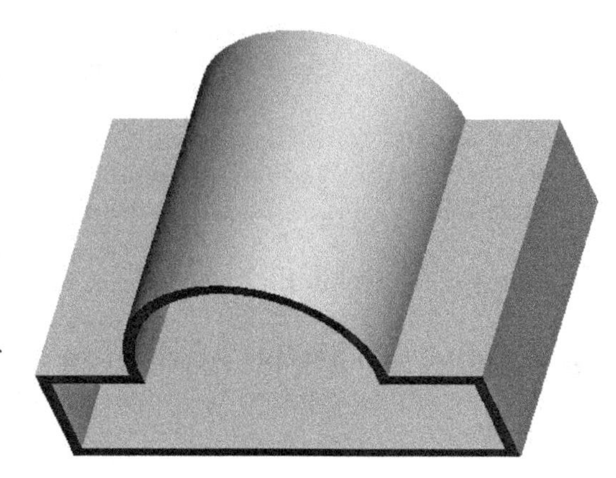

Figure 4-68 The default trimetric view of the solid model

Saving the Model

1. Choose the **Save** option from the **File** menu or choose the **Save** button from the **Quick Access** toolbar; the **Save Object** dialog box is displayed with the name of the object file specified earlier.

2. Choose the **OK** button to save the file.

Closing the Current Window

The given model is completed and is also saved. Now, you need to close the current window.

1. Choose **File > Close** from the menu bar.

Tutorial 4

In this tutorial, you will create the model shown in Figure 4-69. The dimensions of the sketch are shown in Figure 4-70. **(Expected time: 30 min)**

Figure 4-69 The isometric view of the model

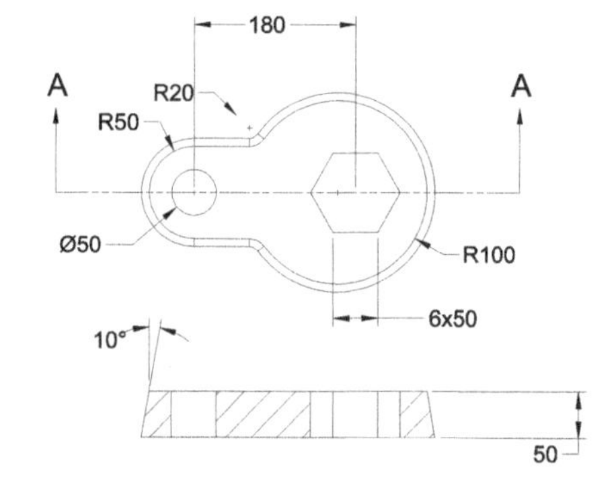

Figure 4-70 The front section and the top views

The following steps are required to complete this tutorial:

a. Create a new object file in the **Part** mode.
b. First examine the model and then determine the type of protrusion for the model. Next, select the sketching plane for the model.
c. Create the sketch for base feature, refer to Figures 4-71 through 4-73.
d. Specify the attributes to create the base feature, refer to Figures 4-74 and 4-75.
e. Create the cut feature by selecting the top face as the sketching plane, refer to Figure 4-76 and 4-77.
f. Specify the attributes to create the cut feature and then save it, refer to Figures 4-78 and 4-79.

The working directory was selected in Tutorial 1, and therefore, you do not need to select the working directory again. But if a new session of PTC Creo Parametric is started, then set the working directory using the Navigator.

Starting a New Object File

1. Open a new object file in the **Part** mode and then name it as *c04tut4*.

If the default datum planes were not turned off in the previous tutorial, they will appear in the drawing area. If the datum planes are not displayed, turn them on by using the **Plane Display** button.

Selecting the Extrude Tool

The desired solid model will be created by extruding the sketch to a distance of 50 units. Therefore, the sketch will be extruded as a solid to create the model.

1. Choose the **Extrude** tool from the **Shapes** group; the **Extrude** tab is displayed in the **Ribbon**.

Most of the attributes needed to create the model will be defined after the sketch is drawn.

Selecting the Sketching Plane

From the isometric view of the model shown in Figure 4-69, it is clear that the direction of extrusion of the solid model is perpendicular to the **TOP** datum plane. Therefore, you need to select the **TOP** datum plane as the sketching plane.

1. Choose the **Placement** tab from the **Extrude** dashboard to display the slide-down panel. Choose the **Define** button from the slide-down panel; the **Sketch** dialog box is displayed.

2. Select the **TOP** datum plane as the sketching plane; the reference plane and its orientation are set automatically.

In the **Sketch** dialog box, the **Reference** collector displays **RIGHT:F1(DATUM PLANE)**. This indicates that the **RIGHT** datum plane is selected as the reference plane. In the **Orientation** drop-down list, the **Right** option is selected by default. As a result, while drawing the sketch, the **RIGHT** datum plane will be on the right.

3. Choose the **Sketch** button from the **Sketch** dialog box to enter the sketching environment.

Drawing the Sketch

Using the sketcher tools, you need to draw the sketch for the thin extruded model.

1. Draw the sketch of the model by using the **Line Chain**, **3-Point / Tangent End**, and **Circular** tools from the **Sketching** group, as shown in Figure 4-71.

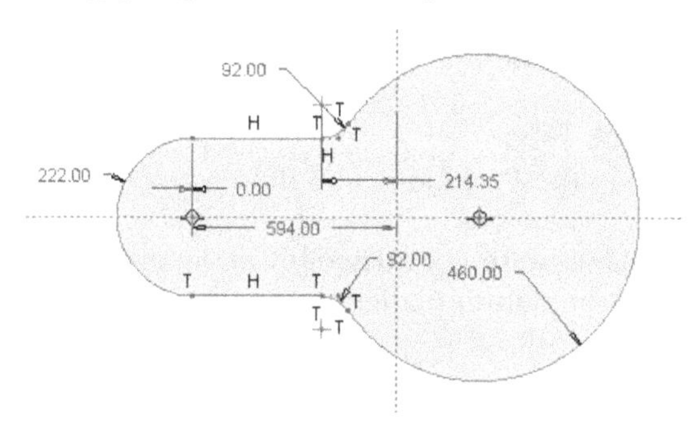

Figure 4-71 First Sketch of the model

Applying Constraints to the Sketch

While drawing the sketch, some weak dimensions are applied to the sketch. These dimensions appear in gray color. Although the weak dimensions are applied, you need to apply other constraints to the sketch.

1. Choose the **Equal** tool from the **Constrain** group and select the two lines and circular fillet to apply the equal length constraint.

2. Choose the **Tangent** tool and select the lines and arcs one by one; the tangent constraint symbol is applied. If these constraints have already been applied, the **Resolve Sketch** dialog box is displayed. Choose **Undo** from this dialog box, the given constraint will be delete.

The sketch after applying constraints is shown in Figure 4-72.

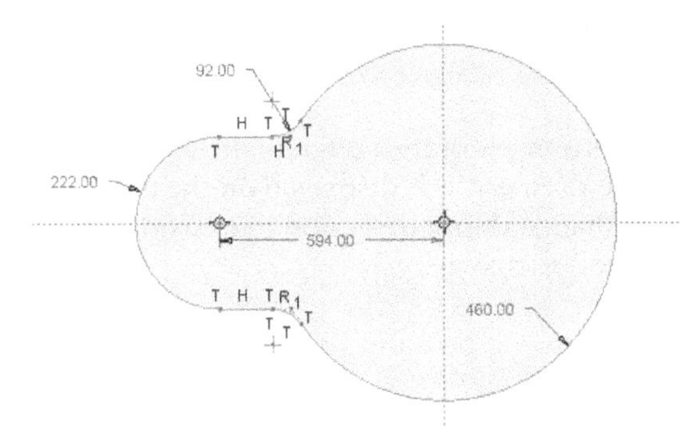

Figure 4-72 Sketch after applying constraints

Modifying Dimensions

You need to modify the dimension values of the weak dimensions.

1. Select the sketch and dimensions by using the CTRL+ALT+A keys.
2. Choose the **Modify** tool from the **Editing** group; the **Modify Dimensions** dialog box is displayed.

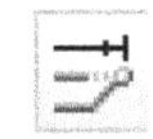

3. In this dialog box, clear the **Regenerate** check box and then modify the values of the dimensions, refer to Figure 4-73. As mentioned earlier, it is recommended that you clear the **Regenerate** check box when multiple dimension have to be modified.

   You will notice that the dimension selected in the **Modify Dimensions** dialog box is enclosed in a blue box in the drawing area.

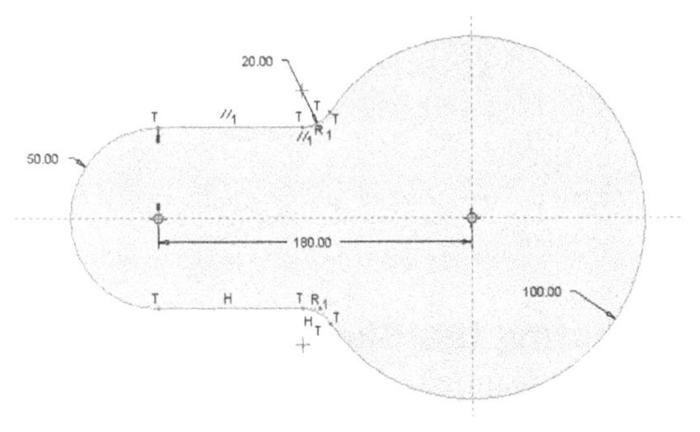

Figure 4-73 Sketch after modifying dimensions

4. After modifying all dimensions, choose the **Regenerate the section and close the dialog** button from the **Modify Dimensions** dialog box; the message **Dimension modifications successfully completed** is displayed in the message area.
5. Choose the **OK** button to exit the sketcher environment.

Specifying the Model Attributes

The attributes required to create the model need to be selected from the **Extrude** dashboard.

1. Choose the **Named View** button from the **Orientation** group in the **View** tab; a flyout is displayed. Choose the **Default Orientation** option from the flyout or

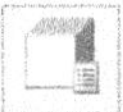

press CTRL+D; the default trimetric view of the model is displayed in the drawing area, as shown in Figure 4-74.

This display gives you a better view of the sketch in the 3D space. The model is displayed in orange. Also, the pink colored arrow is displayed on the model indicating the direction of extrusion. Choose the **Change depth direction of extrude to other side of sketch** button to change the direction of extrusion.

2. Enter **50** in the **Depth** value edit box and press ENTER.

3. Choose the **Options** tab; a flyout displayed. Select the **Add taper** check box from the flyout. Enter the value **10** in the **Add taper** edit box. The base feature is completed as shown in Figure 4-75, and now you need to exit the **Extrude** dashboard.

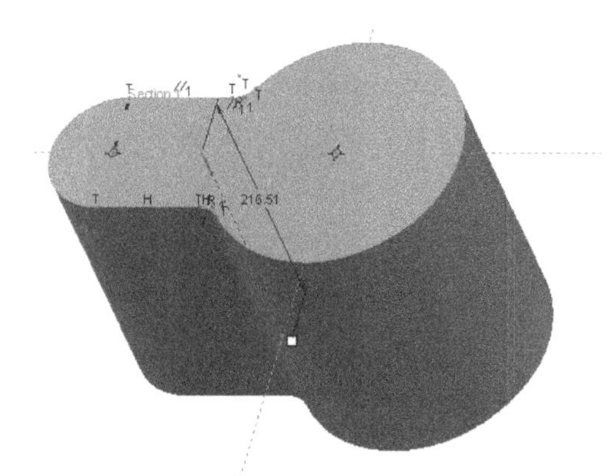

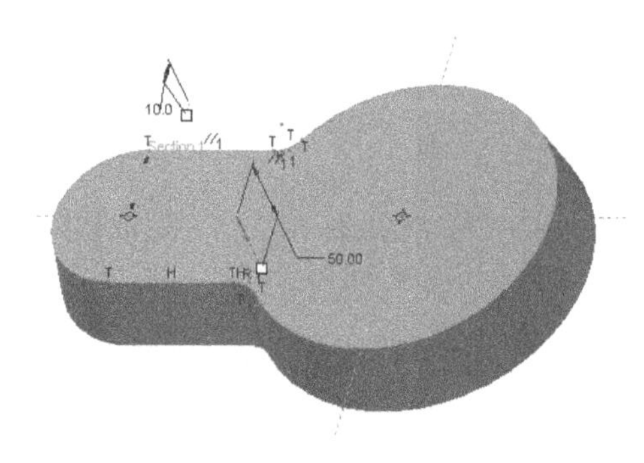

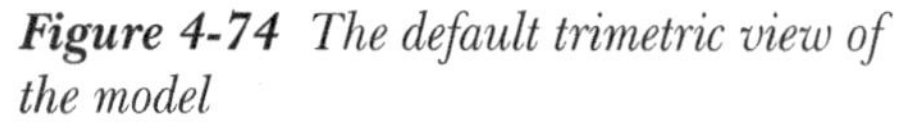

Figure 4-74 The default trimetric view of the model

Figure 4-75 Model after applying taper and depth

Selecting the Sketching Plane for the Cut Feature

Next feature to be created is an extruded cut. This feature will be created on top of the model.

1. Choose the **Extrude** tool from the **Shapes** group; the **Extrude** dashboard is displayed in the drawing area.

2. Choose the **Remove Material** button from the **Extrude** dashboard.

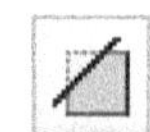

3. Choose the **Placement** tab from the dashboard; a slide-down panel is displayed. Next, choose the **Define** button from it; the **Sketch** dialog box is displayed.

4. Select the top face of the model as the sketching plane, as shown in Figure 4-76.

5. Using the left mouse button, select the **RIGHT** datum plane as the reference plane and then select the **Right** option from the **Orientation** drop-down list.

6. Choose the **Sketch** button; the sketcher environment is displayed.

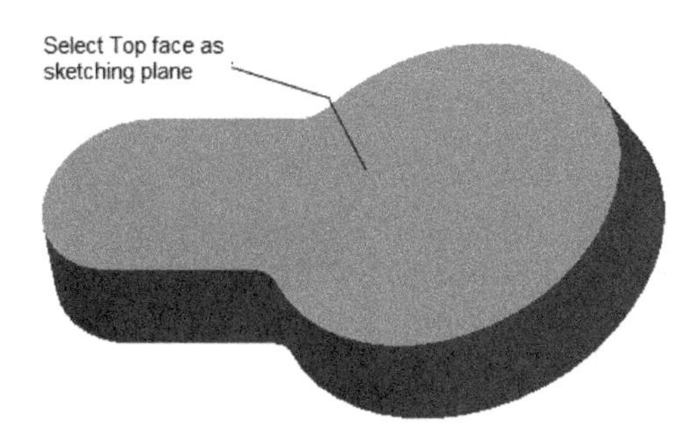

Figure 4-76 *Top face selected as the sketching plane*

Creating the Sketch for the Cut Feature

1. Turn the model display to **No hidden** from the **Display Style** drop-down list in the **Graphics** toolbar. Choose the **Sketch View** button from the **Graphics** toolbar to orient the sketching plane parallel to screen. Draw the sketch of the cut feature and add dimensions to it, as shown in Figure 4-77.

2. Choose the **OK** button and turn the model display to **Shading**. Choose the **Default Orientation** option from the **Named Views** flyout from the **Graphics** toolbar or press CTRL+D; the default trimetric view of the model is displayed in the drawing area as shown in Figure 4-78.

3. Two arrows also appear in the model, refer to Figure 4-78. One arrow indicates the direction of feature creation and the other arrow indicates the direction along which the material will be removed. Click on **Change depth direction of extrude to other side of sketch** to change the direction of material removal.

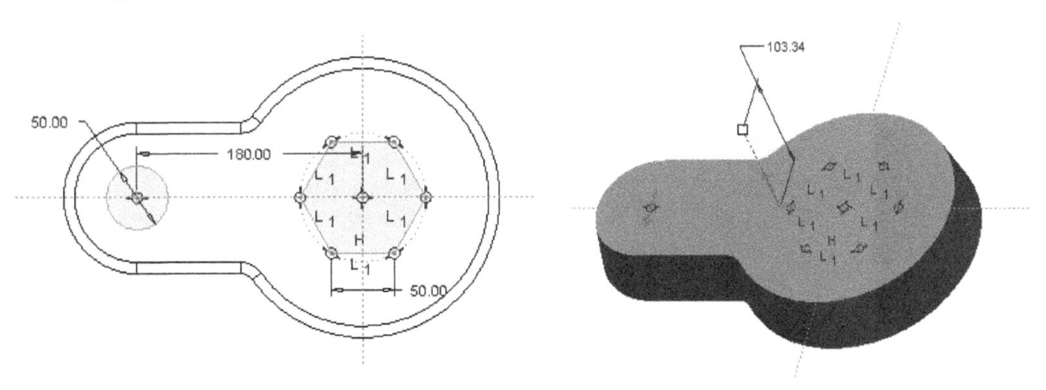

Figure 4-77 *Sketch of the cut feature of the model*

Figure 4-78 *The default trimetric view of the model*

4. Choose the **Build feature** button from the **Extrude** dashboard; the default trimetric view of the model is displayed, as shown in Figure 4-79.

Figure 4-79 The default trimetric view of the solid model

Saving the Model

1. Choose the **Save** option from the **File** menu or choose the **Save** button from the **Quick Access** toolbar; the **Save Object** dialog box is displayed with the name of the object file specified earlier.

2. Choose the **OK** button to save the file.

Closing the Current Window

The given model is completed and is also saved. Now, you need to close the current window.

1. Choose **File > Close** from the menu bar.

Tutorial 5

In this tutorial, you will create the model shown in Figure 4-80. Figure 4-81 shows the front view and dimensions of the model. Also, you will generate the general drawing view on a format file and then print the file. The thickness of the model is 1 unit.

(Expected time: 30 min)

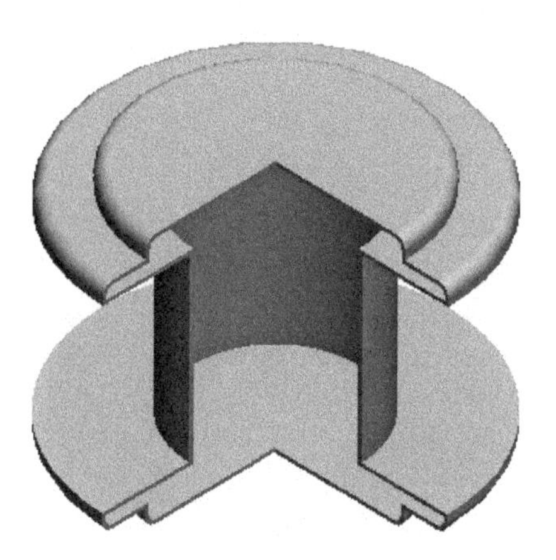

Figure 4-80 The isometric view of the solid model

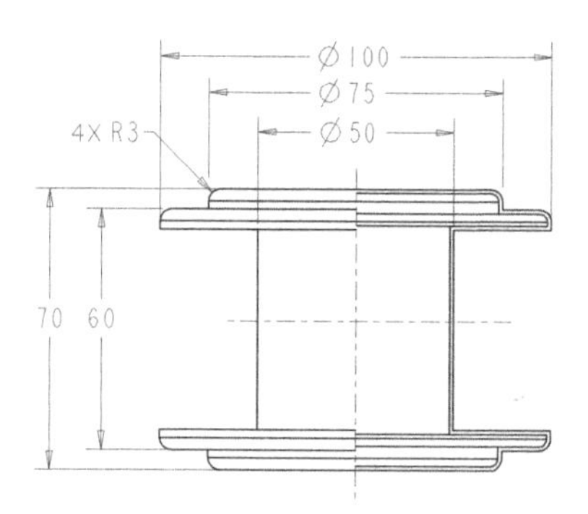

Figure 4-81 The front view of the solid model

The following steps are required to complete this tutorial:

a. Create a new object file in the **Part** mode.
b. First examine the model and then determine the type of protrusion for the model. Next, select the sketching plane for the model.
c. Draw the sketch by using sketching tools, apply dimensions, and modify dimension values, refer to Figures 4-82 through 4-85.
d. Specify the attributes for the model and then save it, refer to Figures 4-86 and 4-87.
e. Insert the isometric view of the model in a format file and then print the file.

The working directory was selected in Tutorial 1, therefore, you do not need to select the directory again. But if a new session of PTC Creo Parametric is started, then set the working directory using the Navigator.

Starting a New Object File

1. Start a new object file in the **Part** mode and then name the file as *c04tut5*.

The default datum planes are displayed, if they are not turned off. If the default datum planes were not turned off in the previous tutorial, they will appear in the drawing area. If the datum planes are not displayed, turn them on by using the **Plane Display** button from the **View** tab.

Selecting the Revolve Tool

Now, you need to create a revolved thin model by revolving the sketch through a given angle. Therefore, you need to use the **Thicken Sketch** button from the **Revolve** dashboard to create the model. The procedure is given next.

1. Choose the **Revolve** tool from the **Shapes** group; the **Revolve** dashboard is displayed on the top of the drawing area.

2. Choose the **Thicken Sketch** button from the dashboard. Make sure that the **Revolve as solid** button is chosen in the dashboard.

In this tutorial, the **Thicken Sketch** button is chosen before drawing the sketch because the sketch to be drawn is an open section. It is not possible to draw an open sketch with the default attributes selected in the **Revolve** dashboard.

3. Choose the **Placement** tab; a slide-down panel is displayed. Choose the **Define** button from the slide-down panel; the **Sketch** dialog box is displayed.

Selecting the Sketch Plane

Note that in Figure 4-80, the imaginary axis of revolution of the revolved feature is normal to the **TOP** datum plane. Therefore, the **TOP** datum plane will be selected to be at the top while drawing the sketch. Now, you need to decide the sketching plane from the **RIGHT** and **FRONT** datum planes. Any of the two datum planes can be selected as the sketching plane. In this case, the **RIGHT** datum plane will be selected as the sketching plane.

1. Select the **RIGHT** datum plane as the sketching plane. As you select the sketching plane, the reference plane and its orientation are set automatically by default.

   In the **Sketch** dialog box, the **Reference** box displays **TOP:F2(DATUM PLANE)**. This indicates that the **TOP** datum plane is selected as the reference plane. But you need to change the orientation of the plane.

2. From the **Orientation** drop-down list, select the **Top** option.

   The **TOP** datum plane will be perpendicular to the sketching plane and the sketching plane will be parallel to the screen.

3. Choose the **Flip** button to change the direction of the pink arrow that appears on the sketching plane; the direction of viewing the sketching plane is reversed.

4. Choose the **Sketch** button in the **Sketch** dialog box.

   Choose the **Sketch View** button from the **Graphics** toolbar to make the sketching plane parallel to the screen.

Drawing the Sketch

You need to draw the sketch of the thin extruded model. The sketch can be a closed loop or an open loop. Here, you need to draw an open sketch.

1. Draw a geometric centerline and then draw the sketch, as shown in Figure 4-82. The geometric centerline is the axis of revolution. To fillet the corners, choose the **Circular** tool from the **Fillet** drop-down in the **Sketching** group.

   While drawing the sketch, weak constraints are applied to the sketch.

Applying Constraints to the Sketch

Weak constraints are applied to the sketch while drawing but you need to apply the constraints to the sketch using the **Constrain** group.

1. Choose the **Equal** tool from the **Constrain** group.

2. Select the two horizontal lines with the constraint symbol $\mathbf{L_1}$ one by one, as shown in Figure 4-83, to apply the equal length constraint.

3. Select the two horizontal lines with the constraint symbol $\mathbf{L_2}$ one by one, as shown in Figure 4-83, to apply the equal length constraint.

4. Select the vertical lines with the constraint symbol $\mathbf{L_3}$, as shown in Figure 4-83, to apply the equal length constraint.

5. Select all fillets to apply equal radii constraint.

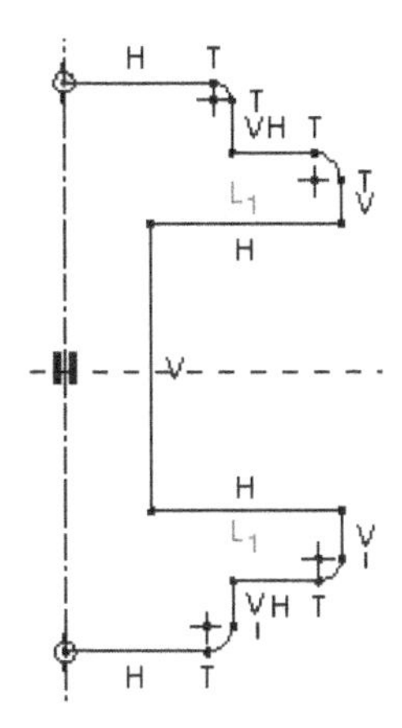

Figure 4-82 Sketch for the revolved feature with the weak dimensions turned off for clarity

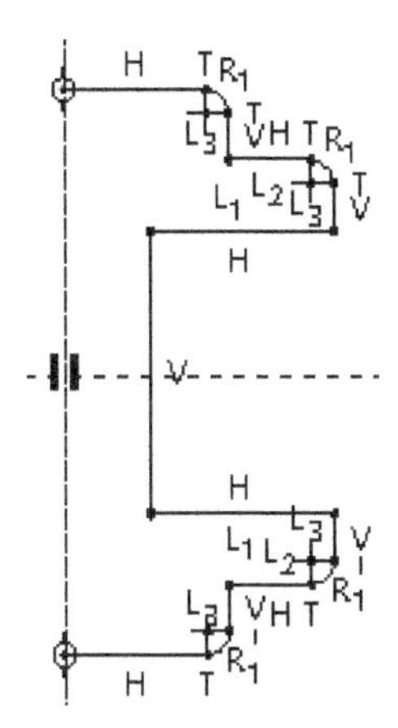

Figure 4-83 Sketch after applying the constraints with the weak dimensions turned off for clarity

Applying Dimensions to the Sketch

The weak dimensions are automatically applied to the sketch. Remember that since the model is a revolved feature, you need to apply the linear diameter dimensions to the sketch manually. You need to apply the linear diameter dimensions by using the centerline drawn in the sketch.

1. Choose the **Normal** tool from the **Dimension** group.

2. Dimension the sketch, as shown in Figure 4-84.

Modifying the Dimensions

You need to modify the dimension values of the weak dimensions.

1. Select the sketch and dimensions by using the CTRL+ALT+A keys.
2. Choose the **Modify** tool from the **Sketching** group; the **Modify Dimensions** dialog box is displayed.

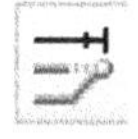

3. In this dialog box, clear the **Regenerate** check box and then modify the values of the dimensions, as shown in Figure 4-85.

   You will notice that the dimension you select in the **Modify Dimensions** dialog box is enclosed in a blue box in the drawing area.

4. After modifying all dimensions, choose the **Regenerate the section and close the dialog** button from the **Modify Dimensions** dialog box; the message **Dimension modifications successfully completed** is displayed in the message area.
5. Choose the **OK** button to exit the sketcher environment.

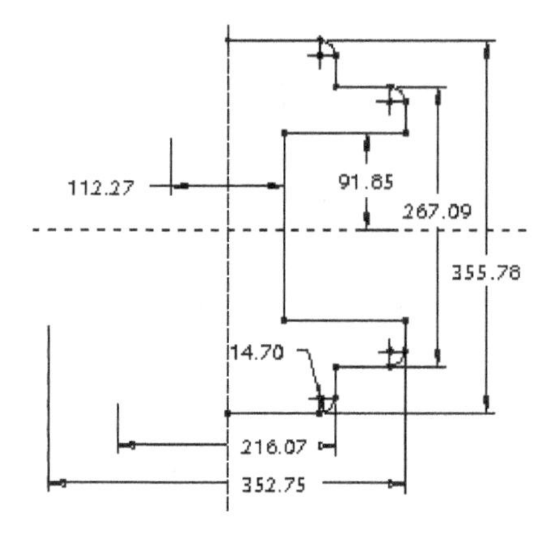

Figure 4-84 Sketch after dimensioning with the constraints turned off for clarity

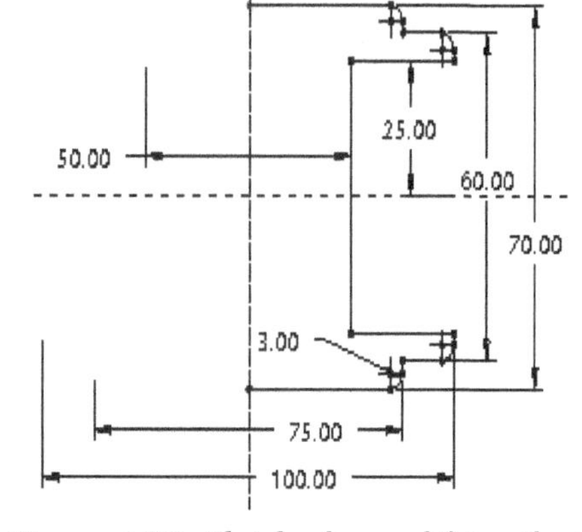

Figure 4-85 Sketch after modifying the dimensions with the constraints turned off for clarity

Specifying the Model Attributes

After completing the sketch, you need to specify the side to which thickness of material will be applied. The thickness can be applied outside the section, inside the section, or symmetrically to both sides of the section boundary. In this model, you need to apply the thickness inside the section.

When you exit the sketcher environment, the model assumes some default attributes and therefore it is not displayed in the drawing area.

1. Enter the thickness value as **1** in the drop-down list available on the **Revolve** dashboard, and press ENTER.

2. Enter **270** in the edit box present on the left of the **Revolve** dashboard and press ENTER. The default value of angle of revolution is displayed as 360.

3. Choose the **Named View** button from the **View** tab to display the flyout.From the flyout, choose the **Default Orientation** option or press CTRL+D; the trimetric view of the model after specifying all attributes is displayed, as shown in Figure 4-86.

4. Choose the **Build feature** button from the **Revolve** dashboard. The default trimetric view of the model is shown in Figure 4-87.

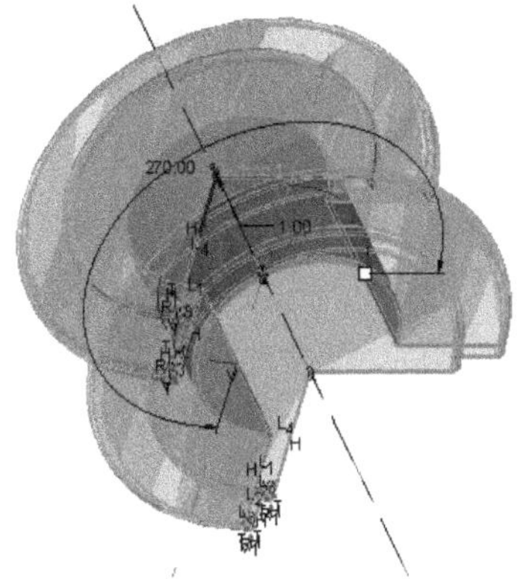
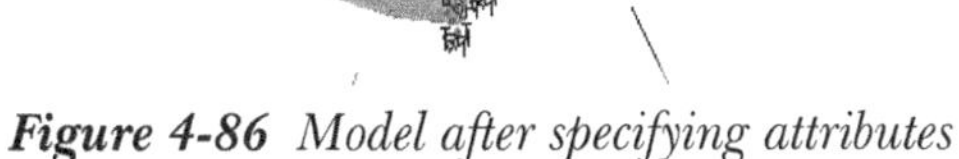

Figure 4-86 *Model after specifying attributes*

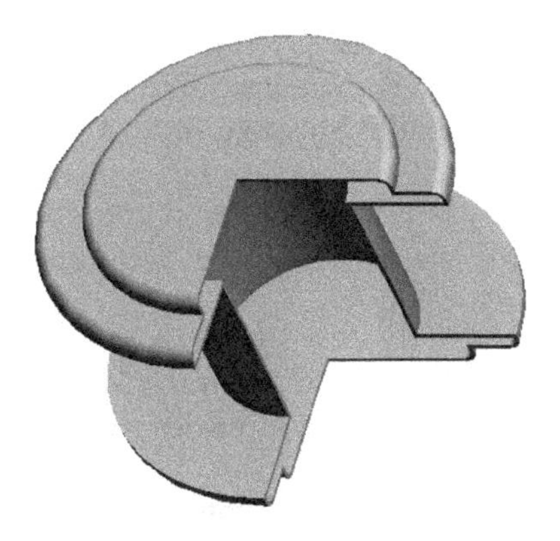

Figure 4-87 *Default trimetric view of the model*

Saving the Model

1. Choose the **Save** option from the **File** menu or choose the **Save** button from the **Quick Access** toolbar; the **Save Object** dialog box is displayed with the name of the object file specified earlier.

2. Choose the **OK** button to save the file.

Starting the Format File

1. Choose the **New** button from the **Quick Access** toolbar; the **New** dialog box is displayed.

2. Select the **Format** radio button from the **Type** area in the **New** dialog box and name the file as *Format1*. Choose **OK** from the **New** dialog box; the **New Format** dialog box is displayed, as shown in Figure 4-88.

3. Select the **Empty** radio button from the **Specify Template** area and the **Landscape** radio button from the **Orientation** area of the **New Format** dialog box, if they are not selected.

4. Select **A** from the **Standard Size** drop-down list in the **Size** area and then choose **OK** to proceed to the **Format** mode. Note that a sheet of the size A is displayed on the screen. This is evident from the text displayed below the sheet on the screen.

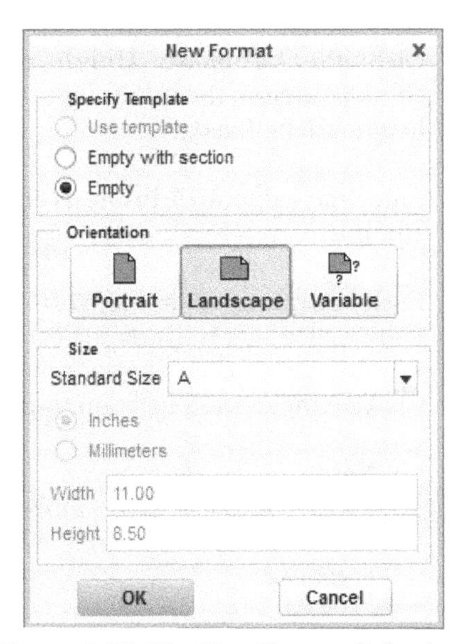

Figure 4-88 *The* ***New Format*** *dialog box*

Creating Format

1. Choose the **Offset Edge** tool from the **Edge** drop-down in the **Sketching** group of the **Sketch** tab in the **Ribbon**; the **OFFSET OPER** menu is displayed. Choose the **Ent Chain**

option from the **OFFSET OPER** menu and then select all four border lines of the format by drawing a window around them. Press the middle mouse button after making the selection; the **Message Input Window** is displayed at top of the drawing area.

An arrow is displayed pointing outward. This arrow displays the direction of the offset. Since you need to offset the lines in the opposite direction, you will specify a negative offset distance.

2. Enter the value **-0.25** in the **Message Input Window** and press ENTER. Now, press the middle mouse button twice to exit the **Menu Manager** dialog box.

3. Choose the **Insert Table** tool from the **Table** drop-down in the **Table** group of the **Table** tab in the **Ribbon**; the **Insert Table** dialog box is displayed, as shown in Figure 4-89.

4. Choose the **Table growth direction: leftward and ascending** button from the **Direction** area of the dialog box.

5. Set the value of the **Number of columns** spinner to **4** and **Number of rows** spinner to **1**.

6. Clear the **Automatic Height Adjustment** check box. Now, the value entered for the row height will be fixed.

7. Enter the value **0.75** in the **Height (INCH)** edit box.

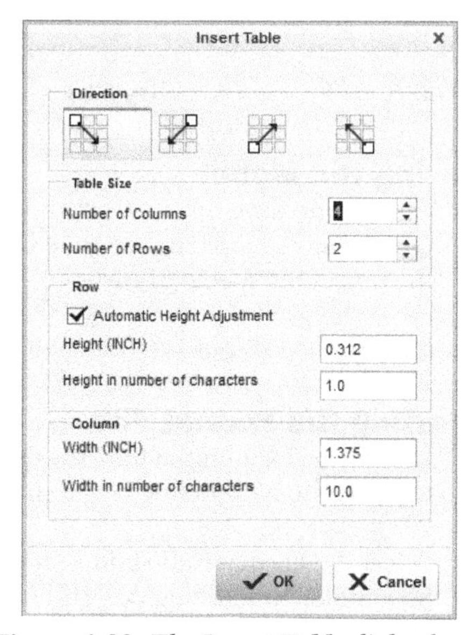

***Figure 4-89** The **Insert Table** dialog box*

8. Enter the value **1.25** in the **Width (INCH)** edit box and choose the **OK** button; the **Select Point** dialog box is displayed.

9. Click on the desired location to place the origin of the table; the table will be created.

Note

*You need to double-click on any of the columns of the table to enter the text in the respective columns, as required. When you double-click on any of the columns, the **Format** tab is displayed in **Ribbon**.*

10. Choose the **Save** button from the **Quick Access** toolbar; the **Save Object** dialog box is displayed with the name of the file specified earlier.

11. Choose the **OK** button from the **Save Object** dialog box to save the file.

12. Choose **File > Close** from the menu bar to close the format file.

Generating the Isometric View in the Format File

Now, you need to create a new drawing file and then generate the isometric view of the model.

1. After saving the model, choose the **New** button from the **File** menu; the **New** dialog box is displayed.

2. Select the **Drawing** radio button from the **Type** area; a default name is displayed in the **Name** edit box.

3. Enter *c04tut5* in the **Name** edit box and choose **OK** from the **New** dialog box; the **New Drawing** dialog box is displayed, as shown in Figure 4-90. The name *c04tut5* is displayed in the **Default Model** edit box and the **Use template** radio button is selected by default in the **Specify Template** area of the dialog box.

4. Select the **Empty with format** radio button from the **Specify Template** area and choose the **Browse** button from the **Format** area; the **Open** dialog box is displayed.

5. Select the *Format1.frm* file that you have created and choose **Open**. Next, choose the **OK** button from the **New Drawing** dialog box; the drawing mode is invoked and the format file is displayed in the drawing area.

6. Choose the **General** tool from the **Model Views** group in the **Layout** tab of the **Ribbon**; the **Select Combined State** dialog box is displayed.

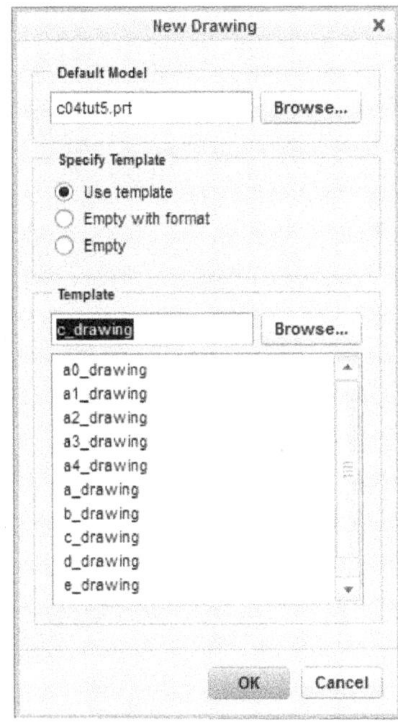

Figure 4-90 *The **New Drawing** dialog box*

7. Choose the **OK** button from this dialog box and click in the drawing area; the default view of the model is displayed in the drawing area and the **Drawing View** dialog box is displayed, as shown in Figure 4-91.

8. Select the **Default Orientation** option from the list below the **Model view names** area. Also, select the **Isometric** option from the **Default orientation** drop-down list and choose **Apply** from the dialog box. The default view of the model is displayed in the drawing area.

9. Now, select the **Scale** option from the **Categories** list in the **Drawing View** dialog box, the dialog box is displayed, as shown in Figure 4-92.

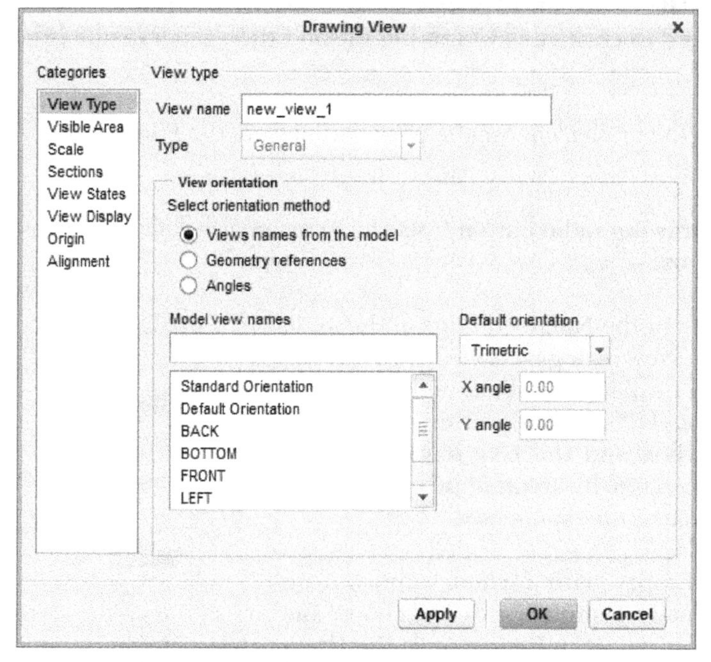

*Figure 4-91 The **Drawing View** dialog box*

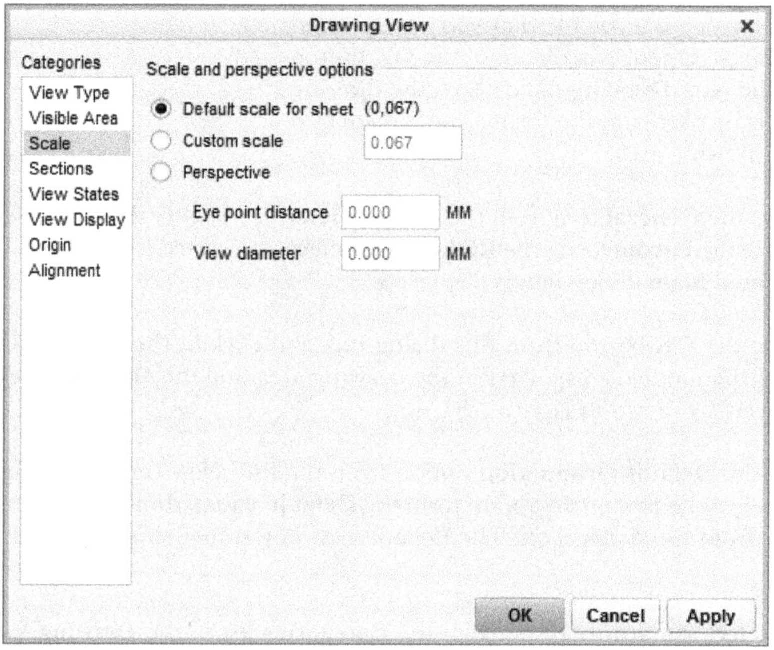

*Figure 4-92 The **Scale** option selected in the **Drawing View** dialog box*

The **Default scale for sheet** radio button is selected in the **Scale and perspective options** area by default and the default scale value for the sheet is displayed.

10. Select the **Custom scale** radio button and enter **0.04** in the edit box displayed next to the **Custom scale** radio button. Now, choose **Apply** from the **Drawing View** dialog box; the scaled model is displayed in the drawing area.

11. Choose **Close** from the **Drawing View** dialog box to close the dialog box and complete the generation of the isometric view of the model.

12. Notice that the isometric view of the model is displayed in the drawing area along with the scale factor mentioned at the bottom of the isometric view. If you want to erase the note displayed, choose the **Annotate** tab in the **Ribbon**. Next, expand the **new_view_1** node and then the **Annotations** node in the drawing tree. Select the draft note node and right-click to invoke a shortcut menu. Choose **Erase** from the shortcut menu to erase the note. The drawing area after erasing the note is similar to the one shown in Figure 4-93.

Saving and Printing the Drawing File

1. Choose **File > Print > Print** from the menu bar; the **Print Preview** tab is displayed in the **Ribbon**.

2. Choose the **Settings** button from the **Settings** group to display the **Printer Configuration** dialog box. Next, select the **MS Printer Manager** option from the dialog box.

Note

*The **MS Printer Manager** option is used to print an active model through the printer configuration installed as the default printer on your computer. It is recommended that you use the laser printer to print the drawing files.*

3. Now, choose the **OK** button from the **Printer Configuration** dialog box.

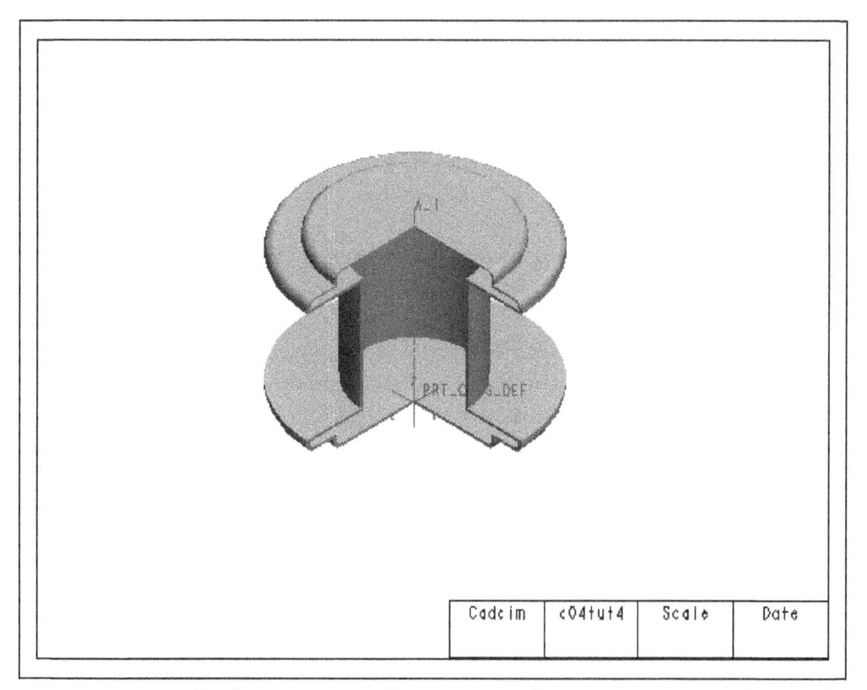

Figure 4-93 The drawing area after inserting the isometric view of the model

4. Choose the **Print** button from the **Finish** group of the **Print Preview** tab; the **Print** dialog box is displayed. Now, choose the **OK** button from the new **Print** dialog box to print the drawing file.

Closing the Current Window

The given model is completed and saved. Now, you can close the current window.

1. Choose **File > Close** from the menu bar; the window is closed.

Self-Evaluation Test

Answer the following questions and then compare them to those given at the end of this chapter:

1. All features created in the **Part** mode are called the base features. (T/F)

2. You can extrude a sketch to both sides of the sketching plane symmetrically. (T/F)

3. When the sketcher environment is invoked, the references required for creating a sketch are automatically selected. (T/F)

4. The **Sketch** dialog box is used to select the sketching plane. (T/F)

5. After selecting the sketching plane, the orange arrow displayed on it shows the direction in which you will view the sketching plane. (T/F)

6. The __________ button is chosen by default in the **Extrude** dashboard.

7. The __________ button in the **Revolve** dashboard is used to create a thin revolved model.

8. The **Orientation** dialog box is displayed when you choose the __________ button.

9. After you exit the sketcher environment, the __________ appear on the model to dynamically modify the extrusion depth or the angle of revolution.

10. The **Revolve** tool revolves the sketched section about the __________ through the specified angle.

Review Questions

Answer the following questions:

1. In which direction does a sketch, by default, revolve about a geometric centerline?

   (a) Clockwise (b) Counterclockwise
   (c) Both (a) and (b) (d) None of these

2. Which of the following groups in the **Model** tab contains the **Extrude** tool?

   (a) **Editing** (b) **Get Data**
   (c) **Shapes** (d) **Operations**

3. How many datum planes are available when you enter the **Part** mode?

   (a) 4 (b) 3
   (c) 2 (d) None of these

4. Which of the following groups is used to turn off the display of datum planes?

   (a) **Orientation** (b) **Window**
   (c) **Model Display** (d) **Show**

5. Which of the following combinations is used to change the orientation of the model to the default orientation?

   (a) CTRL+D (b) CTRL+left mouse button
   (c) CTRL+right mouse button (d) None of these

6. Features created by using the **Blind** option do not have a dimension associated with them. Hence, they cannot be modified by changing the dimension value. (T/F)

7. It is recommended not to use the default datum planes to create the base feature. (T/F)

8. The section drawn for revolving as a solid should be a closed loop. (T/F)

9. The revolved section should have a centerline. (T/F)

10. A revolved section can be drawn on both sides of a centerline. (T/F)

Exercises

Exercise 1

Create the model shown in Figure 4-94. The dimensions of the model are shown in Figure 4-95. **(Expected time: 20 min)**

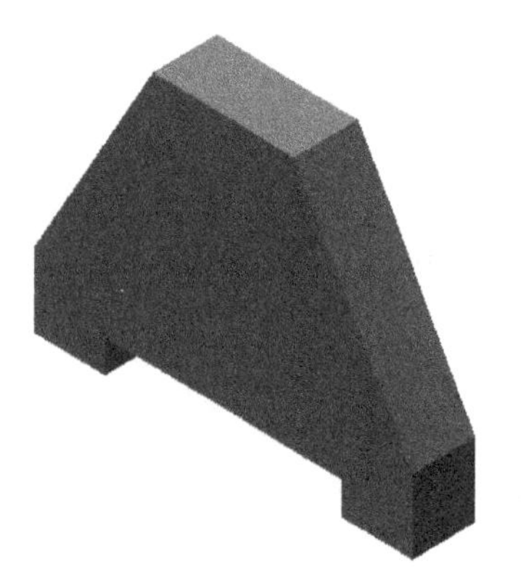

Figure 4-94 The isometric view of the model

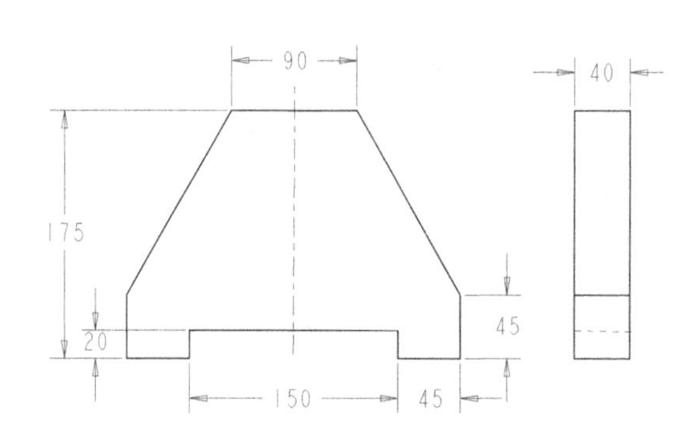

Figure 4-95 The front and right views of the model

Exercise 2

Create the model shown in Figure 4-96. The dimensions of the model are shown in Figure 4-97. **(Expected time: 30 min)**

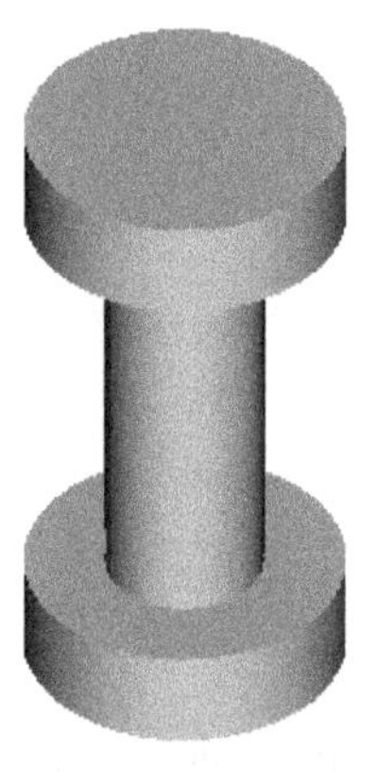

Figure 4-96 The isometric view of the model

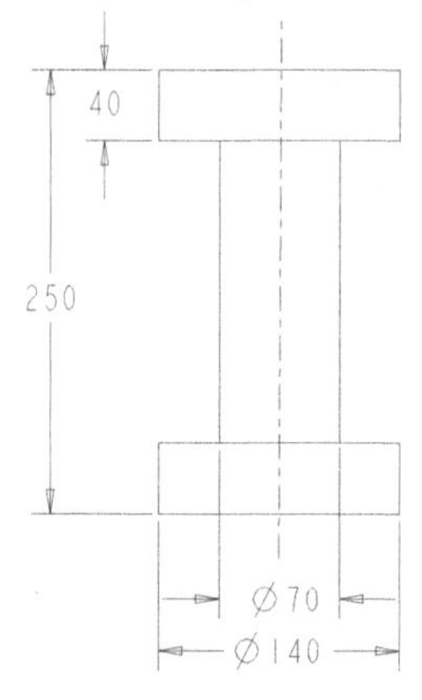

Figure 4-97 The front view of the model

Exercise 3

Create the model shown in Figure 4-98. The dimensions of the model are shown in Figure 4-99. **(Expected time: 30 min)**

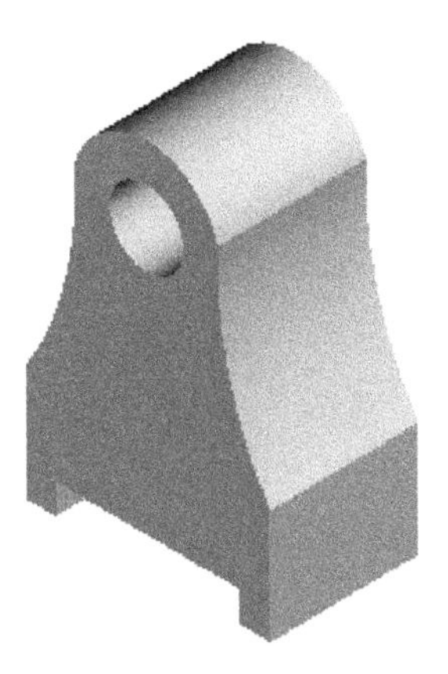

Figure 4-98 *The isometric view of the model*

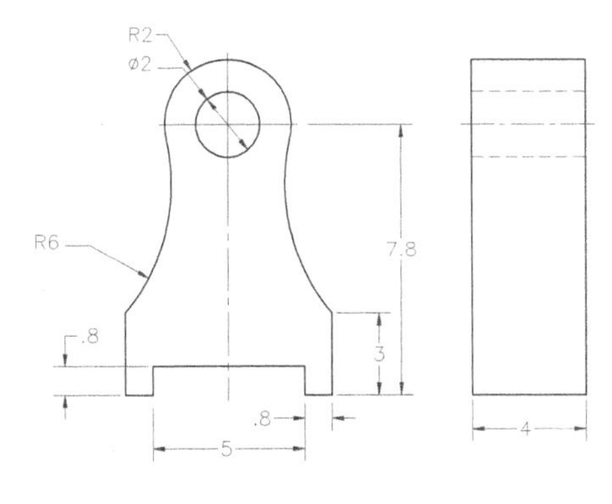

Figure 4-99 *The front view of the model*

Exercise 4

Create the model shown in Figure 4-100. The dimensions of the model are shown in Figure 4-101. **(Expected time: 30 min)**

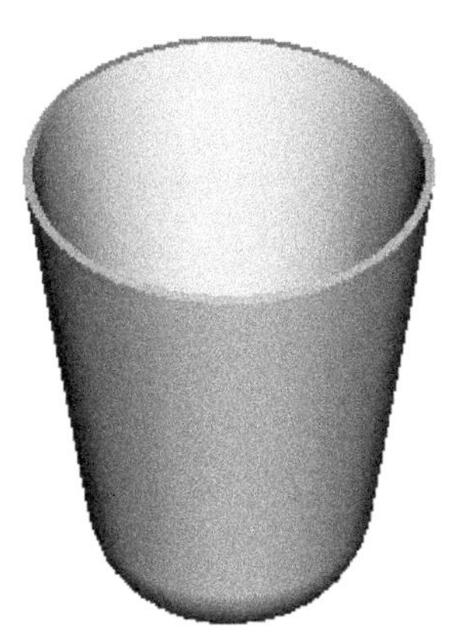

Figure 4-100 *The isometric view of the model*

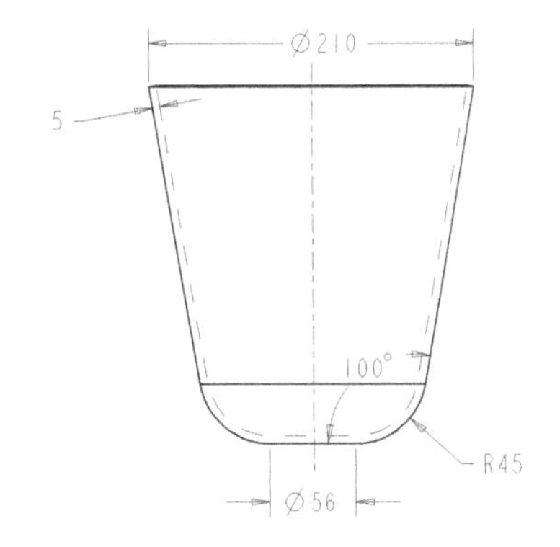

Figure 4-101 *The front view of the model*

Exercise 5

Create the model shown in Figure 4-102. The dimensions of the model are shown in Figure 4-103. **(Expected time: 30 min)**

Figure 4-102 The isometric view of the model

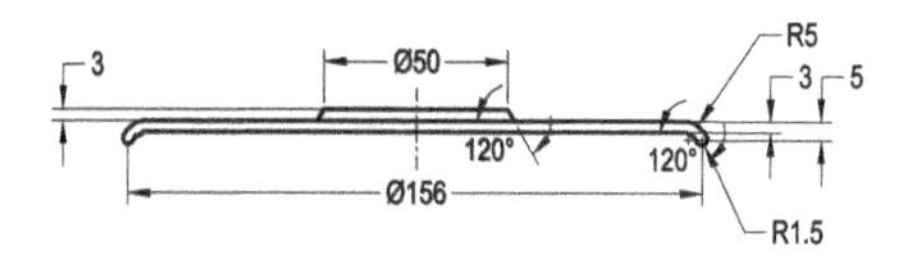

Figure 4-103 The front view of the model

Answers to Self-Evaluation Test

1. F, **2.** T, **3.** T, **4.** T, **5.** T, **6. Extrude as solid**, **7. Thicken Sketch**, **8. Reorient view**, **9.** handles, **10.** Geometric Centerline.

Chapter 5

Datums

Learning Objectives

After completing this chapter, you will be able to:

- *Work with three default datum planes.*
- *Understand selection methods in PTC Creo Parametric.*
- *Create datum planes using different constraints.*
- *Create datum planes on-the-fly.*
- *Create datum axes using different constraints.*
- *Create datum points.*
- *Create datum coordinate system.*

DATUMS

Datums are imaginary features with no mass or volume. Datums help you in creating models. They act as reference for sketching a feature, orienting a model, assembling components, and so on. Remember that datums play a very important role in creating complex models in PTC Creo Parametric; therefore, you must have a good understanding of datums. Datums are considered to be features but not model geometry. In PTC Creo Parametric, datums exist as datum plane, datum curve, datum point, datum coordinate system, datum graph, and so on.

Default Datum Planes

When you enter the **Part** mode or the **Assembly** mode, the three datum planes are displayed by default in the drawing area. These datum planes are known as the default datum planes and they are mutually perpendicular to each other. The only difference between the default datum planes of the **Part** mode and those of the **Assembly** mode lies in the names of the datum planes.

The default datum planes in the **Part** mode are named as **FRONT**, **TOP**, and **RIGHT**. In case of the **Assembly** mode, the default datum planes are named as **ASM_FRONT**, **ASM_TOP**, and **ASM_RIGHT**. However, the names of the default datum planes can be changed if required.

To change the name of a default plane, choose **File > Prepare > Model Properties** from the menu bar; the **Model Properties** window will be displayed. In this window, click on the **change** link in front of the **Names** line under the **Features and Geometry** head; the **Rename** dialog box will be displayed, as shown in Figure 5-1. Next, click on the datum plane you want to rename and enter the desired name to change the default name of the plane. Next, choose **OK** from the **Rename** dialog box and then close the **Model Properties** window.

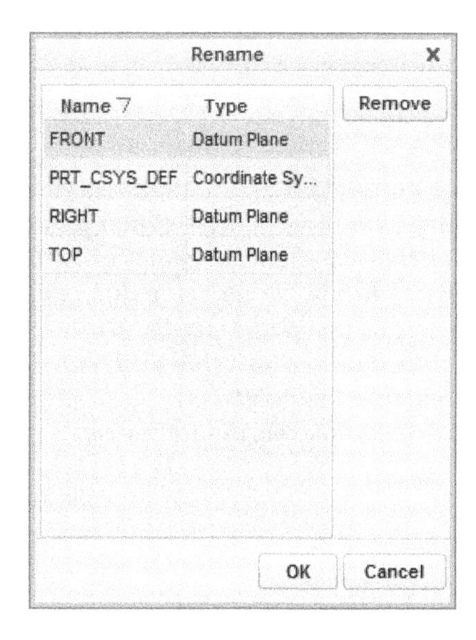

Figure 5-1 *The **Rename** dialog box*

Alternate method to rename a datum plane is to select it and right-click. On doing so, a shortcut menu will be displayed. Choose **Properties** or **Rename** from it and then enter the desired name for it at the appropriate place.

Note

*The **Model Properties** window can also be used to set the material, units, and other parameters related to the model.*

NEED FOR DATUMS IN MODELING

Generally, most of the engineering components or designs consist of more than one feature. First the base feature of the model is created and then other features of the model are added.

Since all features of a model cannot be drawn on a single plane, therefore, to draw rest of the features, sometimes, additional planes have to be created or selected. Also, most of the times, the three default datum planes are not enough to create a complex model having many features. For example, Figure 5-2 shows a simple model that consists of two features that require two different planes.

Tip: *Whenever you come across any solid model, first try to visualize the number of features in that model and then decide which feature is to be considered as the base feature.*

In Figure 5-2, any one of the two features defined on two different planes can be considered as the base feature. However, in this discussion, the feature that is selected as the base feature is shown in Figure 5-3. After creating the base feature, the next feature will be created. For the next feature, a sketching plane has to be defined. Therefore, an additional plane has been created on which you can draw the sketch for the second feature.

Figure 5-2 *Model having two extruded features* ***Figure 5-3*** *Base feature of the model*

As shown in Figure 5-4, the plane that is used to create the base feature is highlighted by a mesh. To create the second feature, a new plane is created, which is shown in Figure 5-5. The sketch of the second feature will be drawn on this plane.

Note

Throughout the book, at some instances, the datum planes are shown by a mesh plane as is evident from Figure 5-4. This view of the datum plane is only for explanation. In PTC Creo Parametric, the datum planes do not appear in the form of mesh.

To create different types of datums, you need to select the references on the model. Based on the selection you make on the model, PTC Creo Parametric applies constraints to create the datum.

SELECTION METHOD IN PTC Creo Parametric

The selection method in PTC Creo Parametric is divided into two types:

1. Selection
2. Collection

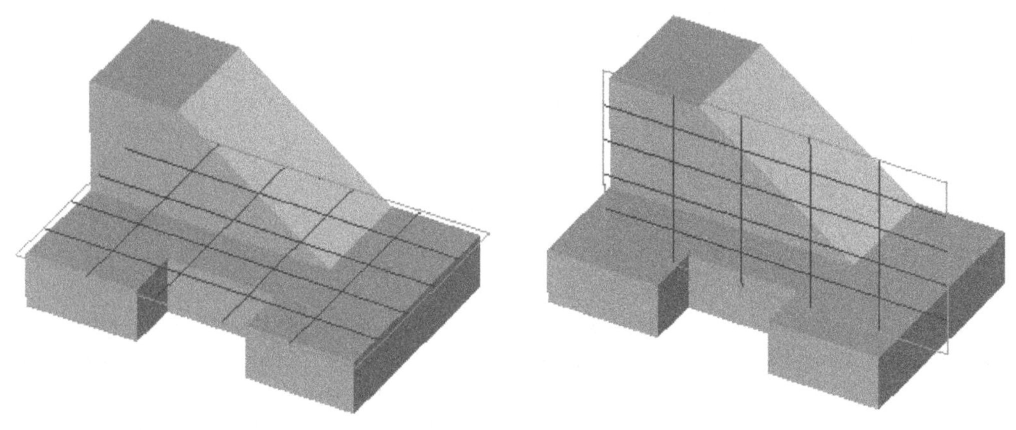

Figure 5-4 Plane selected for the base feature *Figure 5-5 Plane selected for the second feature*

Selection

In PTC Creo Parametric, selection refers to an action in which you first select entities like edge, plane, face, axis, coordinate system, and so on, and then invoke a feature creation tool. This property of a design package is also known as Object Action. To make your selection process easier, filters are available in PTC Creo Parametric. Filters are available in the **Filter** drop-down list located at the right corner of the Status Bar at the bottom. The options in the **Filter** drop-down list change depending upon the mode that is active. The options in this drop-down list, when the **Part** mode is active, are shown in Figure 5-6.

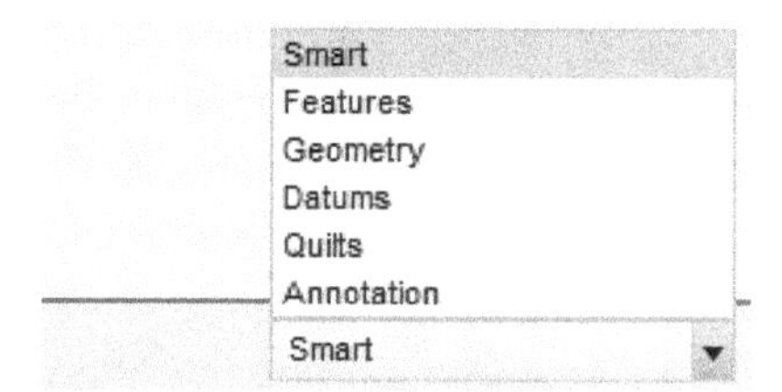

*Figure 5-6 The **Filter** drop-down list*

As you select entities on the model, the numbers of selected entities appear on the left of the **Filter** drop-down list in the Status Bar. When you double-click on the number, the **Selected Items** dialog box will be displayed, as shown in Figure 5-7. All selections that you make on the model are available in this dialog box. You can use the dialog box to remove the selected entities. These selections appear highlighted on the model. To remove a selection on the model, press the CTRL key and then select the highlighted entity.

*Figure 5-7 The **Selected Items** dialog box*

By default, the **Smart** filter is selected in the **Filter** drop-down list. The second filter in the **Filter** drop-down list is **Features**. This filter is used to select the features on the model. To redefine a feature, select this option from the drop-down list. The third filter is **Geometry**. This filter allows you to select only the edges, vertices, and surfaces. The fourth filter is **Datums**. This filter allows you to select the datum features. The fifth filter is **Quilts**. This filter allows you to select surfaces. The sixth filter is **Annotation**. This filter is used to select the notes from the drawing area.

Collection

In PTC Creo Parametric, collection refers to an action in which you select the entities to collect the references for creating a feature. To make your selection process easy, filters can be used. The filters available in the **Filter** drop-down list depend on the feature creation tool that you have invoked.

DATUM OPTIONS

After discussing the default datum planes, which are the first features in the **Part** mode, you must know various other features created using the datum options. Datums are also considered as features having no geometry. Figure 5-8 shows the **Datum** group and various datum types options from the group.

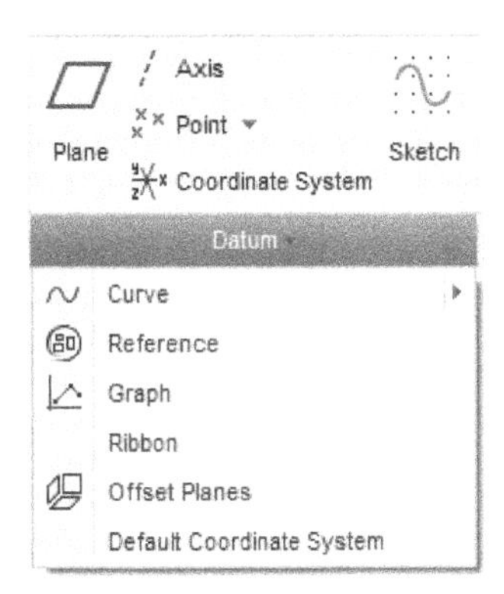

*Figure 5-8 The **Datum** group*

Datum Planes

Ribbon: Model > Datum > Plane

You can create datum planes, other than the three default datum planes by using the **Datum** group. You can create a datum plane even when any other tool is invoked. You can turn on or off the display of the datum planes by using the **Plane Display** button from the **Show** group of **View** tab. Before discussing the procedure to create datum planes, it is important to understand the use of datum planes. Some of the uses of datum planes in PTC Creo Parametric are listed next.

1. Datum planes are used as sketching planes to create sketches for the features of a model.

2. Datum planes are used as reference planes for sketching.

3. Datum planes are used as references for placing holes and for assembling components.

4. Datum planes are used as a reference for mirroring features, copying features, creating a cross-section, and for orientation of references.

   PTC Creo Parametric provides you with various constraints to create additional datum planes. Additional datum planes are created with the help of constraints and the filters available in the **Filter** drop-down list. Datums can also be created while you are in the sketcher environment. When you choose **Datum > Plane** from the **Ribbon**, the **Datum Plane** dialog box will be displayed, as shown in Figure 5-9. Separate buttons for creating datum axis, datum curve, datum point, and datum coordinate system are available in the **Datum** group.

Figure 5-9 shows various options in the **Datum Plane** dialog box to create datum planes. The options in different tabs of the **Datum Plane** dialog box are discussed next.

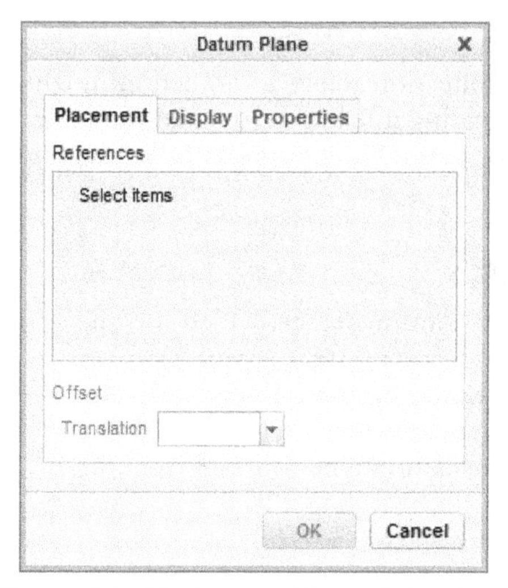

Figure 5-9 *The* ***Datum Plane*** *dialog box with the* ***Placement*** *tab chosen*

Tip: *Generally, the three default datum planes are used for creating the base feature. As the part becomes complex or in other words, as the number of features increases, the need for additional datum planes arises.*

Placement Tab

The **Placement** tab in the **Datum Plane** dialog box is chosen by default. In this tab, the **References** area displays the references selected to create the datum plane. All valid constraint types for these references are displayed in a drop-down list on the right of the selected reference. These constraints are applied automatically based on the references you select. To change a constraint, select the constraint in the dialog box; you will notice that a drop-down list appears in its place. From this drop-down list, select a different constraint. This drop-down list is available only for those constraints in the dialog box that can be substituted by another constraint. The constraints that are used to create a datum plane are discussed later in the chapter.

The **Offset** area will be available only when you use the **Offset** constraint to create a datum plane. The **Translation** edit box in this area is used to specify the offset distance of the new datum plane. This distance can also be set dynamically on the model by using the drag handle. The **Rotation** edit box will be available only when you create a datum plane at an angle with the selected edge or axis. This edit box is used to specify the rotation angle of the new datum plane. You can also set the rotation angle dynamically by using the drag handle displayed on the model.

Display Tab

The **Display** tab of the **Datum Plane** dialog box is shown in Figure 5-10. The **Flip** button in this dialog box is used to change the normal direction of the datum plane being created. The arrow direction and hence the direction of the plane can be changed by using the **Flip** button. By default, the **Adjust Outline** check box is clear. If you select this check box, it allows you to set the size of the datum plane.

Properties Tab

The **Datum Plane** dialog box with the **Properties** tab chosen is shown in Figure 5-11. This tab, when chosen allows you to name the datum plane you are creating. By default, the first datum plane you create is named DTM1 and then other datum planes created are successively numbered.

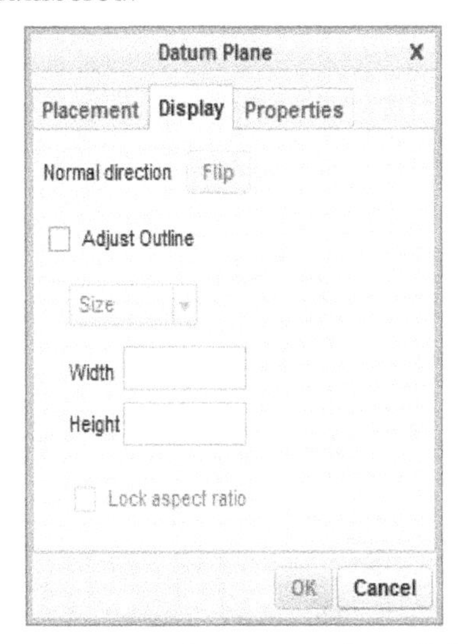

*Figure 5-10 The **Datum Plane** dialog box with the **Display** tab chosen*

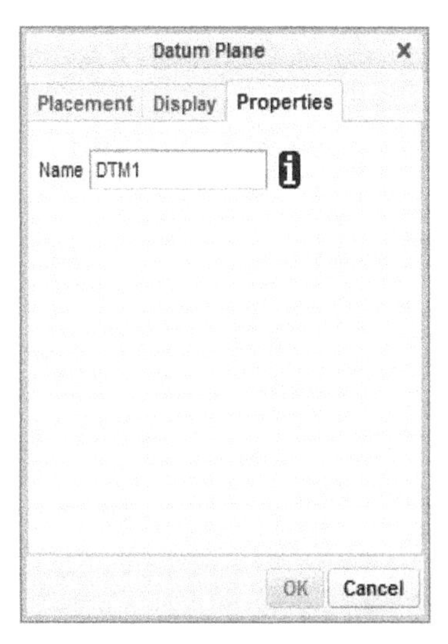

*Figure 5-11 The **Datum Plane** dialog box with the **Properties** tab chosen*

Creating Datum Planes

When you create a datum plane, the constraints are applied automatically based on the selection you make on the model. These constraints appear in the **Datum Plane** dialog box. Sometimes, while applying constraints to define a datum plane, a single constraint is enough to define the datum plane and sometimes, you may need more than one constraint to do the same. When only one constraint is used to create a datum plane, it is called stand-alone constraint. The stand-alone constraints are sufficient by themselves to constrain a datum plane definition. The constraints that are used to create a datum plane are discussed next.

Through Constraint

The **Through** constraint is used to create a datum plane through any specified axis, edge, curve, point/vertex, plane, cylinder, or coordinate system. This constraint can be used in combination with other constraints that are discussed next.

The **Through** constraint can also be used as a stand-alone constraint when you select a plane. Figure 5-12 shows the datum plane constraint combinations using the **Through** constraint. Datum planes can be created using any of the combinations shown in Figure 5-12. The possible combinations of datum plane creation are referred to as **Yes** and the combinations that are not possible are referred to as **No** in the figure.

DATUM PLANE CONSTRAINT COMBINATIONS (USING THROUGH CONSTRAINT)		Through			Normal	Parallel	Angle	Tangent	Standalone Constraints
		Axis/Edge/ Curve	Point/ Vertex	Cylinder	Plane	Plane	Plane	Cylinder	
Through	Axis/Edge/ Curve	Yes	Yes	Yes	Yes	Yes	Yes	Yes	No
	Point/ Vertex	Yes	Yes	Yes	Yes	Yes	No	Yes	No
	Plane	No	No	No	No	No	No	No	Yes
	Cylinder	Yes	Yes	Yes	Yes	Yes	Yes	Yes	No

***Figure 5-12** Datum plane constraint combinations using the **Through** constraint*

While reading the table shown in Figure 5-12, first preference is given to the text written in the first column and then the text in the first row should be read. For example, if you want to create a datum plane that is passing through a cylinder and normal to a plane then look for **Through** in the first column and then for **Cylinder** in the second column. Now, look for **Normal** in the first row and for **Plane** in the second row. After finding all combinations, trace them in the respective column and row till they intersect. If the result is **Yes**, the creation of a datum plane that passes through a cylinder and is normal to a plane is possible. While reading the table shown in Figure 5-12, remember that the constraints that are not stand-alone have to be applied in pairs.

The options in the **Filter** drop-down list shown in Figure 5-13 are used to make a selection on a model. This drop-down list is located at the right corner of the main window. The options in this drop-down list change depending on the operation being performed. For example, when you are creating a datum plane, the options shown in Figure 5-13 will be available. Once you exit the datum plane creation process, the options in this drop-down list change.

All
Datum Point
Vertex
Intent Datum Point
Intent Chain End
Axis
Intent Datum Axis
Edge
Intent Chain
Cable Segment
Curve
Intent Datum Curve
Coordinate System
Intent Coordinate System
Surface
Datum Plane
Intent Datum Plane
Channel
Facet Vertex
Facet Edge
Feature

***Figure 5-13** The **Filter** drop-down list with options*

Figures 5-14 and 5-15 show the cylindrical surface and the default datum plane used to create a datum plane at an angle to the selected default datum plane and passing through the center of the cylindrical surface. When you select the cylindrical surface, it is highlighted in green, indicating that it is selected. To make the second selection, select the **Datum Plane** filter from the **Filter** drop-down list. Now, press the CTRL button and then select the datum plane. This method of selecting by pressing the CTRL button is known as collection. Click on the constraint displayed against the selected plane in the **Datum Plane** dialog box and then select the **Offset** option from the drop-down list. Enter the angle value in the **Rotation** dimension box. Choose the **OK** button to create the datum plane.

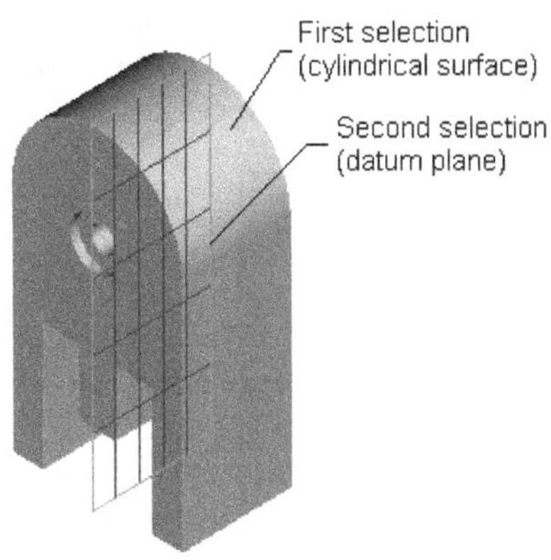

Figure 5-14 Selecting a cylindrical surface and a default datum plane to create a datum plane

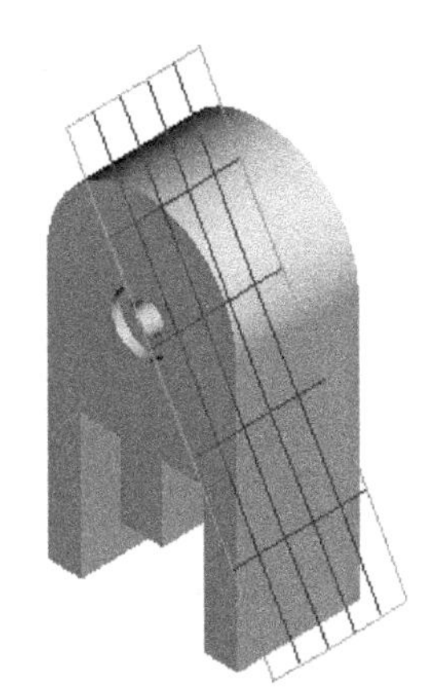

Figure 5-15 Resulting datum plane passing through the center of a cylinder

Note

1. The first reference to constrain a datum plane is selected by using the left mouse button. The second reference is selected by using the CTRL+left mouse button.

*2. Make sure you change the option in the **Filter** drop-down list when you invoke any other tool.*

Normal Constraint

The **Normal** constraint is used to create a datum plane normal to any specified axis, edge, curve, or plane. This constraint is used in combination with other constraints. The **Normal** constraint cannot be used as a stand-alone constraint. Figure 5-16 shows the datum plane constraint combinations that can be used with the **Normal** constraint.

DATUM PLANE CONSTRAINT COMBINATIONS (USING NORMAL CONSTRAINT)		Through			Normal	Parallel	Angle	Tangent	Standalone Constraints
		Axis/Edge/ Curve	Point/ Vertex	Cylinder	Plane	Plane	Plane	Cylinder	
Normal	Axis/Edge/ Curve	Yes	Yes	Yes	No	No	No	No	No
	Plane	Yes	Yes	Yes	Yes	Yes	No	Yes	No

*Figure 5-16 Datum plane constraint combinations using the **Normal** constraint*

The possible combinations of datum plane creation are referred to as **Yes** and the combinations that are not possible are referred to as **No**.

Figure 5-17 shows a planar surface and a cylindrical surface of a solid model. The planar face is selected as the normal surface and the cylindrical face is selected to be tangent to the datum plane. The following steps explain the procedure to create this datum plane:

1. Invoke the **Datum Plane** dialog box by choosing **Plane** tool from the **Datum** group in the **Model** tab. Select the first reference shown in Figure 5-17. On doing so, the **Offset** constraint is automatically applied and will be displayed in the **References** collector.

2. Click on the **Offset** constraint in the collector; a drop-down list appears. Select the **Normal** constraint from this drop-down list.

3. Now, use the CTRL+left mouse button to select the second reference shown in Figure 5-17. On doing so, the **Through** constraint will be displayed in the **References** collector.

4. Change this constraint to **Tangent** from the drop-down list that appears when you click on the constraint in the **References** collector. The datum plane that is created is shown in Figure 5-18.

5. Choose the **OK** button from the **Datum Plane** dialog box.

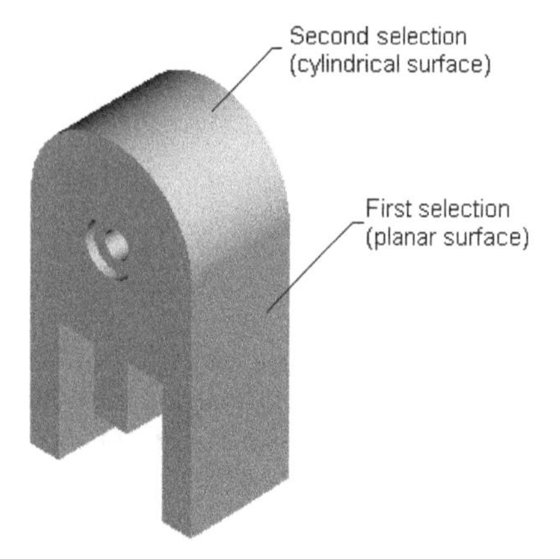

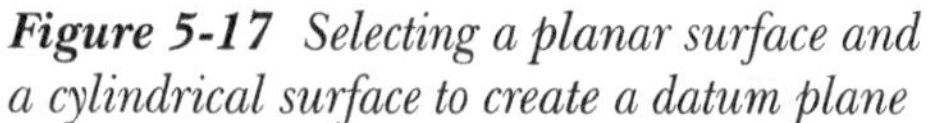

Figure 5-17 *Selecting a planar surface and a cylindrical surface to create a datum plane*

Figure 5-18 *Resulting datum plane*

Parallel Constraint

The **Parallel** constraint is used to create a datum plane parallel to any specified datum plane or planar face. This option is used in combination with other constraints. The **Parallel** constraint cannot be used as a stand-alone constraint. Figure 5-19 shows various datum plane constraint combinations using the **Parallel** constraint. The possible combinations of datum plane creation are referred to as **Yes** and the combinations that are not possible are referred to as **No** in Figure 5-19.

DATUM PLANE CONSTRAINT COMBINATIONS (USING PARALLEL CONSTRAINT)		Through			Normal	Parallel	Angle	Tangent	Standalone Constraints
		Axis/Edge/ Curve	Point/ Vertex	Cylinder	Plane	Plane	Plane	Cylinder	
Parallel	Plane	Yes	Yes	Yes	No	No	No	Yes	No

Figure 5-19 *Datum plane constraint combinations using the* ***Parallel*** *constraint*

Figure 5-20 shows the selection of a default datum plane and an axis to create a datum plane. The resulting datum plane is parallel to the selected datum plane and passes through the axis, as shown in Figure 5-21. The steps explaining the procedure to create this datum plane are given next.

1. Invoke the **Datum Plane** dialog box. Select the first reference refer to in Figure 5-20; the **Offset** constraint is automatically applied and will be displayed in the **References** collector.

2. Click on the **Offset** constraint in the collector; a drop-down list appears. Select the **Parallel** constraint from this drop-down list; the plane is oriented parallel to the selected reference.

3. Now, press CTRL+left mouse button and select the second reference refer to Figure 5-20; the **Through** constraint will be displayed in the **References** collector.

4. Choose the **OK** button from the **Datum Plane** dialog box. The resulting datum plane is shown in Figure 5-21.

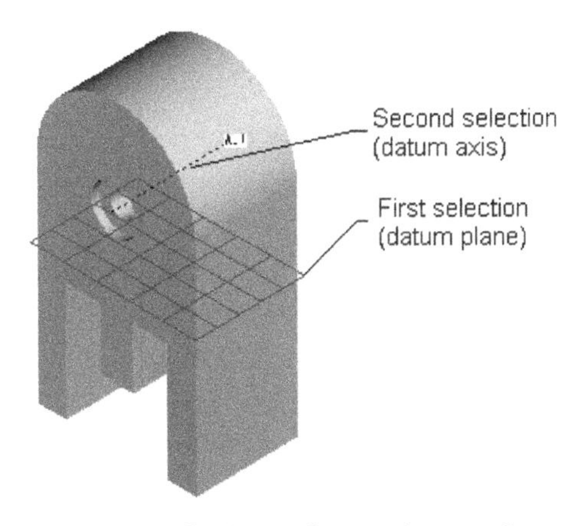

Figure 5-20 *Selecting a datum plane and an axis to create another datum plane*

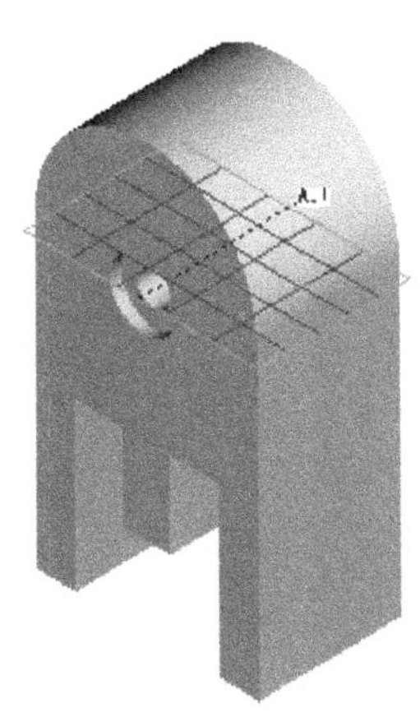

Figure 5-21 *Resulting datum plane*

Offset Constraint

The **Offset** constraint is used to create a datum plane at an offset distance to any specified plane or coordinate system. This option is used in combination with other constraints. However, the **Offset** constraint can be used as a stand-alone constraint when you select a plane to offset from.

When the **Offset** constraint is applied, the **Translation** dimension list box is activated in the **Datum Plane** dialog box. This list box is used to specify the offset distance. In the case of angular planes, the **Translation** dimension box changes to the **Rotation** dimension box in which you need to specify the angle. Also, an arrow appears on the model that shows the positive direction of the offset distance or angle.

Figure 5-22 shows various datum plane constraint combinations using the **Offset** constraint. The possible combinations of datum plane creation are referred to as **Yes** and the combinations that are not possible are referred to as **No** in the Figure 5-22.

DATUM PLANE CONSTRAINT COMBINATIONS (USING OFFSET CONSTRAINT)		Through			Normal	Parallel	Angle	Tangent	Standalone Constraints
		Axis/Edge/ Curve	Point/ Vertex	Cylinder	Plane	Plane	Plane	Cylinder	
Offset	Plane	Yes	No	Yes	No	No	No	No	Yes
	Coord System	No	No	No	No	No	No	No	Yes

***Figure 5-22** Datum plane constraint combinations using the **Offset** constraint*

Figure 5-23 shows the selection of a default datum plane and an edge to define the offset datum plane. The resulting datum plane will be at the specified offset to the selected datum plane and will pass through the edge, as shown in Figure 5-24.

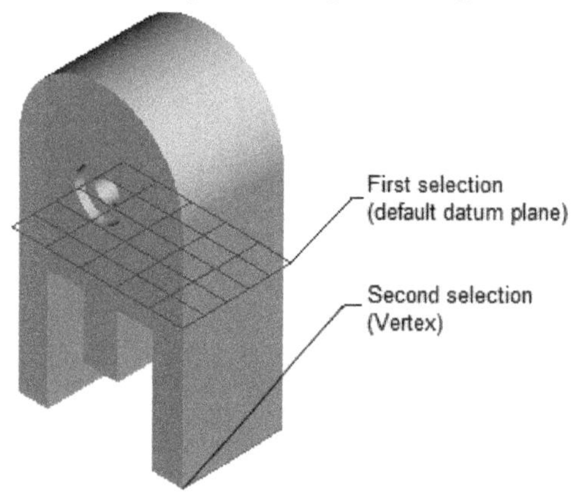

***Figure 5-23** Selecting a datum plane and an edge to create a datum plane*

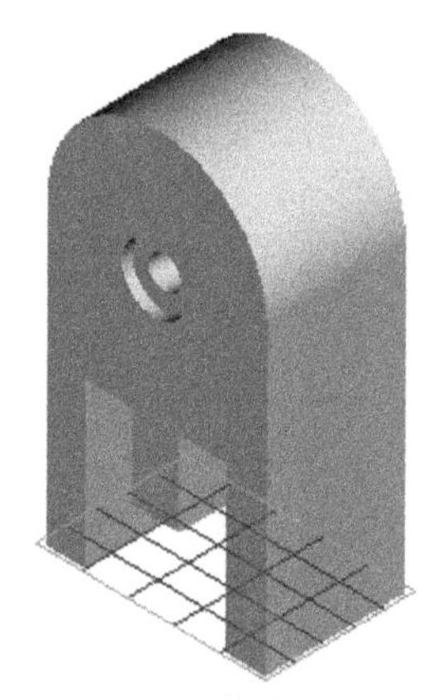

***Figure 5-24** Resulting datum plane*

The following steps explain the procedure to create this datum plane:

1. Invoke the **Datum Plane** dialog box. Select the first reference refer to Figure 5-23. The **Offset** constraint is automatically applied and will be displayed in the **References** collector. Also, the **Translation** dimension box appears.

2. Now, use the CTRL+left mouse button to select the second reference, which is a vertex shown in Figure 5-23; the **Through** constraint will be displayed in the **References** collector.

3. Choose the **OK** button from the **Datum Plane** dialog box; the datum plane that is created is shown Figure 5-24.

Note

*The **OK** button in the **Datum Plane** dialog box will be enabled only when the datum plane you are creating is fully constrained.*

Tangent Constraint

The **Tangent** constraint creates datum planes tangent to cylindrical features. This constraint is also used with other constraints to create various types of datum planes. Figure 5-25 shows the datum plane constraint combinations using the **Tangent** constraint. The possible combinations of datum plane creation are referred to as **Yes** and the combinations that are not possible are referred to as **No** in the figure.

DATUM PLANE CONSTRAINT COMBINATIONS (USING TANGENT CONSTRAINT)		Through			Normal	Parallel	Angle	Tangent	Standalone Constraints
		Axis/Edge/Curve	Point/Vertex	Cylinder	Plane	Plane	Plane	Cylinder	
Tangent	Cylinder	Yes	No	No	Yes	Yes	No	No	No

*Figure 5-25 Datum plane constraint combinations using the **Tangent** constraint*

Note

*If you want to remove the reference that was selected while creating the datum plane, select the reference in the **Datum Plane** dialog box and press the DELETE key. Alternatively, right-click on the reference in the dialog box and then choose **Remove** from the shortcut menu displayed.*

Datum Planes Created On-the-Fly

The term **On-the-Fly** refers to the creation of a datum plane when the system prompts you to select or create a plane. At this step, if you choose the **Plane** tool from the **Datum** group, the datum plane created will be called datum plane on-the-fly. When you create a datum plane on-the-fly, it is neither visible in the drawing area nor it will be displayed under the default display of the **Model Tree** once the feature is completed.

You can create a datum plane on-the-fly by following these steps:

1. Choose the **Extrude** tool; the **Extrude** dashboard will be displayed.

2. In the **Extrude** dashboard, choose the **Placement** tab and then choose the **Define** button to invoke the **Sketch** dialog box.

   Now, you need to select a sketching plane. You can select an existing datum plane, a face as the sketch plane, or you can create a datum plane on-the-fly.

3. Choose the **Plane** tool from the **Datum** group in the dashboard. Select the references and constraints to create the datum plane. This datum plane when created is automatically selected as the sketching plane.

When you need to select the references to orient the sketching plane, at this step also you can create a datum plane on-the-fly. Once the datum plane on-the-fly is created, it cannot be referenced by other features.

The datum plane created on-the-fly does not appear in the **Model Tree** under the default display. But it can be seen in the **Model Tree** when you click the arrow sign (▶) that appears on the left of the feature that you have created. The datum plane and the feature that you have created are grouped and appear as a grouped feature in the **Model Tree**. This means that the datum on-the-fly that was created during feature creation belongs to the feature being created.

Note
You can create a rotational pattern without creating a datum on-the-fly.

Datum Axes

Ribbon: Model > Datum > Axis

Datum axis is an imaginary axis that helps in creating a model. Datum axes can be created manually. They are also created automatically when any cylindrical feature is created. The display of the datum axis can be turned on or off by selecting the **Axis Display** check box from the **Datum Display Filters** drop-down in the **Graphics** toolbar. The uses of datum axes are given next.

1. Datum axes act as reference for feature creation.

2. They are used in creating a datum plane along with different constraint combinations.

3. They are used in placing features co-axially.

4. They are also used to create rotational patterns. You will learn to create patterns in Chapter 7.

In PTC Creo Parametric, datum axes are named by default. The default name of a datum axis is **A_(Number)**, where **Number** represents the number of datum axis. However, the default name of the datum axes can be changed in the same way as that of the datum planes.

When you choose the **Axis** tool from the **Datum** group, the **Datum Axis** dialog box is displayed with the **Placement** tab chosen in it, as shown in Figure 5-26. The options in this dialog box are discussed next.

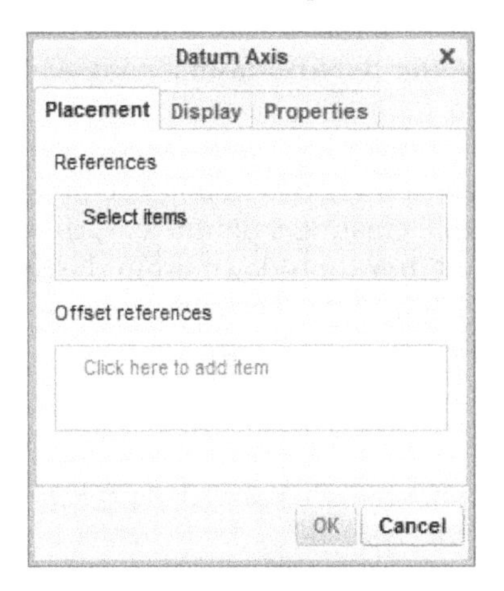

*Figure 5-26 The **Datum Axis** dialog box with the **Placement** tab chosen*

Placement Tab

When you invoke the **Datum Axis** dialog box, the **Placement** tab is chosen by default. Under this tab, the **References** collector allows you to select the references that will be used to create the datum axis. The constraints are displayed on the right of the references. These constraints are applied automatically based on the reference you select. The constraints that are used to create a datum axis are discussed later in this chapter.

The **Offset references** collector will be available only when the **Normal** constraint is used in the **References** collector to define a datum axis. To define the datum axis, you need to select two references. These references can be an edge, datum plane, face, or axis. The references can be specified dynamically on the model by dragging the handles displayed on the datum axis. The handles on the datum axis are displayed only when you select a reference. You can also select the references by using the **Datum Axis** dialog box.

There are three drag handles available on the model, as shown in Figure 5-27. The middle handle is used to move the position of the axis you are creating and it appears like a white square. The other two handles are used to specify references for dimensioning and appear like a green square.

Display Tab

The options in this tab are used to control the length of an axis. The **Datum Axis** dialog box with the **Display** tab chosen is shown in Figure 5-28. Select the **Adjust Outline** check box to activate the **Length** edit box. By default, the **Length** edit box is in the inactive mode. You need to select the **Adjust Outline** check box to activate it. You can enter a value in this edit box or drag the handles displayed at both ends of the datum axis to modify its length.

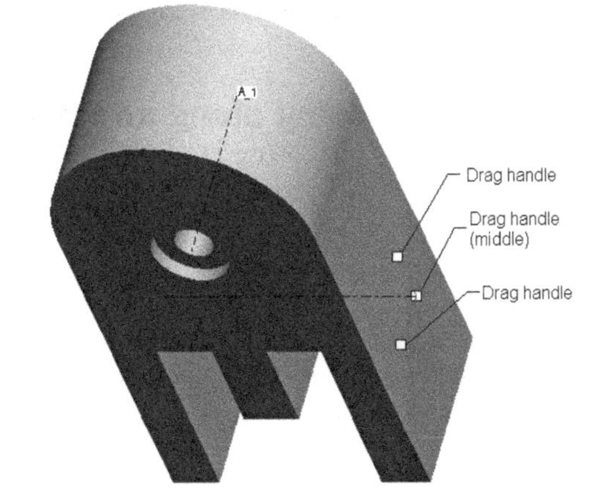

Figure 5-27 Solid model with the datum axis and three drag handles

Properties Tab

The **Properties** tab of the **Datum Axis** dialog box is shown in Figure 5-29. This tab, when chosen, allows you to name the datum axis you are creating. By default, the first datum axis you create is named A_1 and then additional datum axes created are named successively. As mentioned earlier, a datum axis is created using constraints that are applied automatically. While creating a datum axis, the options available in the **Filter** drop-down list in the **Status Bar** are shown in Figure 5-30. The constraints used to create a datum axis are discussed next.

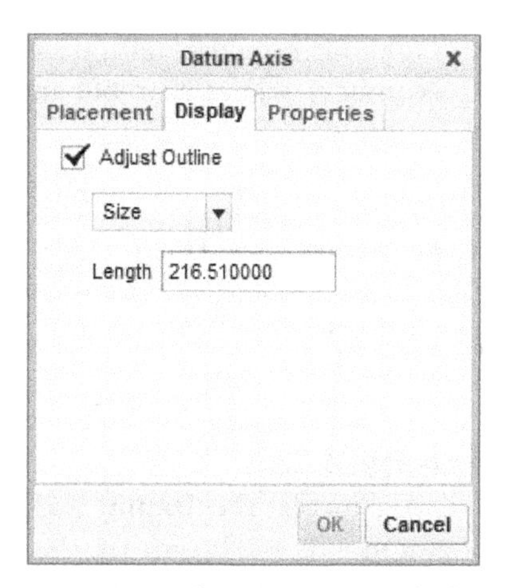

*Figure 5-28 The **Datum Axis** dialog box with the **Display** tab chosen*

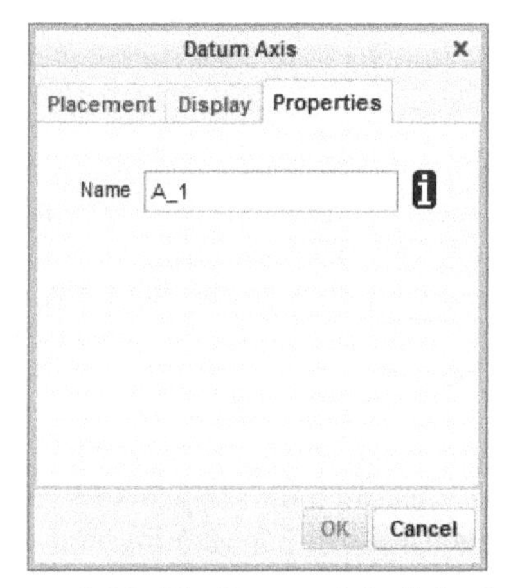

*Figure 5-29 The **Datum Axis** dialog box with the **Properties** tab chosen*

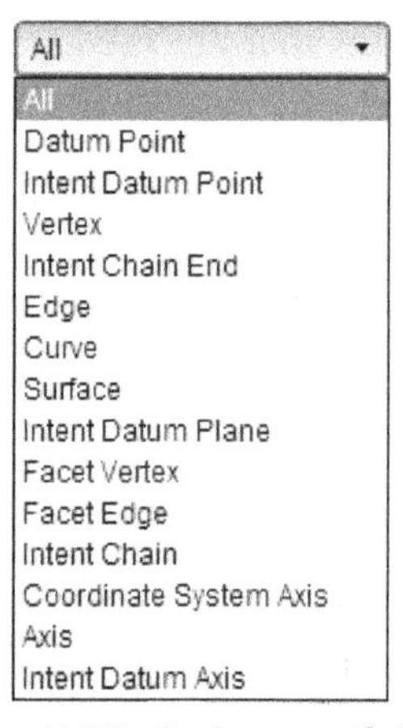

*Figure 5-30 Options available in the **Filter** drop-down list*

Datum Axis Passing along an Edge

The **Through** constraint is used to create a datum axis passing through any selected edge. This constraint will be displayed in the dialog box when you select an edge, as shown in Figure 5-31. In Figure 5-31, **A_1** is the datum axis created using this option.

Note

While creating a datum axis, you may need to create datum points passing through the axis. Therefore, in various cases that are discussed next, you may need to create datum points. Datum points are discussed in detail later in this chapter.

Datum Axis Normal to a Plane

The **Normal** constraint is used to create a datum axis normal to a selected face or datum plane. When you select a face or a datum plane, the **Normal** constraint will be displayed in the dialog box and you are prompted to select two references to place the axis. Notice that three drag handles appear on the model. Use the middle handle to change the location of the axis on the selected face. After you specify its placement location, you need to select two edges, axes, datums, or faces to specify the linear dimension for the placement of the datum axis. Click in the **Offset references** collector; the collector turns green. When you select the first edge for the placement dimensions of the axis, the offset value will be displayed in the **Offset references** collector. You can accept the default dimension or click on it to change its value. Similarly, select the second edge for dimensioning by pressing CTRL+left mouse button and enter the dimension value. Figure 5-32 shows the preview of the datum axis with drag handles. It should be noted that the references for dimensioning can also be selected dynamically using these drag handles.

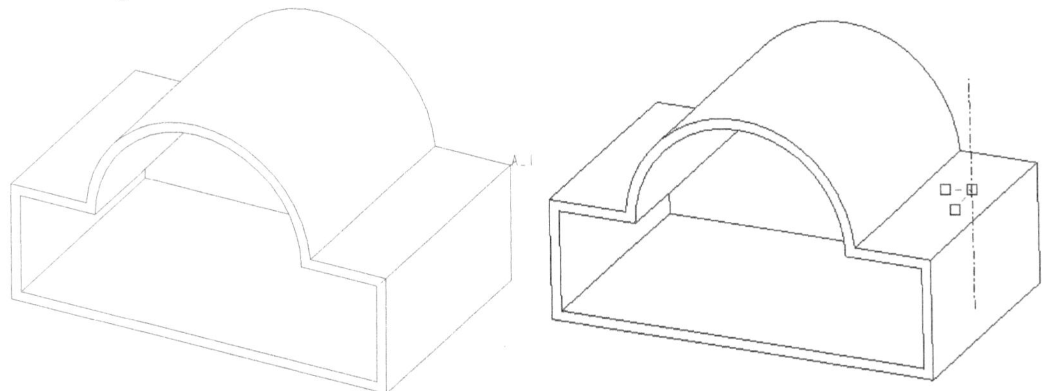

Figure 5-31 *Datum axis created along the edge*

Figure 5-32 *Datum axis created normal to the plane*

Note

When you select a reference for creating the datum axis, you are prompted to select the offset references. For selecting the offset references, you need to activate the ***Offset references*** *collector by clicking on it. On doing so, the area in the* ***Datum Axis*** *dialog box will be highlighted, which indicates that you can select references to place the axis. Until the* ***Offset references*** *collector is activated, you will not be able to select placement references.*

Datum Axis Passing through a Datum Point and Normal to a Plane

The **Normal** constraint creates a datum axis passing through a datum point or vertex and normal to any face or datum plane. When you select a face or a datum plane to which the datum axis will be normal, you will be prompted to select two references to place the axis. Use CTRL+left mouse button to select the datum point or vertex to create an axis passing through it. Choose the **OK** button to exit the **Datum Axis** dialog box. Figure 5-33 shows the preview of the datum axis and the datum point.

Datum Axis Passing through the Center of a Round Surface

The **Through** constraint is used to create a datum axis passing through the center of a cylindrical or round surface. To create this datum axis, select the round surface through which you need to pass the datum axis; an axis will be created automatically and will pass through the center. Figure 5-34 shows the preview of the datum axis. In this figure, the selected cylindrical surface is highlighted.

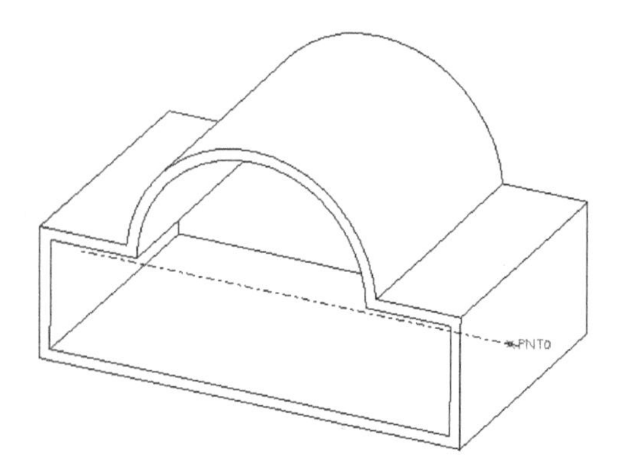

Figure 5-33 *Datum axis passing through the datum point and normal to the plane*

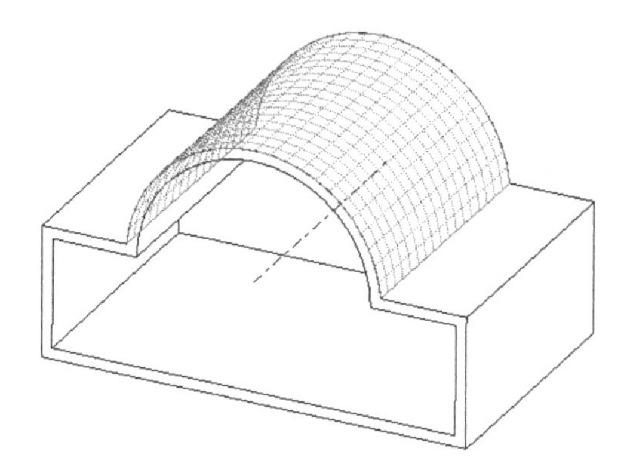

Figure 5-34 *Datum axis passing through the center of a cylinder*

Datum Axis Passing through the Edge Formed by Two Planes

The **Through** constraint is used to create a datum axis passing through the intersection edge of two planar faces or planes. When you select the face, the **Normal** constraint is applied automatically. Now, select the second face using the CTRL+left mouse button; the **Through** constraint is applied to both faces automatically and a datum axis is created passing through the edge formed by the two faces or planes. Figure 5-35 shows the preview of the datum axis. The two faces that were selected are also highlighted in this figure.

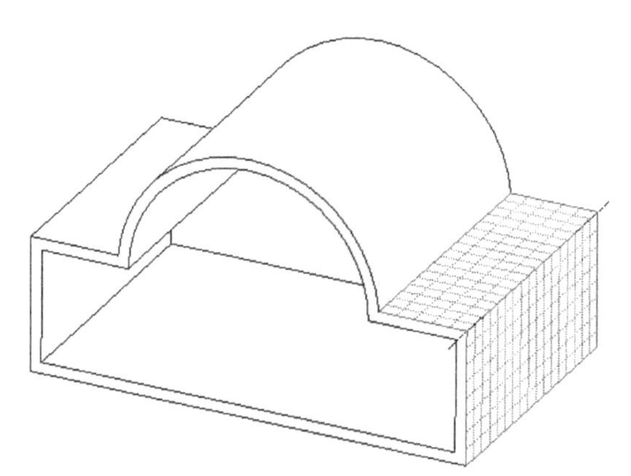

Figure 5-35 *Datum axis created at the edge where the two selected planes meet*

Datum Axis Passing through Two Datum Points or Vertices

The **Through** constraint is used to create a datum axis between two datum points or vertices. To create this datum axis, select the first vertex and then select the second vertex using CTRL+left mouse button; the datum axis will be created along the two selected datum points or vertices. Figure 5-36 shows the preview of the datum axis with the two vertices highlighted.

Datum Axis Tangent to a Curve and Passing through its Vertex

The **Tangent** and **Through** constraints create a datum axis tangent to a curve and passing through one of its vertex. Select the edge of the cylindrical surface as the first selection. Make sure you do not select the cylindrical surface. After you select a curve or an edge, use CTRL+left mouse button to select one vertex of the edge. The datum axis is created tangent to the curve and passing through its selected vertex. Figure 5-37 shows the preview of the datum axis created by using the **Tangent** and **Through** constraints. The curved edge is also highlighted in this figure.

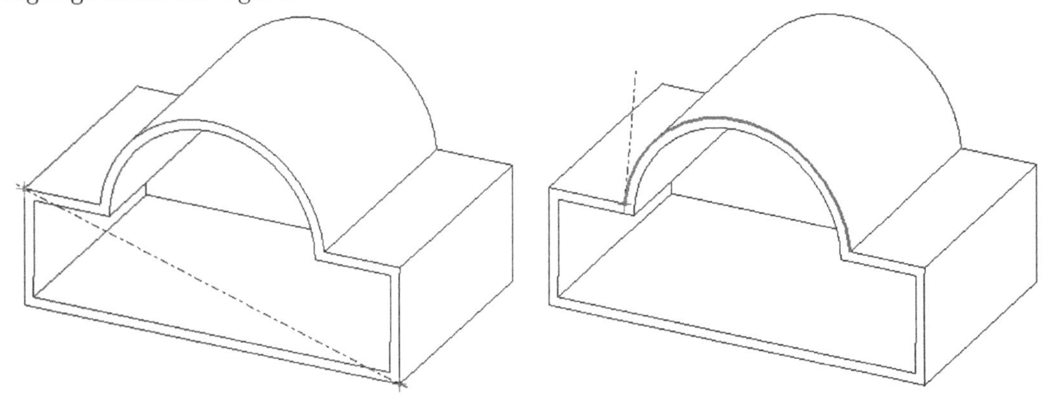

Figure 5-36 *Datum axis created between the two selected vertices*

Figure 5-37 *Datum axis created tangent to the selected curve*

Datum Points

Ribbon: Model > Datum > Point > Point

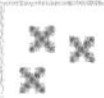

Datum points are imaginary points created to aid in creating models, drawings, analyzing models, and so on. The uses of datum points are as follows:

1. To create datum planes and axes.
2. To associate note in the drawings and attach datum targets.
3. To create coordinate system.
4. To specify point loads for mesh generation.
5. To create pipe features.

The default name associated with a datum point in PTC Creo Parametric is **PNT(Number)** where **Number** indicates the number of datum points created in a particular model. However, you can change the default name associated with the datum points.

When you choose the **Point** tool from **Point** drop-down in the **Datum** group of the **Ribbon**, the **Datum Point** dialog box will displayed with various options to create datum points, as shown in Figure 5-38. This dialog box can be used to create more than one datum point. The options in the **Datum Point** dialog box are discussed next.

Tip: *On selecting references in the model while creating the datum feature, some constraints appear in the **References** area automatically. You need to choose the required constraint from these constraints to constrain the datum feature. In this process, you will notice that the range of available constraints reduces, thereby enabling you to make a faster decision.*

Placement Tab

The **Placement** tab in the **Datum Point** dialog box has the options that change with the reference selected. However, the **References** collector shown in Figure 5-38 is always present.

This area is used to select the references which aid in placing the datum point. The constraints available in the drop-down list in the **References** collector depend on the references you select from the model. This means that the references narrow down the type of constraints you can apply on the datum point that you are creating.

Various methods of creating datum points are discussed next.

*Figure 5-38 The **Datum Point** dialog box with the **Placement** tab chosen*

Datum Point on a Face or a Datum Plane

The **On** constraint is used to create datum points on a face or a datum plane. When you select a face or a datum plane to place the datum point, a white colored point will be displayed at the selected location on the surface with the three drag handles, as shown in Figure 5-39.

Next, click in the **Offset references** collector to make it green. You are prompted to select two references to specify the linear dimensions for the placement of the datum point. Note that the second selection to select the reference should be made by using CTRL+left mouse button. After you select the two planes or edges for the placement of dimension of the point, a default value will be displayed in the **Offset references** collector. You can accept the default value or change it to the required value. After the datum point is located at the desired position on the face or datum plane, choose the **OK** button to exit the dialog box.

Figure 5-39 Datum point on a face and the three drag handles

Datum Point Offset to a Face or a Datum Plane

The **Offset** constraint creates datum points at an offset distance from a specified face or a datum plane in a specified direction. Select a face or a datum plane from where the offset distance for the placement of the datum point will be measured. The **On** constraint is displayed automatically in the **Datum Point** dialog box. From the drop-down list in the **References** collector, change the constraint to **Offset** and make the required settings, as shown in Figure 5-40. Figure 5-41 shows the datum point created at an offset from the sketched face.

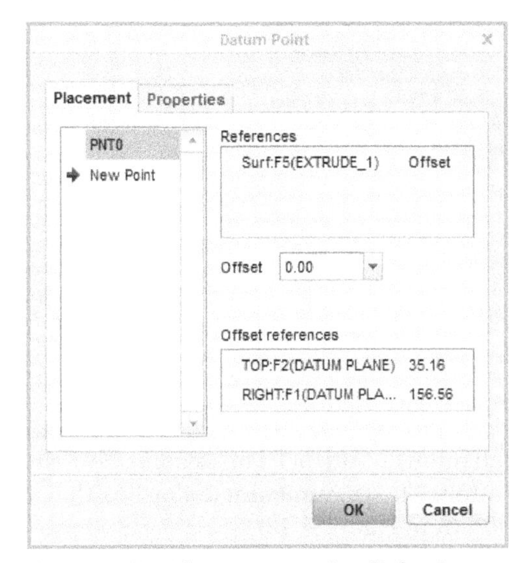

*Figure 5-40 The **Datum Point** dialog box with the **Offset** constraint selected*

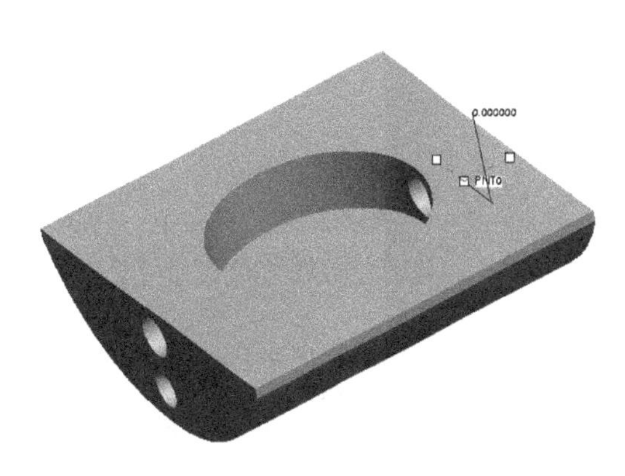

*Figure 5-41 Datum point on the top face with offset **0** and the three drag handles*

In the **Offset** edit box under the **References** collector, enter the offset value; you will be prompted to select the planes or the edges for dimensioning the point. Select two planes or edges for dimensioning and enter the distances from the highlighted references.

You can also specify the offset distance dynamically by using the middle drag handle. The two references for dimensioning the datum point can be specified by using the drag handles or by directly pointing them on the model.

Note

*When you use the **Offset** constraint, the middle drag handle is used to specify the offset distance and when you use the **On** constraint, it is used to specify the location of the datum point on the selected face.*

Datum Point at the Intersection of Three Surfaces

The **On** constraint is used to create a datum point at the intersection of three surfaces. To create this datum point, select the three surfaces; a datum point will be created, as shown in Figure 5-42.

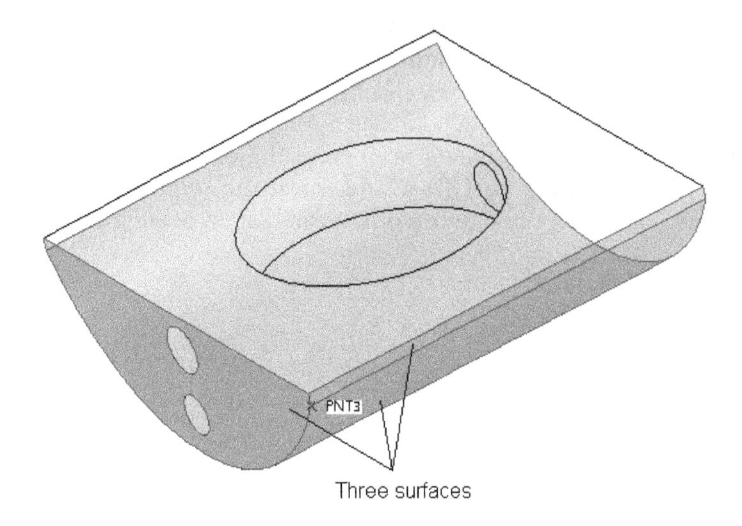

***Figure 5-42** The three highlighted surfaces and the datum point at the intersection of the three surfaces*

The datum point is created on the vertex that is common to the three surfaces. Remember that the first reference is selected using the left mouse button. The second and third references are selected using CTRL+left mouse button. The **Datum Point** dialog box that will be displayed after selecting the three surfaces is shown in Figure 5-43.

Sometimes, more than one intersection point exists with the current selection of references. In such a case, use the **Next Intersection** button in the **Datum Point** dialog box to select the other intersection points to create the datum point on them.

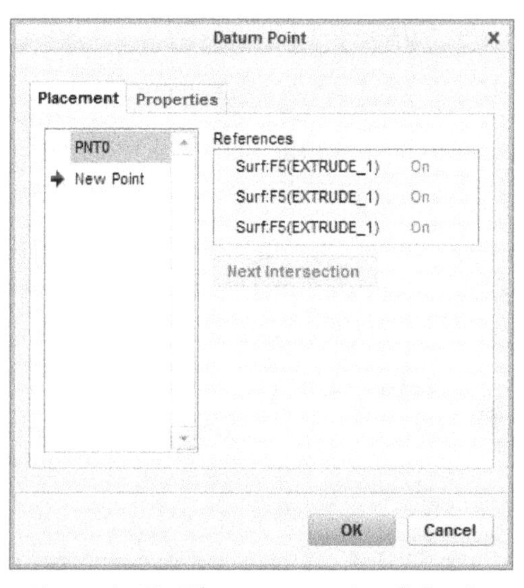

***Figure 5-43** The **Datum Point** dialog box*

Datum Point on a Vertex

The **On** constraint is used to create a datum point on the vertex of a face, an edge, or a datum curve. Invoke the **Datum Point** dialog box and then select a vertex to create the datum point. Choose the **OK** button to exit the dialog box.

Note

You can create more than one datum point by simply selecting the desired vertex.

Datum Point at the Center of a Curved Edge

The **Center** constraint is used to create a datum point at the center of an arc or a curved edge. Invoke the **Datum Point** dialog box and select the curve edge; the **On** constraint is applied by default. To modify the constraint, select the constraint in the **References** collector; a drop-down list appears to the right of the constraint. Select the **Center** constraint from it to create the datum point at the center of the curved edge.

Datum Point on an Edge or a Curve

The **On** constraint is used to create a datum point on an edge or a curve. Invoke the **Datum Point** dialog box, and select an edge or a curve from the drawing area. On doing so, the datum point will be placed on the selected edge, as shown in Figure 5-44, and the **Datum Point** dialog box will be displayed.

In the **Datum Point** dialog box, after the point is placed, the dimension type can be defined. The **Offset** dimension box displays the offset distance of the datum point from one end of the edge. The offset distance is measured as a ratio or a real value from the end of the edge, which can be selected from the drop-down list, refer to Figure 5-45.

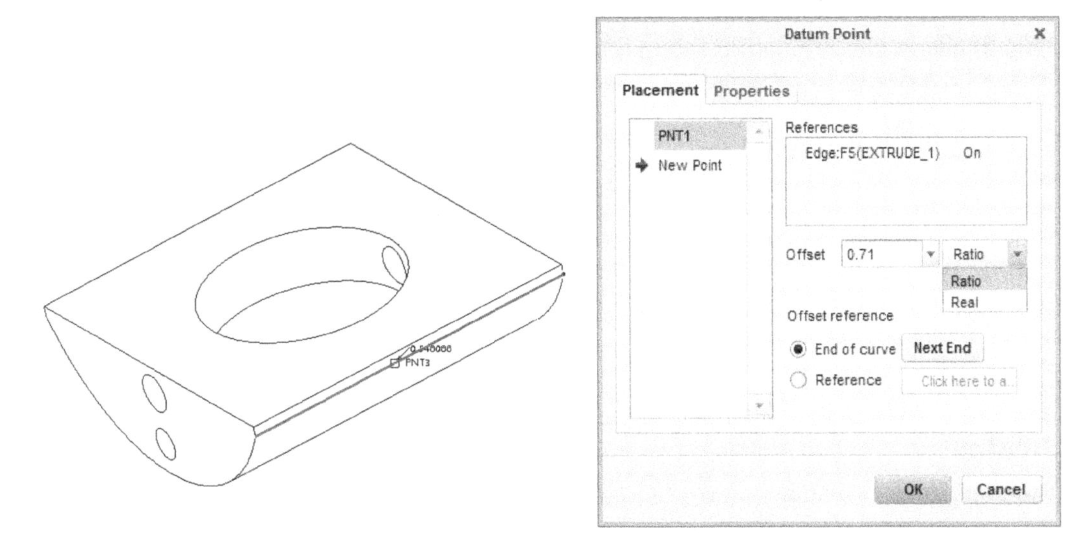

Figure 5-44 The highlighted edge and the datum point at some offset distance from one end of edge

*Figure 5-45 The **Datum Point** dialog box with **Ratio** option selected*

In the **Offset reference** area, the **Next End** button is used to flip the end of the edge from where the offset distance is measured. This button is available only when the **End of curve** radio button is selected. If you select the **Reference** radio button, you need to select a reference from which the datum point will be dimensioned.

Note

The geometry points created in the sketcher environment can also be used as datum points.

Creating an Array of Datum Points from a Coordinate System

Ribbon: Model > Datum > Point > Offset Coordinate System

The **Offset Coordinate System** tool available in the **Point** drop-down of the **Datum** group is used to create an array of datum points at an offset distance from a coordinate system. You can change the array of points by redefining the array. Note that the Datum Coordinate system must be defined before you create an array of datum points.

When you invoke this tool, the **Datum Point** dialog box will be displayed and you will be prompted to select a coordinate system. After selecting a coordinate system, you can enter the values of coordinates for the datum points, refer to Figure 5-46.

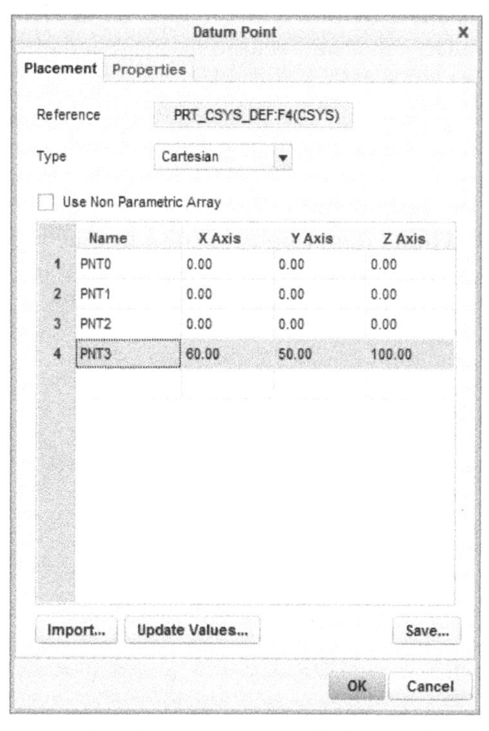

*Figure 5-46 The **Datum Point** dialog box displayed on choosing the **Offset Coordinate System** tool*

To add a point, click under the **Name** column in the list box; the first row gets activated. Click under the **X Axis** column; the edit box appears in which you can enter an offset value. Similarly, enter the offset values for **Y Axis** and **Z Axis.** To add another point click in the next row. From the **Type** drop-down list, select the type of coordinate system: **Cartesian**, **Cylindrical**, or **Spherical**. The options in the list box change according to the coordinate system selected from the **Type** drop-down list. In Figure 5-46, the coordinates of the three datum points are entered in the **Datum Point** dialog box.

Creating a Datum Point by Clicking

Ribbon: Model > Datum > Point > Field

When you choose the **Field** tool from the **Point** drop-down of the **Datum** group, the **Datum Point** dialog box will be displayed, as shown in Figure 5-47. Using this dialog box, you can create a datum point anywhere on the surface or edge of the model. You just need to specify the location of the datum point by using the left mouse button to place a datum point.

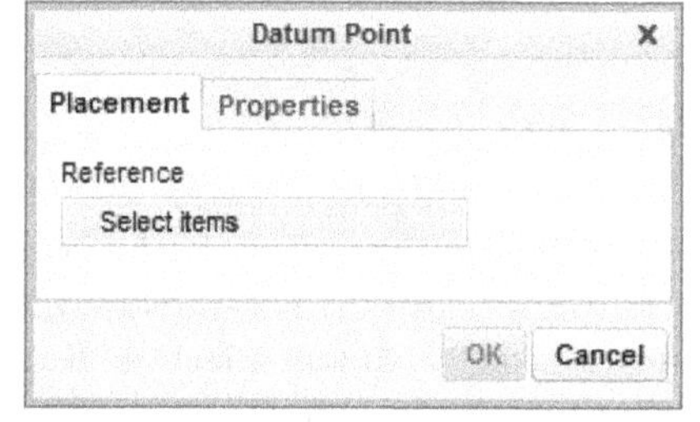

*Figure 5-47 The **Datum Point** dialog box displayed on choosing the **Field** tool*

Datum Coordinate System

Ribbon: Model > Datum > Coordinate System

Coordinate system is used to determine the position of a point or other geometric element. A coordinate system helps you create datum planes and points, set modeling and assembly reference, and define specific location in space.

The default name associated with a coordinate system in PTC Creo Parametric is **CS(Number)**, where **Number** indicates the number of coordinate systems created for a particular model. However, you can change the default name associated with the coordinate system.

For creating a coordinate system, choose the **Coordinate System** tool from the **Datum** group of the **Ribbon**, the **Coordinate System** dialog box will be displayed, as shown in Figure 5-48. The **Properties** tab similar to the one discussed in earlier sections of this chapter. Rest of the options and tabs in the **Coordinate System** dialog box are discussed next.

Origin Tab

The options in this tab change based on the reference selected. However, the **References** collector shown in Figure 5-48 will be always available in this tab. It is used to select the references which help in placing the coordinate system.

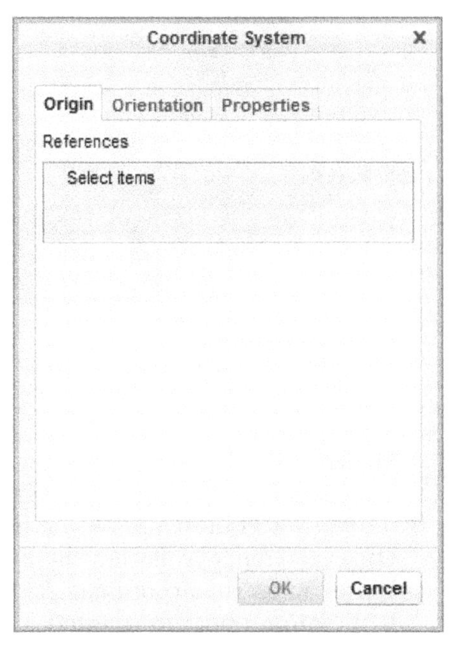

*Figure 5-48 The **Coordinate System** dialog box with the **Origin** tab chosen*

Orientation Tab

This tab is used to orient the coordinate system with respect to the selected reference. In this tab, the **References selection** and **Selected CSYS axes** radio buttons are used to orient the coordinate system.

Various methods of creating a coordinate system are discussed next.

Creating a Coordinate System on a Face or a Datum Plane

To create a coordinate system on a face, invoke the **Coordinate System** dialog box. Next, select the face on which you want to create the coordinate system; a orange colored point will be displayed at the selected location with the three drag handles, refer to Figure 5-49. Also, the options in the **Coordinate System** dialog box will be modified. By default, the **On** constraint is selected in the **References** collector of the **Origin** tab. As a result, the coordinate system will be created on a face or a datum plane.

Next, click on the **Offset references** collector to highlight it; you will be prompted to select two references to specify the linear dimensions for the placement of the coordinate system. Select the first reference and then select the second reference by using CTRL+left mouse button.

After you select the two planes or edges for the placement of dimension of the coordinate system, a default value will be displayed in the **Offset references** collector. You can accept the default value or change it to the required value.

When you select a flat face or a datum plane, a **Type** drop-down list will be displayed. The options in this drop-down list are Linear, Radial, and Diameter, refer to Figure 5-50. The options in this drop-down list are discussed next.

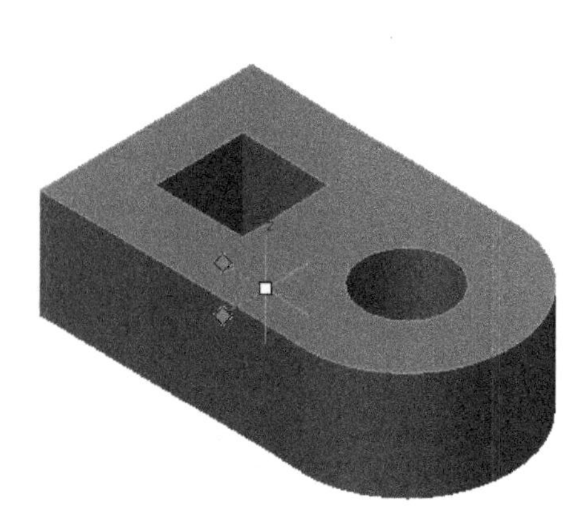

Figure 5-49 The Coordinate System on a face with three drag handles

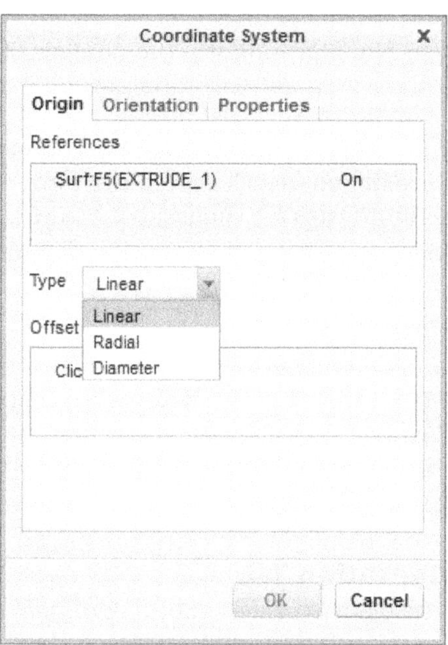

*Figure 5-50 The **Coordinate System** dialog box with the **Type** drop-down list*

Linear

This option is used for linear reference selection. When you select any surface, plane, or edge, the **Linear** option will be selected by default in the **Type** drop-down list. The **Offset references** collector will display two options: **Offset** and **Align**. The **Offset** option is used to create a coordinate system at an offset distance from the selected reference. If the **Align** option is selected, the coordinate system created will be aligned with the selected reference.

Radial

This option is used for radial reference selection. In the radial referencing, one option must be surface or plane for angle reference and other must be edge or axis for radius reference.

Diameter

This option is used for diametrical reference selection. In the diameter referencing, one option must be surface or plane for angle reference and other must be edge or axis for diameter reference.

Note

*When you select any circular surface, only the **Linear** and **Radial** options will be displayed in the **Type** drop-down list.*

After specifying settings in the **Origin** tab, choose the **Orientation** tab. As you have selected a face as a reference, the **References selection** radio button will be automatically selected, as shown in Figure 5-51. Also, the **Add rotation about the first axis** check box will be displayed in this tab. This check box is used to rotate the coordinate system with respect to first offset reference.

*Figure 5-51 The **Orientation** tab with **References selection** radio button selected*

After settings all the parameters in the **Coordinate System** dialog box, choose the **OK** button; the coordinate system will be created on the selected face.

Creating a Coordinate System Using an Existing Coordinate System

To create a coordinate system using an existing coordinate system, invoke the **Coordinate System** dialog box. Next, select the existing coordinate system of the model; a white colored point will be displayed, as shown in Figure 5-52. Also, the options in the **Coordinate System** dialog box will be modified, as shown in Figure 5-53. By default, the **Offset** constraint is selected in the **References** collector of the **Origin** tab. As a result, the coordinate system will be created at an offset distance.

There are four options in the **Offset type** drop-down list to create a coordinate system, refer to Figure 5-53. The options in this drop-down list are discussed next.

Cartesian

When you select this option, the **X**, **Y**, and **Z** edit boxes will be displayed. You can specify the values in these edit boxes to change the location of the coordinate system.

Cylindrical

When you select this option, the **R**, θ, and **Z** edit boxes will be displayed. You can specify the values in these edit boxes to change the location of the coordinate system.

Spherical

When you select this option, the **r**, Φ, and θ edit boxes will be displayed. You can specify the values in these edit boxes to change the location of the coordinate system.

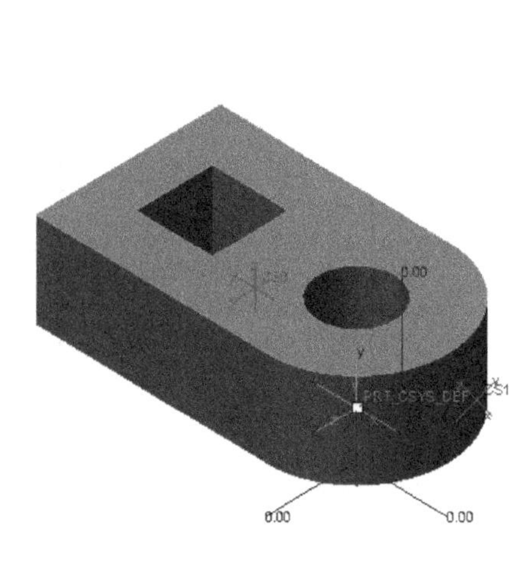

*Figure 5-52 The coordinate system created on an existing coordinate system with offset **0** for all axes*

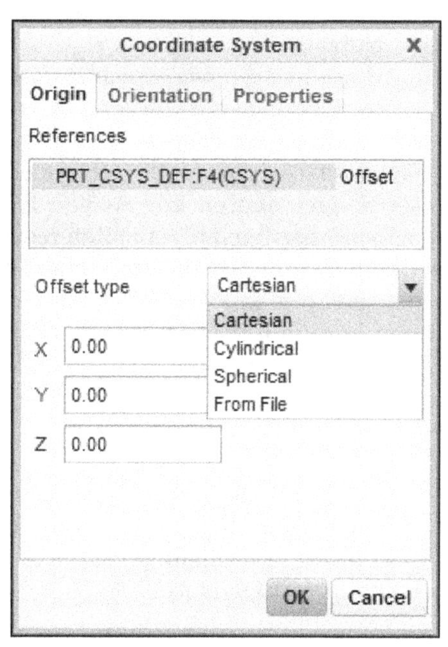

*Figure 5-53 The **Coordinate System** dialog box with the **Offset type** drop down list*

From File
When you select this option, a window will be displayed. In this window, you can browse to the ***.trf*** file (transformation file). This file is used to place the coordinate system.

After specifying settings in the **Origin** tab, choose the **Orientation** tab. As you have selected an existing coordinate system, the **Selected CSYS axes** radio button will be selected by default, refer to Figure 5-54. The orientation can be changed by specifying the values in the **X**, **Y**, and **Z** edit boxes.

After settings all the parameters in the **Coordinate System** dialog box, choose the **OK** button; the coordinate system will be created.

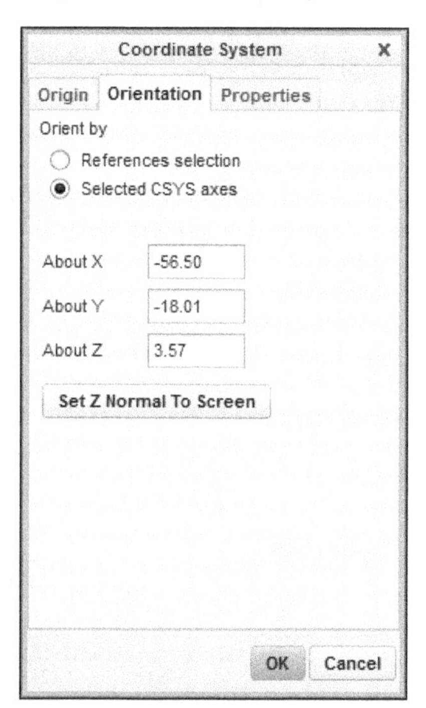

*Figure 5-54 The **Orientation** tab with **Selected CSYS axes** radio button selected*

TUTORIALS

Tutorial 1

In this tutorial, you will create the model shown in Figure 5-55. The front and right views of the solid model are shown in Figure 5-56. **(Expected time: 30 min)**

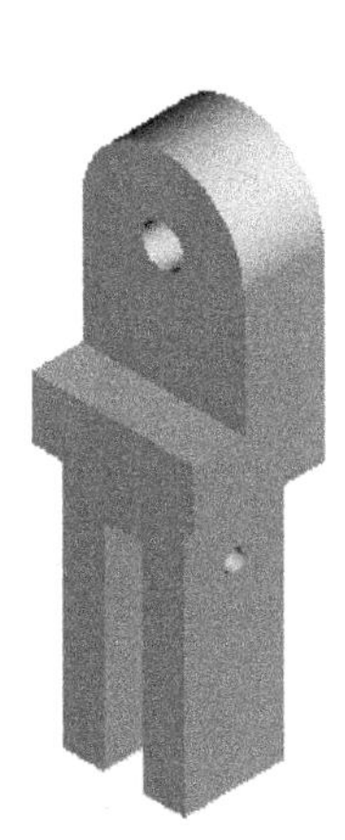

Figure 5-55 Model for Tutorial 1

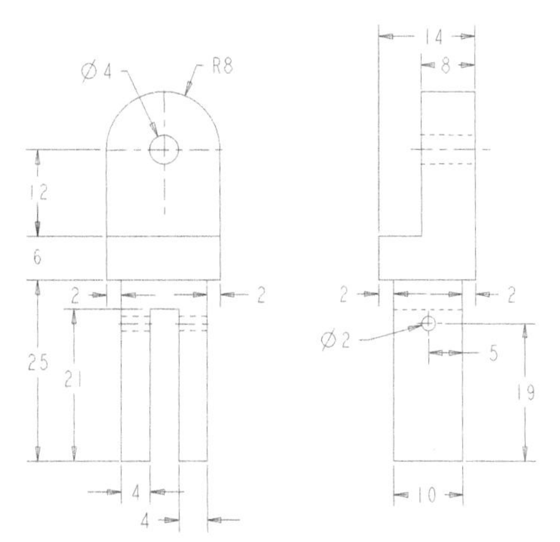

Figure 5-56 Front and right views of the model

The following steps are required to complete this tutorial:

Examine the model and determine the number of features in it, refer to Figure 5-55.

a. Create the base feature, refer to Figures 5-57 and 5-58.
b. Create the second extrude feature, refer to Figures 5-59 through 5-61.
c. Create the third feature on an offset plane, refer to Figures 5-62 through 5-67.
d. Create the circular cut feature, refer to Figures 5-68 through 5-71.

Setting the Working Directory

After starting the PTC Creo Parametric session, the first task is to set the working directory. A working directory is a directory on your system where you can save the work done in the current session of PTC Creo Parametric. You can set any existing directory on your system as the working directory.

1. Choose **Manage Session > Select Working Directory** from the **File** menu; the **Select Working Directory** dialog box is displayed. Select the *C:\Creo-3.0* folder from this dialog box.

2. Choose the **Organize** tab from the **Select Working Directory** dialog box to display the flyout. Next, choose the **New Folder** option from the flyout; the **New Folder** dialog box is displayed.

3. Enter **c05** in the **New Directory** edit box and choose **OK** from the dialog box; a folder with the name *c05* is created at *C:\Creo-3.0*.

4. Next, choose **OK** from the **Select Working Directory** dialog box. The working directory is set to *C:\Creo-3.0\c05*. Also, a message **Successfully changed to C:\Creo-3.0\c05 directory** is displayed in the message area.

Starting a New Object File

1. Choose the **New** button to invoke the **New** dialog box. Next, enter *c05tut1* in the **Name** edit box.

   The three default datum planes are displayed in the drawing area. If required, close the **Model Tree** by choosing the **Show Navigator** button on the bottom left corner of the program window so that the drawing area is increased.

Selecting the Sketching Plane for the Base Feature

To create the sketch for the base feature, you first need to select the sketching plane. In this model, you need to draw the base feature on the **FRONT** datum plane because it is evident from the isometric view of this model that the direction of extrusion for this feature is perpendicular to the **FRONT** datum plane.

Note

You can select any plane as the sketching plane for creating the base feature. The base feature thus created may not have the proper orientation. In this case, the final model will be oriented in the wrong direction. So, you need to be careful while defining the sketching plane for creating the base feature. The desired orientation of the model is shown in Figure 5-55.

1. Choose the **Extrude** tool from the **Shapes** group; the **Extrude** dashboard is displayed above the graphics window.

2. Choose the **Placement** tab from the dashboard; a slide-down panel is displayed. Next, choose the **Define** button from the slide-down panel; the **Sketch** dialog box is displayed.

3. Select the **FRONT** datum plane as the sketching plane; an arrow is displayed on the **FRONT** datum plane, pointing in the direction of view.

4. Select the **TOP** datum plane from the drawing area and then select the **Top** option from the **Orientation** drop-down list.

   The **TOP** datum plane is selected in order to orient the sketching plane.

5. Choose the **Sketch** button from the **Sketch** dialog box to enter into the sketcher environment.

Note

If the sketching plane does not orient parallel to the screen then you need to choose the ***Sketch View*** *button from the* ***Graphics*** *toolbar. You can make the sketching plane parallel to the screen by default. To do so, choose* ***File > Options*** *from the menu bar. Then, choose* ***Sketcher*** *from the list shown in the left area of the* ***PTC Creo Parametric Options*** *dialog box. Now, select the*

Make the sketching plane parallel to the screen check box from the Sketcher startup area and then choose the OK button from the window; the PTC Creo Parametric Options message window will be displayed. Choose the Yes button from this window; the Save As dialog box will be displayed. Choose the OK button to save and exit.

Creating and Dimensioning the Sketch for the Base Feature

The base feature can be created by drawing the sketch and then extruding it to the given distance.

1. Draw the sketch using various sketcher tools and then add required constraints and dimensions to the sketch, as shown in Figure 5-57. When you initially draw the sketch, it is dimensioned automatically and some weak dimensions are assigned to it.

Tip: *It is recommended that you use the **Modify** button to modify weak dimensions. In the **Modify Dimensions** dialog box that appears on choosing this button, clear the **Regenerate** check box and then modify dimensions by using the thumbwheel or the edit boxes. This way the sketch will not be regenerated while editing.*

2. Modify the dimension values, as shown in Figure 5-57.

3. After the sketch is completed, choose the **OK** button. Now, you are out of the sketcher environment.

4. Choose the **Named Views** button from the **Graphics** toolbar; a flyout is displayed. Choose the **Default Orientation** option from the flyout; the model orients in its default orientation, which is the trimetric view, refer to Figure 5-58. The pink colored arrow is displayed on the model, indicating the direction of extrusion.

5. In the dimension box of the **Extrude** dashboard, enter **8**; the model looks like the one shown in Figure 5-58.

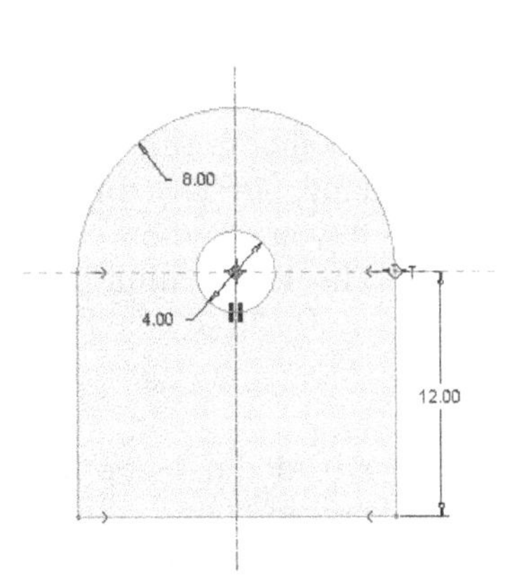

***Figure 5-57** Sketch for the base feature with dimensions and constraints*

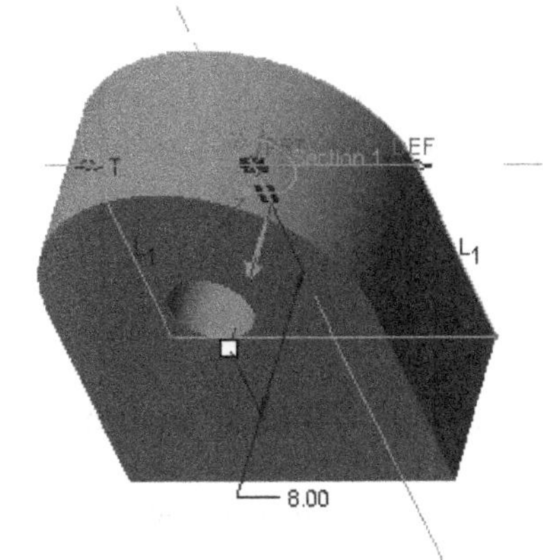

***Figure 5-58** Default orientation of the model and the arrow showing the direction of feature creation*

6. Choose the **Build feature** button from the **Extrude** dashboard; the base feature is created. You can use the middle mouse button to spin the model in order to view it from various directions.

Note

*When you choose the **Default Orientation** option from the **Named Views** drop-down, the resulting orientation of the model is trimetric, not isometric. If you want the model to be displayed in the isometric view whenever you choose the **Default** option, use the **PTC Creo Parametric Options** dialog box. To display this dialog box, choose **Files > Options** from the menu bar. In this dialog box, choose the **Model Display** from the left area and select the **Isometric** option from the **Default model orientation** drop-down list. Next, choose the **OK** button and save these settings. Now, the default orientation will be set to isometric.*

Tip: *It is recommended to check the orientation of the base feature of a model when it is completed. To check whether the plane you specified for sketching was correct, choose the **Named Views** button from the **View** tab; a flyout is displayed. Next, choose the **FRONT** option from the flyout; the base feature will reorient in the drawing area such that you can view the front view of the base feature.*

Selecting the Sketching Plane for the Second Feature

The second feature is an extrude feature. It will be created on the plane that was used to create the base feature.

1. Choose the **Extrude** tool again from the **Shapes** group; the **Extrude** dashboard is displayed above the drawing area.

2. Choose the **Placement** tab from the dashboard. Next, choose the **Define** button from the slide-down panel; the **Sketch** dialog box is displayed.

3. Choose the **Use Previous** button from the **Sketch Plane** area in the **Sketch** dialog box; the **Top** datum plane with default orientation is selected automatically.

   When you choose the **Use Previous** button, the system selects the sketching plane that was used previously to create the base feature. You need to choose this button because the base feature and the second feature are on the same plane, but have different depths of extrusion. In case they had the same depth of extrusion, you could have drawn them on the same plane at the same time as a single feature. The **TOP** datum plane and its orientation are set automatically.

Drawing the Sketch for the Second Feature

The second feature has a rectangular section that will be extruded to a depth of 14 units. To improve the clarity of edges of the base feature, choose the **No hidden** button from the **Display Style** drop-down of the **Model Display** group in the **View** tab before you start sketching the second feature.

1. Draw the sketch, as shown in Figure 5-59.

2. The sketch is automatically constrained and some weak dimensions are assigned to it. Add required constraints and modify weak dimensions, refer to Figure 5-59.

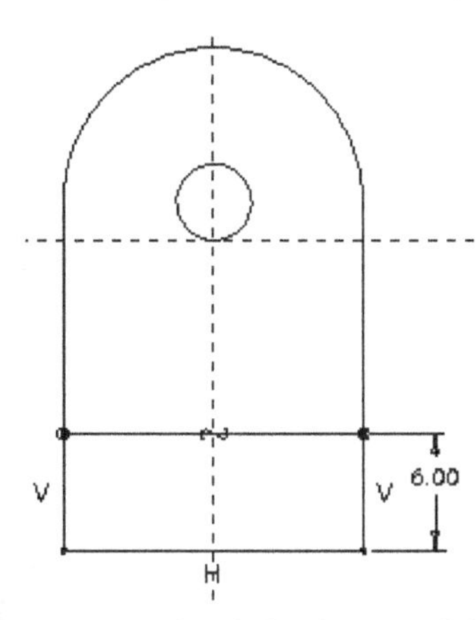

Figure 5-59 *Sketch for the second feature*

3. Choose the **OK** button to exit the sketcher environment; the **Extrude** dashboard is enabled above the drawing area. Choose the **Shading** button from the **Display Style** drop-down in the **Graphics** toolbar to view the shaded model.

Tip: *You can use the **Project** button from the **Sketching** group to use the bottom edge of the base feature for sketching. Else, you need to draw an aligned line on the edge.*

4. Use the middle mouse button to orient the model, as shown in Figure 5-60. This orientation of the model gives you a better view of the sketch in three-dimensional (3D) space. The colored arrow is also displayed on the model, indicating the direction of extrusion.

5. Enter **14** in the dimension box present on the **Extrude** dashboard and press ENTER. The second extruded feature is completed and its preview is displayed in the drawing area.

6. Choose the **Named Views** button from the **Graphics** toolbar; a flyout is displayed. Choose the **Default Orientation** option from the flyout; the model orients in its default orientation, which is the, trimetric view.

7. Now, choose the **Build feature** button from the **Extrude** dashboard; the feature is created and orients on the screen, as shown in Figure 5-61. You can also use the middle mouse button to spin the model.

Creating the Datum Plane for the Third Feature

A new datum plane is required to create the next feature. The datum plane will be created at an offset distance of 2 units from the front face of the second feature. You need to turn on the display of datum planes, if it is off.

1. Choose the **Plane** button from the **Datum** group; the **Datum Plane** dialog box is displayed.

Make sure that before opening the dialog box, no selection is made. This can be confirmed from the Status Bar that is available below the drawing area. If a selection exists, you can remove it by left-clicking in the empty space of the drawing area.

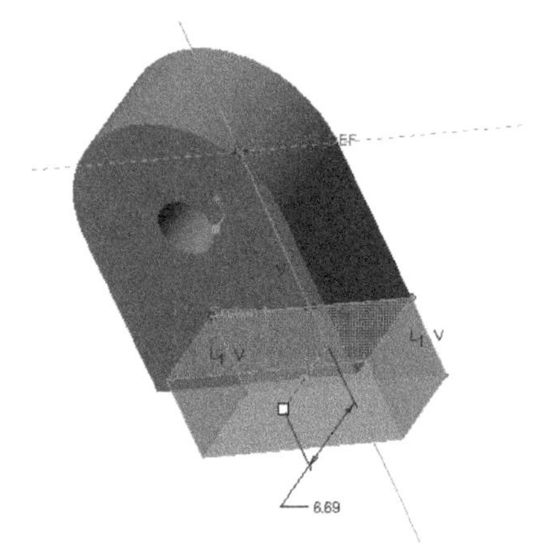

Figure 5-60 *Arrow showing the direction of feature depth*

Figure 5-61 *Second extruded feature with the base feature*

2. Select the front face of the second feature, as shown in Figure 5-62. The boundary of the selected front face is highlighted in green.

   In the **Datum Plane** dialog box, the **Offset** constraint is available under the **References** collector. This means that by the smart selection property of PTC Creo Parametric, the **Offset** constraint is applied automatically to the datum plane that you are going to create.

   Now, you need to specify the offset distance. If you enter a positive value, the datum plane will be created along the direction of arrow and if you enter a negative value then the datum plane will be created in the direction opposite to that shown by the arrow.

3. In the **Translation** edit box under the **Offset** area of the **Datum Plane** dialog box, enter **-2** and press ENTER.

   The negative value is entered because the datum plane has to be created in the direction opposite to that shown by the arrow.

4. Choose the **OK** button; the datum plane named **DTM1** is created, as shown in Figure 5-63.

Creating the Third Feature on DTM1

The datum plane **DTM1** is created and can be seen in the **Model Tree** as well as in the drawing area. The sketch of the next feature that will be extruded has to be created on the datum plane **DTM1**.

1. Choose the **Extrude** tool from the **Shapes** group of the **Model** tab; the **Extrude** dashboard is displayed above the drawing area.

2. Choose the **Placement** tab; a slide-down panel is displayed. Now, choose the **Define** button from the panel; the **Sketch** dialog box is displayed.

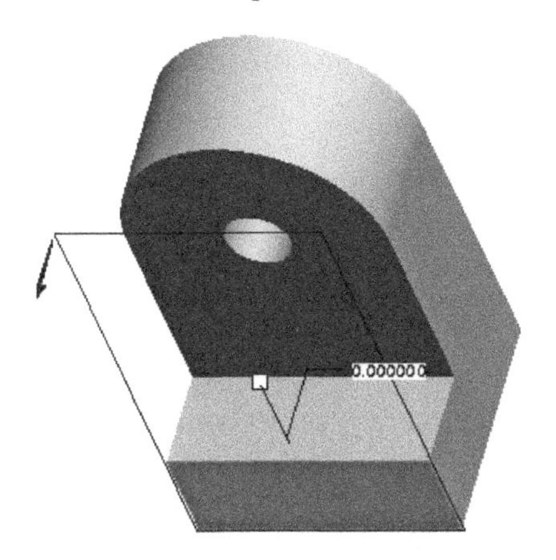

Figure 5-62 The front face of the second feature selected

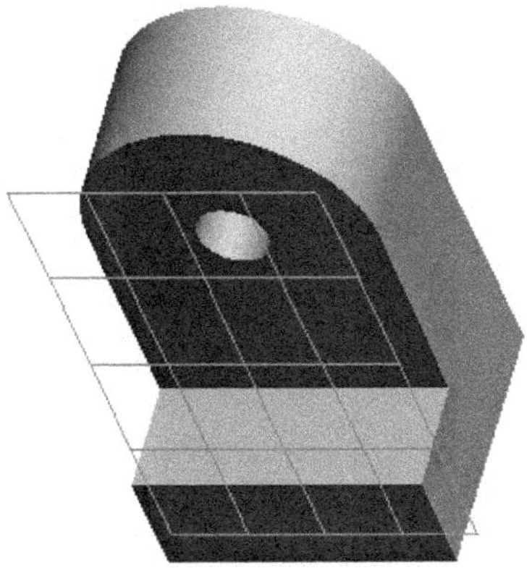

Figure 5-63 The datum plane created

3. Select **DTM1** as the sketching plane for the third feature; a pink arrow is displayed on the selected datum plane, as shown in Figure 5-64. This arrow shows the direction of viewing the sketching plane.

   The **TOP** datum plane and its orientation are selected by default.

4. Choose the **Sketch** button from the **Sketch** dialog box to enter the sketcher environment. Choose the **No hidden** button from the **Display Style** drop-down in the **Graphics** toolbar to view the visible edges of the model.

5. Sketch the section of the third feature of the model and add constraints and dimensions to the sketch, as shown in Figure 5-65.

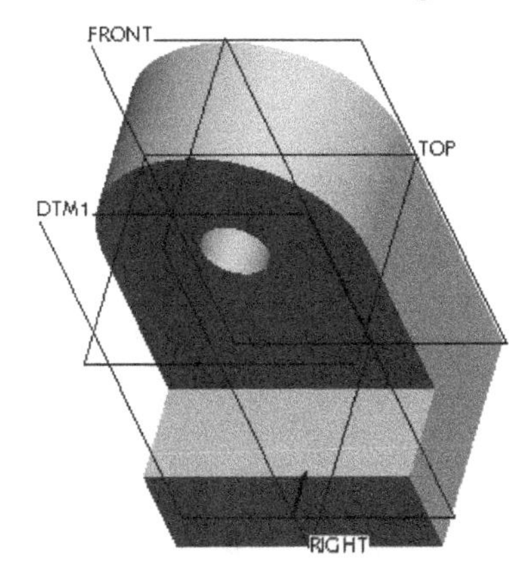

*Figure 5-64 Arrow on **DTM1** showing the direction of viewing the sketching plane*

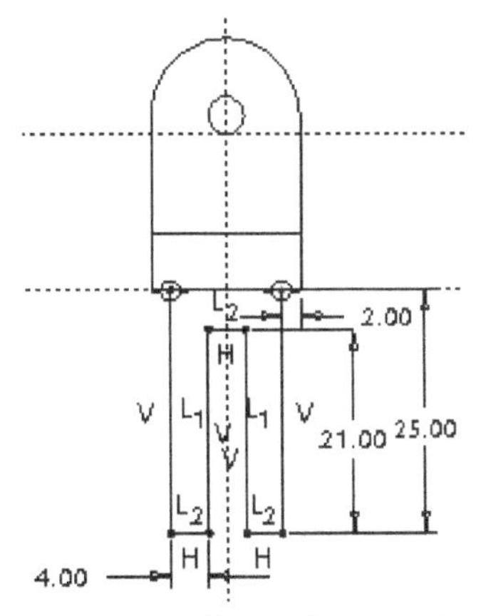

Figure 5-65 Sketch of the third feature with dimensions and constraints

6. Choose the **OK** button to exit the sketcher environment; the **Extrude** dashboard is enabled above the drawing area.

7. Turn the model display to **Shading**. Use the middle mouse button to orient the model, as shown in Figure 5-66. This orientation gives a better view of the model.

   Notice that the arrow is pointing toward the direction of feature creation. But, you need to extrude the sketch in the opposite direction.

8. Choose the **Change depth direction of extrude to other side of sketch** button from the **Extrude** dashboard to change the direction of arrow. 

9. Enter **10** in the dimension box present on the **Extrude** dashboard; the preview of the third feature is displayed in the drawing area.

10. Choose the **Build feature** button from the **Extrude** dashboard to confirm the feature creation.

    Choose the **Named Views** button from the **Graphics** toolbar; a flyout is displayed. Choose the **Default Orientation** option from the flyout; the model orients to its default orientation, the trimetric view, as shown in Figure 5-67.

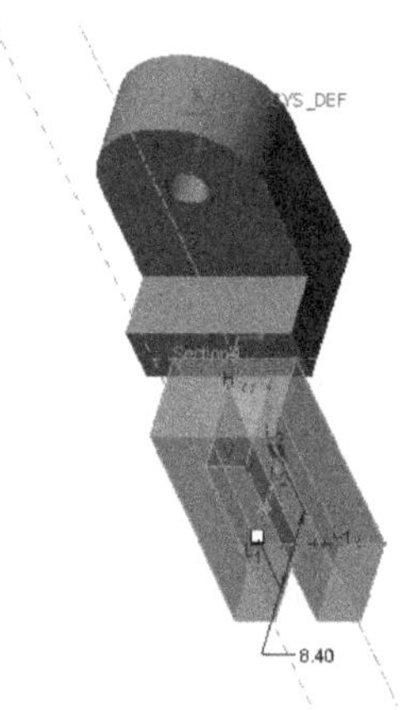

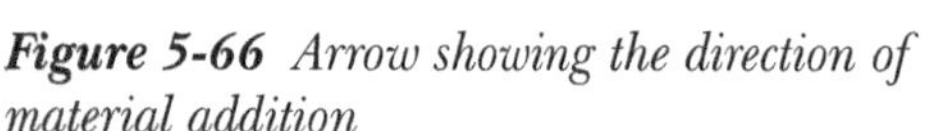

Figure 5-66 *Arrow showing the direction of material addition*

Figure 5-67 *Model after creating the third feature*

Selecting the Sketching Plane for the Cut Feature

You need to create a circular section for circular cut feature. The sketching plane for the cut feature is shown in Figure 5-68.

Note

*The circular cut feature can also be created using the **Hole** tool, which will be discussed in Chapter 6.*

1. Choose the **Extrude** tool from the **Shapes** group; the **Extrude** dashboard is displayed above the drawing area.

2. Choose the **Remove Material** button from the **Extrude** dashboard.

3. Choose the **Placement** tab from the dashboard; a slide-down panel is displayed. Next, choose the **Define** button from it; the **Sketch** dialog box is displayed.

4. Select the face shown in Figure 5-68 for sketching.

5. Using the left mouse button, select the **TOP** datum plane and then select the **Top** option from the **Orientation** drop-down list.

6. Choose the **Sketch** button; the system takes you to the sketcher environment.

Creating the Sketch for the Cut Feature

1. Turn the model display to **No hidden**. Draw the sketch of the cut feature and add dimensions to it, as shown in Figure 5-69.

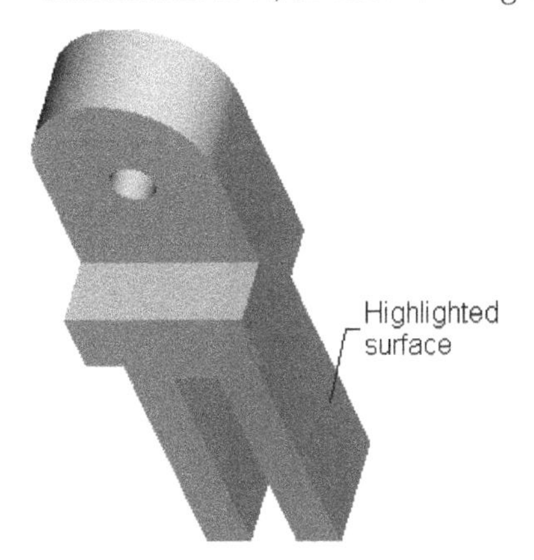

Figure 5-68 *Sketching plane for the cut feature*

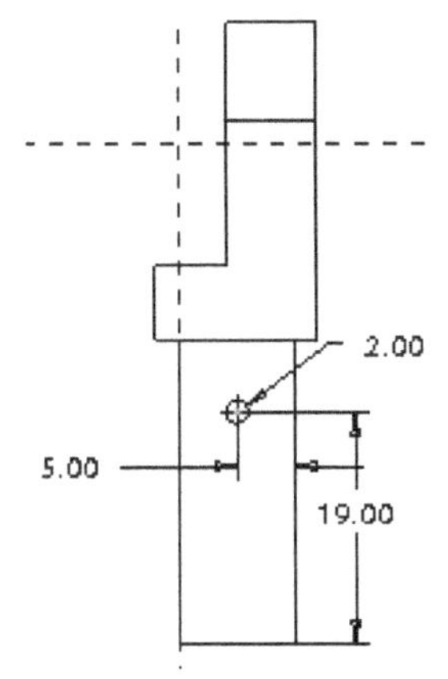

Figure 5-69 *Sketch and dimensions for the cut feature*

2. Choose the **OK** button and turn the model display to **Shading**.

3. Choose the **Named Views** button from the **Graphics** toolbar; a flyout is displayed. Choose the **Default Orientation** option from the flyout; the model orients to its default orientation, the trimetric view, as shown in Figure 5-70. Two arrows also appear on the model. One arrow indicates the direction of feature creation and the other arrow indicates the direction along which the material will be removed. Figure 5-70 shows the direction where the arrows are pointing.

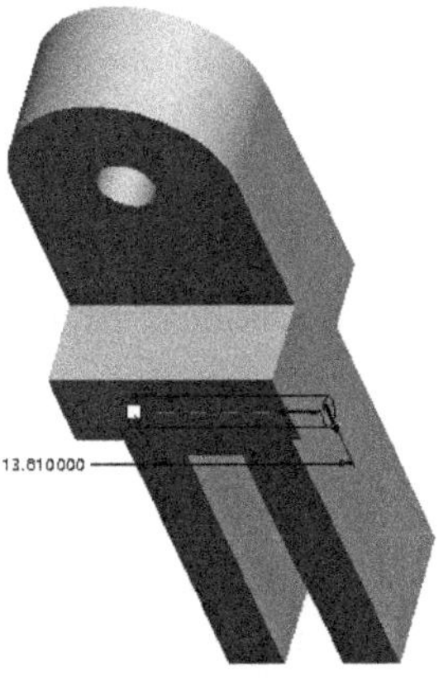

Figure 5-70 *The two arrows on the cut feature*

4. Choose the **Options** tab in the **Extrude** dashboard; the **Depth** slide-down panel is displayed.

5. From the **Side 1** drop-down list, choose the **Through All** option; cut feature is created and can now be previewed in the drawing area.

6. Choose the **Build feature** button from the **Extrude** dashboard to accept the feature creation. The trimetric view of the model is shown in Figure 5-71.

Saving the Model

1. Choose the **Save** button from the **File** menu and save the model.

The order of feature creation can be seen in the **Model Tree** shown in Figure 5-72.

Figure 5-71 *Completed model for Tutorial 1*

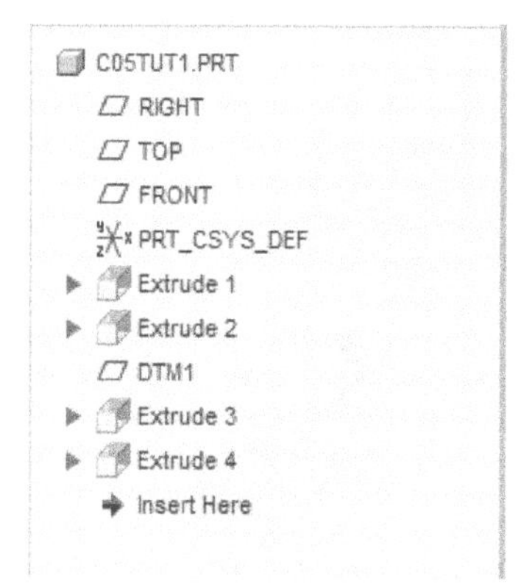

Figure 5-72 *The **Model Tree** for Tutorial 1*

Tutorial 2

In this tutorial, you will create the model shown in Figure 5-73. This figure also shows the front, top, and right views of the solid model. **(Expected time: 30 min)**

The following steps are required to complete this tutorial:

Examine the model and determine the number of features in it, refer to Figure 5-73.

a. Start a new file and create the base feature, refer to Figures 5-74 and 5-75.
b. Create the second feature on the left face of the base feature, refer to Figures 5-76 through 5-80.
c. Create the third and fourth features on the same plane, but with different extrusion depths, refer to Figures 5-81 through 5-88.
d. Save the model.

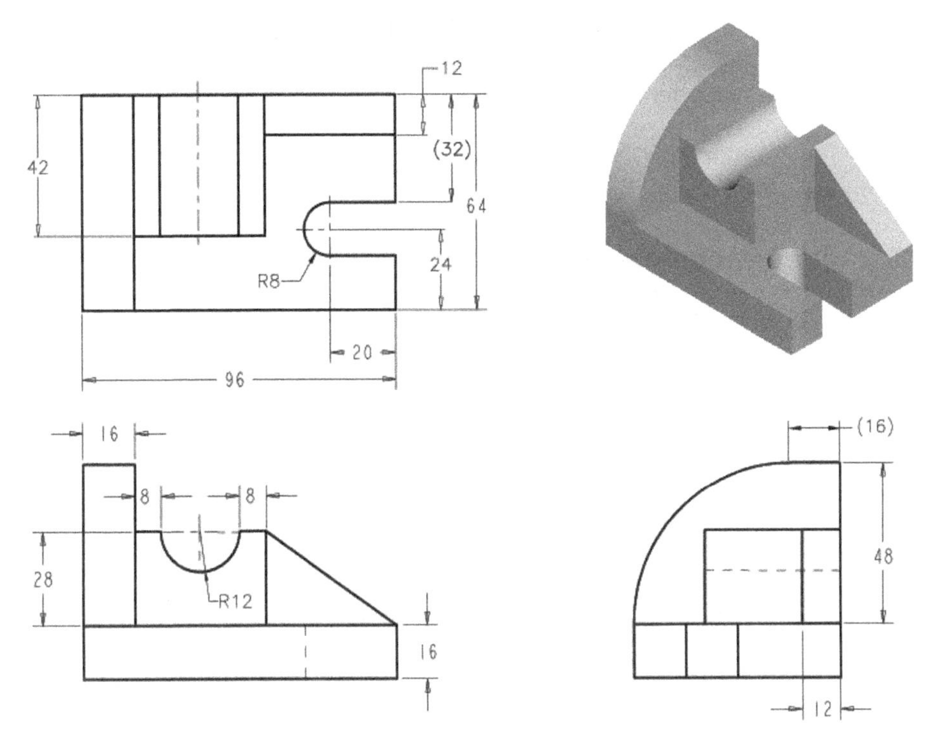

Figure 5-73 The top, front, right, and isometric views of the model

After understanding the procedure for creating the model, you are now ready to create it. Set the working directory, if required.

Starting a New Object File

1. Start a new part file and name it as *c05tut2*.

   Three default datum planes are displayed in the drawing area.

Selecting the Sketching Plane for the Base Feature

To create the sketch for the base feature, you first need to select the sketching plane for the base feature. In this model, you need to draw the base feature on the **TOP** datum plane. This is because the direction of extrusion of the base feature is perpendicular to the **TOP** datum plane.

1. Choose the **Extrude** tool from the **Shapes** group; the **Extrude** dashboard is displayed above the drawing area.

2. Choose the **Placement** tab from the dashboard to display a slide-down panel. Choose the **Define** button from it; the **Sketch** dialog box is displayed.

3. Select the **TOP** datum plane as the sketching plane; an arrow is displayed on the **TOP** datum plane, pointing in the direction of viewing the sketch. The **RIGHT** datum plane and its orientation are set automatically.

4. Choose the **Sketch** button in the **Sketch** dialog box to enter into the sketcher environment.

Creating and Dimensioning the Sketch for the Base Feature

The sketch for the base feature has a rectangular shape with a slot, refer to Figure 5-74. When you extrude this sketch, the base feature will be created with the slot, as shown in Figure 5-75.

1. Draw the sketch using various sketcher tools and add required constraints and dimensions to it. Modify these dimensions, as shown in Figure 5-74.

2. Choose the **OK** button from the dashboard to exit the sketcher environment.

3. Choose the **Named View** button from the **View** tab; a flyout is displayed. Choose the **Default Orientation** option from the flyout; the model appears with its default orientation, which is the trimetric view.

   The default view of the model is displayed, but it does not fit on the screen. You may need to zoom in the model so that it is displayed properly in the drawing area. This gives you a better view of the sketch in the 3D space. A pink arrow is also displayed on the model, indicating the direction of extrusion.

4. Enter **16** in the dimension box present on the **Extrude** dashboard; the preview of the base feature is displayed in the drawing area.

5. Choose the **Build feature** button from the **Extrude** dashboard; the base feature is created, as shown in Figure 5-75. You can use the middle mouse button to spin the object to view it from various directions.

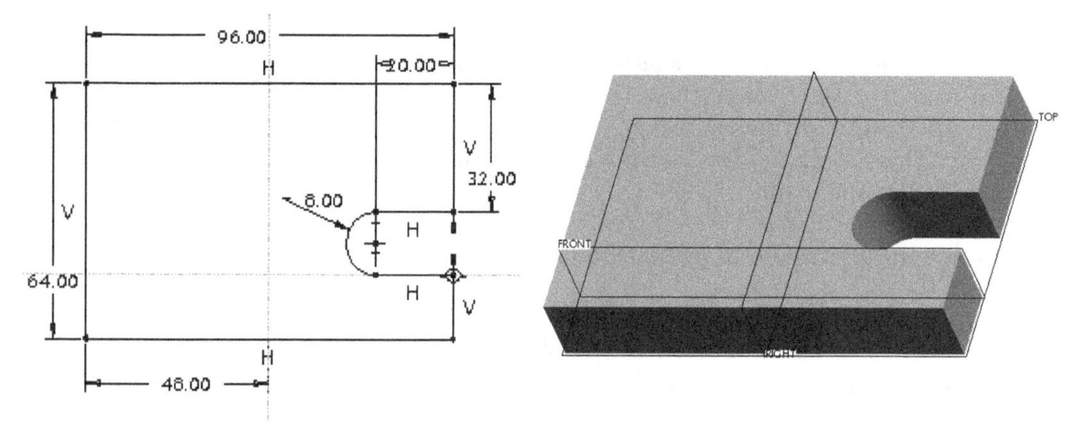

Figure 5-74 Sketch of the base feature with dimensions and constraints

Figure 5-75 Base feature of the model

Selecting the Sketching Plane for the Second Feature

The next feature is an extruded feature and will be created on the left face of the base feature. Therefore, you need to select the left face of the base feature as the sketching plane and then draw the sketch.

1. Choose the **Extrude** tool from the **Shapes** group; the **Extrude** dashboard is displayed, above the drawing area.
2. Choose the **Placement** tab from the dashboard to display a slide-down panel and then choose the **Define** button from it; the **Sketch** dialog box is displayed.
3. Use the middle mouse button to spin the model to get the view shown in Figure 5-76.
4. Now, select the left face of the base feature as the sketching plane; an arrow pointing in the direction of view of the sketching plane appears on the left face of the base feature.
5. Choose the **Flip** button to flip the arrow and to change the direction of viewing the sketching plane, refer to Figure 5-76.
6. Click in the **Reference** collector and select the top face of the base feature, refer to Figure 5-77. Next, select the **Top** option from the **Orientation** drop-down list.

   By selecting the top surface of the base feature, the model will be oriented in such a way that the highlighted planar surface will be at the top while sketching.
7. Choose the **Sketch** button from the **Sketch** dialog box to enter into the sketcher environment.

Figure 5-76 Arrow showing the direction of viewing the model

Figure 5-77 Surface selected to be at top

Creating and Dimensioning the Sketch for the Second Feature

The sketch for the second feature consists of two lines and an arc. The bottom edge of the sketch coincides with the top edge of the base feature. This sketch is extruded to a depth of 16. Before drawing the sketch, turn the model display to **No hidden**.

1. Draw the sketch using various sketcher tools, as shown in Figure 5-78. The sketch is dimensioned automatically and some weak dimensions are assigned to it.

2. Apply constraints and modify weak dimensions to the dimensions shown in Figure 5-78.

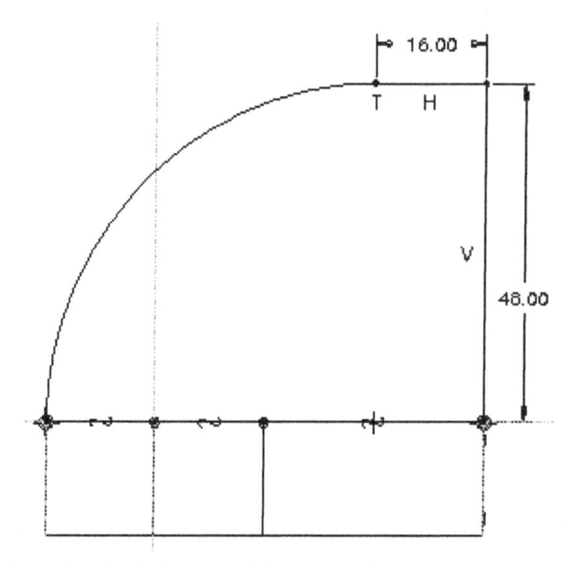

Figure 5-78 *Sketch of the second feature with dimensions and constraints*

3. After the sketch is complete, turn the model display to **Shading** and choose the **OK** button; the **Extrude** dashboard is displayed above the drawing area.

   Use the middle mouse button to orient the model, as shown in Figure 5-79. On doing so, a pink arrow is displayed on the model, indicating the direction of extrusion.

4. Enter **16** in the dimension box present in the **Extrude** dashboard and press ENTER; its preview is displayed on the screen.

5. Choose the **Build feature** button from the **Extrude** dashboard; the second feature is completed and is shown in Figure 5-80. You can use the middle mouse button to spin the model to view it from various directions.

Selecting the Sketching Plane for the Third Feature

The third feature is an extruded feature and will be created on the back planar surface of the base feature. Therefore, you need to define the back face of the base feature as the sketching plane and then draw the sketch.

1. Choose the **Extrude** tool from the **Shapes** group; the **Extrude** dashboard is displayed above the drawing area.

2. Choose the **Placement** tab from the dashboard to display a slide-down panel and then choose the **Define** button from it; the **Sketch** dialog box is displayed.

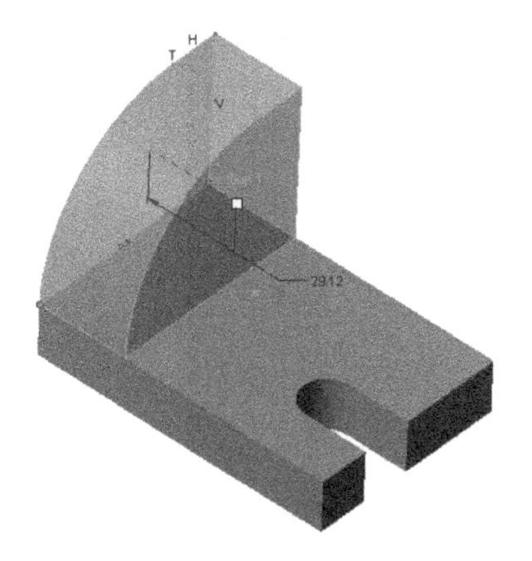

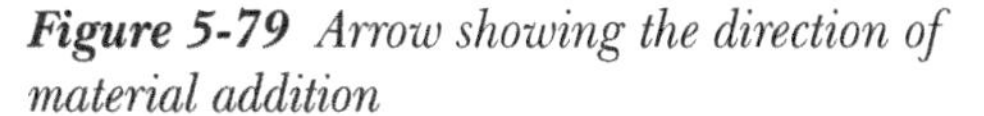

Figure 5-79 *Arrow showing the direction of material addition*

Figure 5-80 *Model with the second extruded feature*

3. Use the middle mouse button to spin the model and then select the face of the base feature shown in Figure 5-81, as the sketching plane; an arrow is displayed on the selected face.

4. Choose the **Flip** button to flip the arrow; the arrow points in the direction of viewing the sketching plane.

5. Select the top face shown in Figure 5-82 and then select the **Top** option from the **Orientation** drop-down list. Now, the top face of the base feature is selected to be at top while drawing the sketch.

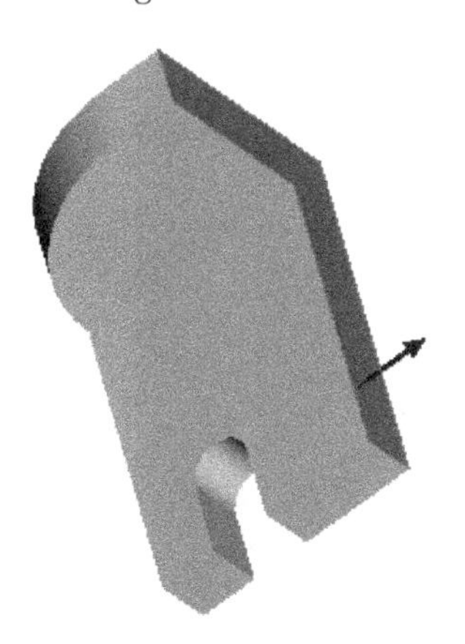

Figure 5-81 *Selected face and arrow indicating the direction of view of the model*

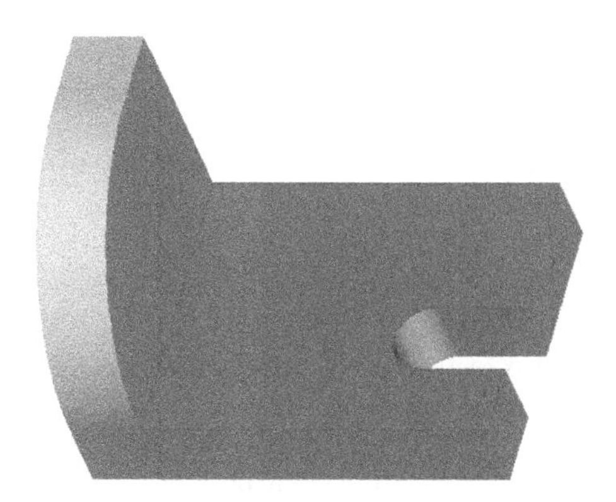

Figure 5-82 *Face selected to be at top*

6. Choose the **Sketch** button to enter the sketcher environment.

Creating and Dimensioning the Sketch for the Third Feature

The sketch for the base feature consists of a rectangular section with a semicircular cut at the top. When this section is extruded, a feature with a semicircular slot will be created. Turn the display to **No hidden**.

1. Draw the sketch using the sketcher tools, as shown in Figure 5-83. The sketch is dimensioned automatically and some weak dimensions are assigned to it.

2. Add required constraints and modify weak dimensions to the dimensions shown in Figure 5-83.

3. Choose the **OK** button to exit the sketcher environment; the **Extrude** dashboard is enabled above the drawing area. Turn the display to **Shading**.

4. Use the middle mouse button to orient the model, as shown in Figure 5-84. This orientation of the model gives a better view of the sketch in 3D space. An arrow is displayed on the model, indicating the direction of extrusion.

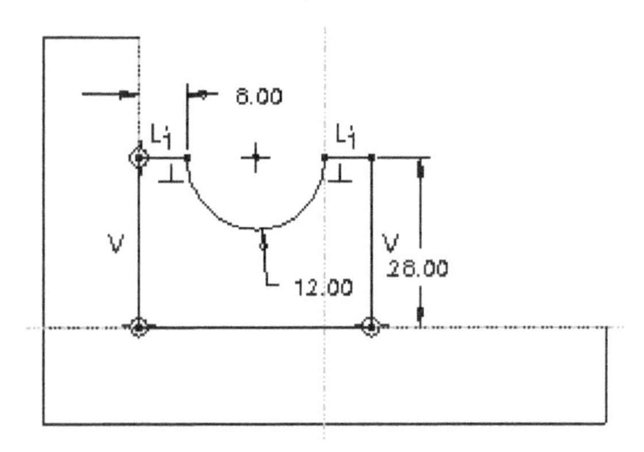

Figure 5-83 Sketch with dimensions and constraints

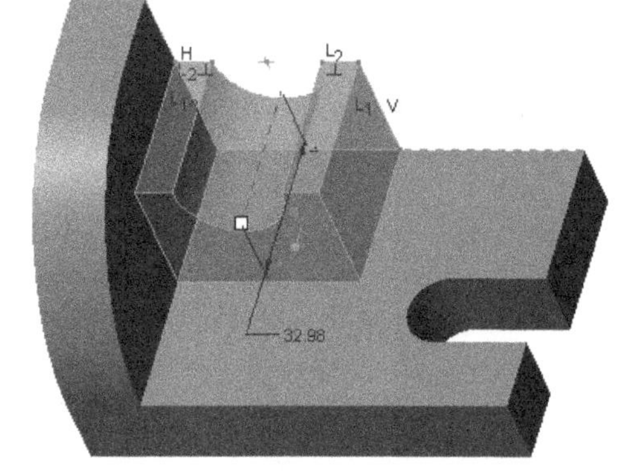

Figure 5-84 Arrow showing the direction of material addition

5. Enter **42** in the dimension box present in the **Extrude** dashboard and press ENTER. This feature is completed and its preview can be seen in the drawing area.

6. Choose the **Build feature** button from the **Extrude** dashboard to confirm the feature creation. Choose the **Named View** button from the **View** tab; a flyout is displayed. Choose the **Default Orientation** option from the flyout; the default view of the model is displayed, as shown in Figure 5-85.

Figure 5-85 Model with the third extruded feature

Selecting the Sketching Plane for the Last Feature

The last feature and the third feature are created on the same plane. But the depth of extrusion is different for both of them. This is the reason they are considered as separate features. Therefore, for sketching this feature, you can use the sketching plane that was used for creating the third feature.

1. Choose the **Extrude** tool and then invoke the **Sketch** dialog box.

2. Choose the **Use Previous** button from the **Sketch Plane** area in the **Sketch** dialog box; the sketcher environment gets activated automatically.

   When you choose the **Use Previous** button, the system selects the previous sketching plane that was used to create the base feature.

Creating and Dimensioning the Sketch for the Last Feature

The sketch of the last feature consists of three lines. The bottom edge of the sketch is aligned with the top edge of the base feature and the left edge is aligned with the right edge of the third feature. When this sketch is extruded, it creates a rib shape.

1. Draw the section sketch by using sketcher tools. The sketch is dimensioned automatically and some weak dimensions are assigned to it. You can add required constraints (**Coincident** constraint) to align the lines and points in the sketch with other features, as shown in Figure 5-86. In this figure, the constraint symbol displayed on the line indicates that the edges of the adjacent features are used to close the section.

2. After the sketch is complete, choose the **OK** button; the **Extrude** dashboard is enabled. Use the middle mouse button to orient the model, as shown in Figure 5-87.

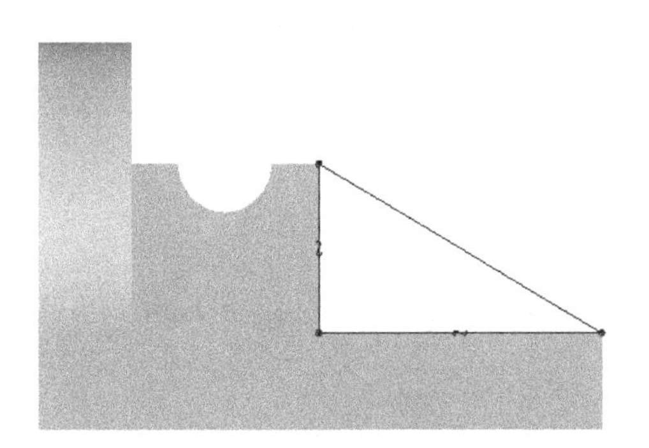

Figure 5-86 *Sketch and constraints for the last feature*

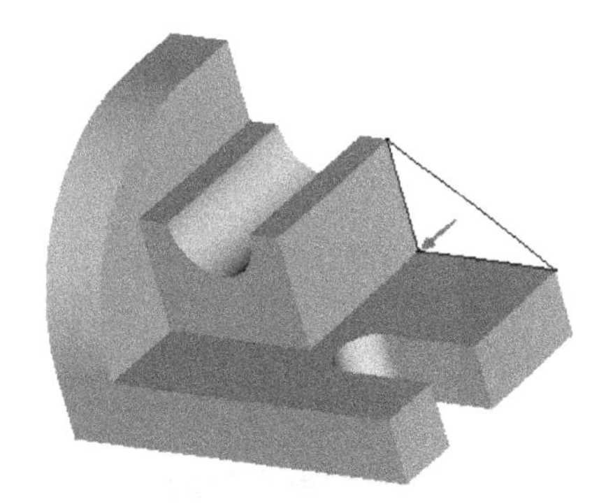

Figure 5-87 *Arrow showing the direction of material addition*

3. Enter the value **12** in the dimension box present in the **Extrude** dashboard; the feature is completed and its preview can be seen in the drawing area.

4. Choose the **Build feature** button in the **Extrude** dashboard.

The default view of the model is shown in Figure 5-88. To invoke the default view, choose the **Named View** button from the **View** tab; a flyout is displayed. Next, choose the **Default Orientation** option from this flyout.

Saving the Model

1. Choose the **Save** button from the **File** menu and save the model. The order of feature creation can be seen from the **Model Tree** shown in Figure 5-89.

Figure 5-88 Final model for Tutorial 2

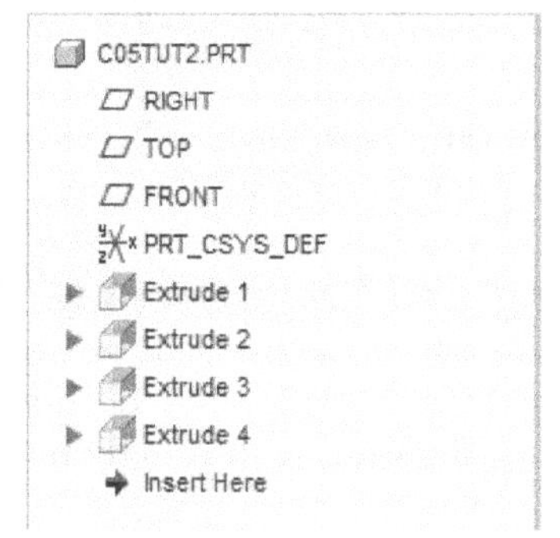

Figure 5-89 ***Model Tree*** *for Tutorial 2*

Tutorial 3

In this tutorial, you will create the model shown in Figure 5-90. This figure also shows the front and top views of the solid model. **(Expected time: 30 min)**

Examine the model and then determine the number of features in it, refer to Figure 5-90.

The following steps are required to complete this tutorial:

a. Start a new file and create the base feature, refer to Figures 5-91 and 5-92.
b. Create the second feature, refer to Figures 5-93 through 5-96.
c. Create the hollow cylindrical feature on an offset datum plane, refer to Figures 5-97 through 5-102.

Starting a New Object File

1. Set the working directory, if required, and then start a new part file with the name *c05tut3*.

The three default datum planes are displayed in the drawing area.

Selecting the Sketching Plane for the Base Feature

To create the sketch for the base feature, you first need to select the sketching plane. In this model, you need to draw the base feature on the **FRONT** datum plane because the direction of extrusion is perpendicular to the **FRONT** datum plane.

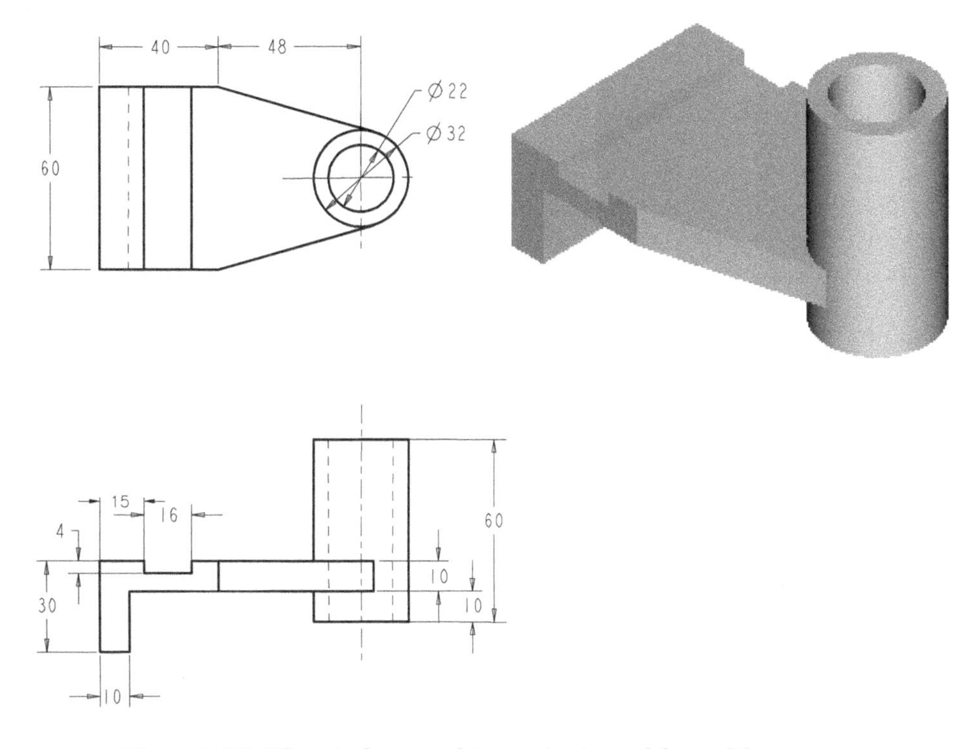

Figure 5-90 The top, front, and isometric views of the model

1. Choose the **Extrude** tool from the **Shapes** group.
2. Choose the **Placement** tab from the dashboard to display a slide-down panel. Next, choose the **Define** button; the **Sketch** dialog box is displayed.
3. Select the **FRONT** datum plane as the sketching plane.

   A pink arrow is displayed on the **FRONT** datum plane and it points in the direction of viewing the sketch plane. The **RIGHT** datum plane and its orientation are selected automatically.
4. Choose the **Sketch** button to enter the sketcher environment.

Creating and Dimensioning the Sketch for the Base Feature

The section to be extruded for the base feature is evident from the model. The section sketch is shown in Figure 5-91. When this sketch is extruded, it will create the base feature.

1. Draw the sketch using various sketcher tools, as shown in Figure 5-91.

   The sketch is dimensioned automatically and some weak dimensions are assigned to it.

2. Add the required constraints and modify weak dimensions to the dimension, as shown in Figure 5-91.

3. Choose the **OK** button; the **Extrude** dashboard is displayed.

4. Choose the **Named Views** button from the **Graphics** toolbar; a flyout is displayed. Choose the **Default Orientation** option from the flyout; the model orients to its default orientation, which is the trimetric view, and an arrow is also displayed on it, indicating the direction of extrusion.

5. Enter **60** in the dimension box that is present on the **Extrude** dashboard and then press ENTER.

6. Choose the **Build feature** button from the **Extrude** dashboard.

   The base feature is completed, as shown in Figure 5-92. You can use the middle mouse button to spin the model to view it from various directions.

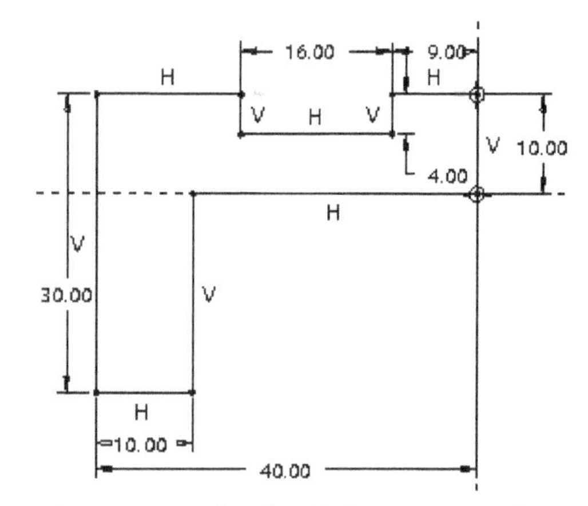

Figure 5-91 Sketch with dimensions and constraints for the base feature

Figure 5-92 Base feature of the model

Selecting the Sketching Plane for the Second Feature

The next feature is an extruded feature. The sketching plane for this feature is the top face of the base feature.

1. Choose the **Extrude** tool from the **Shapes** group.

2. Choose the **Placement** tab from the dashboard; the slide-down panel is displayed. Choose the **Define** button from the slide-down panel; the **Sketch** dialog box is displayed.

3. Select the top face of the base feature as the sketching plane; an arrow pointing in the direction of viewing the sketch is displayed on the top face, refer to Figure 5-93.

4. Choose the **Flip** button to flip the arrow to point in the direction.

5. Select the **RIGHT** datum plane and then select the **Right** option from the **Orientation** drop-down list.

6. Choose the **Sketch** button to enter the sketcher environment.

Creating and Dimensioning the Sketch for the Second Feature

The next feature to be created is an extrude feature. The section for the extrude feature is shown in Figure 5-94. Before drawing the sketch, turn the model display to **No hidden**.

1. Draw the sketch using various sketcher tools. In the sketch, draw a center line passing through the center of the arc, refer to Figure 5-94. This center line helps in dimensioning the sketch. Add required constraints and dimensions to the sketch, refer to Figure 5-94. Before exiting the sketcher environment, turn the model display to **Shading**.

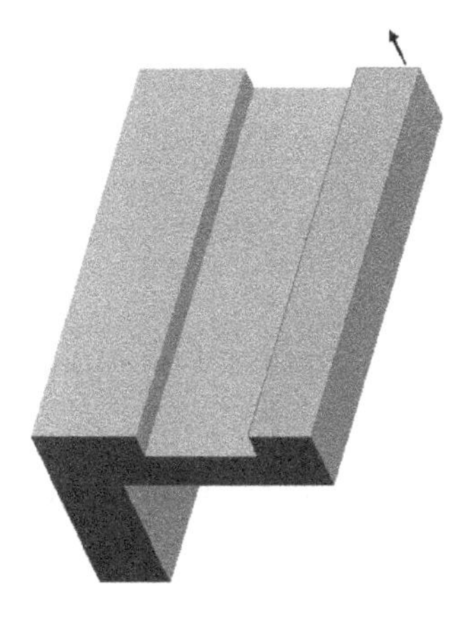

Figure 5-93 Arrow pointing from the sketching plane in the direction of viewing the sketch

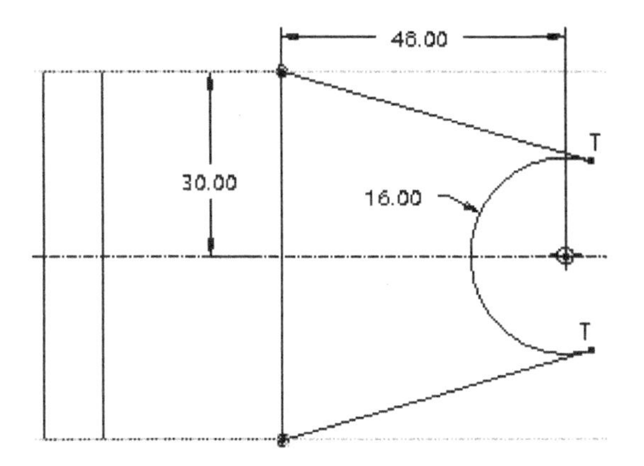

Figure 5-94 Sketch with dimensions and constraints

2. Choose the **OK** button; the **Extrude** dashboard is enabled. Use the middle mouse button to orient the model, refer to Figure 5-95.

3. Enter **10** in the dimension box present on the **Extrude** dashboard and press ENTER.

4. Now, choose the **Build feature** button to confirm the feature creation. The default trimetric view of the extruded feature is shown in Figure 5-96.

Creating a Datum Plane for the Last Feature

To create the hollow cylindrical feature, you need a datum plane. This datum plane will be created at an offset distance of 10 units from the bottom face of the second feature shown in Figure 5-97.

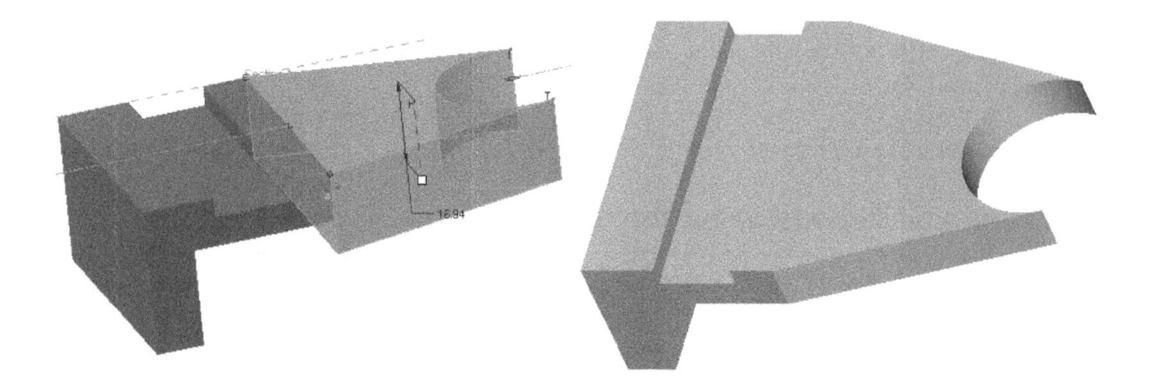

Figure 5-95 Arrow showing the direction of material addition

Figure 5-96 Model with the second extruded feature

Note

The other method to create this feature is to select the top planar surface of the second feature as the sketching plane and extrude the sketch on both sides of the sketching plane. The depth of extrusion will be different on both the sides. If you use this method to create this cylindrical feature, you do not need to create a datum plane.

1. Choose the **Plane** tool from the **Datum** group; the **Datum Plane** dialog box is displayed.

2. Spin the model using the middle mouse button and then select the bottom face of the second feature.

   As you select the face of the second feature, the **Offset** constraint is displayed in the **References** collector of the dialog box.

3. In the **Translation** dimension box, enter **10**.

4. Choose the **OK** button from the **Datum Plane** dialog box; the datum plane **DTM1** is created, as shown in Figure 5-98, and is selected as the sketching plane for creating the sketch.

Selecting the Sketching Plane for the Last Feature

The plane **DTM1** will be selected as the sketching plane and the depth of extrusion will be defined from this plane.

1. Choose the **Extrude** tool from the **Shapes** group.

2. Choose the **Placement** tab from the dashboard; the slide-down panel is displayed. Choose the **Define** button from the slide-down panel; the **Sketch** dialog box is displayed.

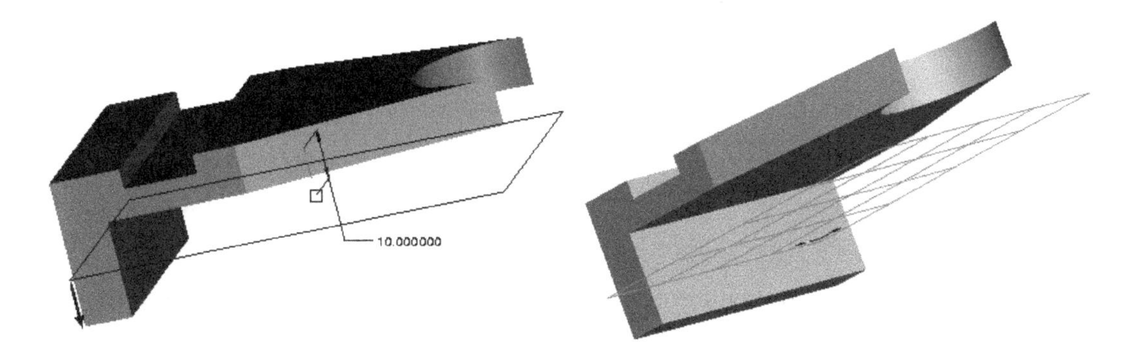

Figure 5-97 *Creating the datum plane*

Figure 5-98 *Model after creating the datum plane*

3. Select **DTM1** as the sketching plane; an arrow appears on the datum plane.

4. Choose the **Flip** button to reverse the direction of viewing the sketch. Figure 5-99 shows the plane with the direction of arrow reversed.

5. Select the **FRONT** datum plane and select the **Bottom** option from the **Orientation** drop-down list.

6. Choose the **Sketch** button to enter into the sketcher environment.

Creating and Dimensioning the Sketch for the Last Feature

The sketch for the hollow cylindrical feature will be drawn on the datum plane **DTM1**. The sketch for the hollow cylindrical feature consists of two concentric circles. Before drawing the sketch, turn the model display to **No hidden**.

1. Draw the sketch using sketcher tools and add required constraints and dimensions to it, as shown in Figure 5-100.

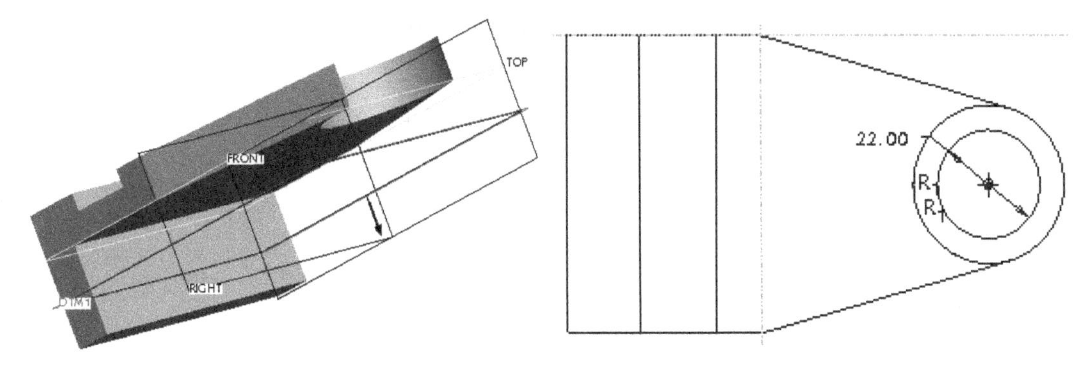

Figure 5-99 *Direction of viewing the sketching plane*

Figure 5-100 *Sketch with dimensions and constraints*

2. Choose the **Done** button; the **Extrude** dashboard is displayed. Now, turn the model display to **Shading**.

   Use the middle mouse button to orient the model, as shown in Figure 5-101. This orientation gives you a better view of the sketch in the 3D space.

3. Enter **60** in the dimension box present on the **Extrude** dashboard and then press ENTER.

4. Choose the **Build feature** button to confirm the feature creation. The default trimetric view of the complete model is shown in Figure 5-102.

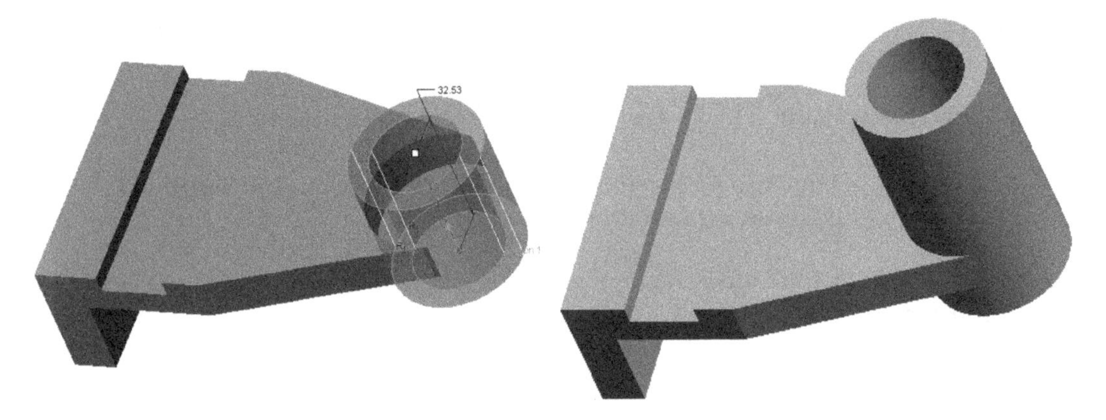

Figure 5-101 *Preview of the feature* ***Figure 5-102*** *Final model of Tutorial 3*

Saving the Model

1. Choose the **Save** button from the **File** menu and save the model.

Self-Evaluation Test

Answer the following questions and then compare them to those given at the end of this chapter:

1. You can change the default names assigned to datum planes. (T/F)

2. Datum points are also used to add notes in drawings and attach datum targets to them. (T/F)

3. The constraint combination for creating an offset plane through a cylinder is a valid combination. (T/F)

4. Generally, all features in a model are created on a single sketching plane. (T/F)

5. A sketching plane can be selected on the existing face of a feature. (T/F)

6. While creating a datum feature, the __________ selected from model helps you choose a constraint from the available constraints to constrain the datum feature.

7. Datum axis is an __________ axis that is created in PTC Creo Parametric.

8. The trimetric view of the model is displayed when you choose the __________ option from the **Named View** drop-down.

9. When you create a new object file, __________ default datum planes are displayed in the drawing area.

10. The default datum planes in the **Part** mode are named as __________, __________, and __________.

Review Questions

Answer the following questions:

1. In PTC Creo Parametric, which one of the following is not a type of datum?

   (a) Axis (b) Plane
   (c) Circle (d) Curve

2. How many methods are available in PTC Creo Parametric to create a datum plane?

   (a) One (b) Two
   (c) Three (d) Four

3. Which one of the following dialog boxes is displayed when you choose the **Point** tool from the **Datum** group while extruding a section sketch?

   (a) **Datum Point** (b) **Datum Plane**
   (c) **Datum Axis** (d) None of these

4. Which one of the following options is used to spin a model in the drawing area?

   (a) CTRL+ALT (b) middle mouse button
   (c) left mouse button (d) CTRL+right mouse button

5. Datum planes are considered as feature geometry and have mass and volume. (T/F)

6. To set the default orientation of a model to isometric, you need to use the **Environment** dialog box. (T/F)

7. Generally, the sketching plane for the base feature of any model is determined after viewing the isometric view or drawing views of the model. (T/F)

8. Datum planes are used as a reference for mirroring features, copying features, creating a cross-section as well as for changing the orientation of the references. (T/F)

9. Unlike the datum planes constraint option, all datum axes constraint options are stand-alone. (T/F)

10. Coordinate system is used to determine the position of a point or other geometric element. (T/F)

Exercises

Exercise 1

Create the model shown in Figure 5-103. The dimensions, front view, and right-side view of the model are shown in Figure 5-104. **(Expected time: 45 min)**

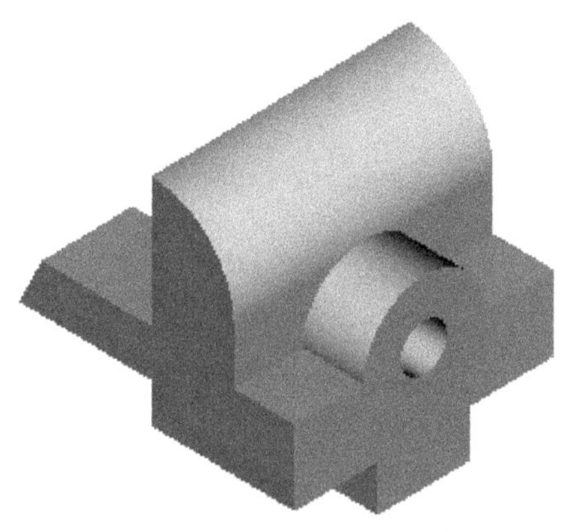

Figure 5-103 *Isometric view of the model*

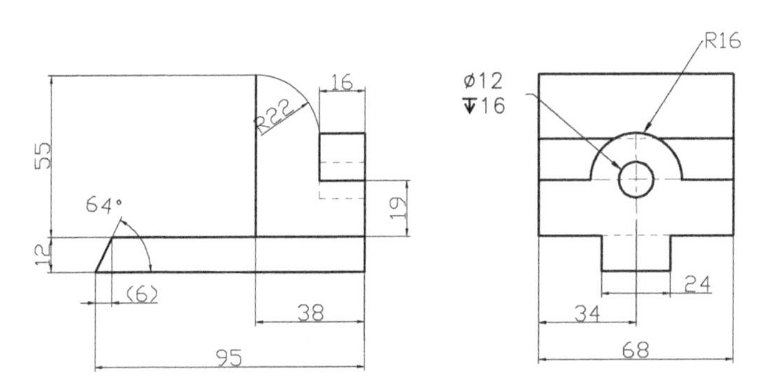

Figure 5-104 *The front and right views of the model*

Exercise 2

Create the model shown in Figure 5-105. The dimensions as well as the front, top, right-side, and isometric views of the model are also shown in this figure. **(Expected time: 45 min)**

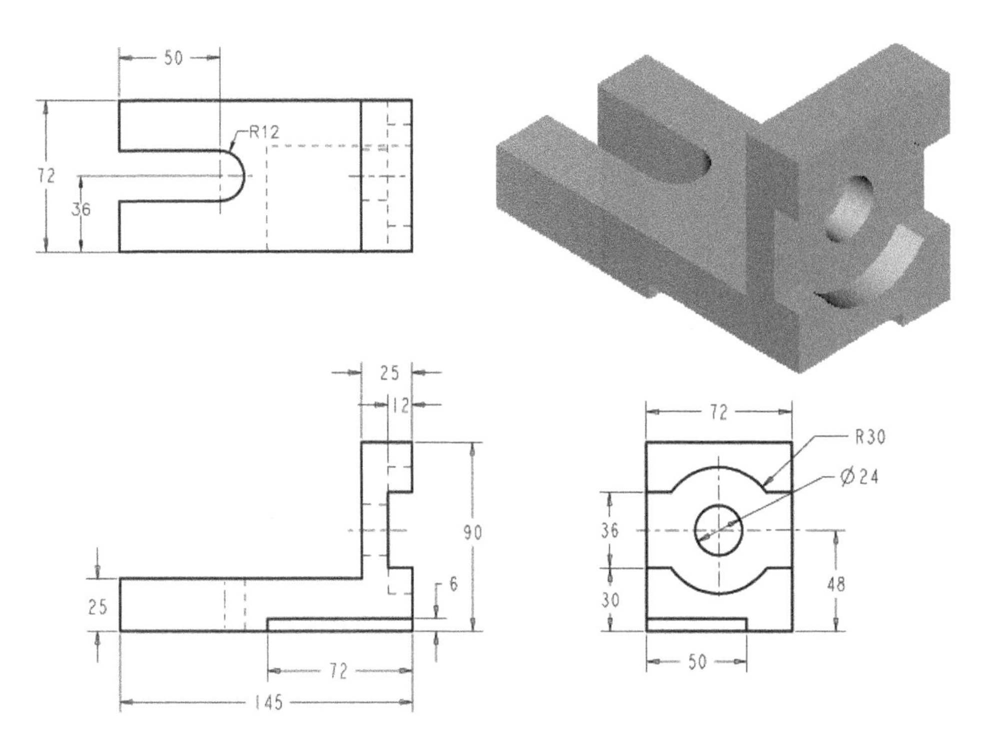

Figure 5-105 *The top, front, right-side, and isometric views of the model*

Exercise 3

Create the model shown in Figure 5-106. The top, front, section and two auxiliary views of the model are shown in Figure 5-107. **(Expected time: 1 hr)**

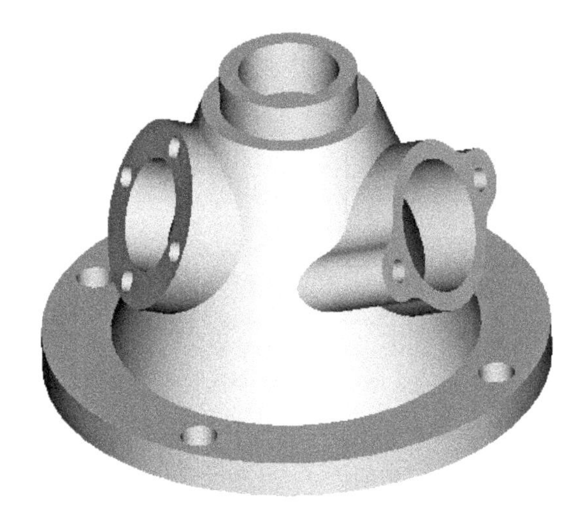

Figure 5-106 *The 3D view of the model*

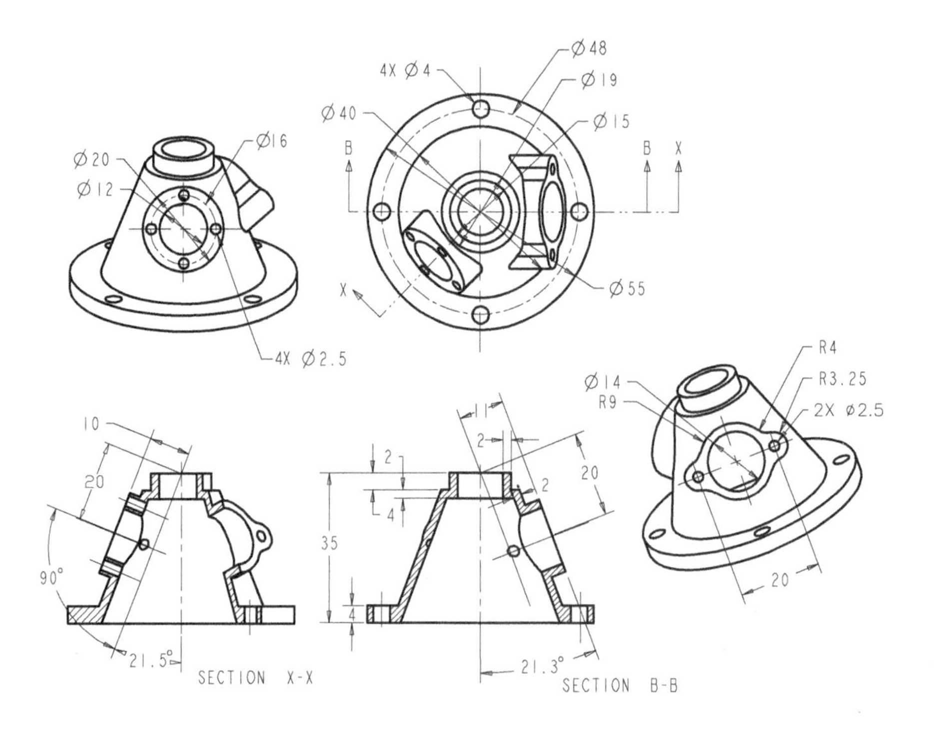

Figure 5-107 *Top, front, section, and two auxiliary views of the model*

Hint to create the two side features in Exercise 3:

Feature on the right

1. Create an axis passing through the top face of the base feature and through the **RIGHT** datum plane.

2. Create a datum plane passing through the axis (created in the previous step) and at an angle of **21.3** degrees from the **RIGHT** datum plane.

3. Now, create an offset plane at a distance of 11 from the datum plane (created in the previous step). Select this datum plane as the sketching plane.

Feature on the left

1. Create a datum plane passing through the axis of revolution of the base feature and at an angle of 45 degrees from the **FRONT** datum plane.

2. Create a datum axis passing through the top face of the base feature and through the datum plane (created in the previous step).

3. Create a datum plane passing through datum axis (created in the previous step) and at an angle of **21.5** degrees from the datum plane (created in step1).

4. Create a datum plane offset to a distance of 10 from the datum plane (created in the previous step).

Exercise 4

Create the model shown in Figure 5-108. The dimensions, the left-side view, auxiliary view, and front view of the model are also shown in Figure 5-109. **(Expected time: 45 min)**

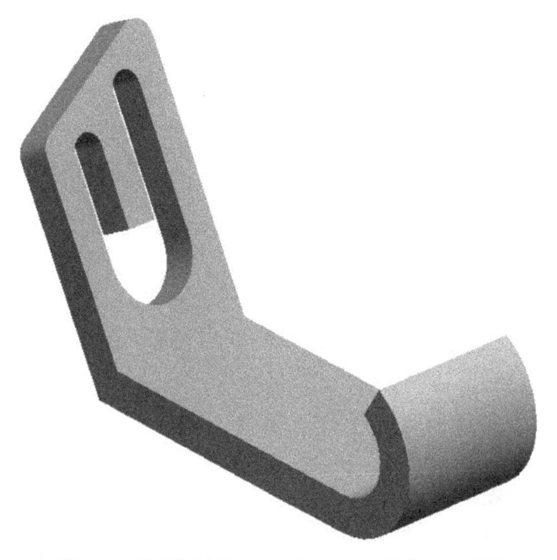

Figure 5-108 *Isometric view of the model*

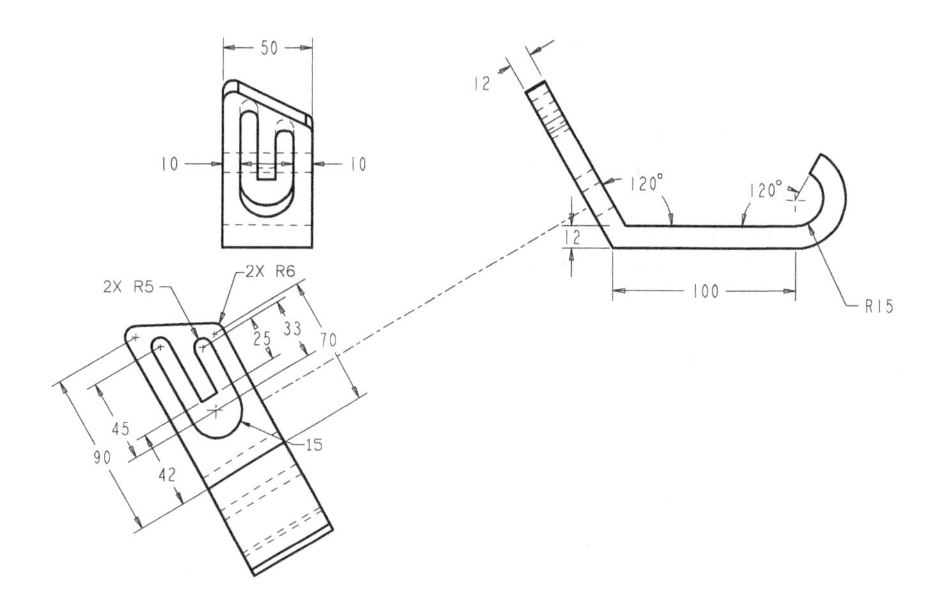

Figure 5-109 Left, auxiliary, and front views of the model

Answers to Self-Evaluation Test

1. T, **2.** T, **3.** T, **4.** F, **5.** T, **6.** geometry, **7.** imaginary, **8. Default Orientation**, **9.** three, **10. TOP, FRONT, RIGHT**

Chapter 6

Options Aiding Construction of Parts-I

Learning Objectives

After completing this chapter, you will be able to:

- *Create holes.*
- *Create Round, Chamfer, and Rib.*
- *Edit features.*
- *Redefine, reroute, and reorder features.*
- *Suppress and delete features.*
- *Modify features.*

OPTIONS AIDING CONSTRUCTION OF PARTS

This chapter explains the feature creation tools provided in PTC Creo Parametric that help in creating a model and editing it. In this chapter, you will also learn to create various types of holes that are required in most of the engineering designs. In previous chapters, you learned to create holes using extrude cut, but in this chapter you will create holes using the **Hole** dashboard. You can create and modify holes in an easier manner using the options in the **Hole** dashboard.

In this chapter you will also learn to create rounds, chamfers, and ribs.

CREATING HOLES

In engineering components, holes can be counterbore, countersink, tapered, or drilled. PTC Creo Parametric allows you to create all such types. PTC Creo Parametric also provides industry standard holes that have standard dimensions. In PTC Creo Parametric, holes are created using the **Hole** dashboard.

The Hole Dashboard

Ribbon: Model > Engineering > Hole

The **Hole** dashboard is displayed when you choose the **Hole** tool from the **Engineering** group in the **Model** tab. You can create two types of holes using the **Hole** dashboard, the **Simple** hole and the **Standard** hole. The **Simple** hole options enable you to create holes using three buttons and these buttons are discussed next.

Creating Simple Holes Using Predefined Rectangle Profile

When you invoke the **Hole** dashboard, the **Use predefined rectangle as drill hole profile** button is chosen by default, as shown in Figure 6-1. The options and the buttons available in this dashboard are discussed next.

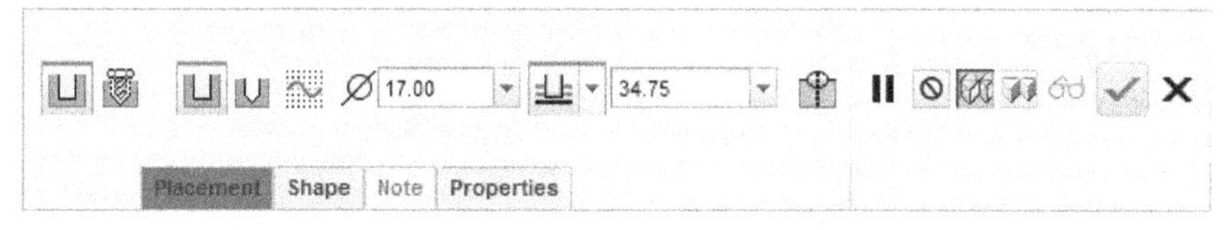

***Figure 6-1** The partial view of the **Hole** dashboard*

Dimension Boxes

The first dimension box from the left in the **Hole** dashboard is used to specify the diameter of the hole.

The second dimension box from the left in the **Hole** dashboard is used to specify the depth of the hole. The options to specify the depth of the hole are the same as those available in the **Extrude** dashboard.

Placement Tab

When you choose the **Placement** tab, a slide-down panel will be displayed, as shown in Figure 6-2. This slide-down panel has the parameters that need to be specified to define a linear hole. The **Placement** collector in the slide-down panel will be displayed. This indicates that PTC Creo Parametric is prompting you to select the placement plane (primary reference). You can select the placement plane only when this collector is highlighted. Once you have selected the primary reference, its name appears in the box. The primary reference can be a plane, axis, or point.

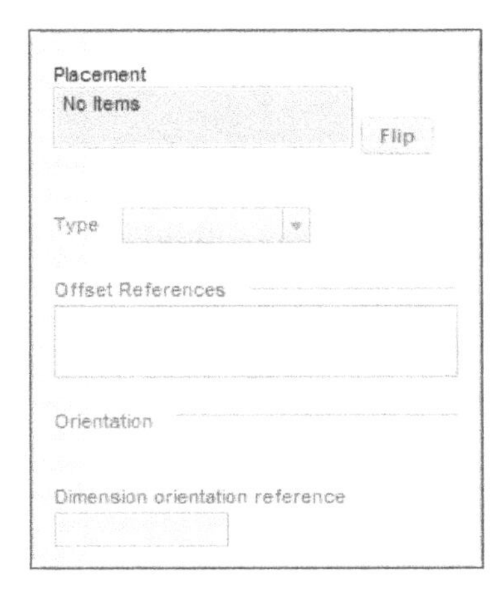

***Figure 6-2** The **Placement** slide-down panel*

The **Flip** button is used to change the direction of the hole creation with reference to the primary reference. When you choose this button, the direction of the hole creation is reversed and you can view its preview in the drawing area. Remember that the hole will be created only if it is possible to create the hole in the specified direction.

The **Type** drop-down list below the **Placement** collector contains the options that determine the type of placement for the hole. You can use the options in this drop-down list after selecting the placement plane or the primary reference. These options are discussed next.

Linear: When you select this option, you need to specify the distance from two linear references. Generally, these linear references are the edges of the planar surface on the model, any two planar surfaces or axes, or a combination of any of these. Figure 6-3 shows a linear hole on the top face of the block. The procedure to create this hole using the **Hole** dashboard is as follows:

1. Invoke the **Hole** dashboard.

2. Choose the **Placement** tab to open a slide-down panel.

3. Select the top face of the model to place the hole; the preview of the hole will be displayed.

4. Click in the **Offset References** collector to make the secondary selections.

5. Select the left face or the top left edge of the model.

6. Use CTRL+left mouse button to select the front face or the top front edge of the model. Now, you have selected the two references that are needed to dimension the position of hole.

7. Specify the exact location of hole in the edit boxes of the **Offset Reference** collector.

8. Specify the diameter of the hole in the dimension box present on the **Hole** dashboard.

9. From the depth flyout, choose the **Drill to intersect with all surfaces** button.

10. Choose the **Build feature** button to create the hole and exit the **Hole** dashboard.

Note

*To select the secondary references, you need to click in the **Offset References** collector and then select the references on the model. Dimensions of the center of the hole from the selected references are displayed in the drawing area as well as in the **Offset References** collector. You can click on the dimensions to modify them.*

Radial: This option is used to create a hole that can be referenced to an axis. When you select this option, you need to select an axial reference and an angular reference to place the hole. The distance from the axis and the angle from the plane are entered in the boxes under the **Offset References** collector. Figure 6-4 shows a radial hole on a curved surface. The following steps explain the procedure to create this hole.

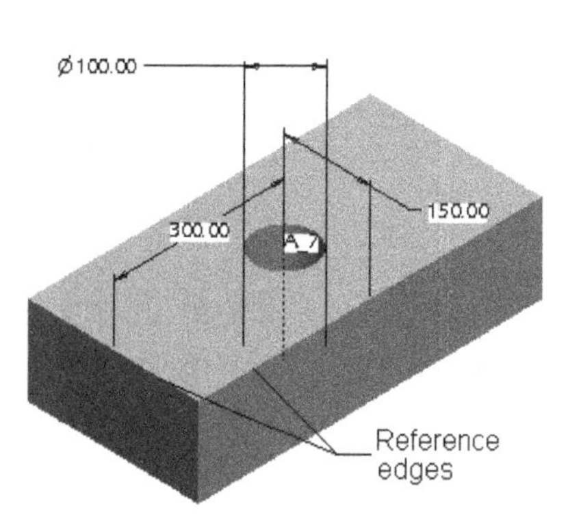

Figure 6-3 Linear hole on the top face of the block

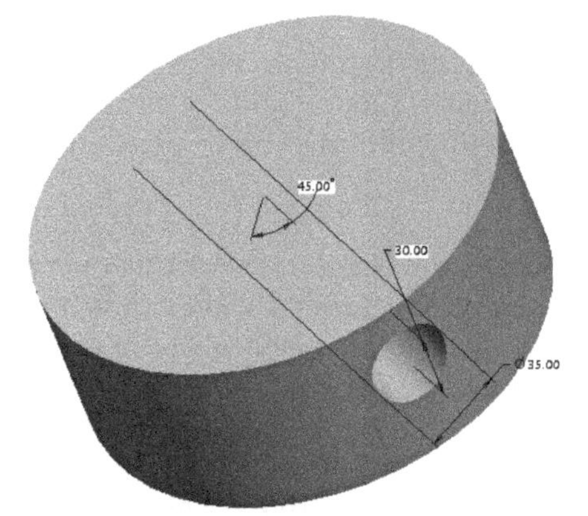

Figure 6-4 Radial hole on the curved surface

Tip: *Note that since you have selected the curved surface as the placement plane for the hole, therefore, you need not to select the axis to specify the radial distance. However, if the placement plane is a plane surface, then you do need to select an axis to create a radial hole at some radius from the selected axis.*

1. Invoke the **Hole** dashboard.

2. Choose the **Placement** tab to open the slide-down panel.

3. Select the curved surface of the model to place the hole; the preview of the hole will be displayed.

4. Click in the **Offset References** collector to make the secondary selections.

5. Select the datum plane that is normal to the curved surface to specify the angular value.

6. Use CTRL+left mouse button to select the top face or the top edge of the model. Now, you have selected the two references that were needed to dimension the hole.

7. Specify the diameter of the hole in the dimension box on the **Hole** dashboard.

8. From the depth flyout, choose the **Drill to intersect with all surfaces** button.

9. Choose the **Build feature** button to create the hole and exit the **Hole** dashboard.

Diameter: This option is used to create a diametrically placed hole. When you select this option, you need to select an axial reference and an angular reference to place it. Figure 6-5 shows a diameter hole on a plane. The following steps explain the procedure to create this hole.

1. Invoke the **Hole** dashboard.

2. Choose the **Placement** tab to open the slide-down panel.

3. Select the top face of the model to place the hole; the preview of the hole will be displayed.

4. From the drop-down list, select the **Diameter** option.

5. Click in the **Offset References** collector to make the secondary selections.

6. Select the face to specify the angular value, refer to Figure 6-5.

7. Use CTRL+left mouse button to select the axis of another reference hole that is on the top face. Now, you have selected the two references that are needed to dimension the hole.

8. Specify the distance for the hole measured from the axis of the reference hole and the angle in the **Offset References** collector.

9. From the depth flyout, choose the **Drill to intersect with all surfaces** button.

10. Choose the **Build feature** button to exit the **Hole** dashboard.

Coaxial: This option is used to create a coaxial hole. When you select this option, you need to select an axis. No dimensions are required to place a coaxial hole. Figure 6-6 shows a coaxial hole on a plane. The procedure to create this hole is given next.

1. Invoke the **Hole** dashboard.

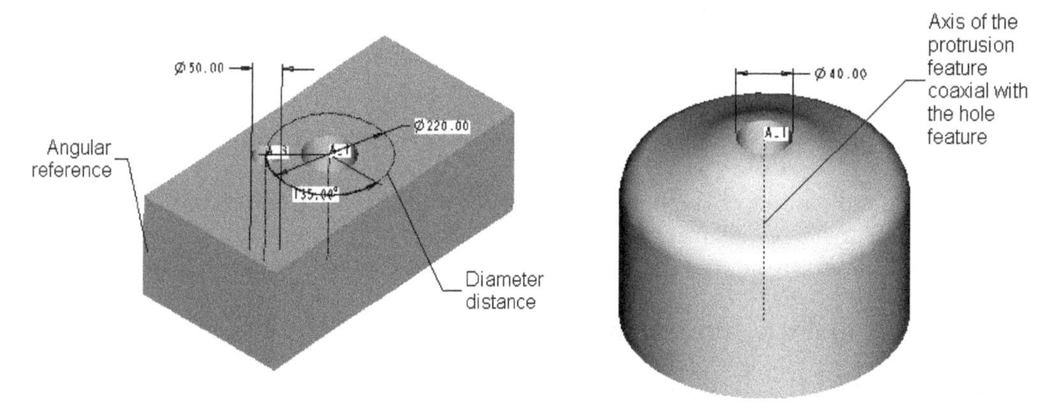

***Figure 6-5** Diameter dimensioning of the hole*

***Figure 6-6** Coaxial hole*

2. Choose the **Placement** tab to open the slide-down panel.
3. Select the axis of the model to place the hole; the preview of the hole will be displayed.
4. In the **Type** drop-down list, the **Coaxial** option is selected by default.
5. Now, press the CTRL key to select the placement plane or the surface where you want to place the base of the hole.
6. Specify the diameter of the hole in the dimension edit box available on the **Hole** dashboard.
7. From the depth flyout, choose the **Drill to intersect with all surfaces** button.
8. Choose the **Build feature** button to exit the **Hole** dashboard.

Tip: *Remember that while placing the linear, diameter, and radial holes using the **Hole** dashboard, you need to define two things. First is the placement plane on which the hole feature will be created and second is the dimensional references. But for the **Coaxial** hole, you need to select an axis and for the **On Point** hole, you need to select a datum point.*

On Point: This option will be available only when you select the datum point as the primary placement reference. Note that the datum point is located either on the surface or at an offset from it. The procedure of creating this hole is explained next.

1. Invoke the **Hole** dashboard.
2. Choose the **Placement** tab to open the slide-down panel.
3. Select the datum point to place the hole; the preview of the hole will be displayed.

In the **Type** drop-down list, the **On Point** option is selected by default.

4. Specify the diameter of the hole in the first dimension edit box available on the **Hole** dashboard.

5. From the depth flyout, choose the **Drill to intersect with all surfaces** button.

6. Choose the **Build feature** button to exit the **Hole** dashboard.

Shape Tab

When you choose the **Shape** tab, a slide-down panel will be displayed, as shown in Figure 6-7. This slide-down panel can be used to specify the depth and the diameter of the hole. The depth of the hole can be specified in two directions with respect to the placement plane. The options to specify the depth of the hole are similar to those that are available in the **Extrude** dashboard. In the slide-down panel, the sketch of the hole, gives an idea of the shape of the hole and its dimensions. The sketch that is shown in the slide-down panel shows the depth of the hole in the **Side 1** direction. The **Side 2** drop-down list is used to specify the depth of the hole in the direction opposite to **Side 1**. This means the hole can be created on both sides of the placement plane. By default, the **None** option is selected in this drop-down list.

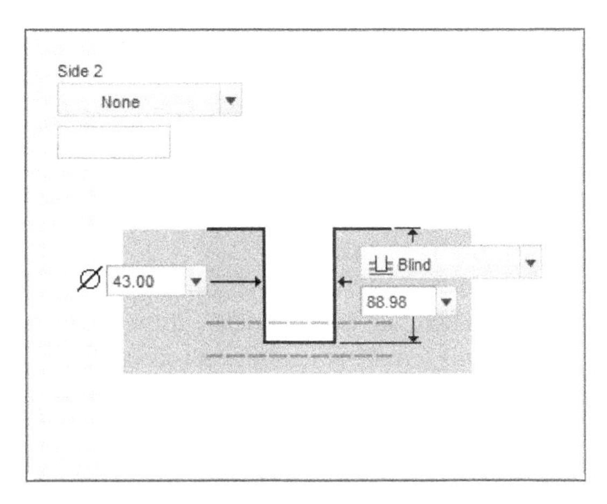

***Figure 6-7** The **Shape** slide-down panel*

Note

*The termination options in the **Shape** slide-down panel are similar to those discussed in the **Extrude** dashboard.*

When you specify the placement plane for a hole and then specify two references in case of a linear hole, you will notice that four drag handles are displayed on the hole. These four drag handles have four different functions to place the hole dynamically on the placement plane. Using these handles, you can set the diameter, depth, and specify two references to place the hole. Remember that these drag handles appear only when you select the primary reference, or in other words, the placement plane.

Tip: *The diameter of the hole and its depth in the **Side 1** direction can be specified without using the slide-down panels. The options to specify the diameter and depth in the **Side 1** direction are available on the **Hole** dashboard. If you want to specify the depth in the **Side 2** direction, open the slide-down panel by choosing the **Shape** tab from the **Hole** dashboard.*

Properties Tab

The **Properties** slide-down panel of the **Hole** dashboard is shown in Figure 6-8. This tab, when chosen, allows you to name the hole you are creating. By default, the first hole you create is named as HOLE_1 and then every hole created is successively numbered.

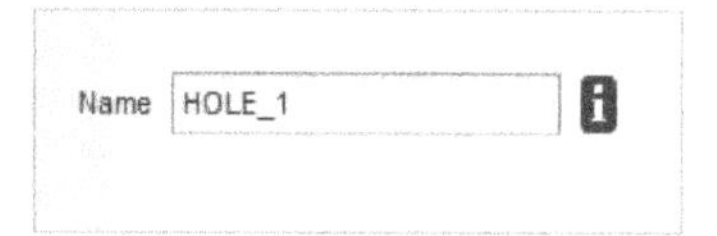

*Figure 6-8 The **Properties** slide-down panel*

Creating Simple Holes Using Standard Hole Profile

You need to choose the **Use standard hole profile as drill hole profile** button from the **Hole** dashboard, as shown in Figure 6-9. On choosing this button, you can specify the countersink, counterbore, and depth of the hole. The options and tool buttons available in the dashboard are discussed next.

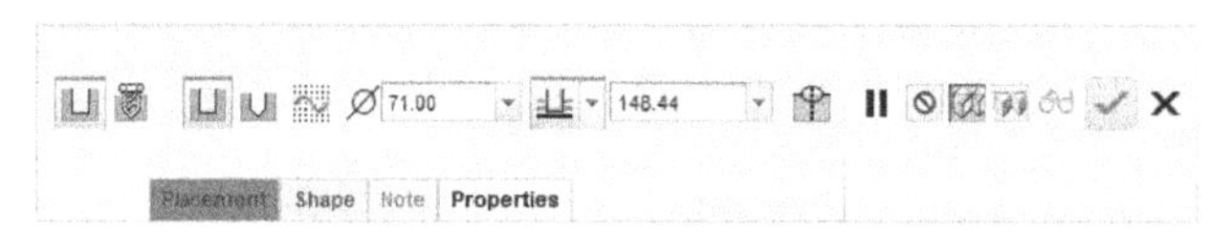

***Figure 6-9** Partial view of the **Hole** dashboard*

Placement Tab

The use of the placement options in this tab is same as you did while creating the **Simple** hole using the **Use predefined rectangle as the drill hole profile** button. These placement options have already been discussed.

Shape Tab

The options in the slide-down panel will be displayed when you choose the **Shape** tab from the **Hole** dashboard. These options are different from the options that were available when you created the **Simple** hole using the **Use predefined rectangle as the drill hole profile** button. When you choose the **Shape** tab, the slide-down panel is displayed. The sketch for the hole in the slide-down panel gives an idea about the shape of the hole and its parameters, as shown in Figure 6-10.

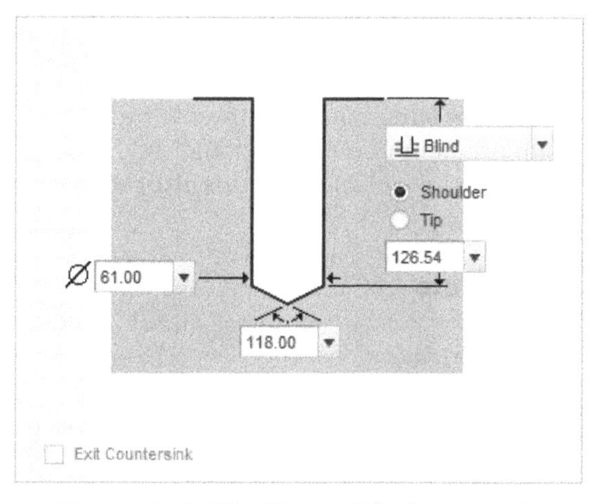

***Figure 6-10** The **Shape** slide-down panel*

Dimension Edit Boxes

The first dimension edit box in the **Hole** dashboard is used to specify the diameter of the hole.

The second dimension edit box in the **Hole** dashboard is used to specify the depth of the hole. The options to specify the depth of the hole are the same as those available in the **Extrude** dashboard.

Adds countersink

Choose this button to create a countersink hole. A countersink hole has two diameters and the transition between the larger diameter and the smaller diameter is in the form of a cone. In the slide-down panel of the **Shape** tab, the preview of the cross-section with parameters of the countersink hole will be displayed, as shown in Figure 6-11. In the preview of the section, the dimensions can be edited as per the requirement.

Adds counterbore

Choose this button to create a counterbore hole. A counterbore hole is a stepped hole, and it has two diameters, a larger diameter and a smaller diameter. The larger diameter is called counter diameter and the smaller diameter is called drill diameter. In the slide-down panel of the **Shape** tab, the preview of the cross-section with parameters of the counterbore hole will be displayed, as shown in Figure 6-12. In the preview of the section, the dimensions can be edited as required. Note that if both the buttons are chosen, then the hole created will be a combination of the counterbore and countersink holes.

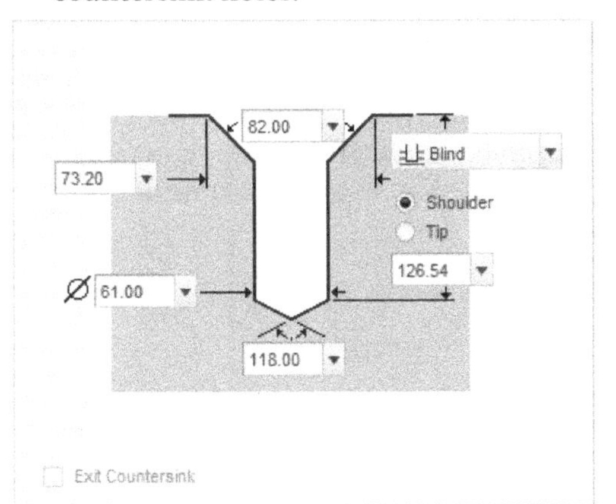

***Figure 6-11** The preview of the cross-section with parameters of countersink hole*

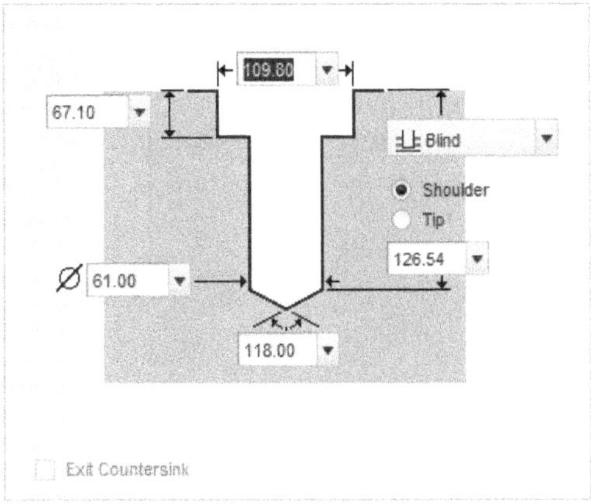

***Figure 6-12** The preview of the cross-section with parameters of the counterbore hole*

Drilled hole shoulder depth

Choose this button to specify the depth of the hole. When you choose the **Drilled hole shoulder depth** button from the **Hole** dashboard, the depth of the hole will be measured upto the shoulder only. In the slide-down panel of the **Shape** tab, the preview of the cross-section with parameters of the hole, whose depth is measured upto the shoulder, will be displayed, as shown in Figure 6-13.

Drilled hole depth

When you choose the **Drilled hole depth** button from the **Hole** dashboard, the depth of the hole is measured upto the tip of the hole. To invoke this button, choose the black arrow on the right of the **Drilled hole shoulder depth** button; a flyout will be displayed. Choose the **Drilled hole depth** button from the flyout. In the slide-down panel of the **Shape** tab, the preview of the cross-section with parameters of the hole, whose depth is measured upto the tip, will be displayed, as shown in Figure 6-14.

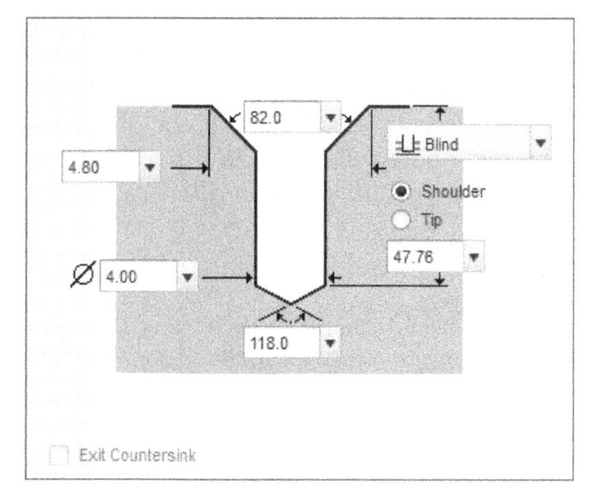

Figure 6-13 Parameters of the hole with depth measured upto the shoulder

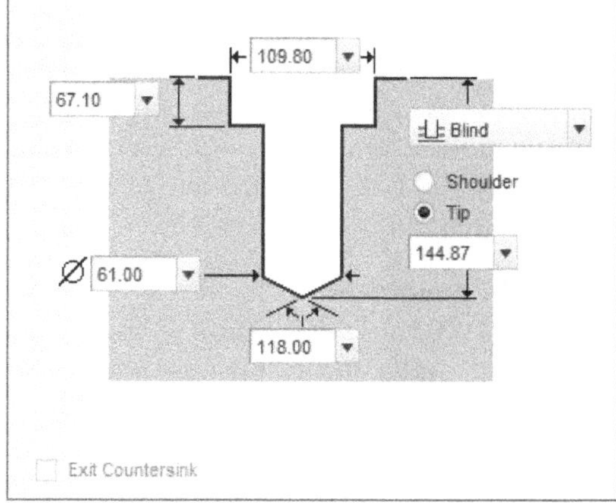

Figure 6-14 Parameters of the hole with depth measured upto the tip

Creating Simple Holes Using Sketched Profile

This option allows you to sketch the cross-section of the hole that is revolved about a center axis. This option is used to draw custom shapes for the hole. When you choose the **Use sketch to define drill hole profile** button from the **Hole** dashboard, the **Hole** dashboard appears, as shown in Figure 6-15. The tools and options in the dashboard are discussed next.

Figure 6-15 Partial view of the ***Hole*** *dashboard*

Activates Sketcher to create section

Choose the **Activates Sketcher to create section** button to open a new window with the sketcher environment. The cross-section for the hole is sketched using the normal sketcher tools available. While drawing the sketch, a center line must be drawn that acts as the axis of revolution for the section of hole. The sketched holes can be blind or through all, depending on the dimensions of its sketch.

When you complete the sketch for the hole and choose the **OK** button to exit the sketcher environment, the **Hole** dashboard will be enabled above the drawing area. Now, you need to specify the placement options to place the hole. The placement options are the same as discussed earlier.

Note

The placement options to place the hole can be specified after drawing the sketch for the hole or before drawing the sketch. The only difference is that when you specify the placement options after drawing the sketch, the preview of the hole is not available in the drawing area. This is because the primary reference is not selected yet. Therefore, it is recommended to specify the placement options before drawing the sketch for the hole.

Opens an existing sketched profile

On the **Hole** dashboard, the **Opens an existing sketched profile** button is used to open an existing sketch that is saved in *.sec* file format. When you choose this button, the **OPEN SECTION** dialog box will be displayed. This dialog box is used to select the *.sec* file to define the shape of the hole.

Placement Tab

The options displayed in the slide-down panel on choosing the **Placement** tab from the **Hole** dashboard are the same as those discussed earlier. You can also use the same type of placement options as discussed earlier.

Shape Tab

The options displayed in the slide-down panel on choosing the **Shape** tab from the **Hole** dashboard are different than the options that were available in case of **Straight** holes. When you choose the **Shape** tab, the slide-down panel displays the preview of the sketch that you have drawn in the sketcher environment. You can pan and zoom the sketch in the slide-down panel.

Note

*If you choose the **Shape** tab before drawing the sketch for the hole, then the slide-down panel becomes blank and does not show a sketch.*

Creating Standard Holes

The standard holes can be created by choosing the **Create standard hole** tool from the **Hole** dashboard. The holes created using this button are based on industry standard fastener tables.

Note

You need to select the required template for the measurement of holes based on the industry standards. This can be done while creating a new file.

When you choose the **Create standard hole** tool from the **Hole** dashboard, the dashboard appears, as shown in Figure 6-16. The tools and options in the dashboard are discussed next.

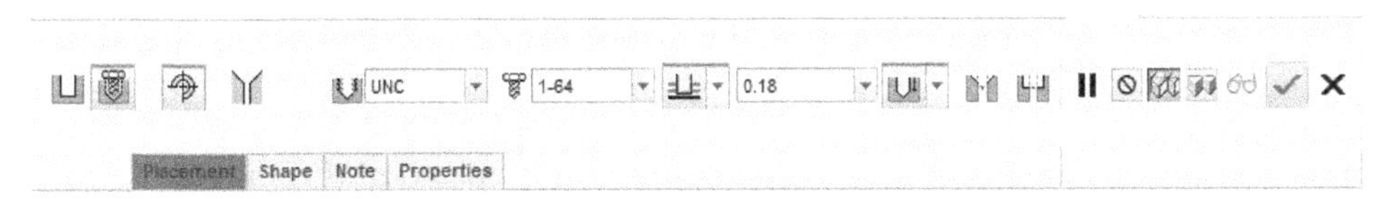

Figure 6-16 *Partial view of the **Hole** dashboard*

Drop-down Lists

The first drop-down list from the left in the dashboard is used to specify the type of thread on the hole. The options in this drop-down list are **ISO**, **UNC**, and **UNF**. The default option selected in this list depend upon the template selected in the **New File Options** dialog box.

The second drop-down list in the **Hole** dashboard is used to specify the size of the screw that corresponds to the hole.

The third drop-down list in the **Hole** dashboard is used to specify the depth of the hole. The options in this drop-down list are similar to those available in the **Extrude** dashboard.

Adds tapping

Choose this button to tap a hole. Tapping is the process of cutting threads in the hole. In PTC Creo Parametric, these threads are known as cosmetic threads. This means that the threads created will not be visible in the hole. In the slide-down panel of the **Shape** tab, the preview of the cross-section with parameters of the tapped hole will be displayed, as shown in Figure 6-17. In the preview of the section, the dimensions can be edited, as required. The dotted lines represent the threading outside the hole. However, on deactivating the **Adds tapping** button, the **Create drilled hole** and the **Create clearance hole** tools will be displayed. Choose the required button and then set the parameters in the dashboard to create the desired hole.

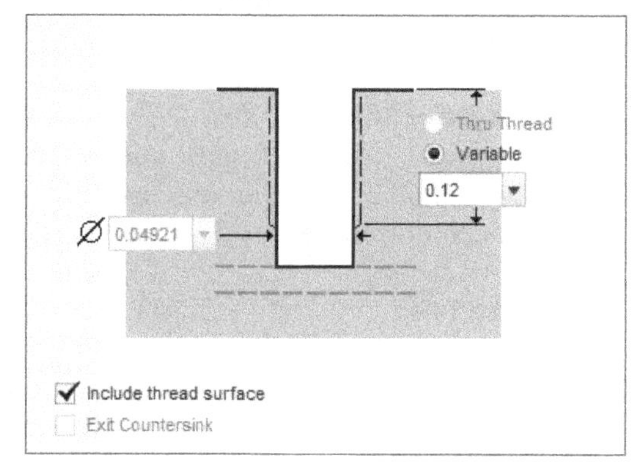

***Figure 6-17** Parameters of the tapped hole*

Create tapered hole

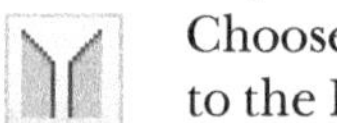

Choose this button to create tapered threaded holes. The sizes of holes can conform to the ISO, JIS, and ANSI standards. There are three standards in the thread type drop-down list, **ISO_7/1**, **NPT**, and **NPTF**. Set the required standard, specify the screw size, and the drill depth value to create the tapered hole. You can also add countersink or counterbore to the tapered hole, if required.

Adds countersink

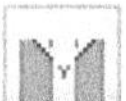

As discussed earlier, this button is used to create a countersink hole. A countersink hole has two diameters and the transition between the bigger diameter and the smaller diameter is in the form of a cone. In the slide-down panel of the **Shape** tab, the preview of the cross-section with threads and parameters of the countersink hole will be displayed, as shown in Figure 6-18. In the preview of the section, the dimensions can be edited as required.

Note that you need to deactivate the **Adds countersink** button and then choose the **Adds counterbore** button to create the counterbore hole. Else, the hole created will be a combination of the countersink and counterbore holes.

Adds counterbore

As discussed earlier, this button is used to create a counterbore hole. A counterbore hole is a stepped hole and has two diameters, a larger one and a smaller one. The larger diameter is called counter diameter and the smaller diameter is called the drill diameter. In the slide-down panel of the **Shape** tab, the preview of the cross-section with threads and parameters of the counterbore hole will be displayed, as shown in Figure 6-19. In the preview of the section, the dimensions can be edited as required.

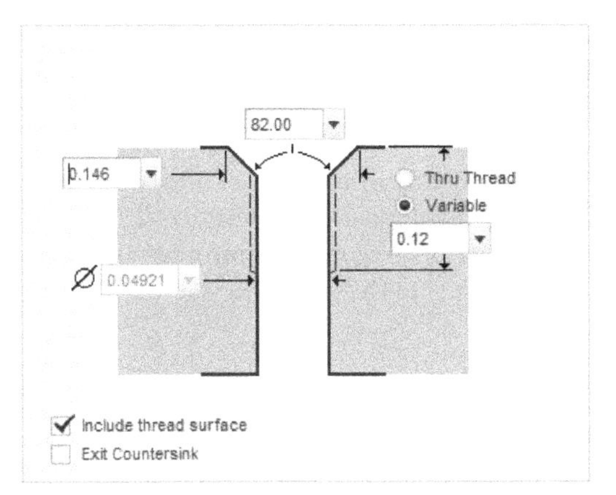

Figure 6-18 The preview of the cross-section with threads and the parameters of the countersink hole

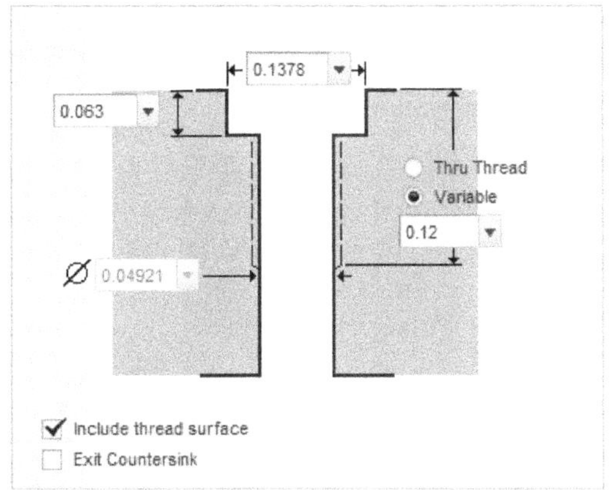

Figure 6-19 The preview of the cross-section with threads and the parameters of the counterbore hole

Placement Tab

The options in the slide-down panel that will be displayed when you choose the **Placement** option from the **Hole** dashboard are the same as those that were available in case of the **Simple** hole. You can use the same type of placement options that were available when you created the **Simple** hole.

Shape Tab

The options in the slide-down panel that will be displayed when you choose the **Shape** tab from the **Hole** dashboard are different from the ones that were available in case of **Simple** hole. When you choose the **Shape** tab, the slide-down panel displays the preview of the hole. You can specify the dimensions for the hole in the respective dimension boxes.

Note Tab

This tab displays the specifications of the hole in the text format. This option is available only when you create a **Standard** hole.

Properties Option

This option displays the properties of the hole that you need to create.

Previewing the Hole

The **Feature Preview** button is used to preview the hole created before confirming its creation. Changes and modifications in the hole parameters can be made once the hole is previewed.

Note

*The holes created using the **Hole** dashboard are parametric in nature and can be modified anytime using the **Model Tree**. The method of modifying the holes using the **Model Tree** is discussed later in this chapter. You can also dynamically edit the parameters of the hole as well as relocate the hole after the hole is created. To do so, select the hole and right-click to display a shortcut menu. Choose **Dynamic Edit** from it; all values of parameters and the dynamic edit handles will be displayed on it. You can double-click on the value and enter a new value or drag the edit handles to the required distance.*

Important Points to Remember While Creating a Hole

The points given next should be remembered while creating a hole.

1. While drawing the sketch of a hole, it should have an axis of revolution and at least one entity normal to it. Now, when you place this hole on the primary reference (**Placement** reference), then this normal entity is aligned with the plane (primary reference). If there are two entities normal to the axis, then the normal entity at the top of the sketch is aligned with the placement plane.

2. While creating a hole, the primary reference for the placement of hole can be selected without choosing any option from the **Hole** dashboard. However, if you need to specify the secondary references (**Offset references**), then choose the **Placement** tab from the **Hole** dashboard; the slide-down panel will be displayed. From the slide-down panel, click in the **Offset References** collector and when the **Placement** collector appears white in color, you can select the secondary references.

   Remember when the first reference is selected by using the left mouse button, then the second reference is selected by using CTRL+left mouse button.

3. It is recommended that if you are creating a **Standard** hole then the units of the model should be in mm. This can be achieved by using the appropriate template at the time of creating a new file or by setting the system of units.

4. An application of the **Sketched** hole is a tapered hole. **Sketched** holes are always blind and have depth in one direction only.

5. You cannot select a convex or concave surface as the placement plane. To create a hole on such a surface, you need to create a datum plane passing through the surface and then select this datum plane as the placement plane.

CREATING ROUNDS

The **Round** tool is used to create a fillet or a smooth rounded transition between two adjacent faces, with a circular or a conic profile. This feature creation tool can be invoked from the **Engineering** group in the **Model** tab. Using the **Round** tool, you can add or remove material, depending on the edge selected. You can select the edges or faces and create a round in a single set or define more than one set.

Figures 6-20 and 6-21 show some examples of rounds. From the figures, it is evident that the geometry of the rounds is tangent to the references selected. These references can be edges or surfaces. In PTC Creo Parametric, rounds are created using the **Round** dashboard.

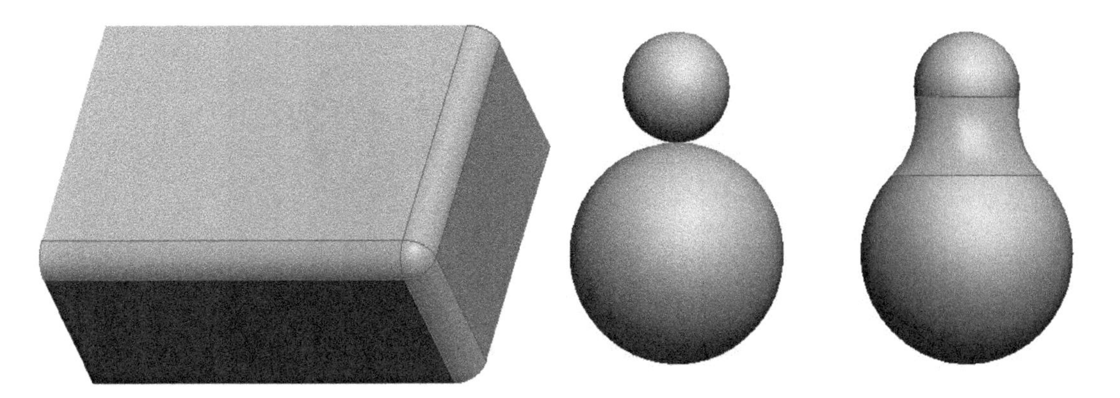

Figure 6-20 Rounds created on the edges *Figure 6-21 Round created on surfaces*

Note

The round created after you select references on the model has a circular cross-section and rolling shape. In PTC Creo Parametric, there is more than one shape that a round can have.

Creating Basic Rounds

Ribbon: Model > Engineering > Round

Choose the **Round** tool from the **Engineering** group; the **Round** dashboard will be displayed, as shown in Figure 6-22. Also, you will be prompted to select an edge or a surface set to create the round. Select an edge; the preview of the round with a default radius value and drag handles will be displayed on the selected edge. Similarly, you can go on selecting edges to add rounds to them. But if you select a surface, you are prompted to select another surface to include in the same set. Use CTRL+left mouse button to select another surface; the preview of the round with a default radius value and drag handles will appear on the model, as shown in Figure 6-23. Note that because the radius of all rounds in a set is equal, the drag handles are available on only one edge. You can drag the handles to dynamically specify the radius of the round or specify the radius in the **Round** dashboard. Once all parameters of the round are specified, choose the **Build feature** button to create it. The various options in the **Round** dashboard are discussed next.

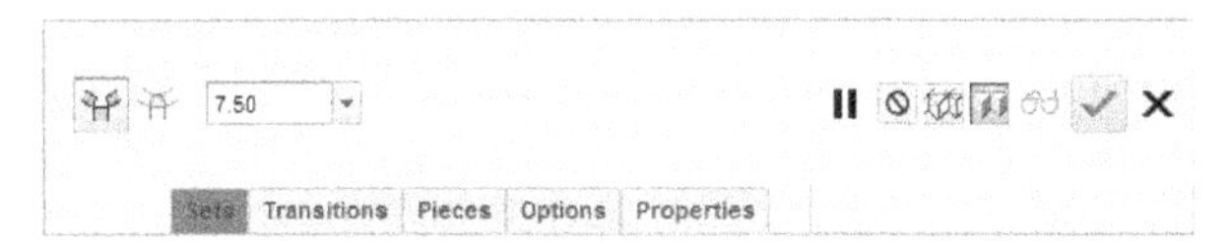

*Figure 6-22 Partial view of the **Round** dashboard*

Switch to set mode

When you invoke the **Round** dashboard, the **Switch to set mode** button is chosen by default and is used to create rounds by defining sets. When this button is chosen, the dimension box is displayed on its right and it is used to specify the radius of round.

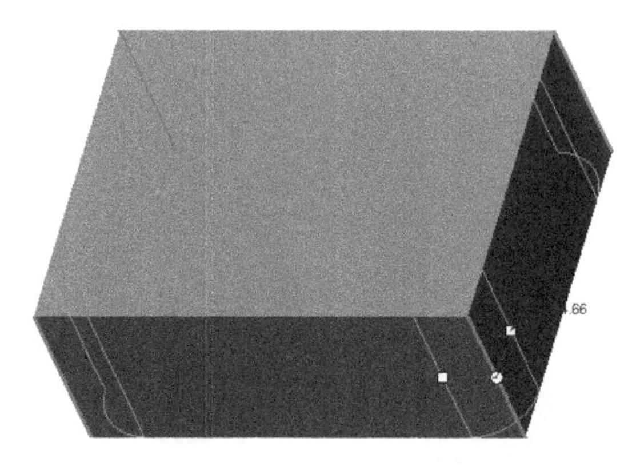

Figure 6-23 Preview of the round feature along with the radius of round and the drag handles

Tip: *You can create a round on an edge without invoking the **Round** tool. To do so, choose the **Geometry** option from the **Filter** drop-down list and then select an edge. Press and hold the right mouse button and choose the **Round** option from the shortcut menu. The **Round** dashboard is invoked and the preview of the round appears on the edge selected.*

Switch to transition mode

The **Switch to transition mode** button will be available only if at least one reference for the round is defined. This button has primarily two applications. The first one is to define the shape of the round that is formed at a vertex and the other is to define a limit up to which the round will be created. Both these applications are discussed next.

Tip: *The need for creating rounds by defining more than one set arises when you want to create rounds that have different radii. The rounds created by using this method will be a single feature and appear as a single feature in the **Model Tree**.*

If the round feature is being created on three intersecting edges, then this button is used to define the shape of the round that is created at the vertex. After choosing the **Round** button, press and hold the CTRL key down and select the three edges, as shown in Figure 6-24. Now, choose the **Switch to transition mode** button from the **Round** dashboard. The various transitions that are possible on the current selection are highlighted on the model, as shown in Figure 6-25. Select the transition that will be displayed at the vertex; a drop-down list will be displayed in the **Round** dashboard. The options available in this dashboard are **Default** (**Round Only 2**), **Intersect**, **Corner Sphere**, **Round Only 1**, and **Patch**. Select the desired option to apply the transition.

The **Intersect** option merges the round in such a way that sharp edges are formed at the vertex, as shown in Figure 6-26.

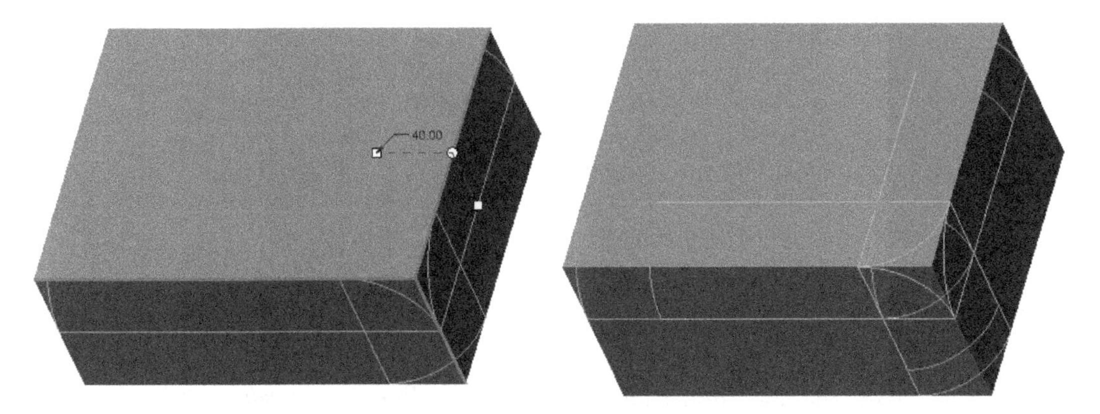

Figure 6-24 *Edges selected for creating rounds* ***Figure 6-25*** *Various possible transitions*

The **Corner Sphere** option results in a smooth spherical merging of rounds at the vertex. If you increase the default radius value of the round feature by dragging the handle while creating the round using the **Corner Sphere** option, handles are displayed on all selected edges. You can drag these handles to dynamically define the length of the fillet along the respective edges. The length values can also be defined in edit boxes that are displayed in the **Round** dashboard. Figure 6-27 shows the round created using the **Corner Sphere** option.

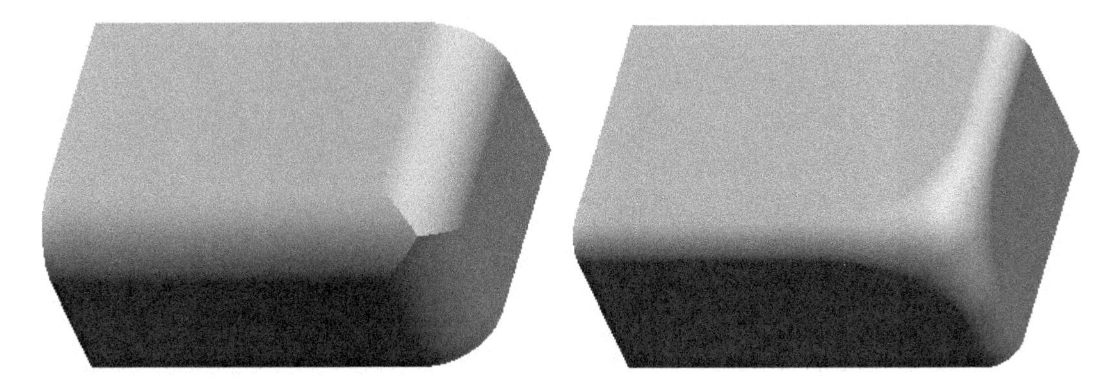

Figure 6-26 *Round created using the **Intersect** option* ***Figure 6-27*** *Round created using the **Corner Sphere** option*

The **Switch to transition mode** button also allows you to select a reference to stop the round on an edge. This means the round will be created on the selected edge up to the selected reference only. After selecting the edge for creating the round, choose the **Switch to transition mode** button to display the two transitions at both the ends. Select any one of them to make it active. Now, select the **Stop at Reference** option from the drop-down list available in the **Round** dashboard. Select a plane that will act as the stopping reference. Note that the edge on the side in which the transition is active will be retained and the round will be created on the other side. Figure 6-28 shows an example of creating a round up to a stopping reference. In this case, four edges have been filleted and the reference plane, which is highlighted by a mesh in the figure, is selected as the stopping reference.

Note

You can select only one transition at a time to make it active. The stopping references selected will be associated with the active transition only. Therefore, if you want to define the stopping reference for more than one edge, then it has to be done individually for each active transition.

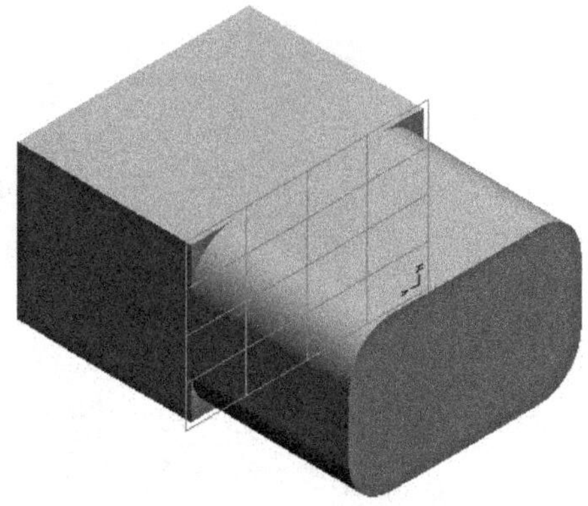

***Figure 6-28** Transition rounds created on four edges and stopped by the highlighted reference plane*

Sets Tab

When you choose the **Switch to set mode** button and then choose the **Sets** tab in the **Round** dashboard, the slide-down panel will be displayed, as shown in Figure 6-29. All references and the geometry of the round are specified in this slide-down panel. The options in this slide-down panel are discussed next.

First Drop-down List

The first drop-down, refer to Figure 6-30, that appears in the slide-down panel is used to specify the geometry of the round you need to create. The options in this drop-down list are discussed next.

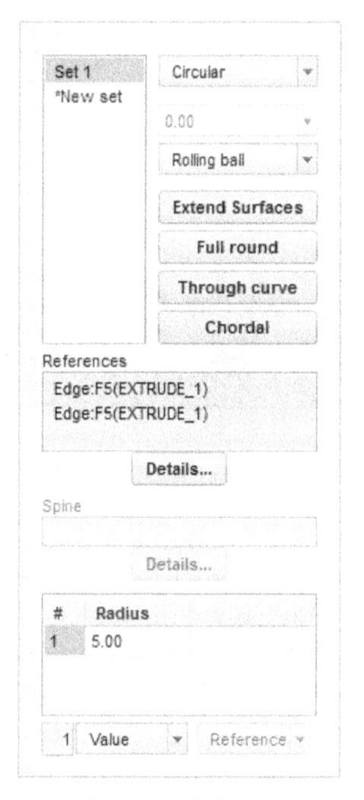

***Figure 6-29** The **Sets** slide-down panel*

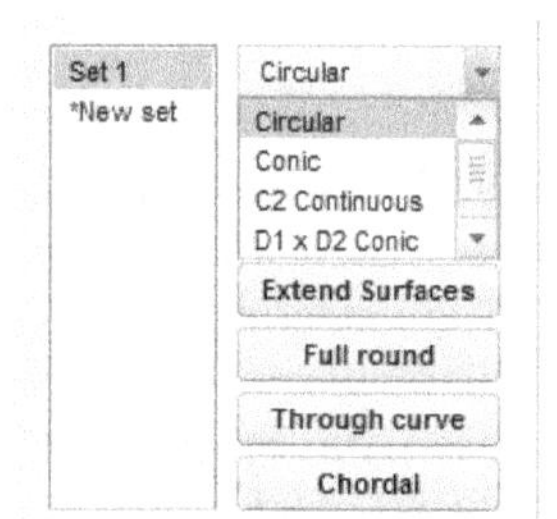

***Figure 6-30** First drop-down list*

Circular: When you choose this option from the drop-down list, the round that is created has a circular cross-section. This is the most widely used geometry type of rounds. The geometry is controlled by a radius value.

Conic: This type of round has geometry of a conic. When you choose this option, the **Conic Parameter** drop-down list appears on the **Round** dashboard. The geometry of a conic round is controlled by two dimensions: the leg dimension and the conic parameter. The value of the conic parameter lies between 0.05 and 0.95.

In Figure 6-31, the round on the model that is on the left is created using a circular geometry and the round on the model that is on the right is created using a conic geometry.

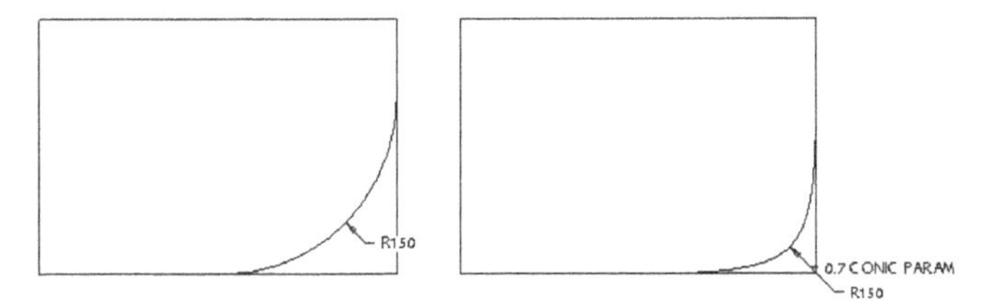

Figure 6-31 *Preview of the round feature along with the radius of round*

C2 Continuous: This type of round has a spline geometry. This type of round maintains the curvature continuity with adjacent surfaces. You can set the sharpness of the spline by specifying a value in the **C2 Shape Factor** edit box in the dashboard. The sharpness value varies between 0.05 - 0.95. You can also set the value of the conic distance in the second edit box in the dashboard. Alternatively, drag the handles on the model preview to adjust the sharpness and conic distance.

D1 x D2 Conic: The round created using this option has a geometry that can be controlled by three dimensions. Figure 6-32 shows the round that is created using the **D1 x D2 Conic** option. This figure shows the two end legs dimensions and the middle dimension is the conic parameter. When you select this option to create a round, the three edit boxes are displayed on the **Round** dashboard. The **Reverse direction of conic distances** button is used to reverse the conic legs dimensions. Figure 6-33 shows the geometry of the conic round and the dimensions of the two legs after choosing this button.

D1xD2 C2: This type of round has spline geometry and it maintains the curvature continuity with adjacent surfaces. In this case also, you can control the geometry by three dimensions as you did in **D1 x D2 Conic.**

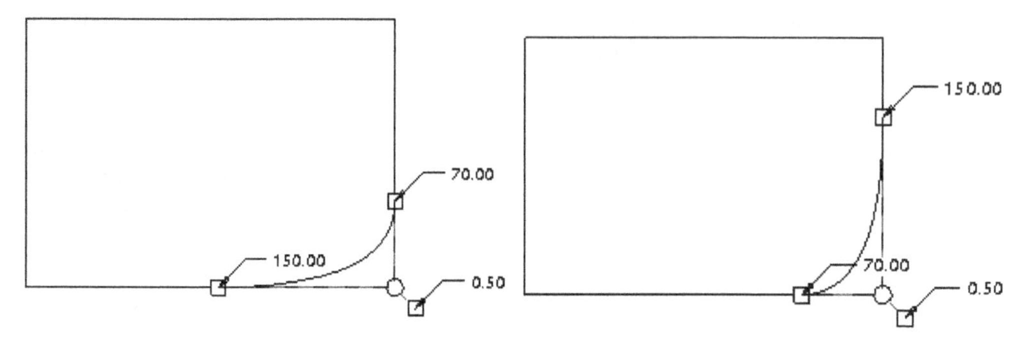

*Figure 6-32 The **D1 x D2 Conic** round*

*Figure 6-33 The **D1 x D2 Conic** round after reversing*

Note

The rounds and chamfers are defined in sets when there is a variation in their dimensions. For example, consider a situation, when you may want to create rounds on two sets of edges and the dimensions for both the rounds are different. In this case, you can define two sets of rounds; one set will define the rounds on one edge with some radius value and the second set will define the rounds on the other edge with a different radius value. The sets are also used in a similar way while creating chamfers.

Dimension Box

The edit box that is present below the first drop-down list is available only when you select the **Conic** or **D1 x D2 Conic** options. This is because this dimension box is used to enter the value for the sharpness of the conic round. The value lies between 0.05 and 0.95.

Second Drop-down List

The second drop-down list that is present in the slide-down panel is used to specify the method that PTC Creo Parametric will use to create the round. The options in this drop-down list are **Rolling ball** and **Normal to spine**. These options are discussed next.

Rolling ball

When you select this option, the round is created by rolling a spherical ball along the surfaces to which it will possibly stay tangent. This is the most common type of round that you will create.

Normal to spine

This option creates a round by sweeping an arc or a conic cross-section normal to the selected spine. Remember that in this case, the cross-section remains normal to the spine.

The area to the left of the drop-down lists in the slide-down panel, lists the number of sets formed for creating rounds. This means you can define more than one set to create the rounds. Each set may have different values and references. To add a set, right-click in this area and choose the **Add** option from the shortcut menu or, you can just click on the **New set**. The set selected in this area is the active set. It is recommended that you fully define one set of round and then define the other set.

Full Round

The **Full Round** button is used to create a complete round between two selected edges or two planar or non-planar surfaces. This button will be available only after selecting two references and will not be available when you choose the **Conic**, **C2 Continuous**, **D1 x D2 Conic**, and **D1 x D2 C2** options or the **Normal to spine** option. The following steps explain the procedure to create a full round by selecting two edges.

1. Invoke the **Round** dashboard.

2. Select the first edge and then use CTRL+left mouse button to select the second edge. Note that the edges selected should have a surface between them that can be converted into a round.

3. Make sure the **Circular** and **Rolling ball** options are selected from the first and third drop-down lists, respectively.

4. Choose the **Full Round** button from the slide-down panel that will be displayed when you choose the **Sets** tab from the **Round** dashboard; the round is created.

   The following steps explain the procedure to create a full round by selecting the two surfaces.

1. Invoke the **Round** dashboard.

2. Select the first surface and then use CTRL+left mouse button to select the second surface. Note that the surfaces selected should have a surface between them that can be converted into a round, for example, the top and bottom faces of a rectangular block.

3. If the **Full round** button is not chosen automatically then choose it from the slide-down panel that will be displayed when you choose the **Sets** tab from the **Round** dashboard. On doing so, you will be prompted to select a surface to replace it with a full round.

Notice that in the slide-down panel, the **Driving surface** display box turns orange in color. This indicates that now you can select the driving surface (surface to replace with a full round). Figure 6-34 shows the surfaces selected for creating full round using the **Full round** option and the preview of the round to be created.

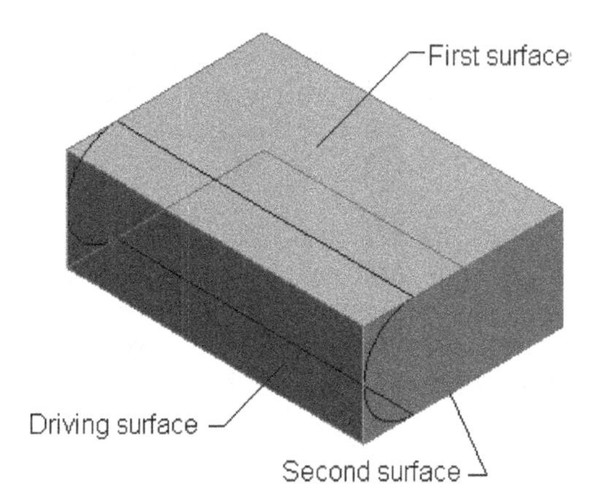

Figure 6-34 Surfaces selected for creating a full round

Through Curve
The **Through Curve** button is used to create a round whose radius is specified by an existing curve.

Chordal
The **Chordal** button is used to create a round with constant chordal length.

References Collector
The **References** collector displays the references selected on the model. These references can be deleted by selecting them in the collector and then holding down the right mouse button to display the shortcut menu. Choose the **Remove** or **Remove All** option from this shortcut menu as per the requirement.

Spine Collector
The **Spine** collector displays the spine selected on the model. This collector will get active only when you select the **Normal to spine** option from the second drop-down list. The **Spine** collector gets modified to the **Driving surface** collector when you select two surfaces and then choose the **Full Round** button. The **Spine** collector is modified to the **Driving Curve** collector when you choose the **Through curve** button.

The area under the **Spine** collector is used to define the dimension values of the rounds.

Transitions Tab

When you choose the **Transitions** tab, a slide-down panel will be displayed, as shown in Figure 6-35. This slide-down panel displays the transitions if they are present in a round. For example, in Figure 6-28, there are four transitions (each at the four edges).

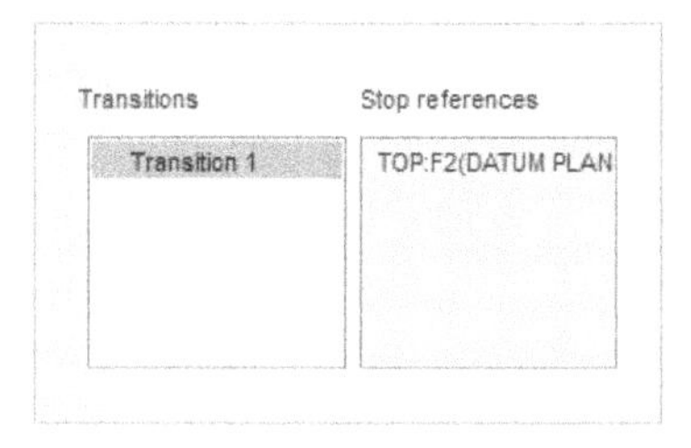

*Figure 6-35 The **Transitions** slide-down panel*

Pieces Tab

When you choose the **Pieces** tab, a slide-down panel will be displayed. It provides the possible locations of the round in the form of pieces. For example, consider a case where you select two faces to fillet. One of the faces has a cut feature such that it has two disjointed surfaces, as shown in Figure 6-36. When you move the cursor over any piece under the **Pieces** area, that piece is highlighted on the model. This tab is useful in resolving ambiguity which arises if multiple placement locations for rounds with the current selection of the references exist. This happens if two surfaces intersect each other at more than one place. From the pieces listed in the **Pieces** area, you can select the pieces to exclude by selecting the **Excluded** option from the drop-down list that will be displayed when you click on the **Included** field, as shown in Figure 6-37.

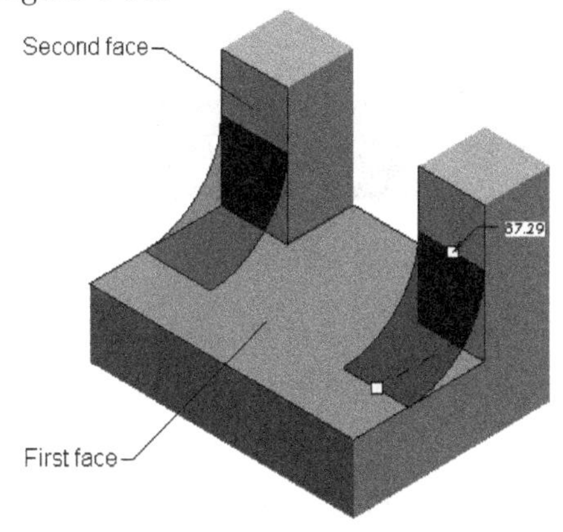

Figure 6-36 Faces selected for creating a round

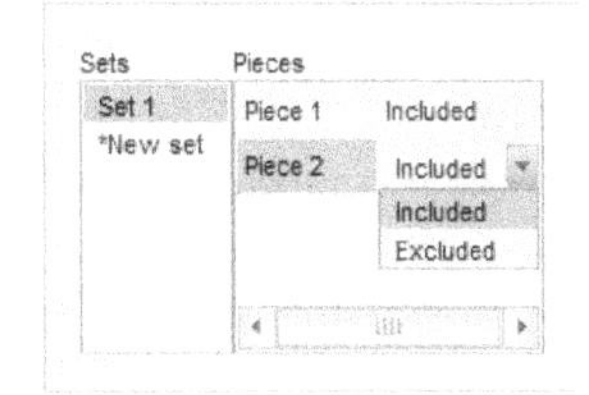

*Figure 6-37 The **Pieces** slide-down panel*

Options Tab

When you choose the **Options** tab, a slide-down panel will be displayed that has two radio buttons. These radio buttons allow you to make the round a surface or a solid. When you select the **Surface** radio button, the **Create end surfaces** check box will be enabled, refer to Figure 6-38. This check box is used to cap all round ends of the model by surfaces.

*Figure 6-38 The **Options** tab*

Properties Tab

When you choose this option, the slide-down panel will be displayed. This slide-down panel displays the feature identity. Click on the **i** button to read the information about the feature. Click on the cross on the top-right of the **Feature info** window to close it.

Creating a Variable Radius Round

A variable radius round is the one in which the radius of the round varies through the length of the edge. For example, refer to Figure 6-39, which shows a constant radius round and Figure 6-40, which shows a variable radius round. To create a variable radius round, choose the **Round** tool and select the edge. When the preview of the round appears on the edge, hold down the right mouse button anywhere in the drawing window to display a shortcut menu. Choose the **Make variable** option from this shortcut menu. Now, in the slide-down panel that will be displayed when you choose the **Sets** tab, the two default values of the radius are available, which can be modified as per requirement. You can drag the handles displayed at both ends of the selected edge to dynamically modify the radius of round.

Figure 6-39 *Constant radius round*

Figure 6-40 *Variable radius round*

Note

You can also create a variable radius round by selecting two surfaces. In this case, the edge that is common to the two surfaces is converted into a round.

You can also add additional points between the two end points and define a variable radius in these additional points. To define additional points, choose the **Sets** tab to invoke the slide-down panel. Bring the cursor to the area in the slide-down panel where the radius value and the location of the points are displayed and right-click to invoke the shortcut menu. Choose the **Add radius** option; one additional point appears in the same area of the slide-down panel and also on the selected edge. This point by default is placed on the selected edge by a ratio value with respect to the end point. You can select the **Reference** option from the drop-down list in this area and select a location on the edge to place the point. The location of the point can also be modified by dragging the circular handle displayed on the edge.

Note

Rounds can be created on one edge, chain of edges, between two surfaces, an edge and a surface, and between two edges.

Points to Remember While Creating Rounds

The following points should be remembered while creating rounds:

1. To avoid conflict due to the parent-child relationship, other features should not be referenced to the edges created by rounds or to the round surface.

2. Preferably, rounds should be added at the end of model creation. This means, the rounds should be the last features in any solid model.

3. The regeneration time of the model increases if the number of round features in it is increased.

4. Use the shortcut menus extensively while creating rounds to reduce the modeling time.

Creating Auto Rounds

Ribbon: Model > Engineering > Round > Auto Round

You can create rounds on all the edges of the model by using the **Auto Round** tool in a single step. To do so, choose the **Auto Round** tool from the **Round** drop-down of the **Engineering** group; the **Auto Round** dashboard will be displayed, as shown in Figure 6-41. There are two check boxes available in the dashboard. By default, these check boxes are selected. The **Selects convex edges to be rounded by Auto-Round** check box is used to create round on the outer edges of the model. The **Selects concave edges to be rounded by Auto-Round** check box is used to create round on the inner edges of the model, if the edges are compatible with the radius value. The tabs available in this dashboard are discussed next.

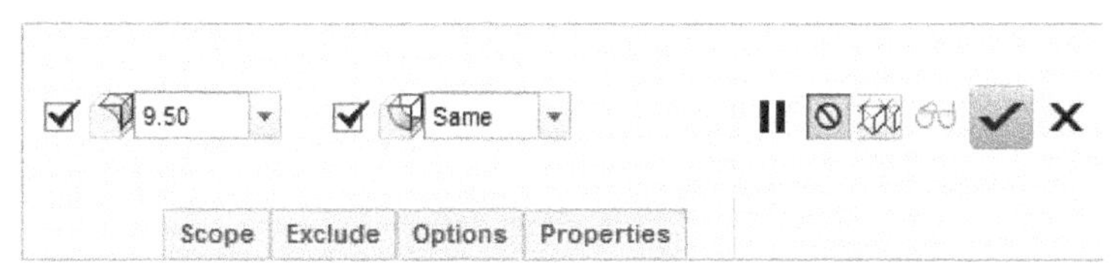

*Figure 6-41 Partial view of the **Auto Round** dashboard*

Scope Tab

When you choose the **Scope** tab, a slide-down panel will be displayed, as shown in Figure 6-42. This slide-down panel displays three radio buttons and two check boxes. These options are discussed next.

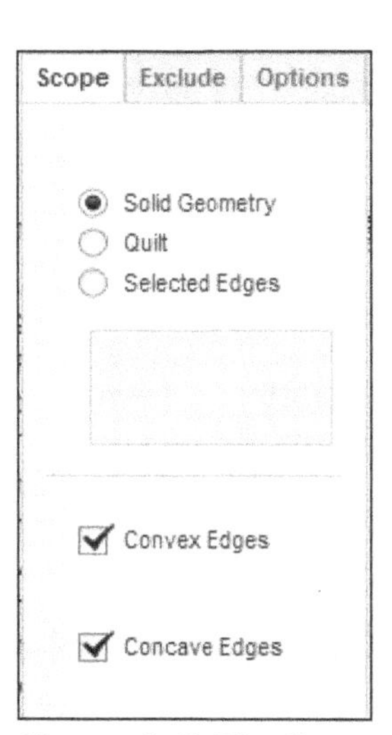

*Figure 6-42 The **Scope** slide-down panel*

Solid Geometry

This radio button is used to create rounds in solid models. The solid models have property to contain weight.

Quilt

This radio button is used to create rounds in surface models. The surface models do not have property to contain weight.

Selected Edges

This radio button is used to create rounds on the selected edges of the model.

Convex Edges

This check box is used to include or exclude the outer edges of the model. Select this check box to include the outside edges.

Concave Edges

This check box is used to include or exclude the inner edges of the model. Select this check box to include the inside edges.

Exclude Tab

When you choose the **Exclude** tab, a slide-down panel will be displayed. It is used to exclude those edges that you do not want to make round. By holding the **CTRL** key, select the edges you want to exclude; the selected edges will be displayed in the **Excluded Edges** area of the slide-down panel.

Options Tab

When you choose the **Options** tab, a slide-down panel will be displayed, as shown in Figure 6-43. It is used to exclude the edges from the model that you do not want to make round. In the **Options** slide-down, the **Create Group of Regular Round Features** check box is available. When you select this check box, the rounds will be created in a group. Note that you can not edit these rounds further using the **Auto Round** tool. You have to edit them individually.

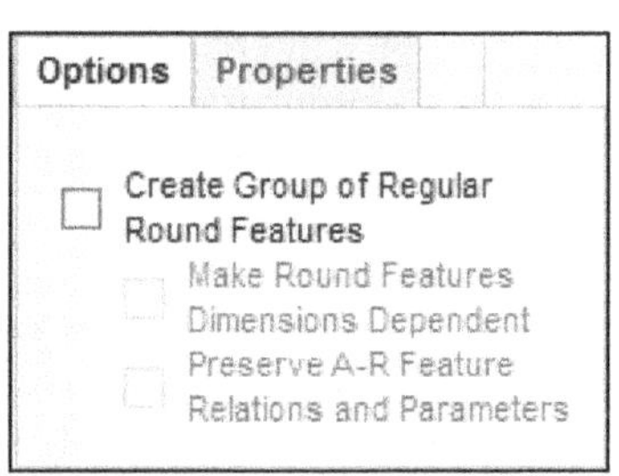

Figure 6-43 The ***Options*** *slide-down panel*

Tip: *Rounds and chamfers are used in components to reduce the stress concentration at the sharp corners and edges. Therefore, they reduce the chances of failure of a component under a specified loading condition.*

CREATING CHAMFERS

The **Chamfer** tool is used to bevel the selected edges and corners as per some specified parameters. In PTC Creo Parametric you can create two types of chamfers. The first is the **Corner** chamfer and the second is the **Edge** chamfer. Figure 6-44 shows two types of chamfers. To create a chamfer on a corner, choose the **Corner Chamfer** tool from the **Chamfer** drop-down in the **Engineering** group. To create a chamfer on an edge, choose **Chamfer > Edge Chamfer** from the **Engineering** group.

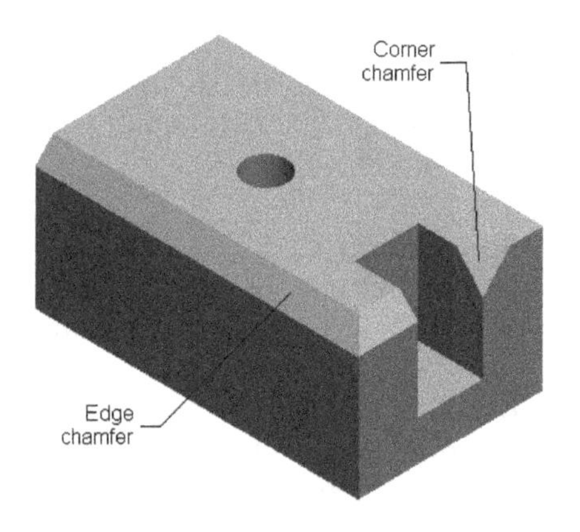

Figure 6-44 *Two types of chamfers*

Corner Chamfer

Ribbon: Model > Engineering > Chamfer > Corner Chamfer

A corner chamfer creates a beveled surface at the intersection of three edges. When you choose the **Corner Chamfer** tool from the **Chamfer** drop-down list of the **Engineering** group, the **Corner Chamfer** dashboard will be displayed, refer to Figure 6-45. Also, you will be prompted to select a corner to chamfer. Select one of the corner points, the preview of chamfer will be displayed. Now, you need to specify the required values of D1, D2, and D3 in their respective field in the dashboard. After specifying the values, choose the **Build Feature** button from the dashboard to create the chamfer.

*Figure 6-45 Partial view of the **Corner Chamfer** dashboard*

Note

*When you are creating the corner chamfer, notice that only the **Vertex** filter is available in the **Filter** drop-down list in the **Status Bar**. As mentioned earlier, the filters in PTC Creo Parametric narrow the entities available for selection. Therefore, you can easily select the reference entities on the model.*

Edge Chamfer

Ribbon: Model > Engineering > Chamfer > Edge Chamfer

An edge chamfer creates a beveled surface along the selected edge. When you choose the **Edge Chamfer** tool from the **Chamfer** drop-down in the **Engineering** group, the **Edge Chamfer** dashboard will be displayed, as shown in Figure 6-46 and you will be prompted to select any number of edges to create a chamfer. The options in this dashboard are discussed next.

Switch to set mode

When you invoke the **Chamfer** dashboard, the **Switch to set mode** button is chosen by default. This button is used to create chamfers by defining either one set or more than one sets. When this button is chosen, the drop-down list to the right of this button in the **Chamfer** dashboard is also displayed. This drop-down list is used to specify the type of chamfer. The options in the drop-down list are discussed next.

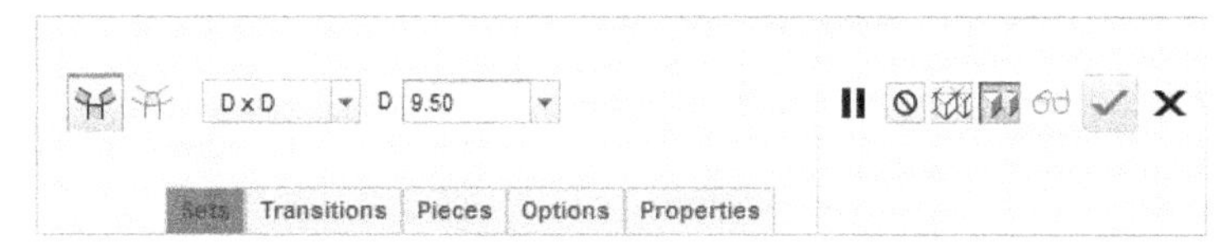

*Figure 6-46 Partial view of the **Edge Chamfer** dashboard*

D x D

The **D x D** option creates a chamfer such that the chamfer is created at an equal distance from the selected edge on both the faces connected by the edge. In Figure 6-47, the dimension value 10 is the value of D, where the distance D is measured from the corner.

D1 x D2

The **D1 x D2** option creates a chamfer at two user-defined distances from the selected edge. In Figure 6-48, the dimension value 10 is the value of D1 and the dimension value 16 is the value of D2. The values for D1 and D2 can be entered from the **Chamfer** dashboard. You can choose the **Interchange distance dimensions of the chamfer** button available on the right of the D2 edit box to interchange the two dimensions.

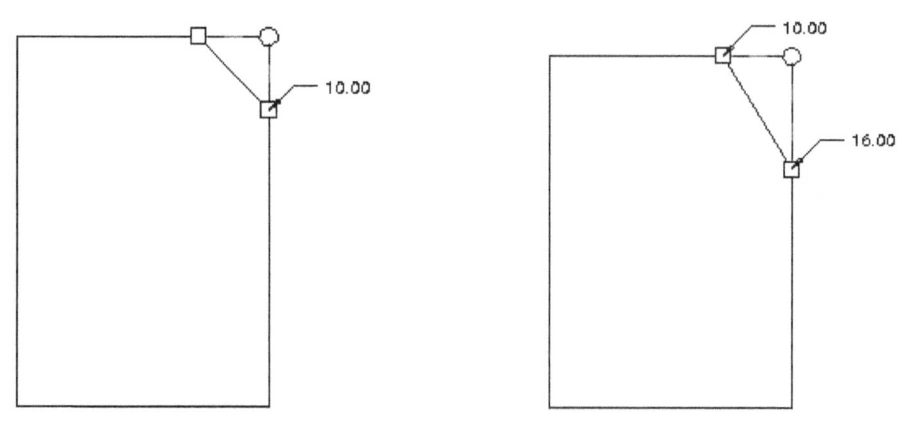

*Figure 6-47 Creating chamfers using the **D x D** option*

*Figure 6-48 Creating chamfers using the **D1 x D2** option*

Angle x D

The **Angle x D** option creates a chamfer at a user-defined distance from the selected edge and also at a user-defined angle measured from a specified surface. In Figure 6-49, the angle is 60 degrees and the value of D is 15.

45 x D

The **45 x D** option creates a chamfer at the intersection of two perpendicular surfaces. The chamfer created is at an angle of 45-degree from both the surfaces and at a distance D from the edge along each surface. Once the chamfer is created, the distance D can be modified. Figure 6-50 shows this type of chamfer where the value of D is 25.

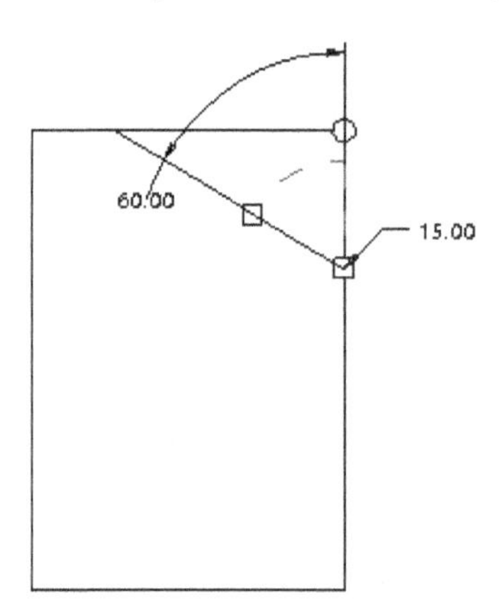

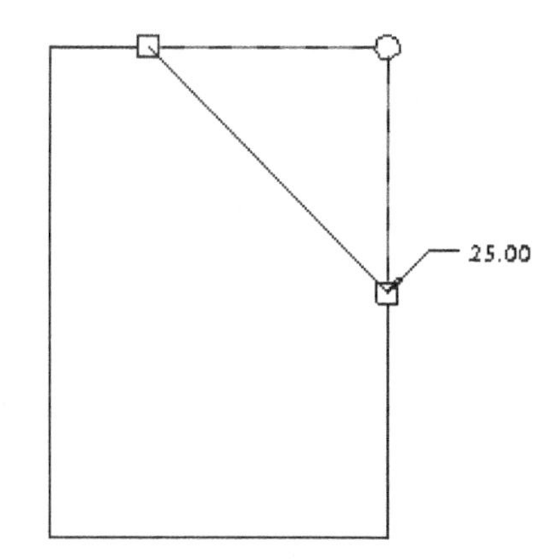

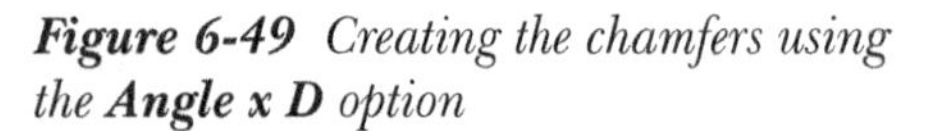

*Figure 6-49 Creating the chamfers using the **Angle x D** option*

*Figure 6-50 Creating the chamfers using the **45 x D** option*

O x O

The **O x O** is used when **Offset Surfaces** is selected from the slide-down panel that is displayed on choosing the **Sets** tab of the **Chamfer** dashboard. In this case, the chamfer is created by offsetting the neighboring surfaces and you need to define the offset distance from the edge.

01 x 02
The chamfer is created by offsetting the neighboring surfaces independently.

Switch to transition mode

The **Switch to transition mode** button in the **Chamfer** dashboard is used in the same way as it is used in the **Round** dashboard. This button will be available only when you have defined at least one reference for the chamfer. Using this button, you can define the shape of the chamfer that is formed at a vertex and also define a limit up to which the chamfer will be created. Both these applications are discussed next.

Note
*After entering into the transition mode, you cannot change the value of **D**. Therefore, it is recommended to set the value of **D** prior to specifying the transition. In case it is required to change the value of **D** after setting the transition, choose the **Switch to set mode** button to return to the set mode and change the value of **D**.*

If the chamfer feature is being created on three intersecting edges, then this button is used to define the shape of the chamfer that is created at the vertex. After choosing the **Edge Chamfer** button, select the three edges. Now choose the **Switch to transition mode** button from the **Edge Chamfer** dashboard; various transitions that are possible on the current selection are highlighted on the model, as shown in Figure 6-51. Select the transition that will be displayed at the vertex. Now from the drop-down list in the **Edge Chamfer** dashboard, select the desired option, which can be **Default (Intersect)**, **Patch**, or **Corner Plane**.

The **Default (Intersect)** option merges the chamfer in such a way that sharp edges are formed at the vertex, as shown in Figure 6-52.

Figure 6-51 *Various possible transitions*

Figure 6-52 *Chamfer created using the **Default (Intersect)** option*

The **Patch** option results in a smooth transition of chamfer by creating a fillet at the vertex with respect to a selected surface. After selecting this option from the drop-down list, the **Optional surface** collector will be displayed to its right. Click once in this collector to bring it into the selection mode. Now, select one of the surfaces that forms the selected edges. The chamfer created by selecting the top surface of the model as the **Optional surface** is shown in Figure 6-53. Note that the fillet radius will be displayed on the selected surface and can be dynamically modified by dragging the handle displayed on the preview or by entering a value in the **Radius** edit box displayed in the **Chamfer** dashboard.

The **Corner Plane** option can be used only when three chamfers converge at a vertex. By using this option, a plane is created at the point of merger, as shown in Figure 6-54. This plane is in the shape of a triangle and its size depends on the specified value of **D**.

Figure 6-53 *Chamfer created using the* ***Patch*** *option*

Figure 6-54 *Chamfer created using the* ***Corner Plane*** *option*

The **Switch to transition mode** button also allows you to select a reference to stop the chamfer on an edge. As discussed earlier, when you choose this button, the transitions that are possible on the current selection are highlighted on the model. The highlighted piece that you select is not chamfered up to the selected reference. The reference to stop the transition can be selected after you select the **Stop at Reference** option from the drop-down list in the **Edge Chamfer** dashboard. This means the chamfer will be created on the selected edge starting from the opposite end and up to the selected reference only, as shown in Figure 6-55. The reference plane is highlighted by a mesh in the figure.

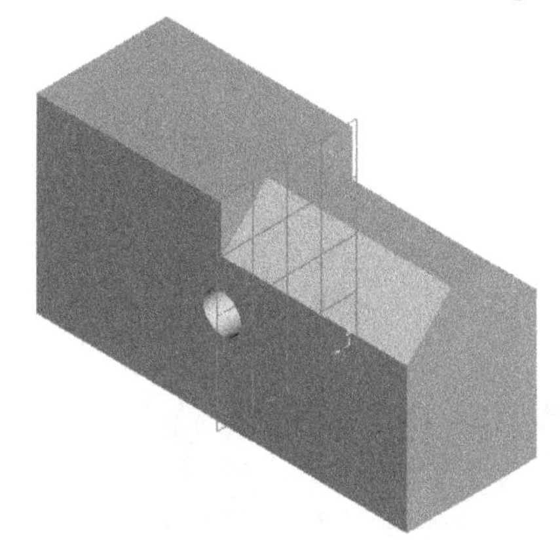

Figure 6-55 *Transition chamfer*

Sets Tab

When you choose the **Sets** tab in the **Chamfer** dashboard, a slide-down panel will be displayed, as shown in Figure 6-56. All references and dimensions of the chamfer can be specified in this slide-down panel. The options in this slide-down panel are discussed next.

Set 1 Display Area

This display area displays the number of sets created. To add a set, right-click to invoke a shortcut menu and choose the **Add** option from it or just click on **New set**. It is recommended that you define an additional set only when you have defined the first set completely. You can switch among the sets by selecting them from this area. The set that is highlighted in this area can be previewed in the drawing area.

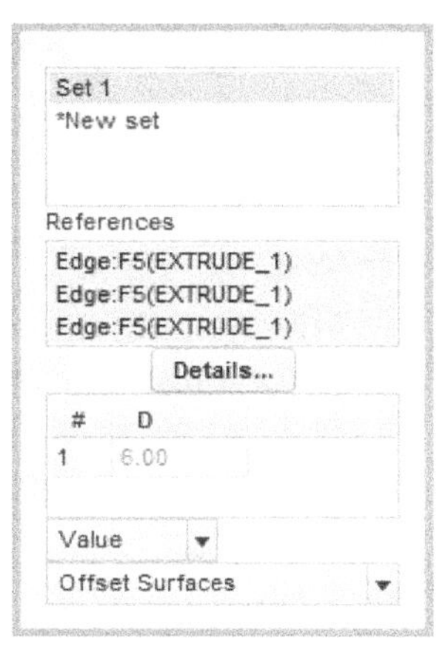

Figure 6-56 The ***Sets*** *slide-down panel*

References collector

The **References** collector displays the edges selected to chamfer. An edge can be removed from the selection set by right-clicking. On doing so, a shortcut menu will be invoked and you can choose the **Remove** option from it.

The area below the **References** collector is used to specify the dimension values. The dimension values can also be specified on the dashboard.

Transitions Tab

When you choose the **Transitions** tab, a slide-down panel will be displayed, as shown in Figure 6-57. This slide-down panel displays the transitions, if they are present in a chamfer. For example, in Figure 6-55, there are two transitions (each at the two edges).

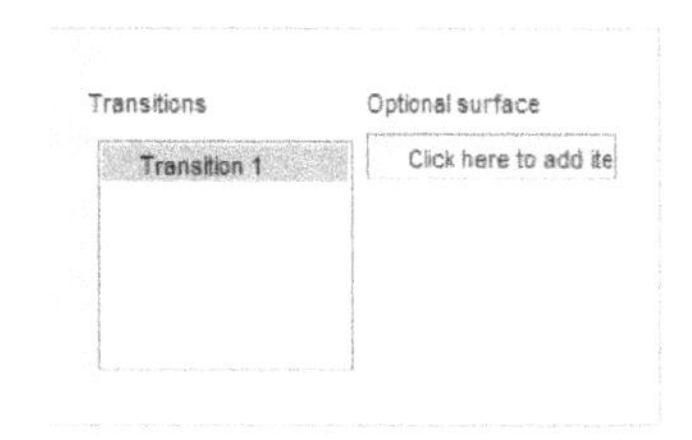

Figure 6-57 The ***Transitions*** *slide-down panel*

Pieces Tab

The use of **Pieces** tab is the same as discussed in the **Round** dashboard. You can use it to exclude selected pieces in case some ambiguity exists while creating chamfers.

Options Tab

When you choose the **Options** tab, the slide-down panel, which has two radio buttons will be displayed. These radio buttons allow you to change the chamfer as a surface or as a solid.

Properties Tab

When you choose this option, the slide-down panel will be displayed, which shows the feature identity.

UNDERSTANDING RIBS

Ribs are defined as thin wall-like structures used to bind the joints together so that they do not fail under an increased load. In PTC Creo Parametric, the section for the rib is sketched as an open section and can be extruded equally in both directions of the sketch plane or on either side. The procedure of creating a rib is similar to that of creating an extrusion.

In PTC Creo Parametric, you can create two types of ribs: Trajectory Ribs and Profile Ribs. The steps to create both the ribs are almost same. These steps are discussed next.

Creating Trajectory Ribs

Ribbon: Model > Engineering > Rib > Trajectory Rib

Adding Trajectory Ribs to a model is a new feature of PTC Creo Parametric. It is mostly used to strengthen plastic parts that include a base and a shell or a hollow area in between the pocket surfaces. To create a trajectory rib, choose the **Trajectory Rib** tool from the **Rib** drop-down in the **Engineering** group. On doing so, the **Trajectory Rib** dashboard will be displayed, as shown in Figure 6-58.

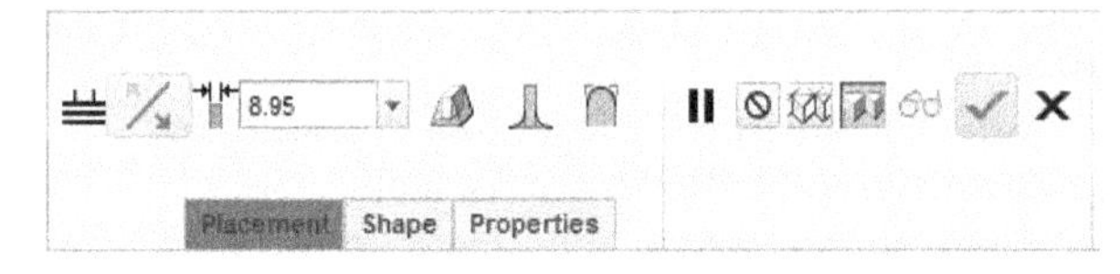

*Figure 6-58 Partial view of the **Trajectory Rib** dashboard*

To create a trajectory rib, first you need to have a shell feature with a solid base, or a hollow area in between the pocket surface. Next, you need to select a sketching plane and then draw the geometry of the rib feature in the sketcher environment. The profile for the rib feature can be open, closed, or intersecting. Also, the profile must be created in between the hollow area and can include any number of segments.

You can now add draft to ribs by using the **Adds draft** button from the dashboard. When you choose this button, a drag handle for adding a draft is displayed. Using this drag handle, you can set the value of draft.

You can also add rounds at the inner edges as well as the exposed edges by using the **Add rounds on internal edges** and **Add rounds on exposed edges** buttons, respectively. Choose the **Add rounds on internal edges** button; the drag handle will be displayed to control radius of round.

The procedure to create a trajectory rib feature is discussed next.

1. Choose the **Trajectory Rib** tool from the **Rib** drop-down in the **Engineering** group of the **Model** tab.

2. Choose the **Placement** tab from the dashboard to display a slide-down panel and then choose the **Define** button from the group; the **Sketch** dialog box will be displayed.

3. Select the sketch plane and specify the reference in the **Sketch** dialog box. The sketch plane will define the top surface of the rib, refer to Figure 6-59.

4. Draw the sketch and exit the sketcher environment; the preview of the rib feature will be displayed with the default width. Also, an arrow will be displayed in the drawing area. This arrow shows the direction of material addition and should point toward the geometry of the model and not away from it. The direction of the arrow can be reversed, if required, by using the **Change depth direction of rib to other side of the sketch** button available in the dashboard.

The width of rib can be dynamically modified by dragging the handles displayed on the rib or by entering a value in the edit box available in the dashboard.

5. After specifying all parameters for the rib feature, choose the **Build feature** button; the rib will be created in between the pocket surfaces with the bottom of rib intersecting the base of solid, as shown in Figure 6-60.

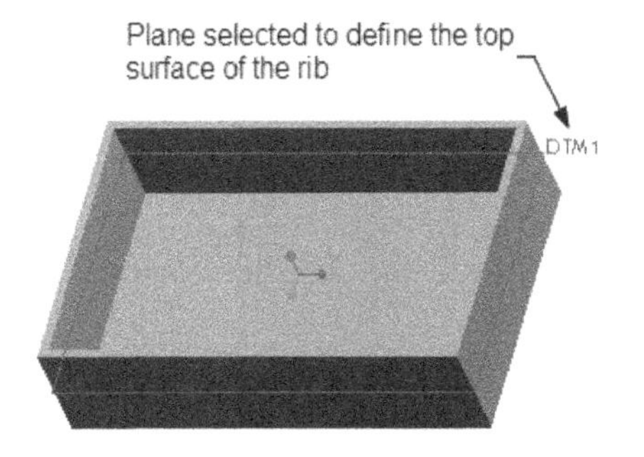

Figure 6-59 *The sketching plane selected*

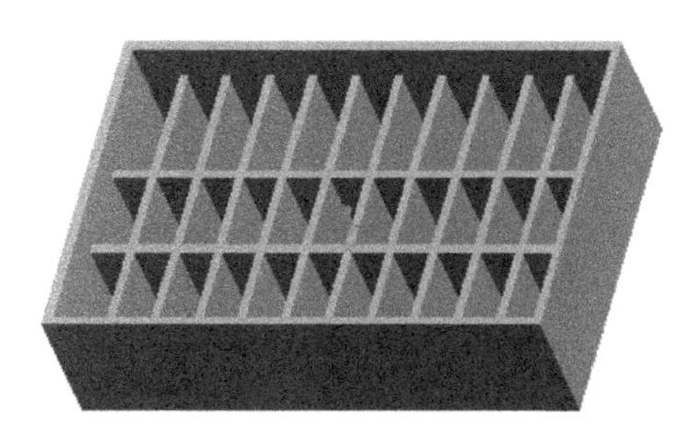

Figure 6-60 *The resultant rib feature*

Tip: *You can externally define the rounds of a Trajectory Rib. To do so, right click on the* ***Trajectory Rib*** *in the* ***Model Tree*** *and select* ***Externalize Rounds;*** *a warning message box will be displayed. Choose* ***OK*** *to accept. The round feature will be added in the* ***Model Tree****. Now you can edit the value of rounds from it.*

Note

You can change the values of rounds and drafts of the rib by using the edit boxes available in the ***Shape*** *tab.*

Creating Profile Ribs

Ribbon: Model > Engineering > Rib > Profile Rib

To create a profile rib, choose the **Profile Rib** tool from the **Rib** drop-down in the **Engineering** group; the **Profile Rib** dashboard will be displayed, as shown in Figure 6-61.

To create a rib, first you need to select a sketching plane and then draw the geometry of the rib feature in the sketcher environment. The sketch for the rib feature should be an open entity.

Figure 6-61 *Partial view of the* ***Profile Rib*** *dashboard*

The options and the tools in the **Rib** dashboard are discussed next.

References Tab

When you choose the **References** tab, a slide-down panel will be displayed, as shown in Figure 6-62. In the slide-down panel, the **Sketch** collector shows **Select 1 item**. This indicates that the sketch for the rib feature is not drawn. To draw the sketch, choose the **Define** button; the sketch dialog box will be displayed using which you can define the sketch plane and the references.

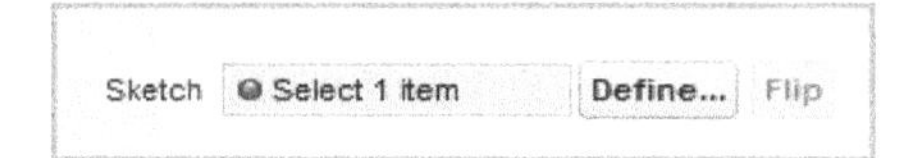

Figure 6-62 *The **References** slide-down panel*

The **Flip** button is used to flip the side where the material should be added. This button will be available only when you draw the sketch and exit the sketcher environment.

Change the thickness option between both sides, side 1, and side 2 Button

This button is available on the **Rib** dashboard only when the sketch of the rib is drawn and you exit the sketcher environment. You have an option to extrude the sketch on either sides or symmetrically on both sides of the sketching plane.

Note

Since the sketch of the rib can be extruded on both sides of the sketching plane, therefore, while creating a rib, you must always select an appropriate sketching plane such that it lies at the center of the rib feature.

The procedure to create a profile rib feature using the **Rib** dashboard is discussed next.

1. Invoke the **Profile Rib** dashboard by choosing the **Rib > Profile Rib** from the **Engineering** group.

2. From the **Profile Rib** dashboard, choose the **References** tab to display the slide-down panel and choose the **Define** button from it.

3. Select the sketch plane and specify the reference in the **Sketch** dialog box. Now, create an open sketch that defines the rib profile. Make sure that the endpoints of the sketch are aligned with the edges of the model.

Tip: *You can use the **References** tool from the **Setup** group of the **Sketch** tab to select the required edges for creating the Profile Rib.*

4. When you exit the sketcher environment, an arrow will be displayed in the drawing area. This arrow shows the direction of material addition and should point toward the geometry of the model and not away from it. The direction of the arrow can be reversed by using the **Flip** button available in the slide-down panel that will be displayed on selecting the **References** tab.

5. If the arrow direction is correct, then you can preview the rib feature on the graphics window. The rib thickness can be dynamically modified by dragging the handles displayed on the rib or by entering a value in the edit box available in the **Rib** dashboard.

6. Now, use the **Change the thickness option between both sides, side 1, and side 2** button to change the direction of extrusion. You can create the rib feature either symmetrically about the sketching plane or on either side of it. Every successive click on this button will change the position of the rib feature with respect to the sketching plane.

7. After specifying all parameters for the rib feature, choose the **Build feature** button.

The profile ribs can be rotational ribs and straight ribs. Rotational ribs are constructed on cylindrical parts and straight ribs are created on planar faces. Figure 6-63 shows a rotational rib and Figure 6-64 shows a straight rib. The options for the creating these ribs are the same. The creation of these ribs depends on the geometry of the feature on which the rib is created.

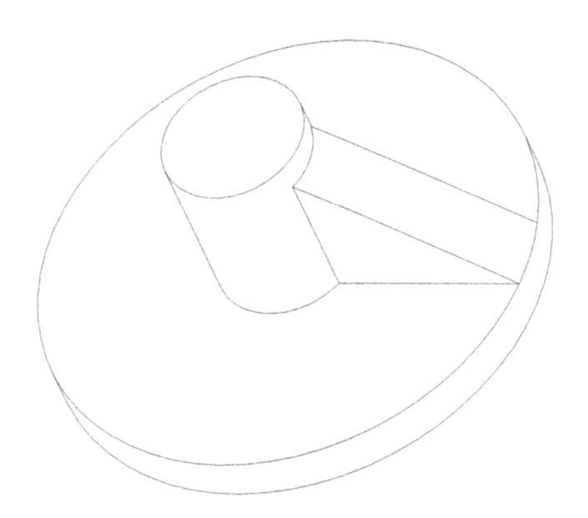

Figure 6-63 *The rotational rib feature*

Figure 6-64 *The straight rib feature*

EDITING FEATURES OF A MODEL

Editing is one of the most important aspects of the product design. Most of the designs require editing during their creation or after they are created. As mentioned earlier, PTC Creo Parametric is a parametric and feature-based solid modeling software. Hence, the features constituting a model can be individually edited. For example, Figure 6-65 shows a cylindrical part that has six counterbore holes created at some pitch circle diameter (PCD).

Now, in case you have to edit the features such that the six holes are to be converted into eight holes and the counterbore holes are to be converted into countersink holes, as shown in Figure 6-66, you just need to use two operations. The first operation will convert six holes into eight holes. To do so, select the pattern of hole from the **Model Tree** and hold down the right mouse button; shortcut menu will be displayed. Choose the **Edit** option from the **Edit Actions** area, specify the number of holes, and then choose **Regenerate** to regenerate the model. Now to convert counterbore holes into countersink holes, select the hole from the **Model Tree** and choose the **Edit** option from the **Edit Actions** area and then change the hole type to countersink.

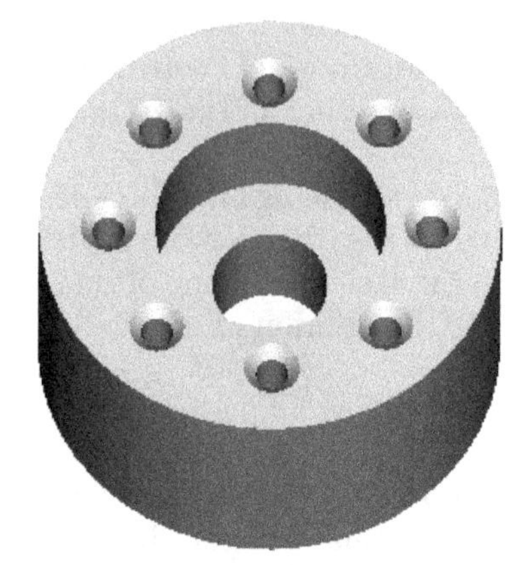

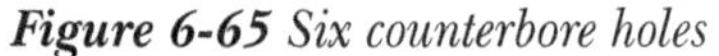

Figure 6-65 *Six counterbore holes*

Figure 6-66 *Eight countersink holes*

Similarly, you can also edit the datums and the features referenced to these datums. As a parent-child relationship exists between the two, therefore the child feature is also modified when the parent feature is modified. For example, if you have created a feature using a datum plane that is at some offset distance, the feature will be automatically repositioned when the offset value of the datum plane is changed. The following methods explain how to edit the features in PTC Creo Parametric.

Editing Definition or Redefining Features

Editing the definition of features allows you to make changes in the parameters that were used to create it. You can also modify the sketches of the sketched features by editing their definitions. There are three methods available in PTC Creo Parametric to edit definition or redefine features. These methods are listed next.

1. Select the feature that has to be redefined from the drawing area; the selected feature is highlighted. Now, hold down the right mouse button; a shortcut menu will be displayed. Choose the **Edit the definition of the selected object** option from the **Edit Actions** area of this menu.

Tip: *Suppose you have created a cube and the display is selected to be* ***Shaded****. Now, you need to select the top face of the cube. By default, in the* ***Filter*** *drop-down list, the* ***Smart*** *option is selected. Select the cube on the top face; the whole cube turns green in color. Again click on the top face; it gets selected and is highlighted in green.*

Sometimes, it becomes difficult to select the desired feature from the model to redefine it. Therefore, you can use the filters in the **Filter** drop-down list in the **Status Bar**.

2. You can also use the **Model Tree** to redefine a feature. Select the feature that has to be redefined from the **Model Tree**; the selected feature is highlighted in the drawing area. Now, right-click on the feature listed in the **Model Tree**; a shortcut menu will be displayed. Choose the **Edit the definition of the selected object** button from the **Edit Actions** area of this menu.

3. Select the feature to redefine from the drawing area or from the **Model Tree** and then choose **Edit Definition** from the **Operations** group. Remember that the **Edit Definition** option in the **Operations** group is available only when a feature has been selected.

Reordering Features

Ribbon: Model > Operations > Reorder

Reordering the features is defined as the process of changing the sequence of feature creation in a model. Sometimes after creating a model, it may be required to change the order in which the features of the model were created. A feature can be placed before or after another feature. For this purpose, the **Model Tree** or the **Reorder** tool is used.

If you use the **Model Tree**, you just have to drag and drop the feature that you want to reorder below or above the destination feature.

Choose the **Reorder** tool from the expanded **Operations** group; the **Feature Reorder** dialog box will be displayed, as shown in Figure 6-67. Click in the **Features to be reordered** collector of the **Feature Reorder** dialog box and select the features to be reordered from the **Model Tree** or from the graphics window. By default, the **After** radio button is selected in the **New location** area. Now select the **Target Feature** collector from the dialog box and then, in the **Model Tree**, select the feature after which the target feature is to be placed.

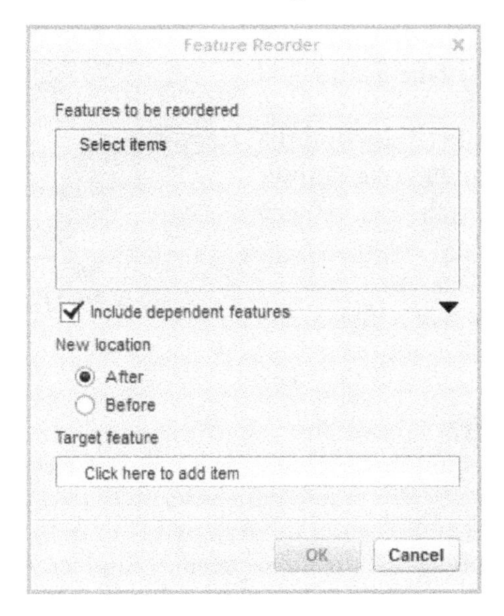

Figure 6-67 *The* ***Feature Reorder*** *dialog box*

In the **New location** area, there are two radio buttons: **After** and **Before**. If you choose the **After** radio button then the feature to be reordered will be placed after the target feature. If you choose the **Before** radio button then the feature to be reordered will be placed before the target feature.

The example that is discussed here will explain why and when the need for reordering of features in a model arises. Consider the model shown in Figure 6-68. It consists of a rectangular pattern of three columns and two rows. Now, a shell feature is created on this model, which removes the top and the front face, as shown in Figure 6-69.

Figure 6-68 *Model with pattern*   ***Figure 6-69*** *Model after creating the shell feature*

In Figure 6-69, it is evident that the material equal to the wall thickness of the shell feature is added around the hole features. If this is not desired, the model has to be modified and here the need for reordering the feature arises. You will reorder the features such that the shell feature is inserted before the hole feature and its pattern. When you reorder the features, all features will be automatically adjusted in the new order. The model after reordering the hole feature is shown in Figure 6-70.

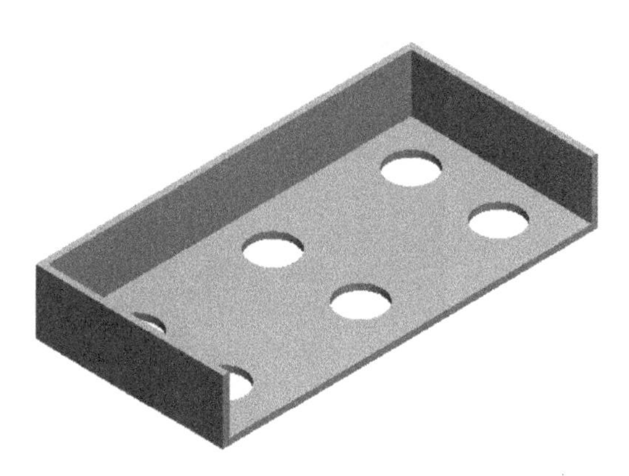

Figure 6-70 *Model after reordering the features*

Note

*Usage of the **Pattern** tool is discussed in the Chapter 7.*

Rerouting Features

Enhanced

Rerouting of features is required when there is a need of changing its sketch plane. While doing so, you can also redefine the references that were selected while creating the sketch. For rerouting a feature, select the required feature from the **Model Tree** or from the drawing area. Now, choose the **Edit References** or **Replace References** option from the **Operations** group. Note that the **Edit References** or **Replace References** option will only be available after a feature has been selected.

To edit the reference, choose the **Edit References** option from the **Operation** group of the **Model** tab; the **Edit References** dialog box will be displayed, as shown in Figure 6-71. Choose the reference to be replaced from the **Original references** collector. Next, click in the **New reference** collector of the dialog box, select the replacement reference, and then choose the **OK** button from the dialog box. To reset the changed reference back to the original reference, click on the **Reset** button of the dialog box.

To replace the reference, choose the **Replace References** option from the **Operation** group of the **Model** tab; the **Replace References** dialog box will be displayed, as shown in Figure 6-72. Choose the reference to be replaced from the **Original references** collector. Now, click in the **New reference** collector in the dialog box and select the replacement reference. Next, choose the **OK** button from the dialog box. To reset the changed reference back to the original reference, choose the **Reset** button from the dialog box.

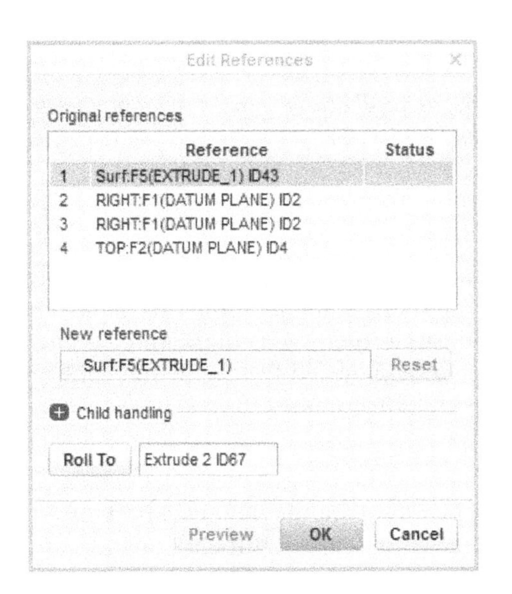

*Figure 6-71 The **Edit References** dialog box*

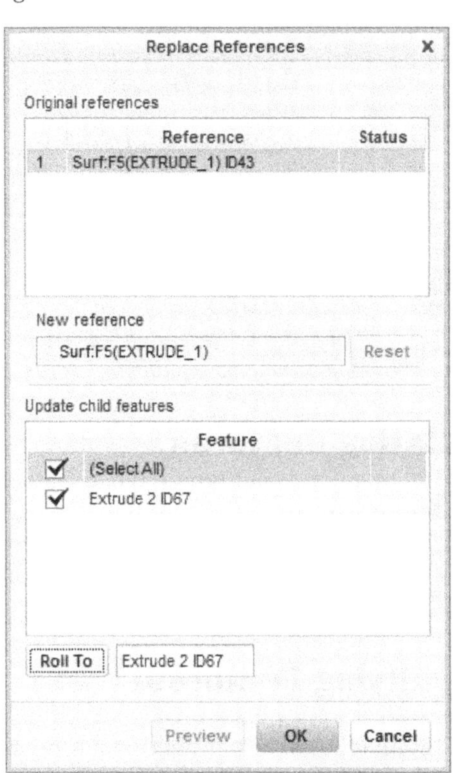

*Figure 6-72 The **Replace References** dialog box*

Note

While rerouting a feature, if you do not want to replace a highlighted surface then select the same surface again as the alternate surface.

Suppressing Features

Once the feature is suppressed, it will neither be displayed in the drawing area nor in the drawing views. Note that on suppressing, the feature is not deleted, only its visibility is turned off. You can resume the feature anytime by unsuppressing it using the **Model Tree** or using the **Resume** option from the **Operations** group. As soon as you unsuppress the feature, it will be displayed in the drawing area and also in the drawing views. When a model has many features, then suppressing some features decreases its regeneration time. To suppress a feature, right-click on it in the **Model Tree** or select it in the drawing area and then press and hold down the right mouse button to display the shortcut menu.

Choose **Suppress** from the shortcut menu to suppress the selected feature; the **Suppress** message box will be displayed prompting you to confirm the suppression of the highlighted features in the drawing area. Confirm the suppression by choosing the **OK** button in this message box. You can also suppress a feature by first selecting it from the **Model Tree** or the drawing area and then choosing **Suppress > Suppress** from the **Operations** group. To unsuppress the feature, right-click in the **Model Tree** or in the drawing area, invoke the shortcut menu, and choose the **Resume** option from it.

Tip: *By default, the suppressed features are not displayed in the* ***Model Tree****. To display the suppressed features, choose* ***Settings > Tree Filters*** *from the* ***Model Tree****; the* ***Model Tree Items*** *dialog box will be displayed. Select the* ***Suppressed Objects*** *check box and choose the* ***OK*** *button. The suppressed features will now be displayed in the* ***Model Tree*** *with a small black box displayed against their names.*

Note

If the feature being suppressed has child features, they will also be suppressed on suppressing the parent feature. However, before suppressing the features, the system prompts you to confirm the suppression of the highlighted features in the drawing area.

Deleting Features

The feature that is not required can be deleted from the model. Right-click on the feature in the **Model Tree** or in the drawing area to invoke the shortcut menu. From this menu, choose the **Delete** option. Once the feature to be deleted is selected, it gets highlighted along with its child features. The system confirms you to delete the selected feature. If you confirm the deletion, the feature will be deleted along with its child features.

Modifying Features

Once a feature is created, you can still modify it by modifying its dimensions. This editing operation reflects the parametric nature of PTC Creo Parametric. Right-click on the feature in the **Model Tree** or in the drawing area to invoke the shortcut menu. From this menu, choose the **Edit** option from the **Edit Actions** area; the dimensions of the feature are displayed in the drawing area. These dimensions are the same as those defined while sketching or creating the feature.

In some cases, for example, rounds, holes, and so on, the dimensions of these features were defined as parameters. Therefore, these dimensions are also displayed when you select them to modify. To modify the dimensions, double-click on the dimension in the drawing area; the dimension is converted into an edit box. Type the required value for the selected dimension and press ENTER. Once you modify a dimension, you also need to regenerate the feature. Choose the **Regenerate** button from the **Quick Access** toolbar or press CTRL+G.

Tip: *You can double-click on a feature in the drawing area to display its dimensions. A dimension can be modified first by double-clicking on it, then entering a new value in the edit box displayed, and finally by regenerating it.*

If you want to modify a pattern, it is recommended that you can select the instance as well as the original model. Patterns are discussed in Chapter 7.

Modifying the Sketch of a Feature Dynamically

If you want to modify the sketch of a feature without entering the sketcher environment, then you need to modify it dynamically. To dynamically modify a sketch, follow these steps:

1. Select the feature by using the **Features** filter from the **Filter** drop-down list.

2. When the feature turns green in color, hold down the right mouse button to invoke the shortcut menu.

3. In the shortcut menu, choose the **Edit** option from the **Edit Actions** area; the dimensions of the sketch appear on the feature.

4. Select the sketch and hold down the right mouse button to invoke the shortcut menu. Choose the **Edit** option from this shortcut menu.

5. Now, bring the cursor close to the sketch. You will notice that the cursor changes to a pointer. The pointer symbol represents that you can select an entity from the sketch and drag it to modify its shape.

6. Modify the shape of the sketch by dragging its entities.

7. Choose the **Regenerate** button from the **Quick Access Toolbar** to apply the modifications.

Modifying the Features Dynamically

Dynamic editing of models is a new feature of PTC Creo Parametric and it has been introduced to simplify the editing process. This feature enables you to modify the model after it has been created and can be performed any time during the creation. To edit a feature dynamically, select the feature from the drawing area or from the **Model Tree** and right-click to invoke a shortcut menu. Next, choose the **Edit** option from the **Edit Actions** area of the shortcut menu. Now you can edit the features dynamically.

The following entities displayed on the feature can be modified:

1. Small square boxes are displayed for the options that were defined in the dashboard while creating the feature. These handles enable you to adjust the values by selecting and dragging them to the desired position.

2. The profile used for the feature is also displayed. You can drag the entities and modify the shape of the sketch to modify the feature.

3. All the dimensions that were defined while sketching the model as well as those that were entered in the dashboard are displayed on the feature. You can double-click on the dimension and enter a new value to modify the feature accordingly.

Note that in the process of dynamic modification, the feature gets modified instantaneously and you need not regenerate the model.

TUTORIALS

Tutorial 1

Create the model shown in Figure 6-73. The dimensions, and the front, top, and left-side views of the model are also shown in the figure. **(Expected time: 45 min)**

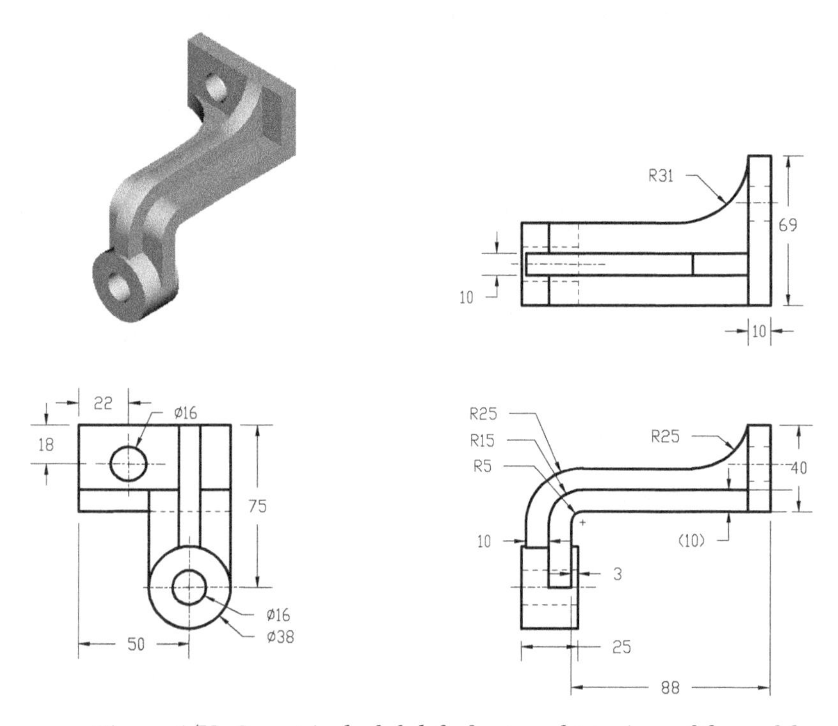

Figure 6-73 *Isometric shaded, left, front, and top views of the model*

The following steps are required to complete this model:

Examine the model and determine the number of features in it, refer to Figure 6-73.

a. Create the base feature on the **FRONT** datum plane, refer to Figures 6-74 and 6-75.
b. Create the second extrude feature on the **TOP** datum plane, refer to Figures 6-76 and 6-77.
c. Create the third extrude feature on the front planar surface of the second feature, refer to Figures 6-78 and 6-79.
d. Create the non symmetrically extruded cylindrical feature on the front face of the third feature, refer to Figures 6-80 and 6-81.
e. Create a hole feature that is coaxial to the cylindrical feature, refer to Figure 6-82.
f. Create the round features, refer to Figures 6-83 and 6-84.
g. Create the last feature, which is the rib, refer to Figures 6-85 and 6-86.

When a PTC Creo Parametric session starts, the first task is to set the working directory. As this is the first tutorial of this chapter, you need to select the Working Directory first. Set *C:\ Creo-3.0\c06* as the Working Directory; the message **Successfully changed to C:\ Creo-3.0\c06 directory** is displayed in the message area.

Starting a New Object File

1. Start a new part file and name it as *c06tut01*.

   The three default datum planes are displayed in the drawing area. Also, the **Model Tree** is displayed on the left of the drawing area.

Creating the Base Feature

To create the sketch of the base feature, you first need to select the sketching plane for it. In this model, you need to draw the base feature on the **FRONT** datum plane because the direction of extrusion of this feature is normal to this plane.

1. Choose the **Extrude** tool from the **Shapes** group; the **Extrude** dashboard is displayed.

2. Choose the **Placement** tab to display the slide-down panel and then choose the **Define** button; the **Sketch** dialog box is displayed.

3. Select the **FRONT** datum plane as the sketching plane; the **RIGHT** datum plane and its orientation are selected by default.

4. Select the **TOP** datum plane and then choose the **Top** option from the **Orientation** drop-down list.

5. Choose the **Sketch** button from the **Sketch** dialog box to enter into the sketcher environment.

6. After you enter into the sketcher environment, create the sketch of the base feature and then apply constraints and dimensions to it, as shown in Figure 6-74. Note that in the sketch, the bottom line of the rectangular section coincides with the **TOP** datum plane.

   As is evident from the sketch of the base feature shown in Figure 6-74 that the **RIGHT** datum plane is located at a distance of 50 from the left edge because later in the tutorial, the rib feature will be created on this plane. Refer to Figure 6-73 for the distance required for **RIGHT** datum plane. As mentioned earlier, that, by default, the sketch for the rib feature is extruded on both sides of the sketching plane. Therefore, when you create the sketch for the rib feature on the **RIGHT** datum plane, it will extrude on both the sides.

7. After the sketch is completed, choose the **OK** button to exit the sketcher environment; the **Extrude** dashboard is enabled and appears above the drawing area. All model attributes that are selected by default are accepted to create the model.

8. Enter **10** as the depth in the dimension box on the **Extrude** dashboard.

9. Choose the **Named Views** button from the **Graphics** toolbar; a flyout is displayed. Choose the **Default Orientation** option from the flyout; the model orients in its default orientation, that is, the trimetric view.

10. Next, choose the **Build feature** button from the **Extrude** dashboard to create the base feature.

Creating the Second Feature

The second feature is also an extruded feature and it will be created on the **TOP** datum plane. Therefore, you need to define the **TOP** datum plane as the sketching plane.

1. Choose the **Extrude** tool from the **Shapes** group; the **Extrude** dashboard is displayed.

2. Choose the **Placement** tab; the slide-down panel is displayed. From the slide-down panel, choose the **Define** button; the **Sketch** dialog box is displayed.

3. Select the **TOP** datum plane as the sketching plane.

4. Select the front face of the base feature shown in Figure 6-75 and then select the **Bottom** option from the **Orientation** drop-down list.

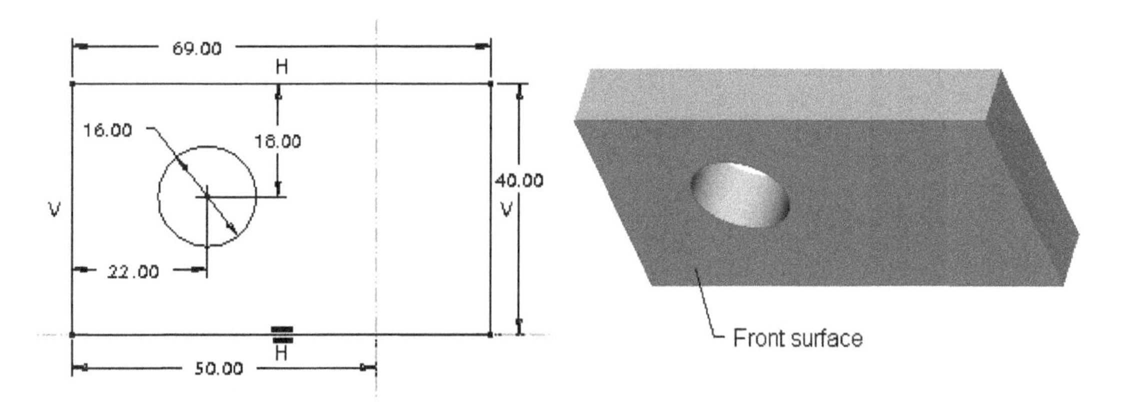

Figure 6-74 *Sketch of the base feature with dimensions and constraints*

Figure 6-75 *Front face of the base feature selected*

5. Choose the **Sketch** button from the **Sketch** dialog box to enter the sketcher environment. After entering the sketcher environment, turn the model display to **No hidden**.

Note

*Before you start sketching a section for a feature, it is recommended that you turn the model display to **No hidden** and turn off the datum plane display using the **Plane Display** button. This is done to improve the clarity of the drawing while sketching.*

6. Create the sketch for the second feature and apply the constraints and dimensions to it, as shown in Figure 6-76.

You can choose the **Project** button from the **Sketching** group to use the edge of the base feature. The edge of the base feature is required to close the section for the second feature. However, if you do not use the edge, you can draw a line to close the sketch. Align it with the bottom edge, and add constraints and dimensions, to the line created refer to Figure 6-73.

Note

If you do not close the section loop by drawing a line or using the edge of the base feature, a arrow is displayed on exiting the sketcher environment. Using this arrow, you can specify the direction in which the material will be added.

7. After completing the sketch, turn the model display to **Shading** and choose the **OK** button; the **Extrude** dashboard is enabled above the drawing area.

8. Enter **10** in the dimension box in the **Extrude** dashboard and press ENTER.

9. Choose the **Build feature** button from the **Extrude** dashboard; the second feature is completed and the shaded default trimetric view is displayed, as shown in Figure 6-77.

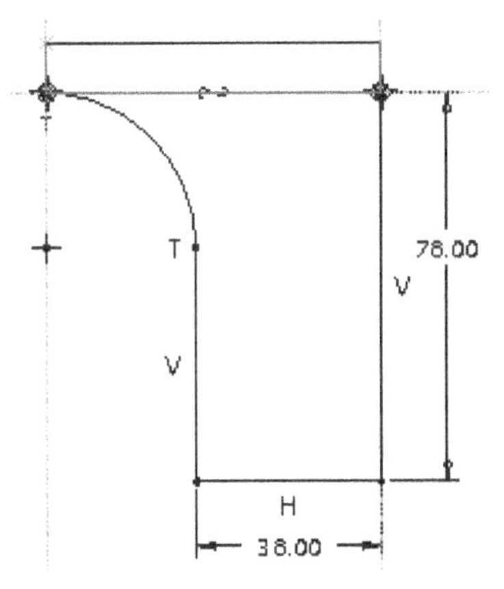

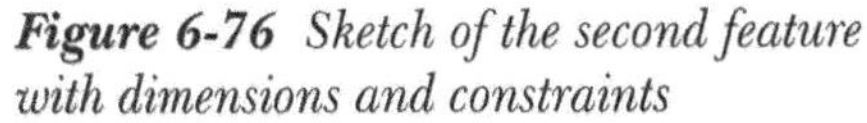

Figure 6-76 *Sketch of the second feature with dimensions and constraints*

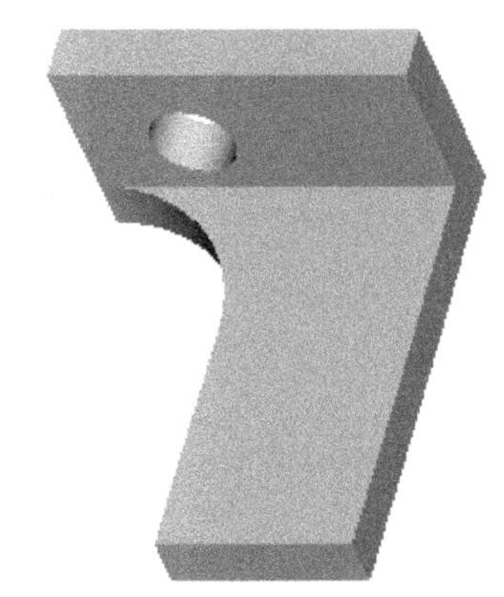

Figure 6-77 *The default trimetric view of the completed second feature and the base feature*

Creating the Third Feature

The sketch of the third feature will be drawn on the front planar surface of the second feature and it will be extruded to the given depth.

1. Choose the **Extrude** tool from the **Shapes** group.

2. Choose the **Placement** tab. Next, from the slide-down panel, choose the **Define** button; the **Sketch** dialog box is displayed.

3. Select the face of the second feature as the sketching plane, as shown in Figure 6-78.

4. Select the top face of the second feature as the reference and then select the **Top** option from the **Orientation** drop-down list, only if it is not selected by default.

5. Choose the **Sketch** button from the **Sketch** dialog box to enter into the sketcher environment.

6. Once you enter in the sketcher environment, turn the model display to **No hidden**. Create the sketch for the third feature and apply constraints and dimensions to it, refer to Figure 6-79.

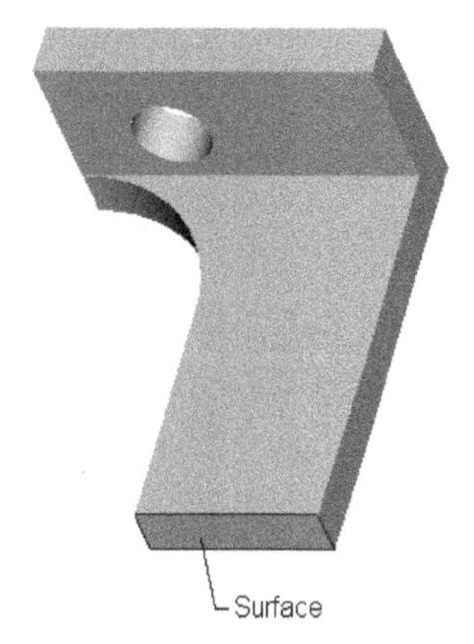

Figure 6-78 *Planar surface selected as the sketching plane for the third feature*

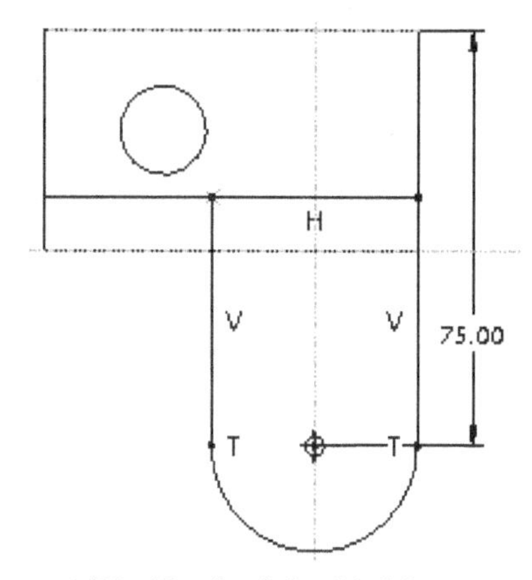

Figure 6-79 *Sketch of the third feature with dimensions and constraints*

Note

While drawing the sketch of any feature, it is recommended that you first apply the required constraints and then dimension the sketch.

7. Choose the **OK** button from the **Close** group; the **Extrude** dashboard is enabled above the drawing area.

8. Enter **10** in the dimension box in the **Extrude** dashboard.

9. Choose the **Build feature** button from the **Extrude** dashboard. Turn the model display to **Shading**. The third feature is completed. You can use the middle mouse button to spin the model to view its different orientations.

Creating the Fourth Feature

The fourth feature of the model is an extruded feature and its sketch is drawn on the front planar surface of the third feature.

1. Choose the **Extrude** tool from the **Shapes** group.

2. Choose the **Placement** tab and then from the slide-down panel, choose the **Define** button; the **Sketch** dialog box is displayed.

3. Select the front planar face of the third feature, shown in Figure 6-80, as the sketching plane. The extrusion of the cylindrical feature will be created on both sides of the front face.

   The attributes of the feature, like extrusion on both sides and the depth of extrusion, will be specified after the sketch of the feature is drawn.

4. Select the top face of the second feature as reference and then select the **Top** option from the **Orientation** drop-down list, if they are not selected by default.

5. Choose the **Sketch** button from the **Sketch** dialog box to enter into the sketcher environment.

6. Once you enter into the sketcher environment, turn the model display to **No hidden**.

7. Choose the **Concentric** button from the **Circle** drop-down in the **Sketching** group to draw the circular section for the fourth feature, refer to Figure 6-81. Select the arc using the left mouse button. As you move the mouse, the rubber-band circle changes its size. Move the cursor close to the arc; the cursor snaps to the arc. Use the left mouse button and select a point on the arc. You will notice that the equal radius constraint is applied to the sketch. Now, exit this tool.

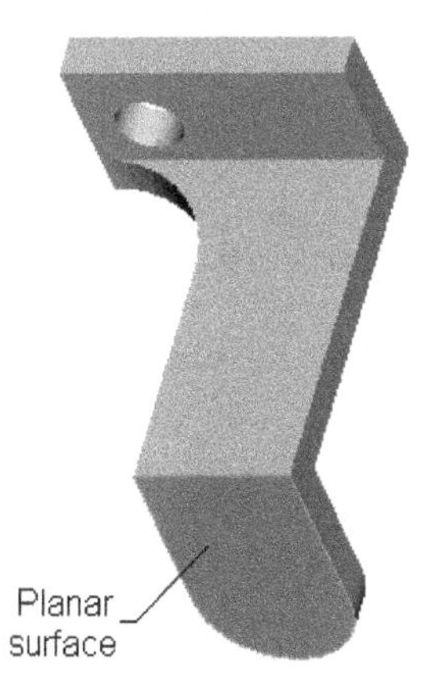

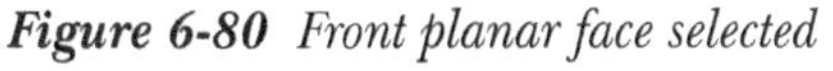
Figure 6-80 *Front planar face selected*

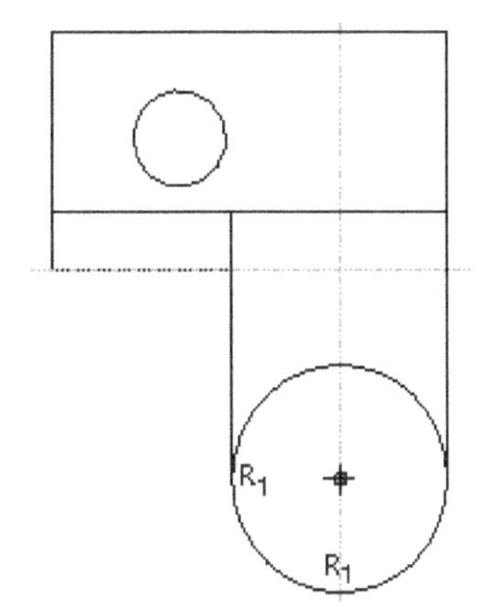

Figure 6-81 *Sketch for the fourth feature*

8. After the sketch is completed, choose the **OK** button from the **Close** group; the **Extrude** dashboard is enabled above the drawing area.

9. Choose the **Options** option from the **Extrude** dashboard. In the slide-down panel, choose the **Blind** option from both the **Side 1** and **Side 2** drop-down lists.

10. Enter **12** in the dimension box on the right of the **Side 1** drop-down list.

11. Enter **13** in the dimension box that is on the right of the **Side 2** drop-down list.

12. Turn the model display to **Shading** and choose the **Build feature** button from the **Extrude** dashboard to create the fourth feature.

Creating the Hole Feature

The hole feature is created using the **Hole** dashboard. The hole will be placed coaxial to the fourth cylindrical feature.

Note

*For creating a coaxial hole, you need to select the axis of the circular feature. To do so, you need to turn on the display of the axis by selecting the **Axis Display** check box from the **Datum Display Filters** drop-down list in the **Graphics** toolbar, if it is not displayed.*

1. Choose the **Hole** tool from the **Engineering** group; the **Hole** dashboard is displayed and the **Create simple hole** tool is chosen by default.

2. Choose the **Placement** tab from the **Hole** dashboard; a slide-down panel is displayed.
3. Select the front face of the cylindrical feature, as shown in Figure 6-82, to place the hole. As you select the front face of the cylindrical feature, the preview of the hole is displayed in the drawing area. Now, you need to specify the reference for placing the hole.
4. Press the CTRL key and select the axis of the cylindrical feature from drawing area.

   Note that the **Coaxial** option is automatically selected in the drop-down list of the slide-down panel of the **Placement** tab.
5. In the diameter dimension box on the **Hole** dashboard, type **16** and press ENTER.
6. In the depth flyout of the **Hole** dashboard, choose the **Drill to intersect with all surfaces** button.

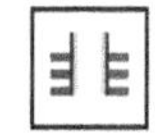

7. Choose the **Build feature** button from the **Hole** dashboard; the hole is created and the trimetric view of the shaded model with the hole is displayed, as shown in Figure 6-83.

Creating Two Round Features

Now, the two round features need to be created. Both the rounds have different radii and will be created by defining two sets.

1. Choose the **Round** button from the **Engineering** group; the **Round** dashboard is displayed.

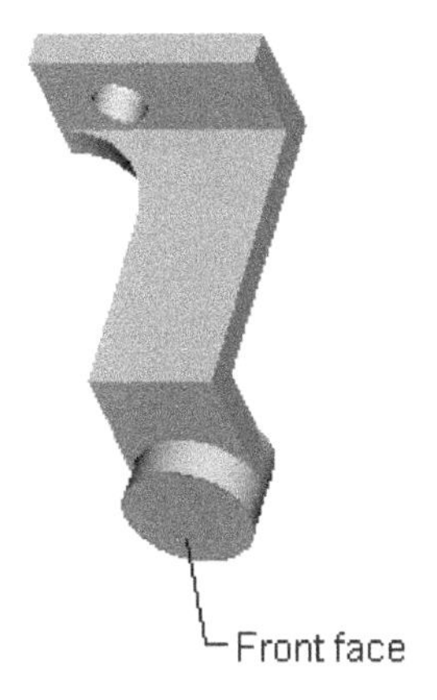

Figure 6-82 Planar face

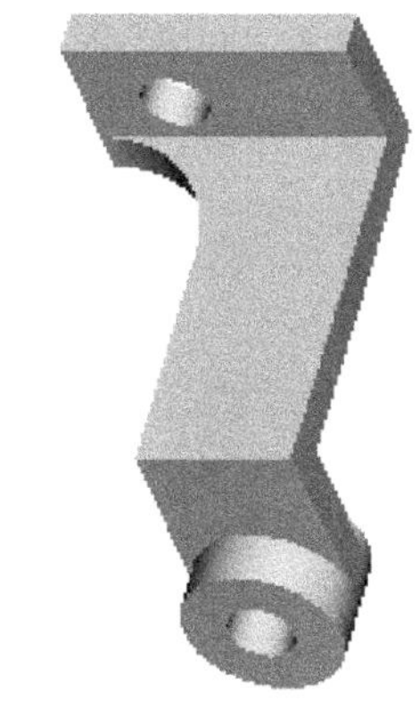

Figure 6-83 The default trimetric view

2. Choose the **Sets** tab to display the slide-down panel. Let this slide-down panel remain open so that you can view the selections made on the model.

3. Spin the model and select one of the faces shown in Figure 6-84. Press the CTRL key and then select the other face.

   The preview of the round is created on the edge that connects the two selected faces; the default value of the radius is also displayed.

4. Double-click on the default radius value that is displayed on the preview of the round. In the edit box that appears, enter the value **5** and press ENTER; the first round is created.

5. After spinning the model, select one of the faces shown in Figure 6-85 and then use CTRL+left mouse button to select the second face. You will notice that in the slide-down panel, **Set 1** and **Set 2** appear. This indicates that the second set has been defined.

   The preview of the round is created on the edge that connects the two selected faces. The default value of the radius is also displayed.

6. Double-click on the default radius value that is displayed in the preview of the second round. In the edit box that appears, type the value **15** and press ENTER.

7. Choose the **Build feature** button from the **Round** dashboard.

   The second round is created. In the **Model Tree**, the two rounds created appear as a single feature.

Creating the Rib Feature

Advance planning done while drawing the sketch for the base feature will help you to draw the rib feature now. The location for the **RIGHT** datum plane was calculated while sketching the base feature. The section for the rib feature will be drawn on the **RIGHT** datum plane.

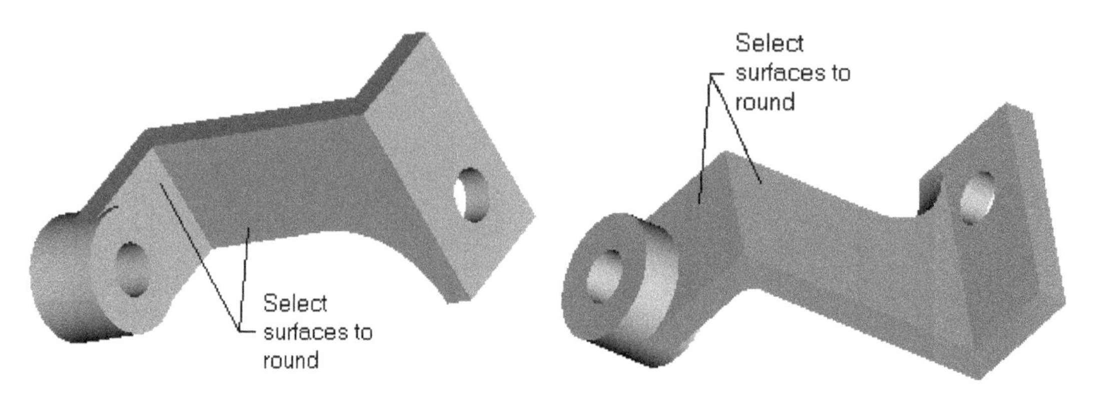

Figure 6-84 *Faces for creating a round of radius* ***5***

Figure 6-85 *Faces for creating a round of radius* ***15***

1. Choose the **Profile Rib** tool from the **Rib** drop-down in the **Engineering** group; the **Profile Rib** dashboard is displayed.

2. A rib feature is always sketched from the side view. Now, turn on the display of the datum planes from the **Datum Display** drop-down list in the **Graphics** toolbar.

3. Choose the **References** tab; the slide-down panel is displayed. Choose the **Define** button; the **Sketch** dialog box is displayed.

4. Select the **RIGHT** datum plane as the sketching plane from the drawing area.

5. Select the **TOP** datum plane as reference and then select the **Top** option from the **Orientation** drop-down list, only if they are not selected by default.

6. Choose the **Sketch** button to enter into the sketcher environment.

   You will need to turn the model display to **No hidden** before drawing the sketch for the rib feature.

7. Draw the open sketch for the rib feature and apply required constraints and dimensions, as shown in Figure 6-86.

8. After the sketch is completed, choose the **OK** button. On doing so, an arrow pointing in the direction of material addition is displayed on the sketch. As the section for the rib feature is open, PTC Creo Parametric allows you to specify the direction where the material should be added. Therefore, you need to change the direction of the arrow toward the model, if it is not already changed.

9. Enter **10** in the dimension box in the **Profile Rib** dashboard.

10. Choose the **Build feature** button from the **Profile Rib** dashboard to complete creating the model and exit this feature creation tool.

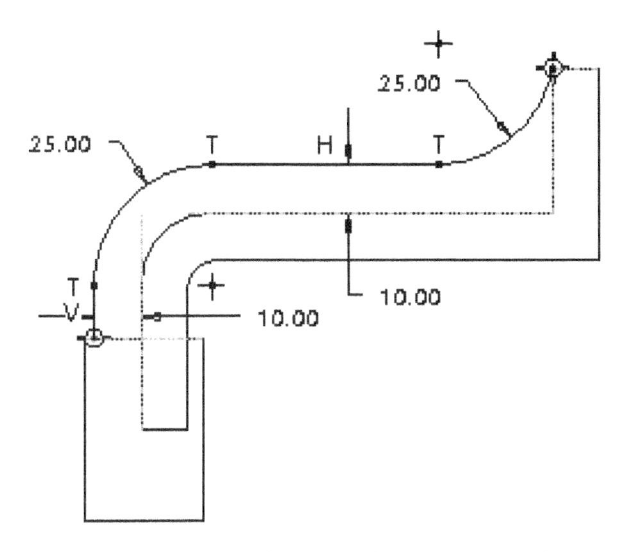

Figure 6-86 *Open sketch of the rib feature with dimensions and constraints*

All features in the model have been created and the model is now complete. The trimetric shaded view of the completed model is shown in Figure 6-87.

Saving the Model

1. Choose the **Save** button from the **File** menu and save the model. The order of feature creation can be seen from the **Model Tree** shown in Figure 6-88.

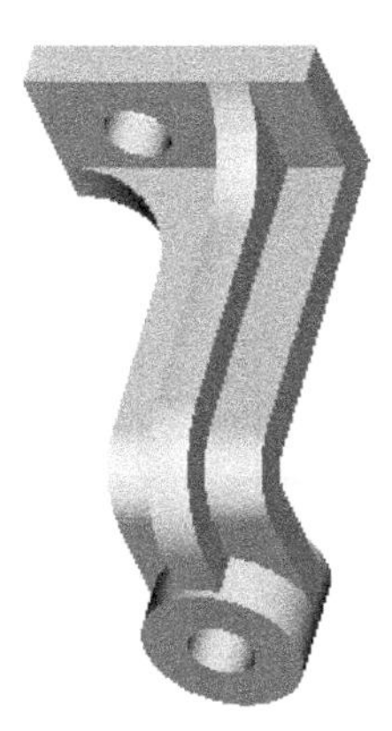

Figure 6-87 *The default trimetric view of the model*

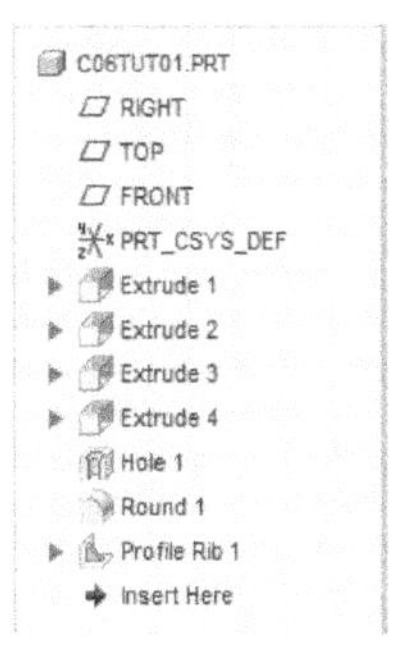

Figure 6-88 *The **Model Tree** for Tutorial 1*

Tutorial 2

In this tutorial, you will create the model shown in Figure 6-89. The dimensions of the model are shown in Figure 6-90. **(Expected time: 30 min)**

The following steps are required to complete this model:

Examine the model and determine the number of features in it, refer to Figure 6-89.

a. Create the base feature, refer to Figures 6-91 and 6-92.
b. Create the cylindrical feature, refer to Figures 6-93 through 6-95.
c. Create the hole feature coaxially on the cylindrical feature, refer to Figure 6-96.
d. Create the rib feature, refer to Figures 6-97 and 6-98.

Figure 6-89 *Isometric view of the solid model*

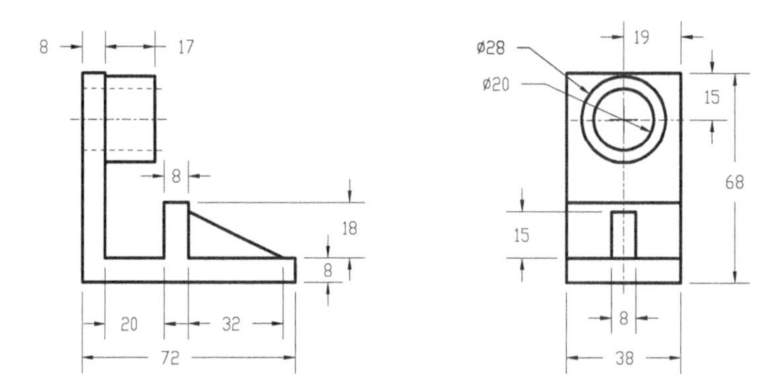

Figure 6-90 *Front and right views of the model for Tutorial 2*

The working directory has already been selected in Tutorial 1 and therefore, you do not need to select it again. However, if you need to change the working directory, choose **File > Select Working Directory** and then select *c06* in the **Select Working Directory** dialog box.

Starting a New Object File

1. Start a new part file and name it as *c06tut02*.

   The three default datum planes and the **Model Tree** appear in the drawing area.

Creating the Base Feature

1. Choose the **Extrude** tool from the **Shapes** group; the **Extrude** dashboard is displayed.
2. Choose the **Placement** tab and then from the slide-down panel, choose the **Define** button; the **Sketch** dialog box is displayed.
3. Select the **FRONT** datum plane for drawing the sketch of the base feature.
4. Select the **TOP** datum plane from the drawing area and then select the **Top** option from the **Orientation** drop-down list.
5. Choose the **Sketch** button from the **Sketch** dialog box to enter into the sketcher environment.
6. In the sketcher environment, create the sketch of the base feature and apply constraints and dimensions to it, as shown in Figure 6-91.
7. After the sketch is completed, choose the **OK** button and exit the sketcher environment; the **Extrude** dashboard is enabled and appears above the drawing area.
8. Enter **38** as the depth in the dimension box that is present on the **Extrude** dashboard; the default trimetric view of the base feature is displayed, as shown in Figure 6-92.
9. Choose the **Build feature** button from the **Extrude** dashboard to exit the feature creation tool.

Note

*In this tutorial, the base feature can also be created by extruding it on both sides of the sketching plane that is the **FRONT** datum plane. This would reduce the step required to create the datum plane, which in turn is used to create the rib feature. However, to familiarize you with creating embedded datum planes, the base feature is extruded on one side of the sketching plane.*

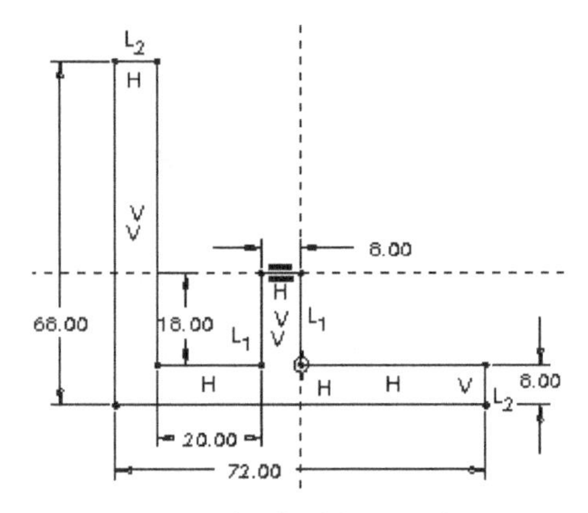

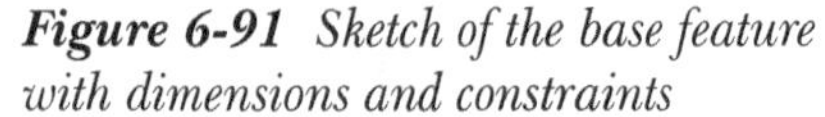
Figure 6-91 Sketch of the base feature with dimensions and constraints

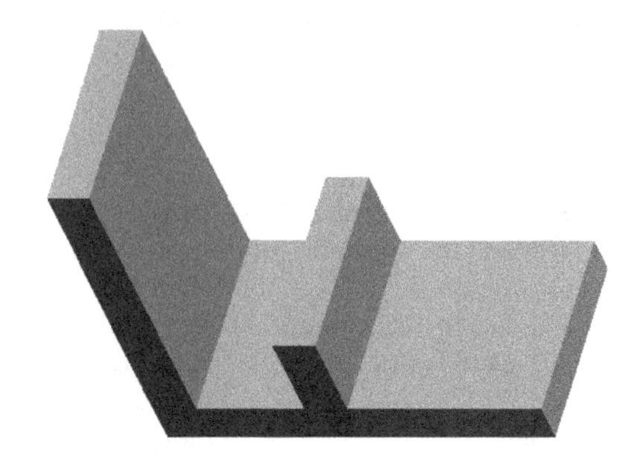
Figure 6-92 The default trimetric view of the base feature

Creating the Second Feature

The second feature is a cylindrical feature that is sketched on the planar surface of the base feature which is shown in Figure 6-93.

1. Choose the **Extrude** tool from the **Shapes** group.

2. Choose the **Placement** tab and then from the slide-down panel, choose the **Define** button; the **Sketch** dialog box is displayed.

3. Select the face of the base feature as the sketching plane, refer to Figure 6-93.

4. Select the **TOP** datum plane from the drawing area and then select the **Top** option from the **Orientation** drop-down list.

5. Choose the **Sketch** button from the **Sketch** dialog box to enter into the sketcher environment.

   After invoking the sketcher environment, turn the model display to **No hidden**.

6. Draw the sketch for the second feature. Apply and modify the dimensions, as shown in Figure 6-94.

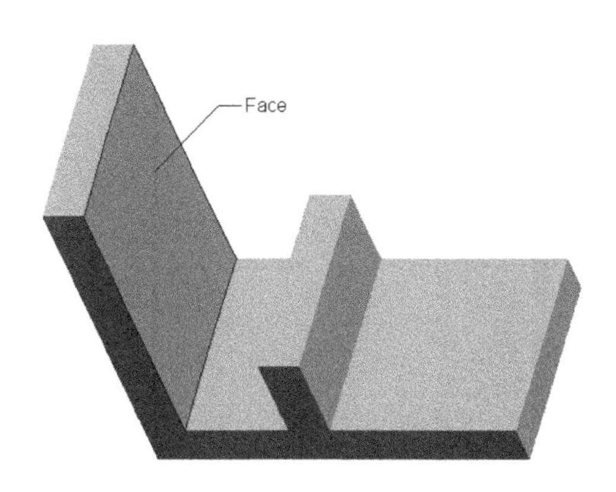

Figure 6-93 The face of the base feature selected as the sketching plane

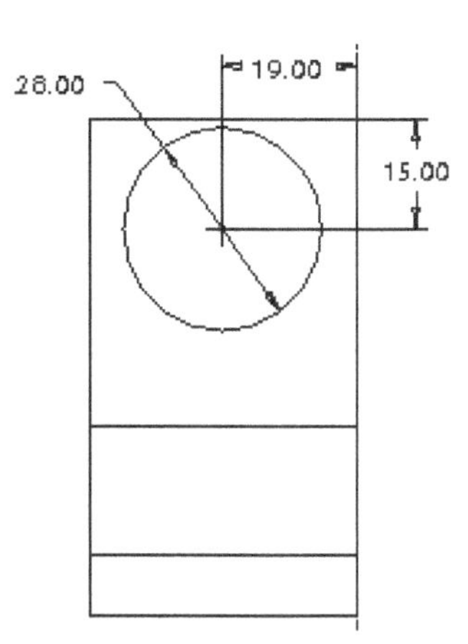

Figure 6-94 Sketch and dimensions of the second feature

7. Turn the display of the model to **Shading**. Exit the sketcher environment by choosing the **OK** button; the **Extrude** dashboard is enabled and appears above the drawing area.

8. Enter **17** in the dimension box in the **Extrude** dashboard.

9. Choose the **Build feature** button from the **Extrude** dashboard to exit the feature creation tool. The trimetric view of the model with the second feature is shown in Figure 6-95.

Creating the Hole Feature

The hole feature will be created using the **Hole** dashboard. The coaxial hole will be created on the cylindrical feature. The axis of the cylindrical feature will be used as the axial reference to create the coaxial hole.

1. Choose the **Hole** tool from the **Engineering** group in the **Ribbon**; the **Hole** dashboard is displayed. The **Create simple hole** tool in the **Hole** dashboard is chosen by default.

2. Choose the **Placement** tab from the **Hole** dashboard; a slide-down panel is displayed.

3. Select the front face of the cylindrical feature to place the hole.

   As you select the front face of the cylindrical feature, the preview of the hole is displayed in the drawing area. Now, you need to specify the reference for the placement of hole.

4. Hold-down the CTRL key and select the axis of the cylindrical feature from the drawing area.

   The **Coaxial** option is selected automatically in the drop-down list of the slide-down panel.

5. In the diameter edit box of the **Hole** dashboard, enter **20** and press ENTER.

6. From the depth flyout on the **Hole** dashboard, choose the **Drill to intersect with all surfaces** button.

7. Choose the **Build feature** button from the **Hole** dashboard; the hole is created and the trimetric view of the shaded model with the hole is displayed, as shown in Figure 6-96.

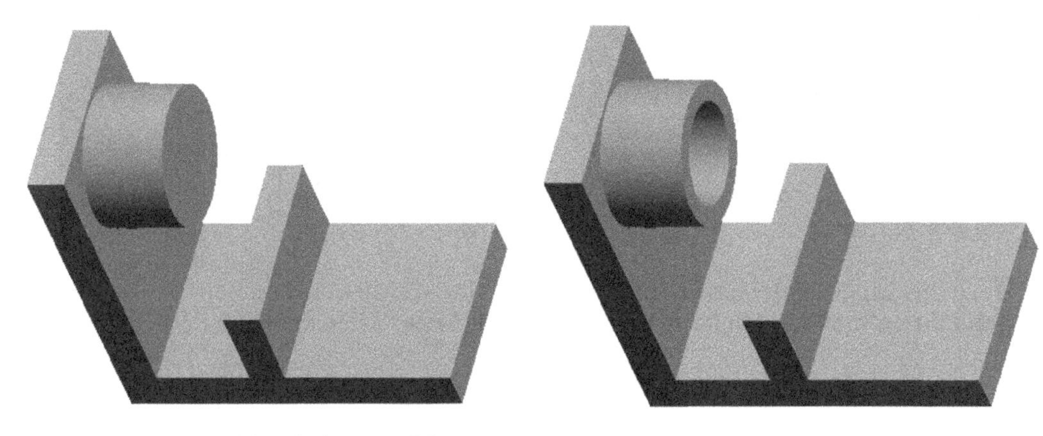

Figure 6-95 *Model with the second feature*

Figure 6-96 *Model with the hole feature*

Creating the Rib Feature

To create the rib feature, a datum plane is created on-the-fly. As mentioned earlier, rib features are always drawn from the side view.

1. Choose the **Profile Rib** tool from the **Rib** drop-down in the **Engineering** group.

2. Choose the **References** tab; a slide-down panel is displayed. Choose the **Define** button; the **Sketch** dialog box is displayed.

3. Choose the **Plane** tool from the **Datum** group of the **Model** tab in the **Profile Rib** dashboard; the **Datum Plane** dialog box is displayed. You may need to move the **Sketch** dialog box to bring the **Plane** button into view.

4. Select the axis of the hole, press the CTRL key, and select the **FRONT** datum plane. You can view your selections in the **References** collector.

5. Choose the **Offset** button in the **References** collector; a drop-down list appears on the right of the reference. Select the **Parallel** option from this drop-down list.

6. Choose **OK** from the **Datum Plane** dialog box to exit it.

   A datum plane that passes through the selected axis and is parallel to the **FRONT** datum plane will be created. This datum plane is selected automatically as the sketching plane. Now, you need to select the reference plane.

7. Select the **TOP** datum plane and then select the **Top** option from the **Orientation** drop-down list.

8. Choose the **Sketch** button; the system takes you to the sketcher environment. Choose the **No Hidden** option from the **Display Style** drop-down in the **Graphics** toolbar.

9. Draw the open sketch of the rib feature and apply dimensions, as shown in Figure 6-97. Exit the sketcher environment by choosing the **OK** button; the **Profile Rib** dashboard is enabled and appears above the drawing area.

10. Choose the **Flip** button on the **References** slide-down panel to flip the direction of the arrow; the preview of the rib is displayed on the model. Choose this button only when the preview of the rib is not displayed. Alternatively, click on the arrow to flip the direction.

11. Enter **8** as the thickness of the rib in the edit box in the **Profile Rib** dashboard. Next, choose the **Build feature** button.

    The trimetric view of the complete model with the rib feature is shown in Figure 6-98.

Saving the Model

1. Choose the **Save** button from the **File** menu to save the model and then close the active window.

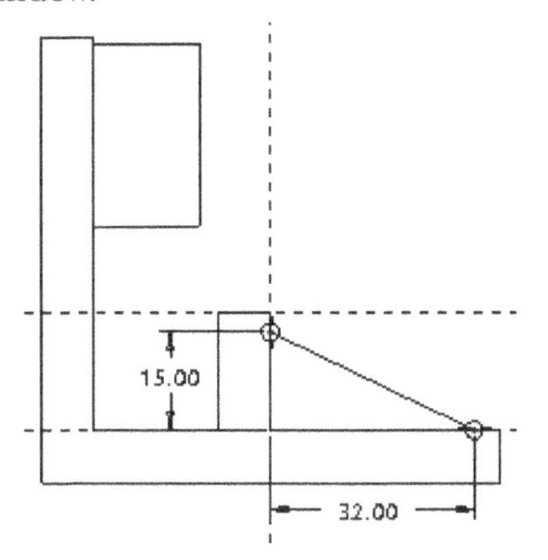

Figure 6-97** Sketch for the rib feature with the model display set to **No hidden

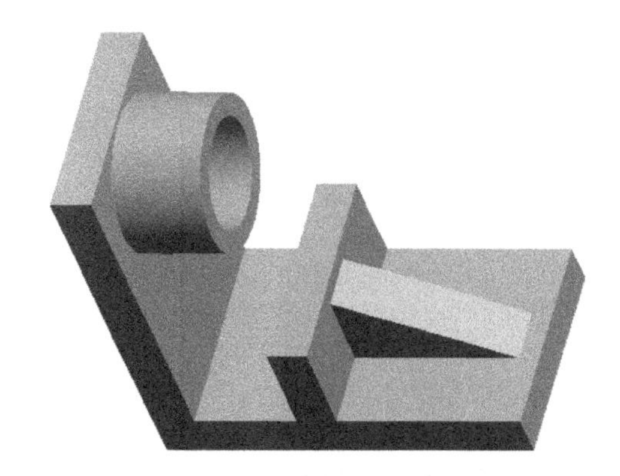

***Figure 6-98** The default trimetric view of the final model*

Tutorial 3

In this tutorial, you will create the model shown in Figure 6-99. The solid model, dimensions, and the front and the right views are also shown in this figure. **(Expected time: 45 min)**

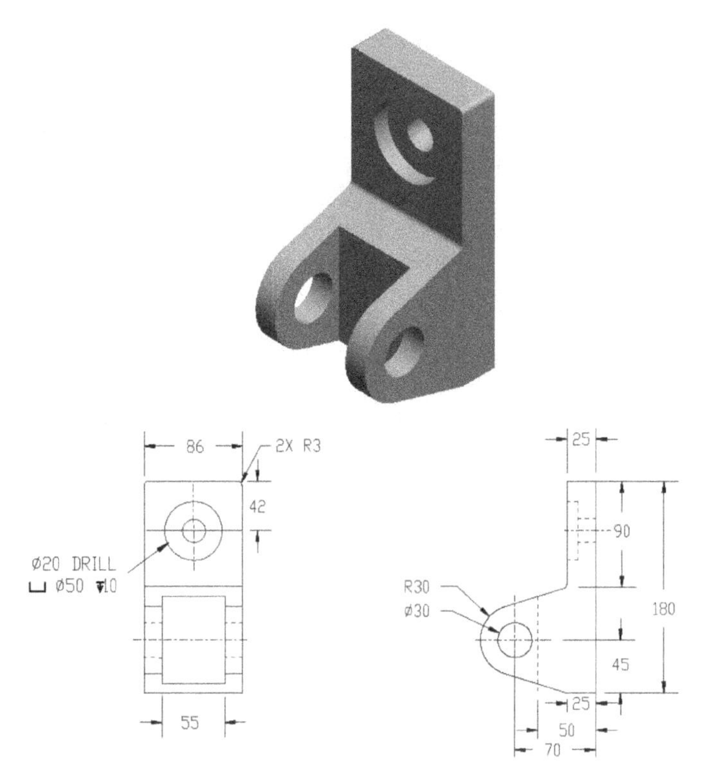

Figure 6-99 The isometric, front, and right views of the model

The following steps are required to complete this model:

Examine the model and determine the number of features in it, refer to Figure 6-99.

a. Create the base feature, refer to Figures 6-100 and 6-101.
b. Create the cut feature, refer to Figures 6-102 and 6-103.
c. Create a counterbore hole, refer to Figures 6-104 through 6-106.
d. Create the rounds, refer to Figures 6-107 and 6-108.

Starting a New Object File

1. Start a new part file and name it *c06tut03*.

   The three default datum planes and the **Model Tree** appear in the drawing area if they were not turned off previously.

Creating the Base Feature

1. Choose the **Extrude** tool from the **Shapes** group.

2. Choose the **Placement** tab; a slide-down panel is displayed. Choose the **Define** button; the **Sketch** dialog box is displayed.

3. Select the **RIGHT** datum plane as the sketching plane.

4. Select the **TOP** datum plane from the drawing area and then select the **Top** option from the **Orientation** drop-down list.

5. Choose the **Sketch** button from the **Extrude** dashboard to enter into the sketcher environment.

6. Once you enter into the sketcher environment, create the sketch of the base feature and apply dimensions and constraints, as shown in Figure 6-100.

7. Exit the sketcher environment by choosing the **OK** button; the **Extrude** dashboard is enabled and appears above the drawing area.

8. Enter **86** in the dimension box in the **Extrude** dashboard.

9. Choose the **Build feature** button to exit the feature creation tool. The base feature is completed. The default trimetric view of the base feature is shown in Figure 6-101.

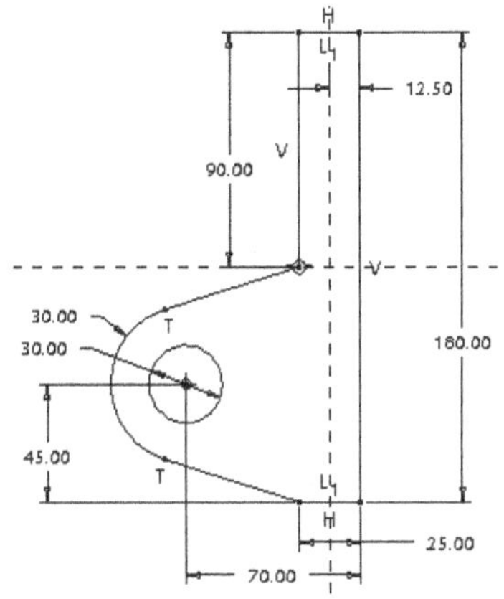

***Figure 6-100** Sketch of the base feature with dimensions and constraints*

Creating the Second Feature

The second feature is an extruded cut feature. This cut feature is created on a datum plane that passes through the center of the base feature.

1. Choose the **Extrude** tool from the **Shapes** group.

2. Choose the **Remove Material** button.

3. In the depth flyout, choose the **Extrude on both sides of sketch plane by half the specified depth value in each direction** button.

4. Choose the **Placement** tab; a slide-down panel is displayed. Choose the **Define** button; the **Sketch** dialog box is displayed.

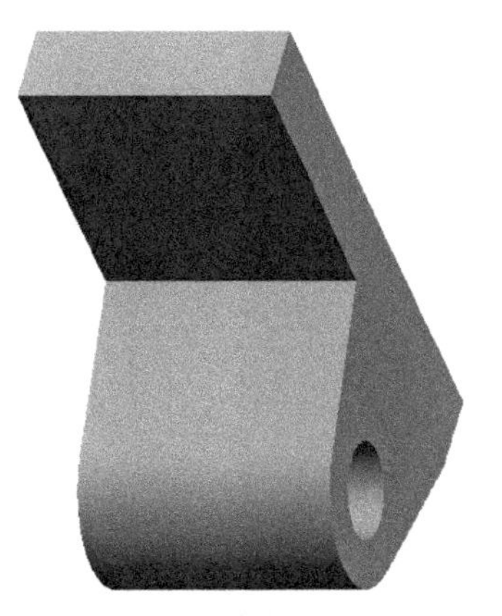

***Figure 6-101** The default trimetric view of the base feature*

5. Choose the **Plane** tool from the **Datum** group in the **Profile Rib** dashboard; the **Datum Plane** dialog box is displayed.

6. Select the **RIGHT** datum plane. In the **Translation** edit box that appears in the dialog box, enter **43**. This value is half of the width of the base feature.

7. Choose **OK** to exit the **Datum Plane** dialog box.

8. Select the **FRONT** datum plane and then select the **Left** option from the **Orientation** drop-down list.

9. Choose the **Sketch** button; you will enter into the sketcher environment.

10. Draw the sketch for the cut feature and apply the required dimensions and constraints, as shown in Figure 6-102.

    In Figure 6-102, some dimensions appear light in color. These dimensions are weak dimensions and it is not important to convert them into strong dimensions. These dimensions are not important for the creation of this feature. However, the geometry for the cut should be similar to that shown in Figure 6-102.

11. After the sketch is completed, exit the sketcher environment; the **Extrude** dashboard is enabled and it appears above the drawing area.

12. Enter a depth value of **55** in the dimension box in the **Extrude** dashboard. The system will accept this depth symmetrical to the sketching plane.

13. Choose the **Build feature** button to exit the feature creation tool. The cut feature is completed now and the default trimetric view of the cut feature along with the base feature is shown in Figure 6-103.

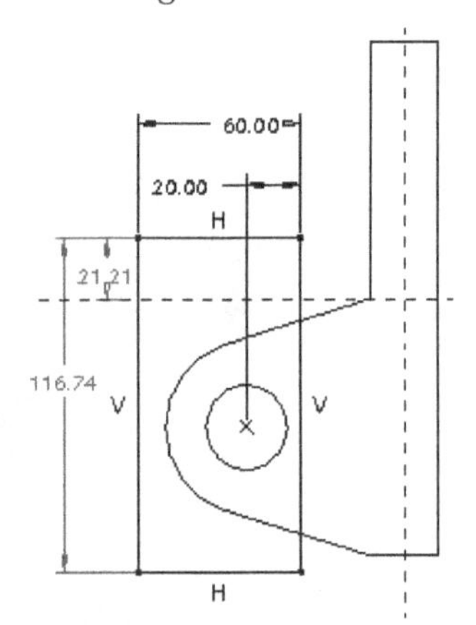

Figure 6-102 Sketch of the cut feature with dimensions and constraints

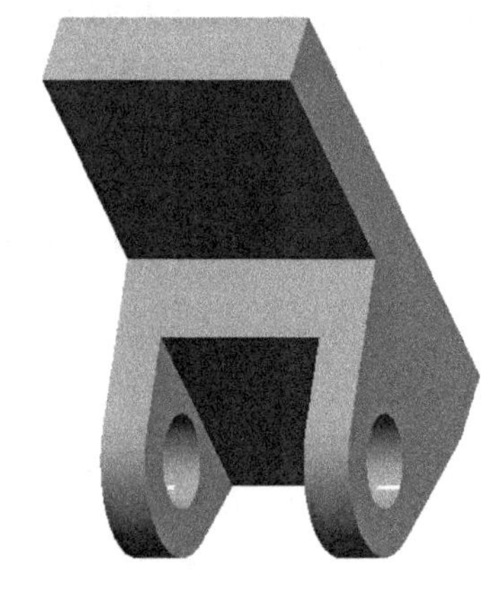

Figure 6-103 The default trimetric view of the model

Creating the Hole Feature

The hole is created using the **Hole** dashboard.

1. Choose the **Hole** tool from the **Engineering** group; the **Hole** dashboard is displayed.

2. In the **Hole** dashboard, choose the **Use sketch to define drill hole profile** button.

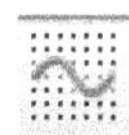

3. Choose the **Activates Sketcher to create section** button to enter into the sketcher environment.

4. In the sketcher environment, draw the sketch of the counterbore hole, as shown in Figure 6-104. Make sure that the center line is a geometric centerline. After drawing the sketch, exit the sketcher environment.

5. Select the front face of the base feature to place the hole; the preview of the hole appears on the selected face.

6. Choose the **Placement** tab from the **Hole** dashboard; a slide-down panel is displayed.

7. Click in the **Offset References** collector; you are prompted to select two references.

8. Select the two references, as shown in Figure 6-105, for dimensioning. Note that the second reference should be selected by using the CTRL key + left mouse button.

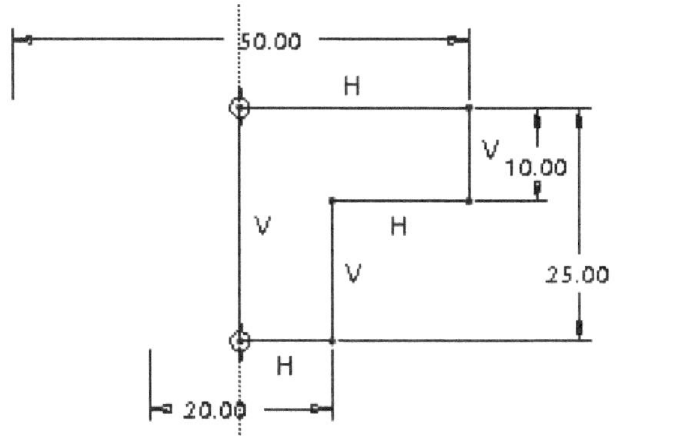

Figure 6-104 Cross-section of the hole with dimensions and constraints

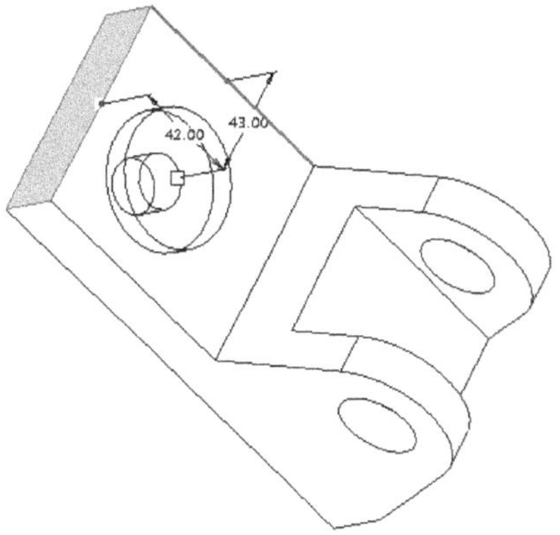

Figure 6-105 Edges to be selected for placing the hole

9. Modify the dimensions of the references by entering suitable values in the **Placement** slide-down panel. The center of hole is at the distance of **42** from the top edge and **43** from the right edge. After viewing the preview of the hole, you may have to specify a negative value for any one of the distances.

10. Choose the **Build feature** button to completely create the hole and Figure 6-106 shows the hole created.

Creating the Round Features

In this section, you will create the round feature using the **Simple** option.

1. Choose the **Round** button from the **Engineering** group; the **Round** dashboard is displayed.

2. Using the CTRL key, select the edges that are shown in Figure 6-107. The selected edges turn green in color.

Figure 6-106 The default trimetric view

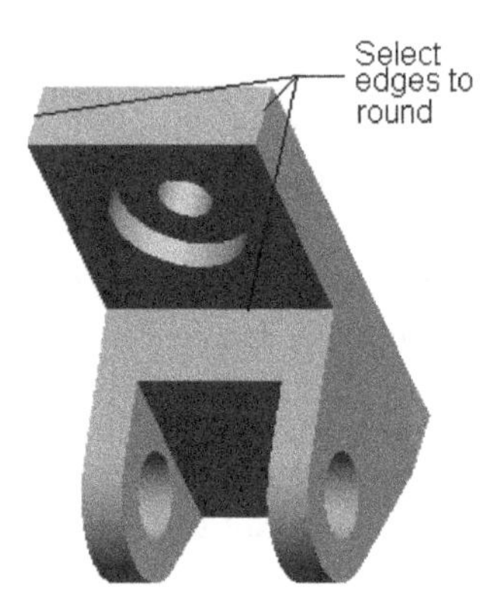

Figure 6-107 Edges selected to create the round feature

3. Enter **3** in the dimension box in the **Round** dashboard or double-click on the default value displayed on the preview of the round and enter the radius value in the edit box displayed. The round is created and its preview appears in the drawing area.

4. Choose the **Build feature** button to exit the round feature creation tool. The trimetric view of the final model with the round feature created is shown in Figure 6-108.

Saving the Model

1. Choose the **Save** button from the **File** menu and save the model.

   The order of feature creation can be seen from the **Model Tree** shown in Figure 6-109.

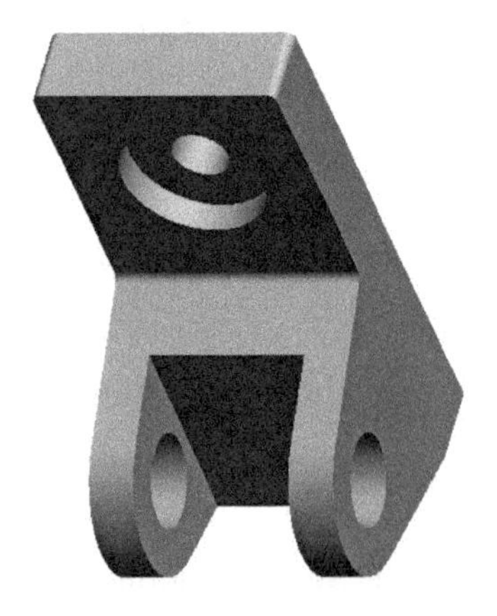

Figure 6-108 Final model for Tutorial 3

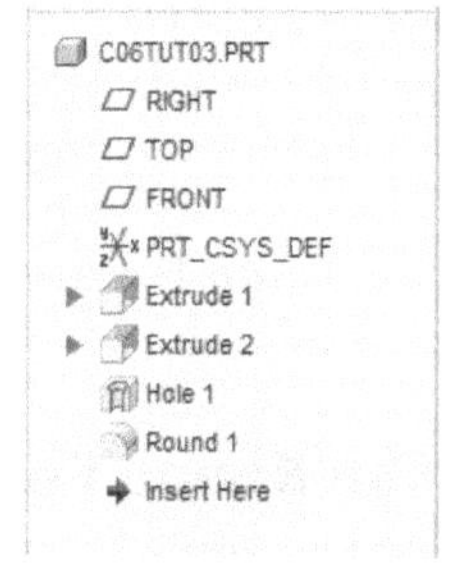

*Figure 6-109 The **Model Tree** for Tutorial 3*

Tutorial 4

Create the model shown in Figure 6-110. The dimensions, top view, and front section view of the model are shown in Figure 6-111. **(Expected time: 45 min)**

The following steps are required to complete this model:

Examine the model and determine the number of features in it, refer to Figure 6-110.

a. Create the base feature, refer to Figures 6-112 and 6-113.
b. Create the cut feature, refer to Figures 6-114 through 6-116.
c. Create the cylindrical feature on an offset datum plane, refer to Figures 6-117 through 6-119.
d. Create a counterbore coaxial hole on the cylindrical feature, refer to Figures 6-120 and 6-121.
e. Create straight holes on the top face of the base feature, refer to Figures 6-122 and 6-123.

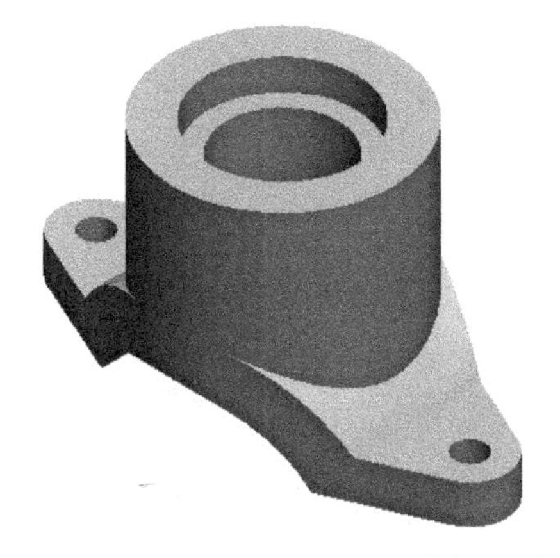

Figure 6-110 *Isometric view of the model*

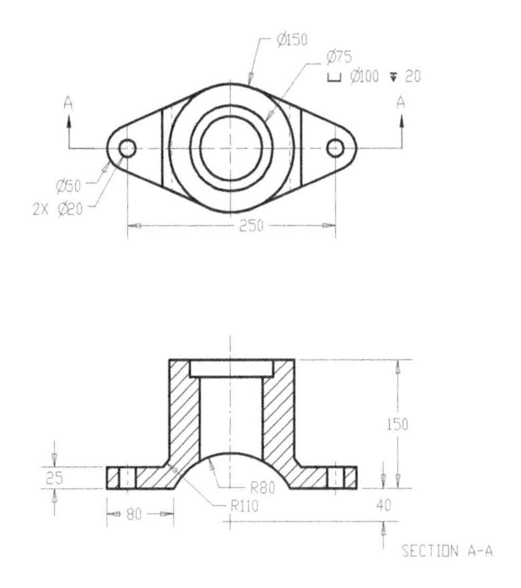

Figure 6-111 Top and front section views of the model

Starting a New Object File

1. Start a new part file and name it as *c06tut04*.

   The three default datum planes are displayed in the drawing area if the **Plane display** button is turned on.

Selecting the Sketching Plane for the Base Feature

In this model, you need to draw the base feature on the **FRONT** datum plane because the direction of extrusion of the base feature is perpendicular to the **FRONT** datum plane.

1. Choose the **Extrude** tool from the **Shapes** group.

2. Choose the **Placement** tab; a slide-down panel is displayed. Choose the **Define** button; the **Sketch** dialog box is displayed.

3. Select the **FRONT** datum plane as the sketching plane. As you select the sketching plane, the **RIGHT** datum plane and its orientation, **Right** are set automatically.

4. Choose the **Sketch** button to enter into the sketcher environment.

Creating the Base Feature

From the model, the section to be extruded for the base feature is not evident. Therefore, you need to visualize the sketch for the base feature. The sketch is shown in Figure 6-112. When this sketch is extruded, it will create the base feature.

1. Draw the sketch using various sketcher tools, as shown in Figure 6-112.

2. The sketch is dimensioned automatically and some weak dimensions are assigned to it. Add the required constraints to it and modify the weak dimensions, as shown in Figure 6-112.

3. Choose the **OK** button; the **Extrude** dashboard is displayed.

4. Choose the **Named Views** button from the **Graphics** toolbar; a flyout is displayed. Choose the **Default Orientation** option from the flyout; the model is set in its default orientation, which is the trimetric view.

   The default view is displayed which gives you a better view of the sketch in the 3D space. Also, an arrow is also displayed on the model indicating the direction of extrusion.

5. Enter **150** in the dimension box that is present on the **Extrude** dashboard.

6. Choose the **Build feature** button from the **Extrude** dashboard.

   The base feature is completed, as shown in Figure 6-113. You can use the middle mouse button to spin the model to view it from various directions.

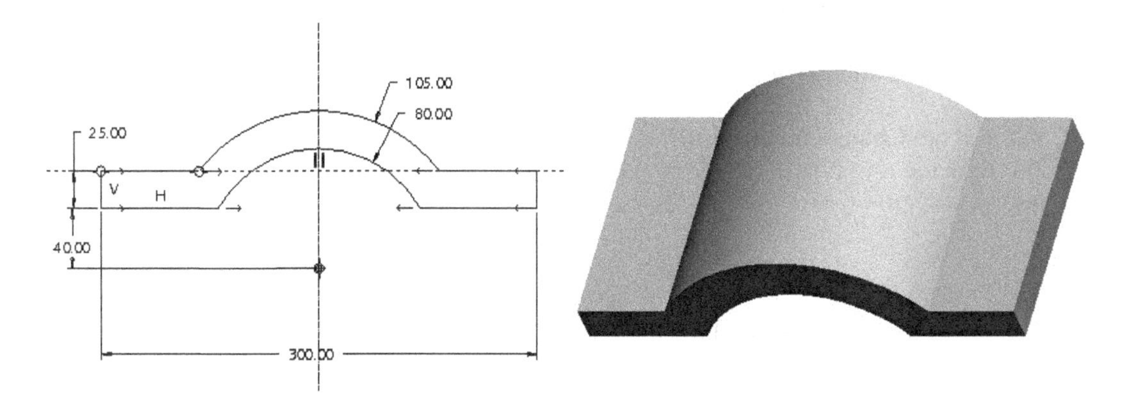

Figure 6-112 Sketch with dimensions and constraints

Figure 6-113 Default view of the base feature

Creating the Second Feature

The next feature is an extruded cut feature. The sketching plane for this feature is the bottom face of the base feature. To get the required shape of the base of the model, you need to cut the base feature in such a way that you get the required shape. This sketch and its dimensions can be referred from the top view of the model shown in Figure 6-111. Before drawing the sketch, change the model display to **No hidden**.

1. Choose the **Extrude** tool from the **Shapes** group.

2. Choose the **Remove Material** button from the **Extrude** dashboard.

3. Choose the **Placement** tab; a slide-down panel is displayed. Choose the **Define** button from the slide-down panel; the **Sketch** dialog box is displayed.

4. Select the bottom face of the base feature as the sketching plane.

5. Select the **RIGHT** datum plane and from the **Orientation** drop-down list, select the **Right** option.

6. Choose the **Sketch** button to enter into the sketcher environment.

7. Draw the sketch of the cut feature, and apply constraints and dimensions to it, as shown in Figure 6-114.

8. Choose the **OK** button to exit the sketcher environment; the **Extrude** dashboard is displayed above the drawing area.

9. Choose the **Shading** button from the **Display Style** drop-down in the **Graphics** toolbar.

10. Choose the **Named Views** button from the **View** tab; a flyout is displayed. Choose the **Default Orientation** option from the flyout; the model orients in its default orientation, which is the trimetric view.

11. Choose the **Change material direction of extrude to other side of sketch** button from the **Extrude** dashboard; an arrow is displayed pointing in the direction of extrusion, as shown in Figure 6-115.

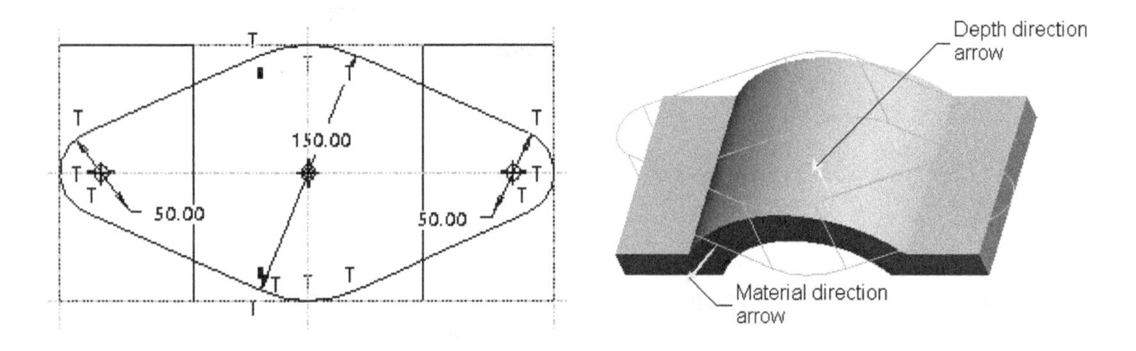

Figure 6-114 Sketch with dimensions and constraints for the cut feature

Figure 6-115 Two arrows and the preview of the cut feature

12. Choose the **Options** tab in the **Extrude** dashboard to display the slide-down panel. From the **Side 1** drop-down list, select the **Through All** option; the model can be previewed in the drawing area.

13. Choose the **Build feature** button from the **Extrude** dashboard to complete the feature creation. The model after creating the cut feature is shown in Figure 6-116.

Creating the Sketching Plane for the Third Feature

You need to create a datum plane to create the third feature. The datum plane will be at a distance of 150 from the bottom face of the model.

1. Choose the **Plane** tool from the **Datum** group; the **Datum Plane** dialog box is displayed.

2. Spin the model using the middle mouse button and then using the left mouse button, select the bottom face of the model.

   When you select the bottom face of the model, the **Offset** constraint is displayed in the **References** collector of the dialog box. Note that the arrow points in the opposite direction. Therefore, you need to enter a negative value of the offset distance.

3. In the **Translation** dimension box, enter **-150** and press ENTER.

4. Choose the **OK** button from the **Datum Plane** dialog box.

   Datum plane **DTM1** is created, as shown in Figure 6-117, and will be selected as the sketching plane for creating the sketch.

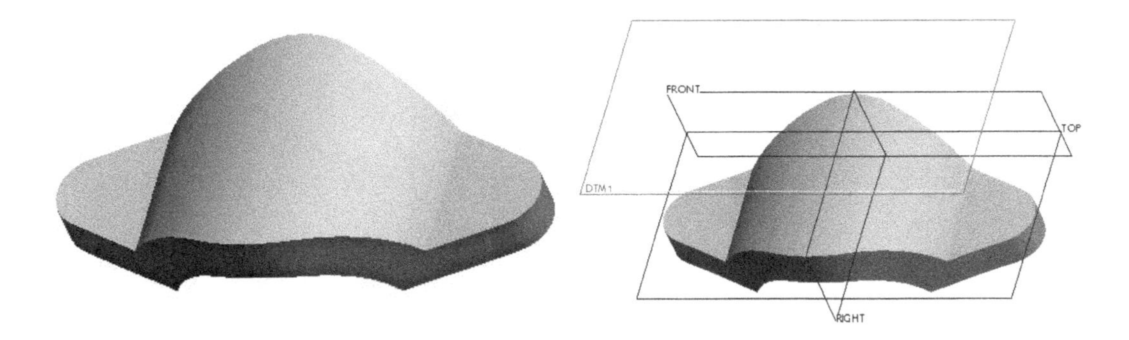

***Figure 6-116** Model after creating the cut feature* ***Figure 6-117** **DTM1** created at an offset distance*

Creating the Third Feature

The plane **DTM1** will be selected as the sketching plane and the depth of extrusion will be given from this plane. The third feature is cylindrical in shape and its outer edge is tangent to the edge of the base feature.

1. Choose the **Extrude** tool from the **Shapes** group; the **Extrude** dashboard is displayed.
2. Choose the **Placement** tab; a slide-down panel is displayed. Choose the **Define** button from the slide-down panel; the **Sketch** dialog box is displayed.
3. Select **DTM1** as the sketching plane; an arrow appears on the datum plane.
4. Select the **FRONT** datum plane, and then from the **Orientation** drop-down list, select the **Top** option.
5. Choose the **Sketch** button to enter into the sketcher environment.
6. Draw a concentric circle for the cylindrical feature, as shown in Figure 6-118.
7. Choose the **OK** button to exit the sketcher environment.
8. Choose the **Named Views** button from the **Graphics** toolbar; a flyout is displayed. Choose the **Default Orientation** option from the flyout; the model orients in its default orientation, which is the trimetric view.
9. Choose the **Options** tab in the **Extrude** dashboard to display the slide-down panel. From the **Side 1** drop-down list, select the **To Next** option. The model can be previewed in the drawing area.
10. Choose the **Build feature** button from the **Extrude** dashboard to complete the feature creation. The model after creating the feature is shown in Figure 6-119.

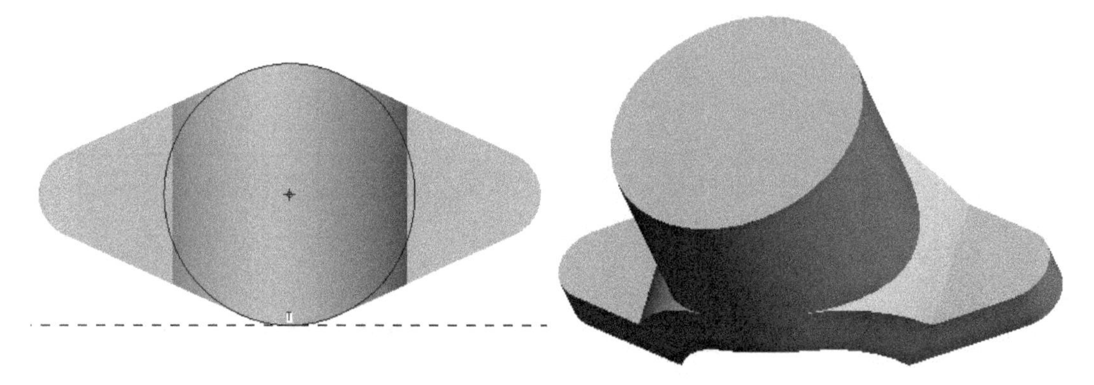

Figure 6-118 *Sketch for the cylindrical feature*

Figure 6-119 *Model after creating the cylindrical feature*

Creating the Counterbore Hole

The fourth feature is a counterbore hole that can be easily created using the **Hole** feature creation tool.

1. Choose the **Hole** tool from the **Engineering** group; the **Hole** dashboard is displayed.

2. Choose the **Create simple hole** and then **Use standard hole profile as drill hole profile** buttons; some buttons are added to the dashboard.

3. Choose the **Adds counterbore** button from the dashboard.

4. Next, choose the **Shape** tab and enter the values, as shown in Figure 6-120.

5. Select the **To Selected** option from the drop-down list and then select the bottom face of the base feature.

6. Choose the **Placement** tab from the **Hole** dashboard; a slide-down panel is displayed.

7. Select the top face of the third feature and then select the axis of the cylindrical feature by pressing the CTRL key; the preview of the hole can be viewed in the model.

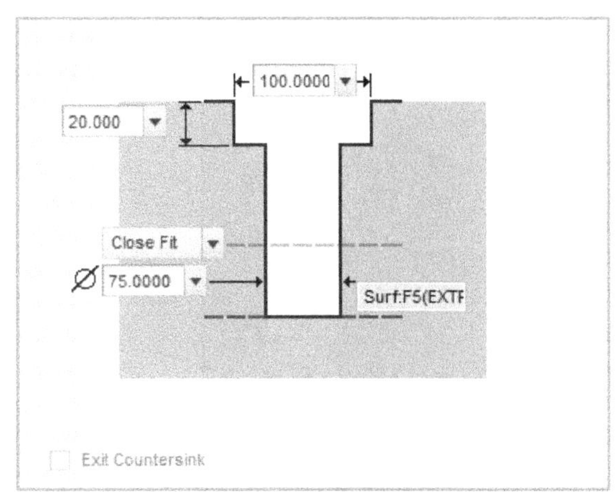

Figure 6-120 *Sketch with dimensions*

The **Coaxial** option is selected automatically in the **Type** drop-down list of the slide-down panel.

8. Choose the **Build feature** button to exit the **Hole** dashboard; the model with the counterbore hole is shown in Figure 6-121.

Creating the Datum Axis for the Hole Features

The last feature is a pair of two holes that is on the top face of the base feature. The two holes will be placed coaxially with the two axes that you need to create.

1. Choose the **Axis** button from the **Datum** group; the **Datum Axis** dialog box is displayed.

2. Select the curved surface shown in Figure 6-122 to create the datum axis.

Figure 6-121 *Model with counterbore hole*

3. Select **Through** from the drop-down list in the **Reference** tab and choose **OK** from the dialog box to accept the settings and exit.

4. Similarly, create a datum axis on the left of the model. Figure 6-123 shows the model after creating the two datum axes.

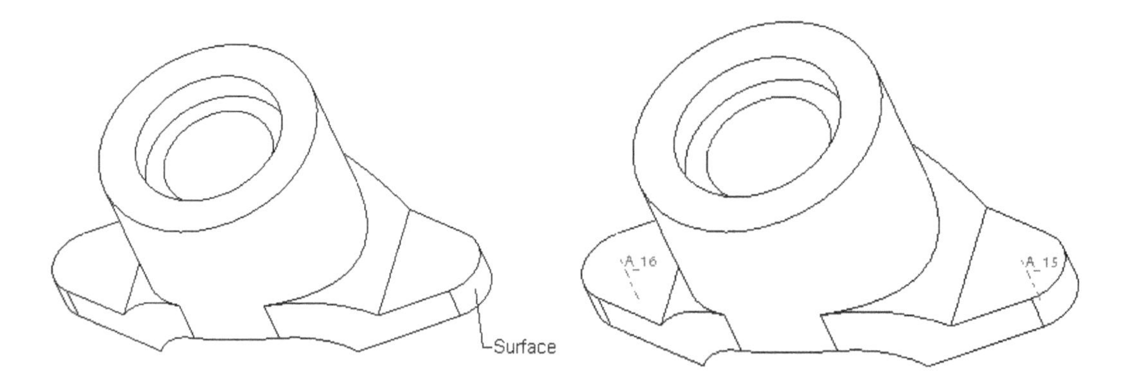

Figure 6-122 Surface to be selected for creating the datum axis

Figure 6-123 The two datum axes

Creating Holes

To create the two coaxial holes, you need to invoke the **Hole** dashboard twice. This is because only one hole can be created at a time by using the **Hole** dashboard.

1. Choose the **Hole** tool from the **Engineering** group; the **Hole** dashboard is displayed.

2. Select the top face of the base feature; the preview of the hole appears on the selected face.

3. Choose the **Placement** tab from the **Hole** dashboard; the slide-down panel is displayed.

4. Select the axis of the feature by pressing the CTRL key.

   The **Coaxial** option is selected automatically in the **Type** drop-down list of the slide-down panel.

5. Enter **20** as the diameter of hole in the dimension box of the **Hole** dashboard and choose the **Drill up to next surface** option from the dashboard.

6. Choose the **Build feature** button to exit the **Hole** dashboard.

7. Again, invoke the **Hole** dashboard and create the second hole similar to the first hole.

   The two holes on the base feature are shown in Figure 6-124.

Saving the Model

1. Choose the **Save** button from the **File** menu and save the model. The order of feature creation can be seen from the **Model Tree** shown in Figure 6-125.

Figure 6-124 *Model after creating the two holes*

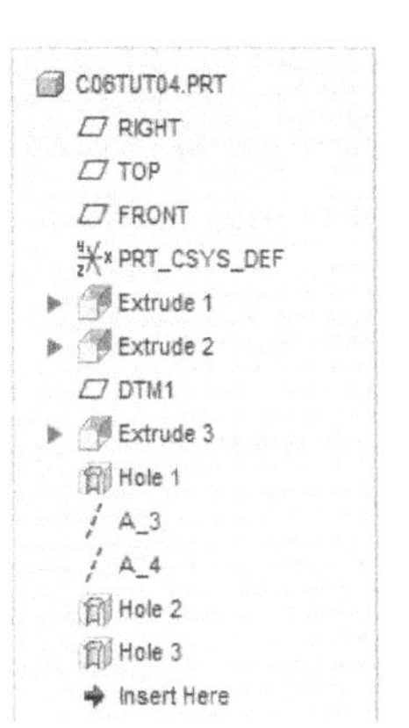

Figure 6-125 *The* **Model Tree** *for Tutorial 4*

Self-Evaluation Test

Answer the following questions and then compare them to those given at the end of this chapter:

1. A hole created using the **Hole** dashboard is parametric in nature. (T/F)
2. A hole cannot be created on both sides of the sketching plane or the placement plane. (T/F)
3. In PTC Creo Parametric, you can sketch profile of a hole. (T/F)
4. The **Full Round** button in the **Sets** slide-down panel is used to enter the radius of the round. (T/F)
5. The **Model Tree** is used extensively in PTC Creo Parametric for editing a model. (T/F)
6. The profile rib feature is always created from the _________ view.
7. A _________ hole is a stepped hole and has two diameters, a larger one and a smaller one.
8. Straight holes are the holes that have a circular cross-section with a _________ diameter throughout the depth.
9. _________ are defined as thin wall-like structures that are used to bind the joints together so that they do not fail under increased load.
10. By default, the suppressed features are not displayed in the _________.

Review Questions

Answer the following questions:

1. Which of the following is a stepped hole with two diameters?

   (a) Counterbore (b) Countersink
   (c) Straight (d) None

2. Which of the following options of the **REROUTE** menu is used to change the sequence in which the features are created?

   (a) **Feature** (b) **Modify**
   (c) **Regenerate** (d) **Reorder**

3. At the intersection of how many edges does a **Corner** chamfer create a beveled surface?

   (a) One (b) Two
   (c) Three (d) None

4. If a feature being suppressed has some child features, then the child features will also be suppressed. (T/F)

5. While in the sketcher mode, the constraints to a sketch should be applied before applying the dimensions. (T/F)

6. The chamfers created in PTC Creo Parametric are parametric in nature. (T/F)

7. The **Model Tree** can be used to redefine a feature. (T/F)

8. When you redefine a rib feature, the __________ dashboard is displayed.

9. The sketch of a rib feature can be extruded to __________ side(s) of the sketching plane.

10. Rounds and chamfers are used in engineering components to reduce the __________ on the corners.

Exercises

Exercise 1

Create the model shown in Figure 6-126. The dimensions and front and top views of the model are shown in Figure 6-127. **(Expected time: 45 min)**

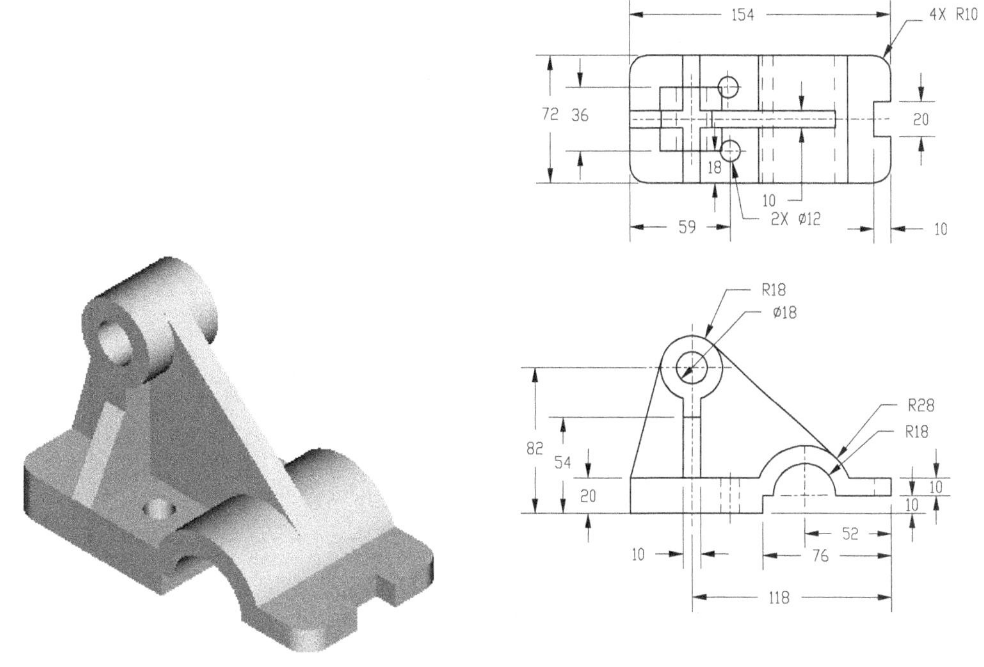

Figure 6-126 *Isometric view of the model*

Figure 6-127 *Top and front views of the model with hidden lines suppressed*

Exercise 2

Create the model shown in Figure 6-128. The dimensions and front and right-side views of the model are shown in Figure 6-129. **(Expected time: 30 min)**

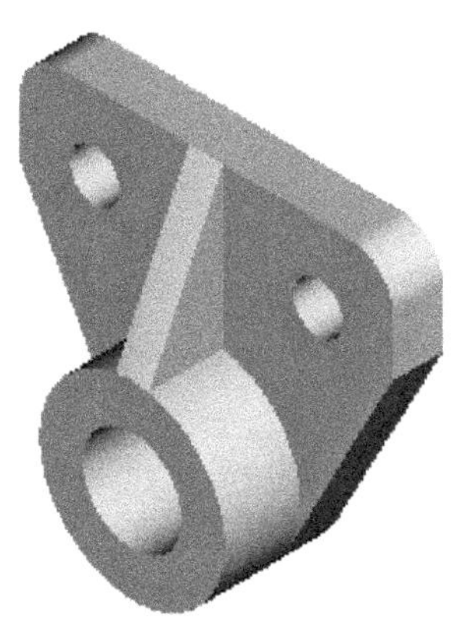

Figure 6-128 *Isometric view of the model for Exercise 2*

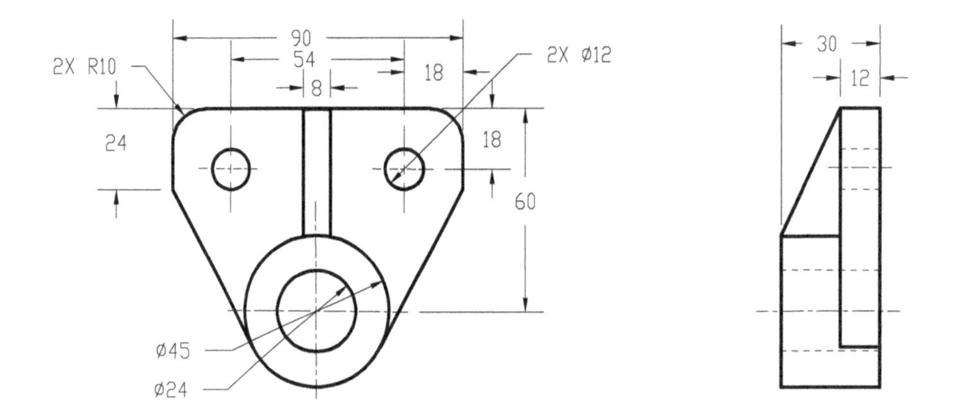

Figure 6-129 *Front and right-side views of the model*

Answers to Self-Evaluation Test

1. T, **2.** F, **3.** T, **4.** F, **5.** T, **6.** side, **7.** counterbore, **8.** constant, **9.** Ribs, **10. Model Tree**

Chapter 7

Options Aiding Construction of Parts-II

Learning Objectives

After completing this chapter, you will be able to:

- *Create a dimension pattern.*
- *Create a direction pattern.*
- *Create an axis pattern.*
- *Create a fill pattern.*
- *Create a reference pattern.*
- *Create a table-driven pattern.*
- *Create a curve-driven pattern.*
- *Create a point pattern.*
- *Control the size of pattern instances using constraints.*
- *Use the Mirror option.*
- *Create a simplified representation of a model.*

INTRODUCTION

In this chapter, you will learn about various methods for duplicating existing features. In PTC Creo Parametric, you can duplicate a feature by using the following methods:

- **Pattern**
- **Copy**
- **Mirror**

You will also learn to create simplified representation in solid models. A solid model is sectioned for viewing the profile of its cross-section and also for creating section drawing views.

CREATING FEATURE PATTERNS

Patterns are used to create incremental array of features in one or two directions from a single feature called the parent feature or the leader. When a pattern is created, the leader also becomes a part of the pattern. When you pattern a feature, you need to specify the total number of features to be created, including the one that is being patterned and the increment in the dimensions, if required.

Uses of Patterns

Patterns are very helpful in solid modeling as they speed up the model creation process. The uses of patterns in solid modeling are discussed next.

1. Patterns are used to create multiple copies of a feature, and therefore, save time that would otherwise be spent in creating the features individually.

2. All instances in a pattern, including the parent feature, act as a single feature. Therefore, they can be easily suppressed or mirrored.

3. All instances in a pattern are related parametrically. Therefore, you can modify the number of instances in a pattern, the spacing between the instances, and other pattern related parameters.

4. If the dimensions of the parent feature are modified, then the dimensions of the child features are also modified.

Creating Patterns

Ribbon: Model > Editing > Pattern > Pattern

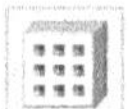

In PTC Creo Parametric, patterns are created by choosing the **Pattern** tool from the **Editing** group. This tool is an object-action tool. This means the **Pattern** tool is enabled in the **Editing** group only when you have selected the feature to be patterned.

To pattern a feature, choose the **Pattern** tool from the **Pattern** drop-down available in the **Editing** group; the **Pattern** dashboard will be displayed, as shown in Figure 7-1. The options and tools in the dashboard depend on the type of pattern you create. In PTC Creo Parametric, you can create eight types of pattern. These patterns are **Dimension**, **Direction**, **Axis**, **Fill**, **Table**, **Reference**, **Curve**, and **Point**. The options and tools in the **Pattern** dashboard are discussed next.

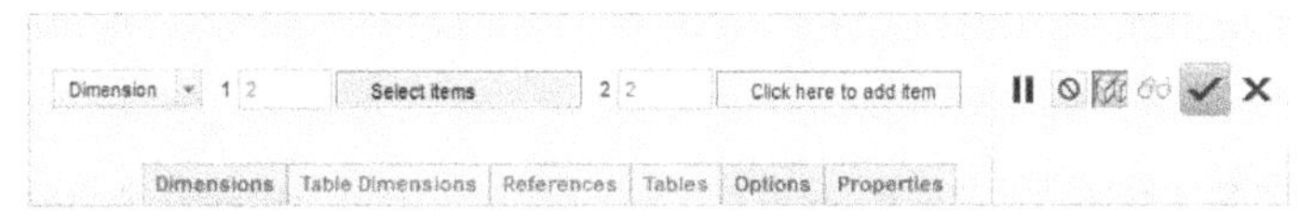

*Figure 7-1 The **Pattern** dashboard*

Drop-down List

In the drop-down list of the **Pattern** dashboard, the **Dimension** option is selected by default. The other options in this drop-down list are **Direction**, **Axis**, **Fill**, **Table**, **Reference**, **Curve**, and **Point**. When you select these options from this drop-down list, the options and tools in the **Pattern** dashboard change accordingly. The types of pattern that can be created in PTC Creo Parametric are discussed next.

Dimension Patterns

In dimension patterns, the existing dimensions of the parent feature are used to create a pattern. This pattern can be created in one direction or in two directions. When you select the option to create the pattern in the second direction, all instances that were created in the first direction can also be created in the second direction. You need to specify the increment value for the instances, which can be positive or negative. Once you have specified the increment value in a direction, the system creates the specified number of instances (including the parent feature) in that direction. Figure 7-2 shows a hole to be patterned and the patterned hole is shown in Figure 7-3.

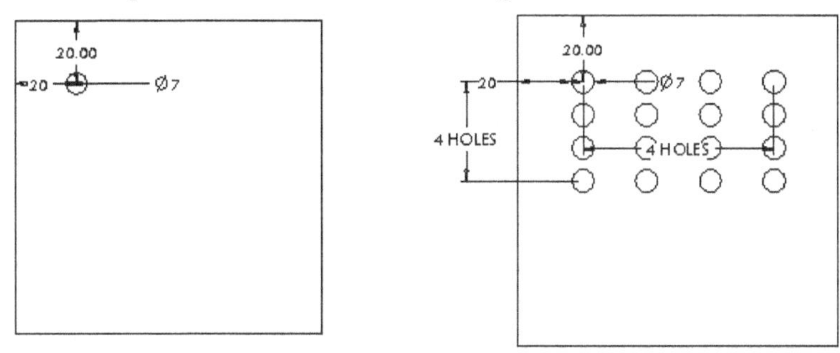

Figure 7-2 *Hole to be patterned* ***Figure 7-3*** *Hole patterned in two directions*

To create a dimension pattern, use the following steps:

1. Select the feature to be patterned and then choose the **Pattern** tool.

2. Click the dimension in the first direction; an edit box will be displayed. Enter the increment in the dimension where you need to place the second instance of the pattern.

3. Enter the number of instances in the **1** edit box of the dashboard. After specifying the instances in the first direction you need to specify the number of instances and the increment in dimension in the second direction. Note that the dimension in the second direction should be selected only if you need to create instances in the second direction.

4. Click on **Click here to add item** in the **2** collector; it will display **Select items**. Click the dimension in the second direction; an edit box will be displayed. Enter the dimension increment in the edit box.

5. In the **2** edit box, enter the number of instances needed in the second direction. On doing so, you can view the black dots where the instances of the pattern will be placed.

6. Choose the **Build feature** button to exit the dashboard and to view the pattern.

Direction Patterns

The **Direction** option is used to create the pattern in the specified direction. The direction is specified by selecting an edge, plane, axis, or linear curve. This pattern is different from the **Dimension** pattern. This pattern does not use the dimensions of the feature to be patterned. When you select this option from the drop-down list, you will be prompted to select the reference for the first direction. After specifying the first direction, an arbitrary dimension that is not the dimension of the parent feature will be displayed in the specified direction. You can specify the second direction to place the instances.

In PTC Creo Parametric, there are three options to define the direction reference, **Translation**, **Rotation**, and **Coordinate System**. These options are available in the **Translation** drop-down lists adjacent to **1** and **2** edit boxes. The drop-down list adjacent to **1** edit box is used for defining direction reference in the first direction and the drop-down list adjacent to **2** edit box is used for defining direction reference in the second direction. The options in these drop-down lists are discussed next.

Translation. When this option is selected in the drop-down list, then you need to select a flat face, axis, plane, linear curve, or coordinate system axis to define the direction reference. When you select a flat face or a plane, then a linear pattern is created in the direction perpendicular to the selected face or plane. When you select an axis, linear curve or a coordinate system axis, then a linear pattern is created in the direction of that curve. Now you need to enter the spacing between two instances in the corresponding edit boxes.

Rotation. When this option is selected in the drop-down list, then you need to select a linear curve, coordinate system axis, or an axis to define the direction reference. When you select an axis, linear curve or a coordinate system axis, then a circular pattern is created around that curve. Next you need to enter the required values in the corresponding edit boxes.

Coordinate System. When this option is selected in the drop-down list, then you need to select a coordinate system to define the direction reference. Select a coordinate system;

the dashboard will be displayed, as shown in Figure 7-4. Specify the required values in **X**, **Y**, and **Z** edit boxes to define the direction vector.

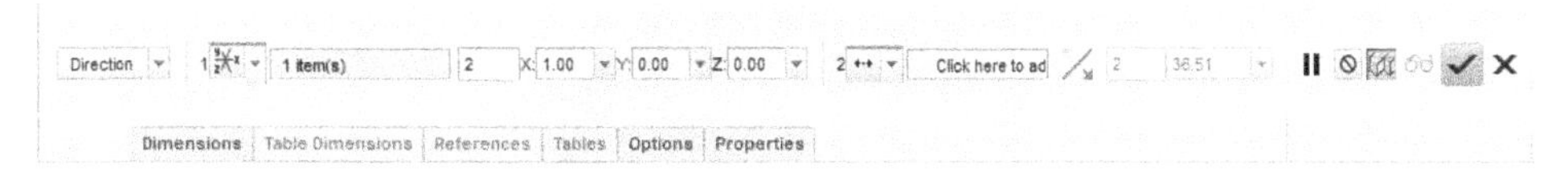

Figure 7-4 *Partial view of the* ***Pattern*** *dashboard with the* ***Coordinate System*** *option selected*

Note that the direction of pattern can be reversed by choosing the **Flip** button in the **Pattern** dashboard. Figure 7-5 shows the linear curve selected to create the direction pattern. The direction is indicated by an arrow. The preview of the pattern instances are shown as black dots and the pattern created is shown in Figure 7-6.

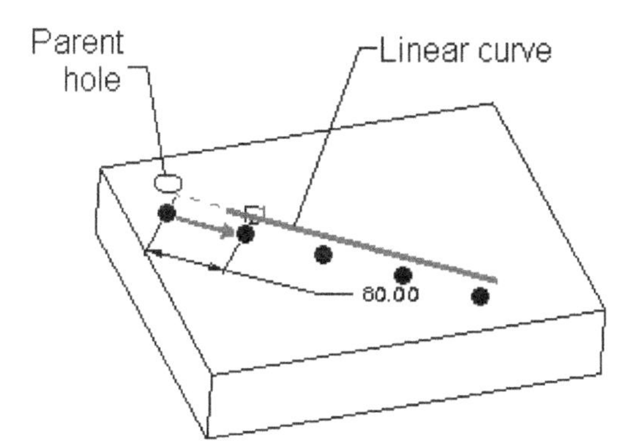

Figure 7-5 *Linear curve selected to create the direction pattern*

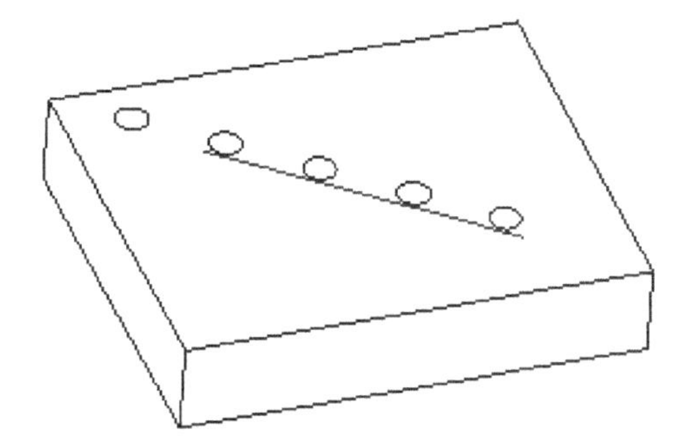

Figure 7-6 *Direction pattern created*

Note

The reference that you select for specifying the direction of a ***Direction*** *pattern should belong to a feature created before the feature to be patterned. Otherwise, the desired reference will not be highlighted for selection.*

Axis Patterns

The **Axis** option is used to create rotational patterns. When you choose this option, the **Pattern** dashboard will be displayed, as shown in Figure 7-7. To create this type of pattern, an axis is required. The axis can be a datum axis or the axis of a feature. For example, an axis formed by a revolve feature, like a hole. The parameters that you need to specify to create an axis pattern are shown in Figure 7-8.

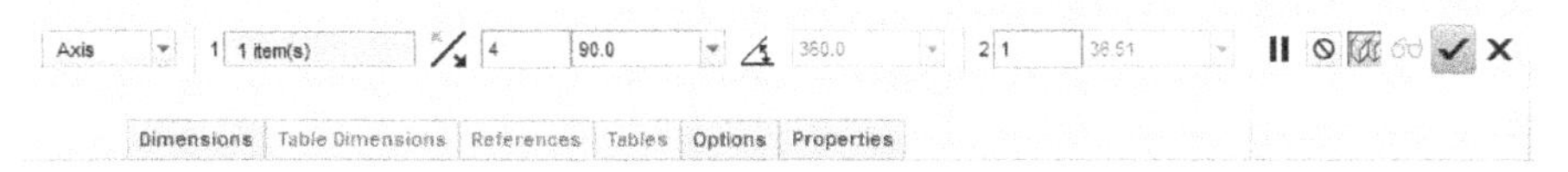

Figure 7-7 *Partial view of the* ***Pattern*** *dashboard with the* ***Axis*** *option selected*

Note

Remember that for creating the axis pattern of a feature, the reference axis must be created before the feature that is to be patterned.

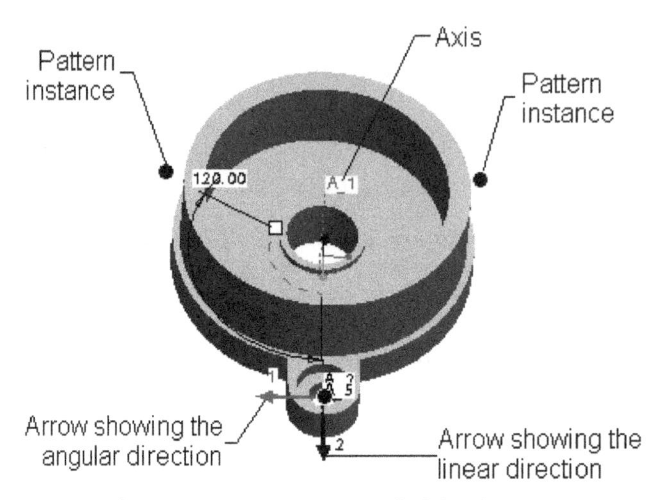

Figure 7-8 *Parameters to be specified for the axis pattern*

To create an axis pattern, use the following steps:

1. Select the feature to be patterned and then choose the **Pattern** tool. Now, select the **Axis** option from the drop-down list in the **Pattern** dashboard; you will be prompted to select an axis about which the pattern will be created.

2. Turn on the display of the axis and then select it.

3. Enter the required angle between each instance and then the number of pattern instances in the edit box. Instead of specifying the angle between each instance, you can specify the total angle of pattern by choosing the **Set the angular extent of the pattern** button from the dashboard. When you enter the value of angular extent of the pattern in the **Enter the angular extent** edit box, the instances will be placed at equal distance on the arc length.

   The instances in the first direction have been specified and now you can specify the number of instances and the increment in dimension in the second direction. Note that the dimension and number of instances in the second direction should be specified only if you need to create instances in the second direction. To create the rotational pattern, as shown in Figure 7-9, you need to create instances in the first direction. The instances in the second direction, along with the instances in the first direction, are shown in Figure 7-10.

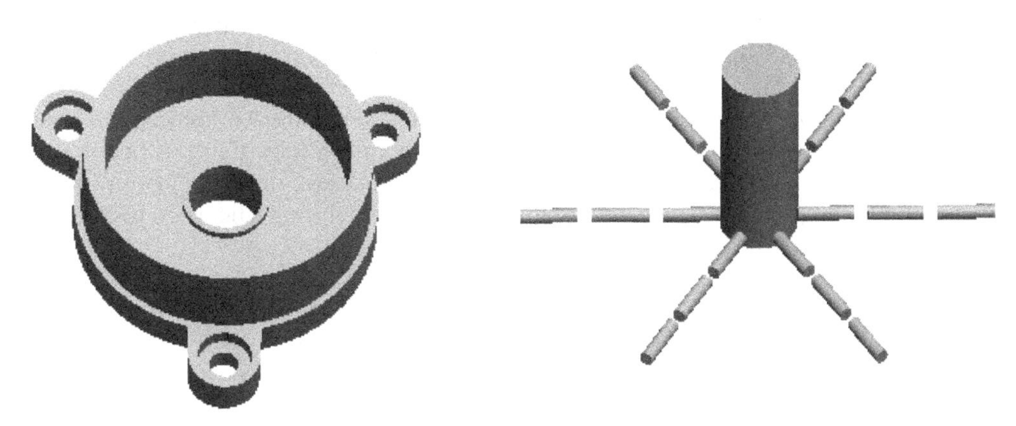

Figure 7-9 *Rotational pattern created using the axis option*

Figure 7-10 *Pattern created both in circular and linear directions*

4. Enter the number of pattern instances in the **2** edit box for the second direction and then enter the radial distance between the pattern instances in the edit box. The pattern is created and now you can view the black dots where the instances will be placed.

5. Choose the **Build feature** button to exit the dashboard and view the pattern.

Fill Patterns

Fill patterns are used to fill the sketched area by a selected feature. This is the easiest and fastest method to create a pattern in the sketched area. When you select the **Fill** option from the drop-down list, the **Pattern** dashboard will appear, as shown in Figure 7-11.

The options and tools in this dashboard are discussed next.

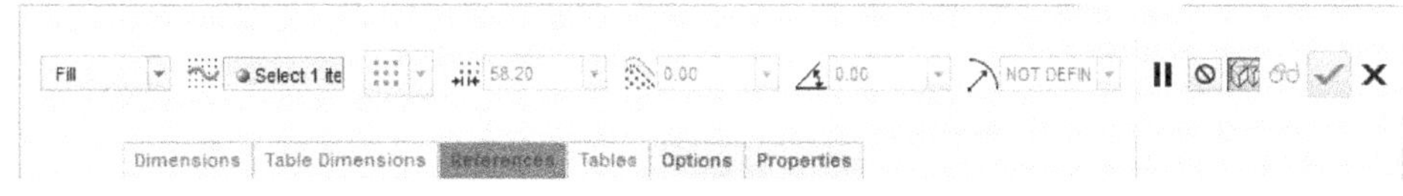

Figure 7-11 *The **Pattern** dashboard with the **Fill** option selected*

References Tab: When you choose this tab, the slide-down panel will be displayed, in which the **Define** button is available. Choose the **Define** button to invoke the **Sketch** dialog box. This dialog box is used to sketch the area that will be filled by the pattern instances. Select the sketching plane and its orientation and then enter the sketcher environment to draw the sketch.

Drop-down list for grid template: This drop-down list is available only when the area to be filled is defined. The options in this drop-down list are used to specify the shape of the fill pattern. Using the options in this drop-down list, you can shape the fill pattern along square pattern, diamond pattern, hexagon pattern, concentric circle pattern, spiral pattern, and along the sketched curves. Generally, it is recommended that the sketch should be similar to the shape of the fill pattern you need. Figures 7-12 through 7-17 show the patterns of various shapes.

Dimension Boxes: The dimension boxes in the dashboard are used to specify the different parameters of the fill pattern. You can specify the gap between the two members of the pattern, the distance of all members (instances) from the sketched boundary, rotation angle of the pattern, and radial distance between instances, in case of circular and spiral shapes.

Options Tab: When you choose this tab, the slide-down panel will be displayed in which four check boxes are available, as shown in Figure 7-18. These check boxes are discussed next.

When you select the **Use alternate origin** check box, you will be prompted to specify the alternate origin point for the pattern. Select a point from where you want to start the pattern.

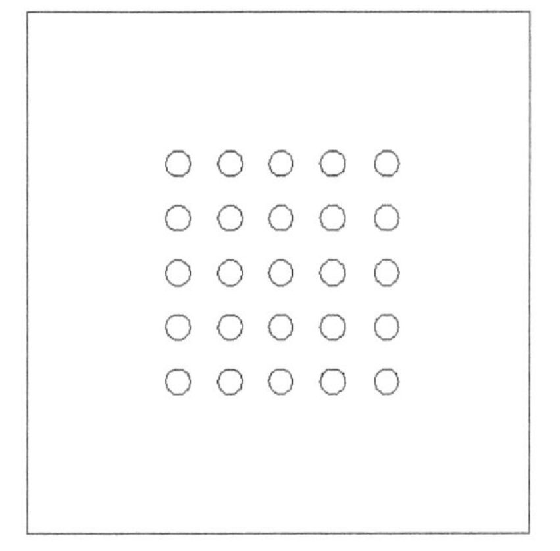

Figure 7-12 *The square pattern fill in the sketched square*

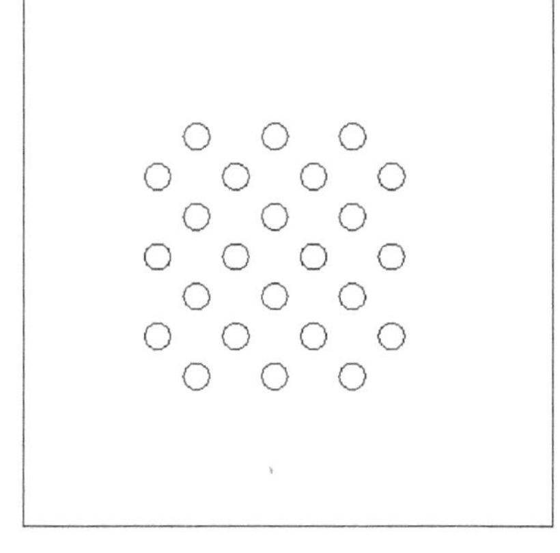

Figure 7-13 *The diamond pattern fill in the sketched square*

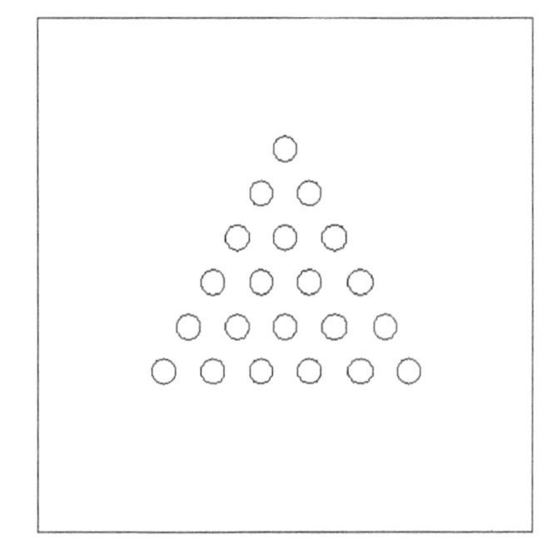

Figure 7-14 *The hexagon pattern fill in the sketched triangle*

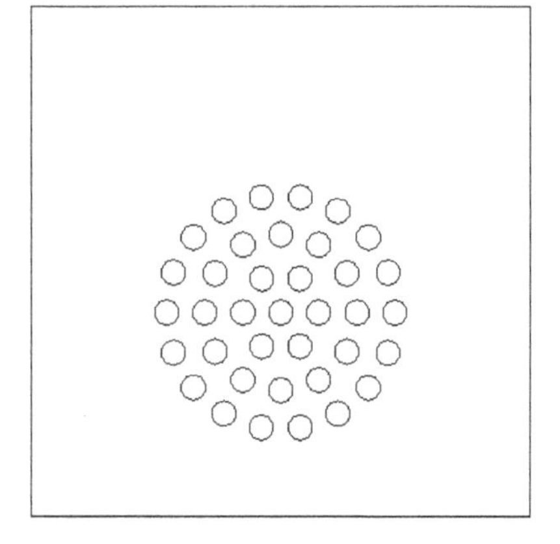

Figure 7-15 *The concentric circle pattern fill in the sketched circle*

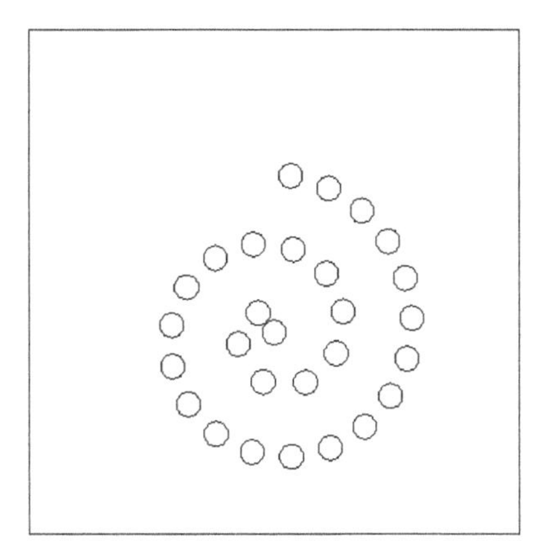

Figure 7-16 The spiral pattern fill in the sketched circle

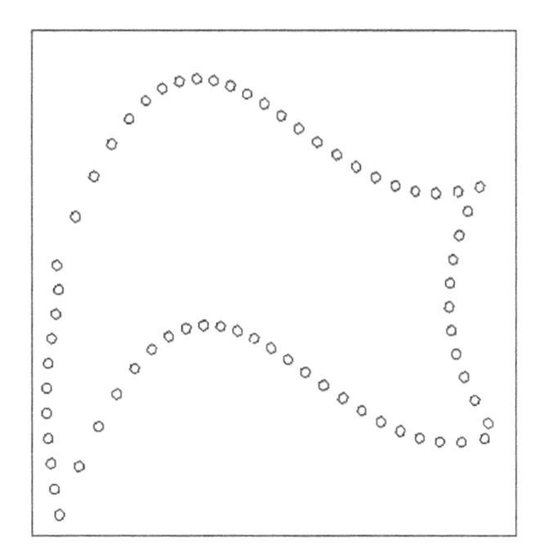

Figure 7-17 Along the sketched curves pattern fill in the sketched curve

The **Follow leader location** check box is selected by default. When you clear this check box, the pattern will be created on the plane which was selected for drawing the area to be filled.

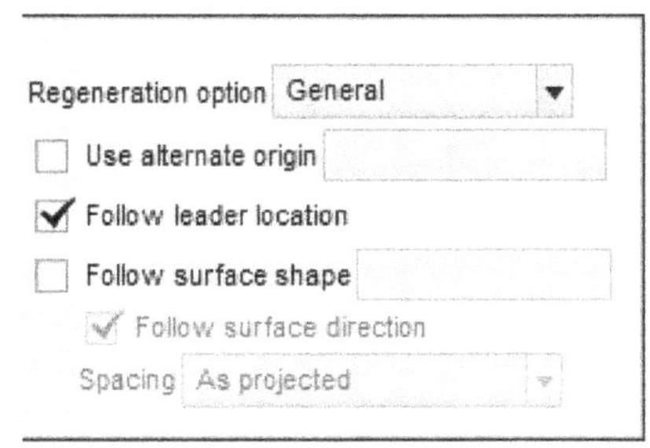

*Figure 7-18 The **Options** slide-down panel*

When you select the **Follow surface shape** check box, you will be prompted to select a surface. Select a surface; the pattern instances will follow the shape of the surface.

The **Follow surface direction** check box is active only when you select the **Follow surface shape** check box. When you select the **Follow surface direction** check box, the **Spacing** drop-down list will get activated. In this drop-down list, three options are available to define the spacing of the pattern instances on the surface. When you select the **As projected** option from the drop-down list, the instances will be projected on the surface opposite to the selected surface. When you select the **Map to Surface Space** option, the instances will be placed on the selected surface but the spacing between them will not be equal. When you select the **Map to Surface UV Space** option, the instances will be equally spaced on the surface.

Tip: *While creating a pattern, you can select the member that you want to remove. This can be done before exiting the **Pattern** dashboard. To remove the member of a pattern, select it in the preview of the pattern. The selected member turns white, suggesting that it is removed. To resume the removed member of the pattern, select it once again.*

Reference Patterns

In the reference patterns, an existing pattern is referenced to create a new pattern. In this pattern, the parent feature of the new pattern should be referenced to the parent feature of the existing pattern.

Figure 7-19 shows a rib feature referenced to the parent hole feature. In this figure, the parent rib feature is created on a plane that was created on-the-fly while creating the rib feature. This plane passes through the axis of the parent hole feature. Therefore, a relationship is built between the parent rib feature and the parent hole feature.

To create the reference pattern of the rib feature, select it from the **Model Tree** and choose the **Pattern** tool from the **Editing** group to invoke the **Pattern** dashboard. The **Reference** option is selected by default in the **Pattern** dashboard. Choose the **Build feature** button; the pattern of the rib feature is created without specifying the increment in dimensions. Figure 7-20 shows the rib feature patterned using the reference of the hole pattern.

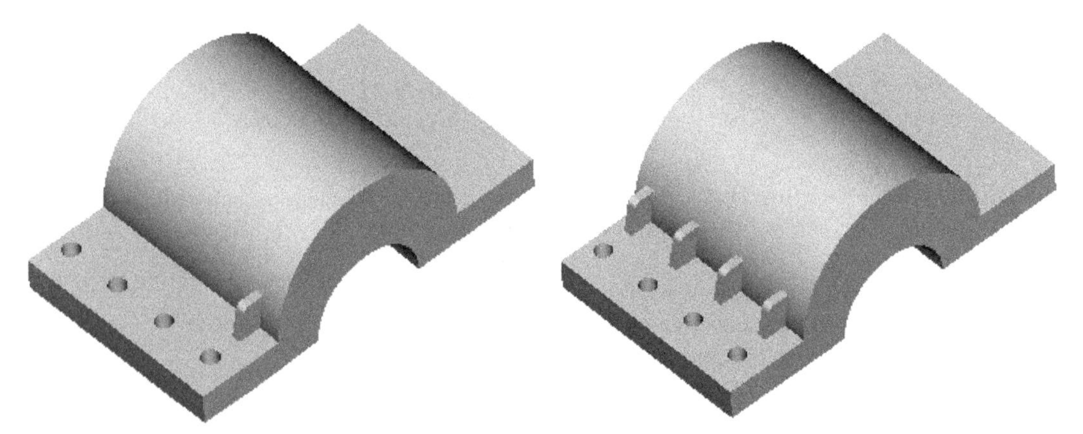

Figure 7-19 *Rib feature referenced to the parent hole feature*

Figure 7-20 *Rib feature patterned using the* ***Reference*** *pattern*

Note

If the rib feature is not created on an embedded datum plane, the relationship between the hole feature and the rib feature will not exist. Therefore, the rib feature cannot be patterned using the ***Reference*** *option. Ensure that the feature which will be patterned using the* ***Reference*** *option, has a relationship with the parent feature, which in this case will be the patterned hole feature.*

Tip: *If you do not want to create an embedded datum plane but still want to create a reference pattern, as shown in Figure 7-20, you need to create a datum plane with a rib feature on it. Group the datum plane and the rib feature. After grouping the two, select the group from the* ***Model Tree*** *and choose the* ***Pattern*** *tool. Now, the reference pattern can be created.*

Table Patterns

Table-driven patterns are created by defining a table. In this table, you need to specify the dimensions of the instances from the edge or faces from where the leader of the pattern is referenced (leader of a pattern is the feature that is selected to create a pattern). When you select the **Table** option from the drop-down list, the **Pattern** dashboard will appear, as shown in Figure 7-21.

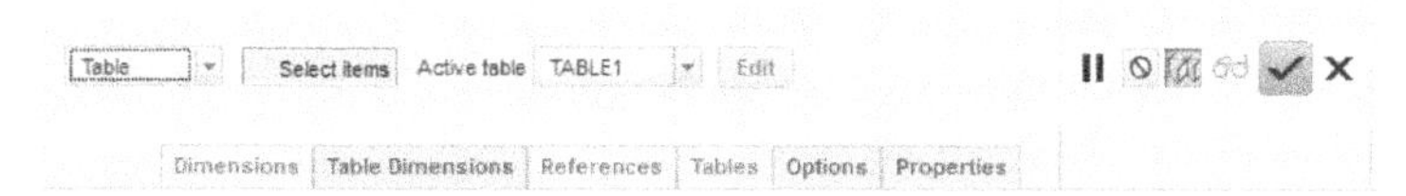

Figure 7-21 *The **Pattern** dashboard with the **Table** option selected*

In Figure 7-21, the **Active table** collector shows the active table. You can create more than one table to create a pattern. You can choose the **Edit** button to display the **Pro/TABLE** window as shown in Figure 7-22. This button will be available only after you have selected at least one dimension.

To create the table-driven pattern shown in Figure 7-23, follow the steps given below:

1. Create a solid protrusion of dimension 120 X 100 X 15.

2. Create a through hole of diameter **5** on the top face of the base feature at a distance of 20 from the left edge and 20 from the bottom edge.

3. Select the hole feature and choose the **Pattern** tool from the **Editing** group; the **Pattern** dashboard will be displayed and the dimensions of the hole feature will appear on the hole.

4. Select the **Table** option from the drop-down list in the dashboard.

5. Select the dimension which is **20** from the left edge and then use CTRL+left mouse button to select the dimension which is **20** from the bottom edge.

6. Choose the **Edit** button from the **Pattern** dashboard to display the **Pro/TABLE**.

7. In column **C1**, click under **idx** (index number). Enter **1** when the cell is highlighted. The value 1 signifies the first instance.

8. In column **C2**, toward the right of **1**, enter the distance along the first dimension and in column **C3**, enter the distance along the second dimension.

9. Similarly, under the rows below **idx**, enter the dimensions of other instances of the holes, as shown in Figure 7-24. Remember that the distance of each hole is measured from where the leader is dimensioned.

10. Choose **File** > **Save**, and then choose the **Close** option to exit the **Pro/TABLE**.

11. Choose the **Build feature** button to exit the feature creation tool.

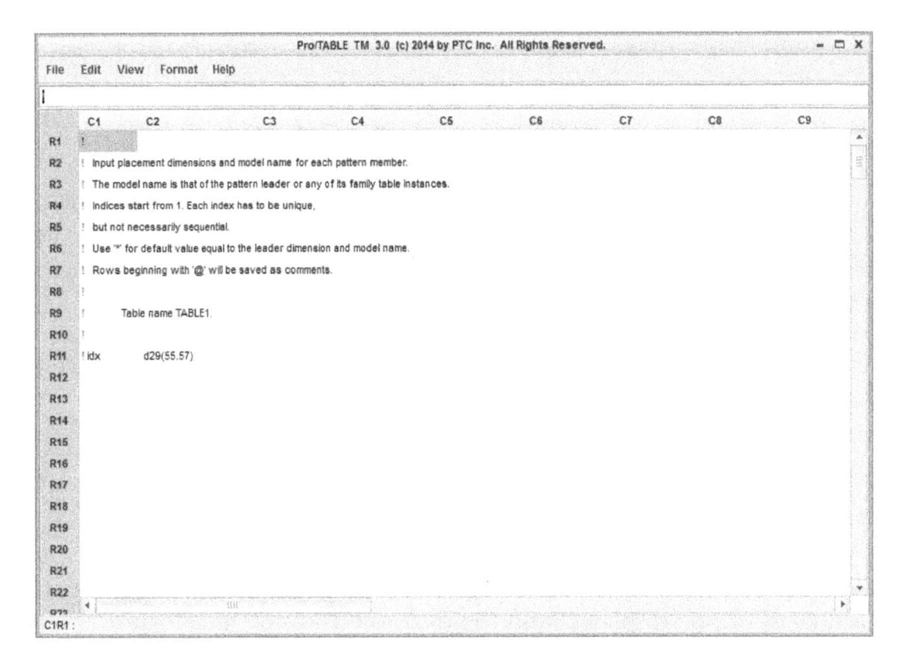

Figure 7-22 The ***Pro/TABLE*** *window*

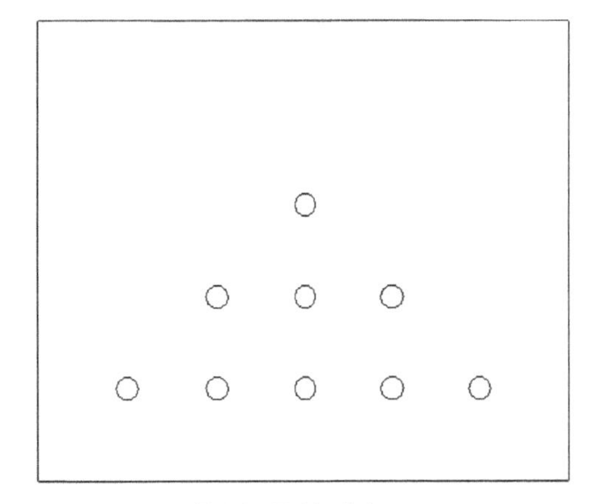

Figure 7-23 *Table-driven pattern*

! idx	d7(20.00)	d8(20.00)
1	40	40
2	40	20
3	60	20
4	60	40
5	60	60
6	80	20
7	80	40
8	100	20

Figure 7-24 *Coordinate values of instances*

Tip: *Pattern tables can be created from scratch by picking dimensions and filling in all values. In most cases, it is easier and less time-consuming to create a dimensional pattern similar to what you need and then convert it into a table.*

Generally, table-driven patterns are used when the incremental distances between the instances of the pattern are nonuniform. This pattern is also useful when the coordinate locations of the instances are known.

Curve Driven Pattern

The **Curve** option is used to create patterns along user-defined curves. When you select the **Curve** option from the drop-down list, the **Pattern** dashboard will appear, as shown in Figure 7-25.

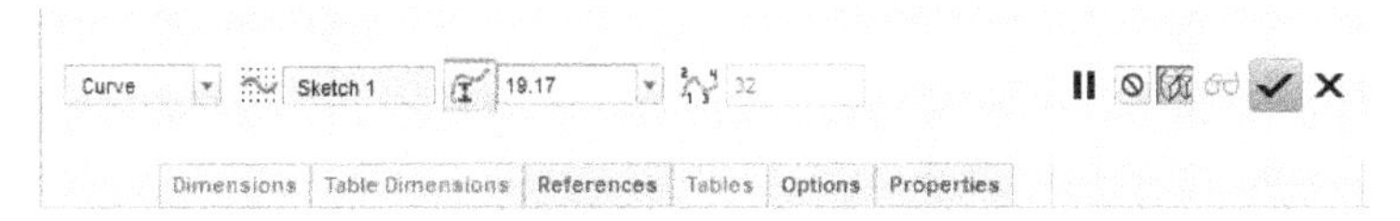

*Figure 7-25 Partial view of the **Pattern** dashboard with the **Curve** option selected*

To create a pattern, first you need to sketch a curve. This curve also specifies the direction for the pattern. All instances created using the **Curve** option are identical in size and geometry. The pattern instances also follow the shape of the sketched curve. After choosing the **Curve** option from the drop-down list, you have to sketch a curve for patterning. Then enter the spacing between the patterned instances in the edit box in the pattern dashboard. The preview of the patterned feature can be seen in the drawing area. Figure 7-26 shows the model with the feature to be patterned along the sketched curve, and Figure 7-27 shows the patterned feature.

Note

The start point of the curve pattern acts as the start point of the curve. The blue arrow on the curve displays the start point of the curve and the direction of the pattern.

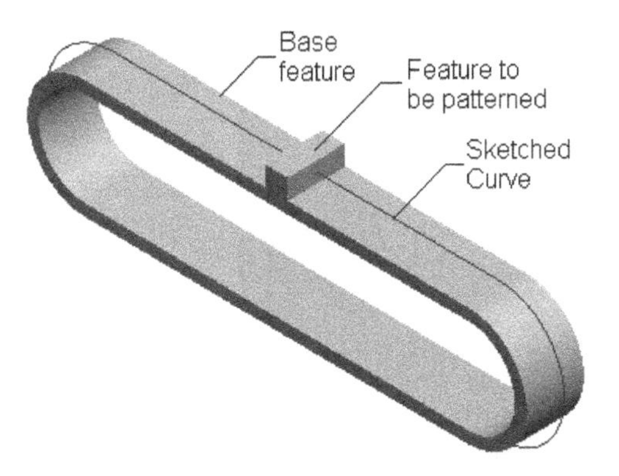

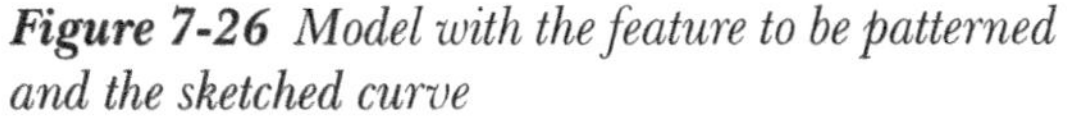

Figure 7-26 Model with the feature to be patterned and the sketched curve

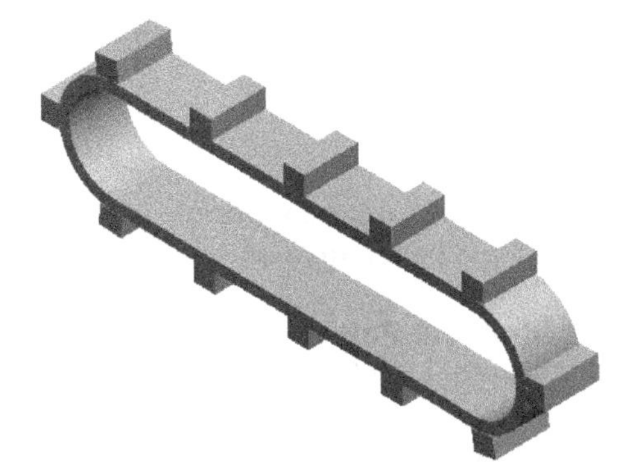

Figure 7-27 The curve-driven pattern

Point Pattern

The **Point** option is used to create patterns at sketched points, datum points, or coordinate systems. To do so, select the feature to be patterned and invoke the **Pattern** tool, as discussed earlier. Select **Point** from the first drop-down list; the dashboard will be displayed, as shown in Figure 7-28.

*Figure 7-28 Partial view of the **Pattern** dashboard with the **Point** option selected*

The two buttons in the dashboard enable you to select the geometry points, geometry coordinate systems, or datum points at which the pattern members will be placed. Select the required entities from the model area and then choose the **Build Feature** button; a pattern will be created at each point or coordinate system, refer to Figures 7-29 and 7-30. Note that the points and the coordinate systems must be drawn first and then you need to invoke the **Pattern** tool. You can also use the geometric points or coordinate systems used in an internal sketch or in an imported feature.

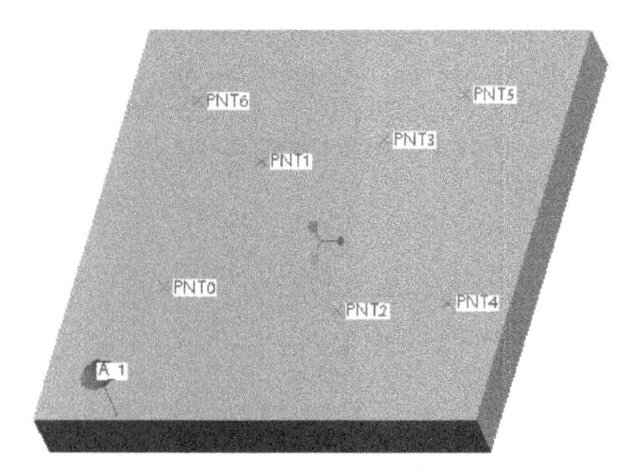

Figure 7-29 Hole on the base feature

Figure 7-30 Hole patterned on the base feature

Options Tab

When you choose the **Options** tab, a slide-down panel will be displayed. The options in this slide-down panel are discussed next.

Identical Pattern

The **Identical** option is used to create an identical pattern. You need to select at least one incremental dimension to pattern the feature. Depending on the incremental dimension selected, the resulting pattern will be linear or rotational. A linear pattern is created when the driving dimension is linear and a rotational pattern is created when the driving dimension is angular. You can enter a positive or a negative value as the increment in a pattern dimension. All instances of a pattern that are created using this option are identical in size and geometry. This is the reason the patterns created using this option are known as Identical patterns. Figure 7-31 shows a hole feature on the base feature and Figure 7-32 shows the holes patterned linearly.

Similarly, Figure 7-33 shows a hole feature on the base feature and Figure 7-34 shows the rotational pattern of the hole feature.

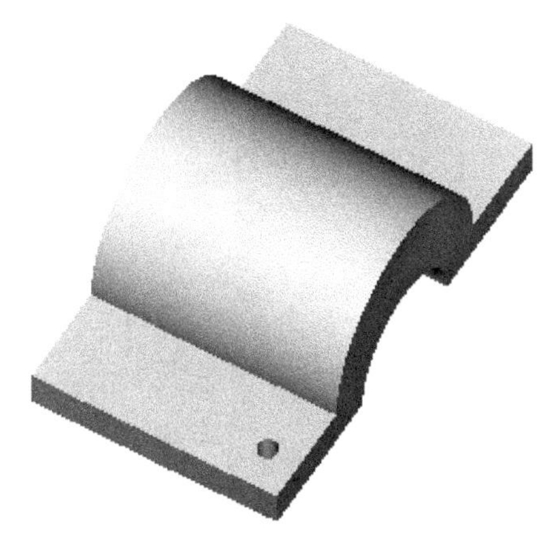

Figure 7-31 *Hole on the base feature*

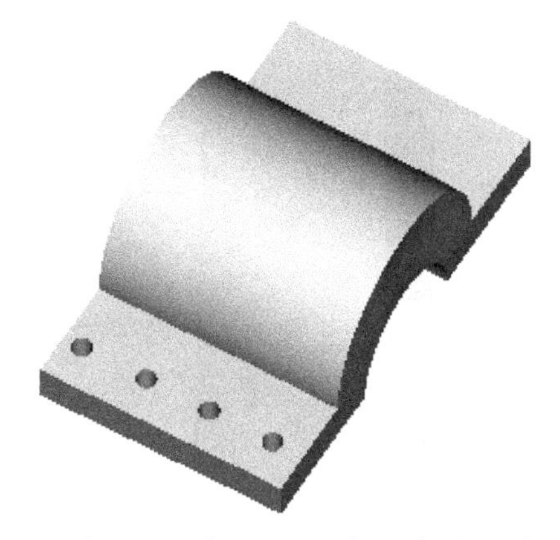

Figure 7-32 *Hole patterned on the base feature*

Figure 7-33 *Hole on the base feature*

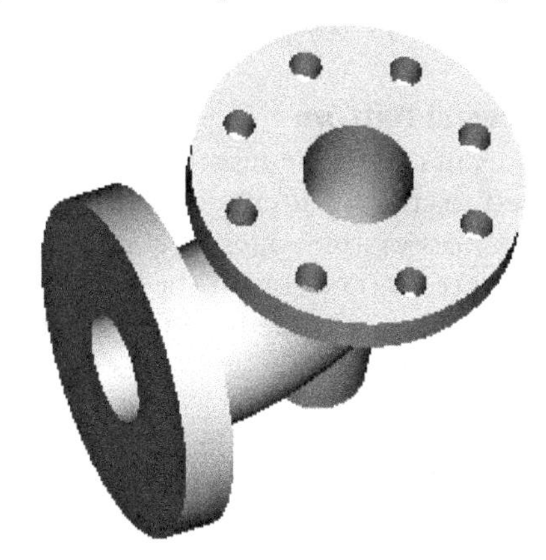

Figure 7-34 *Rotational pattern of the hole feature*

As evident from Figures 7-32 and 7-34, all instances in the identical patterns are placed on the same placement surface and no feature intersects the edges of the placement surface, any other instance, or any other feature other than the placement surface. Note that you cannot pattern the hole feature on the right flap shown in Figure 7-32 by using the **Identical** option. However, you can use the **General** option.

Variable Pattern

The **Variable** pattern is used when instances vary in size. In this type of pattern, the instances can be placed on different surfaces and can also intersect with the edges of the placement surface. The feature, shown in Figure 7-35, is patterned by using the **Variable** option and the patterned feature created by using this option is shown in Figure 7-36. In Figure 7-36, the length and the diameter of the rod vary in all instances. To create the instances that vary in position and size, you need to select the **Variable** option from the **Options** tab and then select the dimension you want to be variable. Next, choose the **Dimensions** tab and specify the increment value of variation in the slide-down panel. You can add more variable dimensions by pressing and holding the CTRL key. Next, specify the number of instances you want in the pattern. Variable pattern can be created only for the **Dimension**, **Direction**, and **Axis** patterns.

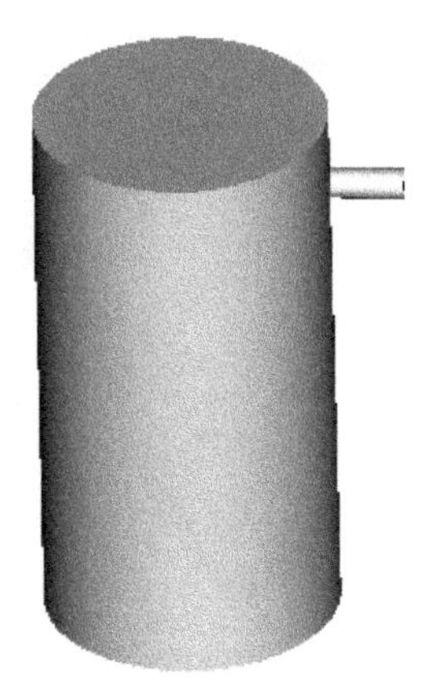

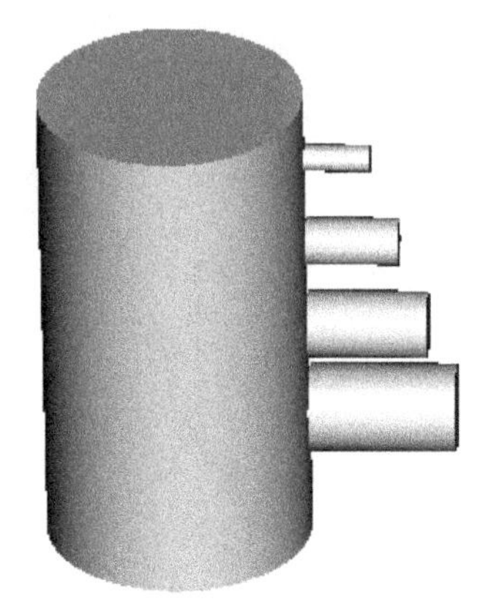

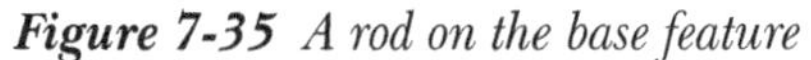

Figure 7-35 *A rod on the base feature* ***Figure 7-36*** *Varying pattern of the rod*

General Pattern

You can create the most complex patterns by using the **General** pattern. This option is used to create patterns in which the instances touch each other and intersect with other instances or the edges of a surface. This option of creating patterns is also used when instances intersect with the base feature and the intersection is not visible. Figure 7-37 shows a hole on the base feature and Figure 7-38 shows the hole pattern created using the **General** option.

Figure 7-37 *Hole on the base feature*

Figure 7-38 *General pattern*

Note

If the features that you need to pattern have formed a group and are listed in the ***Model Tree*** *as a group feature, you need to select the desired feature from the group to pattern it. A feature forms a group when it has used other features to form itself. For example, if you create a datum plane on-the-fly and use it as a sketch plane, then after the feature is created, it is listed in the* ***Model Tree*** *as a group feature. This group comprises feature, datum plane, and sketch of feature.*

Creating Geometry Patterns

Ribbon: Model > Editing > Pattern > Geometry Pattern

In PTC Creo Parametric, the **Geometry Pattern** tool is used to pattern geometries rather than features. In the geometry pattern, the selected features need not be made consecutive.

To create a geometry pattern, select the geometry from the model and then choose the **Geometry Pattern** tool from the **Pattern** drop-down in the **Editing** group; the **Geometry Pattern** dashboard will be displayed. In PTC Creo Parametric, you can create six types of geometry patterns. These patterns are **Direction**, **Axis**, **Fill**, **Table**, **Curve**, and **Point**. The use of these geometry pattern options are similar to **Pattern** tool, as discussed earlier.

Figure 7-39 shows a model for creating geometric pattern. Both the cut features are created by using one sketch. But, we need to pattern only the circular cut feature. Figure 7-40 shows the model after creating the geometric pattern.

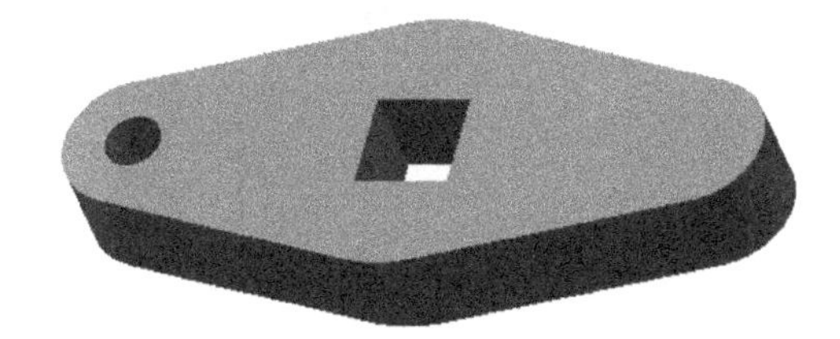

Figure 7-39 *Model with cut feature*

Figure 7-40 *Model after creating the geometry pattern*

Note

*Unlike the **Pattern** tool, the **Geometry Pattern** tool has the ability to pattern the feature more than twice.*

Deleting a Pattern

To delete a pattern select the pattern feature from the **Model Tree** and right-click to invoke the shortcut menu. Choose the **Delete Pattern** option from the shortcut menu. However, note that the parent feature (leader) is not deleted when you delete the pattern by using the **Delete Pattern** option even if it was selected along with other instances for deletion.

If you want to delete the pattern along with the parent feature, choose the **Delete** option from the shortcut menu. The system highlights the pattern in the drawing area and confirms the deletion of the pattern from the user.

MIRRORING A GEOMETRY

Ribbon: Model > Editing > Mirror

PTC Creo Parametric allows you to mirror a feature across a planar surface. To do so, choose the **Mirror** tool from the **Editing** group. This tool is only available after a feature has been selected. Choose this tool; the **Mirror** dashboard will be displayed and you will be prompted to select a datum plane about which the model will be mirrored. Select the required datum plane or a planar surface as the mirroring plane. Next, choose the **Build Feature** button to create the mirrored feature.

The model can be selected by selecting its name available at the top of the **Model Tree**. All features in the mirrored model are related to the parent model. Any modification made in the parent model is reflected in the mirrored model. Figure 7-41 shows a model and a datum plane for mirroring the model and Figure 7-42 shows the resulting mirrored model.

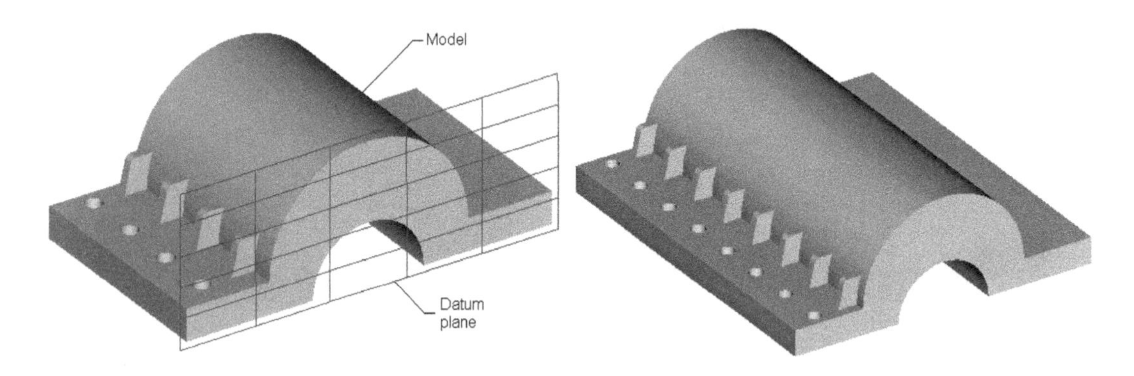

Figure 7-41 *Model and datum plane*

Figure 7-42 *Resulting mirrored model*

CREATING A SIMPLIFIED REPRESENTATION OF A SOLID MODEL

Ribbon: View > Model Display > Manage Views > View Manager

The complex geometry of a model sometimes needs to be simplified so that you can show the inner features which are not visible from outside.

To create the section view, choose the **View Manager** tool from the **Manage Views** drop-down in the **Model Display** group of the **View** tab; the **View Manager** dialog box will be displayed, as shown in Figure 7-43. Choose the **New** button from the dialog box; the **Rep0001** will be displayed in the **Names** display box. This is the default name given to the section that you will create. If required, you can enter another name in the **Names** display box and press ENTER; the **Menu Manager** with the **EDIT METHOD** menu will be displayed, as shown in Figure 7-44.

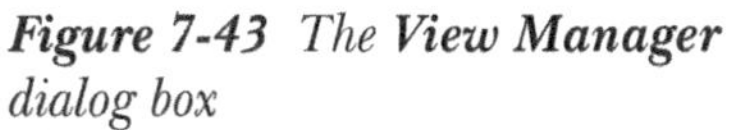
*Figure 7-43 The **View Manager** dialog box*

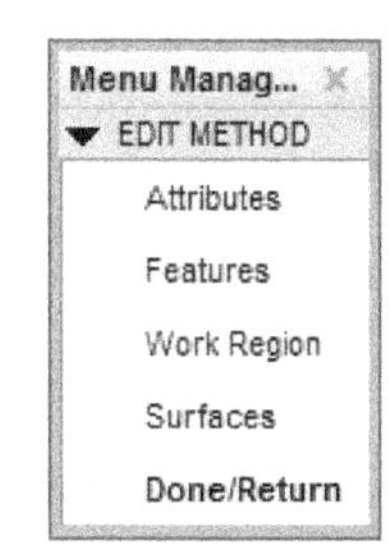

*Figure 7-44 The **Menu Manager** with the **EDIT METHOD** menu*

In this menu, the methods of representing a model are listed. Here, you will learn about the **Work Region** method of representing the model.

Work Region Method

This method of representing a model is very similar to removing material (creating cut) from a model. The following steps explain the procedure to create the section of the model shown in Figure 7-45.

1. Choose the **Work Region** option from the **EDIT METHOD** menu; the **SOLID OPTS** submenu will be displayed, as shown in Figure 7-46.

Figure 7-45 Solid model

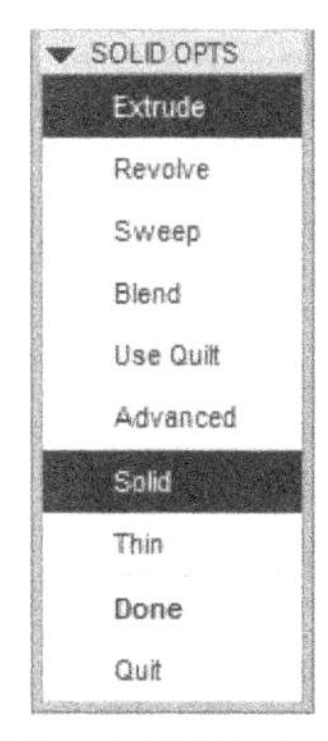

*Figure 7-46 The **SOLID OPTS** submenu*

2. Choose **Extrude > Solid > Done** from the **SOLID OPTS** submenu; the **Extrude** dashboard will be displayed.

3. Choose the **Placement** tab. Next, from the slide-down panel, choose the **Define** button; the **Sketch** dialog box will be displayed.

4. Select the top face of the plate as the sketch plane and select the **RIGHT** datum plane to be at the right while drawing the sketch.

5. Choose the **Sketch** button to enter the sketcher environment. Now you need to draw the section lines, as shown in Figure 7-47. This section is an open loop and the dimensions are not of importance to create this section.

6. Exit the sketcher environment and choose the **Extrude to intersect with all surface** button from the depth flyout; the yellow arrow will point in the reverse direction.

7. Choose the **Change material direction of extrude to other side of sketch** button to reverse the direction of arrow.

8. Choose the **Build feature** button from the **Extrude** dashboard. The sectioned model is shown in Figure 7-48.

9. Choose the **Done/Return** option from the **EDIT METHOD** menu.

To resume the view of the complete model, invoke the **View Manager** dialog box. Right-click on **Master Rep** in the **Names** display box to invoke the shortcut menu and select the **Set Activate** option from it. Close the **View Manager** dialog box when the complete view of the model is resumed.

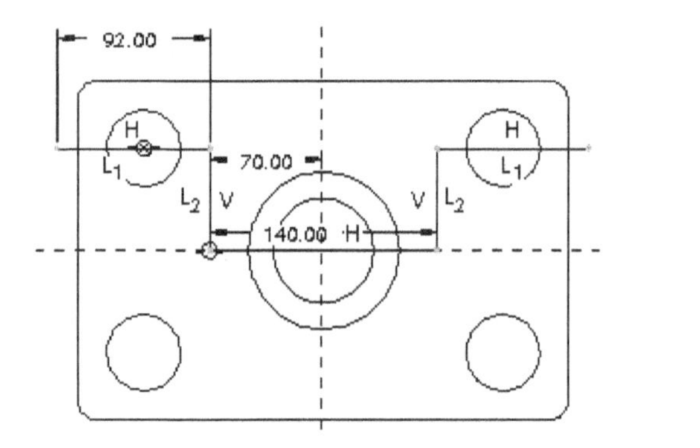

Figure 7-47 *Sketch of section lines*

Figure 7-48 *The sectioned model*

Note

The created section is saved with the name that you entered in the ***Names*** *display box while creating it. You can create multiple representations and save them with different names. To view a particular section, right-click on its name in the* ***Names*** *display box and choose the* ***Activate*** *option from the shortcut menu.*

TUTORIALS

Tutorial 1

In this tutorial, you will create the model shown in Figure 7-49. This figure also shows the top, front, and right views of the model. **(Expected time: 30 min)**

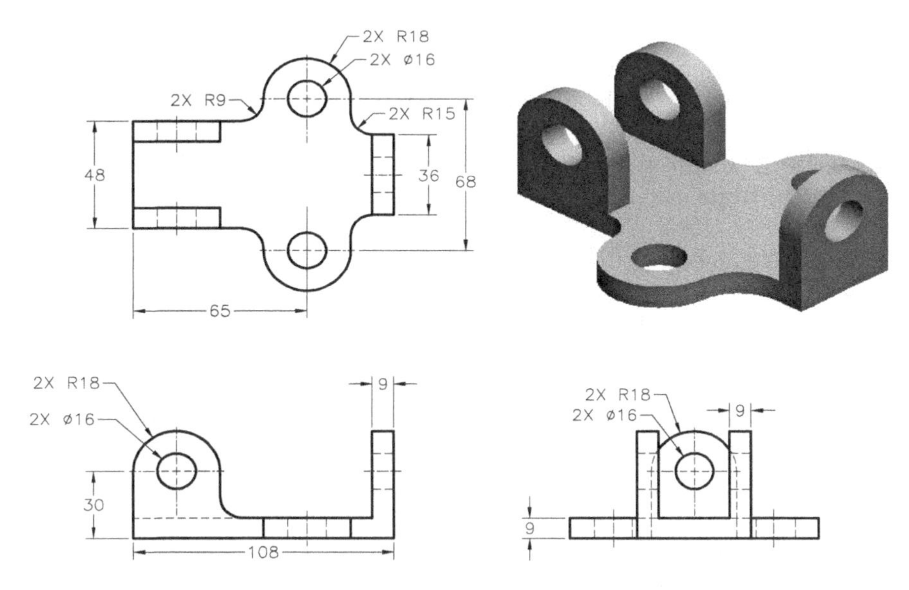

Figure 7-49 *The model and its top, front, and right views*

Examine the model to determine the number of features in it. The model consists of four features, refer to Figure 7-49.

The following steps are required to complete this tutorial:

a. Create the base feature on the **TOP** datum plane, refer to Figures 7-50 and 7-51.
b. Create the second feature on the right face of the base feature, refer to Figures 7-52 through 7-54.
c. Create the third feature, refer to Figures 7-55 and 7-56.
d. Create the fourth feature by mirroring the third feature, refer to Figure 7-57.

After starting PTC Creo Parametric session, the first task is to set the working directory. Since it is the first tutorial of this chapter, you need to create a folder with the name *c07*, if it does not exist, and then set it as working directory.

Starting a New Object File

1. Start a new part file and then name it as *c07tut1*.

The three default datum planes are displayed in the drawing area. Also, the **Model Tree** is displayed in the drawing area.

Creating the Base Feature

To create a sketch for the base feature, you need to select the **TOP** datum plane as the sketching plane.

1. Choose the **Extrude** tool from the **Shapes** group of the **Model** tab; the **Extrude** dashboard is displayed above the drawing area.

2. Choose the **Placement** tab; the slide-down panel is displayed. From the slide-down panel, choose the **Define** button; the **Sketch** dialog box is displayed.

3. Select the **TOP** datum plane as the sketching plane; the **RIGHT** datum plane and its orientation to **Right** are set automatically.

4. Choose the **Sketch** button to enter the sketcher environment.

5. Next, create the sketch of the base feature and apply constraints and dimensions to it, as shown in Figure 7-50.

   Note that in the sketch, the bottom half of the sketch is mirrored to create the top half of the sketch. This is evident from the constraints of symmetry applied to the sketch in Figure 7-50.

6. After the sketch is completed, choose the **OK** button to exit the sketcher environment; the **Extrude** dashboard is enabled and displayed above the drawing area.

7. Enter **9** as the value of depth in the dimension box in the **Extrude** dashboard.

8. Choose the **Build feature** button from the **Extrude** dashboard.

   Now, the base feature is completed and you need to create the second feature. The default trimetric view of the base feature is shown in Figure 7-51.

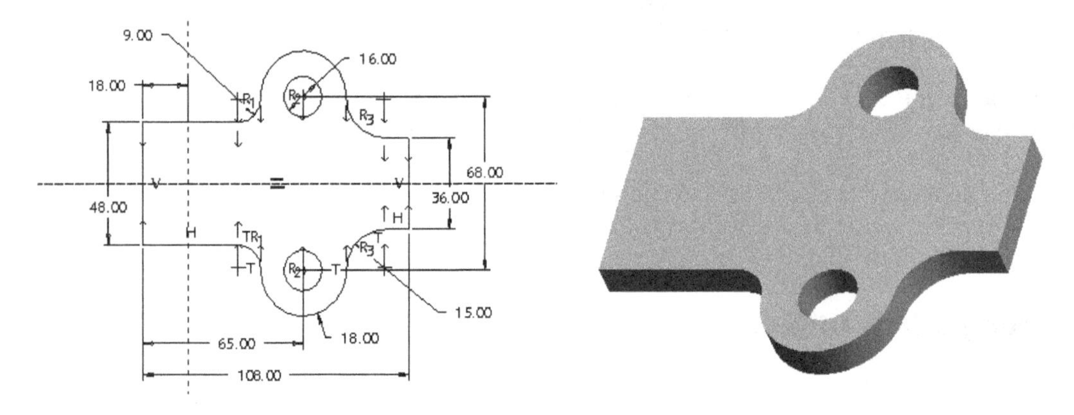

Figure 7-50 Sketch of the base feature

Figure 7-51 The default trimetric view of the base feature

Note

*The two holes are integrated in the base feature. These holes are sketched while drawing the sketch for the base feature. Therefore, the base feature is created as a single feature that includes two holes. The other method is to create the two holes separately on the base feature by using the **Hole** dashboard. When you create the features separately, the total number of features created will be three.*

Creating the Second Feature

The second feature is also an extruded feature. You will create the second feature on the right face of the base feature. Therefore, you need to select the right face as the sketching plane.

1. Choose the **Extrude** tool from the **Shapes** group; the **Extrude** dashboard is displayed above the drawing area.

2. Choose the **Placement** tab to display the slide-down panel. Then, choose the **Define** button from the slide-down panel; the **Sketch** dialog box is displayed.

3. Select the right face of the base feature as the sketching plane.

4. Choose the **Flip** button to reverse the direction of the pink arrow.

5. Select the **TOP** datum plane as reference and then select the **Top** option from the **Orientation** drop-down list.

6. Choose the **Sketch** button from the **Sketch** dialog box to enter the sketcher environment.

   After entering the sketcher environment, turn the model display to **No hidden** to view the wireframe model without the hidden lines.

7. Choose the **Line Chain** tool from the **Sketching** group and draw the right vertical line starting from the point shown in Figure 7-52. Notice that the endpoint of the right vertical line is not aligned with the edge on the top face of the base feature.

8. Next, draw the horizontal line in continuation with the first line, as shown in Figure 7-52. Since the endpoint of the right vertical line is not aligned with the edge on the top face of the base feature, the horizontal line is also not aligned with that edge.

9. Next, draw the left vertical line in continuation with the horizontal line and then complete the sketch by drawing the arc and the circle, see Figure 7-52.

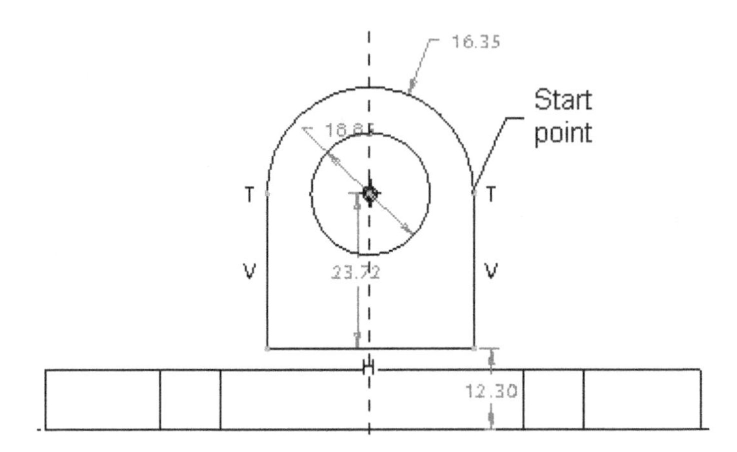

Figure 7-52 Sketch of the second feature with weak dimensions

10. Choose the **Coincident** tool from the **Constrain** group and select the bottom horizontal line of the sketch. Now, select the edge on the top face of the base feature as the second entity to apply the constraint. You will notice that the sketch will extend and the horizontal line gets aligned with the top face.

11. If the center points of the arc and the circle are not aligned with the **FRONT** datum plane, align them.

12. Add dimensions to the sketch of the second feature and then modify them, as shown in Figure 7-53. After completing the sketch, turn the model display to **Shading** and choose the **OK** button from the **Close** group; the **Extrude** dashboard is enabled.

13. Enter **9** as the value of depth in the dimension box in the **Extrude** dashboard.

14. Choose the **Build feature** button from the **Extrude** dashboard; the default shaded trimetric view of the model after creating the second feature is shown in Figure 7-54.

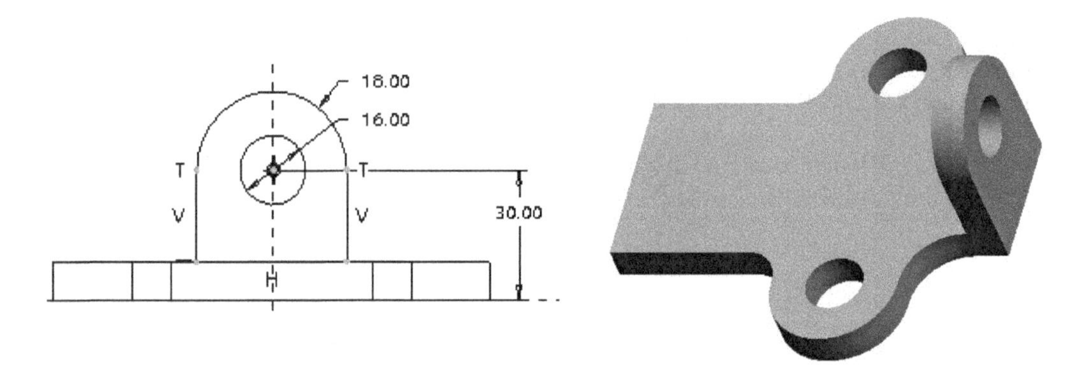

Figure 7-53 Adding dimensions to the second feature

Figure 7-54 Model after creating the second feature

Creating the Third Feature

Now you need to create the third feature.

1. Choose the **Extrude** tool from the **Shapes** group; the **Extrude** dashboard is displayed above the drawing area.

2. Choose the **Placement** tab to display the slide-down panel. Then, choose the **Define** button from the slide-down panel; the **Sketch** dialog box is displayed. In this dialog box, the **Bottom** option from the **Orientation** drop-down list is selected by default.

3. Select the face shown in Figure 7-55 from the drawing area.

4. Choose the **Sketch** button from the **Sketch** dialog box to enter the sketcher environment.

5. Next, create the sketch of the third feature and apply constraints and dimensions to it.

6. After the sketch is completed, choose the **OK** button to exit the sketcher environment; the **Extrude** dashboard is enabled and displayed above the drawing area.

7. Enter **9** as the value of depth in the dimension box in the **Extrude** dashboard and choose the **Change depth direction of extrude to other side of sketch** button from the dashboard.

8. Next, choose the **Build feature** button from the **Extrude** dashboard. The model after creating the third feature is shown in Figure 7-56.

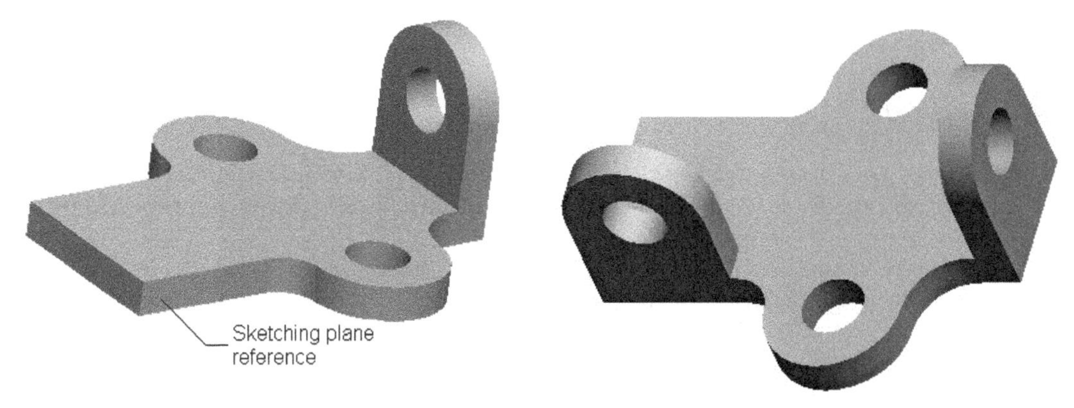

Figure 7-55 *Sketching plane reference for the third feature*

Figure 7-56 *The model after creating the third feature*

Creating the Fourth Feature

The fourth feature can be created by sketching and extruding it to a given depth. You can also create this feature by placing a mirrored copy of the third feature at the required location. In this tutorial, you will use the second method because it consumes less time.

1. Select the third feature from the **Model Tree** and choose the **Mirror** tool from the **Editing** group of the **Model** tab; the **Mirror** dashboard is displayed and you are prompted to select the mirror plane.

2. Select the **FRONT** datum plane as the mirror plane.

3. Choose the **Build feature** button from the **Mirror** dashboard. The third feature is mirrored about the **FRONT** datum plane. The trimetric view of the final model is shown in Figure 7-57.

4. Choose the **Save** button from the **File** menu and save the model. The order of feature creation can be seen from the **Model Tree** shown in Figure 7-58. Note that the feature id numbers in your model may be different from the ones shown in this figure.

Figure 7-57 *Default trimetric view of the model*

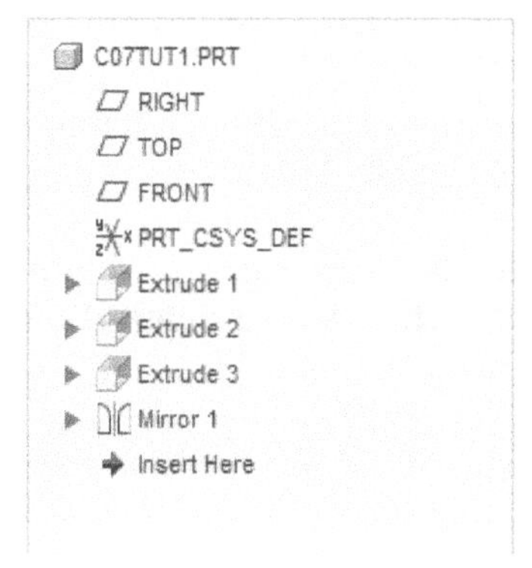

Figure 7-58 *The* ***Model Tree*** *for Tutorial 1*

Tutorial 2

In this tutorial, you will create the model shown in Figure 7-59. The dimensions of the model in the top view and the front section view are shown in Figure 7-60. **(Expected time: 45 min)**

Examine the model and determine the number of features in it. The model is composed of nine features, refer to Figure 7-59.

The following steps are required to complete this tutorial:

a. Create the base feature on the **FRONT** datum plane, refer to Figures 7-61 and 7-62.
b. Create round features, refer to Figures 7-63 and 7-64.
c. Create the hole feature and then pattern it, refer to Figures 7-65 and 7-66.
d. Create the rib feature on the **FRONT** datum plane, refer to Figures 7-67 and 7-68.
e. Mirror the rib feature, refer to Figure 7-69.
f. Create the extrude feature on the top planar face of the base feature, refer to Figures 7-70 and 7-71.

g. Create the cut feature on the bottom planar face of the base feature, refer to Figures 7-72 and 7-73.
h. Create the hole feature on the top face of the model and then copy it, refer to Figures 7-74 and 7-75.

Figure 7-59 *Solid model for Tutorial 2*

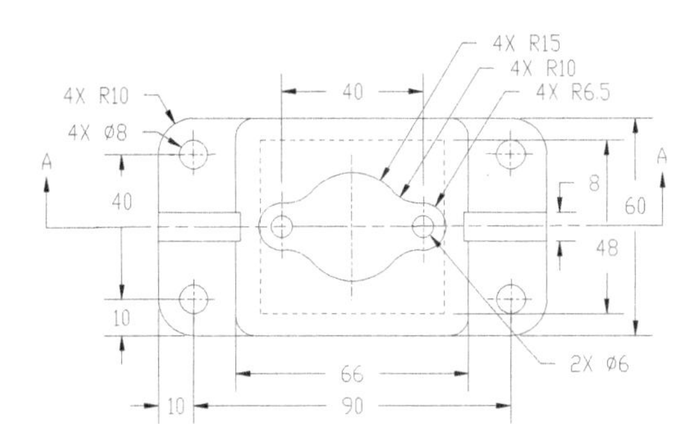

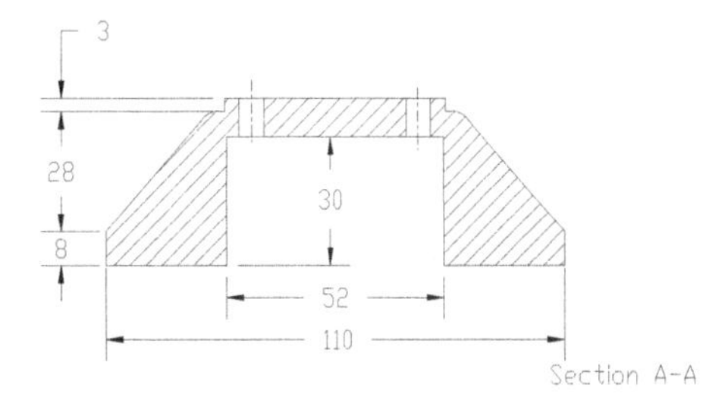

Figure 7-60 *Top view and front section view of the model*

Starting a New Object File

1. If required, set the working directory to the *c07* folder and start a new part file and then name it as *c07tut2*. The three default datum planes are displayed in the drawing area.

Creating the Base Feature

To create the sketch for the base feature, first you need to select the sketching plane for the base feature. In this model, you need to draw the base feature on the **FRONT** datum plane because the direction of extrusion of this feature is perpendicular to it. The base feature will be created symmetric to the **FRONT** datum plane.

1. Choose the **Extrude** tool from the **Shapes** group; the **Extrude** dashboard is displayed above the drawing area.

2. Choose the **Placement** tab; the slide-down panel is displayed. Choose the **Define** button from the panel; the **Sketch** dialog box is displayed.

3. Select the **FRONT** datum plane as the sketch plane.

4. Select the **TOP** datum plane and then select the **Top** option from the **Orientation** drop-down list.

5. Choose the **Sketch** button to enter the sketcher environment.

6. Next, create the sketch of the base feature and apply constraints and dimensions to it, as shown in Figure 7-61.

   In the sketch, note that the base feature is symmetrical; therefore, a vertical center line is drawn and then the right half of the sketch is mirrored to create the left half. This is evident from the constraints of symmetry applied to the sketch in Figure 7-60. These constraints appear as arrow symbols in the sketch.

   As evident from the sketch of the base feature shown in Figure 7-61, the **TOP** datum plane is aligned with the bottom line segment.

7. After the sketch is completed, choose the **OK** button to exit the sketcher environment; the **Extrude** dashboard is enabled and appears above the drawing area. Now you need to extrude the sketch symmetrically to both the sides of the sketching plane in order to create the base feature symmetrical to the **FRONT** datum plane. This is because, later in the tutorial, you need to use the default datum planes as mirror planes for mirroring the features. On extruding symmetrically, you need not create the datum planes.

8. Choose the **Extrude on both sides of sketch plane by half the specified depth value in each direction** button from the depth flyout in the **Extrude** dashboard.

9. Enter **60** as depth value in the dimension box in the **Extrude** dashboard.

10. Choose the **Build feature** button from the **Extrude** dashboard to exit the feature creation tool.

    The base feature is completed and now you need to create the second feature. The default trimetric view of the base feature is shown in Figure 7-62.

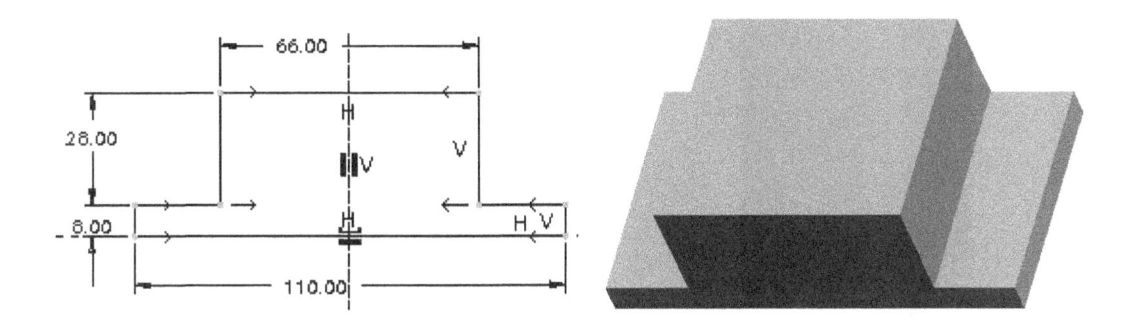

Figure 7-61 *Sketch with dimensions and constraints for the base feature*

Figure 7-62 *Default trimetric view of the base feature*

Creating the Second Feature

The second feature is a round feature of radius 5.

1. Choose the **Round** tool from the **Engineering** group; the **Round** dashboard is displayed. 

2. Choose the **Sets** tab to display the slide-down panel. Let this slide-down panel remain open so that you can view the selections made on the model.

3. Select one of the edges shown in Figure 7-63. To select the second edge and the subsequent edges, press the CTRL key and then select the edges one-by-one.

    The preview of the round is created on the selected edges; a default value of the radius is also displayed.

4. Double-click on the default radius value that is displayed in the preview of the round; an edit box appears. Enter **5** in it. The first set of rounds is created.

    Now you need to create the second set of rounds.

5. After spinning the model, select the four vertical edges of the bottom portion of the base feature to apply round. Remember to use the CTRL key+left mouse button to select the second and subsequent edges.

You will notice that **Set 1** and **Set 2** appear in the slide-down panel. This indicates that the second set is defined. The preview of the round is created on the edges. The default value of the radius is also displayed.

6. Double-click on the default radius value that is displayed along with the preview of the round; an edit box is displayed. Enter **10** in it.

7. Choose the **Build feature** button from the **Round** dashboard to create the round feature; the round feature is completed. The default trimetric view of the round feature is shown in Figure 7-64. In the **Model Tree**, the two rounds created appear as one feature.

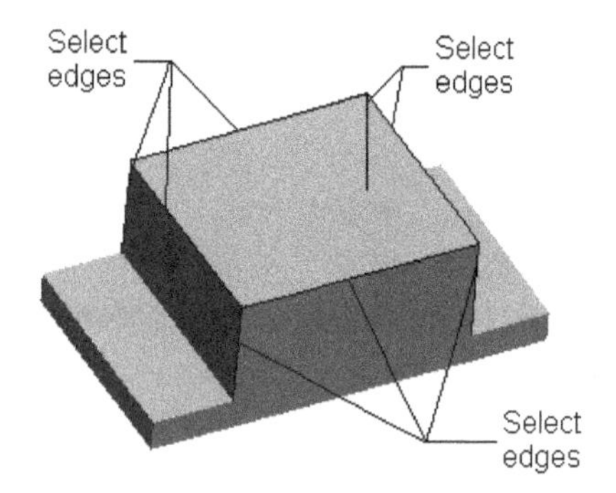

Figure 7-63 Edges selected to create the round features

Figure 7-64 The default trimetric view of the base feature with two sets of round feature

Creating the Third Feature

The third feature is a through hole and needs to be created by using the **Hole** dashboard.

1. Choose the **Hole** tool from the **Engineering** group; the **Hole** dashboard is displayed. By default, the **Create simple hole** button is chosen in it.

2. Create the hole, as shown in Figure 7-65, by specifying the placement parameters. Refer to Figure 7-60 for the placement parameters.

Creating a Pattern of the Hole Feature

As evident from Figure 7-59, you need to create four instances of the hole. The first instance will be created by using the **Hole** dashboard and the remaining three instances will be created by using the **Pattern** tool. You will create a rectangular pattern of the hole feature. You can also create all the holes by using the **Hole** dashboard and specifying the placement parameters for each of them. However, to save time, it is recommended that you create a pattern of the hole.

1. Select the hole feature and then choose the **Pattern** tool from the **Editing** group; the **Pattern** dashboard is displayed. Also, you are prompted to select dimensions to be changed in the first direction.

2. Make sure the **General** option is selected in the **Options** slide-down panel.

   You cannot use the **Identical** option to create the rectangular pattern of the hole feature. This is because when you use the **Identical** option, the pattern cannot intersect the base feature on which the hole is created. If the top portion of the base feature is created as a separate feature, the hole can be patterned by using the **Identical** option.

3. You need to select the dimension value **10** from the drawing area. Since the two dimensions displayed in the drawing area have the dimension value 10, select the dimension that is along the shorter side of the base feature. After you have selected the dimension in the first direction, the edit box is displayed. Enter **40** in it.

4. Hold the right mouse button to display a shortcut menu. Choose the **Direction 2 Dimensions** option from the shortcut menu.

5. Select the dimension value **10** that is along the longer side of the base feature. After you have selected the dimension in the second direction, the edit box is displayed. Enter **90** in the edit box.

   Note that the number of instances specified in the edit boxes present on the dashboard is 2 by default.

6. Choose the **Build feature** button from the **Pattern** dashboard; the rectangular pattern of the hole is displayed.

   You can use the middle mouse button to spin and display the model, as shown in Figure 7-66.

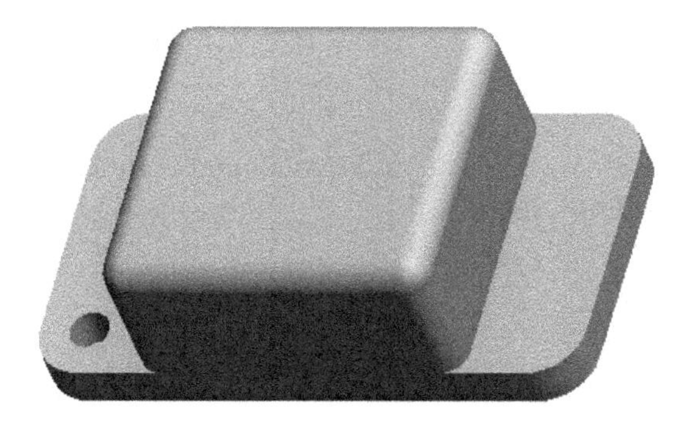

Figure 7-65 *The hole feature on the base feature*

Figure 7-66 *Model after creating the hole pattern*

Note

*If multiple holes need to be created on a model, it is recommended that you create their pattern, if possible. This is because when you assemble bolts in these holes in the **Assembly** mode, it becomes very easy to assemble them using the reference pattern.*

Creating the Rib Feature

The sketch of the rib feature will be created on the **FRONT** datum plane and the required thickness will be applied to it. This is the fourth feature of the model.

1. Choose the **Profile Rib** tool from the **Rib** drop-down in the **Engineering** group; the **Profile Rib** dashboard is displayed.

2. Choose the **References** tab to invoke the slide-down panel. Choose the **Define** button from the slide-down panel; the **Sketch** dialog box is displayed.

3. Select the **FRONT** datum plane as the sketching plane.

4. Select the **TOP** datum plane from the drawing area and then select the **Top** option from the **Orientation** drop-down list.

5. Choose the **Sketch** button to enter the sketcher environment.

6. Draw the sketch of the rib feature, as shown in Figure 7-67, and then exit the sketcher environment.

Note

In Figure 7-67, the top end of the inclined line in the sketch is aligned with the curve and the tangent constraint is applied to the line and the curve. Similarly, the bottom end of the inclined line is also aligned with the two edges. This is the reason, there are no dimensions in the sketch and the sketch for the rib feature is fully constrained.

7. Choose the **Flip** button from the **References** slide-down panel, if the desired direction is not obtained.

8. Specify **8** as the rib thickness in the dimension box in the **Rib** dashboard. Choose the **Build feature** button to create the rib feature.

   The rib feature created is shown in Figure 7-68. You can use the middle mouse button to orient the model to view the correct placement of the rib.

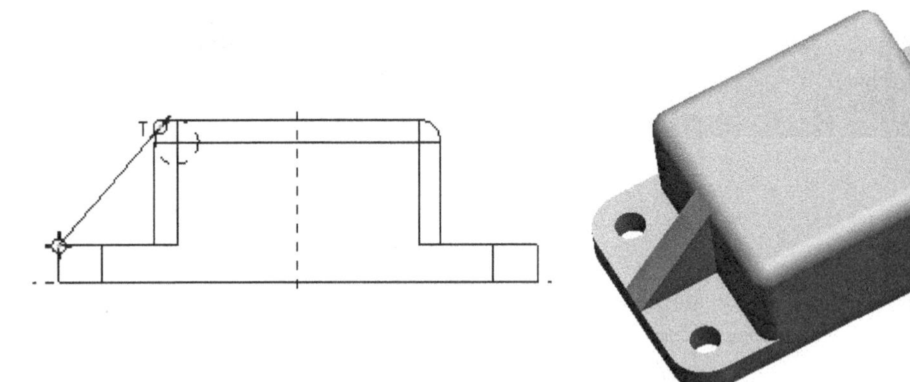

Figure 7-67 *Fully constrained sketch for the rib*

Figure 7-68 *Model after creating the rib feature*

Mirroring the Rib Feature

Now, you need to create another rib feature, as shown in Figure 7-69. You can create this feature either by mirroring or by creating the sketch of the rib feature on the sketching plane. Here, you will create the rib feature by mirroring. The rib feature will be mirrored about the **RIGHT** datum plane.

Figure 7-69 Model after mirroring the rib feature

1. Select the rib feature that you need to mirror. Then, choose the **Mirror** tool from the **Editing** group of the **Model** tab; the **Mirror** dashboard is displayed.

2. Choose the **References** tab to invoke the slide-down panel. In this panel, the **Mirror Plane** collector is selected by default. Select the **RIGHT** datum plane as the mirror plane.

3. Choose the **Build feature** button from the dashboard; the selected feature is mirrored about the **RIGHT** datum plane, refer to Figure 7-69.

Creating the Protrusion Feature

The sixth feature is an extruded feature. You need to create the extruded feature on the top face of the base feature.

1. Choose the **Extrude** tool from the **Shapes** group; the **Extrude** dashboard is displayed.

2. Invoke the **Sketch** dialog box.

3. Select the top face of the base feature as the sketching plane.

4. Select the **RIGHT** datum plane and then select the **Right** option from the **Orientation** drop-down list.

5. Choose the **Sketch** button to enter the sketcher environment.

6. Next, draw the sketch of the extruded feature, as shown in Figure 7-70. The tangent and equal radii constraints are applied to the sketch. Also, the center of the top arc and the bottom arc coincide with the intersection of the two datum planes.

7. After the sketch is complete, choose the **OK** button to exit the sketcher environment; the **Extrude** dashboard is enabled and it appears above the drawing area.

8. Enter **3** as the depth value in the dimension box in the **Extrude** dashboard. Choose the **Build feature** button from the **Extrude** dashboard; the extruded feature is completed and the default trimetric view is shown in Figure 7-71.

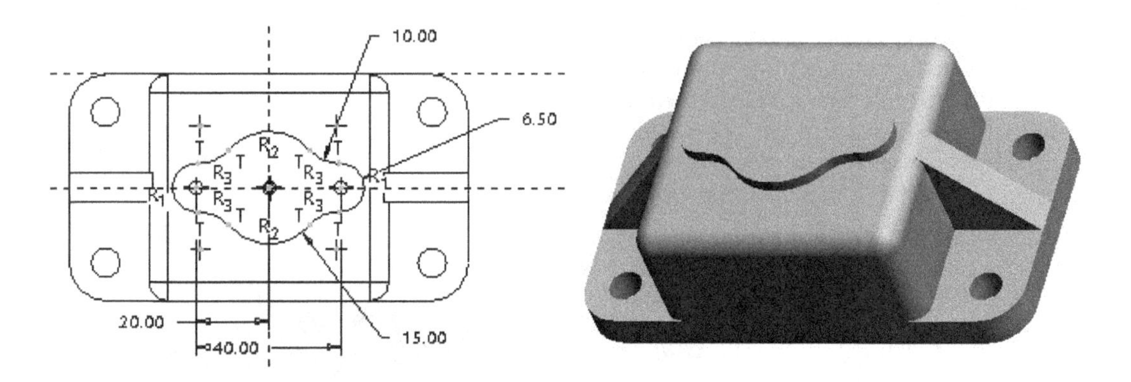

Figure 7-70 Sketch with dimensions and constraints for the extruded feature

Figure 7-71 The default trimetric view after creating the extruded feature

Creating the Cut Feature

You need to create an extruded cut on the bottom planar surface of the base feature. This is the seventh feature of the model.

1. Choose the **Extrude** tool from the **Shapes** group.
2. Select the **Remove Material** button from the **Extrude** dashboard.
3. Choose the **Placement** tab to invoke the slide-down panel. Choose the **Define** button from the slide-down panel; the **Sketch** dialog box is displayed.
4. Select the bottom face of the base feature as the sketching plane.
5. Select the **RIGHT** datum plane and then select the **Right** option from the **Orientation** drop-down list in the **Sketch** dialog box.
6. Choose the **Sketch** button to enter the sketcher environment.
7. Next, draw the sketch of the cut feature and dimension it, as shown in Figure 7-72.
8. After completing the sketch, choose the **OK** button; the **Extrude** dashboard is enabled and it appears above the drawing area.
9. Enter the value **30** in the dimension box of the **Extrude** dashboard.
10. Choose the **Build feature** button from the dashboard to create the cut feature. You can spin the model by using the middle mouse button. The cut feature is shown in Figure 7-73.

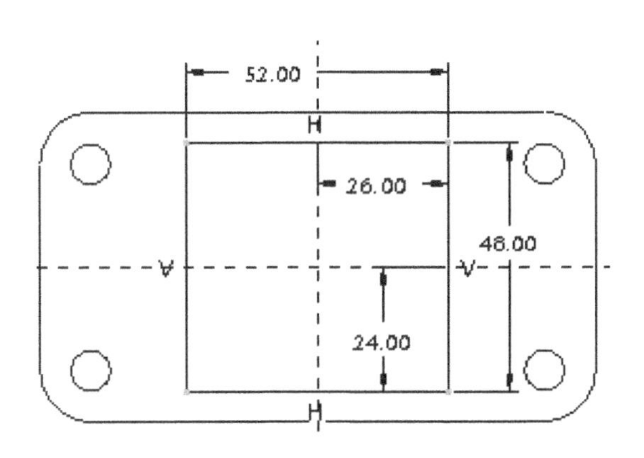

Figure 7-72 Sketch of the cut feature with dimensions

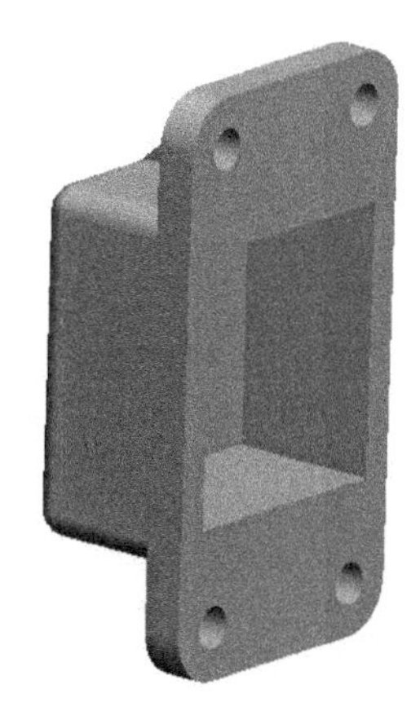

Figure 7-73 Model after creating the cut feature

Creating the Hole

Next, you need to create a through hole on the top extruded feature by using the **Hole** dashboard.

1. Invoke the **Hole** dashboard; the **Create simple hole** button is chosen by default in it. Specify the placement parameters of the hole as given in Figure 7-60 and create the hole. The model after creating the hole is shown in Figure 7-74.

Mirroring the Hole Feature

Now, you need to mirror the hole feature, as shown in Figure 7-75. This is the ninth feature of the model. You will mirror the hole feature about the **RIGHT** datum plane.

1. Select the hole feature to be mirrored. Then, choose the **Mirror** tool from the **Editing** group of the **Model** tab; the **Mirror** dashboard is displayed.

2. Choose the **References** tab to invoke the slide-down panel. Next, select the **RIGHT** datum plane as the mirror plane.

3. Choose the **Build feature** button from the dashboard; the selected feature is mirrored about the **RIGHT** datum plane, as shown in Figure 7-75.

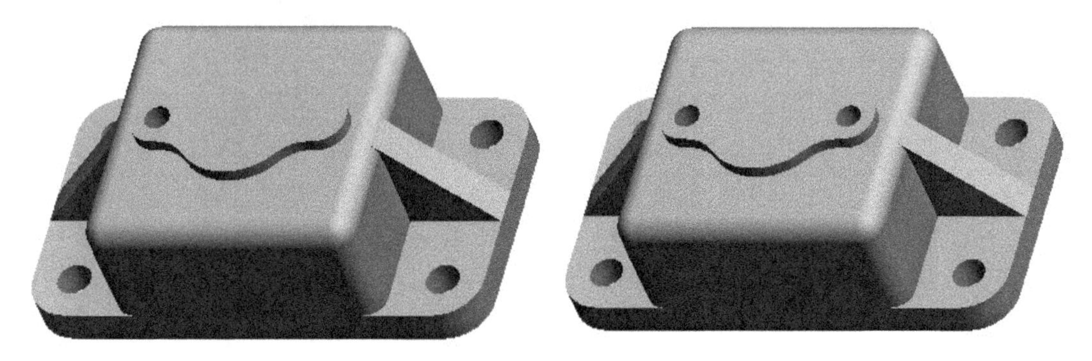

Figure 7-74 The model after creating holes *Figure 7-75 The copied hole feature*

Saving the Model

You need to save the model to use it later.

1. Choose the **Save** button from the **File** menu and save the model.

   The order of feature creation of the model can be seen in the **Model Tree**, as shown in Figure 7-76.

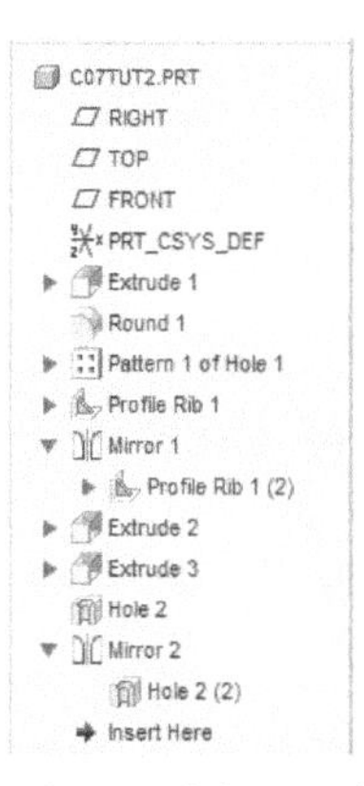

Figure 7-76 The ***Model Tree*** *for Tutorial 2*

Tutorial 3

In this tutorial, you will create the model shown in Figure 7-77. This figure also shows the top and front views of the model. **(Expected time: 30 min)**

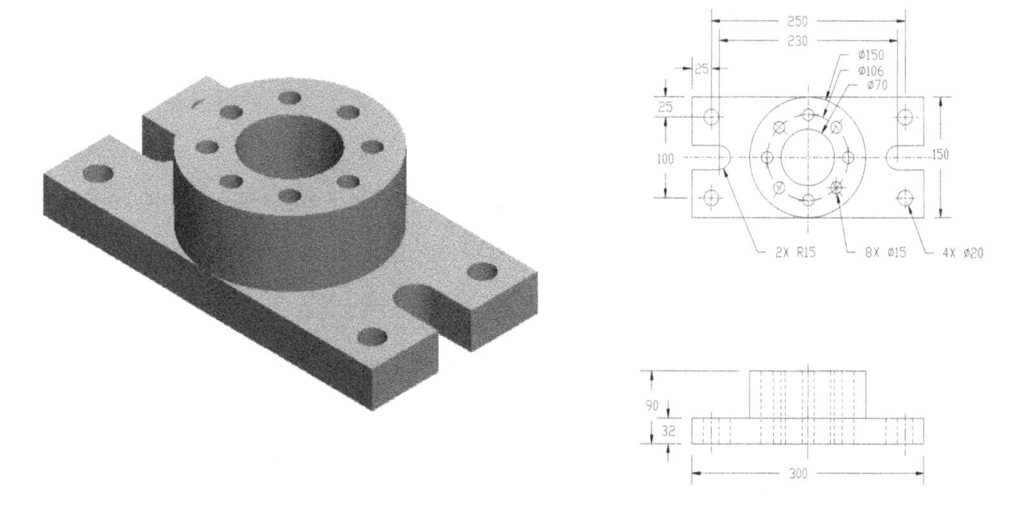

***Figure 7-77** The solid model and its top and front views*

Examine the model and determine the number of features in it. The model consists of five features, refer to Figure 7-77.

The following steps are required to complete this tutorial:

a. Create the base feature on the **TOP** datum plane, refer to Figures 7-78 and 7-79.
b. Create the cylindrical extrude feature, refer to Figures 7-80 and 7-81.
c. Create the hole feature coaxial with the cylindrical feature, refer to Figure 7-82.
d. Create the hole feature on the top planar surface of the base feature and then pattern it, refer to Figures 7-83 and 7-84.
e. Create the hole feature on the top planar surface of the cylindrical feature, refer to Figure 7-85.
f. Create a rotational pattern of this hole, refer to Figure 7-86.

If required, set the working directory to the *c07* folder.

Starting a New Object File

1. Start a new part file and name it as *c07tut3*.

The three default datum planes are displayed in the drawing area.

Creating the Base Feature

To create the sketch for the base feature, you first need to select the sketching plane for the base feature. In this model, you need to draw the base feature on the **TOP** datum plane. This is because the direction of extrusion of this feature is perpendicular to the **TOP** datum plane.

1. Choose the **Extrude** tool from the **Shapes** group; the **Extrude** dashboard is displayed.

2. Invoke the **Sketch** dialog box by using the **Extrude** dashboard.

3. Select the **TOP** datum plane as the sketching plane.

4. Select the **RIGHT** datum plane and then select the **Right** option from the **Orientation** drop-down list, if it is not selected by default.

5. Choose the **Sketch** button to enter the sketcher environment.

6. Create the sketch of the base feature and apply the required constraints and dimensions to it, as shown in Figure 7-78.

7. After the sketch is complete, choose the **OK** button to exit the sketcher environment; the **Extrude** dashboard is enabled and displayed above the drawing area.

8. Enter **32** as the depth value in the dimension box of the **Extrude** dashboard and then choose the **Build feature** button from the dashboard.

   The base feature is completed and now you need to create the second feature. The default trimetric view of the base feature is shown in Figure 7-79.

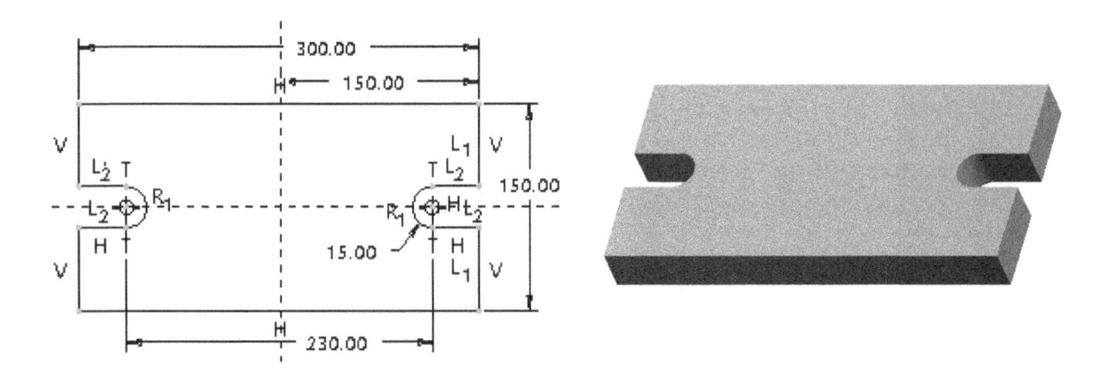

Figure 7-78 *Sketch of the base feature with dimensions and constraints*

Figure 7-79 *Default trimetric view of the base feature*

Creating the Second Feature

The second feature is also an extruded feature. You need to create this feature on the top face of the base feature. Therefore, you need to define the top face as the sketching plane for the second feature.

1. Choose the **Extrude** tool from the **Shapes** group; the **Extrude** dashboard is displayed.

2. Invoke the **Sketch** dialog box by using the **Extrude** dashboard.
3. Select the top face of the base feature as the sketching plane.
4. Select the **RIGHT** datum plane and then select the **Right** option from the **Orientation** drop-down list, if it is not selected by default.
5. Choose the **Sketch** button to enter the sketcher environment.
6. Create the sketch of the second feature and then dimension it, as shown in Figure 7-80.
7. After creating the sketch, choose the **OK** button; the **Extrude** dashboard is enabled.
8. Enter **58** in the dimension box of the **Extrude** dashboard and choose the **Build feature** button. The model, similar to the one shown in Figure 7-81, is displayed in the drawing area.

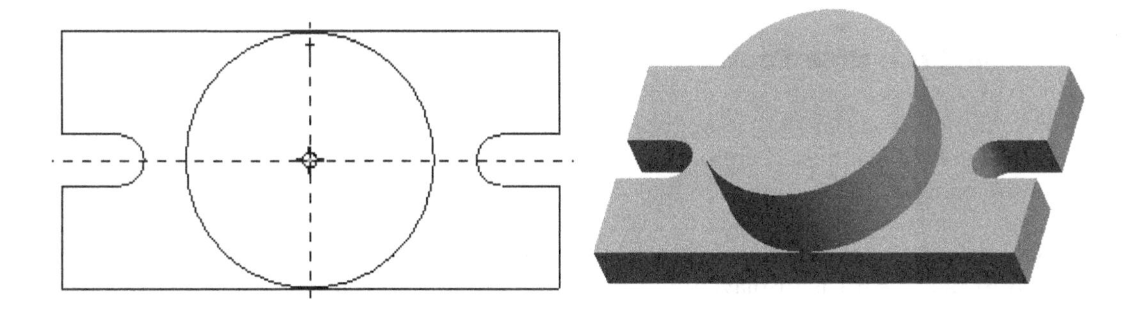

Figure 7-80 *Sketch of the cylindrical feature with diameter of the cylinder*

Figure 7-81 *Default trimetric view of the cylindrical feature*

Creating the Third Feature

The third feature is a through hole that is coaxial to the cylindrical feature. You need to create the hole feature by using the **Hole** dashboard.

1. Choose the **Hole** tool from the **Engineering** group; the **Hole** dashboard is displayed. By default, the **Create simple hole** button is chosen in the **Hole** dashboard.
2. Create a hole of diameter 70, as shown in Figure 7-82. Refer to Figure 7-77 for specifying the placement parameters.

Creating the Fourth Feature

The fourth feature is a through hole. You need to create this feature on the top planar surface of the base feature by using the **Hole** dashboard.

1. Choose the **Hole** tool from the **Engineering** group; the **Hole** dashboard is displayed. By default, the **Create simple hole** button is chosen in the **Hole** dashboard.

2. Create the hole, as shown in Figure 7-83, by specifying the placement parameters (refer to Figure 7-77).

Figure 7-82 *Coaxial hole on the cylindrical feature*

Figure 7-83 *Hole on the base feature*

Patterning the Hole Feature

Next you need to create a rectangular pattern of the hole feature that is created on the base feature. You can also create individual holes but it is time-consuming and increases the number of features. Therefore, it is recommended that you create a rectangular pattern of the hole feature.

1. Select the hole feature and then choose the **Pattern** tool from the **Editing** group; the **Pattern** dashboard is displayed and you are prompted to select the dimensions to be changed in the first direction.

2. Select the **Identical** option from the **Options** slide-down panel.

   Here you need to use the **Identical** option because the feature on which the pattern is created does not intersect the pattern.

3. Select the dimension value **25** from the drawing area. Since both the dimensions displayed in the drawing area have the value **25**, select the dimension **25** that is along the shorter side of the base feature. After you have selected the dimension in the first direction, an edit box is displayed. Enter **100** in the edit box.

4. Hold down the right mouse button to display the shortcut menu. Choose the **Direction 2 Dimensions** option from the shortcut menu.

5. Select the dimension value **25** that is along the longer side of the base feature. After you have selected the dimension in the second direction, an edit box is displayed.

6. Enter **250** in the edit box. Note that the number of instances, **2**, is specified by default, in the instances edit boxes of the dashboard.

7. Choose the **Build feature** button from the **Pattern** dashboard; a rectangular pattern of the hole feature is displayed, as shown in Figure 7-84.

Creating a Hole on the Cylindrical Feature

You need to create the hole on the cylindrical feature diametrically by using the **Hole** dashboard.

1. Choose the **Hole** tool from the **Engineering** group; the **Hole** dashboard is displayed. By default, the **Create simple Hole** tool is chosen in the **Hole** dashboard.

2. Choose the **Placement** tab from the **Hole** dashboard to display the slide-down panel.

3. Select the top face of the cylindrical feature as the placement plane.

4. From the drop-down list in the slide-down panel, select the **Diameter** option.

5. Click in the **Offset References** collector and select the axis of the cylindrical feature.

6. Enter **106** in the second dimension box on the right of the **Diameter** option in the **Offset References** collector.

7. Use the CTRL key+left mouse button and select the **FRONT** datum plane from the drawing area. Enter the value **90** in the dimension box of the **Offset References** collector.

8. Enter **15** as the diameter of the hole in the diameter dimension box in the **Hole** dashboard.

9. Choose the **Drill to intersect with all surfaces** option from the depth flyout in the **Hole** dashboard.

10. Choose the **Build feature** button from the **Hole** dashboard; the hole is created, as shown in Figure 7-85.

Figure 7-84 *Rectangular pattern of the hole feature*

Figure 7-85 *Diametrical hole on the cylindrical feature*

Creating the Rotational Pattern of the Hole Feature

As the creation of the remaining holes individually on the cylindrical feature is time-consuming, you need to create a rotational pattern of the hole feature.

1. Select the hole feature and then choose the **Pattern** tool from the **Editing** group; the **Pattern** dashboard is displayed. Also, the dimensions of the hole feature are displayed in the drawing area and you are prompted to select dimensions to vary in the first direction.

2. Select the **Identical** option from the **Options** slide-down panel.

3. Select the angular dimension **90** from the model; an edit box is displayed.

4. Enter **45** in the edit box and press ENTER. Now you need to specify the number of instances of the hole feature in the pattern.

5. Enter **8** in the **1** edit box and press ENTER. Choose the **Build feature** button from the dashboard. The rotational pattern is created and the model is completed, as shown in Figure 7-86. You can see the order of feature creation in the **Model Tree**, as shown in Figure 7-87.

Figure 7-86 The complete model

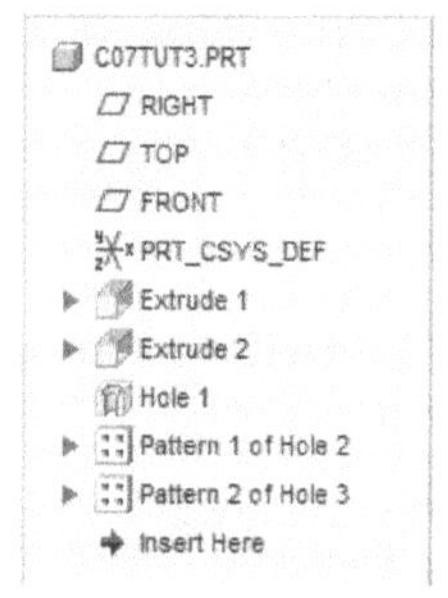

*Figure 7-87 The **Model Tree** for Tutorial 3*

Saving the Model

You need to save the model because you may need it later.

1. Choose the **Save** button from the **File** menu and save the model.

Tutorial 4

In this tutorial, you will create the model of the cylinder head shown in Figure 7-88. Figure 7-89 shows the top view and the front section view of the model. **(Expected time: 45 min)**

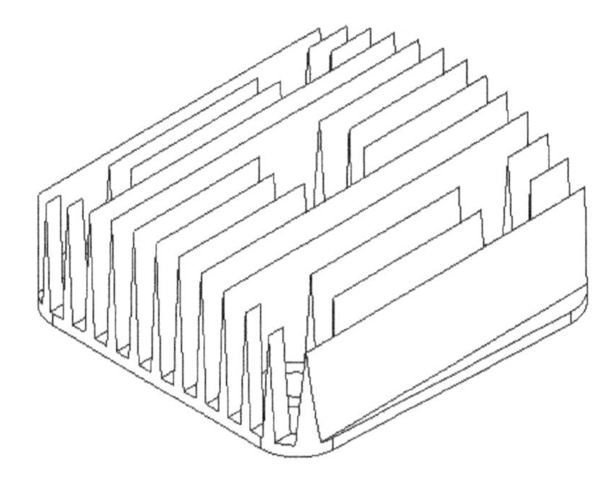

Figure 7-88 *Isometric view of the model*

Examine the model and determine the number of features in it. The model consists of twelve features, refer to Figure 7-88.

The following steps are required to complete this tutorial:

a. Create the base feature on the **TOP** datum plane, refer to Figures 7-90 and 7-91.
b. Create the round features on the vertical edges of the base feature, refer to Figure 7-92.
c. Create the cylindrical feature on the bottom face of the base feature, refer to Figure 7-93.
d. Create the revolve cut feature on a plane passing through the center of the cylindrical feature, refer to Figures 7-94 and 7-95.
e. Create the fifth feature of the fin that will be patterned later on, refer to Figures 7-96 through 7-98.
f. Create the cut feature that will remove the protrusions of the fins projecting out of the base feature, refer to Figure 7-99.
g. Create a circular cut feature on the top face of the base feature, refer to Figures 7-100 and 7-101.
h. Create a cylindrical feature on the top face of the base feature, refer to Figure 7-102.
i. Create the ninth feature, which is a cut feature, refer to Figures 7-103 and 7-104. This feature is reference patterned to create its other instances, refer to Figure 7-105.
j. Create the tenth feature as a cylindrical feature on the top face of the base feature, refer to Figure 7-106.
k. Create the referenced pattern of the tenth feature, refer to Figure 7-107.
l. The eleventh feature is a coaxial hole. After creating the feature, create a reference pattern of this hole, refer to Figure 7-108.
m. Create the last feature as the coaxial hole that will be created on the eighth feature, refer to Figure 7-109.

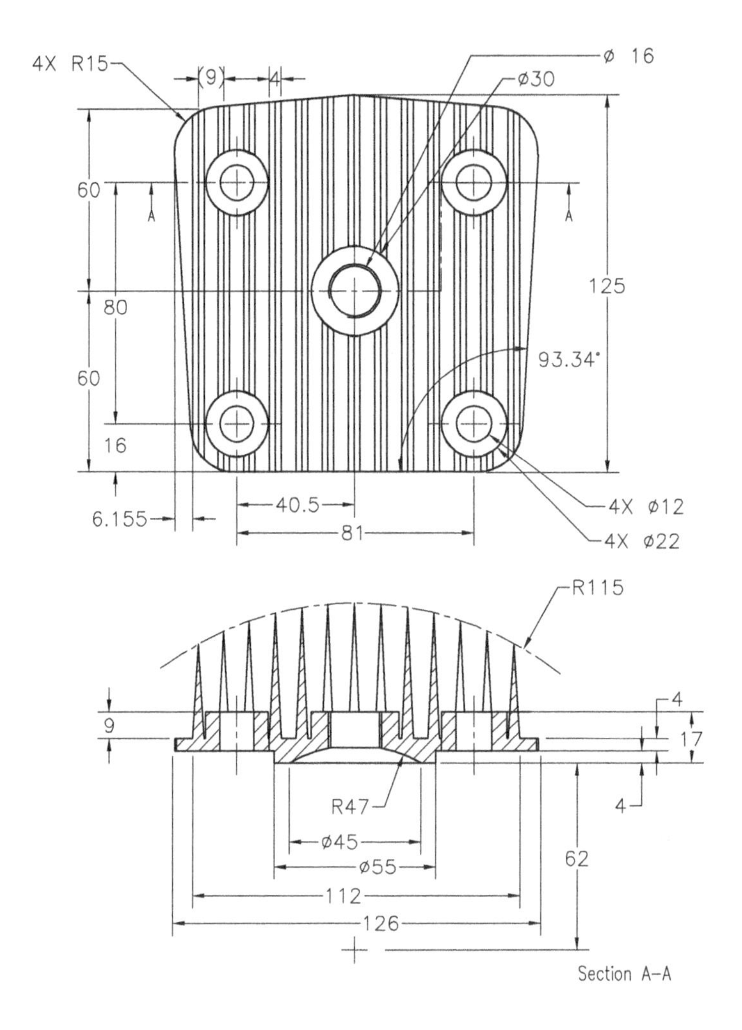

Figure 7-89 *Top and front section views of the model*

If required, set the working directory to the *c07* folder.

Starting a New Object File

1. Start a new part file and name it as *c07tut4*.

   The three default datum planes are displayed in the drawing area.

Creating the Base Feature

In this model, you need to draw the sketch of the base feature on the **TOP** datum plane. The sketch is a polygon. First you need to draw the right half of the polygon and then mirror it about the centerline to create the complete polygon.

1. Choose the **Extrude** tool from the **Shapes** group.

2. Choose the **Placement** tab to display a slide-down panel. From this panel, choose the **Define** button; the **Sketch** dialog box is displayed.

3. Select the **TOP** datum plane as the sketching plane.

4. Select the **RIGHT** datum plane and then select the **Right** option from the **Orientation** drop-down list, if it is not selected automatically.

5. Choose the **Sketch** button to enter the sketcher environment.

6. Next, create the sketch of the base feature and apply the constraints and dimensions to it, as shown in Figure 7-90.

7. After the sketch is complete, choose the **OK** button to exit the sketcher environment. The **Extrude** dashboard is enabled and it appears above the drawing area.

8. Enter **4** as the depth value in the dimension box at the **Extrude** dashboard and then choose the **Build feature** button from the dashboard.

   The base feature is completed and now you need to create the second feature. The default trimetric view of the base feature is shown in Figure 7-91.

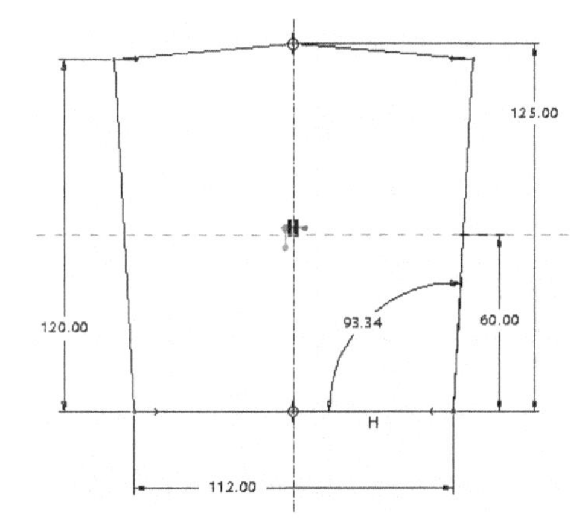

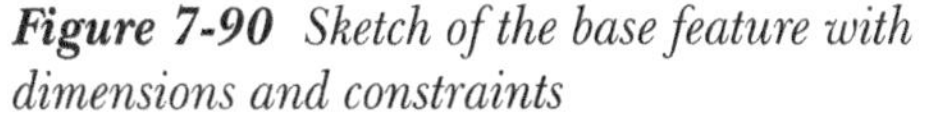

Figure 7-90 *Sketch of the base feature with dimensions and constraints*

Figure 7-91 *Default trimetric view of the base feature*

Creating the Second Feature

The second feature is a round feature of radius 15.

1. Choose the **Round** tool from the **Engineering** group; the **Round** dashboard is displayed.

2. Choose the **Sets** tab to display the slide-down panel.

3. Select one of the four vertical edges at the corners of the base feature. To select the second edge and the subsequent edges, press the CTRL key and then select the edge.

   The preview of the round is created on the selected edges. Also, a default value of the radius is displayed.

4. Double-click on the default radius value displayed along with the preview of the round; an edit box appears. Enter **15** in the edit box. You can also enter the radius value in the dimension box of the **Round** dashboard.

5. Choose the **Build feature** button to create the round feature.

   The default trimetric view of the model after creating the round feature is shown in Figure 7-92.

Creating the Third Feature

The third feature is a cylindrical feature. You need to create the feature on the bottom face of the base feature.

1. Invoke the **Extrude** dashboard and using the **Sketch** dialog box, select the bottom face of the base feature as the sketching plane.

2. Enter the sketcher environment and draw the sketch of the cylindrical feature. Refer to Figure 7-89 for drawing the sketch.

3. Exit the sketcher environment and enter **4** as the depth of extrusion in the dashboard.

4. Exit the **Extrude** dashboard.

   The cylindrical feature is created and is shown in Figure 7-93.

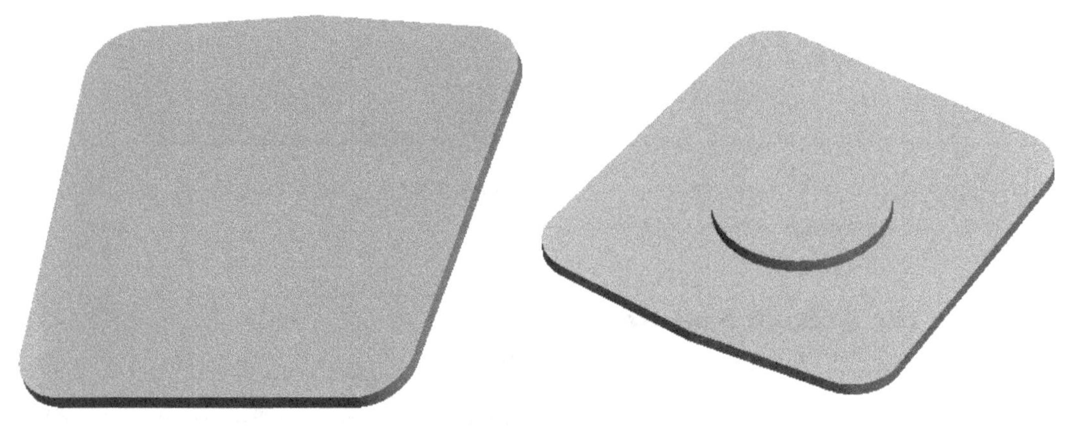

Figure 7-92 *Model with round*

Figure 7-93 *Model after creating the cylindrical feature on the bottom face*

Creating the Fourth Feature

The fourth feature is a revolve cut that will be created on the datum plane passing through the center of the cylindrical feature.

1. Invoke the **Revolve** dashboard and choose the **Remove Material** button.

2. Choose the **Placement** tab; the slide-down panel is displayed. Choose the **Define** button from the slide-down panel; the **Sketch** dialog box is displayed. Select the **FRONT** datum plane as the sketching plane.

3. Select the **RIGHT** datum plane and then select the **Right** option from the **Orientation** drop-down list. Next, choose the **Sketch** button.

4. Once you enter the sketcher environment, create the sketch of the revolve cut feature, draw a vertical geometric centerline, and apply constraints and dimensions to it, as shown in Figure 7-94.

5. After completing the sketch, exit the sketcher environment.

   The **Revolve** dashboard is enabled and displayed above the drawing area. The value for the angle of revolution is 360 by default.

6. Choose the **Build feature** button from the dashboard.

   The revolved cut feature is created and shown in Figure 7-95.

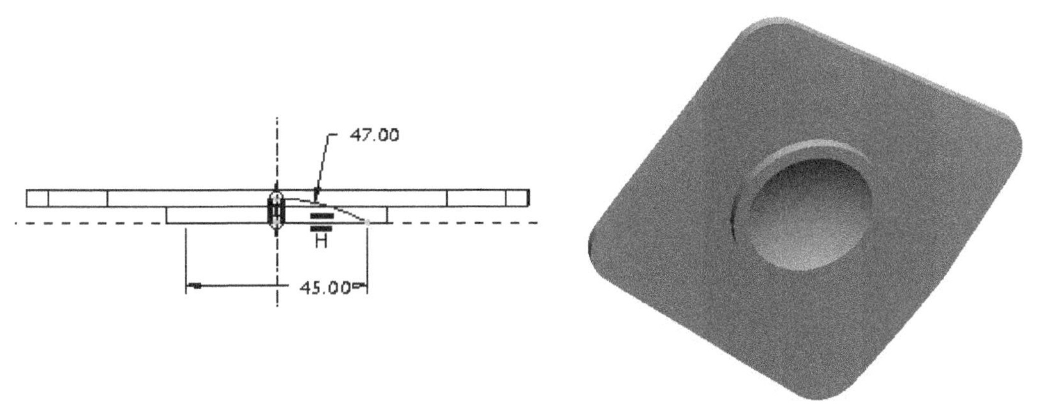

Figure 7-94 *Sketch of the revolved feature with dimensions and constraints*

Figure 7-95 *Revolved cut feature*

Creating the Fifth Feature

The fifth feature is the fin and its pattern. You need to create the sketch of the fin on the **FRONT** datum plane. The pattern of the fin feature will be created by controlling the dimensions of the sketch and constraining it.

This type of pattern can only be created, if dimensions of the feature to be patterned are proper. In the sketch of the feature, a construction entity must exist. This construction entity can be a circle, an arc, or a line. The construction entity is used as the driving entity for the pattern to be created later. The pattern instances are created in such a way that they follow the geometry of the construction entity.

1. Invoke the **Extrude** dashboard and using the **Sketch** dialog box, select the **FRONT** datum plane as the sketching plane.

2. Select the **RIGHT** datum plane and then select the **Right** option from the **Orientation** drop-down list, if it is not selected by default.

3. Choose the **Sketch** button to enter the sketcher environment.

4. Next, draw the arc, refer to Figure 7-96, by using the **Center and Ends** button.

Note

In the sketch of the fin feature, you need to draw the sketch such that the size of the instances of the pattern can be controlled later. You will use the dimension 6.155 to create the pattern. Also, the vertical height of the fin feature is not specified in the sketch because in that case, the fin feature will not slide along the construction arc but will be locked with the top construction arc.

5. Select the arc and then hold down the right mouse button to invoke the shortcut menu. From the shortcut menu, choose the **Construction** option; the arc is converted into a construction arc. Next, create the sketch of the fin feature and apply the required constraints and dimensions to it, as shown in Figure 7-96.

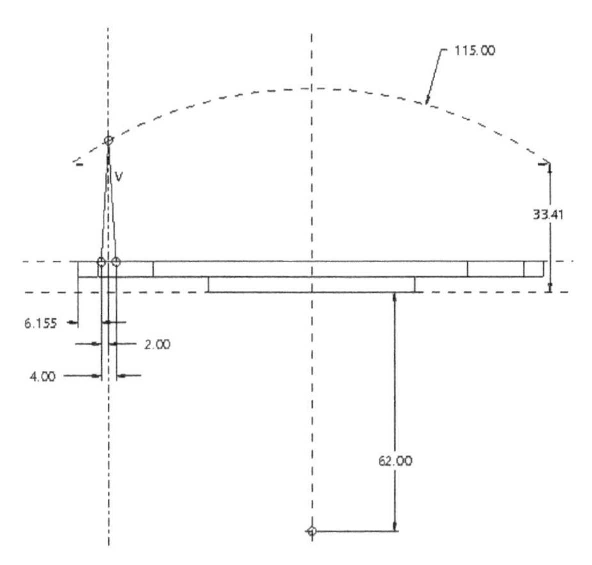

Figure 7-96 Sketch for the fin feature with two weak dimensions

6. After the sketch is completed, exit the sketcher environment. Note that the sketch must be closed.

7. Choose the **Extrude on both sides of sketch plane by half the specified depth value in each direction** button from the depth flyout in the **Extrude** dashboard.

8. Enter **135** in the dimension box of the dashboard.

9. Choose the **Build feature** button from the dashboard.

   The default trimetric view of the fin feature is shown in Figure 7-97. Now you need to pattern the fin feature to create the remaining 12 fins.

10. Select the fin and invoke the **Pattern** dashboard; the dimensions of the fin feature are displayed in the drawing area.

11. Select the dimension value **6.155**; an edit box appears. Enter **9** in this edit box.

12. Enter **13** in the dimension box of the dashboard. These are the number of instances of the fin feature.

13. Choose the **Build feature** button to exit the **Pattern** dashboard. The model after patterning the fin feature is shown in Figure 7-98.

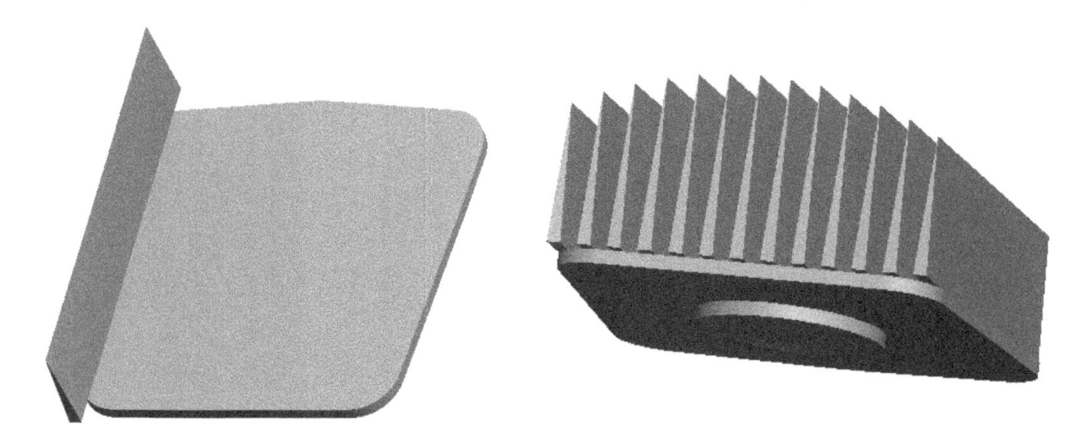

Figure 7-97 Fin created on the base feature

Figure 7-98 Pattern of the fin feature

Creating the Sixth Feature

The sixth feature is an extrude cut. You need to create this feature on the bottom face of the base feature. This cut will remove the fins that are projecting out of the base feature.

1. Invoke the **Extrude** dashboard and then choose the **Remove Material** button.
2. Select the bottom face of the base feature as the sketching plane.
3. Select the **RIGHT** datum plane and then select the **Right** option from the **Orientation** drop-down list, if they are not selected by default.
4. Once you enter the sketcher environment, use the edge of the base feature to create the sketch that will be extruded later to create the cut.
5. After the sketch is complete, exit the sketcher environment. The **Extrude** dashboard is enabled and it appears above the drawing area.
6. Choose the **Extrude to intersect with all surface** button from the depth flyout.
7. Choose the **Change depth direction of extrude to other side of sketch** button from the dashboard to reverse the direction of the pink arrow. This arrow points in the direction from where the material will be removed. Note that the arrows displayed on the preview should be pointing upward and outward. You can click on the arrows to change the direction of the cut.
8. Choose the **Build feature** button from the dashboard. The default trimetric view of the model after creating the cut feature is shown in Figure 7-99.

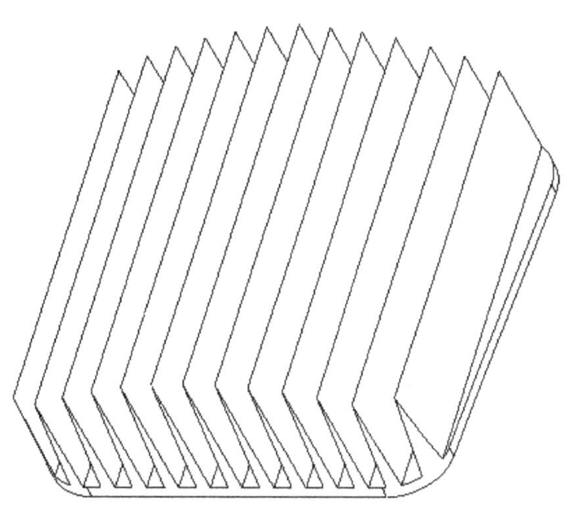

Figure 7-99 *Model after creating the cut feature*

Creating the Seventh Feature

The seventh feature is an extrude cut feature. You need to create the feature on the top face of the base feature.

1. Invoke the **Extrude** dashboard and then choose the **Remove Material** button from it.

2. Select the top face of the base feature as the sketching plane.

3. Select the **RIGHT** datum plane and then select the **Right** option from the **Orientation** drop-down list, if they are not selected by default. Next, choose the **Sketch** button.

4. Draw the sketch of the circular cut feature, as shown in Figure 7-100.

5. After the sketch is completed, exit the sketcher environment.

   The **Extrude** dashboard is enabled and it appears above the drawing area.

6. Choose the **Extrude to intersect with all surface** button from the depth flyout.

7. If required, change the depth direction of extrude such that the fins of the model are cut.

8. Choose the **Build feature** button from the dashboard. The top view of the model after creating the circular cut is shown in Figure 7-101.

Creating the Eighth Feature

The eighth feature is a protrusion feature. You need to create this feature on the top face of the base feature.

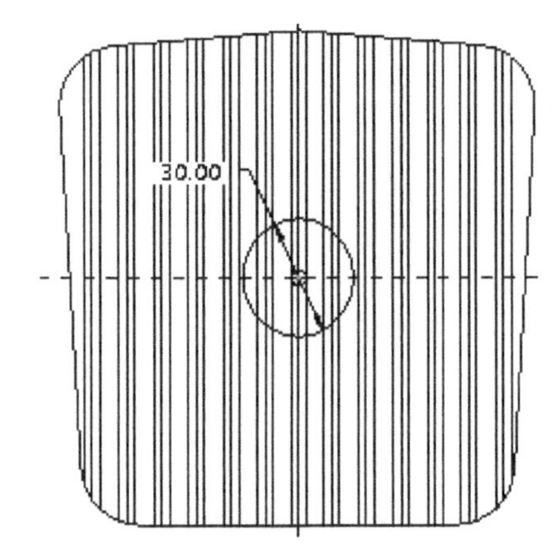

Figure 7-100 Sketch of the circular cut feature

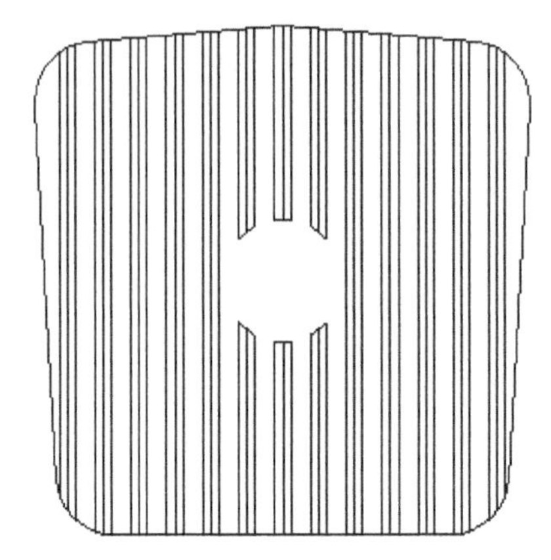

Figure 7-101 Model after creating the circular cut

1. Invoke the **Extrude** dashboard and select the top face of the base feature as the sketching plane.

2. Select the **RIGHT** datum plane and then select the **Right** option from the **Orientation** drop-down list, if they are not selected by default.

3. Once you enter the sketcher environment, use the edge of the cut feature to create the sketch of the protrusion feature.

4. After the sketch is completed, exit the sketcher environment.

5. Enter **9** in the dimension box of the **Extrude** dashboard.

6. Choose the **Build feature** button from the dashboard. The model after creating the cylindrical feature is shown in Figure 7-102.

Creating the Ninth Feature

The ninth feature is a cut feature. You need to pattern this feature to create its other instances. The cut feature is created on the top face of the base feature.

1. Invoke the **Extrude** dashboard and then choose the **Remove Material** button.

2. Select the top face of the base feature as the sketching plane.

3. Select the **RIGHT** datum plane and then select the **Right** option from the **Orientation** drop-down list.

4. Once you enter the sketcher environment, draw the sketch of the circular cut feature, as shown in Figure 7-103.

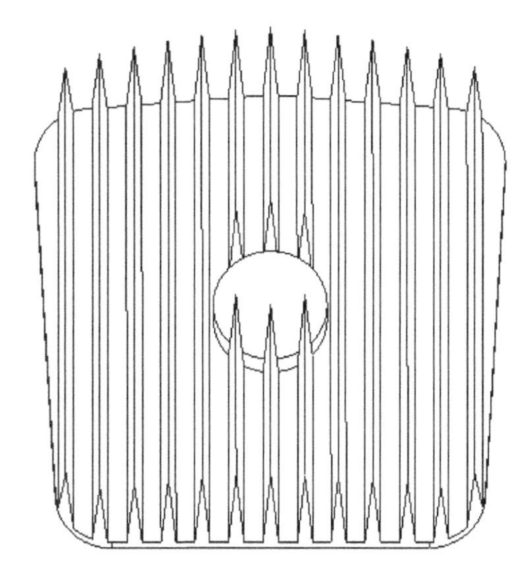

Figure 7-102 *Model after creating the cylindrical feature*

5. After the sketch is completed, exit the sketcher environment. The **Extrude** dashboard is enabled and it appears above the drawing area.
6. Choose the **Extrude to intersect with all surface** button from the depth flyout.
7. Change the depth direction of extrude such that the fins of the model are cut.
8. Choose the **Build feature** button from the dashboard. The model after creating the circular cut is shown in Figure 7-104.

Now you need to pattern the cut feature to create its other instances.

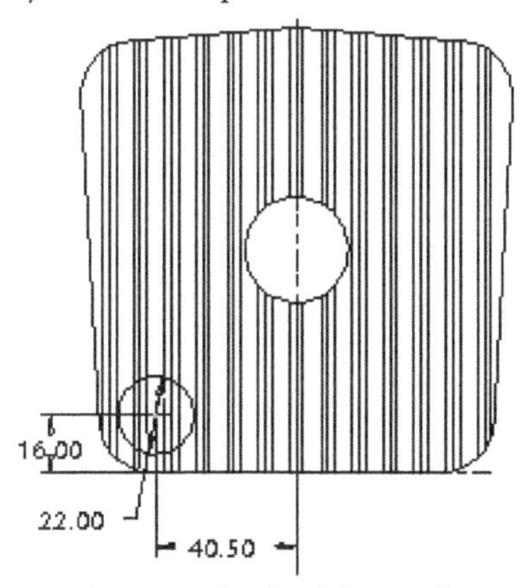

Figure 7-103 *Sketch of the cut feature*

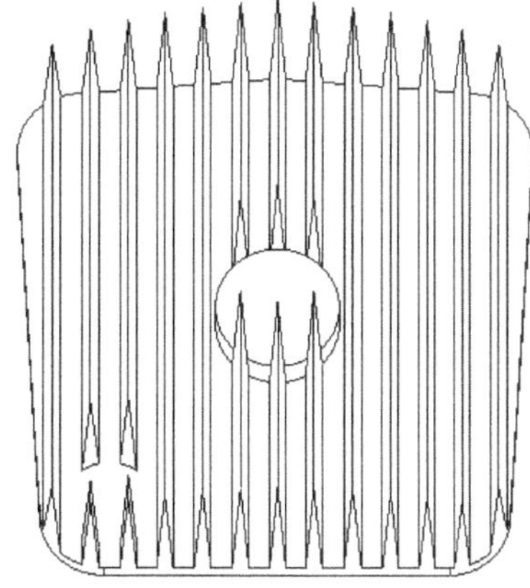

Figure 7-104 *Model after creating circular cut*

9. Select the cut feature from the **Model Tree** and then choose the **Pattern** tool from the **Editing** group. Select the **Dimension** option from the drop-down list, if it is not selected by default. The **Pattern** dashboard is displayed and you are prompted to select dimensions to be changed in the first direction.

10. Select the **General** option from the **Options** slide-down panel, if it is not selected.

11. Select the dimension value **40.5** from the drawing area. After you have selected the dimension in the first direction, an edit box is displayed.

12. Enter **-81** in the edit box.

13. Hold the right mouse button to display the shortcut menu. Choose the **Direction 2 Dimensions** option from the shortcut menu.

14. Select the dimension **16**. After you have selected the dimension in the second direction, an edit box is displayed.

15. Enter **80** in the edit box. Note that in the instances edit boxes in the dashboard, the number of instances, **2**, is specified by default.

16. Choose the **Build feature** button from the **Pattern** dashboard. The rectangular pattern of the cut feature is displayed, as shown in Figure 7-105.

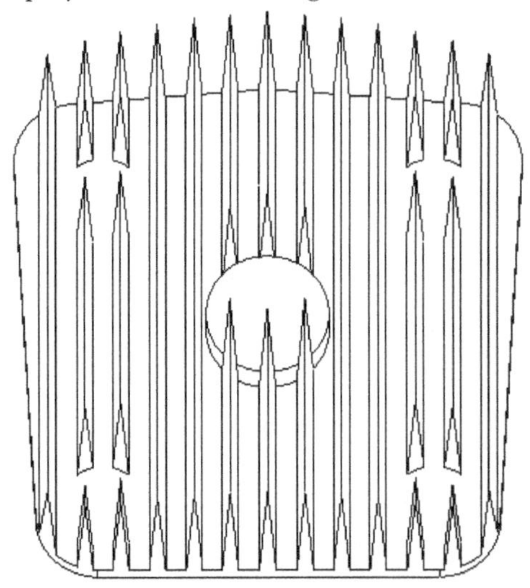

Figure 7-105 *Model after patterning the cut feature*

Creating the Tenth Feature

The tenth feature is an extrude feature created on the top face of the base feature. You need to pattern this feature to create other instances.

1. Invoke the **Extrude** dashboard.

2. Select the top face of the base feature as the sketching plane.
3. Select the **RIGHT** datum plane and then select the **Right** option from the **Orientation** drop-down list, if they are not selected by default.
4. Once you enter the sketcher environment, draw the sketch of the protrusion feature using the edges of the cut feature.
5. After the sketch is completed, exit the sketcher environment.
6. Enter **9** in the dimension box of the dashboard.
7. Choose the **Build feature** button from the dashboard. The model after creating the protrusion feature is shown in Figure 7-106.

   Now you need to create a reference pattern of the protrusion feature to create its other instances.

8. Select the extrude feature from the **Model Tree** and then choose the **Pattern** tool from the **Editing** group.

   The **Reference** option is selected by default in the drop-down list of the **Pattern** dashboard.

9. Choose the **Build feature** button from the **Pattern** dashboard. The model after creating the pattern is shown in Figure 7-107.

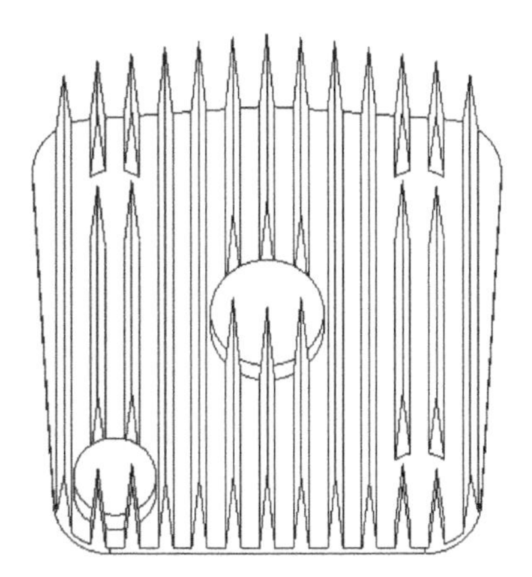

Figure 7-106 *Model after creating the cylindrical feature*

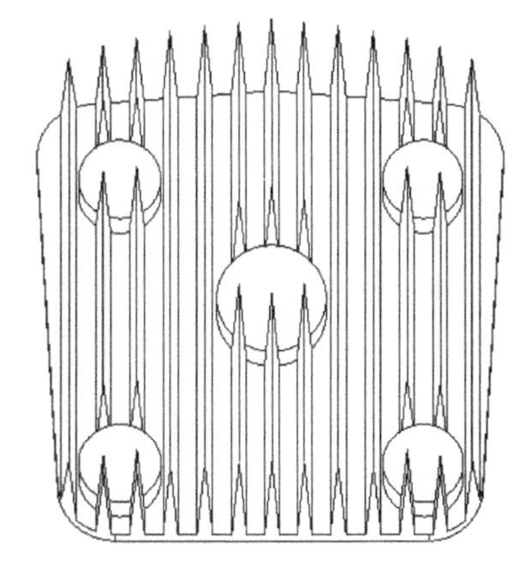

Figure 7-107 *Model after creating the reference pattern of the cylindrical feature*

Creating the Eleventh Feature

The eleventh feature is a coaxial hole and is created on the top face of the tenth feature and later it is reference-patterned to create its other instances.

1. Create a hole of diameter 12 on the top face of the protrusion feature created at the lower left corner. Make sure the bottom face of the base feature is selected. You must create a hole that passes through all faces.

Note

While creating the coaxial hole on the top face of the tenth feature, make sure that the axis of the coaxial hole and the axis of the tenth feature are same.

2. Pattern the feature by using the **Reference** option. The model after creating the pattern of the hole feature is shown in Figure 7-108.

Creating the Last Feature

The last feature is a hole of diameter **16** and is created on the top face of the eighth feature. This hole is coaxial with the cylindrical feature on which it is being created. The model after creating the hole feature is shown in Figure 7-109.

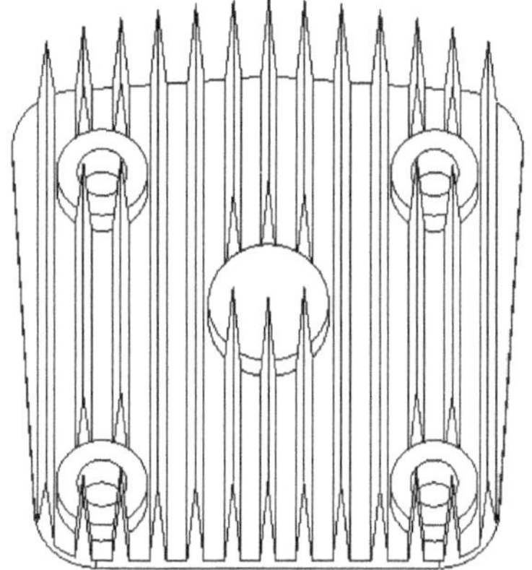

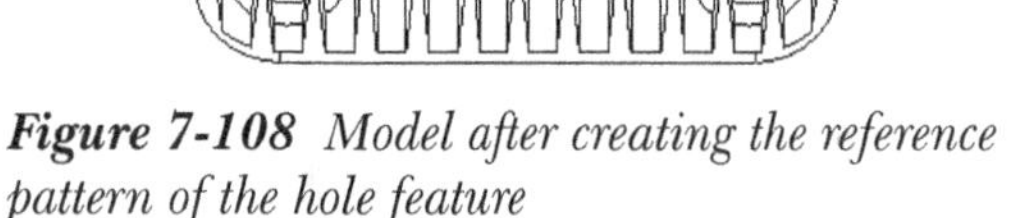

Figure 7-108 *Model after creating the reference pattern of the hole feature*

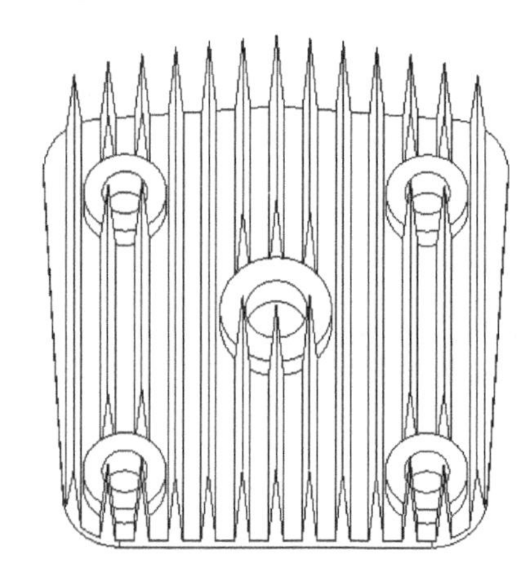

Figure 7-109 *Model after creating the hole on the cylindrical feature*

Saving the Model

You need to save the model because you may need it later.

1. Choose the **Save** button from the **File** menu and then save the model. The order of feature creation can be seen from the **Model Tree** shown in Figure 7-110.

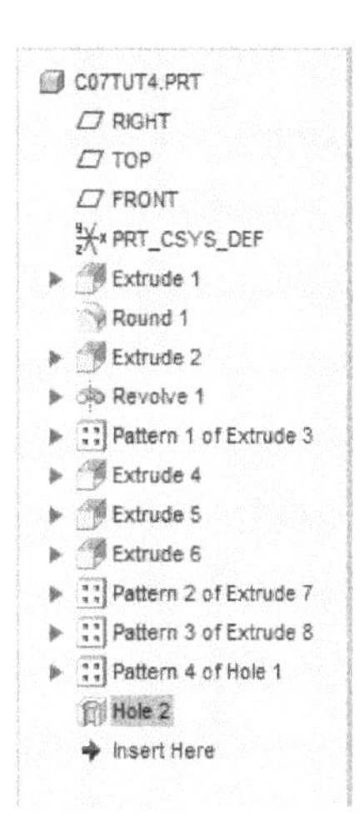

Figure 7-110 *The **Model Tree** for Tutorial 4*

Creating a Simplified Representation View of the Model

1. Choose **Manage Views > View Manager** from the **Model Display** group of the **View** tab; the **View Manager** dialog box is displayed.

2. Choose the **New** button from the dialog box; **Rep0001** is displayed in the **Names** display box.

3. Press ENTER; the dialog box disappears, and the **Menu Manager** with the **EDIT METHOD** menu is displayed.

4. Choose the **Work Region** option from the **EDIT METHOD** menu; the **SOLID OPTS** menu is displayed.

5. Choose **Extrude > Solid > Done** from the **SOLID OPTS** menu; the **Extrude** dashboard is displayed. Next, invoke the **Sketch** dialog box from the **Extrude** dashboard.

6. Select the bottom most planar face of the model as the sketching plane and then select the **RIGHT** datum plane to be at the right while drawing the sketch.

7. If required, choose the **Flip** button to reverse the direction of the pink arrow. You need to draw the sketch from the side of the sketch plane where the fins are facing. This is to ensure that the section lines do not cut the fins while you draw the sketch.

8. While entering the sketcher environment, you need to draw section lines, as shown in Figure 7-111. This section is an open loop and dimensions are not of importance to create it.

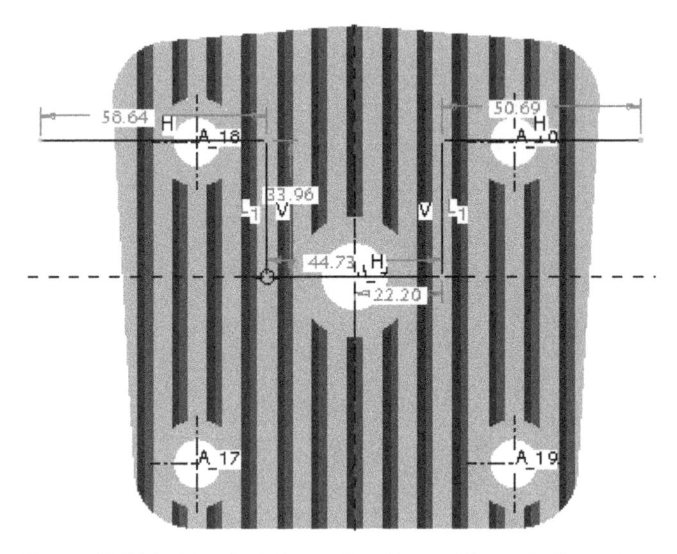

Figure 7-111 *Sketch of the section lines with weak dimensions*

9. Exit the sketcher environment and then choose the **Extrude to intersect with all surface** button from the depth flyout; the arrow will point in the reverse direction.

10. Choose the **Change material direction of extrude to other side of sketch** button to reverse the direction of the arrow.

11. Choose the **Change depth direction of extrude to other side of sketch** button to reverse the direction of the extrude.

12. Choose the **Build feature** button from the **Extrude** dashboard. The sectioned model is shown in Figure 7-112.

13. Choose the **Done/Return** option from the **EDIT METHOD** menu and then close the **View Manager**.

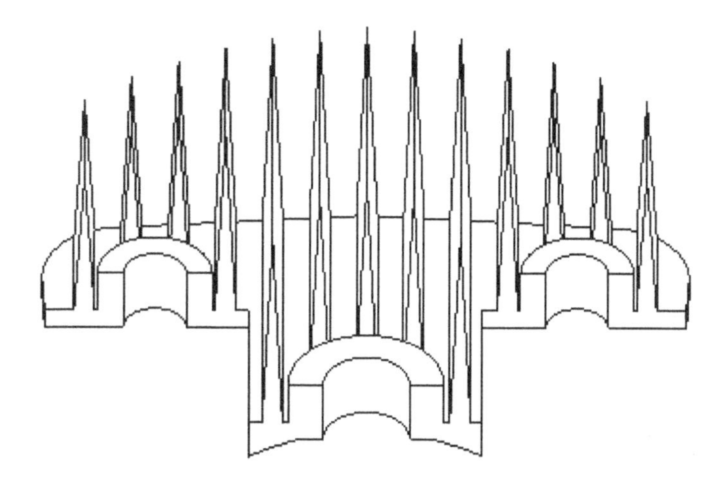

Figure 7-112 *Simplified representation of the model*

Self-Evaluation Test

Answer the following questions and then compare them to those given at the end of this chapter:

1. When a pattern is created, the leader or the parent feature also becomes a part of the pattern. (T/F)

2. Once a pattern is created, all instances in the pattern, including the parent feature, act as a single feature. (T/F)

3. In the **Reference** pattern, an existing pattern is referenced to create a new pattern. (T/F)

4. Using the **Pattern** dashboard, you can only create linear patterns. (T/F)

5. If you select a pattern feature from the **Model Tree** and right-click to display the shortcut menu and then choose the **Delete** option, the whole pattern is deleted including the leader. (T/F)

6. The _________ option is used to delete a pattern except the leader feature.

7. The _________ tool is used to mirror the entire geometry about a plane.

8. The _________ option from the **Pattern** dashboard is used to create a pattern in which the dimensions of instances can be varied.

9. The first feature in a pattern is called _________.

10. The _________ menu is used to select the dimensions that you want to vary from the leader while copying.

Review Questions

Answer the following questions:

1. Which of the following options in the **Pattern** dashboard is used to create patterns in which instances touch each other and intersect with other instances or edges of a surface?

   (a) **Identical** (b) **Variable**
   (c) **General** (d) None of these

2. Which of the following options in the **Pattern** dashboard is used to create a pattern on the specified path?

   (a) **Direction** (b) **Curve**
   (c) **Axis** (d) **Point**

3. Which of the following options in the **Pattern** dashboard cannot be used to create a pattern that intersects an edge of a feature on which the pattern has to be created?

   (a) **Identical** (b) **Variable**
   (c) **General** (d) None of these

4. Which of the following options in the **Pattern** dashboard is used to create a pattern that has all instances of different sizes?

   (a) **Identical** (b) **Variable**
   (c) **General** (d) None of these

5. You can mirror features by using datum planes or planar surfaces. (T/F)

6. To create a rotational pattern, you should specify an angular increment. (T/F)

7. The dimension pattern can be created in two direction. (T/F)

8. In the **Pattern** dashboard, the **Dimension** option is selected by default. (T/F)

9. The **Direction** option in the **Pattern** dashboard is used to create the pattern in the specified direction. (T/F)

10. Fill pattern is used to fill the sketched area by a selected feature. (T/F)

Exercises

Exercise 1

Create the model shown in Figure 7-113. The top, front section, right, detailed, and sectioned views of the model are shown in Figure 7-114. **(Expected time: 45 min)**

Figure 7-113 *Solid model for Exercise 1*

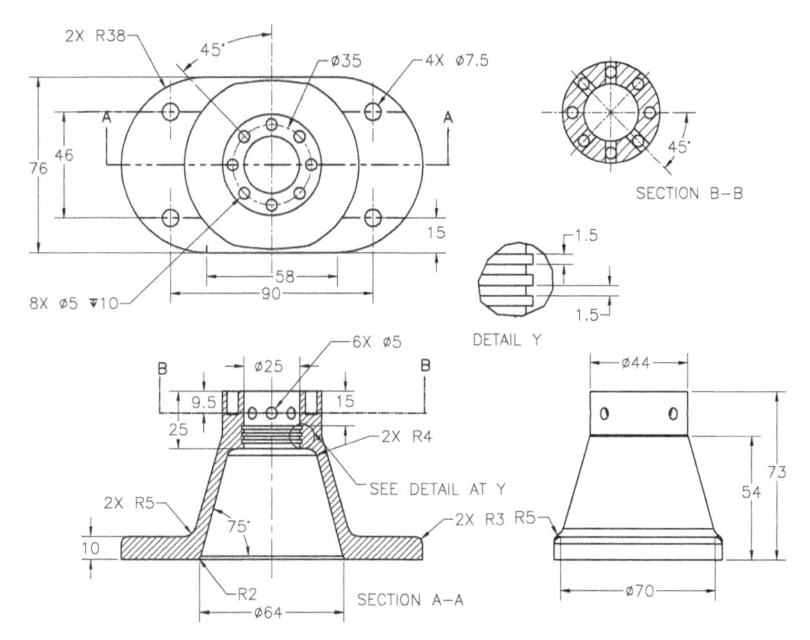

Figure 7-114 *Top, front section, right, detailed, and sectioned views of the model*

Exercise 2

Create the model shown in Figure 7-115. This figure also shows the top, front, and right views of the model. **(Expected time: 30 min)**

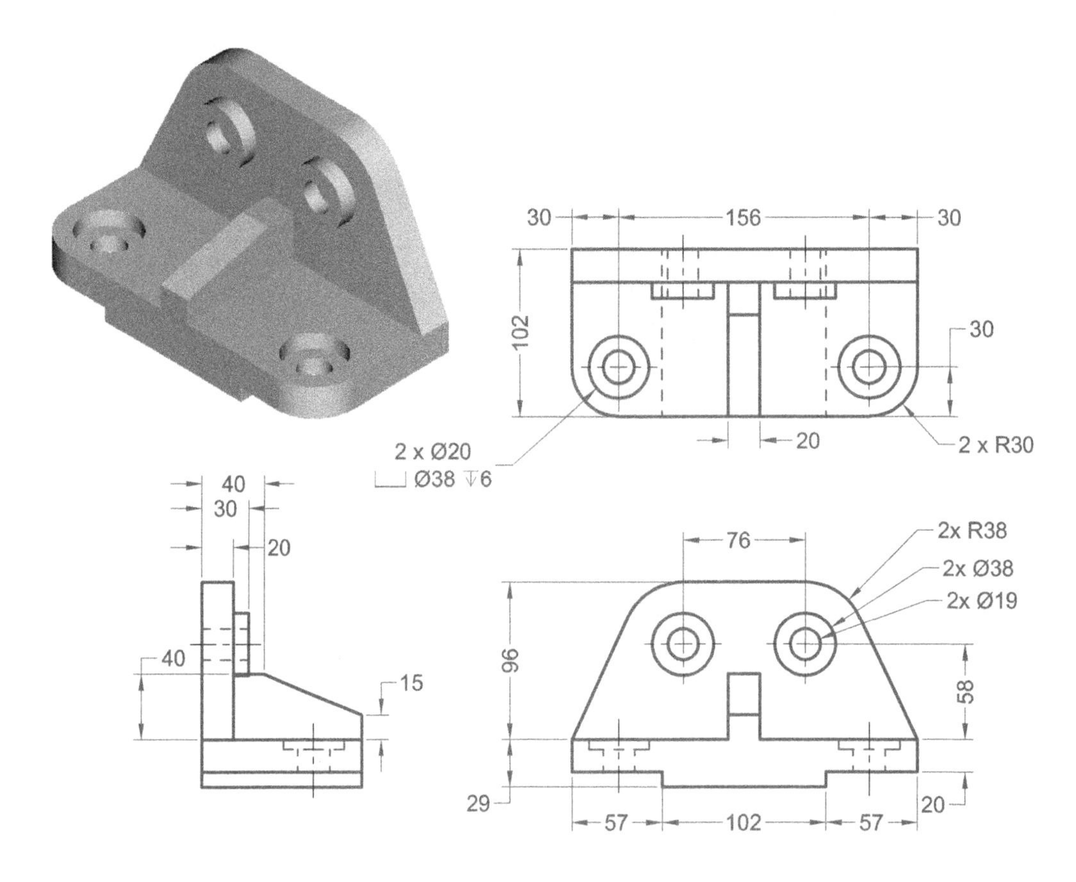

Figure 7-115 The model and its top, front, and right views

Exercise 3

Create the model shown in Figure 7-116. The top, front, and isometric views is shown in the same figure. **(Expected time: 30 min)**

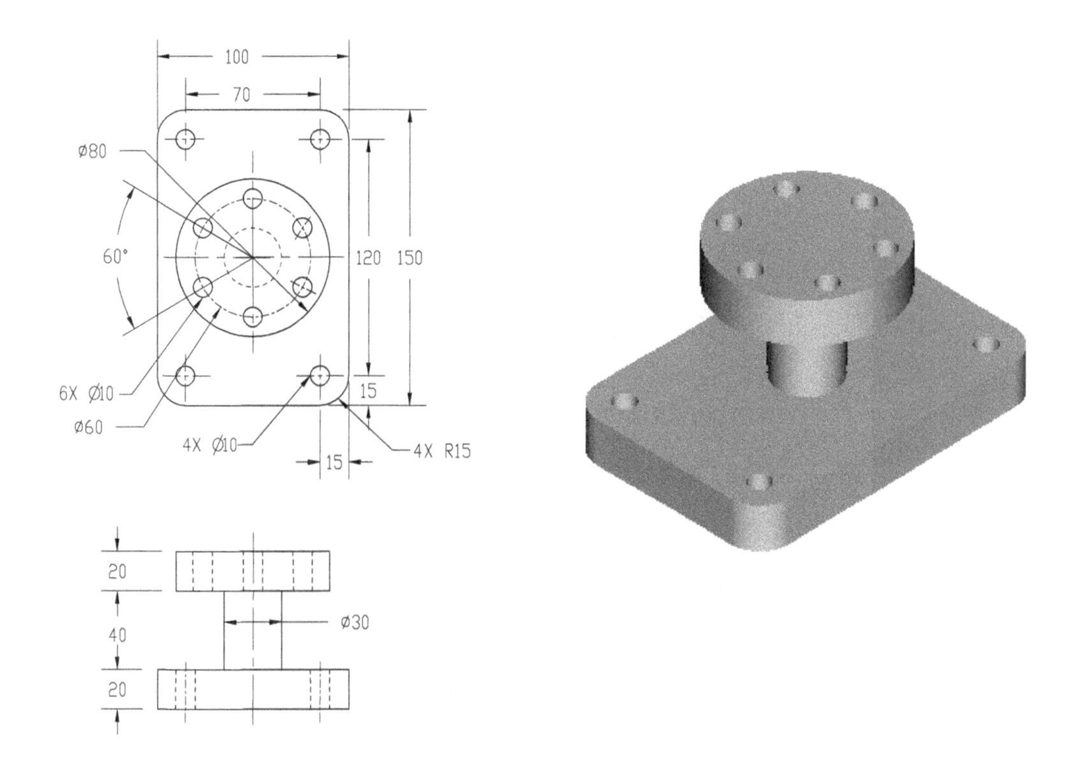

Figure 7-116 *The top, front, and isometric views*

Exercise 4

Create the model shown in Figure 7-117. The top and sectioned right views are shown in Figure 7-118. **(Expected time: 1hr)**

Figure 7-117 Solid model for Exercise 4

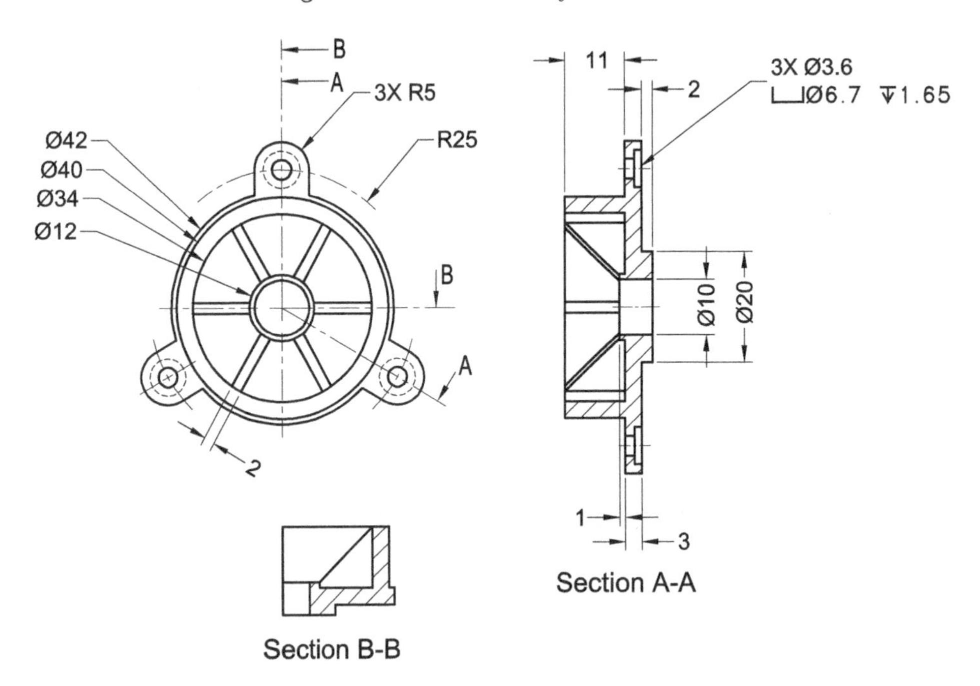

Figure 7-118 Top view and the sectioned views of the model

Answers to Self-Evaluation Test

1. T, **2.** T, **3.** T, **4.** F, **5.** T, **6. Delete Pattern**, **7. Mirror**, **8. Variable**, **9.** leader, **10. GP VAR DIMS**.

Chapter 8

Advanced Modeling Tools-I

Learning Objectives

After completing this chapter, you will be able to:

- *Create sweep features.*
- *Create features by using sweep cut.*
- *Create blend and rotational blend.*
- *Use blend vertex in blend features.*
- *Create shell features.*
- *Create datum curves.*
- *Create draft features.*

OTHER PROTRUSION OPTIONS

In chapter 4, you learned how to use the **Extrude** and **Revolve** options. In this chapter, you will learn how to use the **Sweep** and **Blend** options. As mentioned in previous chapters, **Protrusion** and **Cut** are the two basic options available in PTC Creo Parametric that are used to create a feature.

Note

All options available for creating a Cut feature are similar to those available for creating a Protrusion. Remember that a cut can be created on an an existing feature only and therefore the options related to it will be available only when a base feature exists in the drawing area.

You will also learn about the tools of solid modeling that are used to create a complex model easily. In the next section, you will learn about sweep features.

SWEEP FEATURES

The **Sweep** option extrudes a section along a defined trajectory. The order of operation is to first create a trajectory and then a section. Trajectory is a path along which a section is swept. The trajectory for a sweep feature can be either sketched or selected. The **Sweep** option of protrusion is similar to the **Extrude** option. The only difference is that in the case of the **Extrude** option, the feature is extruded in a direction normal to the sketching plane, but in the case of the **Sweep** option, the section is swept along the sketched or selected trajectory. The trajectory can be open or closed. Normal sketching tools are used for sketching the trajectory. The cross section of the sweep feature remains constant throughout the sweep.

Note

Some important points to remember while drawing a trajectory and a section for a sweep feature are discussed later in this chapter.

Creating Sweep Protrusions

Ribbon: Model > Shapes > Sweep > Sweep

The **Sweep** option can be used for adding material (protrusion) as well as for removing material (cut). To add material on a specified path, choose the **Sweep** tool from the **Sweep** drop-down in the **Shapes** group of the **Model** tab in the **Ribbon**; the **Sweep** dashboard is displayed, as shown in Figure 8-1. To create a sweep protrusion, you need to specify a trajectory and a section. In PTC Creo Parametric 3.0, you need to create a trajectory using the **Sketch** tool before invoking the **Sweep** tool. The trajectory can be closed or open. After creating the trajectory, invoke the **Sweep** tool from the **Shapes** group; the **Sweep** dashboard will be displayed. Now, choose the **Reference** tab from the dashboard; the **Reference** slide-down panel will be displayed. Select the trajectory created earlier; the slide-down panel will be modified, refer to Figure 8-2, and a crosshair will be displayed to locate the center for the section. Also, you will be prompted to sketch the cross-section. Figure 8-3 shows how the section is swept along the sketched trajectory and Figure 8-4 shows the shaded image of the sweep feature.

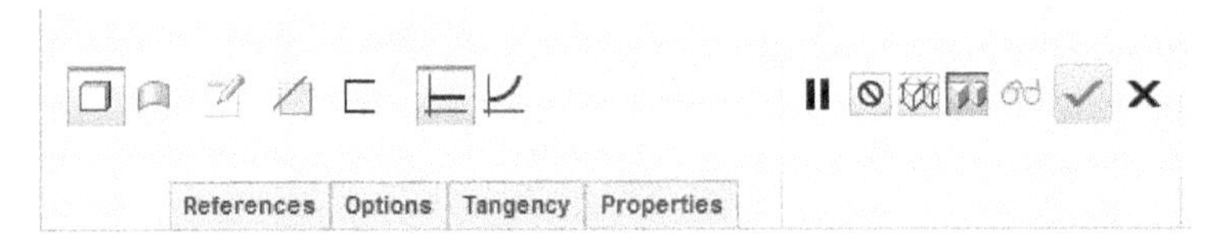

***Figure 8-1** The partial view of the **Sweep** dashboard*

The following points specify the combinations of trajectories and sections that can/cannot be used together to create the sweep feature:

1. Open section and open trajectory are possible for surface sweep feature.
2. Open section and closed trajectory are possible for surface sweep feature.
3. Closed section and open trajectory are possible.
4. Closed section and closed trajectory are possible.

You can also use the edges of an existing base feature as trajectory.

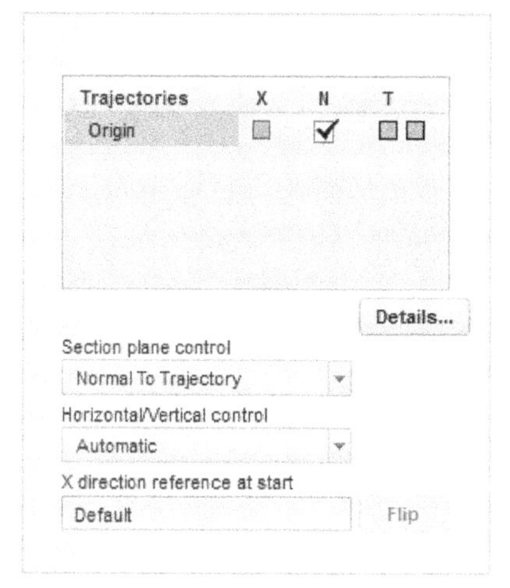

*Figure 8-2 The **Reference** slide-down panel*

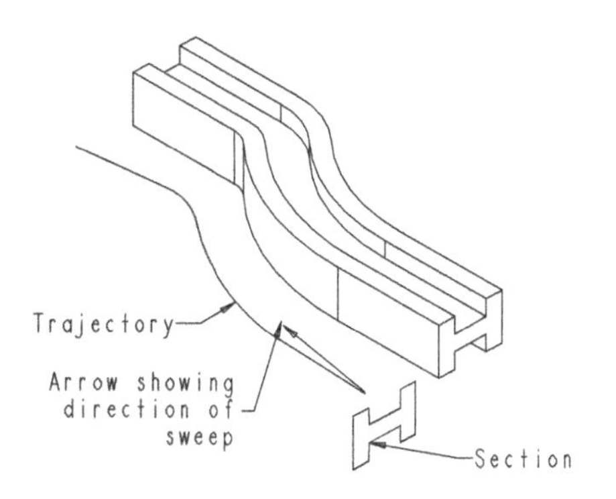

Figure 8-3 Section swept along the sketched trajectory

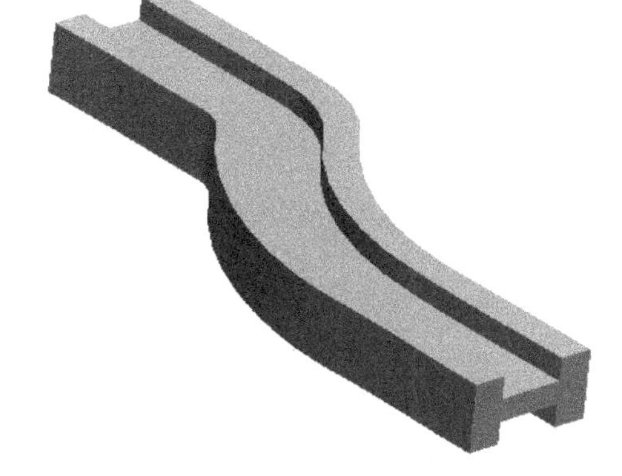

Figure 8-4 Shaded image of the sweep feature

Aligning a Sketched Trajectory to an Existing Geometry

You can align a sketched trajectory to the adjacent geometry of the existing feature by selecting the **Merge ends** check box available in the **Options** tab. This check box is available in the **Options** tab that is displayed after you select the sketch of the trajectory.

Creating a Thin Sweep Protrusion

The **Create a thin feature** tool is used to create a thin sweep feature with a specified thickness. This tool is available in the **Sweep** dashboard. This tool is similar to the **Extrude** tool that was discussed in Chapter 4. In case of thin features, a certain thickness needs to be specified. You can specify the thickness in the drop-down list next to the **Create a thin feature**. The thickness is specified on one side of the section or symmetrically to both the sides of the section. The resulting sweep feature will be a hollow sweep feature. Figure 8-5 shows the sections that can be used to create the model shown in Figure 8-6.

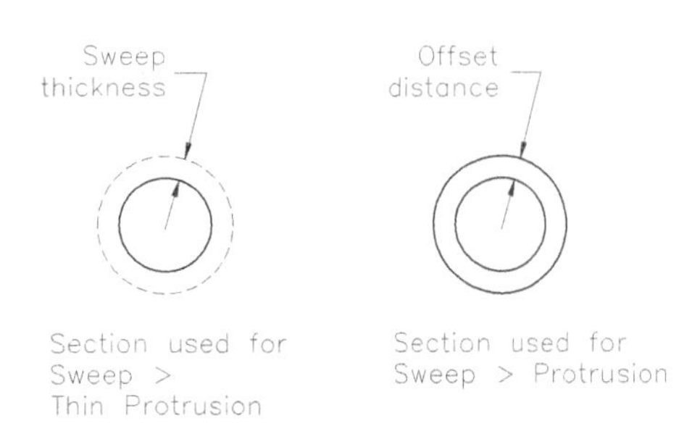

Figure 8-5 *Two possible sections to create the same model*

Figure 8-6 *The model created by using different sections*

Tip: *The following points should be remembered while creating a sweep feature:*

1. *Similar to other sketched features, the trajectory of the sweep feature is also sketched after selecting a sketching plane.*

2. *At bends in a trajectory, the radius of the bend should be proportionate to the cross-section to be swept to avoid overlapping. If the section size is large and the radius of the curve or bend is small, overlapping takes place and the sweep feature will not be created. Therefore, make sure that the ratio of the size of the section to the size of the trajectory is appropriate.*

Creating a Sweep Cut

The procedure to create a sweep cut feature is similar to that of creating the sweep protrusion. The only difference is that in case of cut features, the material is removed from an existing feature. The **Cut** option can be invoked by choosing the **Remove Material** button from the **Sweep** dashboard. The cut can be a solid sweep cut or thin sweep cut. Figure 8-7 shows trajectories for the sweep cut feature. Figure 8-8 shows the shaded model with the open and closed trajectory sweep cuts.

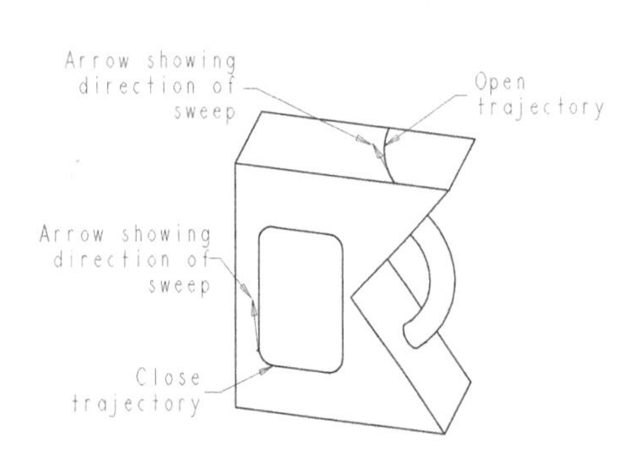

Figure 8-7 *Trajectories for the sweep cut feature*

Figure 8-8 *Shaded model with the open and close trajectory sweep cuts*

BLEND FEATURES

Ribbon: Model > Shapes > Blend

Blend features consists of two or more sections joined through transitional faces at their edges so as to form a continuous feature. The number of entities in each section that are used to create the blend feature should be the same. For example, you cannot blend a circle with a rectangle. This is because a rectangle is composed of four entities and a circle of one entity. It can be achieved only if the circle is divided into four entities.

In PTC Creo Parametric, the blend feature is mainly of two types, Protrusion and Cut. The **Blend** option is used where the feature to be created has varying cross sections. To invoke this option, choose **Blend** from the expanded **Shapes** group; the **Blend** dashboard will be displayed, as shown in Figure 8-9. Options and tabs available in the dashboard are discussed next.

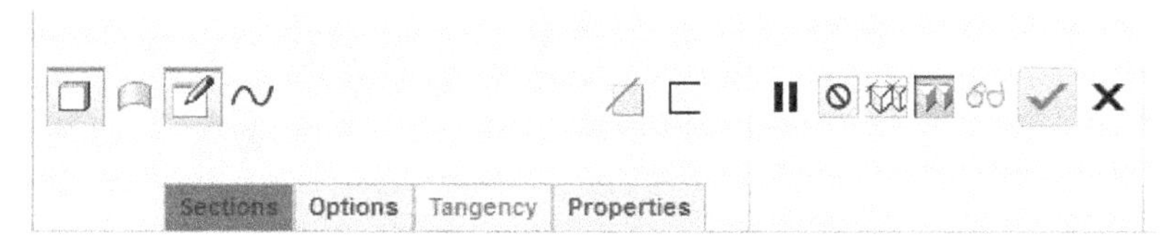

*Figure 8-9 The **Blend** dashboard*

Blend as solid

The **Blend as solid** button is chosen by default and is used to blend two or more sketches in a solid form and create solid feature. Note that the sketch that is drawn for creating solid feature need to be a closed loop.

Blend as surface

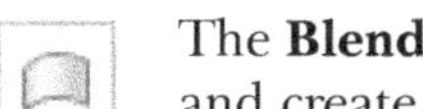

The **Blend as surface** button is used to blend two or more sketches in a surface form and create surface feature. The sketch that is drawn for the surface model need not to be a closed loop.

Blend with sketched sections

The **Blend with sketched sections** button is used to create sketches in the drawing area to form the blend. The options for creating sketches are available in the **Section** tab that is discussed further.

Blend with selected sections

The **Blend with selected sections** button is used to select sketches available in the drawing area to form the blend. The options for inserting sketches are available in the **Section** tab.

Remove Material

The **Remove Material** button is displayed in the **Blend** dashboard only after the base feature is created. This is because this button is used to remove the material from an existing feature.

Creating a Thin Feature

The **Thicken Sketch** button is used to create thin parts with specified thickness. When you choose this button, the **Change direction to thicken between one side, other side, or both sides of sketch** button and a dimension edit box appear on the right side of this button.

Sections Tab

When you choose the **Sections** tab from the dashboard, the **Sections** slide-down panel will be displayed, as shown in Figure 8-10. This panel has two radio buttons that are discussed next.

Figure 8-10 *The* ***Sections*** *slide-down panel*

Sketched sections

This radio button is used to create sketches in the drawing area to form the blend. When you select this radio button, the **Define** button will be displayed in panel which is used to enter the sketcher environment. You can also use the **Insert** button, displayed in panel to add more sketches to the blend.

Selected sections

This radio button is used to select sketches from the drawing area to form the blend. You can use the **Insert** button, displayed in panel to add more sketches to the blend.

Options Tab

The **Options** tab contains the parameters related to the blend feature. The options available in this tab are discussed next.

Straight

The **Straight** radio button is available in the **Blended surfaces** area of this tab. It is used to connect the vertices of all sections in a blend feature with straight lines.

Smooth

The **Smooth** radio button is available in the **Blended surfaces** area. It is used to connect the vertices of all sections in a blend feature with smooth curves.

Capped Ends

The **Capped Ends** check box is available in the **Start and End Sections** area and is used to close the sketch ends. This check box gets activated when you choose the **Blend as surface** option from the dashboard.

Tangency Tab

The **Tangency** tab will only be available when you select the **Smooth** radio button from the **Options** tab. This tab is used to set the boundary of created or selected sketches in the **Free**, **Tangent**, or **Normal** condition.

The steps required to create a blend feature are as follows:

1. Invoke the **Blend** tool by choosing **Model > Shapes > Blend** from the **Ribbon**; the **Blend** dashboard will be displayed.

2. Choose **Blend as solid** from the dashboard and then choose the **Sections** tab to open the slide-down panel. Next, select the **Sketched sections** radio button and then choose the **Define** button to define the sketching plane.

3. Define the required sketching plane and its orientation to draw the sections.

4. Draw the sketch of the first section. Choose **OK** to exit the sketcher environment. An offset value edit box with a default value will be available in the dashboard and in the **Sections** tab. Specify the desired offset value for the second section and draw the second section. Choose the **Insert** button from the **Sections** tab, if you want to add another section.

Note

All sections in a blend feature must have equal number of entities. However, you can blend a point with any section irrespective of the number of entities.

5. Choose the **Build feature** button to exit the **Blend** tool.

Figures 8-11 and 8-12 show the blend features created with straight edges and smooth edges, respectively.

Figure 8-11 *Blend with straight edges*

Figure 8-12 *Blend with smooth edges*

Note

While drawing a section, the start point of all sections should be in the same direction in order to avoid twisted blend features. By default, the start point of an entity drawn to define a section is considered to be the start point of the section. To change the start point of a section, select the point to be defined as the start point and hold the right mouse button to display the shortcut menu. Choose the start point option from this shortcut menu to change the start point.

ROTATIONAL BLEND FEATURES

Ribbon: Model > Shapes > Rotational Blend

The rotational blends are created by using the sections that are rotated about the Y-axis up to a maximum of 120 degrees. To create such features, an angle called rotational blend angle has to be defined between each section. Each section in rotational blend is rotated about a geometric centerline drawn in the first sketch.

The following steps briefly explain the procedure of creating the blend feature shown in Figure 8-13. Figure 8-14 shows the sections and the rotational blend angle.

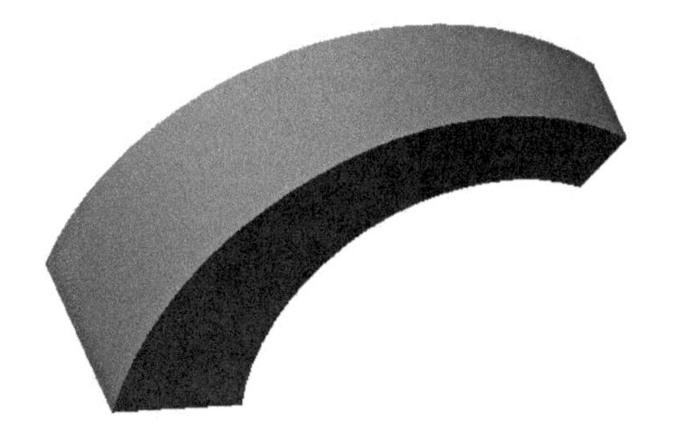

Figure 8-13 *Shaded model of a rotational open blend feature*

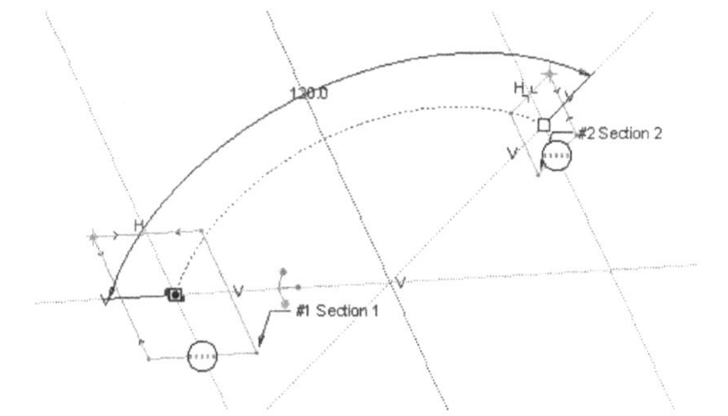

Figure 8-14 *The two sections used to create the blend feature shown in Figure 8-13*

1. Invoke the **Rotational Blend** tool by choosing **Model > Shapes > Rotational Blend** from the **Ribbon**; the **Rotational Blend** dashboard will be displayed.

2. Choose **Blend as solid** and click on the **Sections** tab from the dashboard to open the slide-down panel. Select the **Sketched sections** radio button and choose the **Define** button to define the sketching plane.

3. Define the required sketching plane and its orientation to draw the sections.

4. Draw the sketch of the first section and place the geometric centerline on one side of the sketch. Choose **OK** to exit the sketcher environment. An angle offset value edit box with a default value will be available in the dashboard and in the **Sections** tab. Change

the desired angle offset value for the second section. Choose the **Insert** button from the **Sections** tab, if you want to add another section.

5. Choose the **Build feature** button to exit the **Blend** tool.

Note

*You can select the **Connect end and start sections** check box in the **Options** tab to create a closed blend feature. But for creating a closed blend feature, at least three sections are required.*

USING BLEND VERTEX

As mentioned earlier, each section of the blend feature must have equal number of entities. However, you can use the **Blend Vertex** option if the numbers of entities in all sections are not equal. For example, to create a blending between a square and a triangle, add blend vertex on a point other than the start point of the triangle. The two vertices of the triangle will be blended with the two vertices of the rectangle and the blend vertex in the triangle will be blended with the remaining two vertices in the rectangle, as shown in Figure 8-15.

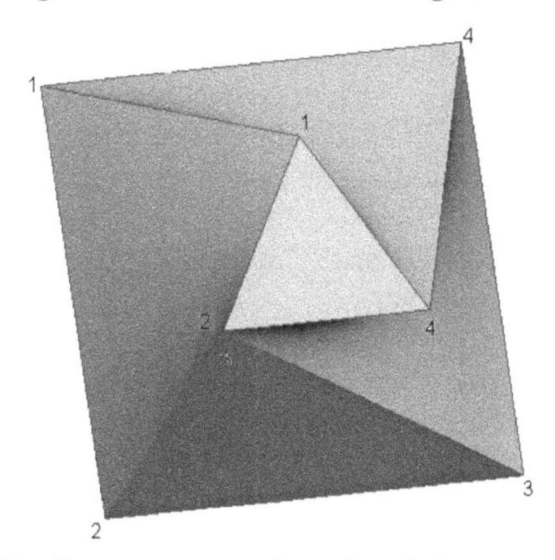

Figure 8-15 *Blending a square with a triangle by using the blend vertex*

To add a blend vertex in a sketch, select the point where you want to place the blend vertex. The selected point is highlighted in red. Choose the **Blend Vertex** option from the **Feature Tools** flyout in the **Setup** group of the **Ribbon**. The blend vertex is placed at the selected point.

Note

*The **Blend Vertex** option can be used either in the first or in the last section of a blend feature.*

SHELL FEATURE

Ribbon: Model > Engineering > Shell

The **Shell** option scoops out the material from the model and at the same time removes the selected faces, leaving behind a thin model with some specified wall thickness. The **Shell** option can be invoked by choosing the **Shell** tool from the **Engineering**

group of the **Model** tab in the **Ribbon**. When you choose this tool; the **Shell** dashboard will be displayed, as shown in Figure 8-16.

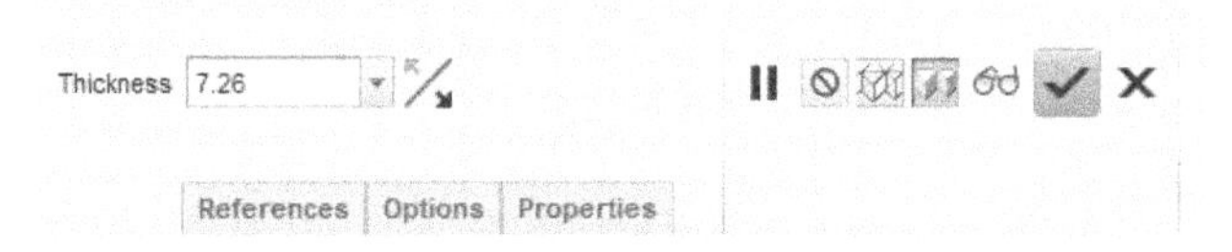

***Figure 8-16** Partial view of the **Shell** dashboard*

Options Tab

When you choose the **Options** tab from the **Shell** dashboard, a slide-down panel will be displayed, as shown in Figure 8-17.

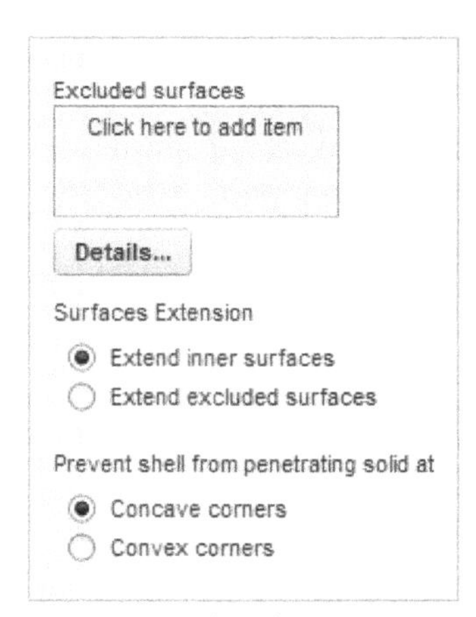

***Figure 8-17** The **Options** slide-down panel*

The **Excluded surfaces** collector lists the surfaces excluded from being removed. If you do not select any surface from being excluded, the entire part will be shelled.

The following steps explain briefly the procedure of creating a shell feature by using the **Excluded surfaces** collector:

1. Invoke the **Extrude** dashboard and then create the extrude feature, refer to Figure 8-18.

2. Once the **Extrude** feature is created, invoke the **Shell** dashboard by choosing the **Shell** tool from the **Engineering** group.

3. Select the top face of the model to remove it.

4. Choose the **Options** tab; the slide-down panel is displayed. Activate the **Excluded surfaces** collector by clicking on it.

5. Select the surfaces to be excluded by pressing the CTRL key, refer to Figure 8-18.

6. Choose the **Build feature** button from the **Shell** dashboard to exit it; the shell created by excluding the specified surfaces is shown in Figure 8-19.

Note

*The **Shell** option is used on existing models and hence, this option is available only when a model exists in the drawing area.*

Using the **Shell** dashboard, you can create the two types of shell: Constant thickness shell and Variable thickness shell.

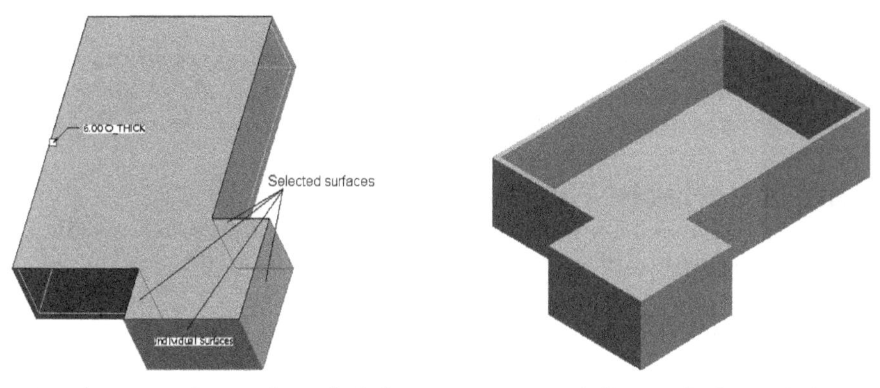

Figure 8-18 *Selecting surfaces to be excluded* ***Figure 8-19*** *Shell created after excluding surfaces*

Creating a Constant Thickness Shell

The constant thickness shell is a shell that has a uniform thickness on all faces of the model. The following steps explain the procedure to create a constant thickness shell.

1. Invoke the **Shell** dashboard.

2. Select the top face of the model to remove it, as shown in Figure 8-20. This is the face that will be removed from the final model, leaving the specified thickness from the boundary of the selected face.

3. Enter the thickness value of the shell in the **Thickness** edit box present on the dashboard.

4. Choose the **Feature Preview** button from the dashboard to preview the shell feature. 

5. Choose the **Build feature** button from the **Shell** dashboard to exit it.

The shell is created on the selected face, as shown in Figure 8-21. The thickness of the shell is uniform on all faces of the model.

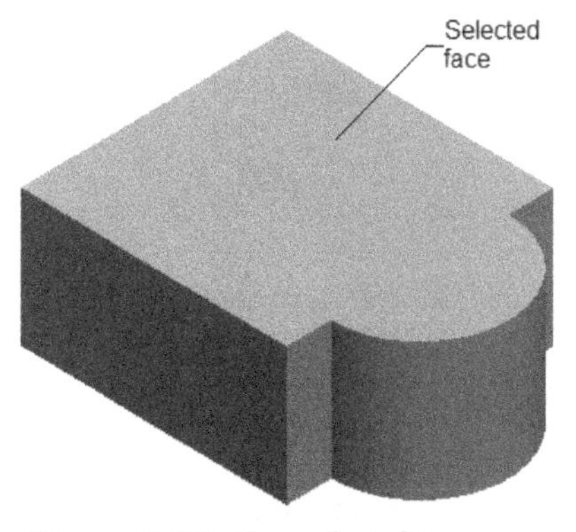

Figure 8-20 *Face selected to remove*

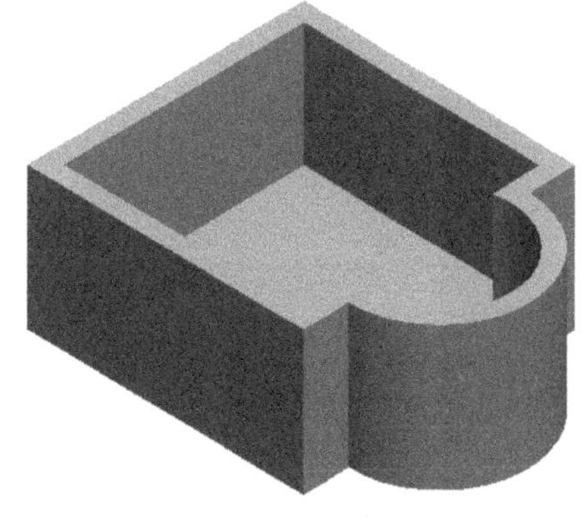

Figure 8-21 *Constant thickness shell feature created on the selected face*

Note

The thickness value entered can be positive or negative. If the value entered is positive, the material is removed, leaving the shell thickness inside the boundary of the selected face. But when the value entered is negative, the shell thickness is added outside the boundary of the selected face.

Creating a Variable Thickness Shell

The variable thickness shell is a shell that has different thickness values assigned to adjacent faces. The following steps explain the procedure to create a variable thickness shell.

1. Invoke the **Shell** dashboard.
2. Select the top face of the model to remove it, as shown in Figure 8-22.
3. Choose the **References** tab to invoke the slide-down panel. In the slide-down panel, there are two collectors: **Removed surfaces** and **Non-default thickness**. The **Removed surfaces** collector shows the surface id of the face that you have selected to remove. Whereas, the **Non-default thickness** collector shows the surfaces that you will select to create the variable thickness shell.
4. Click in the **Non-default thickness** collector and then select the adjacent faces shown in Figure 8-22. To select the second and successive faces, you need to use CTRL+left mouse button. Now, you need to specify different thickness values with reference to the selected face.
5. Enter thickness values in the dimension edit boxes present on the right of the surfaces in the **Non-default thickness** collector or drag the white square to specify the thickness. Alternatively, click on the thickness dimension and enter the new values in the edit boxes displayed.

The shell is created on the selected faces with the thickness values assigned to them, refer to Figure 8-23.

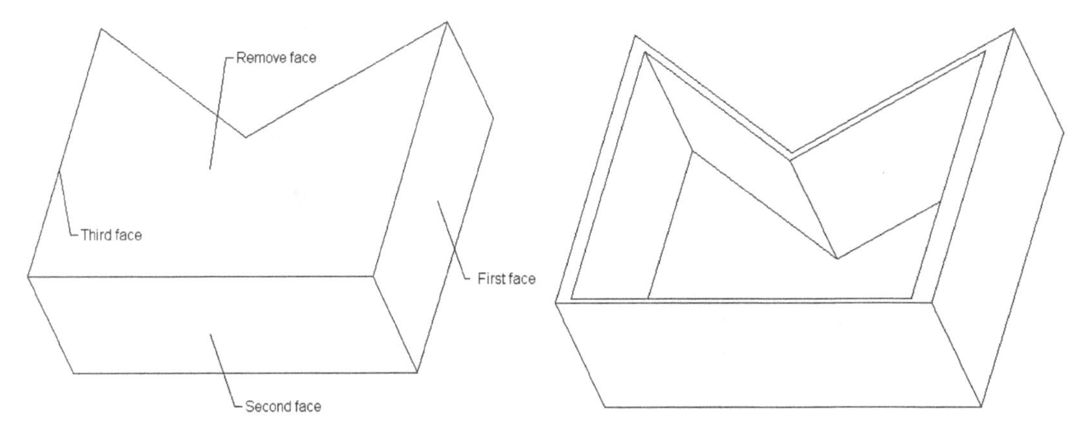

Figure 8-22 *Faces of the model selected for variable thickness and the top face to remove*

Figure 8-23 *Variable thickness shell*

Figure 8-24 shows the model whose top surface is selected to be removed in order to create a shell. Figure 8-25 shows the model after shelling. Notice that the shell thickness is left on the selected face and the remaining material is removed.

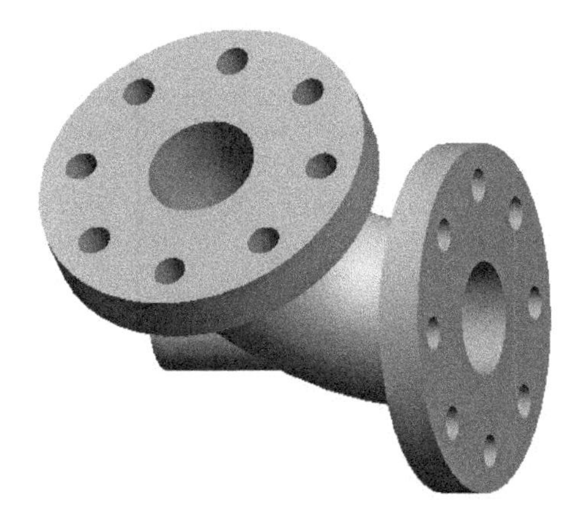

Figure 8-24 *Solid model*

Figure 8-25 *Model after shelling the top face*

DATUM CURVES

Datum curves are useful in creation of advanced solid and surface features such as the sweep trajectories to create a sweep feature. A datum curve is considered as a feature and is displayed in the **Model Tree**. There are various tools and options to create a datum curve. You can choose any of the tools available in the **Curve** drop-down from the **Datum** group to create a datum curve. Various methods of creating curves are discussed next.

Creating a Datum Curve by Using the Curve Drop-down

The **Curve** drop-down is used to create datum curves. There are three options available in this drop-down, as shown in Figure 8-26. Some of the important options in this drop-down are discussed next.

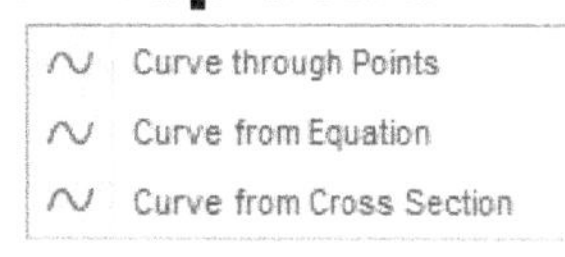

Figure 8-26 *The* ***Curve*** *drop-down*

Curve through Points

Ribbon: Model > Datum > Curve > Curve through Points

The **Curve through Points** option is used to create a datum curve by selecting the existing datum points or vertices. When you choose this option; the **CURVE: Through Points** dashboard will be displayed and you will be prompted to select the datum points. When you select the datum points, an orange colored curve will be displayed in the preview. Choose the **Done** button from the dashboard to create the datum curve. The resulting datum curve may be a spline curve or can be an arc.

There are two radio buttons available in the **Placement** tab to control the trajectory of curve. These buttons are discussed next.

Spline

Using the **Spline** radio button, you can create a datum curve in the form of a spline that passes through the selected datum points or vertices. The curve created between the selected point and the previous point is in the form of a spline.

Straight line

Using the **Straight line** radio button, you can create a datum curve in the form of a straight line between the specified point. You can add a fillet at the selected point by using the **Add fillet** check box. When you select this check box, the **Radius** edit box will be displayed. You can enter the value of the radius in this edit box. You can specify the radius value for each point individually.

Curve from Equation

Ribbon: Model > Datum > Curve > Curve from Equation

The **Curve from Equation** option is used to create datum curves by defining equations using the coordinate systems. When you choose this option, the **CURVE: From Equation** dashboard will be displayed. Select the type of coordinate system from the **Cartesian** drop-down list displayed at the left of the dashboard; you will be prompted to select the coordinate system.

You can choose the type of coordinate system from this dashboard. Choose the coordinate system depending on the equation you want to use to create a datum curve. Then choose the **Equation** button located next to the **Cartesian** drop-down list; the **Equation** window will be displayed, as shown in Figure 8-27.

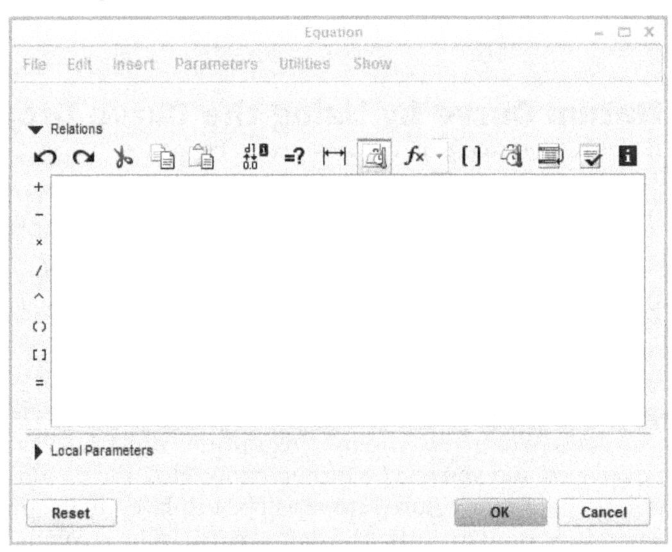

Figure 8-27 The ***Equation*** *window*

Using this window, you can define the equations. The window shows the instructions that should be followed while writing an equation. These instructions vary and depend on the type of coordinate system you have selected. After you have entered the equations in the window, save the file and then exit the notepad.

Creating a Datum Curve in Spiral Shape

The following steps explain the procedure to create a spiral-shaped datum curve:

1. Select the **Cylindrical** option from the **Click an option to set the coordinate system type** drop-down list and choose the **Equation** button; the **Equation** window will be displayed. In the window, the parametric equations that are used to create a circle are given as follows:
   r = 4
   theta = t * 360
   z = 0

   In the above equations, the variable t varies from 0 to 1, r is the radius of the circle, and z is the third equation that is set equal to 0. Now, to understand the given equation of the circle, notice that the radius of the circle is given. The only value that is unknown is theta. The value of theta depends on the variable t. Therefore,
   when t = 0, theta = 0
   when t = 1/2, theta = 180 (semicircle)
   and when t = 1, theta = 360 (circle)

2. Enter the following parametric equation of the spiral curve below the dashed line in the window:
   IR = 8
   OR = 80
   TURNS = 10
   r = IR + t * (OR - IR)
   theta = t * 360 * TURNS
   z = 0

   In the above equations, the value of r is selected to vary because in the spiral curve, the value of r always increases from center (at center of spiral, r = 0). The internal radius of the spiral IR = 8, outer radius OR = 80, and number of turns TURNS = 10 are given.

3. Choose **File > Save Configuration As** from the menu bar to save this file for later use and then exit the window.

4. Choose the **Reference** tab; a slide-down panel is displayed. Now, select the coordinate system from the drawing area.

5. Choose the **Preview** button from the **CURVE: From Equation** dashboard to preview the datum curve.

6. Choose **OK** from this dashboard; the spiral-shaped datum curve appears in the drawing area, as shown in Figure 8-28.

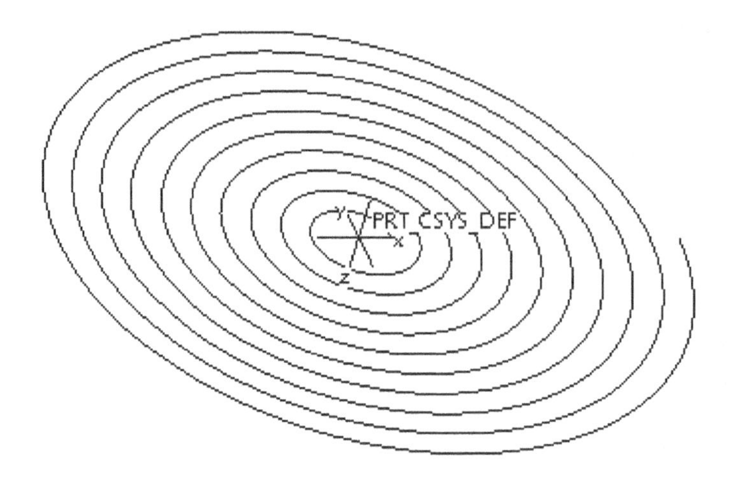

Figure 8-28 Spiral-shaped datum curve

Creating a Datum Curve by Sketching

The **Sketch** button is used to sketch a datum curve using the sketcher tools. This is the most commonly used option to create a datum curve and is drawn in the sketching environment. The sketched curves can consist of one or more sketched entities and of one or more open or closed loops, whereas the datum curves are restricted to a single curve of an open or closed loop.

Creating a Curve Using the Intersect Option

Ribbon: Model > Editing > Intersect

The **Intersect** option can be invoked by choosing the **Intersect** tool from the **Editing** group. This option is an object action tool, therefore you need to select a datum plane and then invoke this option. The **Intersect** option creates a datum curve at the intersection of a face of the model and a datum plane, intersection of a face of the model and a quilt surface, intersection of a quilt surface and a datum plane. Note that you cannot create a datum curve at the intersection of two datum planes, or two model faces using this option.

When you choose this option, the **Intersect** dashboard appears, as shown in Figure 8-29.

*Figure 8-29 The partial view of the **Intersect** dashboard*

To create the ring on the circumference of the cylindrical feature, as shown in Figure 8-30, you need to create a datum curve that is on the circumferential surface of the cylindrical feature.

Figure 8-30 *Ring on the circumference of the cylindrical feature*

To create the datum curve that lies on the outer surface, follow these steps:

1. Select the datum plane that is intersecting with the circumferential surface of the cylinder.
2. Choose the **Intersect** tool from the **Editing** group; the **Intersect** dashboard is displayed.
3. Choose the **References** tab to open the slide-down panel. In the **Surfaces** collector, the selected datum plane is displayed.
4. Use CTRL+left mouse button to select the top or bottom circumferential surfaces of the cylindrical feature, as shown in Figure 8-31.
5. Choose the **Build feature** button from the dashboard to exit the feature creation tool. The datum curve is created on the circumferential surface of the cylinder, as shown in Figure 8-32.

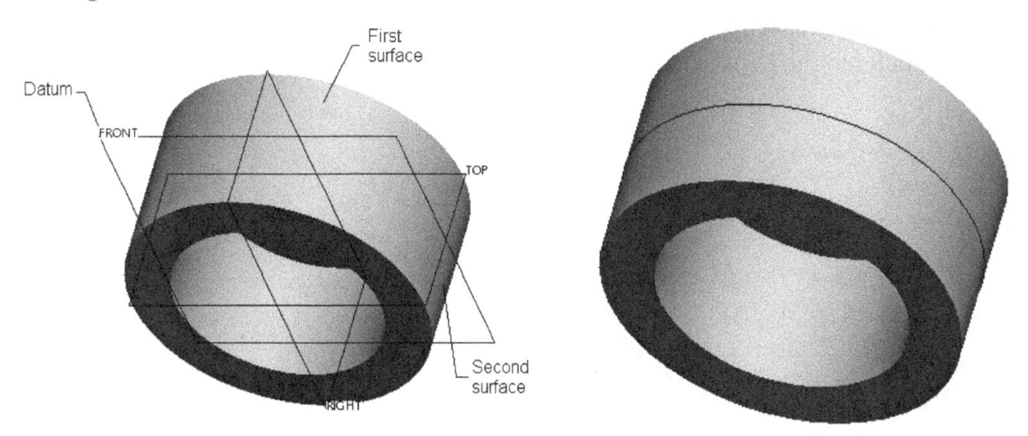

Figure 8-31 *Selections made on the model*

Figure 8-32 *Datum curve created on the circumferential surface of the cylindrical feature*

Creating a Curve Using the Project Tool

Ribbon: Model > Editing > Project

The **Project** tool is used to project a selected or sketched entity on one or more planar or non-planar surfaces or datum planes. The projected datum curve forms a true projection of the selected or sketched entity on the specified surfaces. The dimensions of the original entity may distort while projecting.

Choose the **Project** tool from the **Editing** group; the **Projected Curve** dashboard is displayed, as shown in Figure 8-33. The tools and options in this dashboard are discussed next.

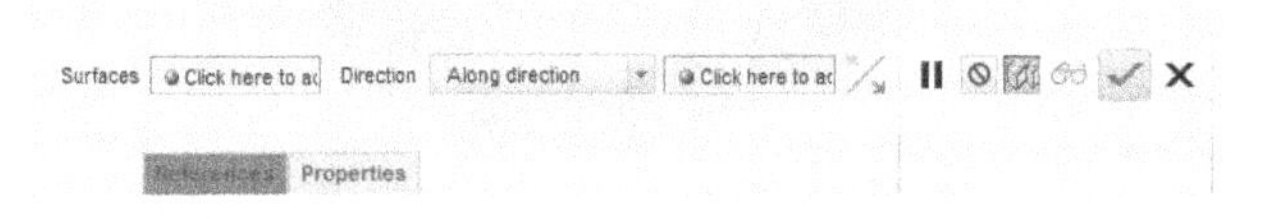

Figure 8-33 The ***Project Curve*** *dashboard*

Surfaces Collector

The **Surfaces** collector is used to select the surface on which you need to project the sketched or selected datum curve.

Direction Drop-down List

The **Direction** drop-down list is used to specify the method of projection of the datum curve on the receiving surface or plane. The two options available in this drop-down list are: Along direction and Normal to surface, and are discussed next.

Along direction

If you use this option, then the datum curve is projected in the direction indicated by the pink arrow. To specify the direction of projection, select the datum plane, edge, or surface. Figure 8-34 shows the top view of the projected datum curves that are overlapping. The datum curve that is selected to project on the receiving surface and the datum curve after projection are of same geometry. This is because, using the **Along direction** option, the true geometry is obtained after projection. Figure 8-35 shows the sketched datum curve after it has been projected on a receiving surface. The curve is sketched on a datum plane that is parallel to the bottom face of the model.

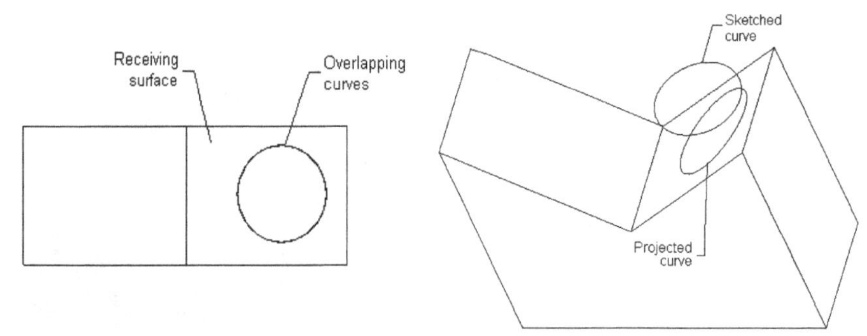

Figure 8-34 *Top view of the projected datum curve*

Figure 8-35 *Projecting a datum curve along the specified direction*

Normal to surface

This option of projecting a datum curve projects the datum curve normal to the receiving surface. Figure 8-36 shows a datum curve selected to be projected on a receiving surface and the datum curve after projection. Figure 8-37 shows the sketched datum curve after the projection on the receiving surface.

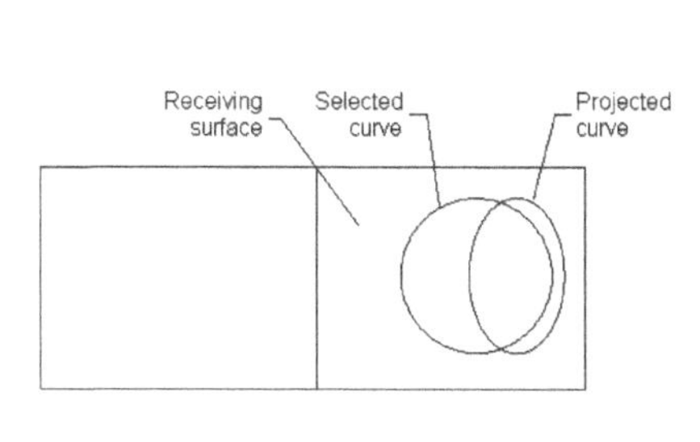

Figure 8-36 *Top view of the projected datum curve*

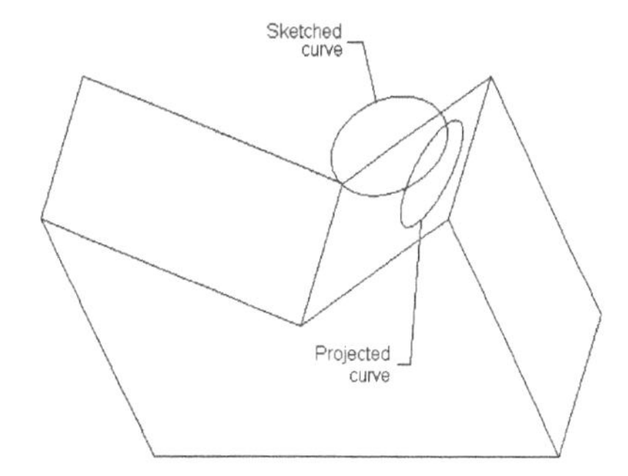

Figure 8-37 *Projecting a datum curve*

References Tab

This tab is used to define the references for projection. If you choose the **References** tab, a slide-down panel is displayed, as shown in Figure 8-38. This slide-down panel is used to select the datum curve to be projected or sketch a datum curve to project. The drop-down list in the slide-down panel lists three options: **Project chains**, **Project a sketch**, and **Project a cosmetic sketch**. Using these options, you can project multiple chains or a sketch on a surface or face. These options are discussed next.

Project chains

The **Project chains** option is used when the datum curve to be projected exists. Click in the **Chains** collector to make it active and then select the datum curve that you need to project. After selecting the datum curve, click in the **Surfaces** collector and select the receiving surface (plane or surface to project on to). Then you need to specify the direction of projection. Remember that if you are using the **Normal to surface** option, then you do not need to specify the direction of projection. The **Direction Reference** collector is displayed while selecting the **Along direction** option from the dashboard, whereas it does not appear while selecting the **Normal to surface** option.

Project a sketch

When you select the **Project a sketch** option from the drop-down list in the **References** slide-down panel, a slide-down panel appears, as shown in Figure 8-39.

Click on the **Sketch** collector and then select a curve or sketch to be projected or create a sketch that can be projected on the sketch. To create a new sketch, choose the **Define** button to invoke the **Sketch** dialog box and to select the sketch plane for the curve to be projected. After sketching, exit the sketcher environment and select the surface or plane on which the curve will be projected. If you are using the **Along direction** option to project the curve then you need to click in the **Direction Reference** collector to specify the direction of projection.

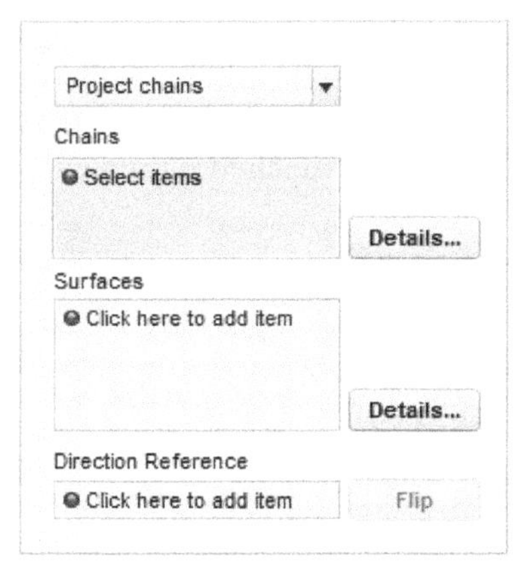

*Figure 8-38 The **References** slide-down panel with the **Project chains** option selected*

*Figure 8-39 The **References** slide-down panel with the **Project a sketch** option selected*

Project a cosmetic sketch

When you select the **Project a cosmetic sketch** option from the drop-down list of the **References** slide-down panel, a slide-down panel will be displayed which is similar to the **Project a sketch** slide-down panel. Click on the **Sketch** collector and then select a cosmetic sketch to be projected or create a sketch that can be projected on the sketch. To create a new cosmetic sketch, choose the **Define** button to invoke the **Sketch** dialog box and then select the sketch plane for sketching the curve. After sketching, exit the sketcher environment and select the surface or plane on which the curve needs to be projected.

Creating a Curve Using the Wrap Option

Ribbon: Model > Editing > Wrap

This option is used to create a datum curve by wrapping a sketched entity around a solid or a quilt. Choose the **Wrap** tool from the **Editing** group, the **Wrap** dashboard will be displayed, as shown in Figure 8-40. The tools and options in this dashboard are discussed next.

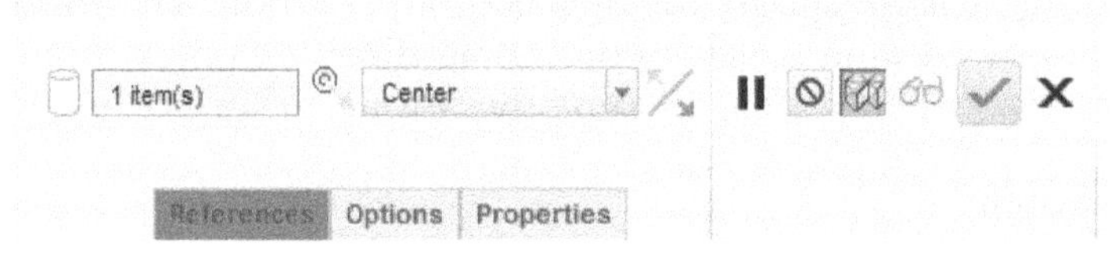

*Figure 8-40 The partial view of the **Wrap** dashboard*

References Tab

When you choose the **References** tab, a slide-down panel will be displayed, as shown in Figure 8-41. The **Define** button available in this slide-down panel is used to invoke the **Sketch** dialog box that can be to select the sketch plane and enter the sketcher environment to draw the curve.

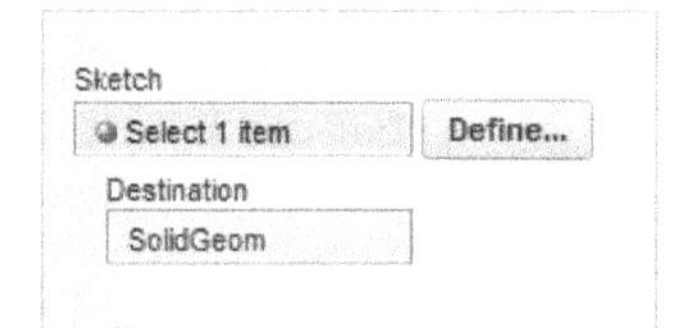

*Figure 8-41 The **References** slide-down panel*

The **Destination** collector in the slide-down panel is used to select the object on which you need to wrap the curve. Generally, if there is a single feature in the drawing area, then you do not need to select the object to wrap on. PTC Creo Parametric automatically wraps the selected curve or the sketched curve on the object. However, if you want to select a different object to wrap on then you can click in this collector to make it active and then select the object.

Options Tab

On choosing the **Options** tab, a slide-down panel will be displayed, as shown in Figure 8-42. There are two options available in this panel. The **Ignore intersection surface** check box when selected, ignores intersection surface, if any and wraps the selected curve on the destination object.

The **Trim at boundary** check box when selected, trims the extra portion of the curve that is beyond the boundary of the destination object.

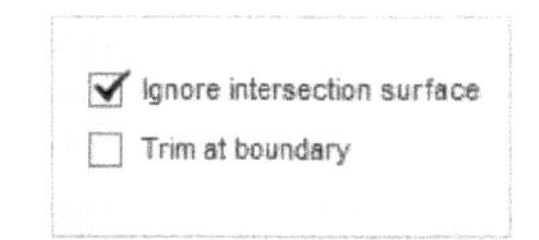

Figure 8-42 *The* ***Options*** *slide-down panel*

Follow the steps given below to sketch and wrap a curve on the rectangular block, as shown in Figure 8-43. It is assumed that the rectangular block of dimension 20x20x50 exists.

1. Choose **Wrap** tool from the **Editing** group to invoke the **Wrap** dashboard.

2. Choose the **References** tab to invoke the slide-down panel.

3. Choose the **Define** button to invoke the **Sketch** dialog box.

4. Choose the datum plane that is passing through the center of the rectangular block as the sketching plane and then choose a reference for orienting the sketching plane.

5. After entering the sketcher environment, draw a line that starts from the bottom of the rectangular block. Its start point is aligned with the center of the rectangular block and with the bottom edge. The endpoint of the line is at a distance of 1000 and at a height of 50 from the bottom of the rectangular block, refer to Figure 8-44.

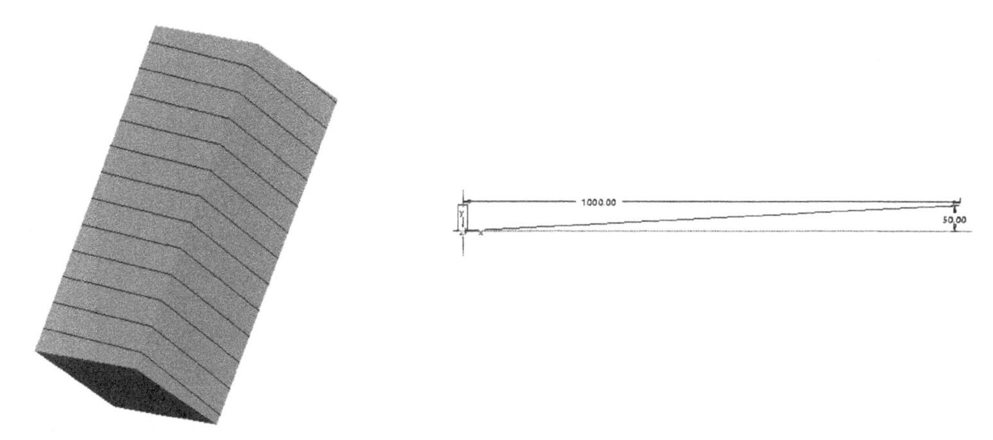

Figure 8-43 *Curve wrapped on the rectangular block*

Figure 8-44 *Sketch of the curve to be wrapped*

6. Place a user-defined coordinate system at the start point of the line.

7. Exit the sketcher environment; the sketched curve will be automatically wrapped on the circumference of the cylinder.

8. Choose the **Build feature** button to exit the dashboard.

CREATING DRAFT FEATURES

Ribbon: Model > Engineering > Draft > Draft

In PTC Creo Parametric, drafts are created on existing surfaces. They are created by rotating the selected surface by a certain angle. One of the applications of draft features is in moulds and castings where a taper is required to separate the casting from the mould or vice versa. To create a draft, choose the **Draft** tool from the **Draft** drop-down available in the **Engineering** group; the **Draft** dashboard will be displayed, as shown in Figure 8-45. The options in this dashboard are discussed next.

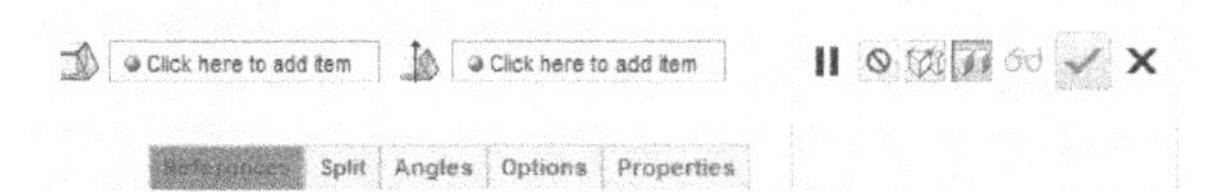

***Figure 8-45** Partial view of the **Draft** dashboard*

References Tab

On choosing the **References** tab, a slide-down panel will be displayed, as shown in Figure 8-46. The options in this slide-down panel are discussed next.

Draft surfaces Collector

The **Draft surfaces** collector is selected by default. If this collector is not selected by default, click in this collector and then select the surface to which you need to add a draft angle. The maximum draft angle that can be added to a surface is 89.9 degrees.

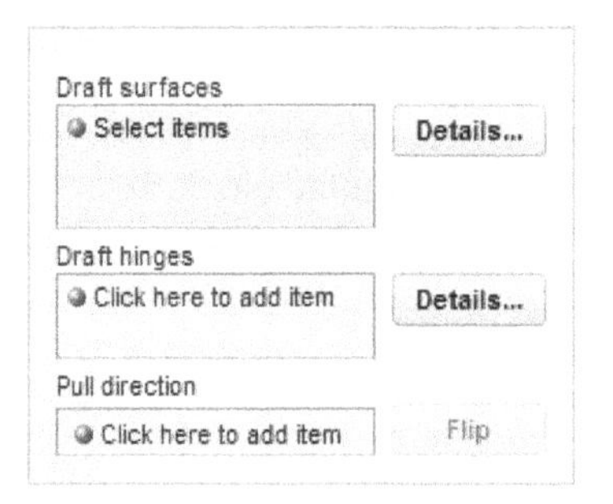

***Figure 8-46** The **References** slide-down panel*

Draft hinges Collector

The **Draft hinges** collector is used to select the hinge of the draft surface. The hinge that you select can be an edge, a surface, an axis, or a datum plane. The draft surface is pivoted on the hinge that you select. In other words, the draft surface is rotated about the hinge. The hinge that you select need not intersect the draft surface. Figure 8-47 shows the draft surface and the hinge selected to create the draft and Figure 8-48 shows the resultant draft surface.

Pull direction Collector

The **Pull direction** collector is used to specify the direction of rotation of the draft surface. The direction of pull is shown by the direction of the pink arrow. Generally, when you select the hinge, the pull direction is selected by default. If the pull direction shown by the pink arrow is not as desired, click in the **Pull direction** collector and then select the pull direction. You can change the direction of the pink arrow by choosing the **Flip** button.

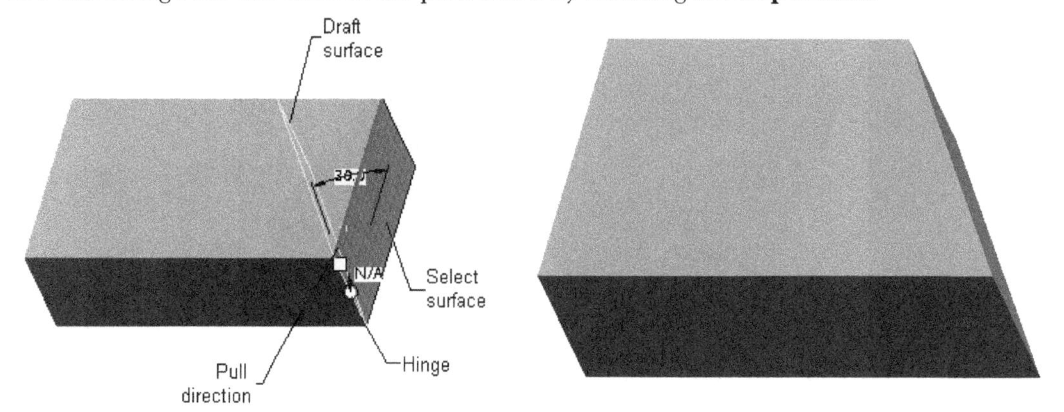

Figure 8-47 Parameters needed to create a basic draft feature

Figure 8-48 Resultant draft feature

Split Tab

If you choose the **Split** tab, a slide-down panel will be displayed, as shown in Figure 8-49. The options in this slide-down panel are discussed next.

Split options Drop-down List

The options in this drop-down list are used to split the surface selected to draft. Using these options, the selected surface gets split into two surfaces and different draft angles can be applied to both the surfaces. The options available in the **Split options** drop-down list are discussed next.

No split

This option is used when you do not want to split the surface selected to give the draft angle.

*Figure 8-49 The **Split** slide-down panel*

Split by draft hinge

This option is available only when you have selected the hinge for the draft surface. When you select this option, the selected surface is divided into two surfaces at the location on the surface where the hinge intersects it. Figure 8-50 shows the surface that is split into two surfaces at the hinge.

Split by split object

This option, when chosen activates the **Split object** collector and the **Define** button. This option is used to split the surface selected to draft by drawing a sketch. In this type of split, surface gets split into two surfaces and the sketch defines the profile of the split. Figure 8-51 shows the parameters that you need to define to create a draft by using the split by **Split object** option. Figure 8-52 shows the draft created on the cylindrical surface.

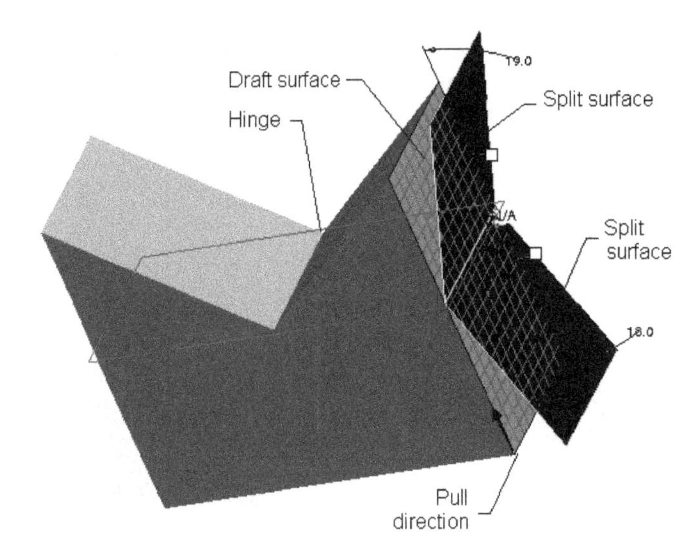

Figure 8-50 Split at draft hinge

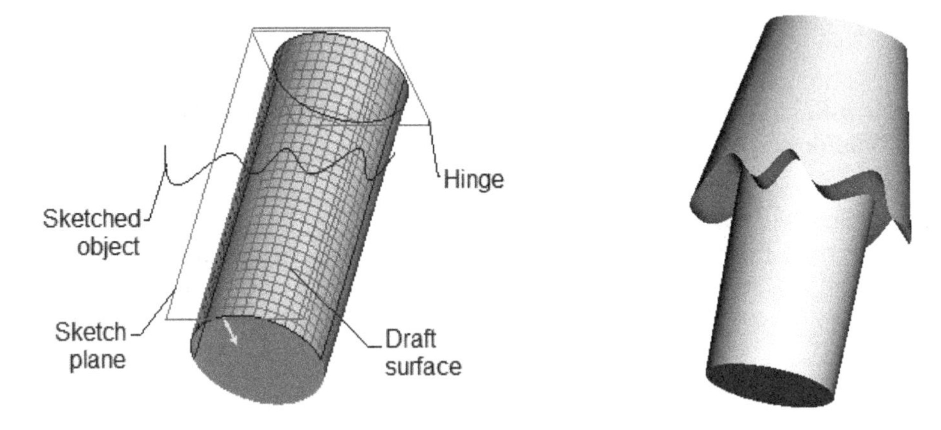

Figure 8-51 Split by split object

Figure 8-52 Resultant model

Side options Drop-down List

The options in this drop-down list are used to specify how to apply the draft once the surface has been split. The options in this drop-down list are discussed next.

Draft sides independently

The two sides of the draft surface that are formed when the surface is split can be given different draft angles. When you select this option, the edit boxes appear on the **Draft** dashboard. You can enter the values of the draft angle in these edit boxes.

Draft sides dependently

Using this option, you can apply same draft angle on both sides of the split surface.

Draft first side only

Using this option, you can apply the draft angle only to the first side of the split surface.

Draft second side only

Using this option, you can apply the draft angle only to the second side of the split surface.

Angles Tab

Choose the **Angles** tab from the **Draft** dashboard after you have specified some angle for draft surface. On choosing this option, a slide-down panel will be displayed, as shown in Figure 8-53.

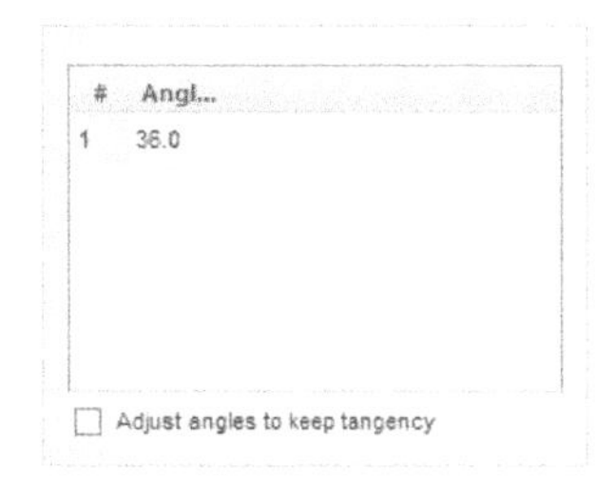

Figure 8-53 *The* ***Angles*** *slide-down panel*

In PTC Creo Parametric, you can create constant as well as variable angle drafts.

Constant Angle Drafts

By default, PTC Creo Parametric creates a constant angle draft feature. This means, the same draft angle is applied to the surface that you have selected.

Variable Angle Drafts

When you apply more than one value of the draft angle to the selected surface, it is called variable angle draft. You need to apply the variable angle draft after you have selected the hinge, the pull direction, and the draft surface.

To create the draft surface shown in Figure 8-54, follow the steps given below. It is assumed that you have created the base feature of the model, as shown in Figure 8-55.

Figure 8-54 *Draft feature on the model*

Figure 8-55 *Base feature*

1. Invoke the **Draft** dashboard and click in the first collector from the left.

2. Select the top face of the base feature. Notice that in the **Pull direction** collector, the direction of pull is selected automatically. Reverse the pull direction by choosing the **Reverse pull direction** button available on the right of the second collector.

3. Hold the right mouse button to invoke the shortcut menu shown in Figure 8-56.

4. Choose the **Draft Surfaces** option from the shortcut menu displayed and select the cylindrical surface of the base feature to add the draft angle. The drag handle appears on the base feature; you can use it to change the draft angle. The draft angle value of **1** also appears in the drawing area. Also notice the white ball that appears on the edge of the top face (face selected as hinge).

5. Bring the cursor on the white ball and hold the right mouse button to invoke the shortcut menu shown in Figure 8-57.

Figure 8-56 *Shortcut menu* ***Figure 8-57*** *Shortcut menu*

6. Choose the **Add Angle** option from the shortcut menu. Notice that now there are two white balls on the edge of the top face. On the first ball, the value of **0.5** appears. This value varies from one end of the face to the other end. This value represents the location of the point that you need to use for reference in order to apply the draft angles.

7. Again bring the cursor to any one of the two white balls and invoke the shortcut menu.

8. Choose the **Add Angle** option from the shortcut menu displayed. Another point is added with a location value on the edge.

9. Add eleven such points for locations varying from **0** to **1**.

10. After adding eleven points, double-click on any of the location value that is present in the drawing area; an edit box appears. Enter **0** in this edit box. Similarly, locate all remaining ten points and increment them by 0.1. Figure 8-58 shows the model after relocating all eleven points on the edge.

11. Double-click on the value of the angle that corresponds to the location **0**. The edit box appears, enter a value in it as **15**.

12. Double-click on the value of the angle that corresponds to the location **0.1** (these values appear above the model and have a default value of 1.0). The edit box appears; enter a value of **5**. Similarly, vary the angle at other locations. Every alternate location should have an angle of **15**. Figure 8-59 shows the preview of the model after modifying the values of the angle at all eleven locations.

13. After modifying the values of the angle at all locations, choose the **Feature Preview** button from the **Draft** dashboard; the model appears, as shown in Figure 8-60.

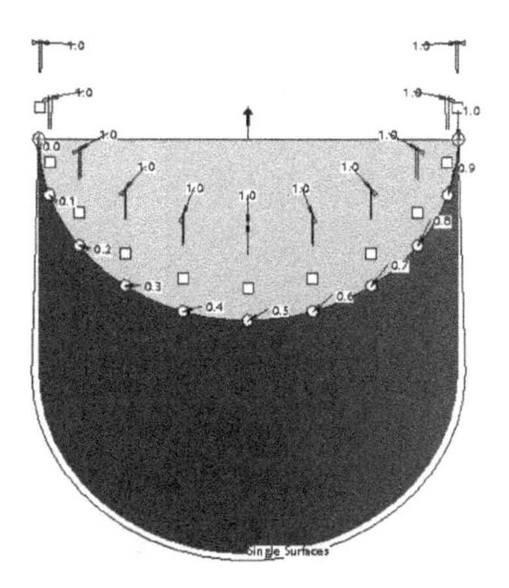

Figure 8-58 *Location points on the edge*

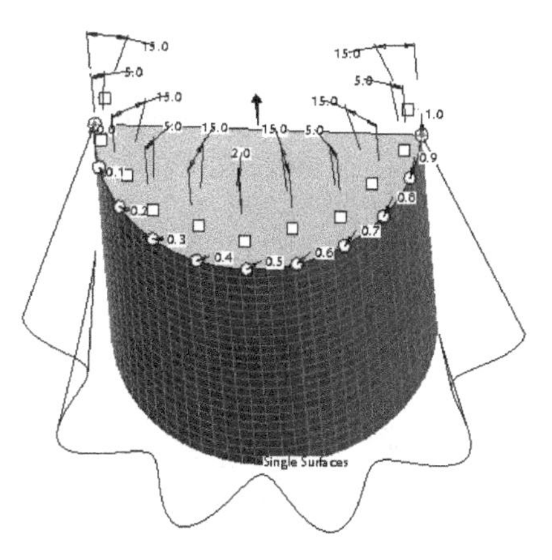

Figure 8-59 *Location points with modified angle values*

14. The model after mirroring the geometry and then shelling with the top and bottom faces removed is shown in Figure 8-61.

Figure 8-60 *Model with the draft feature*

Figure 8-61 *Model of the lamp shade*

Options Tab

On choosing this tab, a slide-down panel will be displayed, as shown in Figure 8-62. The options in this slide-down panel are discussed next.

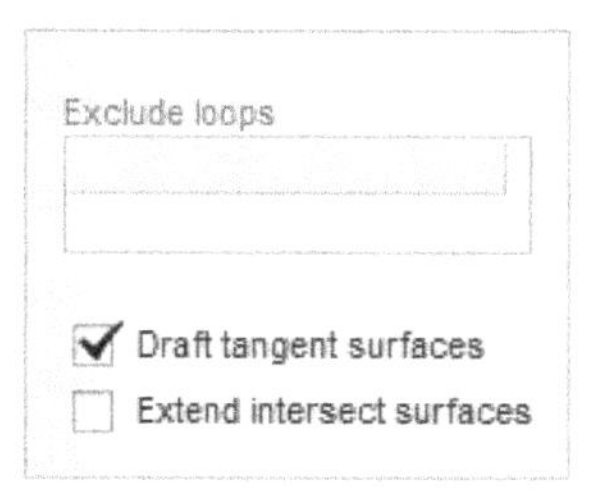

Figure 8-62 *The* ***Options*** *slide-down panel*

Exclude loops Collector

This collector is used to select a surface to which you do not want to add a draft angle. When you select a surface to add a draft angle, all loops that are on the surface are applied the same draft angle. However, using this collector you can select the loop to which you do not want to add a draft angle.

Figure 8-63 shows the surface selected to add a draft angle. You will notice that the surface of the cylindrical feature is selected. If you continue with the draft feature creation and exit the **Draft** dashboard, the draft is created, as shown in Figure 8-64.

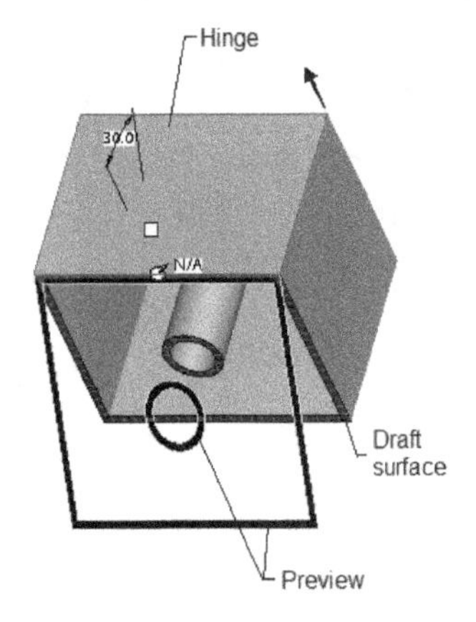

Figure 8-63 *Surface for draft*

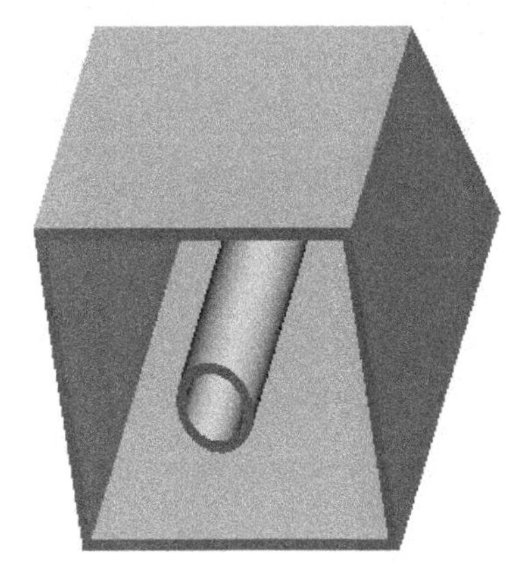

Figure 8-64 *The draft created*

To exclude the face of the cylindrical feature from the loop, click in the **Exclude loops** collector to activate it. Now, bring the cursor close to the face of the cylindrical feature. The boundary of the face is highlighted in cyan. Select the face to exclude it from the loop. Figure 8-65 shows the draft surface after excluding the face of the cylindrical feature from the loop. Note that the model should be a single feature then only the **Exclude loops** collector will be active.

Draft tangent surfaces Check Box

When the **Draft tangent surfaces** check box is selected, the draft is applied to the surfaces tangent to the selected surface. In Figure 8-66, the surface shown is selected to add the draft angle. Since the **Draft tangent surfaces** check box is selected, all surfaces tangent to the selected surface and other surfaces are automatically selected. Figure 8-67 shows the resultant model with the draft feature.

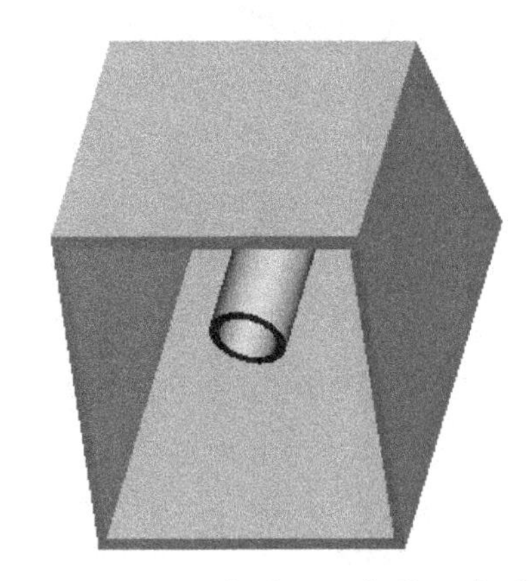

Figure 8-65 *Draft after excluding the face of the cylindrical feature*

Extend intersect surfaces Check Box

When this check box is selected, the draft surface extends in the direction of the feature it intersects. Figure 8-68 shows the example of the draft that intersects with the adjacent feature and extends outside the edge of the feature at the bottom. Figure 8-69 shows the feature when the draft surface gets extended.

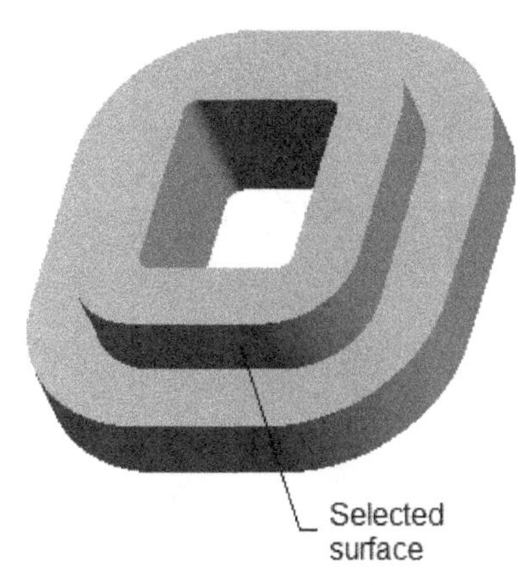

Figure 8-66 *Single surface selected to draft*

Figure 8-67 *The resultant model with draft feature*

Figure 8-68 *Draft created with the* ***Extend intersect surfaces*** *check box cleared*

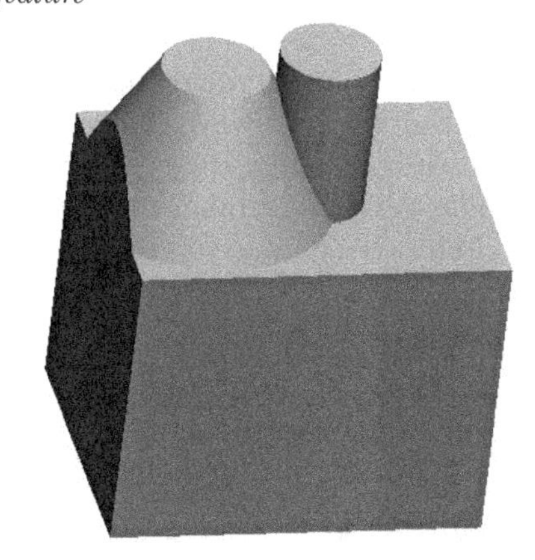

Figure 8-69 *Draft created with the* ***Extend intersect surfaces*** *check box selected*

Figure 8-70 shows another example of extended draft feature. The following steps explain in brief the procedure to create the pencil-shaped model. It is assumed that you have created the model shown in Figure 8-71. The initial model shown is an octagon on the top of which a cylinder is created.

1. Invoke the **Draft** dashboard and select the cylindrical surface to apply draft to it.

2. Select the top face of the cylindrical feature as the hinge. The arrow points in the upward direction.

3. Choose the **Options** tab; a slide-down panel is displayed. Select the **Extend intersect surfaces** check box in it. Also, make sure that the **Draft tangent surfaces** check box is selected.

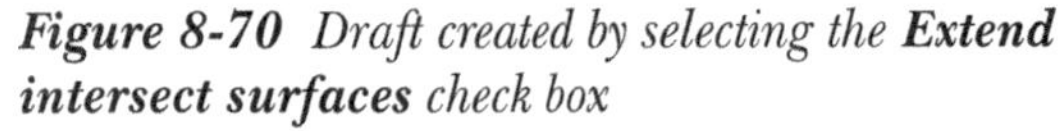

Figure 8-70 *Draft created by selecting the* ***Extend intersect surfaces*** *check box*

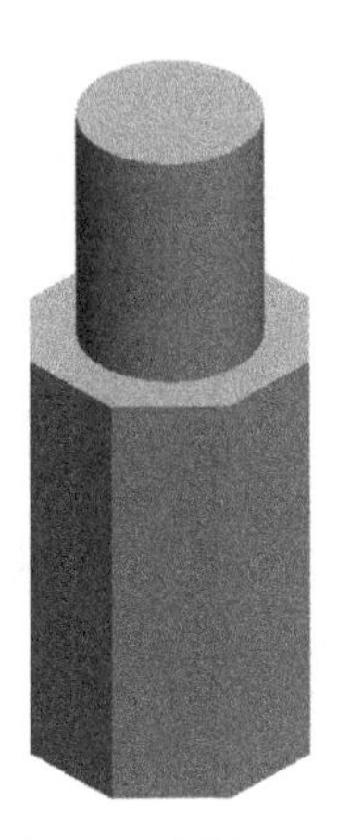

Figure 8-71 *Model initially created*

4. Use the drag handle and increase the draft angle up to the vertex of the octagon. This can be easily done by viewing the preview of the draft surface as you increase the angle. You may even have to use the edit box present on the dashboard to enter the small increase in angle value. Remember that you will get the desired shape of the intersect only when the draft surface intersects the vertices of the octagonal feature.

Properties Tab

When you choose the **Properties** tab, a slide-down panel is displayed, showing the feature id of the feature you have created.

TUTORIALS

Tutorial 1

In this tutorial, you will create the model shown in Figure 8-72. This figure also shows the top, sectioned front, and right views of the solid model with dimensions. The hidden lines are suppressed for clarity. **(Expected time: 45 min)**

Examine the model and determine the number of features in it. The model consists of six features, refer to Figure 8-72.

The following steps are required to complete this tutorial:

a. Start a new file and create the base feature, refer to Figures 8-73 through 8-75.
b. Create the shell feature of given thickness, refer to Figure 8-76.
c. Create the third and fourth extrude features on the two ends of the sweep feature respectively, refer to Figures 8-77 and 8-78.
d. Create the counterbore hole on the third and fourth features, refer to Figures 8-79 and 8-80.
e. Pattern the counterbore holes, refer to Figure 8-81.

When the PTC Creo Parametric session starts, the first task is to set the working directory. As this is the first tutorial of the chapter, you need to create a folder *c08* if it has not been created earlier and set it as the Working Directory.

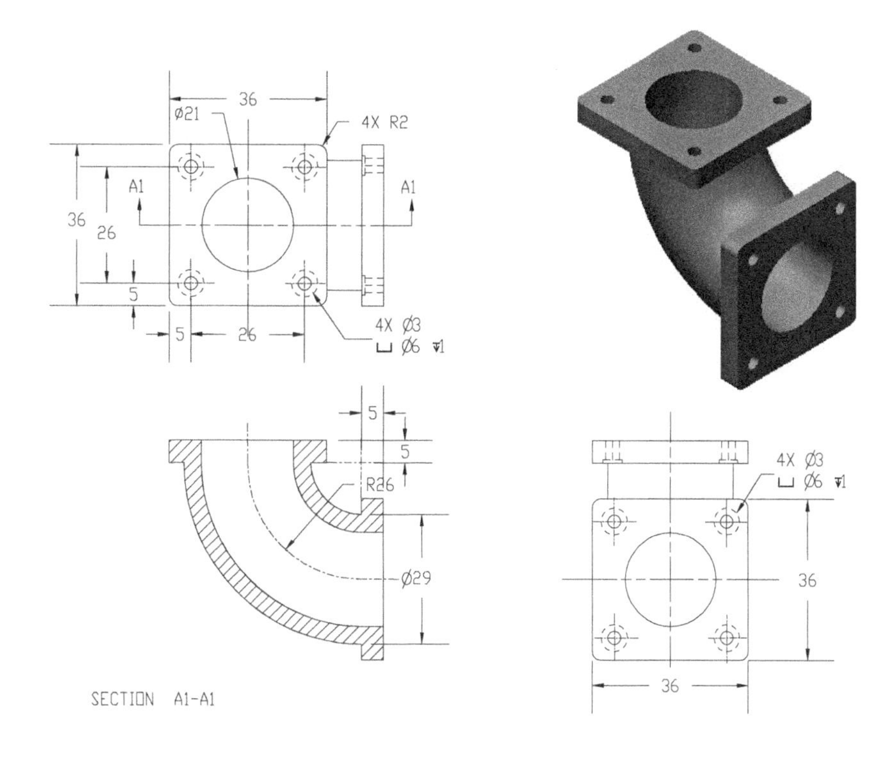

Figure 8-72 *Top, sectioned front, right, and isometric views of the model*

Starting a New Object File

1. Start a new part file and name it as *c08tut1*.

   The three default datum planes as well as the **Model Tree** are displayed in the drawing area.

Sketching the Trajectory

You need to sketch the trajectory of the sweep feature on the **FRONT** datum plane.

1. Invoke the **Sketch** tool from the **Datum** group. Next, select the **FRONT** datum plane as the sketching plane.

2. Select the **Top** datum plane as the reference plane from the **Reference** collector.

3. Select the **Top** option from the **Orientation** drop-down list and then choose the **Sketch** button from the **Sketch** dialog box.

After you have selected the planes for orientation, the system takes you to the sketcher environment.

4. Choose **Center and Ends** from the **Arc** drop-down in the **Sketching** group.

5. Draw an arc such that the center of the arc lies at the intersection of the **TOP** and **RIGHT** datum planes, as shown in Figure 8-73. As you specify the center of the arc, the cursor snaps to the point of intersection of the two planes. Now, draw the arc and exit this tool.

Tip: *To change the start point on the trajectory, click on the point where you want the start point. When the point is highlighted in red color, press and hold the right mouse button to invoke a shortcut menu. From the shortcut menu, choose the* ***Start Point*** *option.*

Modifying the Dimensions of the Trajectory

While drawing the arc, a weak radial dimension is assigned to it automatically. You need to modify the dimension as per your requirement.

1. Double-click on the dimension and modify the radial dimension to **26**, as shown in Figure 8-74. You will notice that the sketch refits on the screen.

2. Choose the **OK** button.

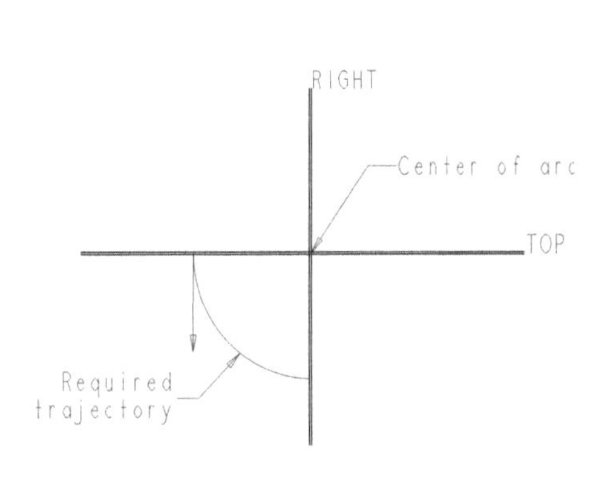

Figure 8-73 *The sketch of the required trajectory*

Figure 8-74 *Dimension for the arc*

Invoking the Sweep Tool

You need to use the **Shapes** group from the **Ribbon** to invoke the **Sweep** tool.

1. Choose the **Sweep** tool from the **Sweep** drop-down in the **Shapes** group of the **Ribbon**; the **Sweep** dashboard is displayed.

2. Choose the **Reference** tab to select the trajectory for the sweep feature; you are prompted to define the sketch.

The endpoints of the arc are automatically aligned with the **TOP** and **RIGHT** datum planes. You will notice that an arrow is attached to the start point of the trajectory. This arrow points in the direction of sweep.

Drawing the Section for the Base Feature

1. Choose the **Create or edit sweep section** button from the dashboard; your sketching plane will be automatically oriented. You are also prompted to draw the cross-section for the sweep feature.

2. Choose the **Center and Point** button from the **Sketch** dashboard and create a circle such that the center of the circle lies at the intersection of the two infinite perpendicular lines.

   Note that, when you draw the circle, the cursor snaps to the intersection point of the cross.

Modifying the Dimensions of the Section

1. Choose the **One-by-One** tool from the **Select** drop-down in the **Operations** group. Next, double-click on the dimension and modify the diameter dimension to **29**.

   The sketch gets modified and refits on the screen.

2. Choose the **OK** button.

Creating the Base Feature

Now, you need to create the base feature, which is a sweep feature. You can preview the sweep feature before creating it.

1. Choose the **Preview** button from the **Sweep** dashboard.

2. Choose the **Named Views** button from the **Graphics** toolbar; a flyout is displayed. Next, choose the **Default Orientation** option from the flyout; the model orients in the drawing area, as shown in Figure 8-75.

   You can use the middle mouse button to change the orientation of the model.

3. Now, choose the **Build Feature** button from the **Sweep** dashboard to exit it.

Tip: *When you enter the sketcher environment to define a section for the sweep trajectory, a pink colored cross is displayed. This cross determines the orientation of the section with respect to the available trajectory.*

Creating the Shell Feature

After creating the sweep feature, you need to create the next feature, called the shell feature.

Note

*Instead of using the **Shell** option, two concentric circles can be drawn while drawing the section for the sweep feature in order to obtain the desired hollow feature. Alternatively, the **Sweep > Thin Protrusion** option can be used to obtain the same hollow feature. In this tutorial, you will use the **Shell** option.*

1. Choose the **Shell** tool from the **Engineering** group; the **Shell** dashboard is displayed. Also, you are prompted to select the surfaces to be removed.

2. Select one end surface of the sweep feature and then by using CTRL+left mouse button, select the other end surface. The two selected surfaces are highlighted in green.
3. Enter **4** as the thickness value of the shell in the dimension edit box present on the **Shell** dashboard and then press ENTER.
4. Choose the **Build feature** button to create the shell feature and exit the **Shell** dashboard.

The default isometric view of the shell feature is shown in Figure 8-76. You can use the middle mouse button to view the model from various directions.

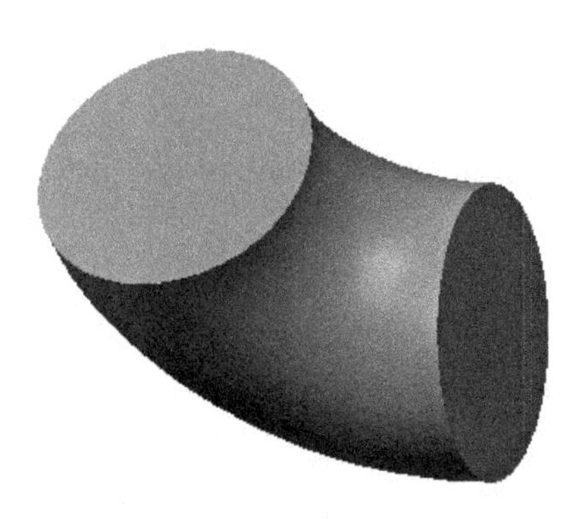

Figure 8-75 *The default view of the model*

Figure 8-76 *Shell feature*

Creating Extrude Features

The next feature is a protrusion feature with a depth of 5 and need to be created at both ends of the sweep feature. While drawing the circle for the sketch of the extrude feature, remember to use the edge of the shell in order to create a hole in the extruded feature as well.

1. Choose the **Extrude** tool from the **Shapes** group; the **Extrude** dashboard is displayed.
2. Choose the **Placement** tab from the dashboard to display the slide-down panel and invoke the **Sketch** dialog box. Select the top face of the sweep feature as the sketching plane.

3. Select the **RIGHT** datum plane and then select the **Right** option from the **Orientation** drop-down list. Then, enter the sketcher environment.

4. Draw the sketch of the extrude feature and then apply constraints and dimensions to it, as shown in Figure 8-77.

5. Create the extruded feature having an extrusion depth of 5. Similarly, create the next extruded feature at the other end of the sweep feature.

   The protrusion features created at both ends of the sweep feature are shown in Figure 8-78.

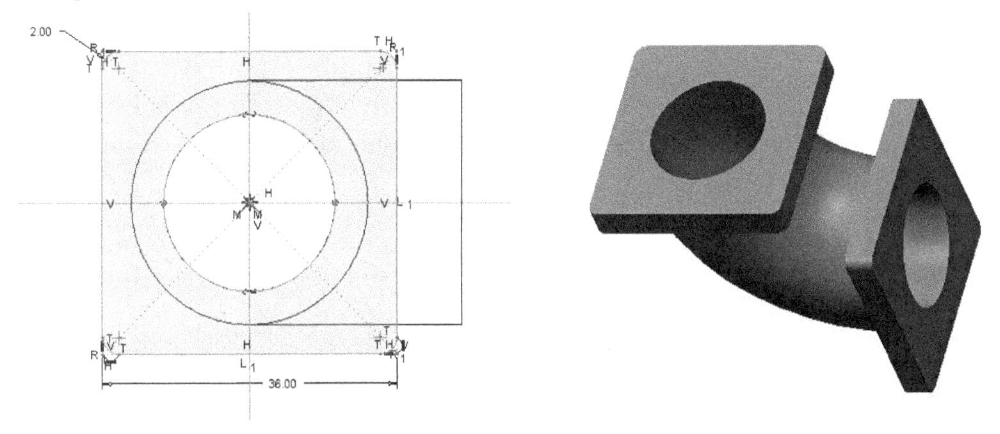

Figure 8-77 *Sketch with dimensions and constraints*

Figure 8-78 *Two extruded features created at both the ends of the sweep feature*

Creating the Hole Feature

After creating the extruded features at both ends of the sweep feature, you need to create the counterbore holes. One hole is to be created on each extruded surface and then they will be patterned on individual planes separately to create the remaining three instances.

1. Choose the **Hole** tool from the **Engineering** group; the **Hole** dashboard is displayed.

2. Choose the **Use standard hole profile as drill hole profile** button from the **Hole** dashboard; two buttons are added to the dashboard.

3. Next, choose the **Adds counterbore** button from the dashboard to make the hole counterbore. Also, choose **Drill up to next surface** from the depth flyout.

4. Choose the **Shape** tab from the dashboard to enter the counterbore parameters of the hole, as shown in Figure 8-79.

5. Select the back face of the third extruded feature for the placement of hole. The preview of the hole appears in the drawing area.

Now, you need to specify the placement parameters for the hole. For this purpose, refer to Figure 8-72 and look for dimensions that can help in placing the hole. The two edges are used to dimension the hole.

6. Choose the **Placement** tab and click in the **Offset References** collector to turn it green.

7. Select the front edge of the back face and then use CTRL+left mouse button to select the left edge of the third feature for specifying linear references. The hole is at a distance of 5 from both the edges. The default dimensions appear on the hole.

   You can also drag the green handles and place them on the two edges.

8. Double-click on the dimensions and modify the linear distance to 5.

9. Choose the **Build feature** button from the **Hole** dashboard; a hole is created on the selected face.

10. Create another hole on the fourth feature by following the same procedure as discussed earlier.

Tip: 1. *You can also create a counterbore hole by sketching its profile. To do so, choose the* ***Use sketch to define drill hole profile*** *button from the dashboard. Next, choose the* ***Activates Sketcher to create section*** *button to activate the sketcher environment.*

2. The same sketch can be used for the second hole from the ***In Session*** *folder.*

The trimetric view of the model completed thus far is shown in Figure 8-80.

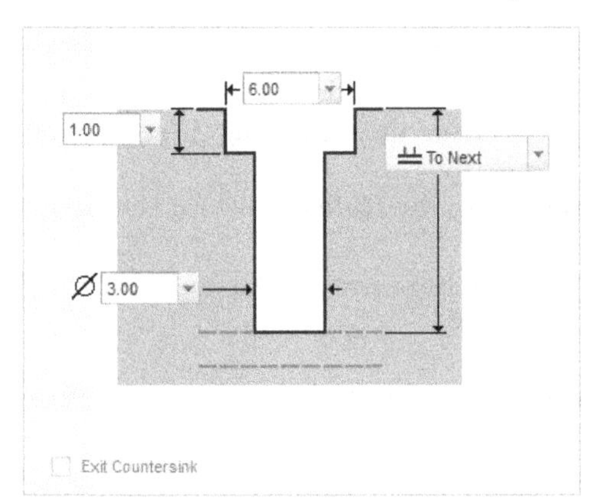

Figure 8-79 *Parameters for the counterbore hole set under the* ***Shape*** *tab*

Figure 8-80 *One hole each on the two extruded features*

Creating the Pattern of Holes

After one hole is placed on each of the two faces, you need to pattern the holes.

1. Select the hole feature from the **Model Tree** or from the drawing area and hold the right mouse button to invoke the shortcut menu. Choose the **Pattern** option from the shortcut menu; the **Pattern** dashboard is displayed.

2. Specify the required parameters in the **Pattern** dashboard and create the pattern of the hole.

Similarly, create the pattern of the hole on the other extruded feature as well. The default trimetric view of the complete model is shown in Figure 8-81.

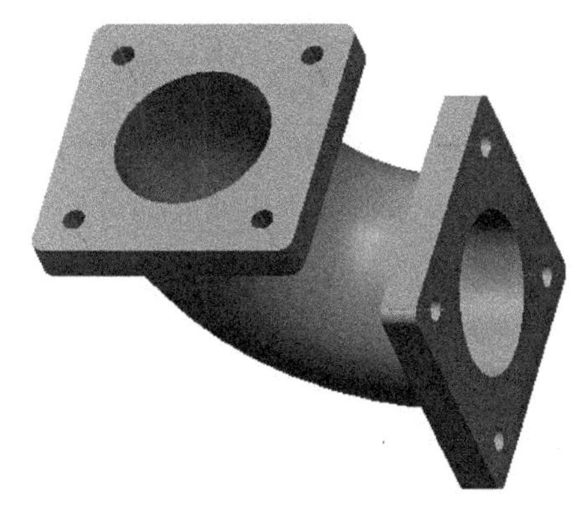

Figure 8-81 *The trimetric view of the complete model*

Saving the Model

1. Choose the **Save** option from the **File** menu and then save the model.

Note

*As discussed in the earlier chapters, the **Model Tree** is used to get an idea about the order of feature creation. Therefore, while creating a model, the order of feature creation will remain same but the id numbers of the features may be different.*

Tutorial 2

In this tutorial, you will create the model of the pencil shown in Figure 8-82 and then change the color of the pencil. Figure 8-83 shows the front, section, and the detail views of the model. **(Expected time: 30 min)**

Figure 8-82 Solid model of the pencil

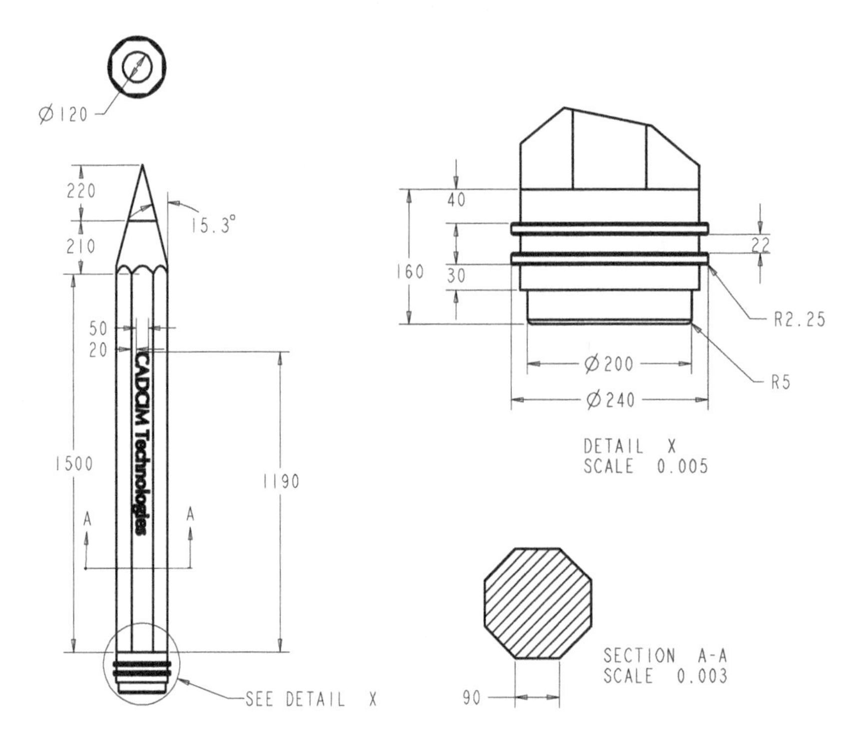

Figure 8-83 Top, front, sectioned, and detail views of the model

Examine the model and determine the number of features in it. The model is composed of six features, refer to Figure 8-82.

The following steps are required to complete this tutorial:

a. Create the base feature, refer to Figures 8-84 and 8-85.
b. Create the second feature that is a cylinder on the top face of the base feature, refer to Figure 8-86.
c. Create the draft feature, refer to Figure 8-87.
d. Create the tip of the pencil that is a revolved feature, refer to Figures 8-88 and 8-89.
e. Create the tail end of the pencil that is again a revolved feature, refer to Figures 8-90 and 8-91.
f. Write the text on the pencil by using the sketched datum curves, refer to Figures 8-92 and 8-93.

Make sure that the *c08* folder is the current Working Directory.

Starting a New Object File

1. Start a new part file and name it as *c08tut2*.

The three default datum planes are displayed in the drawing area.

Creating the Base Feature

The base feature of the pencil is a protrusion feature in which an octagon is extruded to a depth of 1500.

1. Choose the **Extrude** tool from the **Shapes** group to display the **Extrude** dashboard and invoke the **Sketch** dialog box.

2. Select the **TOP** datum plane as the sketching plane.

3. Select the **RIGHT** datum plane from the drawing area and then select the **Right** option from the **Orientation** drop-down list, only if these options are not selected by default.

4. Draw the sketch of the octagon by using the **Palette** button and then apply constraints and dimensions to it, as shown in Figure 8-84.

5. Exit the sketcher environment and extrude the sketch to a depth of 1500. The base feature thus created is shown in Figure 8-85.

Creating the Second Feature

The second feature is a cylindrical feature. You need to draw the feature on the top face of the base feature.

1. Choose the **Extrude** tool from the **Shapes** group; the **Extrude** dashboard is displayed. Next, invoke the **Sketch** dialog box and select the top face of the base feature as the sketching plane.

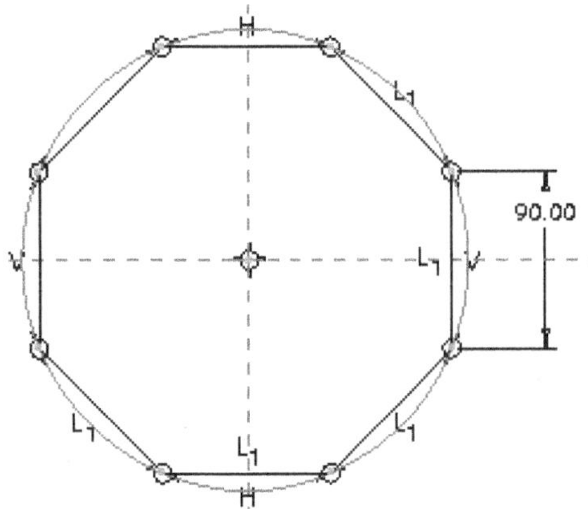

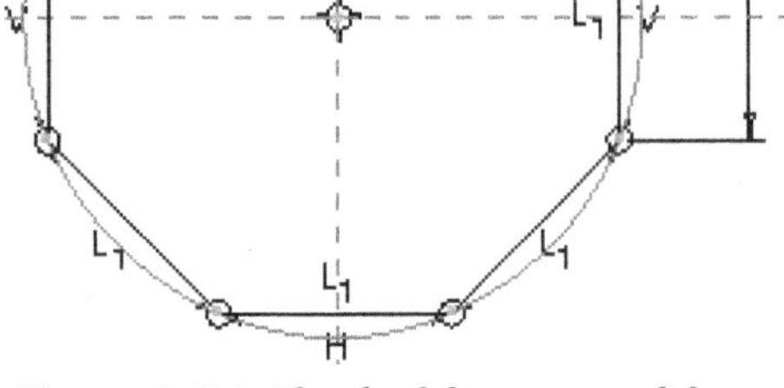

Figure 8-84 *Sketch of the octagonal feature*

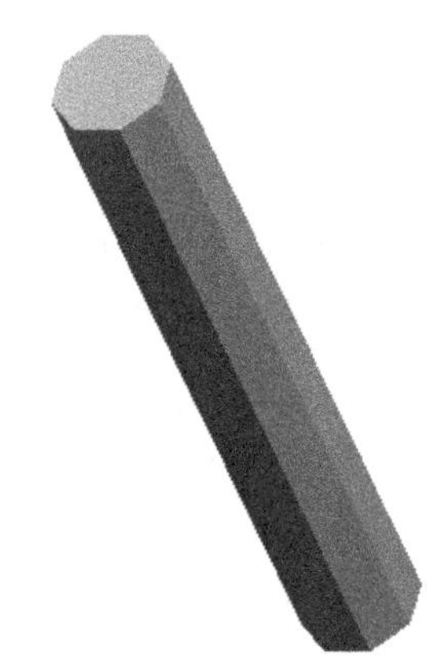

Figure 8-85 *Base feature created by extruding the octagonal sketch*

2. Enter the sketcher environment and draw the sketch of the cylinder. Note that the diameter of the circle is 120. Next, exit the sketching mode.

3. After exiting the sketcher environment, extrude the circle to a depth of 210. The model after creating the cylindrical feature is shown in Figure 8-86.

Creating the Draft Feature

You need to create the draft feature on the cylindrical surface and the draft surface to intersect the base feature. This is because when the draft surface intersects the base feature and is extended, the required shape is obtained at the intersecting edge.

1. Choose the **Draft** tool from the **Engineering** group; the **Draft** dashboard is displayed.

2. Select the cylindrical surface of the cylinder to apply draft to it.

3. Choose the **References** tab to invoke a slide-down panel. Click in the **Draft hinges** collector and then select the top face of the cylindrical feature as the hinge; a pink arrow is displayed pointing upward.

4. Choose the **Flip** button on the right of the **Pull direction** collector to change the direction of the arrow so that it points downward.

5. Choose the **Options** tab to invoke a slide-down panel and then select the **Extend intersect surfaces** check box.

6. Enter **15.3** as the angle value in the edit box present in the **Draft** dashboard.

7. Choose the **Build feature** button from the **Draft** dashboard to complete the draft feature, as shown in Figure 8-87.

Figure 8-86 Model after creating the cylindrical feature

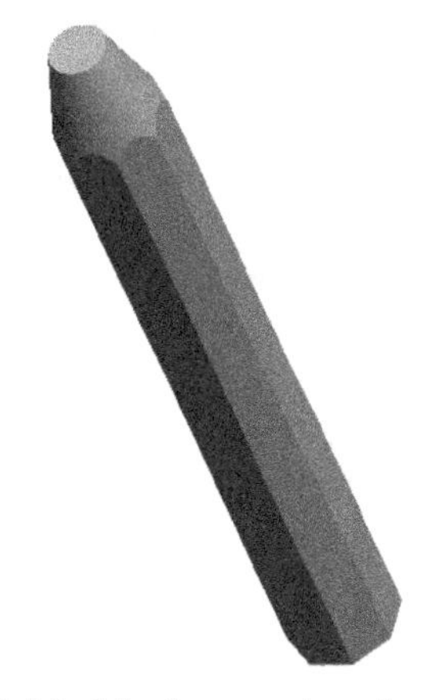

Figure 8-87 Model after creating the draft feature

Creating the Fourth Feature

The fourth feature is the tip of the pencil, which is a revolved feature.

1. Choose the **Revolve** tool from the **Shapes** group; the **Revolve** dashboard is displayed.
2. Invoke the **Sketch** dialog box and select the **FRONT** datum plane as the sketching plane and select the top face as reference.
3. After entering the sketcher environment, draw the sketch and then apply constraints and dimensions to it, as shown in Figure 8-88. Make sure that the geometry centerline is drawn. Exit the sketcher environment.

   The angle of revolution is 360 degree by default.

4. Choose the **Build feature** button to complete the feature creation. The default trimetric view of the model after creating the fourth feature is shown in Figure 8-89.

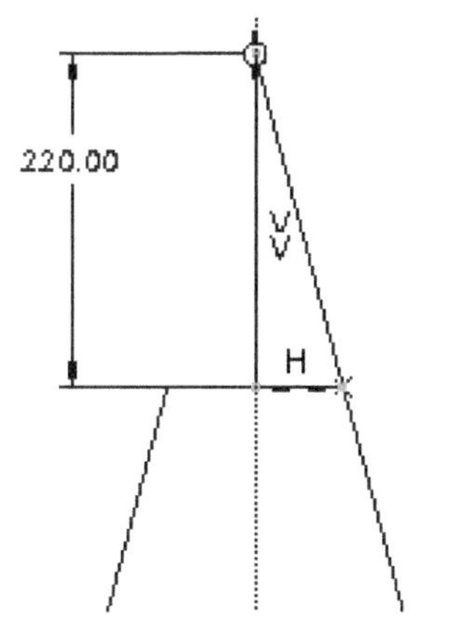

Figure 8-88 Sketch for the tip of the pencil

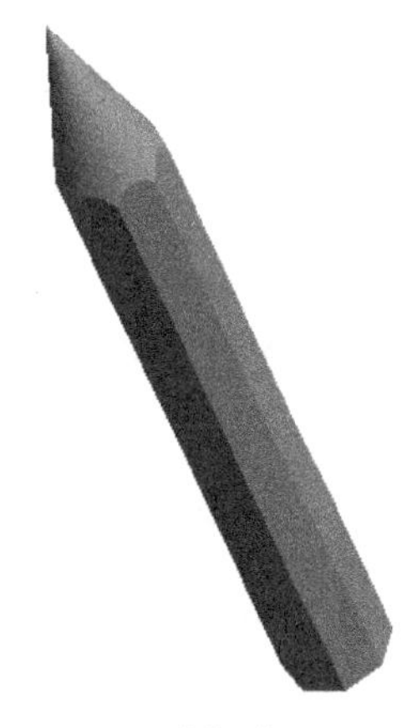

Figure 8-89 Model after creating the tip

Creating the Fifth Feature

The fifth feature, which is a revolve feature, is the bottom portion of the pencil.

1. Choose the **Revolve** tool from the **Shapes** group; the **Revolve** dashboard is displayed.
2. Invoke the **Sketch** dialog box and then select the **FRONT** datum plane as the sketching plane.
3. After entering the sketcher environment, draw the sketch and then apply constraints and dimensions to it, as shown in Figure 8-90. Make sure that the geometry centerline is also drawn.
4. Exit the sketcher environment; the angle of revolution is set to 360-degree by default.
5. Choose the **Build feature** button to complete the feature creation. The default trimetric view of the model after creating the fifth feature is shown in Figure 8-91.

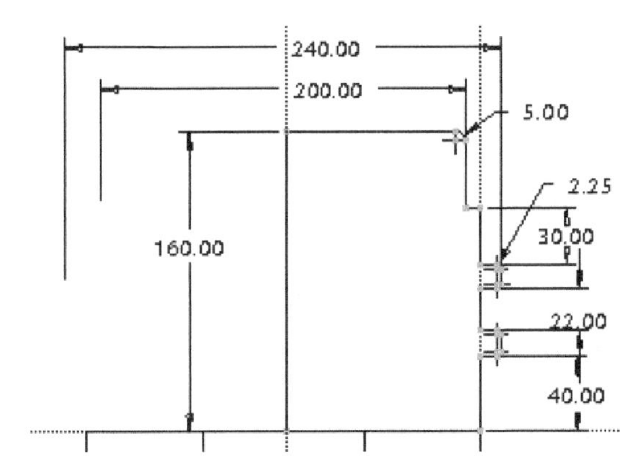

Figure 8-90 *Sketch with dimensions and constraints*

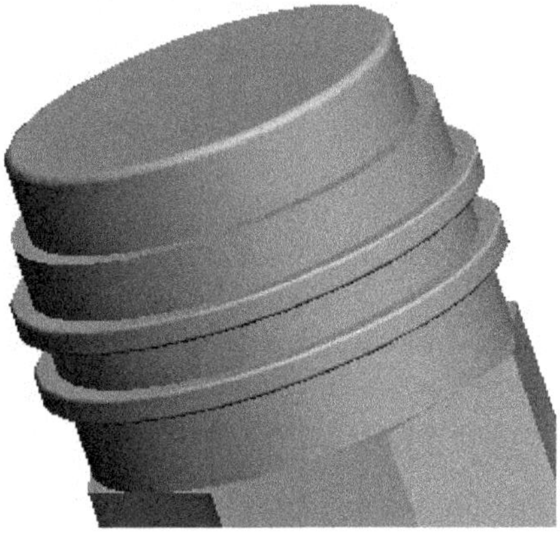

Figure 8-91 *Model after creating the top of the pencil*

Writing the Text

1. Choose the **Sketch** button from the **Datum** group; the **Sketch** dialog box is displayed.
2. Select the front face of the base feature as the sketching plane.
3. After entering the sketcher environment, create the text by using the **Text** button. Figure 8-92 shows the text written with dimensions.
4. Choose the **OK** button to exit the sketcher environment.

The model after creating the text is shown in Figure 8-93. **Model Tree** of the final model is shown in Figure 8-94.

Changing the Colors

You can apply colors and appearances to the surfaces in PTC Creo Parametric. Remember that the colors you apply are saved with the model and remain on the model until you clear them.

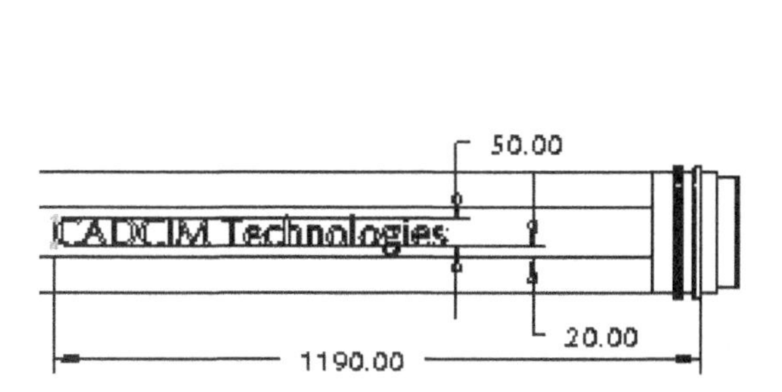

Figure 8-92 Text with dimensions

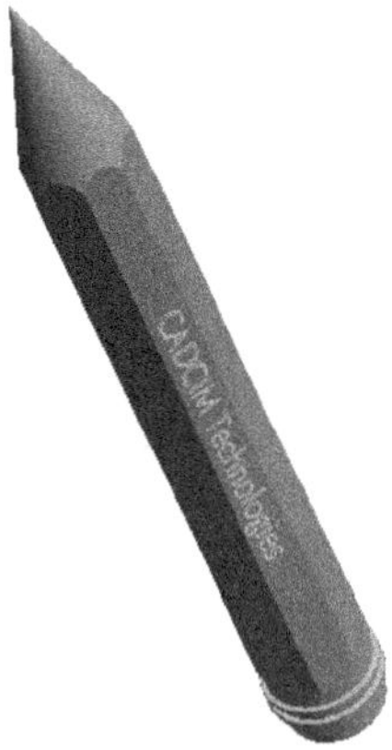

Figure 8-93 Model after creating the text

1. Choose the down arrow on the right of the **Appearance Gallery** button from the **Model Display** group in the **View** tab; a menu is displayed.

2. Choose the **More Appearances** option from the menu; the **Appearance Editor** window is displayed. Next, enter **Lead** in the **Name** edit box in this window.

3. Choose the **Color Editor** button in front of **Color** in the **Properties** area; the **Color Editor** is displayed. Set the RGB values to **0**. The color will be black. You need to apply this color to the tip of the pencil. You can also set the values of **Shine**, **Highlight**, **Reflection**, and **Transparency** to **0**.

Note

*You can also select an appearance icon from the **My Appearances** or **Library** area. Next, invoke the **Appearance Manager** to adjust the values as required. You can also add or remove textures to colors.*

4. Choose the **OK** button from the **Appearance Editor**; the cursor is modified to a paint brush. Also, the **Select** message box is displayed and you are prompted to select surfaces to apply colors.

5. Select the **Surface** option from the **Filters** drop-down list in the Status Bar.

6. Select the tip of the pencil and press the middle mouse button; the tip of the pencil turns black.

7. Similarly, assign different colors to other entities.

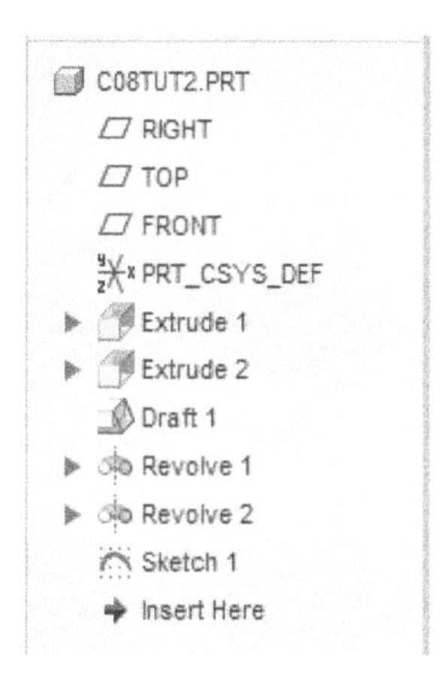

***Figure 8-94 Model tree** for Tutorial 2*

Saving the Model

1. Choose the **Save** button from the **File** menu and save the model.

Note

Colors are not features of the model and therefore, they will not appear in the ***Model Tree****.*

Tutorial 3

In this tutorial, you will create the blend feature shown in Figure 8-95. The two views of the blend feature are shown in Figure 8-96 with dimensions. After creating the model, you will redefine it such that the straight blending is changed into a smooth blending.

(Expected time: 30 min)

Figure 8-95 *Isometric view of the model*

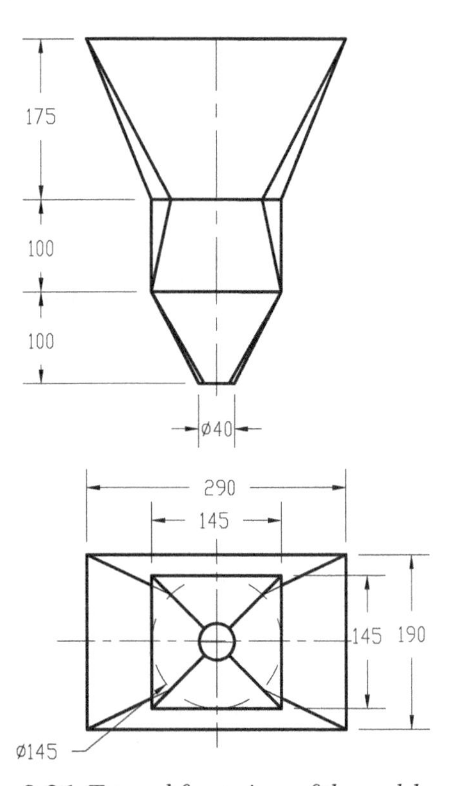

Figure 8-96 *Top and front views of the model*

Examine the blend feature and determine the number of sections in this feature. The blend consists of four sections, refer to Figure 8-96.

The following steps are required to complete this tutorial:

a. Create the sketch for the first section of the blend feature, refer to Figure 8-97.

b. Create the sketch for the second section of the blend feature, refer to Figure 8-98. Define the depth values between section numbers 1 and 2.
c. Create the sketch for the third section of the blend feature and define the depth values between section numbers 2 and 3.
d. Create the sketch for the fourth section, refer to Figure 8-99. Define the depth values between section numbers 3 and 4.
e. Redefine the model to change the straight blending into a smooth blending, refer to Figure 8-101.

Make sure that the *c08* folder is the current Working Directory.

Starting a New Object File

1. Start a new part file and name it as *c08tut3*. The three default datum planes are displayed in the drawing area.

Invoking the Blend Option

1. Choose the **Blend** tool from the expanded **Shapes** group of the **Model** tab; the **Blend** dashboard is displayed.

2. Choose the **Blend as solid** button.

3. Choose the **Sections** tab from the dashboard to open the slide-down panel. Next, select the **Sketched sections** radio button and choose the **Define** button to define the sketching plane.

Selecting the Sketching Plane

1. Select the **FRONT** datum plane as the sketching plane.

2. Select the **RIGHT** datum plane as the reference plane and then select the **Right** option from the **Orientation** drop-down list, only if these options are not selected by default.

Drawing the First Section

1. The first section is a rectangle of 290x190 units. Draw the sketch of the rectangular section and then add constraints and dimensions to it, as shown in Figure 8-97.

2. Choose **OK** to exit the sketcher environment. Next, choose the **Options** tab and select the **Straight** radio button; an offset value edit box with a default value gets displayed in the dashboard and in the **Sections** tab. Enter **175** in the offset value edit box for the second section.

3. Choose the **Sketch** button from the **Sections** tab to enter the sketcher environment.

Note

While drawing the sections for the blend feature, the start point is very important. The start point should be similar to those shown in Figures 8-97, 8-98, and 8-99. If the start point is not at the desired point then select the point where you need the start point. Then hold the right mouse button to invoke the shortcut menu and choose the ***Start Point*** *option from it.*

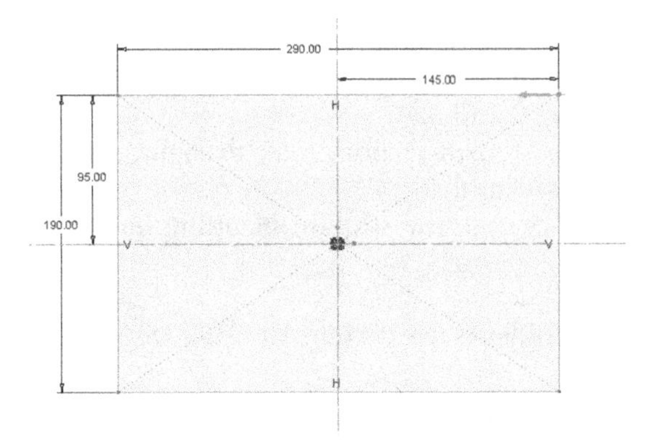

Figure 8-97 First rectangular section with dimensions

Drawing the Next Section

The next section is a circle.

1. Draw the sketch of the circular section, refer to Figure 8-98. Modify the diameter of the circle to **145**. As discussed earlier, the number of entities per section must be equal in a blend feature. Since a circle is a single entity, it should be divided at four points.

Dividing the Circular Section

The circular section should be divided at four points because the rectangle and square each have four entities. When you divide a circle at four points, the number of entities becomes four.

1. Choose the **Divide** button from the **Editing** group.

2. Select four points on the circle, refer to Figure 8-98.

   As you select the points on the circle to divide it, some weak dimensions appear on the circle. Next, you need to apply constraints on the four points that were selected to divide the circle.

Applying Constraints on the Four Points

1. Choose the **Vertical** button from the **Constrain** group and select the two divisional points on the left of the circle to lie in a vertical line. Similarly, select the two points on the right to apply the constraint.

2. Now, choose the **Horizontal** button from the **Constrain** group of the **Sketch** tab and select the two division points on the upper half and the lower half to lie in a horizontal line.

3. Modify the vertical dimension of the upper right division point, as shown in Figure 8-98. After the section is completed, the two sections with dimensions will be similar to the sections shown in Figure 8-98.

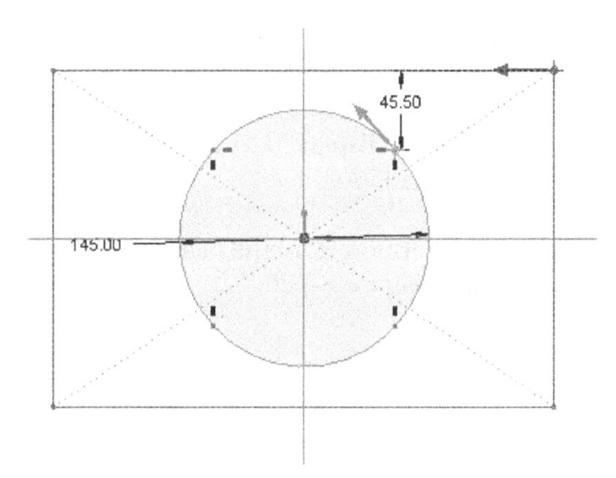

Figure 8-98 *Two completed sections*

4. Choose **OK** to exit the sketcher environment. Now, choose the **Insert** button from the **Sections** tab to add the third section. Enter **100** in offset value edit box to create depth between second and third sections. The next section to be drawn is a square of **145** units.

5. After drawing the square section, choose the **Insert** button to add the fourth section. Enter **100** in the offset value edit box to create depth between third and fourth sections. Draw the circular section. Divide the circular section into four entities similar to second section and then constrain and dimension it. Figure 8-99 shows all completed sections. Choose **OK** to exit the sketcher environment.

Note

In Figure 8-99, note that the direction of the start points is indicated by arrows. These arrows are important as they help you to avoid creating a twisted feature.

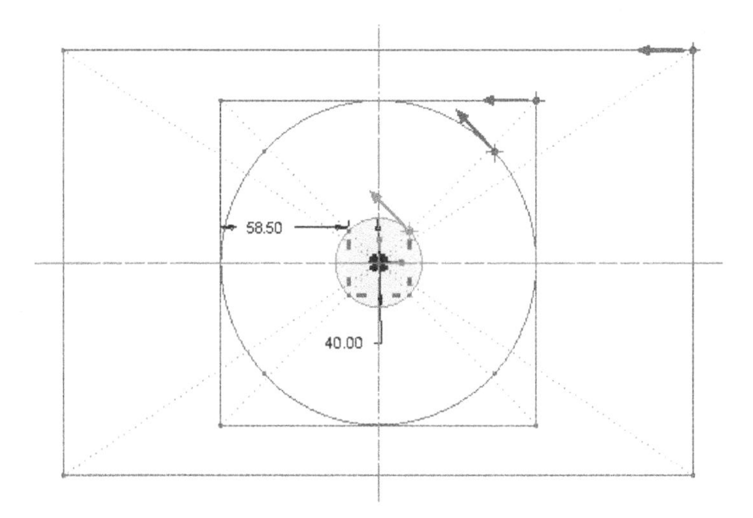

Figure 8-99 *All four completed sections*

Creating the Blend Feature

Now, you need to create the blend feature.

1. Choose the **Preview** button that is displayed at the top right corner of the window to preview the blend feature before creating it.

2. Choose the **Default Orientation** option from the **Named Views** flyout; the model orients in the drawing area, as shown in Figure 8-100.

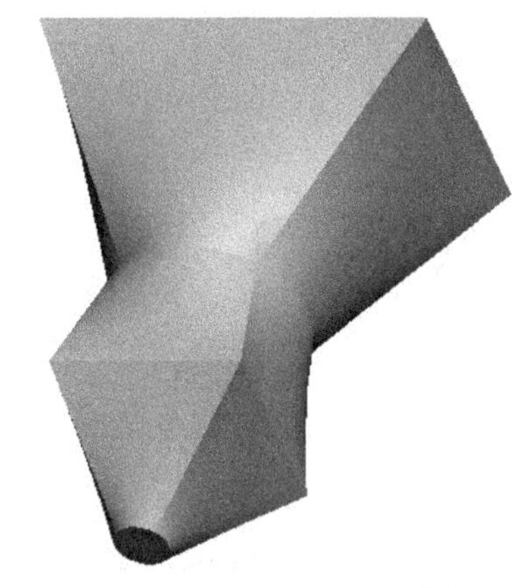

Figure 8-100 *Trimetric view of the model*

3. Now, choose the **OK** button from the dashboard; the trimetric view of the model is displayed in the drawing area.

Saving the Model

1. Choose the **Save** button from the **File** menu and save the model.

Redefining the Blend Feature

After saving the straight blend feature, you need to redefine this feature so that it can be converted to a smooth blend.

1. Select the model in the drawing area; the edges of the model turn green in color.

2. Press and hold the right mouse button in the drawing area until a shortcut menu is displayed.

3. Choose **Edit the definition of the selected object** from the **Edit Actions** area of the shortcut menu; the **Blend** dashboard is displayed.

4. Choose the **Options** tab and then select the **Smooth** radio button.

5. Now, choose the **OK** button; the straight blend feature is converted into a smooth blend, as shown in Figure 8-101.

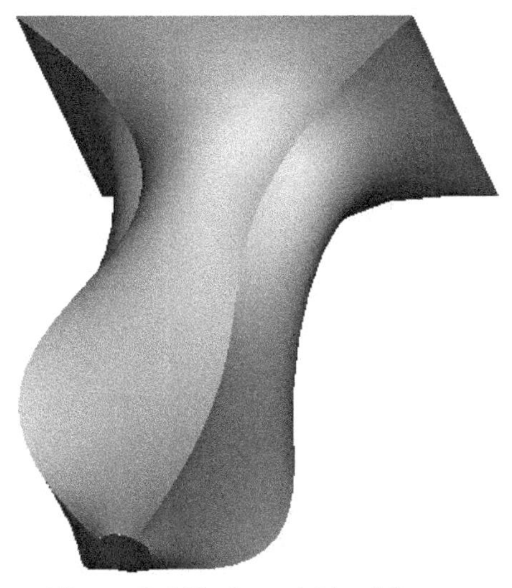

Figure 8-101 Smooth blend feature

Tutorial 4

In this tutorial, you will create the model of a tap by using the views shown in Figure 8-102. All dimensions for the three sections are shown in the figure. Note that the model will be created by using the **Rotational Blend**. **(Expected time: 30 min)**

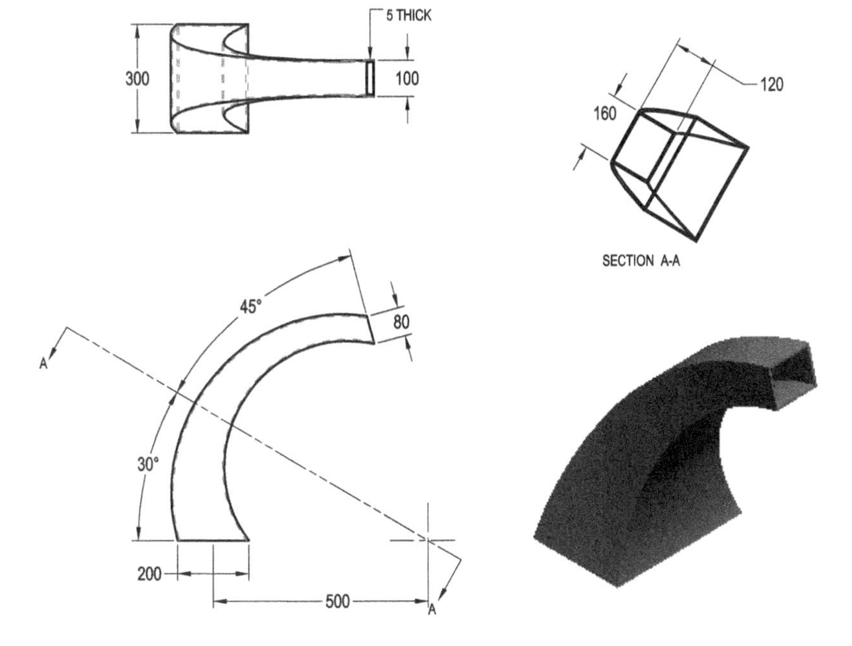

Figure 8-102 Sectioned, front, top and isometric views of the model

Examine the model and determine the number of sections in the blend feature, refer to Figure 8-102.

The following steps are required to complete this tutorial:

a. Start a new file and create the first section for the blend feature, refer to Figure 8-103.
b. Create the second section for the blend feature, refer to Figure 8-104.
c. Create the third section for the blend feature, refer to Figure 8-105.
d. Create the Shell feature, refer to Figure 8-107.

Starting a New Object File

1. Start a new part file and name it as *c08tut4*. The three default datum planes appear in the drawing area if they were not turned off in the previous tutorial.

Invoking the Rotational Blend Option

The first sketch of the rotational blend is 300x200 rectangle.

1. Choose **Rotational Blend** from the expanded **Shapes** group of the **Model** tab; the **Rotational Blend** dashboard is displayed.

2. Choose **Blend as solid** and choose the **Options** tab to select the **Smooth** radio button, if it is not selected by default. Choose the **Sections** tab from the dashboard to open the slide-down panel. Select the **Sketched sections** radio button and choose the **Define** button to define the sketching plane.

Selecting the Sketching Plane

1. Select the **TOP** datum plane as the sketching plane.

2. Select the **FRONT** datum plane as the reference plane and then select the **Bottom** option from the **Orientation** drop-down list. Choose the **Sketch** button to enter the sketching environment.

Drawing the Sketch of the First Section of the Rotational Blend

The blend consists of three sections.

1. Draw the sketch of the first section. That is a rectangle of 300x200 dimension. Also, draw the geometric centerline on the right side of sketch.

2. Apply constraints and modify the dimensions, as shown in Figure 8-103.

3. Choose the **OK** button from the dashboard; an angle offset value edit box with a default value becomes available in the dashboard and in the **Sections** tab. Change the angle offset value to **30** for second section and then choose the **Sketch** button from the **Section** tab to enter the sketching environment.

The sketcher environment is invoked and you are allowed to draw the sketch for the second section. Notice that the second section does not need geometric centerline since it will rotate around the geometric centerline drawn in the first sketch.

Drawing the Sketch for the Second Section

1. Draw the sketch for the second section, apply constraints and dimensions to the sketch, and modify the dimensions, as shown in Figure 8-104.

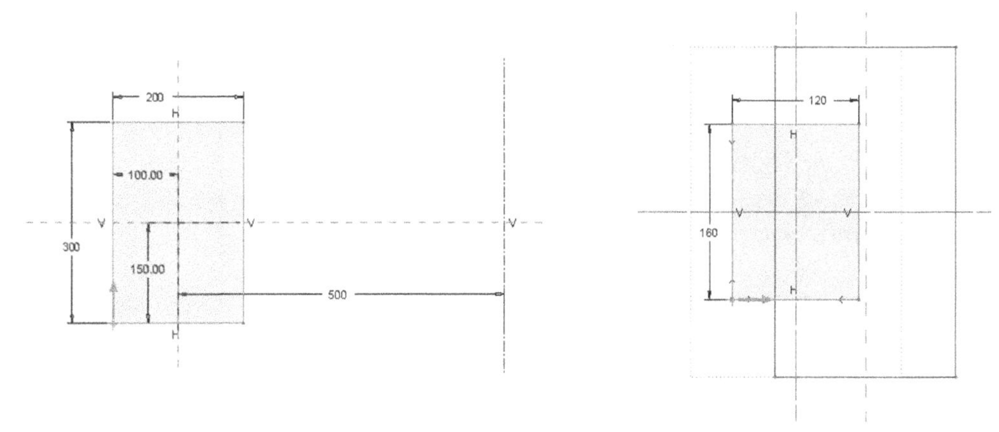

Figure 8-103 Sketch with dimensions and constraints for the first section

Figure 8-104 Sketch with dimensions and constraints for the second section

2. After completing the sketch, choose the **OK** button; the preview of the rotational blend feature is displayed. Now, choose the **Insert** button from the **Sections** tab to add the third section. Enter **45** in angle offset value box for the third section and choose the **Sketch** button from the **Sections** tab.

Drawing the Sketch of the Third Section

1. Draw the sketch for the third section, apply constraints and dimensions to the sketch, and modify the dimensions, as shown in Figure 8-105.

2. After completing the sketch, choose the **OK** button; the preview of the rotational blend feature is displayed.

3. If the preview is twisted at any sectioned sketch. Choose the sketch from the **Sections** tab and choose the **Sketch** button to enter the sketching environment. If the start point is not at the desired point, then select the point where you need the start point. Hold the right mouse button to invoke a shortcut menu, and then choose the **Start Point** option from it. Next, choose the **OK** button to exit the sketching environment.

4. Choose the **Build feature** button to complete the feature creation. The default trimetric view of the model is shown in Figure 8-106.

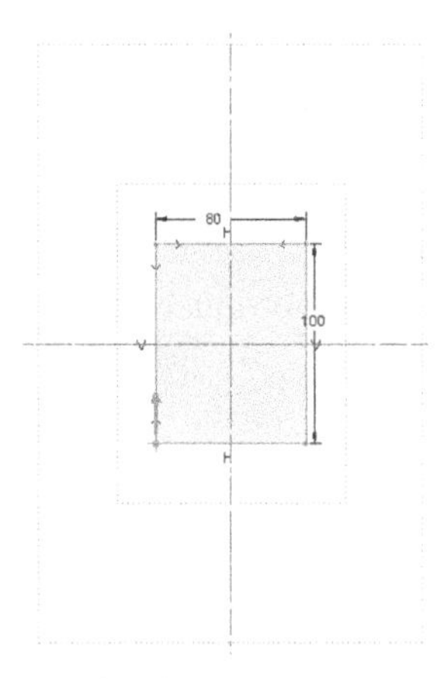

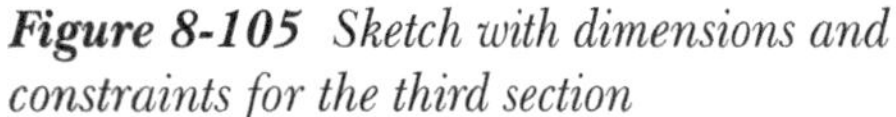

Figure 8-105 *Sketch with dimensions and constraints for the third section*

Figure 8-106 *The default trimetric view of the blend feature*

Creating the Shell Feature

1. Choose the **Shell** tool from the **Engineering** group; the **Shell** dashboard is displayed and you are prompted to select the surfaces to be removed from the part.

2. Select the one end surface of the sweep feature by using the left mouse button and then select the other end surface using the CTRL+left mouse button. The two selected surfaces are highlighted.

3. Enter **5** as the thickness value of the shell in the dimension edit box present on the **Shell** dashboard and press ENTER.

4. Choose the **Build feature** button to create the shell feature.

   The default trimetric view of the model is shown in Figure 8-107. You can use the middle mouse button to change the orientation of the model.

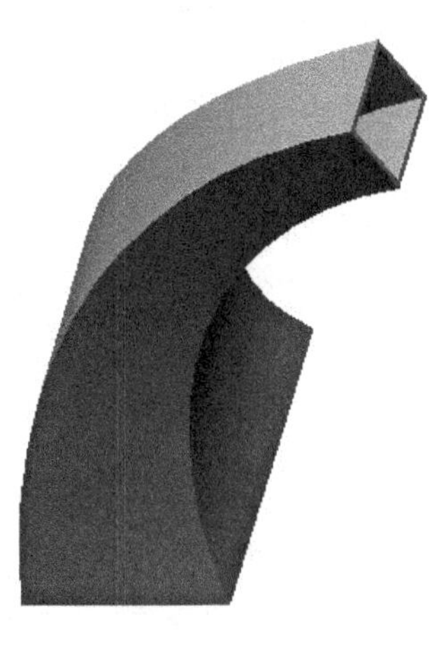

Figure 8-107 *Trimetric view of the model for Tutorial 4*

Saving the Model

1. Choose the **Save** button from the **File** menu and save the model.

Self-Evaluation Test

Answer the following questions and then compare them to those given at the end of this chapter:

1. The types of protrusion in PTC Creo Parametric are different from those available for Cut. (T/F)

2. The **Cut** option is available only if a base feature exists. (T/F)

3. If the value of shell thickness is negative then the shell thickness is added outside the boundary of the face selected for shelling. (T/F)

4. You can create a sweep feature using a closed section and a closed trajectory. (T/F)

5. The procedure to create a sweep cut feature is same as that of a sweep protrusion. (T/F)

6. The **Sweep** option extrudes a section along a __________.

7. The cross-section of a sweep feature remains __________ throughout the sweep.

8. The sketching plane that you select becomes __________ to the screen when you draw the trajectory.

9. An __________ section and an open trajectory are not possible.

10. A quilt is a __________ feature.

Review Questions

Answer the following questions:

1. What is the maximum permissible angle for the rotation of sections in a **Rotational Blend**?

   (a) 120 (b) 90
   (c) 180 (d) 45

2. What is the maximum possible draft angle that can be applied to a solid in PTC Creo Parametric?

   (a) 10 degree (b) 89.9 degree
   (c) 60 degree (d) 90 degree

3. What is the minimum number of sections required for a blend feature?

   (a) one (b) two
   (c) three (d) None of these

4. The **Draft** feature can also be applied to a quilt feature. (T/F)

5. A trajectory of a sweep feature can be modified independent of the geometry of the section. (T/F)

6. You can create a cut feature by using the **Sweep** option. (T/F)

7. You need to create a geometric centerline while creating the **Rotational Blend** feature. (T/F)

8. In **Rotational Blend** feature, the section is rotated about the y-axis of the coordinate system. (T/F)

9. You need to have at least three sections for the **Connect end and start sections** in rotational blend. (T/F)

10. Each section must have equal number of vertices for blending. (T/F)

Exercises

Exercise 1

Create the foundation bolt shown in Figure 8-108. The dimensions of the foundation bolt are shown in Figure 8-109. **(Expected time: 30 min)**

Figure 8-108 Model for Exercise 1

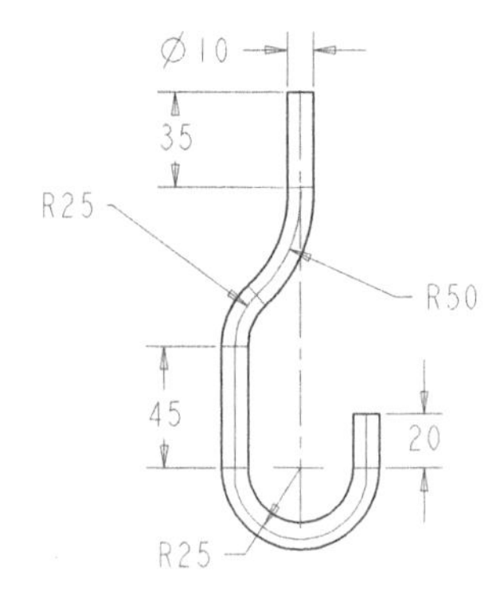

Figure 8-109 Dimensions for Exercise 1

Exercise 2

In this exercise, you will create the model shown in Figure 8-110. Figure 8-111 shows the top, sectioned front, and the right views of the solid model with dimensions. Note that the hidden lines are not displayed for clarity. **(Expected time: 45 min)**

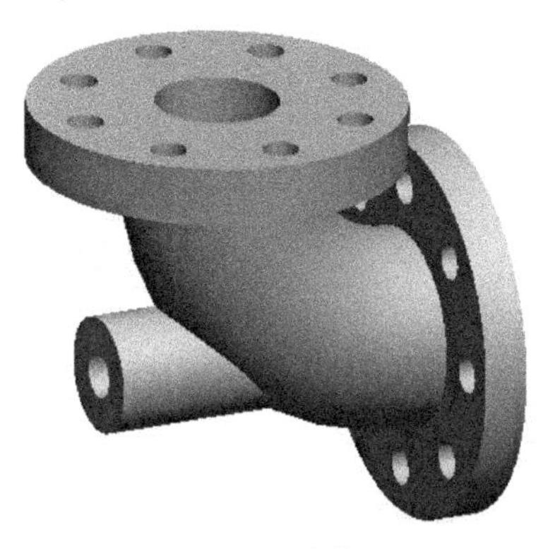

Figure 8-110 *Model for Exercise 2*

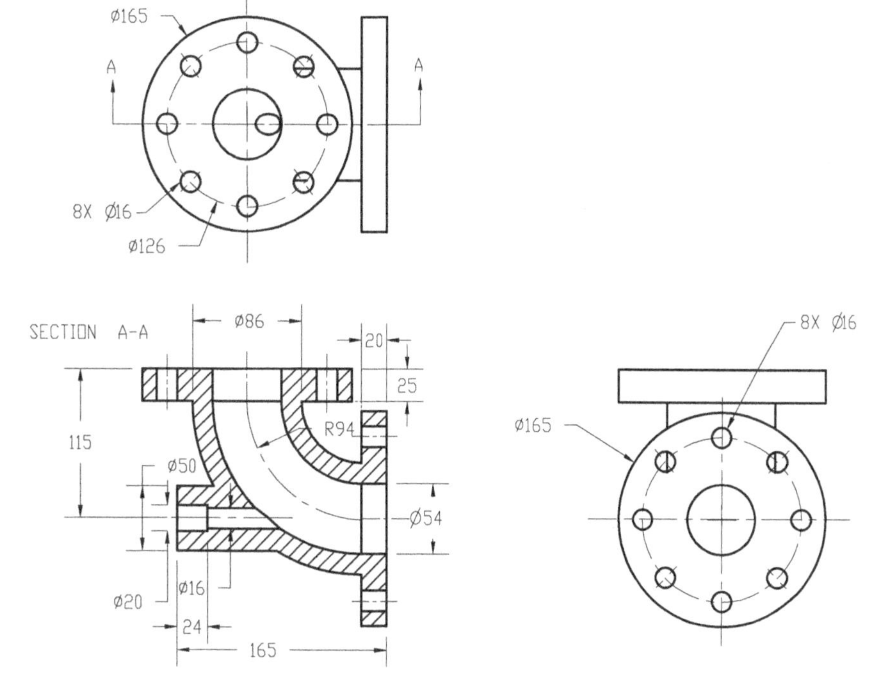

Figure 8-111 *Top, front, and right views of the model*

Exercise 3

In this exercise, you will create the model of a soap case shown in Figure 8-112. Figure 8-113 shows the top, sectioned front, right, and detail views with dimensions.

(Expected time: 45 min)

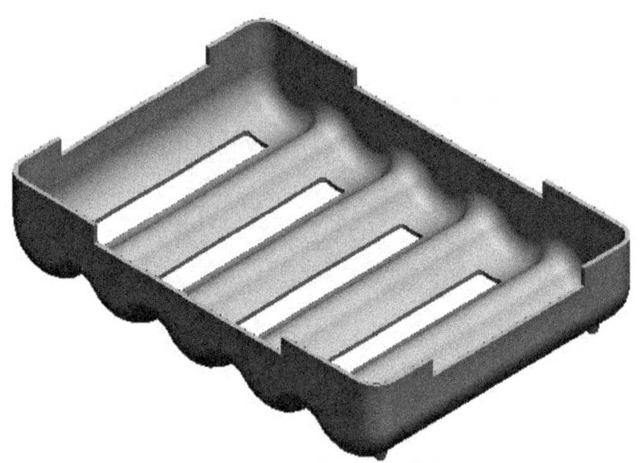

Figure 8-112 Isometric view of the soap case

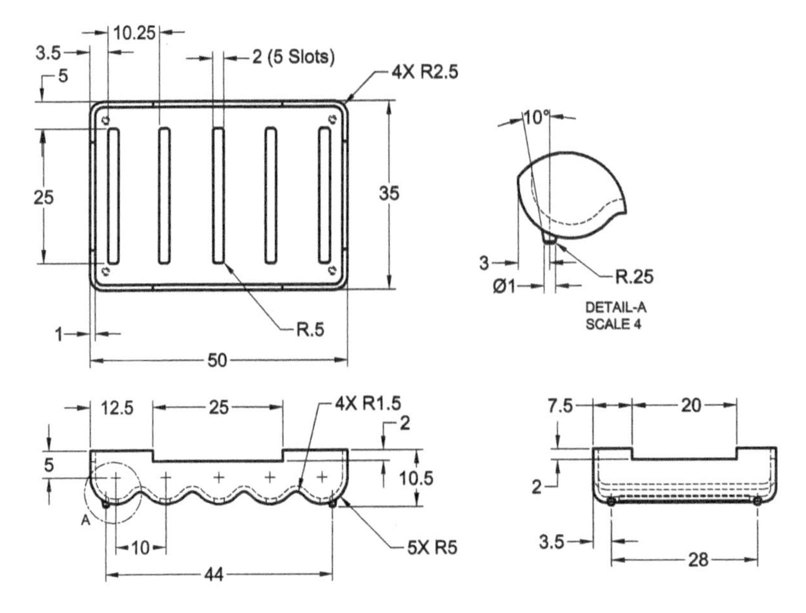

Figure 8-113 Top, sectioned front, right, and detailed views of the soap case

Exercise 4

In this exercise, you will create the model of a carburetor cover shown in Figure 8-114. Figure 8-115 shows the sectioned top view, sectioned front view, sectioned right view, and the sectioned bottom view with dimensions. **(Expected time: 45 min)**

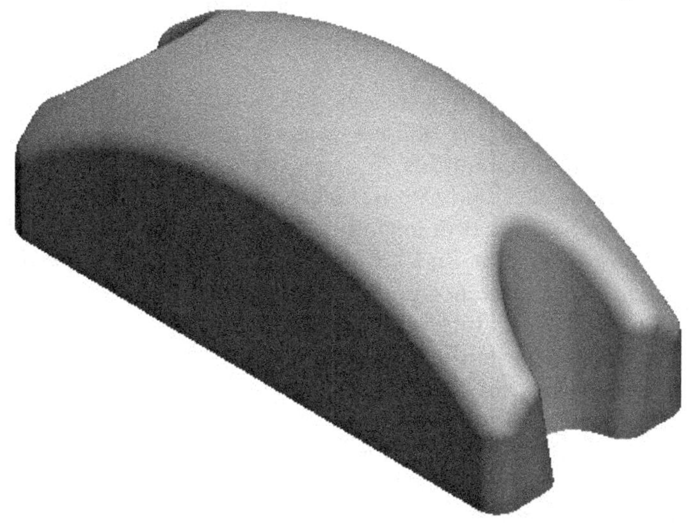

Figure 8-114 *Isometric view of the carburetor cover*

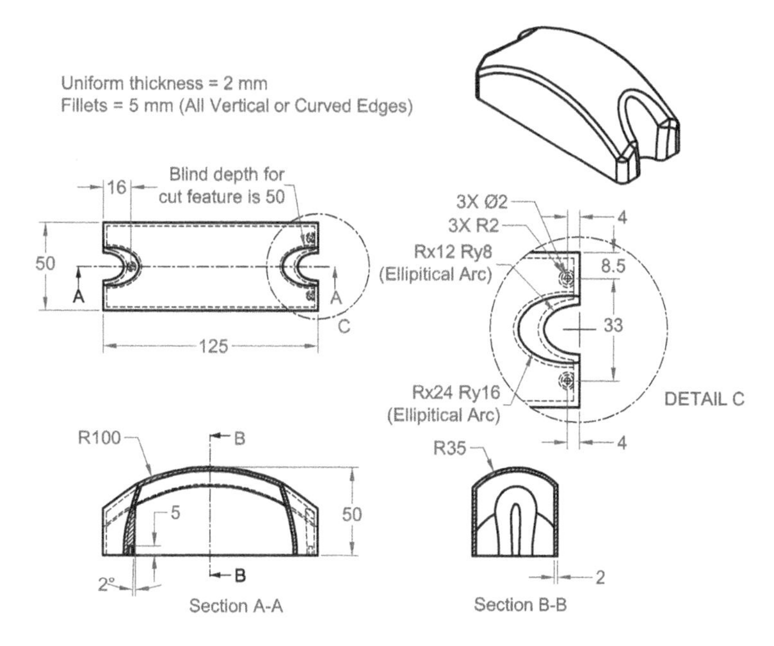

Figure 8-115 *Top, sectioned front, sectioned right, and sectioned bottom views of the carburetor cover*

Hints

1. Create the sketch of the base feature that includes a rectangle of 125x50 and then extrude it.
2. Invoke the sketcher environment, select the **FRONT** datum plane as the sketching plane for sketching a trajectory.
3. Sketch a trajectory by using the **3-Point / Tangent End** button from the **Sketching** group. The start point and the endpoint of the arc should be at a distance of 28 units from the bottom of the base feature, and the radius of the arc should be 100.
4. Exit the sketcher environment by using the **OK** button.
5. Choose the **Sweep** tool from the **Shapes** group of the **Ribbon**, select the trajectory, and choose the **Remove Material** button.
6. Select the **Merge Ends** check box from the **Options** tab.
7. Choose the **Create or edit a sweep section** to create the section for the sweep feature. Choose the **3-Point / Tangent End** button from the **Sketching** group. The arc created should be tangent to the reference lines, and the endpoints of the arc need to be aligned with the edges of the base feature with a radius of 35.
8. Choose the **Done** button from the **Sweep** dashboard.
9. Choose the **Blend** tool in the **Shapes** group of the **Ribbon** and choose **Remove Material** button from dashboard.
10. Choose the **Sections** tab from the dashboard. Choose the **Sketched sections** radio button and click on the **Define** button to define the sketching plane.
11. Select the bottom face of the base feature as the sketching plane.
12. Select the **RIGHT** datum plane as the reference plane and then select the **Right** option from the **Orientation** drop-down list.
13. Create an ellipse of Rx12 and Ry8 by using the **Ellipse** button.
14. Choose **OK** to exit the sketcher environment. Enter **50** in offset value edit box for second section.
15. Create another ellipse of Rx24 and Ry16.
16. Exit the sketcher environment and choose the **Build feature** button to complete the feature creation.
17. Mirror the cut feature about the **RIGHT** datum plane.
18. Create a round feature on all edges except the edges enclosing the bottom planar surface of the base feature.
19. Choose the **Shell** tool from the **Engineering** group. Remove the bottom face of the base feature.
20. Using the bottom face of the base feature as the sketching plane, create three protrusion features that are the supporting structures for the screws. Extrude these features by using the **Extrude up to next surface** button.
21. Use the **Round** tool to round the edges.

Answers to Self-Evaluation Test

1. F, **2.** T, **3.** T, **4.** T, **5.** T, **6.** trajectory, **7.** constant, **8.** parallel, **9.** open, **10.** surface

Chapter 9

Advanced Modeling Tools-II

Learning Objectives

After completing this chapter, you will be able to:

- *Use various advanced options for creating complex models.*
- *Create Variable Section Sweep using the Sweep option.*
- *Create features using the Swept Blend option.*
- *Create features using the Helical Sweep option.*
- *Create features using the Section to Surfaces option.*
- *Create features using the Surfaces to Surfaces option.*

ADVANCED FEATURE CREATION TOOLS

Sometimes a design is too complex and requires advanced options for its completion. PTC Creo Parametric has come up with a design process that can be used to create these complex features with greater ease. All these options are available in the **Shapes** group present in the **Ribbon**.

Variable Section Sweep Using the Sweep Tool

Ribbon: Model > Shapes > Sweep > Sweep

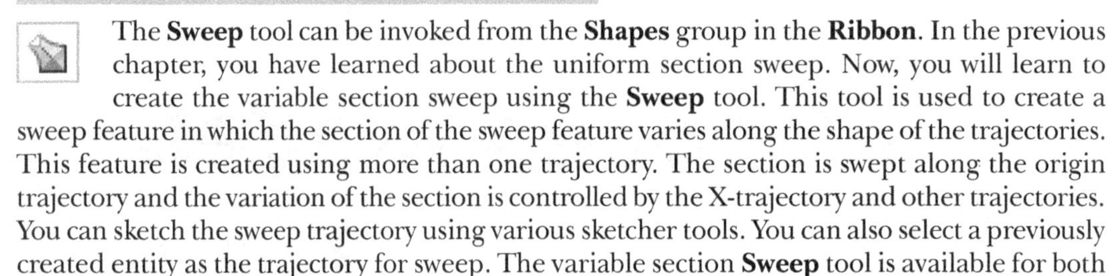

The **Sweep** tool can be invoked from the **Shapes** group in the **Ribbon**. In the previous chapter, you have learned about the uniform section sweep. Now, you will learn to create the variable section sweep using the **Sweep** tool. This tool is used to create a sweep feature in which the section of the sweep feature varies along the shape of the trajectories. This feature is created using more than one trajectory. The section is swept along the origin trajectory and the variation of the section is controlled by the X-trajectory and other trajectories. You can sketch the sweep trajectory using various sketcher tools. You can also select a previously created entity as the trajectory for sweep. The variable section **Sweep** tool is available for both protrusion and cut options.

The following points should be remembered while creating a variable section sweep feature:

1. You should define the origin trajectory to which the section is normal.

2. After creating the origin trajectory, you have to create an X-trajectory that defines the horizontal vector of the section.

3. You can also define additional trajectories to facilitate the creation of a complex profile.

4. When you draw the section for the variable section sweep feature, the section should be aligned to the endpoints of the X-trajectory and the other trajectories. If the section is not aligned, then the feature will be created without following the trajectory path and it will be a uniform section sweep.

When you choose this tool, the **Sweep** dashboard will be displayed, as shown in Figure 9-1. Remember that you should have the trajectories present in the drawing area so that they can be selected for the creation of the variable section sweep feature. This is because you cannot sketch the trajectories by using the **Sweep** dashboard. However, you can draw the trajectories using the **Sketch** button. As mentioned earlier, the process of creating a datum feature when a feature creation tool is active is called datum-on-the-fly. The options in the **Sweep** dashboard are discussed next.

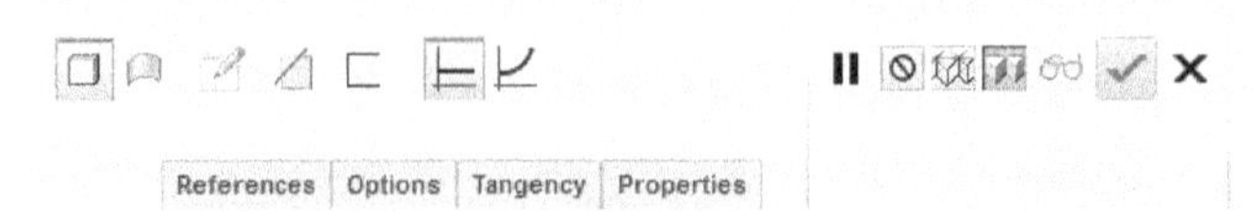

Figure 9-1 *Partial view of the* ***Sweep*** *dashboard*

References Tab

When you choose the **References** tab, the **References** slide-down panel will be displayed, as shown in Figure 9-2. The options available in this slide-down panel are discussed next.

Trajectories Collector

This collector is used for selecting the trajectories by clicking in the collector to activate it. It also lists the selected trajectories. The first trajectory that you select is, by default, listed as the origin trajectory.

*Figure 9-2 The **References** slide-down panel*

Note that you need to first select the normal and X trajectories and then draw the sketch of the section. This is because the sketch plane will be oriented only if you have defined the normal trajectory. Also, since the sketch needs to pass through the endpoints of the X-trajectory, it is better to define the X-trajectory before drawing the section. This way you will have the reference point from which the sketch needs to pass.

This collector has three columns for each selected trajectory listed in the **Trajectories** column. The first column is **X**. When a check box in this column is selected, the trajectory corresponding to that check box is converted into a X trajectory. The section is swept on the X trajectory. The second column is **N**. When a check box in this column is selected, the trajectory corresponding to that check box is converted to the normal trajectory. Remember that only one trajectory can be a normal trajectory and only one trajectory can be an X trajectory. The third column is **T**. There are two check boxes available in this column for each row. All these check boxes are gray in color. When you select any of these check boxes, the name of the trajectory is filled in that cell and the value T=0 appears on the trajectory in the drawing area. This value is displayed on both the ends of the trajectory. Initially, this value is 0. When you enter a value, the point of alignment of the section with the trajectory varies.

Section plane control Drop-down List: The options in this list are used to specify the direction of the section in which it starts sweeping.

The **Normal To Trajectory** option is used to create the variable sweep, the section is normal to the origin trajectory and is swept along the X trajectory. The section should be aligned with the X trajectory to sweep along it. If the section is not aligned with the X trajectory then this option works similar to the **Sweep** option and the section is swept along the origin trajectory only. Figure 9-3 shows the two trajectories and the section used to create the variable section sweep feature. Figure 9-4 shows the resultant variable section sweep feature.

Now, in Figure 9-5, the X trajectory and the origin trajectory are interchanged. The resultant variable section sweep feature is shown in Figure 9-6. Note the difference in the sweep features. In Figure 9-5, the section is normal to the selected origin trajectory and the curve is selected as the origin trajectory.

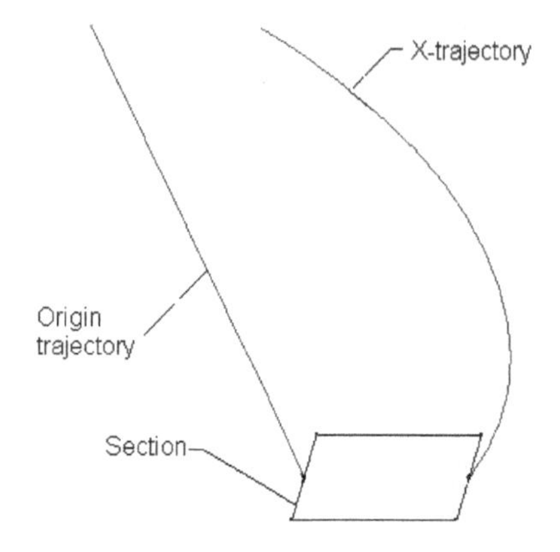

Figure 9-3 *Two trajectories and a section*

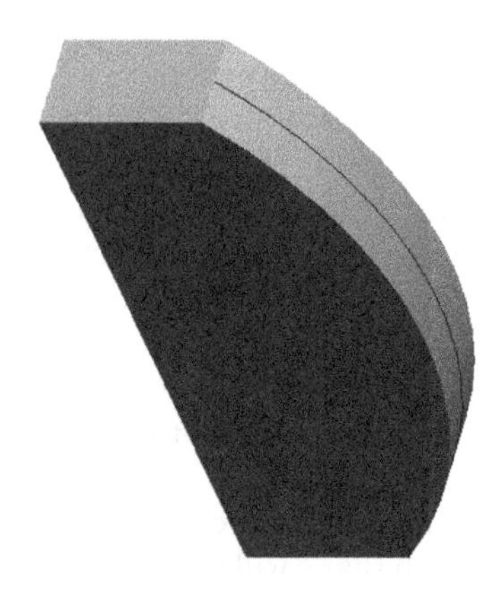

Figure 9-4 *Variable section sweep feature*

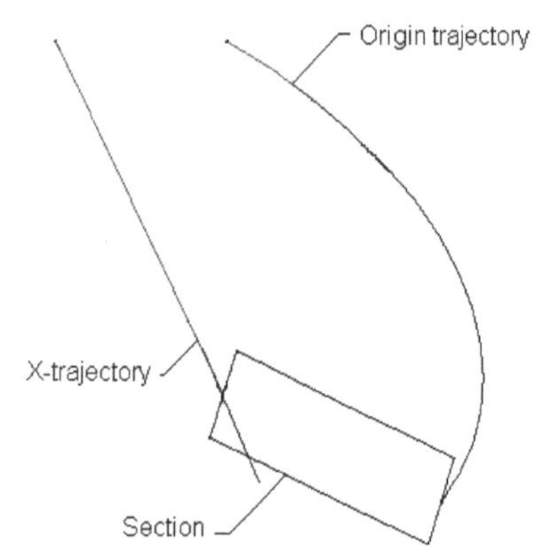

Figure 9-5 *Two trajectories and a section*

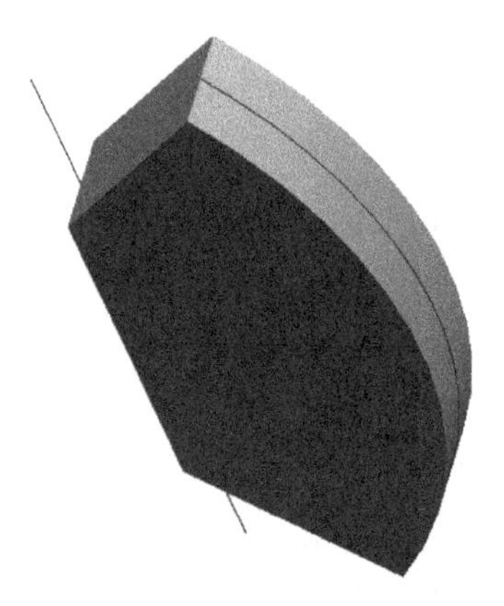

Figure 9-6 *Resultant variable section sweep feature*

The **Normal To Projection** option is used to create a variable section sweep feature in which the section is normal to the two-dimensional (2D) projection of trajectories. The swept section is normal to the origin trajectory and the variation in the section is controlled by the X-trajectory and the other trajectories. When you use this option, you need to select or sketch an origin trajectory, an X-trajectory, other trajectories if required, and the direction of projection. The direction of projection can be specified by using the default coordinate system.

Figure 9-7 shows the origin trajectory and the X trajectory. These two trajectories are used to create the variable section sweep feature shown in Figure 9-8.

The **Constant Normal Direction** option is used to create a variable section sweep in which the section to be swept is normal to the selected face or plane.

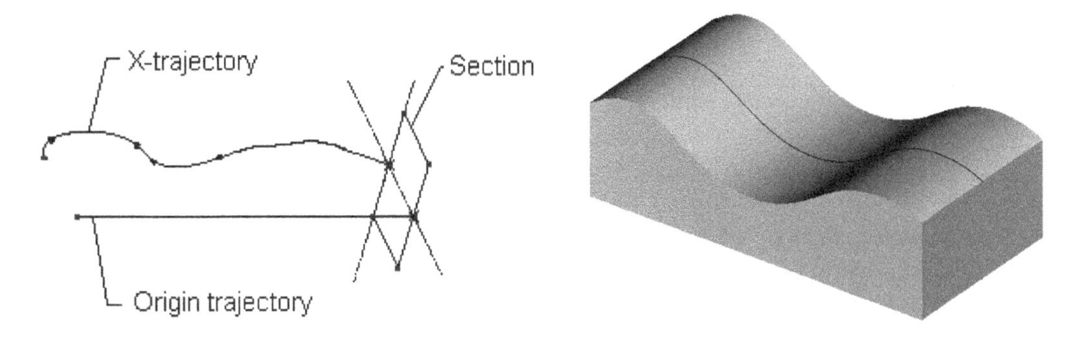

Figure 9-7 *Trajectories and section*

Figure 9-8 *Resultant variable section sweep feature*

When you select the origin trajectory and then select the **Constant Normal Direction** option, you are required to select or create a plane, an edge, a curve, an axis, or a coordinate system to define the constant normal direction. Select a datum plane and then sketch the section. The system takes you to the sketcher environment and automatically orients the sketch plane such that it is normal to the selected direction. Figure 9-9 shows the origin trajectory, the section, and the normal vector of the datum plane to which the section is perpendicular. Figure 9-10 shows the section that is normal to the origin trajectory. This figure explains the difference between the **Normal To Trajectory** option and the **Constant Normal Direction** option.

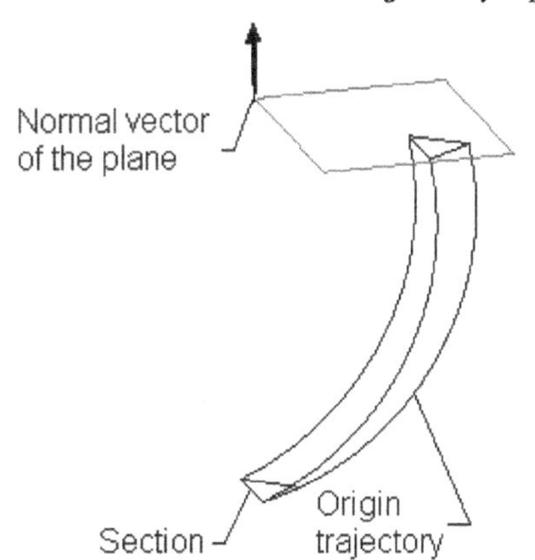

Figure 9-9 *Sweep feature with section perpendicular to the datum plane*

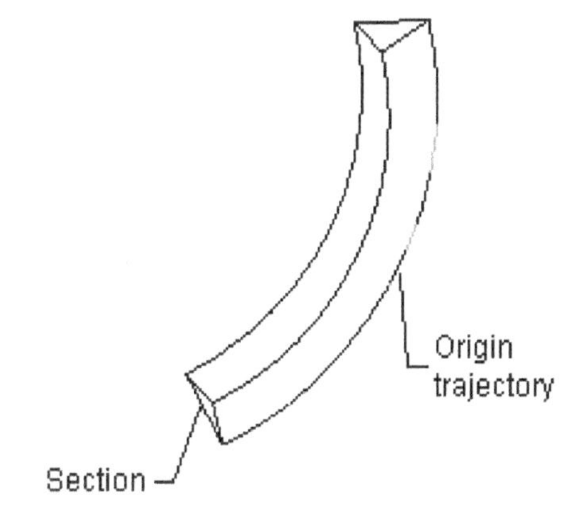

Figure 9-10 *Sweep feature with section normal to the origin trajectory*

Options Tab

When you choose the **Options** tab, the **Options** slide-down panel is displayed, as shown in Figure 9-11. The options in this slide-down panel are discussed next.

Cap ends

On selecting this check box, you can create a surface having closed ends. Note that, this check box is activated only when you choose the **Sweep as surface** button from the dashboard.

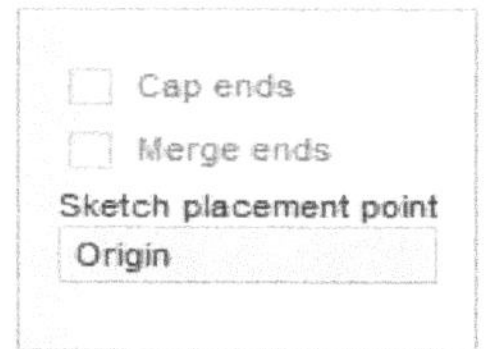

*Figure 9-11 The **Options** slide-down panel*

Merge ends
This check box will be activated only when you create a sweep feature that is attached to other protrusions. This check box is used to fill the gap between the sweep feature to be created and the nearby solid protrusion.

Sketch placement point Collector
By default, **Origin** is selected as the start point of the section. This is the reason you draw the section at the start point of the trajectory. However, if you click in this collector to activate it and then select a datum point that lies on the origin trajectory, the section will start from that point. This means that the selected datum point becomes the start point of the trajectory.

Variable section

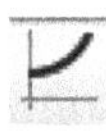

This toggle button is available in the dashboard. It is used to create a sweep feature whose section varies along the trajectory.

Constant section

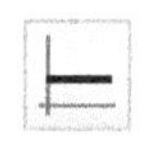

This toggle button is available in the dashboard. It is used to create a sweep feature whose section is constant throughout the trajectory.

Note
*The options available on the **Sweep** dashboard are the same as those available in the **Extrude** dashboard. Using these options, you can specify the type of feature that you want to create such as solid, surface, or thin feature.*

Swept Blend

Ribbon: Model > Shapes > Swept Blend

The **Swept Blend** tool is a combination of sweep and blend features. To create a swept blend feature, choose the **Swept Blend** tool from the **Shapes** group; the **Swept Blend** dashboard will be displayed, as shown in Figure 9-12. Note that the trajectories should be present in the drawing area so that they can be selected for the creation of the swept blend feature. This is because using the options available in the **Swept Blend** dashboard, you cannot sketch the trajectories. However, you can draw curves that can be used as trajectories by using the **Sketch** button when the **Swept Blend** dashboard is active.

Similarly, to create a parallel blend, you need to have more than one section, which have the same number of entities. These sections are then blended, which result in the blend feature. The combination of both these features results in a swept blend feature. The options in the **Swept Blend** dashboard are discussed next.

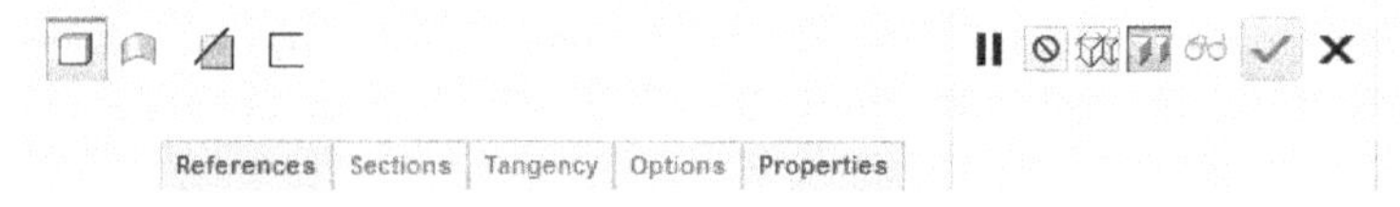

*Figure 9-12 Partial view of the **Swept Blend** dashboard*

References Tab

When you choose the **References** tab, the **References** slide-down panel will be displayed, as shown in Figure 9-13. The options available in slide-down panel are discussed next.

Trajectories Collector

This collector is used for selecting the trajectories. The trajectories that you select will be listed in this collector. The trajectory that you select is listed as the origin trajectory by default (also referred to as normal trajectory). Note that you need to select the origin trajectory and then draw the sketch of the section. This is because the sketch plane will be oriented only if you have defined the origin trajectory. This way you will have the reference point through which the sketch needs to pass. This collector has two columns and the selected trajectories are listed in rows. The first column is **X**. The second column is **N**. Remember that only one trajectory can be a normal trajectory.

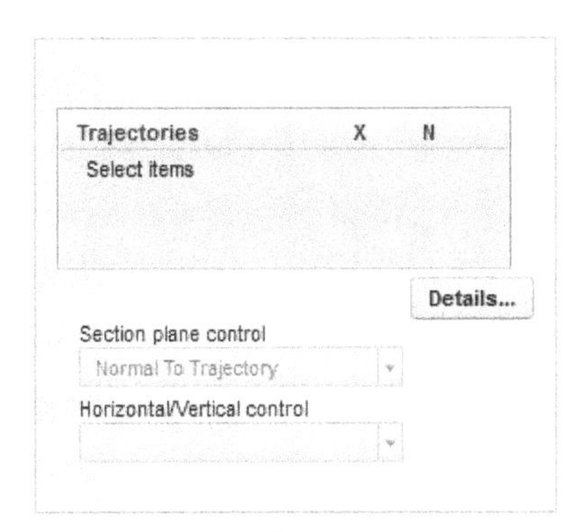

Figure 9-13 The ***References*** *slide-down panel*

Section plane control Drop-down List

The options in this list are used to specify the direction of the section in which the sketch starts sweeping.

Normal To Trajectory Option: This option is used to sweep the section normal to the origin throughout the trajectory.

Normal To Projection Option: The **Normal To Projection** option is used to create a swept blend in which the section to be swept is normal to the 2D projection of the origin trajectory. On selecting this option, you will be prompted to select the direction to which the section is perpendicular. Select the required reference; the selected reference will be displayed in the **Direction reference** collector.

Constant Normal Direction Option: The **Constant Normal Direction** option is used to create a swept blend in which the section is perpendicular to the normal trajectory and is swept along the origin trajectory.

Figure 9-14 shows three rectangular sections. These three sections are connected to each other at their vertices through a trajectory. The shaded view of the model after shelling is shown in Figure 9-15.

Figure 9-16 shows two sections and a trajectory that are used to create the feature shown in Figure 9-17. Note that the number of entities should be same in all the sections.

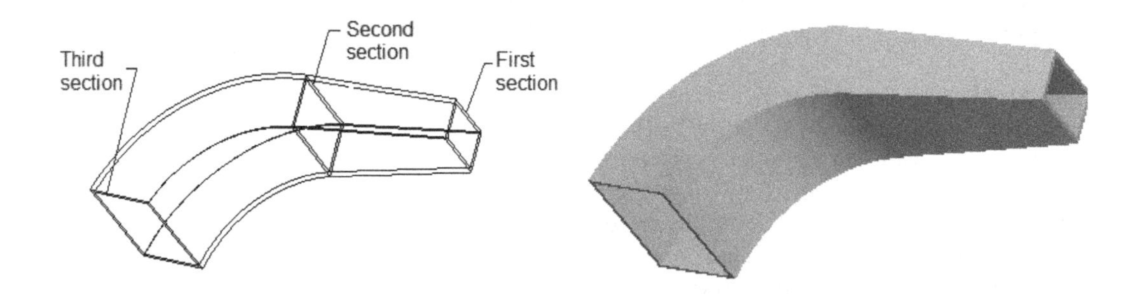

Figure 9-14 Three sections in the swept blend *Figure 9-15 Shaded view of the feature after shelling*

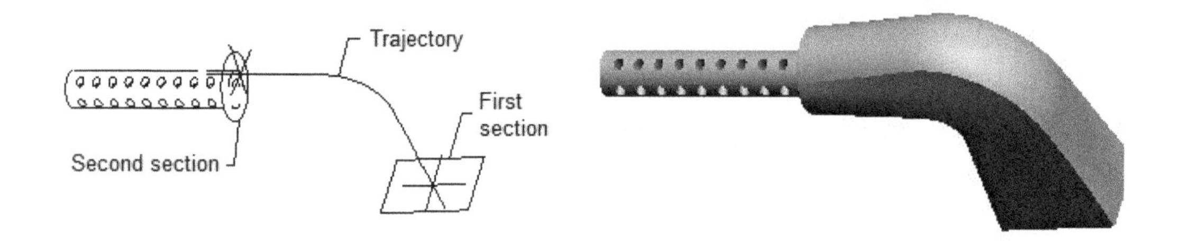

Figure 9-16 Two sections and a trajectory *Figure 9-17 Resultant feature*

Sections Tab

When you choose the **Sections** tab, the **Sections** slide-down panel will be displayed, as shown in Figure 9-18. The options in the slide-down panel are discussed next.

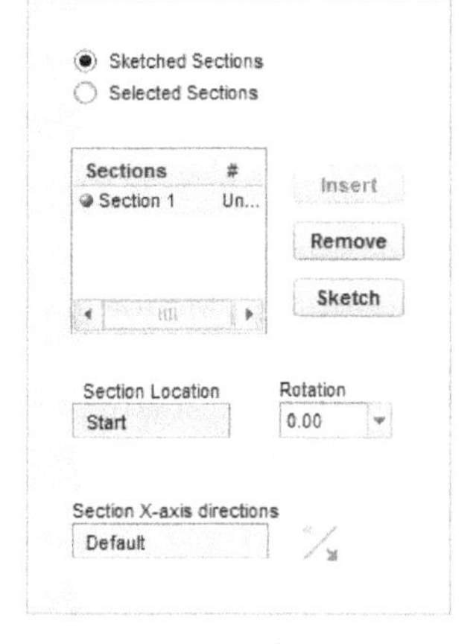

Figure 9-18 The ***Sections*** *slide-down panel*

Sketched Sections

This radio button is used to sketch the section that is used to create the swept blend.

Selected Sections

This radio button is used to select the section that is used to create the swept blend.

Options Tab

When you choose the **Options** tab, the **Options** slide-down panel will be displayed, as shown in Figure 9-19. The **Cap ends** check box, if selected, caps both the end faces of the feature. By default, the **Adjust to keep tangency** check box is selected and is used to maintain the tangency of the surfaces created between two blended sections. The **No blend control** radio button is selected by default. This radio button along with the **Set perimeter control** radio button allows you to control the related parameters of the swept blend feature.

***Figure 9-19** The **Options** slide-down panel*

Helical Sweep

Ribbon: Model > Shapes > Sweep > Helical Sweep

The **Helical Sweep** tool is used to create helical sweep features. You have to define a trajectory that will specify the shape and height of the helix, a pitch value, and a cross-section to create a helical feature using this option. The distance of the trajectory from the center line defines the radius of the helical path and the length of the trajectory defines the length of the sweep feature.

The main use of this tool is to create the helical springs and threads. When you choose the **Helical Sweep** tool from the **Sweep** drop-down in the **Shapes** group, the **Helical Sweep** dashboard is displayed, as shown in Figure 9-20. Figure 9-21 shows the parameters to be defined for the helical sweep and Figure 9-22 shows the resultant helical sweep feature.

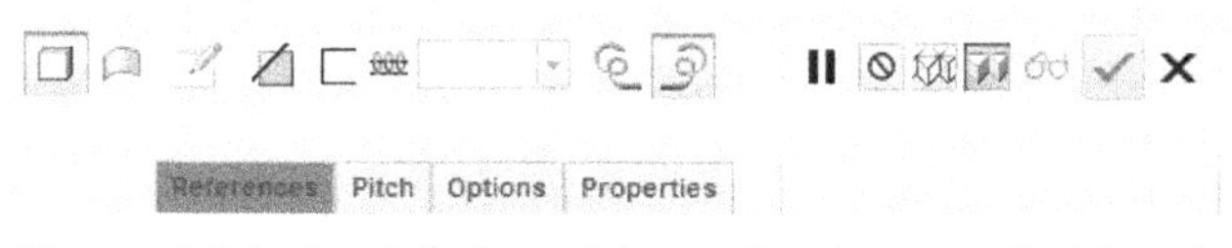

***Figure 9-20** Partial view of the **Helical Sweep** dashboard*

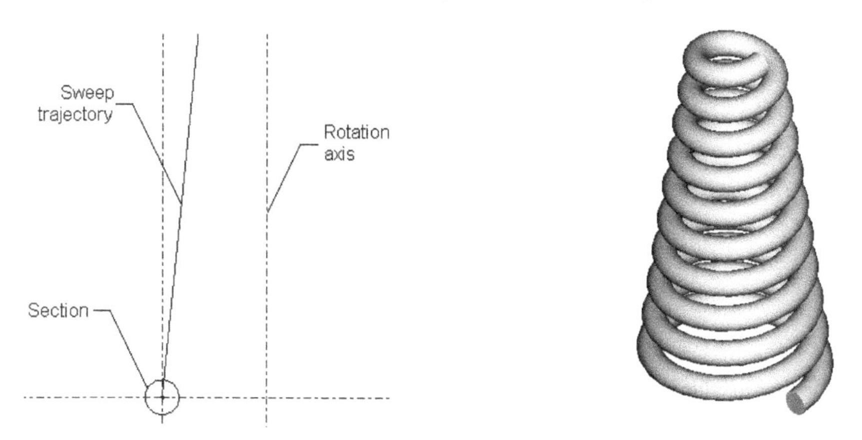

***Figure 9-21** Sweep trajectory, sweep section, and axis of rotation to create a helical feature*

***Figure 9-22** The resultant helical sweep feature*

The options in this dashboard are discussed next.

Keep constant section

The **Keep constant section** radio button is used to create a helical feature with constant pitch, as shown in Figure 9-23. This radio button is available in the **Options** tab.

Vary section

The **Vary section** radio button is also available in the **Options** tab. This radio button is used to create a helical feature of varying pitch. While using this radio button, you have to define the start pitch and the end pitch. You can also define a pitch between the start point and the endpoint of the helix. This pitch value between the endpoint and the start point is specified by creating points on the sweep trajectory. The helical sweep feature with variable pitch is shown in Figure 9-24.

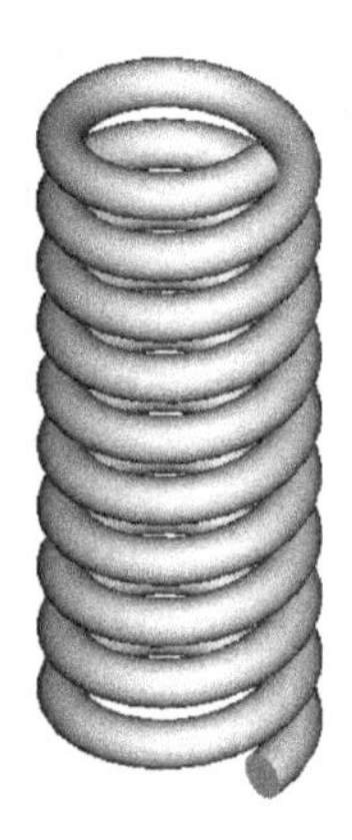

Figure 9-23 *Helical sweep feature with constant pitch*

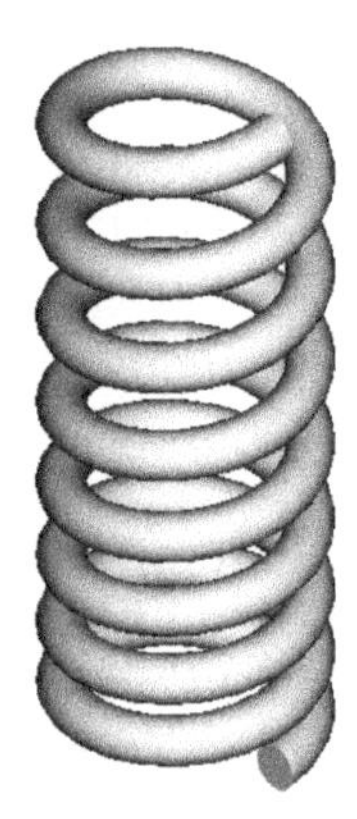

Figure 9-24 *Helical sweep feature with variable pitch*

The brief procedure to create a variable pitch spring is discussed next.

1. Choose **Helical Sweep** from the **Sweep** drop-down in the **Shapes** group of the **Ribbon** and then choose the **Right Handed** button from the dashboard.

2. Select the **Through axis of revolution** radio button from the **Reference** tab.

3. Choose the **Define** button adjacent to the **Helix sweep profile** collector from the slide-down panel of the **Reference** tab.

4. Select the sketching plane and then orient it. Draw the sketch of the sweep trajectory and the center axis.

5. Exit the sketcher environment and specify the value of the pitch in the **Pitch** edit box of the **Pitch** tab in the dashboard.

6. After entering the pitch value at the start point, choose the **Add Pitch** option in the **Pitch** tab to specify the pitch value at the end point.

7. After specifying the pitch value at the two ends, choose the **Add Pitch** option again; a drag-handle is displayed on the centerline of trajectory. Specify the value of location and the pitch value of that point.

8. Similarly, specify the pitch values at other points by using the **Add Pitch** button.

9. After all the pitch values are entered, enter the sketcher environment to draw the section of the spring.

10. Draw the section of the spring at the intersection of the two infinite dotted lines.

11. Exit the sketcher environment and choose **OK** from the **Close** group in the dashboard.

Note

You can use any geometric entity such as spline, line, or arc to create the sweep trajectory.

Through axis of revolution

The **Through axis of revolution** radio button is used to create a helical feature around an axis. This radio button is available in the **Reference** slide-down panel.

Normal to trajectory

The **Normal to trajectory** radio button is used to create a helical feature perpendicular to the sketched trajectory. This radio button is available in the **Reference** slide-down panel.

Right Handed

The **Right Handed** option is used to create a helical feature in which the section is swept in the counterclockwise direction from the start sketch.

Left Handed

The **Left Handed** option is used to create a helical feature in which the section is swept in the clockwise direction from the start sketch.

Blend Section To Surfaces

Enhanced

The **Blend Section To Surfaces** tool is used to blend a selected set of tangential faces with a sketched contour. This option can be used to create a surface or a solid feature. By default, this tool is not available in the **Ribbon**. For adding this tool in the **Ribbon**, you need to customize the **Ribbon**, as discussed next.

To add the **Blend Section To Surfaces** tool, choose **Options** from the **File** menu to invoke the **PTC Creo Parametric Options** dialog box. Now, choose the **Configuration Editor** option from the left pane of the dialog box. Next, choose the **Find** button available at the bottom of the dialog box; the **Find Option** dialog box will be displayed. Now, write **enable_obsoleted_features** in the **Type Keyword** text box and choose the **Find now** button. Now, select the **Yes** option from the **Set Value** drop-down list and choose the **Add/Change** button from the dialog box. Next, close the **Find Option** dialog box and choose the **OK** button from the **PTC Creo Parametric Options** dialog box.

Next, you need to create a new group/tab in the **Ribbon** where this tool will be added. After creating the new group/tab, select the **Commands Not in the Ribbon** option from the **Choose commands from** drop-down list in the dialog box; a list of commands that are not available in the **Ribbon** will be displayed. Choose **Blend Section To Surfaces** tool from the list and then choose the **Add** button; the **Blend Section To Surfaces** tool will be added to the new group/

tab. To create a solid blend feature, choose **Blend Section To Surfaces > Protrusion** from the newly added group/tab in the **Ribbon**; the **PROTRUSION: Section to Surfaces Blend** dialog box will be displayed and you will be prompted to select the surface. Select the surface that you want to blend; you will be prompted to create a sketch. Create a sketch and then choose the **Preview** button from the **PROTRUSION: Section to Surfaces Blend** dialog box; the preview of the blended surface will be displayed. Choose the **OK** button from the dialog box to create the feature. Note that for creating the feature using this option, the sketch must be closed.

Figure 9-25 shows the section to be blended with the selected set of tangential surfaces and Figure 9-26 shows the resultant blended feature.

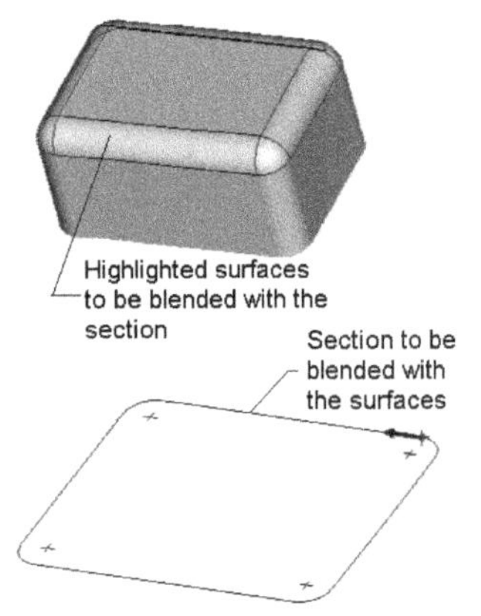

Figure 9-25 *Section to be blended with the selected set of tangential surfaces*

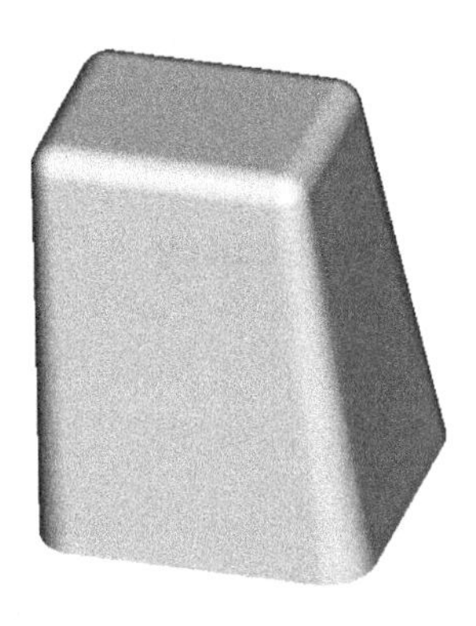

Figure 9-26 *Resultant blended feature*

Blend Between Surfaces

The **Blend Between Surfaces** tool is used to create a blend between two curved quilts, two curved solid surfaces, or a combination of both. The curved surfaces may be revolved features or spheres. This tool will be available in the **Ribbon** only when you customize the **Ribbon**, as discussed in the **Blend Section to Surface** section of this chapter. To create a blend between two surfaces, choose **Blended Between Surfaces** > **Protrusion** from the newly added group/tab in the **Ribbon**; the **PROTRUSION: Surface to Surface Blend** dialog box will be displayed and you will be prompted to select the first surface. Select the first surface that you want to blend; you will be prompted to select the second surface. Select the second surface and then choose the **Preview** button from the **PROTRUSION: Surface to Surface Blend** dialog box; the preview of the blended surface will be displayed. Choose the **OK** button from the dialog box to create the feature. Figure 9-27 shows the two surface spheres to be blended and Figure 9-28 shows the resultant blended feature created using this option.

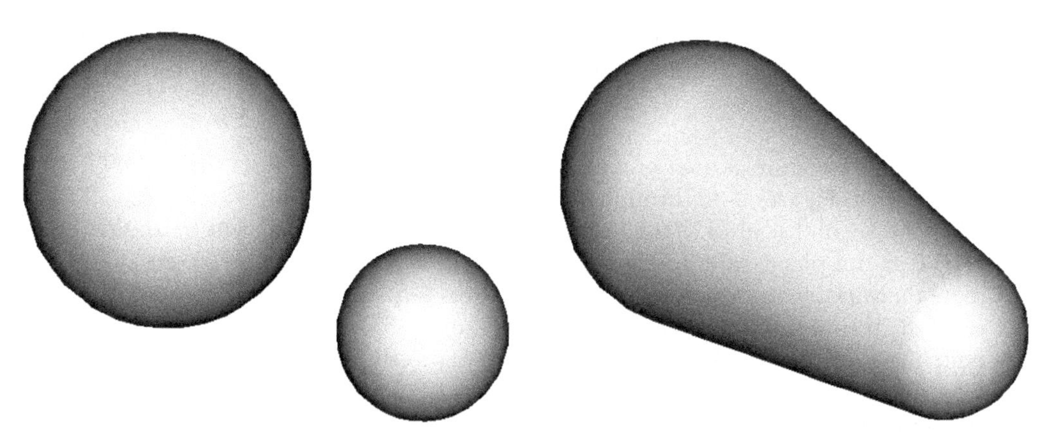

Figure 9-27 The two spheres to be blended

Figure 9-28 Resultant blended feature

TUTORIALS

Tutorial 1

In this tutorial, you will create the model shown in Figure 9-29. This figure also shows the sectioned top, front, right, and isometric views of the model. **(Expected time: 45 min)**

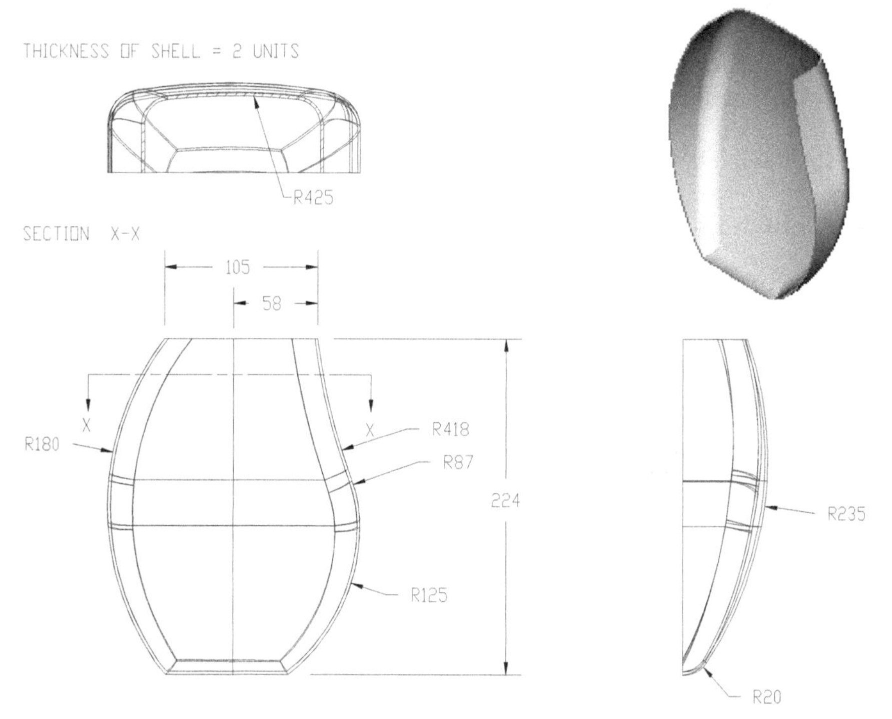

Figure 9-29 Sectioned-top, front, right, and isometric views of the model

Examine the model and determine the number of features in it. The model is composed of five features, refer to Figure 9-29. The following steps are required to complete this tutorial:

a. Create the first feature as a set of datum curves that will be drawn on the **FRONT** datum plane, refer to Figures 9-30 through 9-32.
b. Create the second feature. This feature is the datum curve that will be drawn on the **RIGHT** datum plane, refer to Figure 9-33.
c. Create the variable section sweep, refer to Figures 9-34 through 9-36.
d. Create the round feature, refer to Figures 9-37 and 9-38.
e. Create the shell feature of thickness 2, refer to Figures 9-39 and 9-40.

Make sure that the required working directory is selected.

Starting a New Object File

1. Start a new part file and name it as *c09tut1*.

The three default datum planes are displayed in the drawing area.

Creating the Trajectories

In this section, you need to draw the datum curves that will be selected as trajectories to create the variable section sweep feature. After selecting the sketching plane, draw the sketch for the origin trajectory and dimension it. Then, you need to draw the sketch for the X trajectory on the same plane and dimension it. The X trajectory defines the horizontal vector of a section. If the section is aligned with the X trajectory, the section varies along the path of the X trajectory. Similarly, two additional trajectories will be drawn to guide the section throughout the sweep.

1. Choose the **Sketch** button from the **Datum** group; the **Sketch** dialog box is displayed.

2. Select the **FRONT** datum plane from the drawing as the sketching plane.

3. Select the **TOP** datum plane from the drawing area and then select the **Top** option from the **Orientation** drop-down list. Next, choose the **Sketch** button from the **Sketch** dialog box.

4. Once you enter the sketcher environment, create the sketch of the origin trajectory. The origin trajectory is a straight line segment aligned to the **RIGHT** datum plane and its start point is aligned to the **TOP** datum plane. The sketch of the origin trajectory is shown in Figure 9-30.

Note

1. In the sketcher environment, you will draw all the three datum curves because they are on the same datum plane.

*2. You are drawing the datum curves as individual features because you cannot sketch the trajectories using the tools available in the **Sweep** dashboard.*

5. After the sketch of the origin trajectory is complete, draw the sketch of the X trajectory. The X trajectory is drawn using three arcs. It is recommended that you start the sketch from the top arc. After drawing the first arc, dimension it and then modify its dimension.

6. Draw the second arc and then modify the dimension, as shown in Figure 9-31.

7. Next, draw the third arc. As evident from the sketch of the X trajectory, the center points of the bottom and the second arcs are aligned horizontally.

8. Create the sketch of the third trajectory, as shown in Figure 9-32. This trajectory is created using a single arc.

9. After the sketch is complete, choose the **OK** button to exit the sketcher environment.

10. Similarly, create another trajectory on the **RIGHT** datum plane and keep the **TOP** datum plane at the left. The sketch of the trajectory is shown in Figure 9-33.

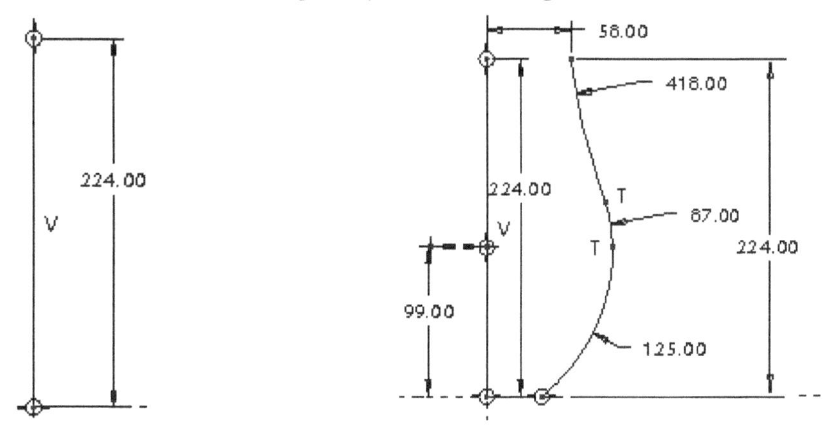

***Figure 9-30** Sketch with dimension of the origin trajectory*

***Figure 9-31** Sketch with dimensions of the X trajectory*

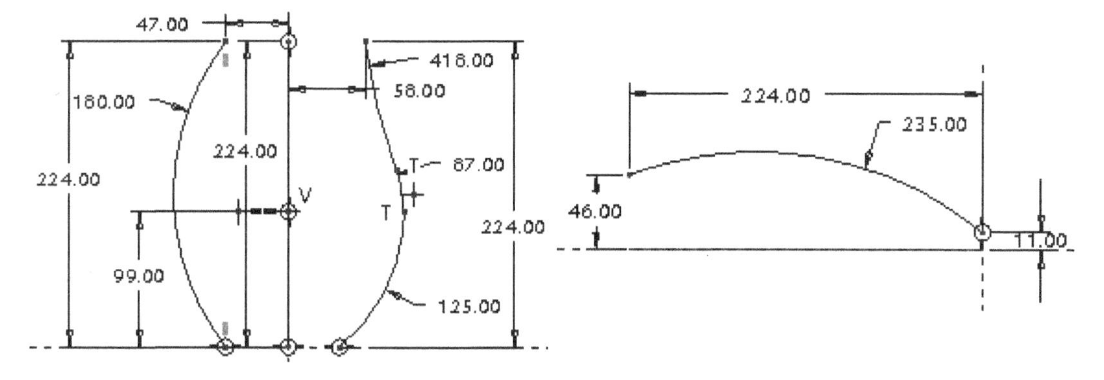

***Figure 9-32** Sketch with dimensions of the third trajectory*

***Figure 9-33** Sketch with dimensions of the additional trajectory*

11. After the sketch is complete, choose the **OK** button to exit the sketcher environment.

12. Choose the **Default Orientation** option from the **Named Views** flyout to view the trajectories that you have created, as shown in Figure 9-34. This view gives a better display of trajectories in 3D space.

Creating the Variable Section Sweep Using Sweep Feature

The four trajectories that will be used to create the variable section sweep have been created. Now, you need to invoke the **Sweep** tool and select these trajectories to create the feature.

1. Choose the **Sweep** tool from the **Shapes** group; the **Sweep** dashboard is displayed.

2. Choose the **References** tab to invoke the slide-down panel. Select the origin trajectory that is the first datum curve drawn.

3. Use CTRL+left mouse button to select the X trajectory (second datum curve). Note that as you are selecting the trajectories, they are being listed in the **Trajectories** collector.

4. Use CTRL+left mouse button to select the remaining two trajectories. Select the check box under the X column for the **Chain 1** row. This makes the second selected trajectory in the row as the X trajectory.

5. Now, choose the **Create or edit sweep section** button from the dashboard to enter the sketcher environment.

6. Draw the closed sketch of the section and apply the dimensions, as shown in Figure 9-35.

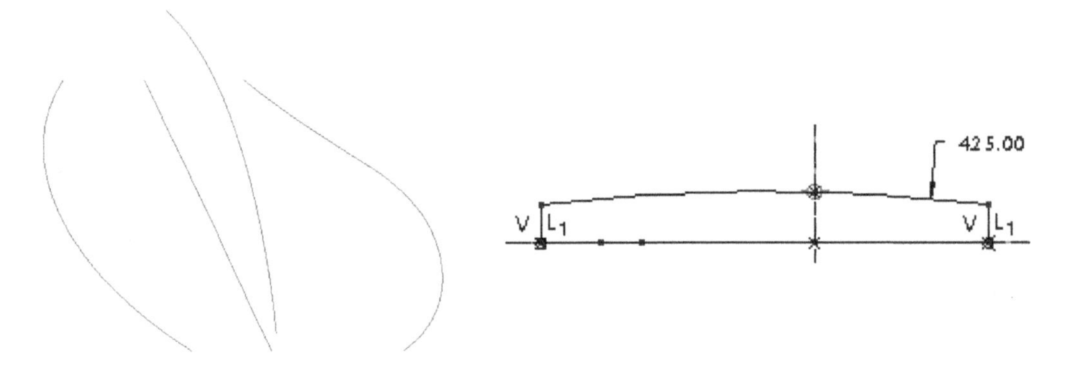

Figure 9-34 *Default view of the four trajectories* *Figure 9-35* *Sketch of the section with dimensions*

Note that the section is aligned with the start points of all the trajectories that are displayed by crosses. The cursor will automatically snap to these points as you move it close to these points. In order to vary the section with the trajectories, it is necessary to align the section with the trajectories.

7. After the sketch is complete, choose the **OK** button to exit the sketcher environment.

   The preview of the variable section sweep feature is displayed in the drawing area.

8. Choose the **Sweep as solid** button from the dashboard, if the preview shows a surface feature.

9. Choose the **Build feature** button from the dashboard.

   Now, you need to hide the datum curves.

10. Choose the **Layer Tree** option from the **Show** flyout of the **Model Tree**. In the Navigator, the layers of the current model are displayed. Right-click on **PRT_ALL_CURVES** to invoke the shortcut menu.

11. Choose the **Hide** option from the shortcut menu. Regenerate the model, if needed. All the datum curves in the model will disappear.

    The variable section sweep feature is completed, as shown in Figure 9-36. You can use the middle mouse button to orient the feature as shown in the figure.

Creating the Round Feature

Now, you need to create round on the edges of the base feature.

1. Choose the **Round** tool from the **Engineering** group; the **Round** dashboard is displayed.

2. Select the edges shown in Figure 9-37. Remember that you need to use CTRL+left mouse button to select more than one edge.

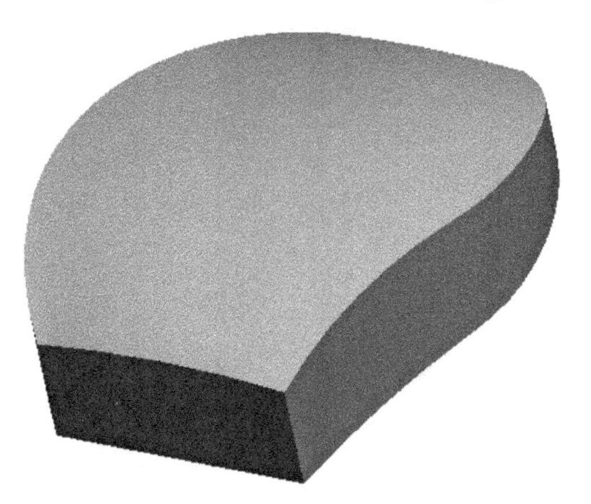

Figure 9-36 Variable section sweep feature

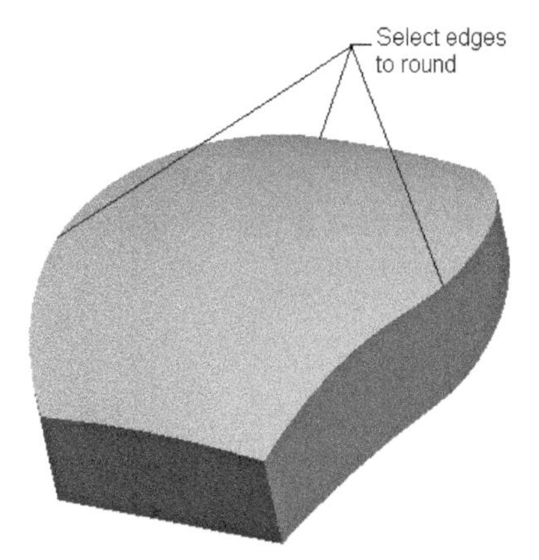

Figure 9-37 Edges to be selected to create round

3. Enter **20** in the dimension edit box present on the dashboard. Choose **Build feature** from the dashboard. The base feature after creating the round is shown in Figure 9-38.

Creating the Shell Feature

Now, you need to create the shell and remove the front and bottom faces of the base feature.

1. Choose the **Shell** tool from the **Engineering** group; you are prompted to select the surfaces to be removed.

2. Select the surfaces, as shown in Figure 9-39. Remember that you need to use the CTRL+left mouse button to select more than one face.

3. Enter **2** in the **Thickness** edit box. Then, choose the **Build Feature** button from the dashboard.

   The model after creating the shell feature is shown in Figure 9-40.

Figure 9-38 Model after creating the round

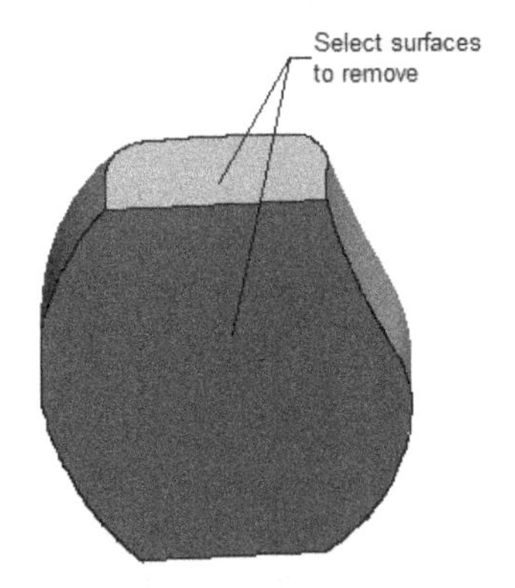

Figure 9-39 Surfaces selected to be removed

Saving the Model

Next, you need to save the model.

1. Choose the **Save** button from the **File** menu and save the model.

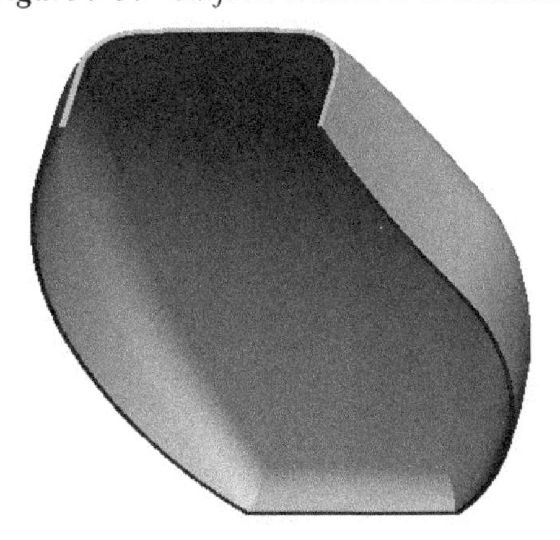

Figure 9-40 Final model after shelling

Tutorial 2

In this tutorial, you will create a solid model of the Upper Housing of a motor blower assembly shown in Figure 9-41. Figure 9-42 shows the left view, top view, front view, and the sectioned left view of the model. **(Expected time: 1 hr)**

Examine the model and determine the number of features in it. The model is composed of ten features, refer to Figures 9-41 and 9-42.

The following steps are required to complete this tutorial:

a. Create the base feature, refer to Figures 9-43 and 9-44.
b. Create the swept blend feature in which the section is normal to the origin trajectory, refer to Figures 9-45 through 9-49.
c. Create the round feature of radius 1.5, refer to Figures 9-50 and 9-51.
d. Create the round of radius 0.5, refer to Figures 9-52 and 9-53.
e. Create the shell feature of thickness 0.25, refer to Figure 9-54.
f. Create an extruded cut, refer to Figures 9-55 and 9-56.
g. Create another extruded cut, refer to Figures 9-57 and 9-58.
h. Create the eighth feature as an extruded feature, refer to Figures 9-59 and 9-60.
i. Create a mirror copy of the eighth feature, refer to Figure 9-61.
j. Create the hole feature, refer to Figure 9-62, and pattern it, refer to Figure 9-63.

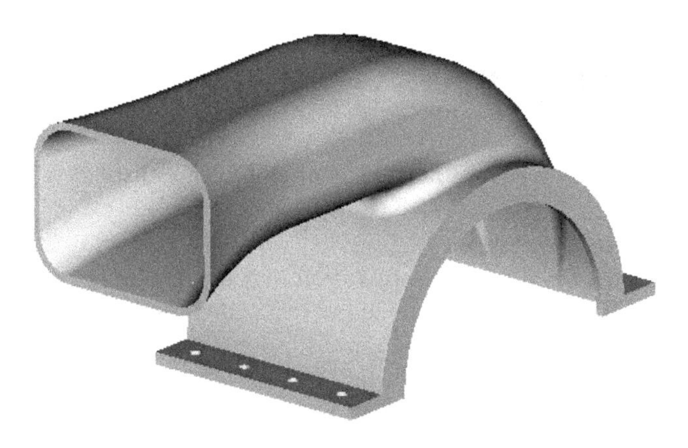

Figure 9-41 *Solid model of the Upper Housing*

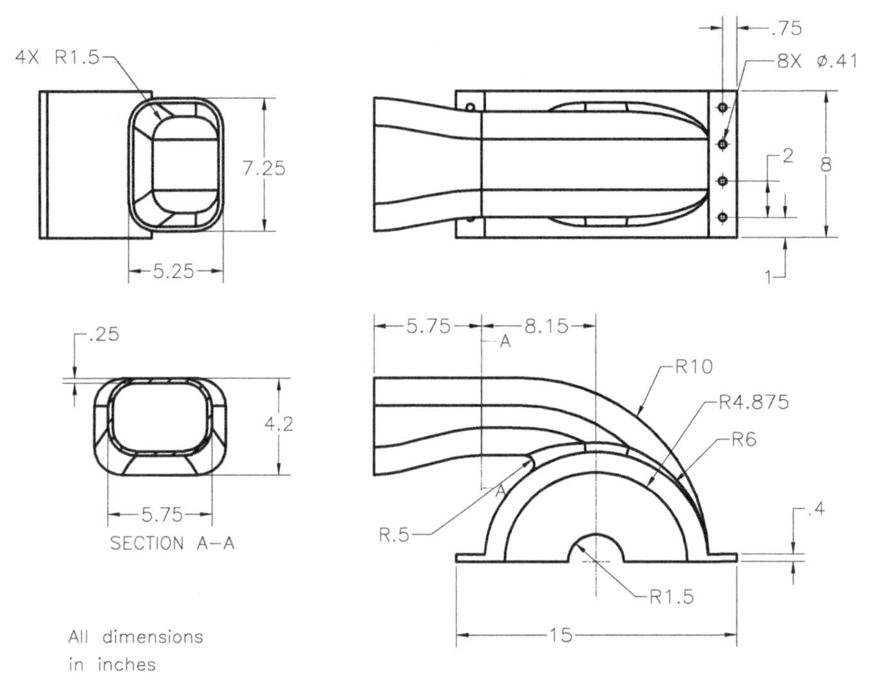

Figure 9-42 *Left view, top view, front view, and the sectioned left view of the model*

Starting a New Object File

1. Start a new part file and name it as *c09tut2*. Also, set the template to **inlbs_part_solid**.

   The three default datum planes and the **Model Tree** are displayed in the drawing area.

Creating the Base Feature

To create the sketch for the base feature, you first need to select the sketching plane for the base feature. In this model, you need to draw the base feature on the **RIGHT** datum plane.

1. Choose the **Extrude** tool from the **Shapes** group. Next, invoke the **Sketch** dialog box.

2. Select the **RIGHT** datum plane as the sketching plane.

3. Select the **TOP** datum plane from the drawing area and then select the **Top** option from the **Orientation** drop-down list.

4. Choose the **Sketch** button from the **Sketch** dialog box to enter the sketcher environment.

5. Once you enter the sketcher environment, create the sketch of the base feature and apply

the required dimensions, as shown in Figure 9-43.

6. After the sketch is complete, choose the **OK** button to exit the sketcher environment.

   The **Extrude** dashboard gets activated and the **Extrude from sketch plane by specified depth value** button is chosen by default in the dashboard.

7. Choose the **Extrude on both sides of sketch plane by half the specified depth value in each direction** button from the depth flyout.

8. Enter the depth as **8** in the **dimension** edit box present in the **Extrude** dashboard. Choose the **Build feature** button from the **Extrude** dashboard.

   The base feature is completed. The default trimetric view of the base feature is shown in Figure 9-44.

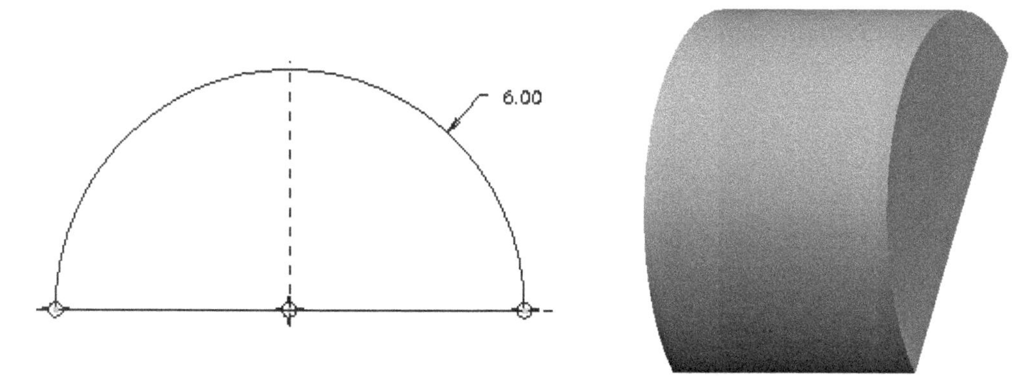

Figure 9-43 *Sketch of the base feature*

Figure 9-44 *Default trimetric view of the base feature*

Creating the Trajectories for the Swept Blend

You can use the **Sketch** button to create the sketch for the trajectories of the swept blend feature before or after invoking the **Swept Blend** dashboard. However, it is recommended that you create the trajectory before invoking the **Swept Blend** dashboard.

1. Choose the **Sketch** tool from the **Datum** group; the **Sketch** dialog box is displayed.

2. Choose the **Use Previous** button from the **Sketch** dialog box to enter the sketcher environment.

3. Draw the sketch, as shown in Figure 9-45, and exit the sketcher environment.

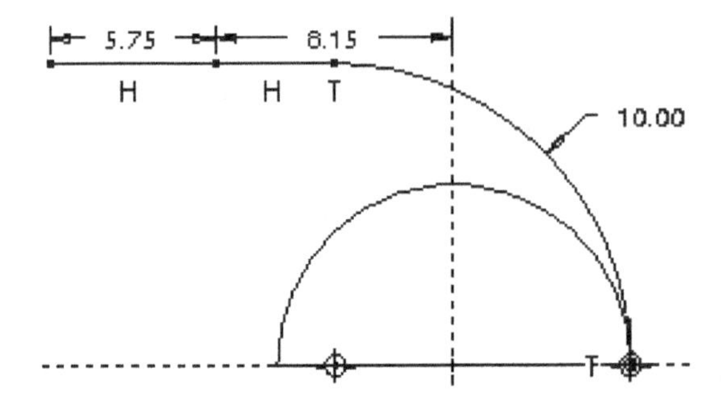

Figure 9-45 Sketch of the origin trajectory

Creating the Swept Blend Feature

1. Choose **Swept Blend** from the **Shapes** group; the **Swept Blend** dashboard is displayed.

2. Choose the **References** tab to display the slide-down panel.

3. Select the trajectory from the drawing area. Note that as you select the trajectory from the drawing area, it gets added to the **Trajectories** collector. Also note that the first trajectory you select from the drawing area is considered as the normal trajectory.

4. Next, choose the **Sections** tab to display the slide-down panel. Select the start point of the arc as the start point of the first cross-section to be created; the **Sketch** button is enabled in the **Sections** slide-down panel. Choose the **Sketch** button to enter the sketcher environment.

5. After entering the sketcher environment, create the sketch for the first cross-section, as shown in Figure 9-46.

6. Choose the **OK** button to exit the sketcher environment.

7. The **Swept Blend** dashboard is enabled and is displayed above the drawing area. Now, choose the **Insert** button from the **Sections** slide-down panel and select the second point on the trajectory which is at a distance of 8.15 from the **FRONT** datum plane. Choose the **Sketch** button from the **Sections** slide-down panel and create the sketch for the second cross-section, as shown in Figure 9-47 and exit the sketcher environment.

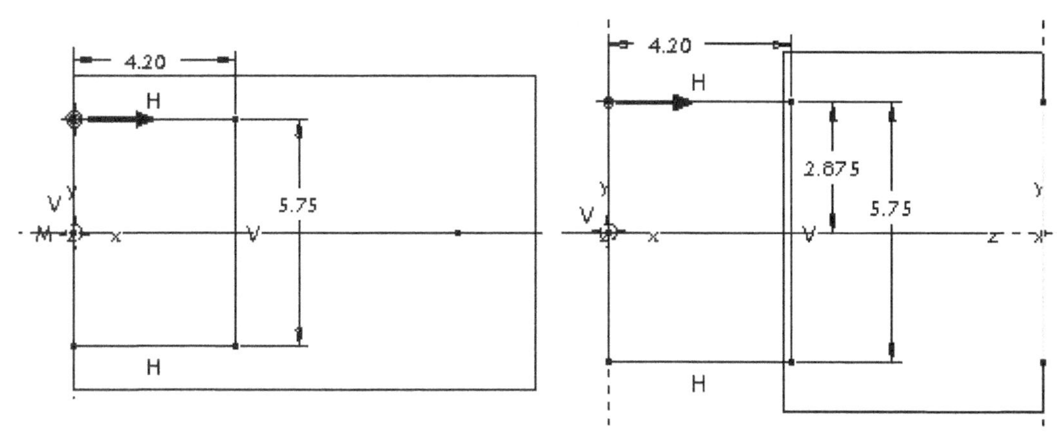

Figure 9-46 *Sketch for the first cross-section with dimensions*

Figure 9-47 *Sketch for the second cross-section with dimensions*

8. Now, choose the **Insert** button and select the other end point of the trajectory to create the third cross-section, as shown in Figure 9-48.

9. Exit the sketcher environment after sketching the third cross-section; the **Swept Blend** dashboard is enabled and displayed above the drawing area. Choose the **Create a Solid** button from the **Swept Blend** dashboard; the preview of the feature is displayed in the drawing area. Choose the **Build feature** button from the **Swept Blend** dashboard to exit the feature creation tool.

   The model, similar to the one shown in Figure 9-49, is displayed in the drawing area.

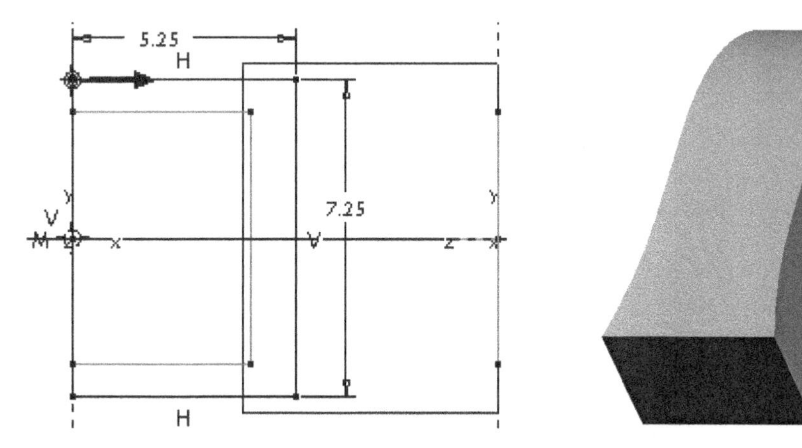

Figure 9-48 *Sketch of the third cross-section with dimensions*

Figure 9-49 *Default view of the swept blend feature*

Creating the Round Features

In this section, you need to create two round features. These features must be created before shelling because the rounds are also shelled along with the faces of the model. The third feature and the fourth feature consist of round feature of radius 1.5 and 0.5, respectively.

1. Choose the **Round** tool from the **Engineering** group; the **Round** dashboard is displayed.

2. Select all four longer edges of the swept feature, as shown in Figure 9-50 to round them. Remember that the first edge can be selected by using the left mouse button and the remaining edges can be selected by using the CTRL+left mouse button. The preview of the round is displayed in the drawing area along with the default radius value.

3. Click on the radius value, enter **1.5** in the edit box that appears, and then press ENTER.

4. Choose the **Build feature** button from the **Round** dashboard. The default view of the model after creating the round feature of radius 1.5 is shown in Figure 9-51.

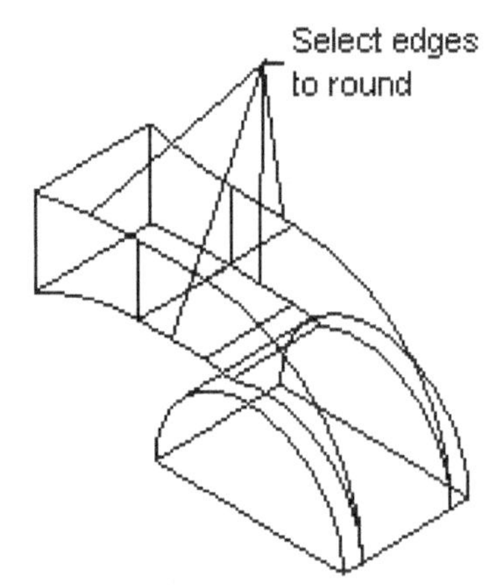

Figure 9-50 *Edges to round*

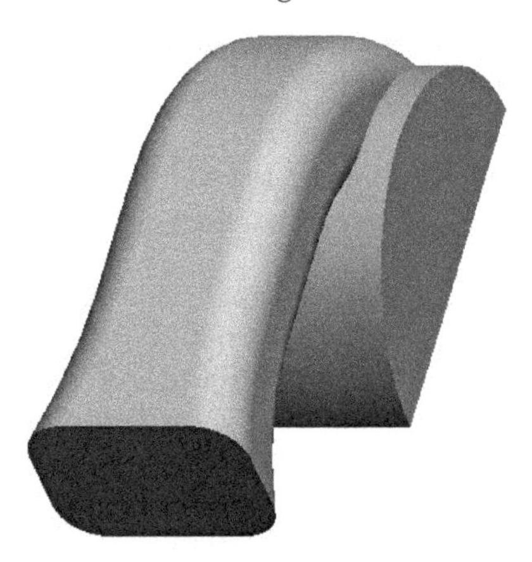

Figure 9-51 *Default view of the model after creating the round feature of radius 1.5*

Note

A round of radius 0.5 cannot be created by adding a round set. This is because this round needs to be created on the geometry of the previous round. Therefore, until the first round with the radius 1.5 is created, you cannot create the second round.

The third feature of the model is completed and now you need to create the fourth feature, which is also a round feature of radius 0.5. To do so, you need to invoke the **Round** dashboard.

5. Choose the **Round** tool from the **Engineering** group; the **Round** dashboard is displayed.

6. Select the edge where the base feature and the swept feature join, as shown in Figure 9-52.

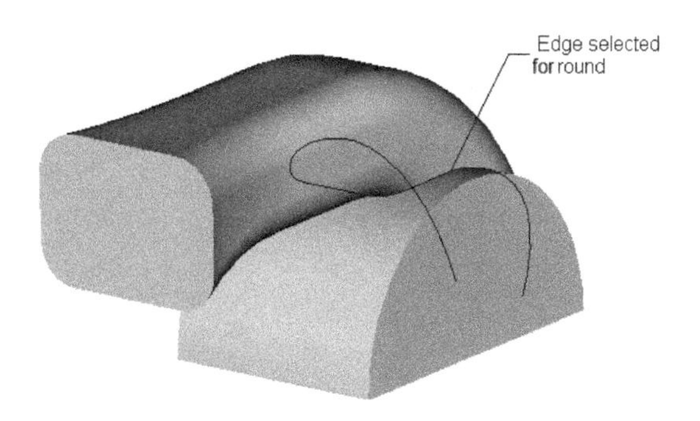

Figure 9-52 *Edge selected to round*

The preview of the round is displayed in the drawing area along with the default radius value.

7. Click on the radius value, enter **0.5** in the edit box that appears, and then press ENTER.

8. Choose the **Build feature** button from the **Round** dashboard. The default view of the model after creating the round feature of radius 0.5 is shown in Figure 9-53.

Creating the Shell Feature

In this section, you need to create a shell of thickness 0.25 on the swept blend feature.

1. Choose the **Shell** tool from the **Engineering** group; the **Shell** dashboard is displayed and you are prompted to select the surface to be removed.

2. Using the left mouse button, select the front face of the swept blend feature and then use the CTRL+left mouse button to select the bottom face of the base feature.

3. Enter **0.25** in the dimension edit box available on the **Shell** dashboard and then press ENTER.

4. Choose the **Build feature** button from the **Shell** dashboard. The shell feature is created as shown in Figure 9-54.

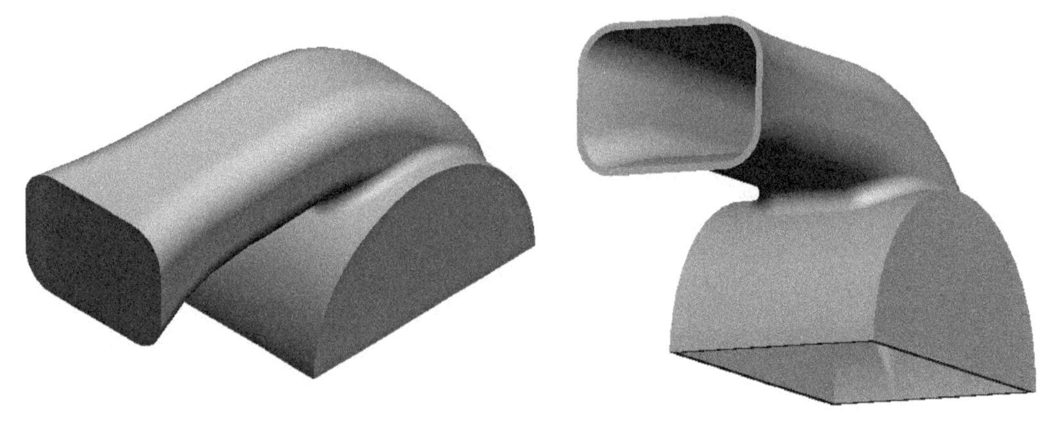

Figure 9-53 Model after creating a round feature of radius 0.5

Figure 9-54 The model after shelling

Creating the Extruded Cut

The extruded cut is the sixth feature of this model. You need to create the cut on the front face of the base feature.

1. Choose the **Extrude** tool from the **Shapes** group; the **Extrude** dashboard is displayed.
2. Choose the **Remove Material** button from the dashboard.
3. Select the front face of the base feature as the sketching plane.
4. Select the **TOP** datum plane from the drawing area and then select the **Top** option from the **Orientation** drop-down list.
5. Choose the **Sketch** button to enter the sketcher environment.
6. Draw the sketch of the cut feature and apply dimension to it, as shown in Figure 9-55. You can use the **Concentric** tool from the **Circle** drop-down in the **Sketching** group to draw the sketch.
7. After the sketch is complete, choose the **OK** button to exit the sketcher environment; the **Extrude** dashboard reappears above the drawing area. The **Extrude from sketch plane by specified depth value** button is chosen by default.
8. Choose the **Extrude upto next surface** button from the depth flyout.
9. Choose the **Build feature** button from the **Extrude** dashboard. The cut feature is created, as shown in Figure 9-56.

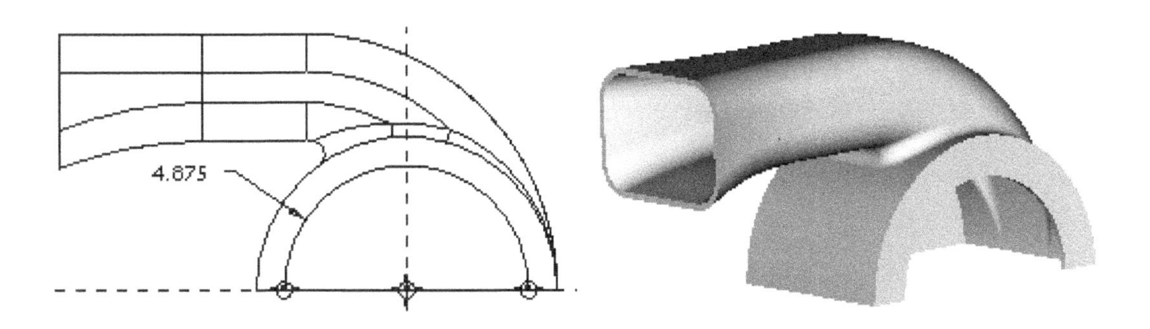

Figure 9-55 Sketch with dimension for cut feature

Figure 9-56 Model with cut feature at the front face

Similarly, create the next extruded cut feature at the back face of the base feature. Select the sketch plane, draw the sketch, apply dimension, and extrude the sketch using the **Extrude upto next surface** button. The sketch for the cut feature is shown in Figure 9-57.

The seventh feature of the model, which is a cut feature, is created, as shown in Figure 9-58.

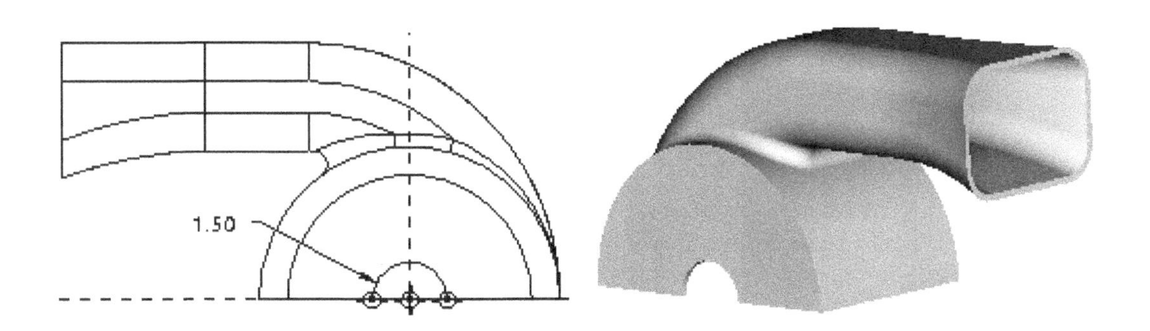

Figure 9-57 Sketch with dimension for cut feature

Figure 9-58 Model with cut feature at the back face

Creating the Extruded Feature

The eighth feature of the model is an extruded feature and it will be created on the front face of the base feature.

1. Invoke the **Extrude** tool and select the front face of the base feature as the sketching plane for the extrude feature. Once you enter the sketcher environment, draw the sketch and then apply constraints and dimensions to it, as shown in Figure 9-59.

2. After the sketch is complete, choose the **OK** button to exit the sketcher environment.

3. From the depth flyout, choose the **Extrude to selected point, curve, plane or surface** button and then select the back face of the base feature.

4. Choose the **Build feature** button from the **Extrude** dashboard. The extruded feature is completed and is shown in Figure 9-60.

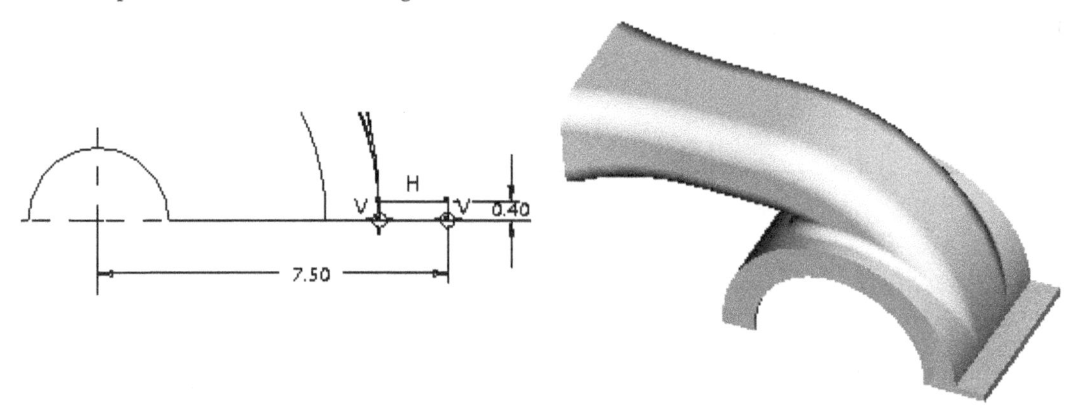

Figure 9-59 Sketch for the extruded feature

Figure 9-60 Model after creating the extruded feature

Creating a Mirror Copy of the Eighth Feature

You need to create a mirror copy of the extruded feature, as shown in Figure 9-61 and this will be the ninth feature of the model. The extruded feature will be mirrored about the **FRONT** datum plane.

1. Select the eighth feature of the model and then choose the **Mirror** tool from the **Editing** group of the **Model** tab; the **Mirror** dashboard is displayed.

2. Select the **FRONT** datum plane as the mirror plane from the **Mirror plane** collector and choose the **Build Feature** button from the dashboard; the extruded feature is mirrored about this datum plane, as shown in Figure 9-61.

Creating the Tenth Feature

The tenth feature is a through hole and it will be created using the **Hole** dashboard.

1. Choose the **Hole** tool from the **Engineering** group; the **Hole** dashboard is displayed in which the **Create simple hole** button is chosen by default.

   Create the hole, as shown in Figure 9-62, by specifying the placement parameters. For placement parameters, refer to Figure 9-42.

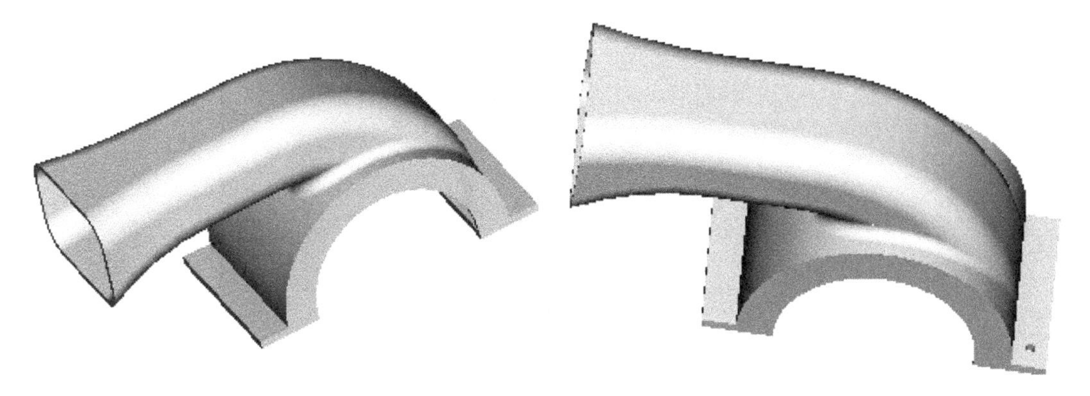

Figure 9-61 Model after mirroring the extruded feature

Figure 9-62 Model after creating the hole

Creating a Pattern of the Hole Feature

As evident from Figure 9-41, you need to create eight instances of the hole. The first instance is created by using the **Hole** dashboard and you can use the **Pattern** option to create the remaining seven instances. You need to create a rectangular pattern of the hole feature. You can also create the holes by using the **Hole** dashboard and by specifying the placement parameters for each of them. But to save time, you should create a pattern of the hole.

1. Select the hole feature from the **Model Tree** or from the model. Hold down the right mouse button to invoke the shortcut menu.

2. Choose the **Pattern** option from the shortcut menu; the **Pattern** dashboard is displayed and the dimensions are displayed on the hole feature.

3. Select the **General** option from the **Regeneration option** drop-down list in the **Options** slide-down panel.

4. Click on the dimension value **1** from the drawing area. Enter **2** in the edit box and press ENTER.

5. Enter **4** in the **1** edit box present on the dashboard and press ENTER. Now, you need to enter dimension increment in the second direction.

6. Click on the collector that displays **Click here to add item** on the right of the **2** edit box. Next, click on the dimension value **0.75** from the drawing area. Enter the value **13.5** in the edit box that appears, and press ENTER.

7. Enter **2** in the **2** edit box present on the dashboard and press ENTER.

8. Choose the **Build feature** button from the dashboard. You can use the middle mouse button to spin the model and display it, as shown in Figure 9-63.

Figure 9-63 *Trimetric view of the completed model*

Saving the Model

1. Choose the **Save** button from the **File** menu and save the model.

Tutorial 3

In this tutorial, you will create the model of the wheel of a car, as shown in Figure 9-64. Figure 9-65 shows the top view, sectioned front view, sectioned right view, detail view, and two blend sections with dimensions. **(Expected time: 45 min)**

Figure 9-64 *Solid model of wheel*

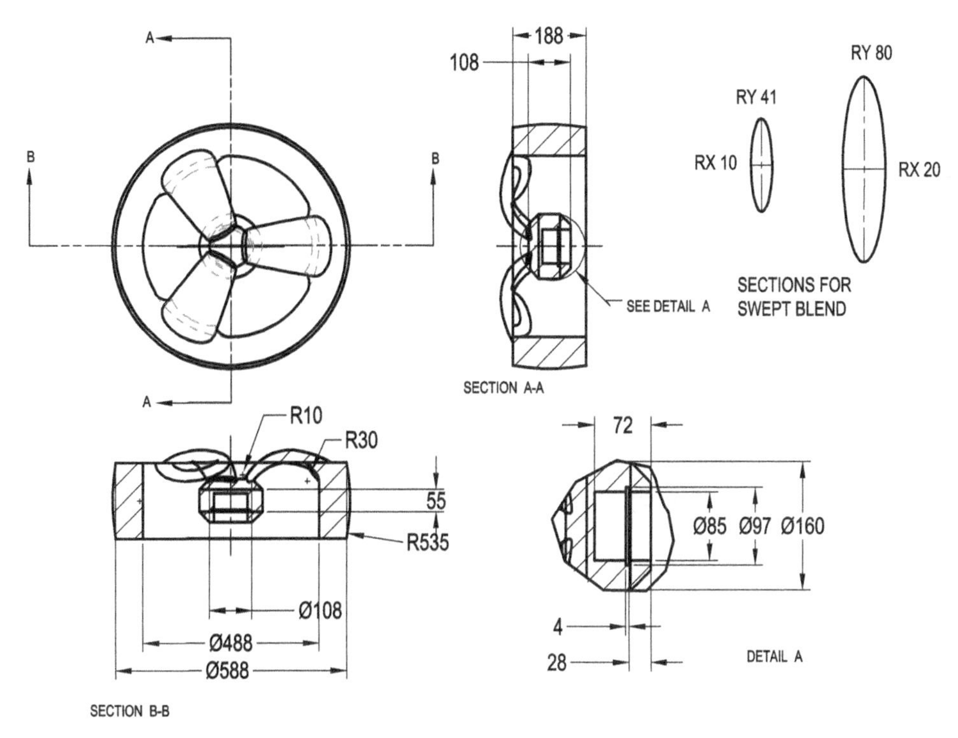

Figure 9-65 *Top view, sectioned front view, sectioned right view, detail view, and two blend sections with dimensions*

Examine the model and determine the number of features in it. The model is composed of four features, refer to Figure 9-64.

The following steps are required to complete this tutorial:

a. Create the base feature as a revolved feature, refer to Figures 9-66 and 9-67.
b. Create the swept blend feature in which the section is normal to the origin trajectory, refer to Figures 9-68 through 9-71.
c. Pattern the swept blend feature, refer to Figure 9-72.
d. Create the round of radius 30 and radius 10, refer to Figures 9-73 through 9-75, respectively.
e. Create the revolved cut, refer to Figures 9-76 and 9-77.

The working directory was already selected in Tutorial 1, and therefore, you do not need to select it again.

Starting a New Object File

1. Start a new part file and name it as *c09tut3*.

The three default datum planes and the **Model Tree** are displayed in the drawing area.

Creating the Base Feature

To create the sketch for the base feature, you first need to select the sketching plane. In this model, you need to draw the base feature on the **FRONT** datum plane.

1. Choose the **Revolve** tool from the **Shapes** group; the **Revolve** dashboard is displayed.
2. Select the **FRONT** datum plane as the sketching plane.
3. Select the **TOP** datum plane from the drawing area and then select the **Top** option from the **Orientation** drop-down list.
4. Once you enter the sketcher environment, create the sketch of the base feature and then apply constraints and dimensions to it, as shown in Figure 9-66. The sections must be closed and a geometry centerline must be drawn for revolution.
5. After the sketch is complete, choose the **OK** button to exit the sketcher environment; the **Revolve** dashboard is displayed.
6. Choose the **Build feature** button from the **Revolve** dashboard to exit it.

The base feature is completed. The default trimetric view of the feature is shown in Figure 9-67.

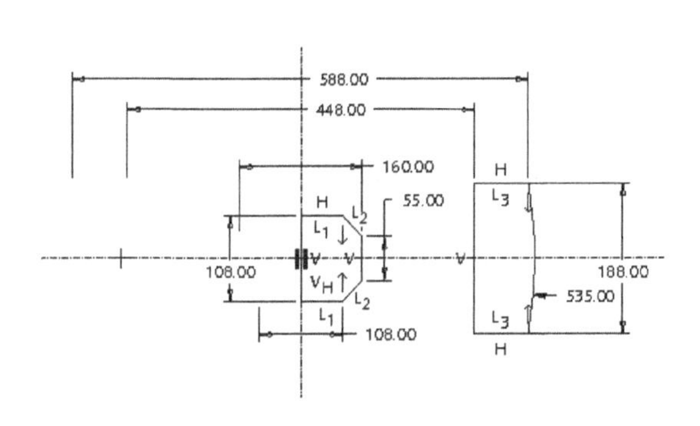

Figure 9-66 *Sketch with constraints and dimensions for the base feature*

Figure 9-67 *Default trimetric view of the base feature*

Creating the Trajectory for the Swept Blend

1. Choose the **Sketch** tool from the **Datum** group; the **Sketch** dialog box is displayed.
2. Select the **FRONT** datum plane as the sketching plane.
3. Next, select the **TOP** datum plane from the drawing area and then select the **TOP** option from the **Orientation** drop-down list.
4. Once you enter the sketcher environment, you need to create the sketch for the origin trajectory using the **Spline** tool and then apply constraints and dimensions to it, as shown

in Figure 9-68.

5. After the sketch is complete, choose the **OK** button to exit the sketcher environment.

Creating the Swept Blend Feature

The second feature of the model is a swept blend. This feature will require the trajectory which has been drawn previously.

1. Choose **Swept Blend** from the **Shapes** group; the **Swept Blend** dashboard is displayed.

2. Choose the **References** tab to invoke the slide-down panel. Select the origin trajectory that was drawn previously.

3. Now, choose the **Sections** tab to invoke the slide-down panel. Ensure that the **Sketched sections** option is selected in the slide-down panel. Select one of the end points on the trajectory and then choose the **Sketch** button from the **Sections** slide-down panel to enter the sketcher environment.

4. Draw the sketch for the first section and add dimensions to it, as shown in Figure 9-69.

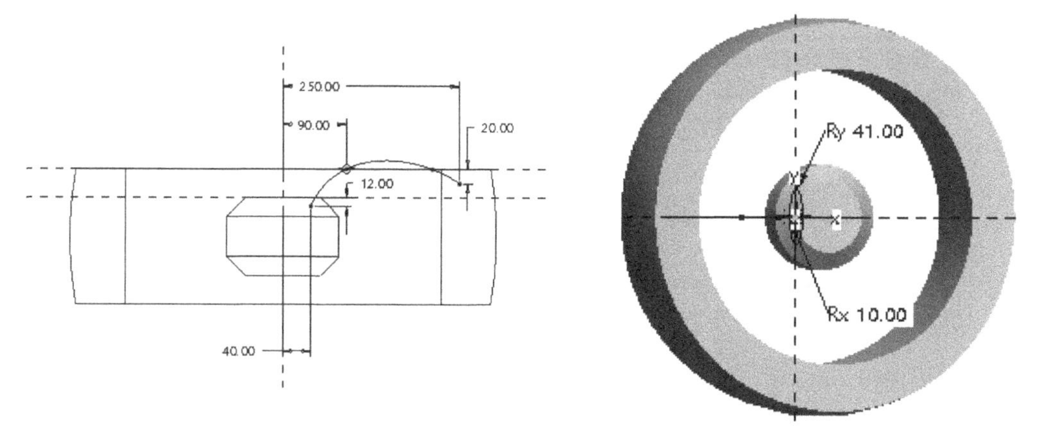

Figure 9-68 *Sketch with dimensions for the origin trajectory*

Figure 9-69 *Sketch for section 1 with dimensions*

5. After the sketch is complete, choose the **OK** button to exit the sketcher environment.

6. After you exit from the sketcher environment, the **Sections** tab is displayed again. Choose the **Insert** button from the slide-down panel. Now, select the other end point on the trajectory and choose the **Sketch** button to enter the sketcher environment.

7. Draw the sketch for the second section and add dimensions to it, as shown in Figure 9-70. The second section is also elliptical in shape. Now, choose the **OK** button to exit the sketcher environment.

8. After the sketch is complete, choose the **Create a Solid** button from the dashboard.

9. Choose the **Build feature** button from the dashboard. The default trimetric view of the swept blend feature with the base feature is shown in Figure 9-71.

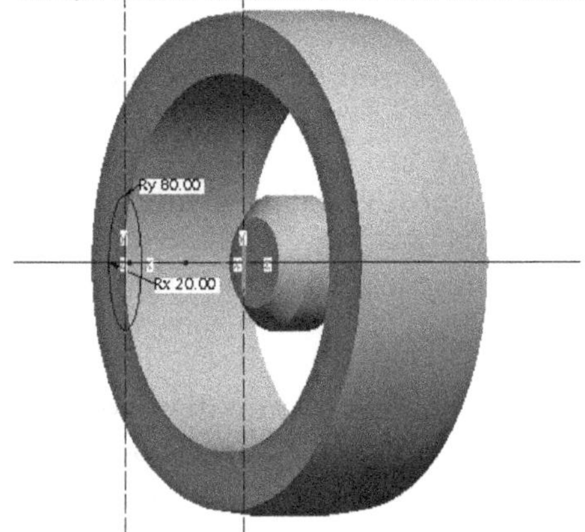

Figure 9-70 Sketch for section 2 with dimensions

Figure 9-71 The default trimetric view of the swept blend feature

Creating the Pattern of the Swept Blend Feature

Creating the swept blend features individually on the base feature will consume a lot of time. Therefore, you need to create a rotational pattern of the swept blend feature.

1. Select the swept blend feature from the **Model Tree** or from the model. Hold the right mouse button; a shortcut menu is displayed.

2. Choose the **Pattern** option from the shortcut menu; the **Pattern** dashboard is displayed and the dimensions are displayed on the selected feature.

3. Select the **Axis** option from the **Dimension** drop-down list; you are prompted to select an axis for creating the pattern feature. Select the axis of the base feature; the preview of the pattern feature is displayed in the drawing area.

4. Select the **General** option from the **Options** tab in the slide-down panel.

5. Enter **3** as the number of instances in the first direction and **120** as the angle between the pattern instances.

6. Choose the **Build feature** button from the dashboard; the rotational pattern is created, as shown in Figure 9-72.

Creating the Round Feature

The third feature to be created is a round feature of radius 30.

1. Choose the **Round** tool from the **Engineering** group; the **Round** dashboard is displayed.

2. Select the edges shown in Figure 9-73 to round. You may need to spin the model to select the inner edges of the swept blend feature. Remember that the first edge can be selected by using the left mouse button and the remaining edges can be selected by using the CTRL+left mouse button.

Figure 9-72 *Model after creating rotational pattern*

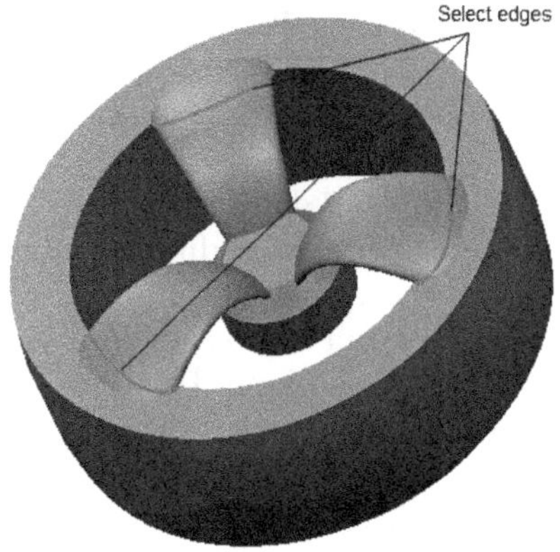

Figure 9-73 *Edges selected to round*

The preview of the round is displayed in the drawing area along with the default radius value.

3. Click on the radius value, enter **30** in the edit box that appears, and then press ENTER. Similarly, select upper edges and apply radius of a value **10**.

4. Choose the **Build feature** button from the **Round** dashboard.

The round feature is completed. The default trimetric view of the round feature is shown in Figure 9-74. Add another set of round of radius 10 as highlighted in Figure 9-75, using the same options.

Figure 9-74 *Model after creating rounds*

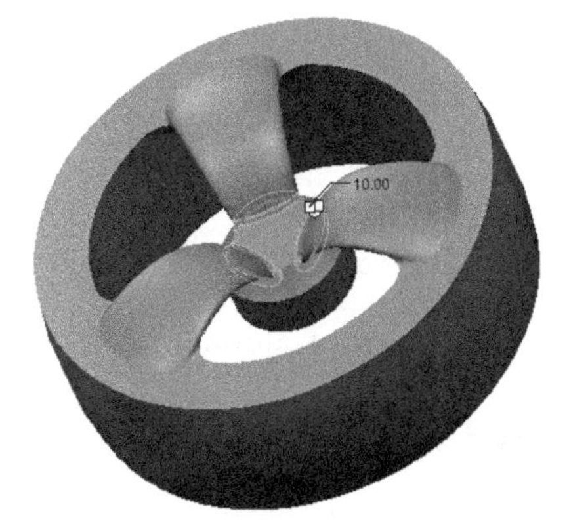

Figure 9-75 *Model highlighting the round feature of radius 10*

Creating the Revolved Cut Feature

You need to first select a sketching plane to draw the sketch of the cut feature and then specify the angle of revolution.

1. Choose the **Revolve** tool from the **Shapes** group; the **Revolve** dashboard is displayed. Next, choose the **Remove Material** button.

2. Select the **FRONT** datum plane as the sketching plane.

3. Select the **TOP** datum plane from the drawing area and then select the **Bottom** option from the **Orientation** drop-down list.

4. Once you enter the sketcher environment, draw the geometry centerline aligned to the **RIGHT** datum plane. Draw the sketch of the revolved cut feature and add dimensions and constraints to it, as shown in Figure 9-76.

5. After the sketch is complete, choose the **OK** button to exit the sketcher environment; the **Revolve** dashboard is enabled again.

6. Choose the **Build feature** button from the **Revolve** dashboard to exit it. The revolve cut feature is shown in Figure 9-77.

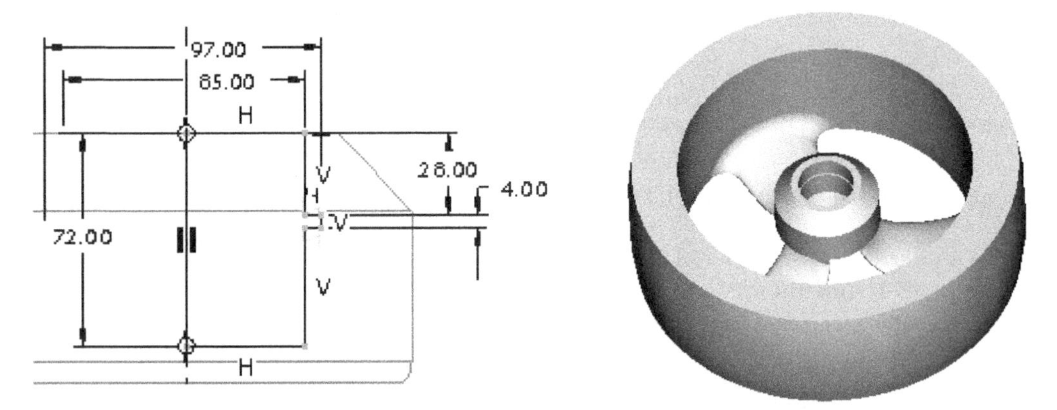

Figure 9-76 Sketch for the revolved cut feature *Figure 9-77 The model with the revolved cut feature*

Saving the Model

You need to save the model because you may need it later.

1. Choose the **Save** button from the **File** menu and save the model.

Tutorial 4

In this tutorial, you will create the spring shown in Figure 9-78. Figure 9-79 shows the front view of the spring with its dimensions. **(Expected time: 15 min)**

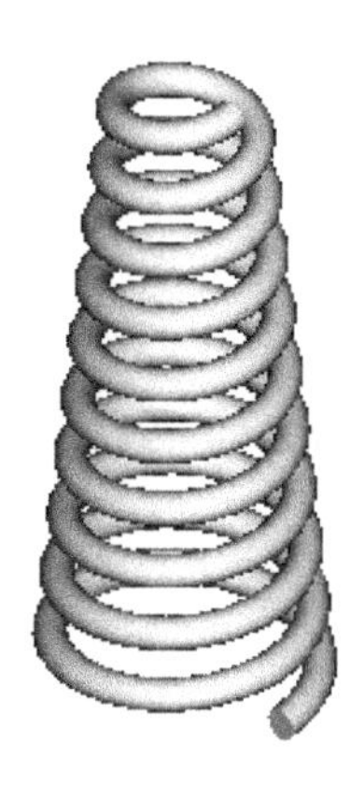

Figure 9-78 Isometric view of the spring

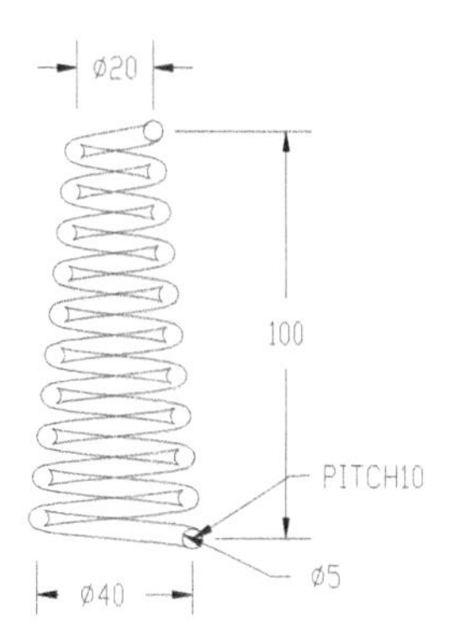

Figure 9-79 Front view with dimensions of the spring

Examine the spring and determine the specifications of the spring, refer to Figure 9-78. The following steps are required to complete this tutorial:

a. Draw the trajectory using the sketcher tools and apply dimensions to it, refer to Figures 9-79 and 9-80.
b. Specify the pitch of the spring.
c. Draw the section of the spring, refer to Figure 9-81.

Make sure that the required working directory is selected.

Starting a New Object File

1. Start a new part file and name it as *c09tut4*.

The three default datum planes and the **Model Tree** are displayed in the drawing area.

Creating the Helical Sweep Feature

The spring that you have to create is a right-handed spring of constant pitch.

1. Choose **Helical Sweep** from the **Sweep** drop-down of the **Shapes** group; the **Helical Sweep** dashboard is displayed. Select the **Through axis of revolution** radio button from the **References** tab.

2. Choose the **Use right handed rule** button from the dashboard.

3. Select the **Keep constant section** radio button from the **Options** tab.

4. Select the **Define** radio button from the **Reference** group; the **Sketch** dialog box is displayed and you are prompted to select a sketching plane.

5. Select the **FRONT** datum plane from the drawing area and then choose the **Sketch** button from the **Sketch** dialog box.

6. Once you enter the sketcher environment, draw the sketch of the trajectory and dimension it, as shown in Figure 9-80. As is evident from Figure 9-79, the endpoint of the trajectory is aligned to the **TOP** datum plane.

   You need to draw a geometry center line in the sketch about which the spring will be rotated. This is the axis of the spring.

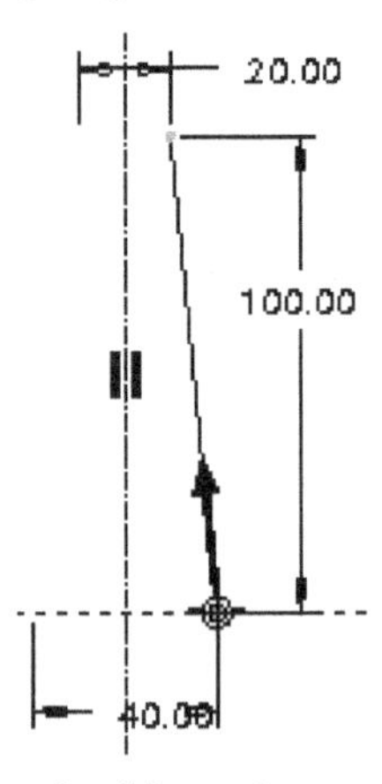

Figure 9-80 *Sketch of the trajectory with dimensions*

7. After you complete the sketch of the trajectory, choose the **OK** button from the dashboard; the **Helical Sweep** dashboard is displayed and a default pitch value of spring is displayed on the trajectory.

8. Enter **10** in the **Pitch** edit box and choose the **Create or edit sweep section** button from the dashboard. You will notice that two pink lines of infinite length crossing each other perpendicularly appear on the screen. The intersection point of these lines is the start point of the trajectory.

9. Draw the section of the spring such that the center of the circle coincides with the intersection of the two perpendicular lines. Then, dimension it, as shown in Figure 9-81.

10. After completing the sketch, choose the **OK** button to exit the sketcher environment.

11. Choose the **Build Feature** button from the **Helical Sweep** dashboard; the spring is created, as shown in Figure 9-82.

Saving the Model

1. Choose the **Save** button from the **File** menu and save the model.

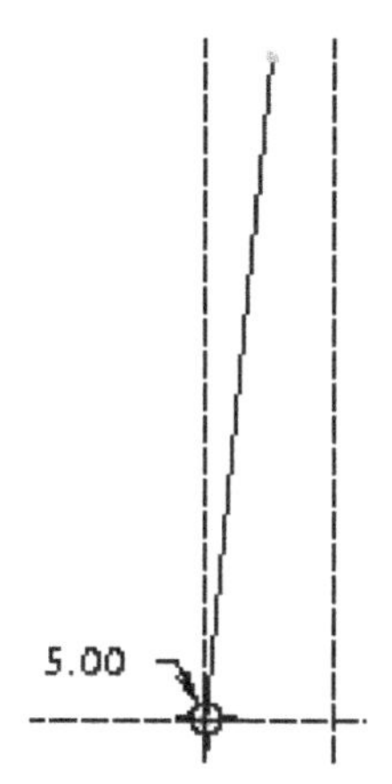

Figure 9-81 *Sketch of the section with dimensions*

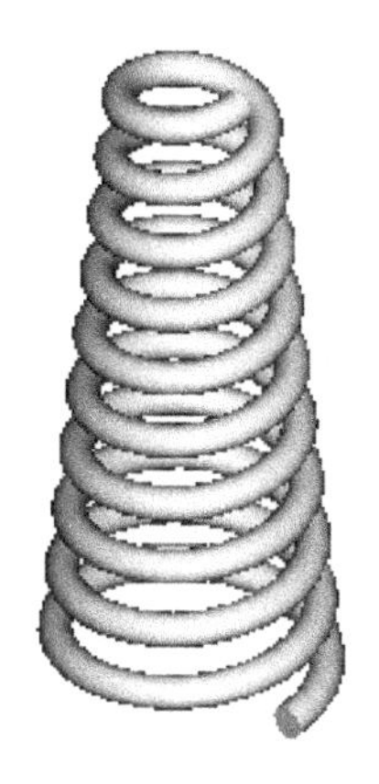

Figure 9-82 *Isometric view of the spring*

Self-Evaluation Test

Answer the following questions and then compare them to those given at the end of this chapter:

1. You can create helical sweep features with a constant as well as a variable pitch. (T/F)
2. The **Helical Sweep** option is used to create cut as well as protrusion features. (T/F)
3. You can create a helical feature having 40 units as pitch and 50 units as the diameter of the circular cross-section. (T/F)
4. You can add as well as delete points while specifying the pitch values during the creation of a variable pitch helical sweep. (T/F)
5. The **Sweep** option is available for both the protrusion and cut options. (T/F)
6. You can sketch as well as select existing sketches present in the drawing area to create a Swept Blend feature. (T/F)
7. While using the **Sweep** option, the contour of the sweep feature depends on the shape of the __________.
8. While using the **Sweep** option, when you draw the section for the sweep feature, the section should be __________ to the endpoints of the trajectories.
9. When you use the **Normal To Trajectory** option to create the variable sweep, the section is created__________ to the origin trajectory and is swept along the X trajectory.
10. The **Swept Blend** option is used to create a model that is a combination of __________ and __________ features.

Review Questions

Answer the following questions:

1. Which of the following dashboards is displayed when you choose the **Helical Sweep** option from the **Shapes** group in the **Ribbon**?

   (a) **Sweep** (b) **Revolve**
   (c) **Swept Blend** (d) None of these

2. Which of the following options is used to create a helical feature with constant pitch throughout the sweep?

   (a) **Right Handed** (b) **Constant**
   (c) **Thru Axis** (d) None of these

3. Which of the following options in the **Swept Blend** tool is used to create a sweep in which the swept section is normal to the trajectory defined as the origin trajectory?

   (a) **Constant Normal Direction** (b) **Normal To Projection**
   (c) **Normal To Trajectory** (d) None of these

4. You can use the **Helical Sweep** tool to create a spring design. (T/F)
5. The **Swept Blend** tool is also used to create a cut. (T/F)
6. The **Sweep** option can be used to create a sweep in which the section of the sweep is constant throughout the sweep. (T/F)
7. The X trajectory is used to guide the section of the sweep feature along it. (T/F)
8. The first trajectory chosen while using the **Sweep** tool is selected as the **Origin** trajectory. (T/F)
9. For creating a Swept Blend, there are three options available in **Tangency** tab. These options are **Free**, **Tangent**, and **Normal**. (T/F)
10. The **Helical Sweep** option is also used to create a cut. (T/F)

Exercises

Exercise 1

Create the model shown in Figure 9-83. Its front view is shown in Figure 9-84.

(Expected time: 15 min)

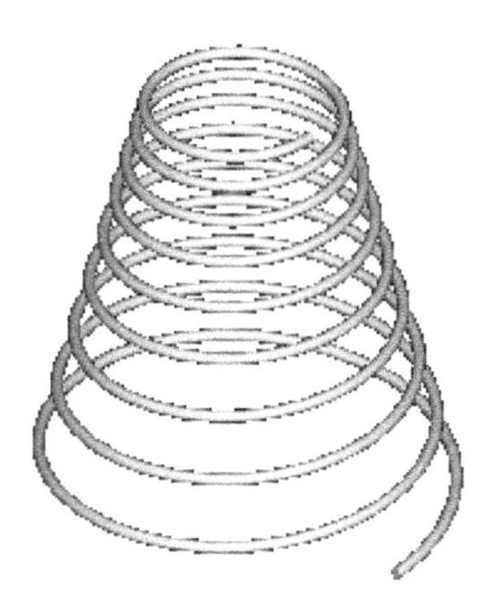

Figure 9-83 Isometric view of the spring

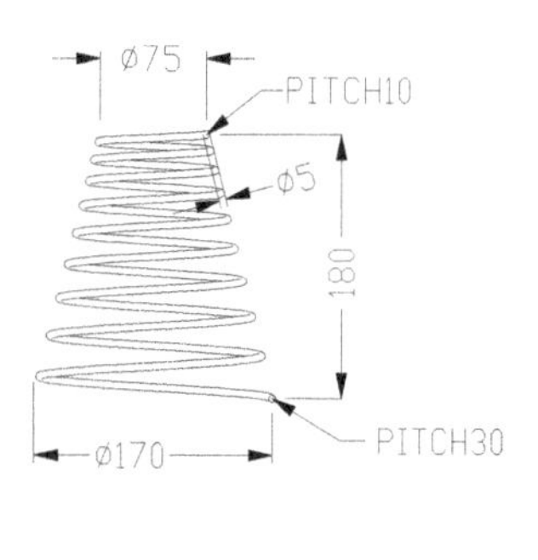

Figure 9-84 Front view of the spring

Exercise 2

Create the model shown in Figure 9-85. The front and top views of the model are shown in Figure 9-86. **(Expected time: 30 min)**

Hints to create the swept blend feature:

1. Select the **FRONT** datum plane as the sketching plane and select the top face of the base feature to be at the top while sketching.
2. Draw an arc of radius 50 and locate its center from the base feature.
3. Exit the sketcher environment by choosing the **OK** button.
4. Choose **Shapes > Swept Blend** from the **Ribbon**; the **Swept Blend** dashboard is displayed.
5. Choose the **Extrude as Solid** button from the dashboard.
6. Choose the **Reference** tab from the dashboard and select the arc drawn.
7. Choose the **Sections** tab and select the endpoint of the arc lying on the base feature, Choose the **Sketch** button from the **Sections** slide-down panel to enter the sketcher environment. Next Draw a rectangle of dimension 15 x 75. Exit the sketcher environment.
8. Choose the **Insert** button from the **Sections** slide-down panel and then select the other end point of the arc. Then, choose the **Sketch** button from the slide-down panel; this will again take you to the sketcher environment. Draw another rectangle with the dimensions 15 x 30. Exit the sketcher environment by choosing the **OK** button.

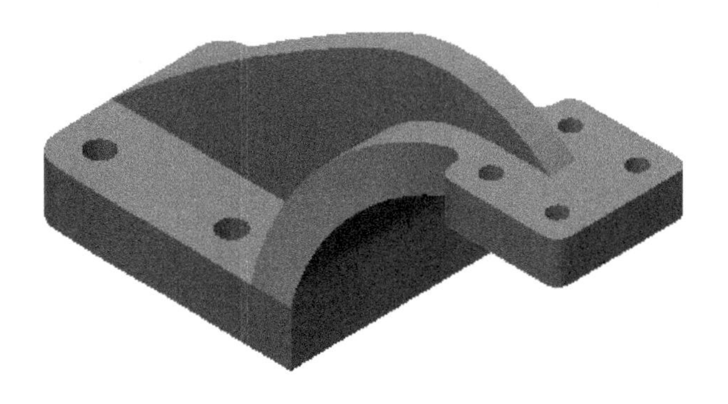

Figure 9-85 *Solid model for Exercise 2*

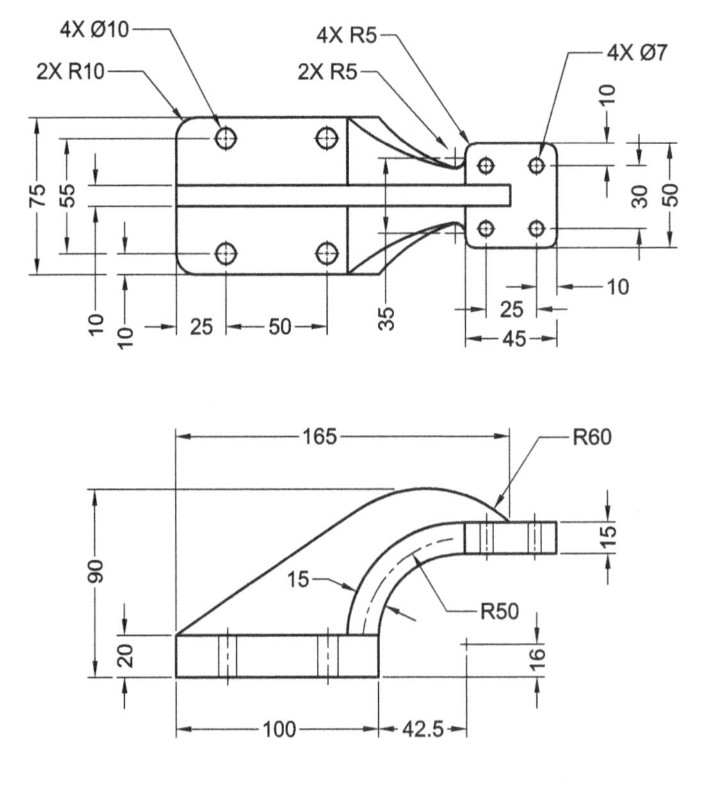

Figure 9-86 *Front and top views of the model*

Exercise 3

Create the model of the Helical Gear shown in Figure 9-87. Its views and dimensions are shown in Figure 9-88. **(Expected time: 45 min)**

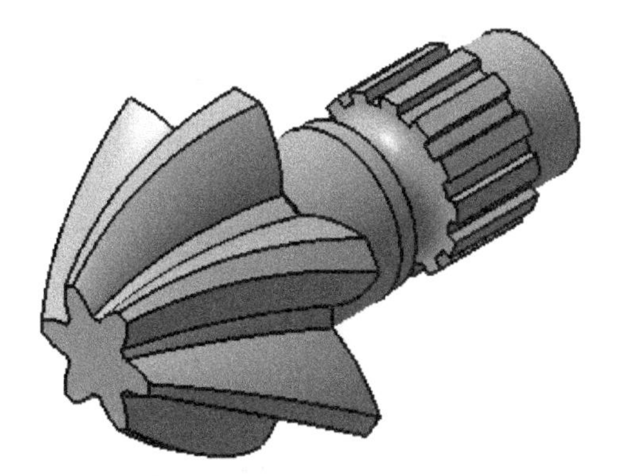

Figure 9-87 Model of the Helical Gear

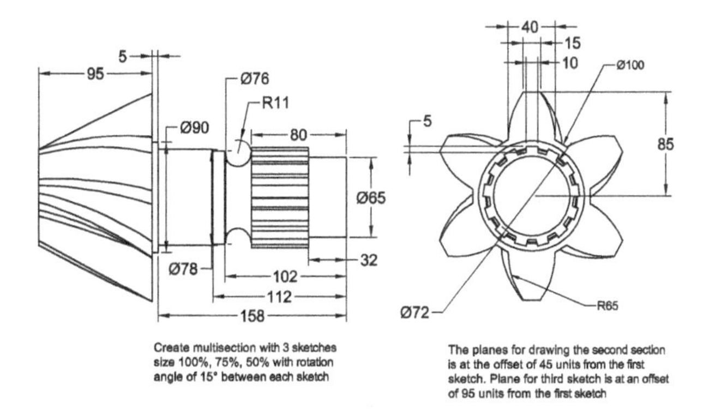

Figure 9-88 Views and Dimensions for the model

Hints to create a Helical Gear is discussed next:

1. Choose the **Blend** tool from the **Shapes** group of the **Model** tab in the **Ribbon** and draw the sketch, as shown in Figure 9-89. Next, choose **OK** to exit the sketching environment.

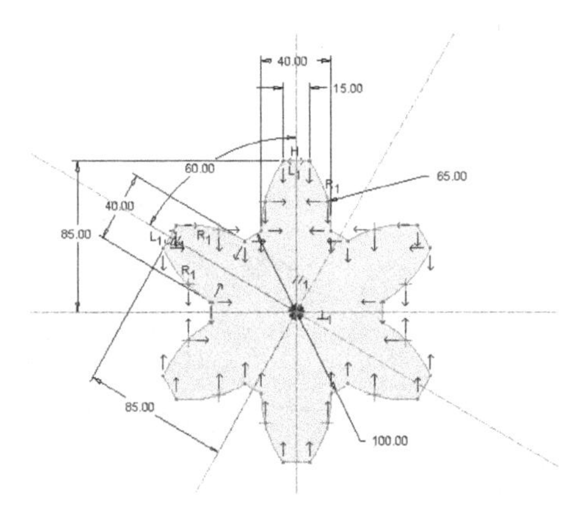

Figure 9-89 *Sketches for the Swept Blend feature*

2. Choose the **Sections** tab and select the **Sketched Sections** radio button from the **Sections** slide-down panel.
3. Now, choose the **Insert** button and enter **45** in the **Offset from** edit box for the second section and choose the **Sketch** button.
4. Next, choose the **Project** tool from the **Sketching** group and choose the **Loop** option from the **Type** menu. Next, select **Section 1** from the drawing area; the section will be projected.
5. Next, select the projected section and choose the **Rotate Resize** tool from the **Editing** group and scale the selected sketch by 75% and rotate it by 15 degrees.
6. Similarly, draw the third sketch at an offset distance of 50 and scale the sketch by 50% and rotate it by 15 degrees from the second sketch.
7. Now, choose the **Build Feature** button; a blend feature is created, as shown in Figure 9-90.

Figure 9-90 *Preview of the Blend feature*

Exercise 4

Create the model of the Hook shown in Figure 9-91. Its views and dimensions are shown in Figure 9-92. **(Expected time: 45 min)**

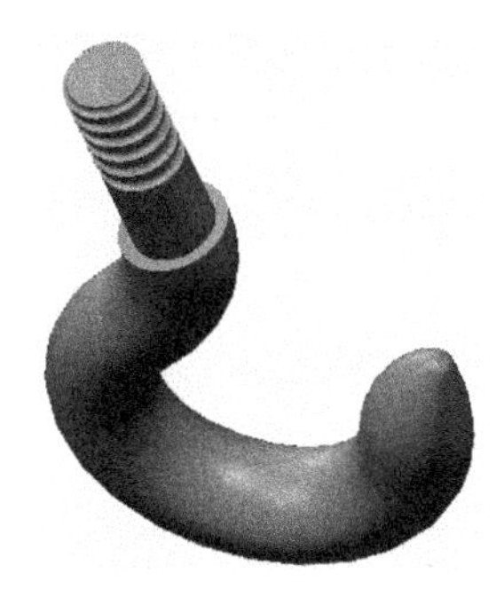

Figure 9-91 *The model of the Hook*

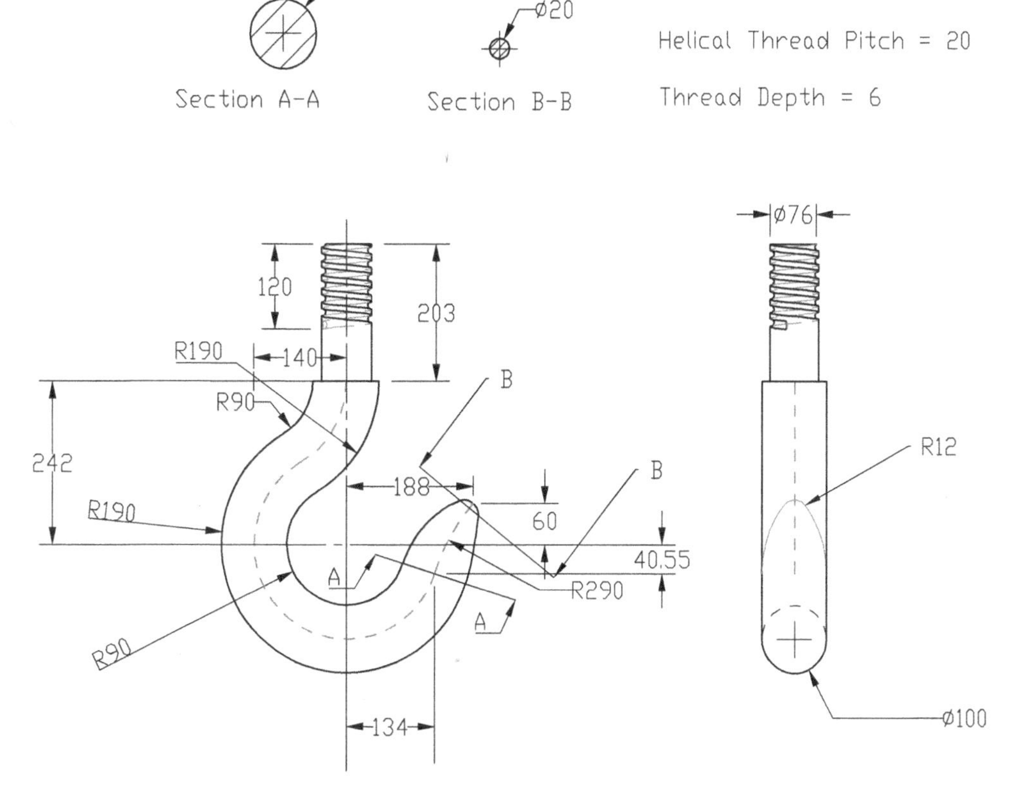

Figure 9-92 *Views and dimensions of the model*

Answers to Self-Evaluation Test

1. T, **2.** T, **3.** F, **4.** T, **5.** T, **6.** T, **7.** X trajectory, **8.** aligned, **9.** normal, **10.** sweep, blend

Chapter 10

Advanced Modeling Tools-III

Learning Objectives

After completing this chapter, you will be able to:

- *Create toroidal bend.*
- *Create spinal bend.*
- *Create warp transformation.*

ADVANCED MODELING TOOLS

In this chapter, you will learn to create models using advanced modeling tools such as **Toroidal Bend**, **Spinal Bend**, and **Warp**. These tools will help you analyze and create complex models with greater ease.

TOROIDAL BEND

Ribbon: Model > Engineering > Toroidal Bend

The **Toroidal Bend** tool is used to create features with curved surfaces, or the models which generally have a cut profile on a curved surface. For example, if you want to create a feature shown in Figure 10-1, you will first create a rectangular plate and cut the profile on it, as shown in Figure 10-2. Then, using the **Toroidal Bend** tool, you will bend the rectangular plate through an angle of 360 degrees, refer to Figure 10-1.

Figure 10-1 The plate after creating the toroidal bend

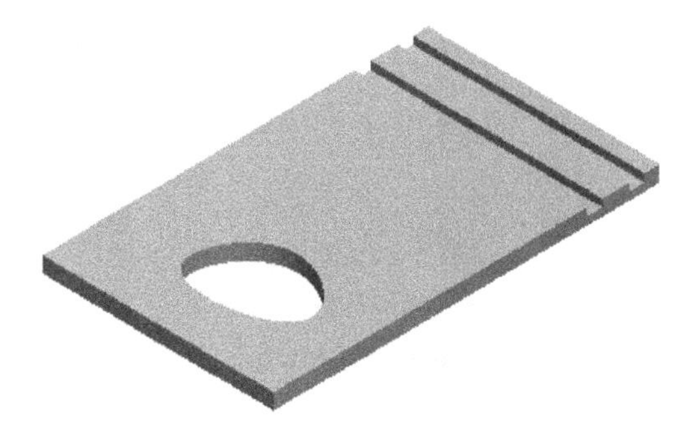

Figure 10-2 Rectangular plate with a cut profile

The following steps explain the procedure to create the model, refer to Figure 10-1. It is assumed that you have created the rectangular plate of dimensions 300 X 500 X 20, refer to Figure 10-2.

1. Choose the **Toroidal Bend** tool from the expanded **Engineering** group to invoke the **Toroidal Bend** option; the **Toroidal Bend** dashboard will be displayed, as shown in Figure 10-3. Next, choose the **References** tab from the dashboard; a slide-down panel will be displayed. Now, choose the **Define** button from the **Profile Section** area.

*Figure 10-3 Partial view of the **Toroidal Bend** dashboard*

2. Select the right face of the base feature (which is the longer face of the base plate) as the sketching plane and orient the sketch plane.

3. After entering the sketcher environment, draw the sketch, as shown in Figure 10-4. The sketch comprises of a line that is drawn from the left edge to the right edge. Also, a user-defined geometric coordinate system is placed at the midpoint of the line.

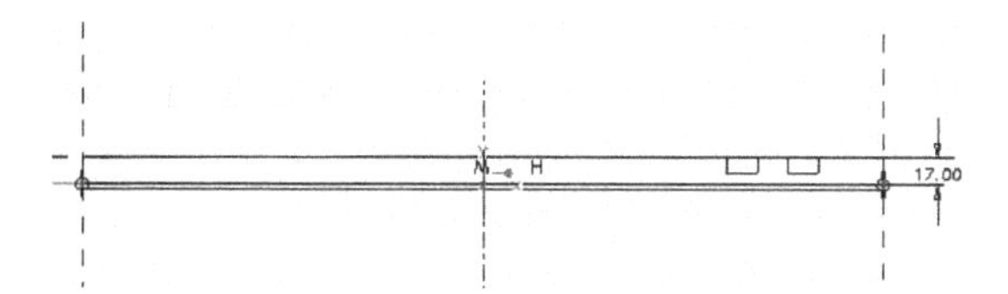

Figure 10-4 *Sketch along with the reference coordinate system*

4. Next, exit the sketcher environment.

Note

The sketch, shown in Figure 10-4, controls the bend profile of the ***Toroidal Bend*** *feature. If the sketch is a straight line, the side profile of the* ***Toroidal Bend*** *will be a line; else it will be an arc. The user-defined coordinate system acts as the neutral point in the bend.*

5. Choose the **References** tab; a slide-down panel will be displayed. Select the **Solid geometry** check box from it.

6. Next, choose the **Options** tab to display a slide-down panel. Select the **Standard** radio button, if it is not already selected.

7. Next, choose the **360 degrees Bend** option from the **Bend Radius** drop-down list; you will be prompted to select the two parallel planes to define the length of the bend.

8. Select the face that was selected as the sketching plane and then select the longer face which is parallel to it; the preview of the toroidal bend will be displayed.

9. Choose the **Build Feature** button from the dashboard; the **Toroidal Bend** will be created and the model similar to the one shown in Figure 10-1 will be displayed in the drawing area.

SPINAL BEND

Ribbon: Model > Engineering > Spinal Bend

The **Spinal Bend** tool is used to bend solids or quilts about a curved spine by continuously repositioning cross sections along the curve. Cross-sections, which are perpendicular to an axis, are repositioned perpendicular to the sketched spine without any distortion. Earlier, the **Spinal Bend** tool was used to bend entire the selected geometry equal to length of the spine but now you have an option to retain the original length of the geometry to be bent. To invoke the **Spinal Bend** dashboard, as shown in Figure 10-5, choose the **Spinal Bend** tool from the expanded the **Engineering** group.

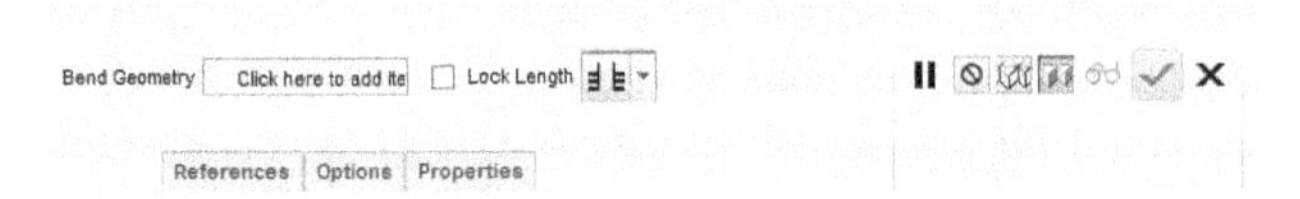

Figure 10-5 *Partial view of the* ***Spinal Bend*** *dashboard*

Figure 10-6 shows the base feature for creating the spinal bend. The procedure to create the model, shown in Figure 10-7, using the **Spinal Bend** tool is discussed next.

Note

It is assumed that you have created the base feature before invoking the ***Spinal Bend*** *tool. Also, a datum plane has to be created on the top face of the model.*

Figure 10-6 *Model for creating the spinal bend*

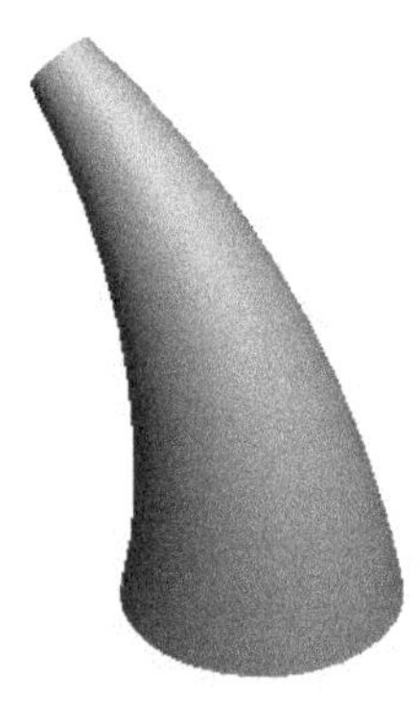

Figure 10-7 *Model showing the spinal bend*

1. Choose the **Spinal Bend** tool from the **Engineering** group to invoke the **Spinal Bend** dashboard and then click in the **Bend Geometry** collector of the dashboard.

2. Select the solid geometry from the drawing area. Next, choose the **References** tab; a slide-down panel will be displayed.

3. Click on the **Spine** collector in the slide-down panel; you will be prompted to select spine.

Note

Note that you have created the trajectory before selection. The trajectory sketched for the spinal bend should be tangent to the solid geometry, as shown in Figure 10-8.

4. Select the spine from the drawing area. Now, you can retain the original length of the selected geometry by selecting the **Lock Length** check box. You can also define the length of the selected geometry to be bent by selecting the required option from the **Define the extent of the selected geometry that will be bent** drop-down list of the dashboard.

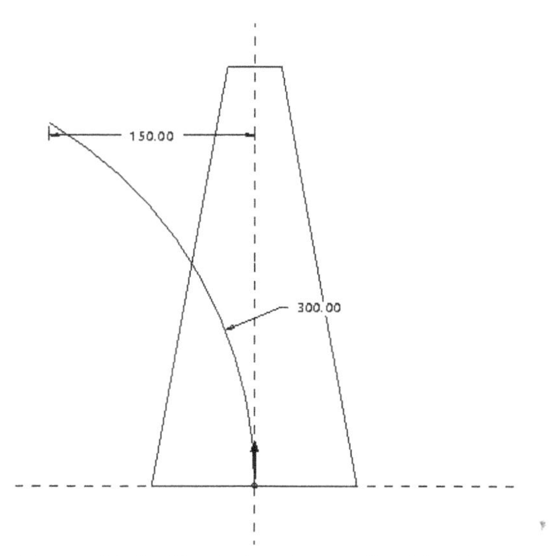

Figure 10-8 Sketched trajectory for spinal bend

WARP

Ribbon: Model > Editing > Warp

The **Warp** tool helps you study the design variations during the conceptual stage. With the **Warp** tool, you can visualize the form and shape of solids, quilts, facets, and curves. The **Warp** tool can be used only in the **Part** mode. Choose the **Warp** tool from the **Editing** group; the **Warp** dashboard will be displayed, as shown in Figure 10-9. The tabs and buttons of the **Warp** dashboard are discussed next.

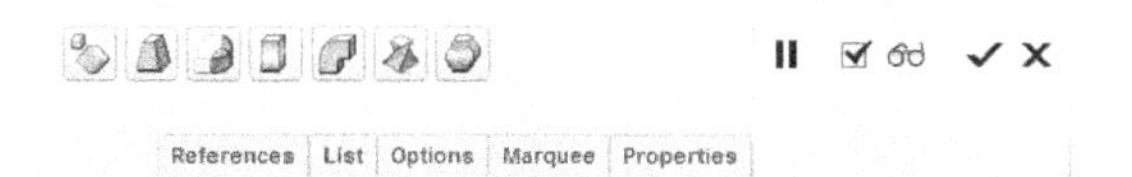

*Figure 10-9 Partial view of the **Warp** dashboard*

References Tab

When the **References** tab is chosen from the **Warp** dashboard, a slide-down panel will be displayed, as shown in Figure 10-10. The **References** area indicates the geometry and direction chosen to create the **Warp** transformation. If the **Hide Original** check box is selected, the original entity selected for the creation of the **Warp** transformation is not displayed in the drawing area. The **Copy Original** check box, if selected, will keep a copy of the original entity chosen for creating the **Warp** transformation.

List Tab

On choosing the **List** tab, a slide-down panel is displayed showng the list of all the **Warp** transformations performed on the selected entity. You can select the required operation from the slide-down panel of the **List** tab, and then edit, preview, or delete it. The **List** slide-down panel is shown in Figure 10-11.

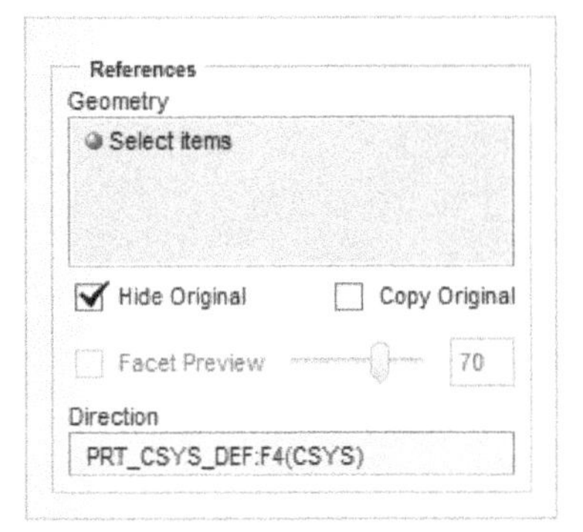

*Figure 10-10 The **References** slide-down panel*

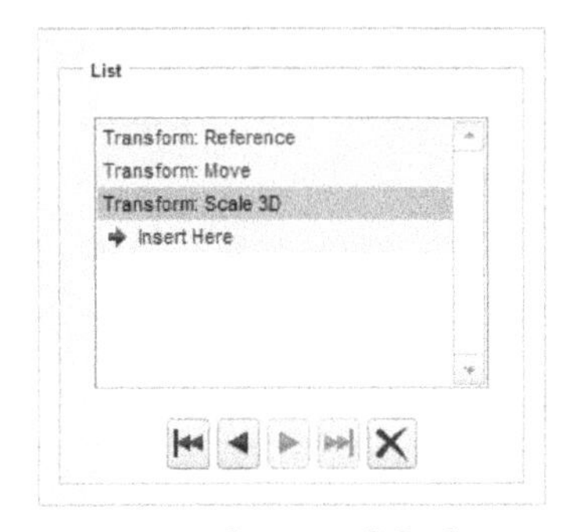

*Figure 10-11 The **List** slide-down panel*

Options Tab

The **Options** tab is used to input various parameters for the **Warp** transformation. The slide-down panel, which will be displayed when the **Options** tab is chosen after invoking the **Transform** tool, is shown in Figure 10-12. The uses of **Transform** tool will be discussed later in this chapter.

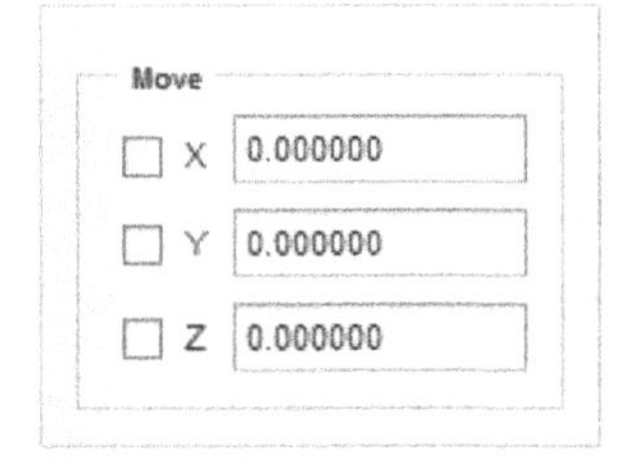

*Figure 10-12 The **Move** area of the **Options** slide-down panel*

This slide-down panel lists the various parameters which control a particular **Warp** transformation. In this slide-down panel, you can enter the values through which the geometry has to be transformed in the X, Y, and Z axes. Similarly, the various parameters which control the particular **Warp** transformation can be edited using the **Options** tab.

Marquee Tab

The **Marquee** tab is used to create the **Warp** transformation. When the **Marquee** tab is chosen after invoking the **Spine** tool, the slide-down panel will be displayed, as shown in Figure 10-13. The **Marquee** tab provides you with the options to control the warping operation of an entity through a curve or a spine sketched before the **Warp** transformation is performed.

You can choose points from the grid in the drawing area and then drag them to get the required shape. Alternatively, you can enter the values for creating the **Warp** transformation in the slide-down panel of the **Marquee** tab.

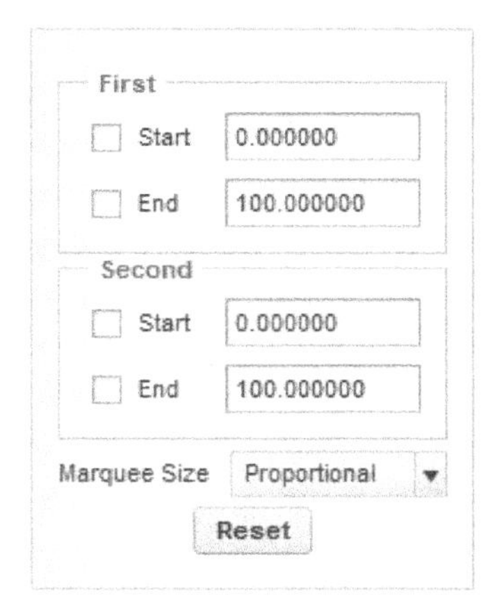

***Figure 10-13** The **Marquee Size** slide-down panel*

Transform Tool

The **Transform** tool can be used for various transformation operations such as translating, rotating, and scaling a geometrical entity. The translate option translates the geometry in the required direction. The rotate option rotates the geometry based on the position of the **Jack** (a blue color coordinate system which appears at the center of the drawing area). The scale option scales the geometry to the required scale factor.

The following steps explain the procedure to translate a geometry using the **Transform** tool, refer to Figure 10-14.

1. Choose the **Warp** tool from the expanded **Editing** group to invoke the **Warp** dashboard.

2. Choose the **References** tab to display the slide-down panel. Select the solid geometry from the drawing area. Now, click once in the **Direction** collector and select the **FRONT** datum plane as the direction reference.

3. Choose the **Transform** tool from the **Warp** dashboard.

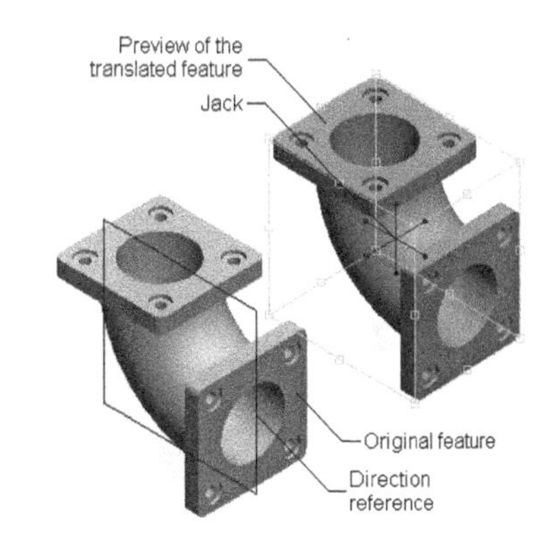

***Figure 10-14** The preview of the translated feature along with the original feature*

4. Click anywhere in the drawing area and drag the mouse to place the geometry at the desired location. Choose the **Build feature** button from the **Warp** dashboard to complete the **Warp** transformation.

The following steps explain the procedure to rotate geometry using the **Transform** tool.

1. Choose the **Warp** tool from the expanded **Editing** group to invoke the **Warp** dashboard.

2. Choose the **References** tab to display the slide-down panel. Select the solid geometry from the drawing area. Now, click once in the **Direction** collector and select the **FRONT** datum plane as the direction reference.

3. Choose the **Transform** tool from the **Warp** dashboard; the **Jack** will be displayed at the center of the drawing area.

4. Click on the handles of the **Jack** and drag the mouse to rotate the geometry to the required orientation.

5. Alternatively, click on the handles of the **Jack** and then choose the **Options** tab from the **Warp** dashboard. Enter the angle of rotation in the **Angle** edit box and choose the **Build feature** button from the **Warp** dashboard to complete the **Warp** transformation.

   Figure 10-15 shows the model before rotating the geometry and Figure 10-16 shows the model after rotating the geometry.

Note

*The **Jack** can be moved independent of the solid geometry. To do so, select **Jack** in the **Transform** drop-down list in the dashboard. Next, click on the **Jack** in the drawing area and drag to place it at the required location. Right-click to invoke the shortcut menu. Choose the **Center Jack** option from the shortcut menu; click and drag on the active handles of the **Jack** to rotate the geometry to the required orientation.*

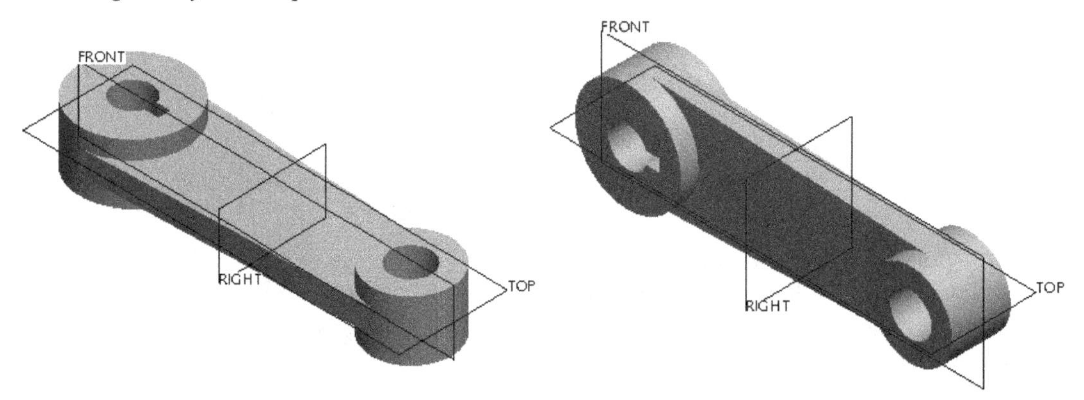

Figure 10-15 *Model before rotating the geometry* ***Figure 10-16*** *Model after rotating the geometry*

The following steps explain the procedure to scale a geometry using the **Transform** tool.

1. Choose the **Warp** tool from the **Editing** group to invoke the **Warp** dashboard.

2. Click on the **References** tab to display the slide-down panel. Select the solid geometry from the drawing area. Now, click once in the **Direction** collector and the **TOP** datum plane as the direction reference.

3. Choose the **Transform** tool from the **Warp** dashboard.

4. Click on any corner of the marquee and choose the **Options** tab from the **Warp** dashboard; the **Options** slide-down panel will be displayed. The **Scale** area of the slide-down panel is displayed, as shown in Figure 10-17.

5. Enter the scale factor in the **Scale** edit box. The two options in the **Toward** drop-down list are **Opposite** and **Center**. The **Opposite** option scales the corner opposite to the selected reference corner and the **Center** option scales the geometry with respect to the center of the geometry.

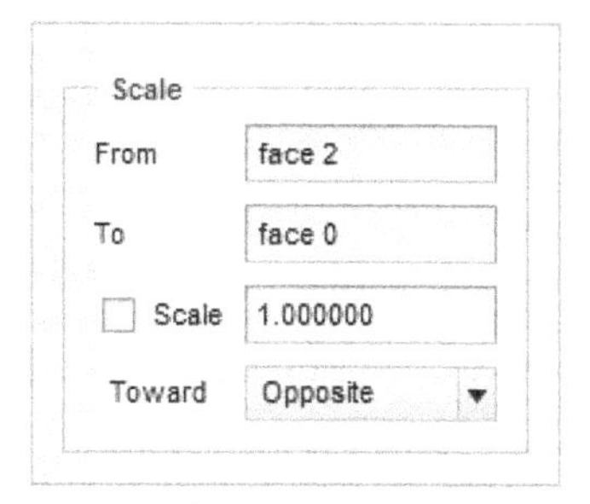

***Figure 10-17** The **Scale** area of the **Options** slide-down panel*

Figure 10-18 shows a scaled feature previewed along with the original feature.

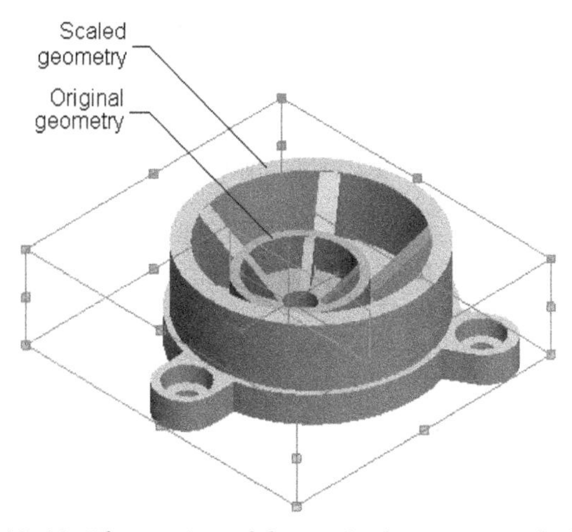

***Figure 10-18** The preview of the original geometry with the scaled geometry*

Warp Tool

The **Warp** tool can be used for changing the shape of the geometry. Some of the operations that are possible using the **Warp** tool are tapering the top or the base of the geometry, or dragging a corner or an edge toward or away from the center.

The following steps explain the procedure to apply wrap transformation using the **Warp** tool.

1. Choose the **Warp** tool from the expanded **Editing** group to invoke the **Warp** dashboard.

2. Choose the **References** tab to display a slide-down panel. Select the solid geometry from the drawing area. Now, click once in the **Direction** collector and select the **TOP** datum plane as the direction reference.

3. Choose the **Warp** tool from the **Warp** dashboard.

4. Move the cursor toward one of the marquee points of the solid geometry to display the arrows. Click on the arrow that is perpendicular to the edge and then choose the **Options** tab from the **Warp** dashboard; the **Options** slide-down panel will be displayed. The **Edge** area of the slide-down panel is displayed, as shown in Figure 10-19.

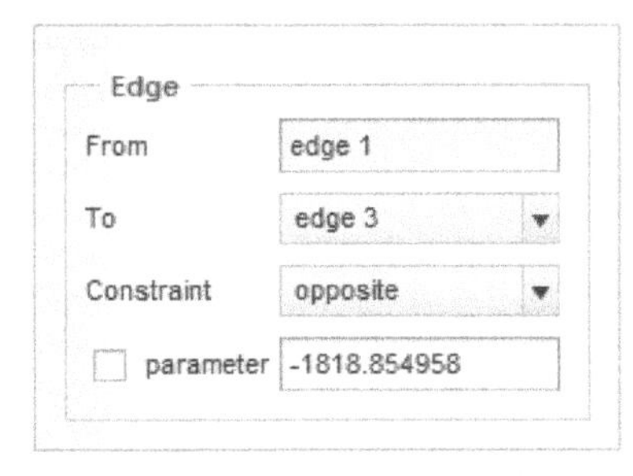

Figure 10-19 *The **Edge** area of the **Options** slide-down panel*

5. Select the **center** option from the **Constraint** drop-down list.

6. Enter the warping length in the **parameter** edit box; the preview of the warped feature appears in the drawing area. Choose the **Build feature** button from the **Warp** dashboard to complete the feature creation.

7. Alternatively, move the cursor to any of the control points to display the arrows. Select one of the arrows and drag it when the arrow on the control point changes its color to green.

   Figure 10-20 shows the model for applying the warp transformation and Figure 10-21 shows the model after applying the warp transformation.

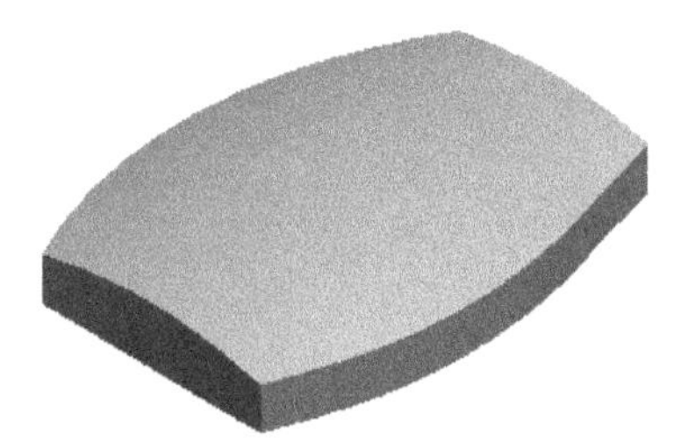

Figure 10-20 *Model for applying the warp transformation*

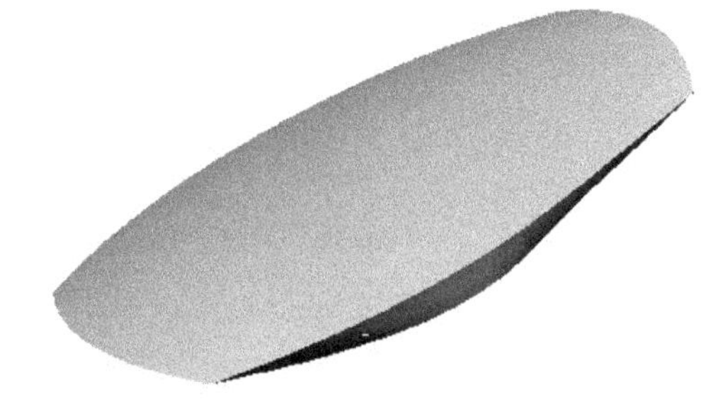

Figure 10-21 *Model after applying the warp transformation*

Note

When you move the cursor onto a control point of the marquee, six arrows will be displayed; three in purple color and the other three in orange color. The arrows in orange are used to warp the model along the selected edge and the arrows in purple are used to warp the model by moving the selected corner within the selected marquee face. In this process, the other two corners follow symmetricity to maintain the original corner angles. If you do not require the other two corners to follow symmetricity, press and hold the ALT key and then drag the selected corner.

If you press and hold the SHIFT and ALT keys together while dragging the marquee points, the warp will take place symmetrically from both the corners.

Spine Tool

The **Spine** tool is used to reshape the geometry by manipulating the defining points of a curve (also referred to as spine). The transformation using the **Spine** tool can be created in the linear or radial direction.

The following steps explain the procedure to apply a warp transformation using the **Spine** tool.

1. The transformation using the **Spine** tool will be applied with reference to a curve (also referred as a spine). You need to create a curve before invoking the **Spine** tool. The transformation will then be applied by dragging the control points of the spine to the required location.

Note

Remember that the spine you create needs to be sketched parallel to the plane on which you apply the warp transformation.

2. Choose **Warp** from the expanded **Editing** group to invoke the **Warp** dashboard.

3. Choose the **References** tab to display the slide-down panel. Select the solid geometry from the drawing area. Now, click once in the **Direction** collector and select the plane, which was the sketching plane for the spine, as the direction reference.

4. Choose the **Spine** tool from the **Warp** dashboard; the **Reference** slide-down panel will be displayed, as shown in Figure 10-22. Now, click once in the **Spine** collector and select the sketched spine from the drawing area.

5. Click and drag any of the control points displayed on the curve (spine) to the desired location. Choose the **Build feature** button from the **Warp** dashboard to complete the creation of the feature.

*Figure 10-22 The **References** slide-down panel*

Stretch Tool

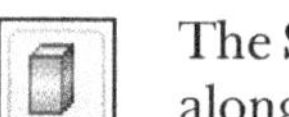

The **Stretch** tool can be used to stretch geometry along the required direction. The range and scale of stretch can be controlled using the various options in the **Options** tab of the **Warp** dashboard.

The following steps explain the procedure to apply a warp transformation using the **Stretch** tool.

1. Choose **Warp** from the expanded **Editing** group to invoke the **Warp** dashboard.

2. Choose the **References** tab to display a slide-down panel. Select the solid geometry from

the drawing area. Next, click once in the **Direction** collector and select the **FRONT** datum plane as the direction reference.

3. Choose the **Stretch** tool from the **Warp** dashboard. You can apply the transformation by moving the cursor onto the small white rectangle which will be displayed in the drawing area and then dragging it to the required location.

4. Drag the rectangle to stretch the model. Alternatively, enter a desired value of stretch in the **Scale** edit box; the model will be stretched in the direction of axis displayed and the preview of the stretched feature will be displayed in the drawing area.

5. To stretch the model in the other direction, choose the **Switch to next axis** button from the dashboard; the axis and a small white rectangle will be displayed in the other direction. Now, you can drag the rectangle to stretch the model in that direction. You can reverse the direction of stretch by choosing the **Reverse axis direction** button.

6. To set the length of the axis, invoke the **Marquee** tab; a slide-down panel will be displayed, as shown in Figure 10-23. Enter the start and end points of the axis in the **Start** and **End** edit boxes, respectively.

7. Choose the **Build feature** button from the **Warp** dashboard to complete the creation of the feature.

Figure 10-24 shows the model before the stretching transformation and Figure 10-25 shows the model after the stretching transformation.

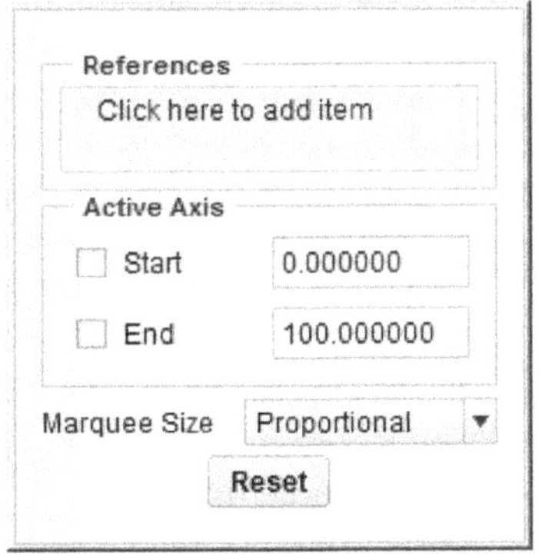

Figure 10-23 *The* ***Marquee*** *slide-down panel*

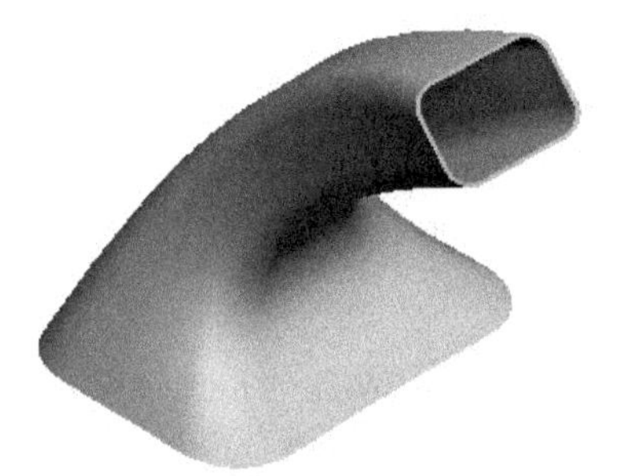

Figure 10-24 *Model before stretching*

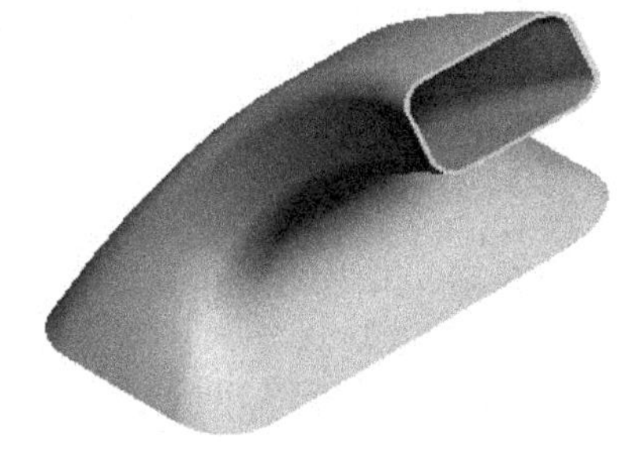

Figure 10-25 *Model after stretching*

Note

The ***Proportional*** *marquee size is selected by default in the* ***Marquee*** *slide-down panel, thereby stretching the model relative to its previous position. You can select* ***Absolute*** *to stretch the model to the specified length. To specify the stretch from an offset distance, select the* ***Offset*** *marquee size.*

Bend Tool

The **Bend** tool is used to bend the geometry to the required orientation. The following steps explain the procedure to create a bend using the **Bend** tool.

1. Choose the **Warp** tool from the expanded **Editing** group to invoke the **Warp** dashboard.

2. Choose the **References** tab to display the slide-down panel. Select the solid geometry from the drawing area. Now, click in the **Direction** collector once and select the **FRONT** datum plane as the direction reference.

3. Choose the **Bend** tool from the **Warp** dashboard. The model will be displayed in the drawing area with marquee and a orange line with a small rectangle attached at one end.

4. Move the cursor onto the small white rectangle and drag it until the desired orientation is achieved. Alternatively, enter the desired orientation value in the **Angle** edit box available in the dashboard; the preview of the bend will be displayed in the model area.

5. Next, choose the **Options** tab; a slide-down panel will be displayed, as shown in Figure 10-26.

   The **Pivot** edit box in the **Bend** area of this panel is used to specify the fixed point of the bend and the **Tilt** edit box is used to specify the angle of the bend. Enter the desired values in the respective edit boxes. You can also set the range of bend, similar to stretch, by invoking the **Marquee** tab in the dashboard.

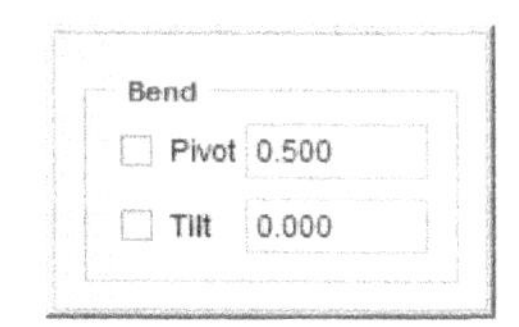

Figure 10-26 *The* ***Bend*** *area of the* ***Options*** *slide-down panel*

Figure 10-27 shows the geometry before creating the bend feature and Figure 10-28 shows the geometry after creating the bend feature.

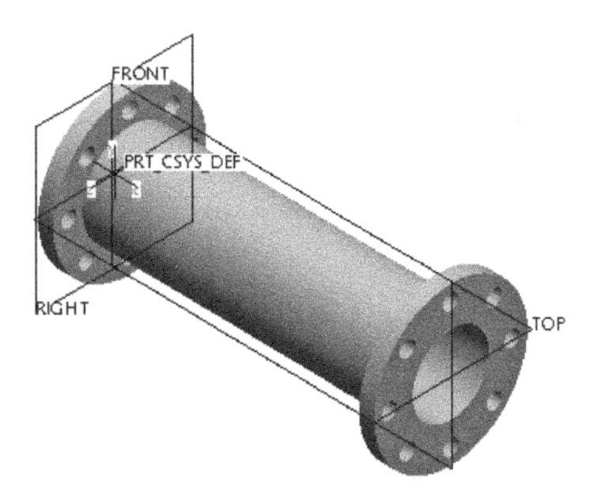

Figure 10-27 *Model before bending*

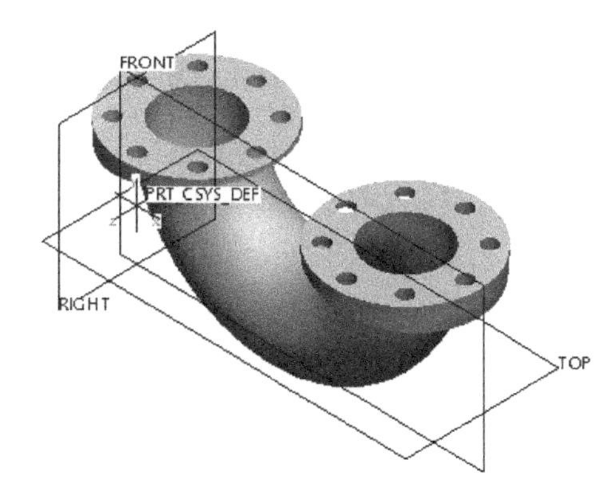

Figure 10-28 *Model after bending*

The bend operation can also be controlled by using the buttons in the **Warp** dashboard.

Switch to next axis button

 This button switches the axis of bend from one plane to another.

Reverse axis direction button

 This button reverses the direction of the axis of the bend.

Increase tilt by 90 degrees button

 This button increases the tilt or the bending angle by 90 degrees from the earlier applied bend angle.

Twist Tool

 The **Twist** tool is used to twist the geometry around an axis by the user specified angle.

The following steps explain the procedure to twist a geometry using the **Twist** tool.

1. Choose the **Warp** tool from the expanded **Editing** group to invoke the **Warp** dashboard.

2. Choose the **References** tab to display a slide-down panel. Select the solid geometry from the drawing area. Now, click once in the **Direction** collector and select the **TOP** datum plane as the direction reference.

3. Choose the **Twist** tool from the **Warp** dashboard; the model will be displayed in the drawing area along with the marquee and a small white rectangle.

4. Move the cursor onto the small white rectangle and drag to twist the model to the required orientation. Alternatively, enter the angle of twist in the **Angle** edit box in the dashboard; the preview of twist is displayed in the drawing area.

5. You can define the range of the twist feature similar to **Stretch** and **Bend**. To do so, choose the **Marquee** tab; a slide-down panel will be displayed, as shown in Figure 10-29. The **Start** and **End** edit boxes in this slide down panel are used to define the range of twist.

6. Choose the **Build feature** button from the **Warp** dashboard to complete the creation of the feature.

   Figures 10-30 shows the model before twisting and Figure 10-31 shows the model after twisting.

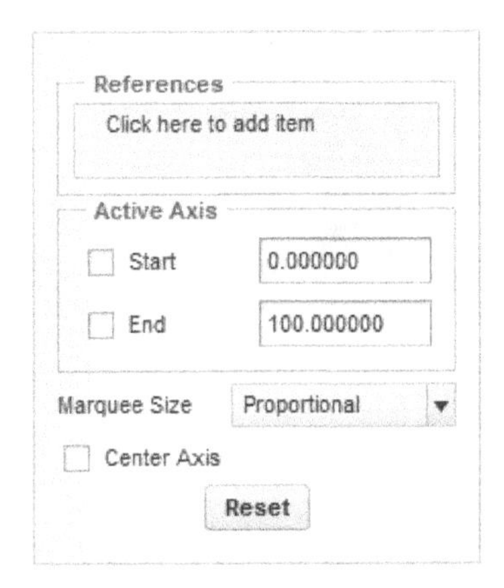

Figure 10-29 *The* ***Marquee*** *slide-down panel*

Note

*The direction of twist and the axis of twist can be reversed or switched by using the **Reverse axis direction** and **Switch to next axis** buttons, respectively in the **Warp** dashboard.*

Figure 10-30 *Model before twisting*

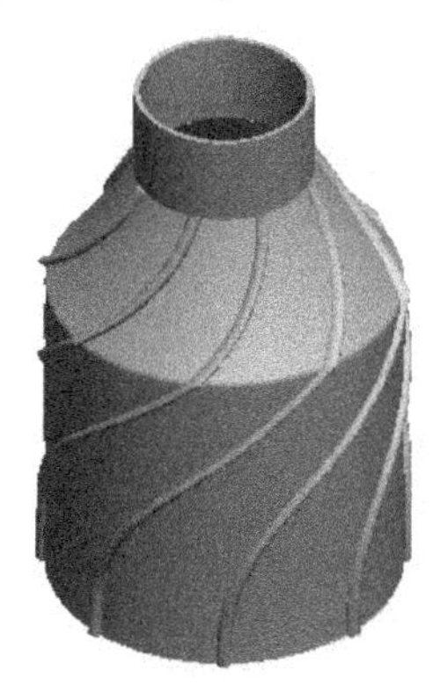

Figure 10-31 *Model after twisting*

Sculpt Tool

The **Sculpt** tool allows you to deform the geometry through a mesh. The following steps explain the procedure to create a sculpt transformation.

1. Choose the **Warp** tool from the expanded **Editing** group to invoke the **Warp** dashboard.

2. Click on the **References** tab to display a slide-down panel. Select the solid geometry from the drawing area. Now, click once in the **Direction** collector and select the TOP datum plane as the direction reference.

3. Choose the **Sculpt** tool from the **Warp** dashboard.

4. Choose the **Options** tab from the **Warp** dashboard; a slide-down panel will be displayed, as shown in Figure 10-32.

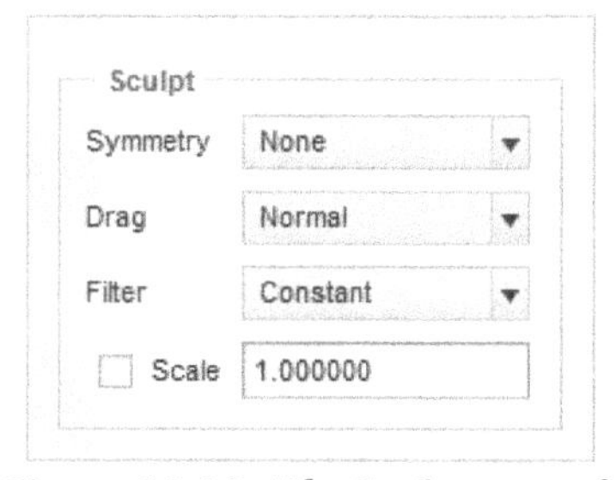

Figure 10-32 *The **Sculpt** area of the **Options** slide-down panel*

5. Select the type of symmetry from the **Symmetry** drop-down list.

Note

*The **Symmetry** drop-down list specifies the direction in which the selected group of points move while sculpting. The **None** option is selected by default in the **Symmetry** drop-down list. If the **Horizontal** option is selected from the drop-down list, then the points move symmetric to the centerline of the face parallel to the selected reference plane.*

6. Select the type of drag from the **Drag** drop-down list.

Note

*The **Drag** drop-down list specifies the constraints on the group of points being sculpted. The **Normal** option is selected by default in this drop-down list so the points move normal to the selected face when the transformation is created. If the **Free** option is selected from the **Drag** drop-down list, then the points selected for sculpting move freely in the drawing area. The **Along Row/Column** option will be used to move the points selected for sculpting towards the neighboring rows or columns.*

7. Select the type of filter required from the **Filter** drop-down list.

Note

*The **Filter** drop-down list specifies the behavior of the points being sculpted. The **Constant** filter is selected by default so the points selected for sculpting move to the same distance as the dragged point. If the **Linear** filter type is selected, then the selected points drop-off linearly with respect to the dragged point. The **Smooth** filter drops off the selected points smoothly with respect to the dragged point.*

8. Select one or more curves or points from the marquee displayed (hold down the CTRL button for multiple point selections). Drag the selected points to the required location. Choose the **Build feature** button from the **Warp** dashboard to complete the **Warp** transformation.

   Figure 10-33 shows the model before applying the **Sculpt** transformation and Figure 10-34 shows the model after applying the **Sculpt** transformation.

Figure 10-33 *Model before applying the **Sculpt** transformation*

Figure 10-34 *Model after applying the **Sculpt** transformation*

You may also specify how the geometry is sculpted by choosing the buttons that will become available when the **Sculpt** tool is chosen in the **Warp** dashboard. These buttons are discussed next.

Switch mesh orientation to the next marquee face

This button helps you to align the mesh in the direction in which the geometry will be sculpted.

Apply to one side of selected items

This button sculpts the selected face in only one direction and is chosen by default.

Apply to both sides of selected items

This button applies the motion of selected face to the opposite face.

Apply symmetrically to both sides of selected items

This button, if chosen, allows the geometry to be symmetric from its center.

Note

The number of rows and columns in the marquee can be controlled by specifying the required value in the ***Rows*** *and* ***Columns*** *edit boxes, respectively in the dashboard.*

TUTORIALS

Tutorial 1

In this tutorial, you will create the model shown in Figure 10-35. **(Expected time: 30 min)**

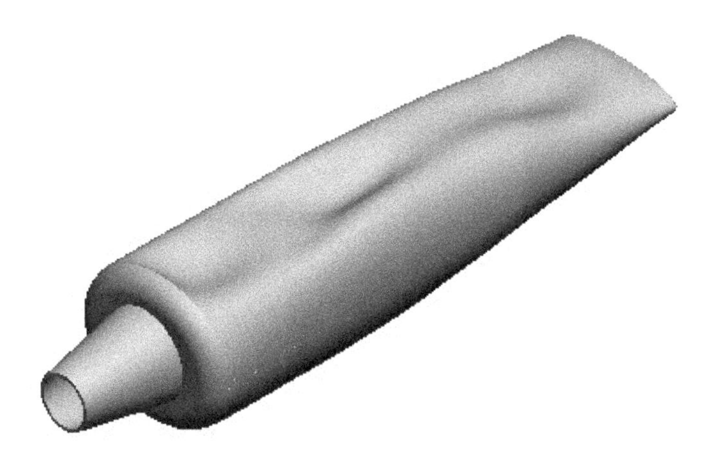

Figure 10-35 *Model for Tutorial 1*

The following steps are required to complete this tutorial:

a. Create the blend feature, refer to Figure 10-36.
b. Create the revolve feature, refer to Figures 10-37 and 10-38.
c. Create the round feature on the two ends of the blend feature, refer to Figure 10-39.
d. Create the shell feature by removing the front face of the model, refer to Figure 10-40.
e. Create the warp features using the **Spine** tool, refer to Figures 10-41 through 10-44.

When you start the PTC Creo Parametric session, the first task is to set the working directory. Make sure the required working directory is selected.

Starting a New Object File

Start a new part file and name it as *c10tut1*.

Creating the Blend feature

1. Choose the **Blend** tool from the expanded **Shapes** group; the **Blend** dashboard is displayed.

2. Choose **Blend as solid button** from the dashboard. Choose the **Options** tab and select the **Straight** radio button. Next, choose the **Sections** tab from the dashboard; the slide down panel is displayed. Choose the **Sketched sections** radio button and click on the **Define** button to define the sketching plane.

3. Select the **FRONT** datum plane as the sketching plane.

4. Select the **RIGHT** datum plane as the reference plane. Next, select the **Right** option from the **Orientation** drop-down list if these options are not selected by default.

   Next, you need to create three elliptical sections for the blend feature.

5. Choose the **Center and Axis Ellipse** tool from the **Ellipse** drop-down in the **Sketching** group and draw an ellipse with the **Rx** dimension as **20** and **Ry** dimension as **5**.

6. Choose **OK** to exit the sketcher environment. Now, choose the **Insert** button from the **Sections** slide-down panel to add the second section. Enter **15** in offset value edit box as depth between the first and the second sections. Next, choose the **Sketch** button from the **Sections** tab to enter the sketcher environment.

7. Draw the second section of the blend feature which is an ellipse with the **Rx** dimension as **20** and the **Ry** dimension as **10**.

8. Choose **OK** to exit the sketcher environment. Now, choose the **Insert** button from the **Sections** slide-down panel to add the third section. Enter **150** in the offset value edit box for specifying depth between the second and the third sections. Next, choose the **Sketch** button for the third section.

9. Draw the third section which is also an ellipse with the **Rx** dimension as **25** and the **Ry** dimension as **20**.

10. Choose the **OK** button to exit the sketcher environment. Choose the **Build feature** button from dashboard to exit it. The model similar to the one shown in Figure 10-36 is displayed in the drawing area.

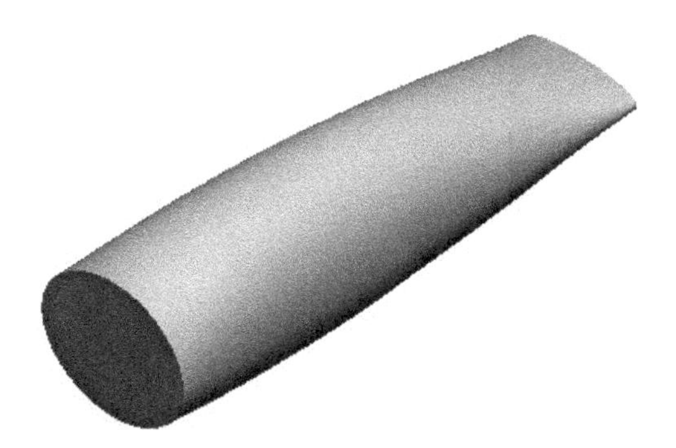

Figure 10-36 Isometric view of the Blend feature

Creating the Revolve Feature

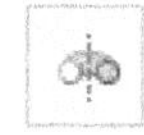

1. Choose the **Revolve** tool from the **Shapes** group and then create the sketch for the revolve feature, as shown in Figure 10-37. Next, revolve the sketch through an angle of 360 degrees. The model after creating the revolve feature is shown in Figure 10-38.

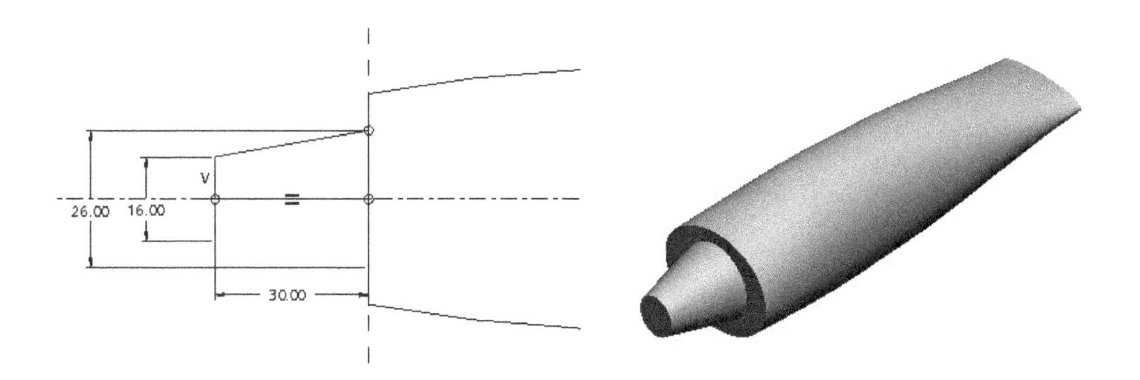

Figure 10-37 Sketch for the revolve feature *Figure 10-38 Model after creating the revolve feature*

Creating the Round Feature

1. Choose the **Round** tool from the **Shapes** group; you are prompted to select an edge or a surface to create the round. Create a round feature of radius **1.5** at the rear edge of the model and then create a round feature of radius **6** at the front-end of the model. The model similar to the one shown in Figure 10-39 is displayed in the drawing area.

Creating the Shell Feature

1. Choose the **Shell** tool, specify the wall thickness value as **1**, and then select the front face of the revolved feature.

The model after creating the shell feature is shown in Figure 10-40.

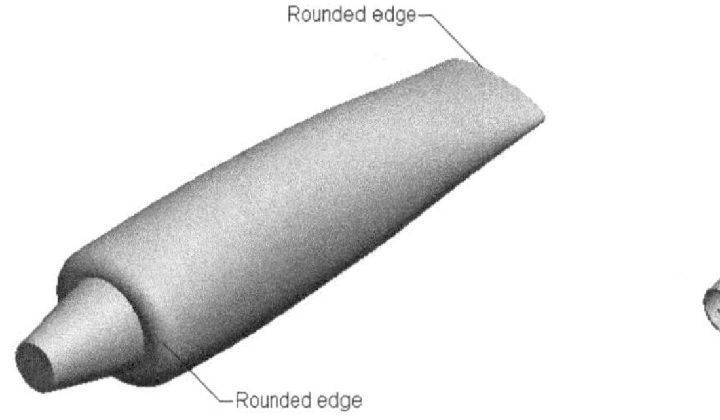

Figure 10-39 Model after creating the round feature

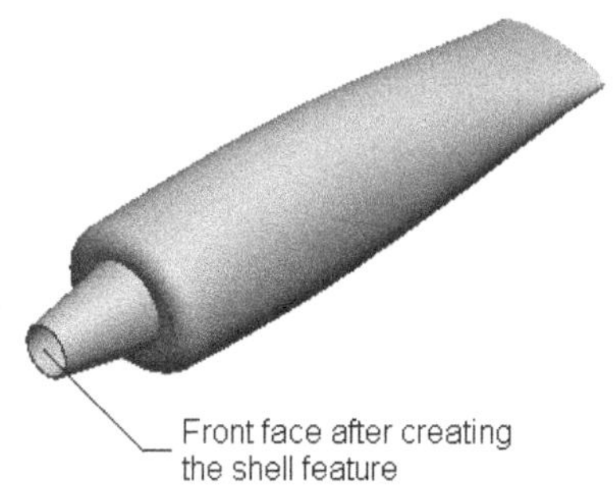

Figure 10-40 Model after creating the shell feature

Creating the Sketch for the Warp Feature

You need to use the **Spline** tool to create deformation in the tube. Therefore, you first need to create the curve which will be used to create the warp feature.

1. Choose the **Sketch** tool from the **Datum** group; the **Sketch** dialog box is displayed.

2. Choose the **RIGHT** plane as the sketching plane and select the **Top** option from the drop-down list in the **Orientation** tab. Next, choose the **Sketch** button to enter the sketching environment.

3. Choose the **Spline** tool from the **Sketching** group and draw the spline curve shown in Figure 10-41. Note that the dimensions of the spline are not important as this sketch will be used as reference for deciding the shape after deformation.

4. Choose the **OK** button to exit the sketcher environment.

Applying the Warp Transformation by Using the Spine Tool

1. Choose the **Warp** tool from the expanded **Editing** group; the **Warp** dashboard is displayed.

2. Choose the **References** tab; the **Geometry** collector is enabled by default. Select the model from the drawing area.

3. Click in the **Direction** selection box; you are prompted to select a plane or a coordinate system to define the direction of the warp.

4. Select the **TOP** plane from the drawing area; you are prompted to select the deformation tool.

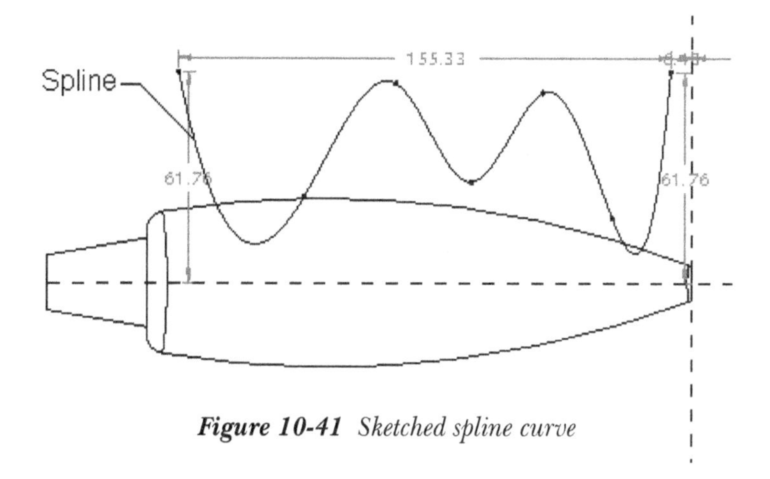

Figure 10-41 Sketched spline curve

5. Choose the **Spine** tool from the **Warp** dashboard; you are prompted to select a curve to define the deformation.

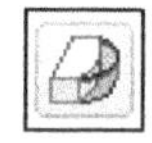

6. Select the sketched spine as the reference curve for the warp transformation. The model similar to the one shown in Figure 10-42 is displayed in the drawing area.

7. Select the control points on the reference curve and drag them until the required spinal deformation is obtained. The position of the control points after carrying out the deformation is shown in Figure 10-43.

Note

The control points for creating the deformation may be randomly selected based on the profile required.

8. Choose the **Build feature** button to create the feature. Figure 10-44 shows the final model after applying the warp transformation and hiding the spline created.

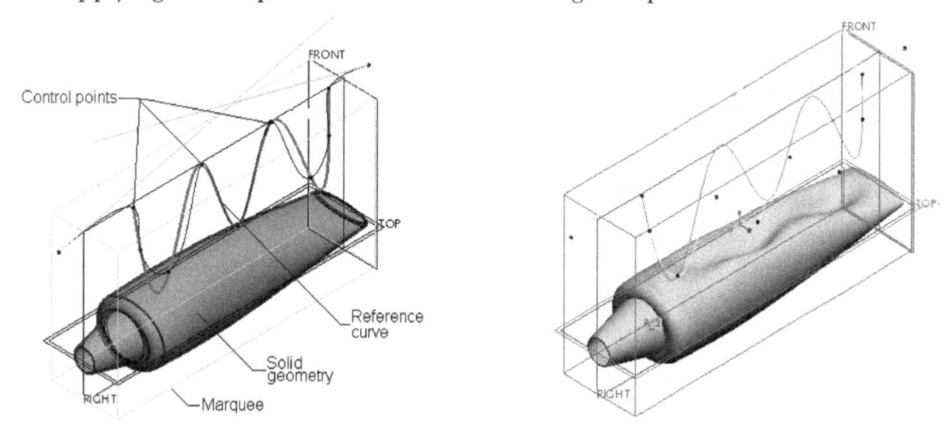

Figure 10-42 *Figure showing the parameters for the warp transformation*

Figure 10-43 *Figure showing the deformed model*

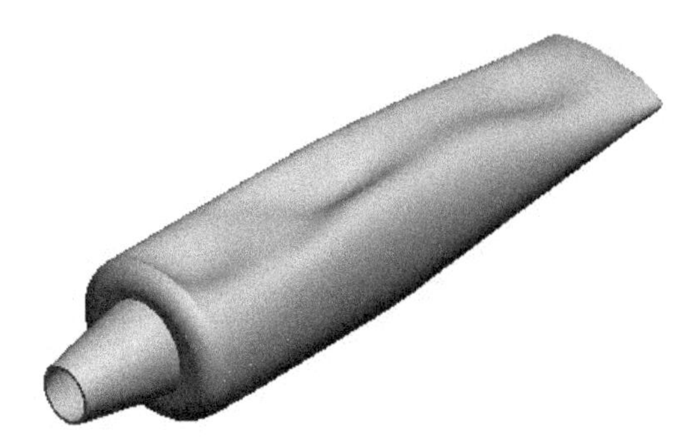

Figure 10-44 Final model after applying the warp transformation

Saving the Model

1. Choose the **Save** button from the **File** menu to save the model.

Tutorial 2

In this tutorial, you will create the model shown in Figure 10-45. **(Expected time: 30 min)**

The following steps are required to complete this tutorial:

a. Create the revolve feature, refer to Figure 10-47.
b. Create the shell feature with thickness as 8, refer to Figure 10-48.
c. Create the extrude feature, refer to Figure 10-49, and then pattern the extruded feature, refer to Figure 10-50.
d. Create the round feature, refer to Figure 10-51.
e. Create the warp transformations, refer to Figure 10-52.

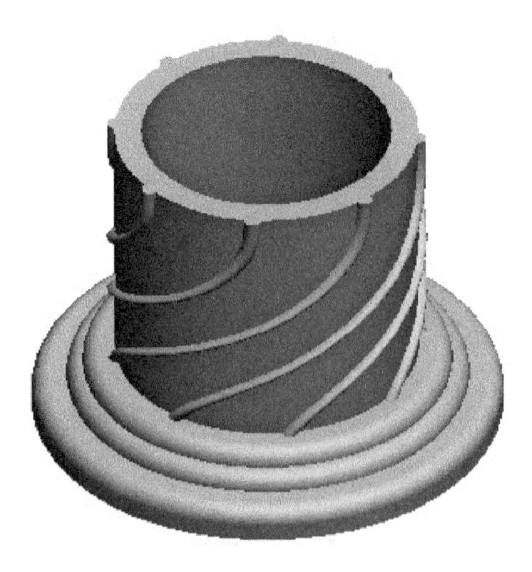

Figure 10-45 Model for Tutorial 2

Starting a New Object File

Start a new part file and name it as *c10tut2*.

Creating the Revolve Feature

1. Draw the sketch of the revolved feature in the **FRONT** datum plane, as shown in Figure 10-46.

2. Revolve the sketch through 360 degrees to create the base feature, as shown in Figure 10-47.

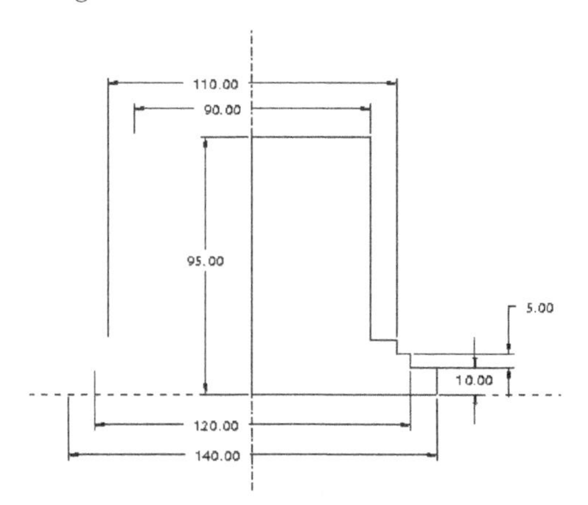

Figure 10-46 *Sketch for the revolved feature*

Figure 10-47 *Model showing the base feature*

Creating the Shell Feature

1. Create the shell feature with a wall thickness of value **8** and select the top face of the base feature. The model after creating the shell feature shown in Figure 10-48.

Creating the Extrude Feature and its Pattern

1. Create the extrude feature similar to the one shown in Figure 10-49.

Figure 10-48 *Model after creating the shell feature*

Figure 10-49 *Model after creating the extrude feature*

2. Create the axis pattern of the extrude feature, refer to Figure 10-50.

Creating Round Features

1. Choose the **Round** tool from the **Engineering** group; you are prompted to select the edges to create the round.

2. Choose the edges, as shown in Figure 10-51.

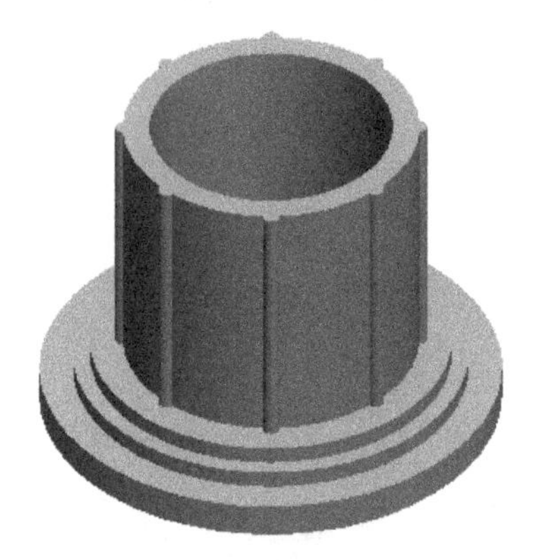

Figure 10-50 Model after creating the pattern

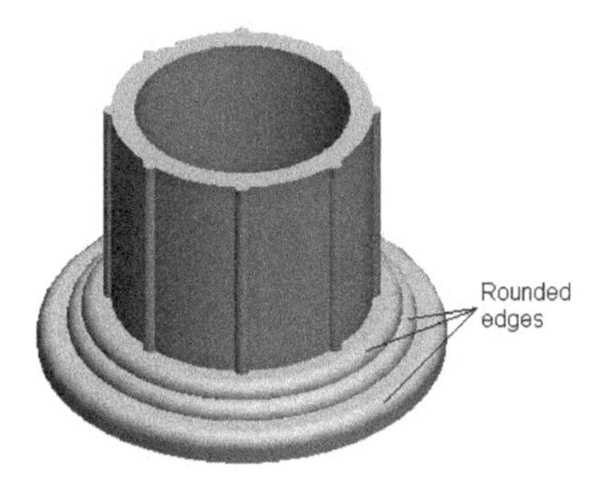

Figure 10-51 Figure showing the round feature

Note
Hold the CTRL key to select multiple edges.

3. Enter **5** in the edit box available in the **Round** dashboard and then choose the **Build feature** button to exit the **Round** dashboard.

   The model after creating the round feature is shown in Figure 10-51.

Applying the Warp Transformation by Using the Twist Tool

1. Choose the **Warp** tool from the expanded **Editing** group; the **Warp** dashboard is displayed and you are prompted to select the solids or quilts to warp.

2. Select the model in the drawing area and then choose the **References** tab from the **Warp** dashboard. Click in the selection area for the **Direction** collector; you are prompted to select a plane or a coordinate system to select the direction of the warp.

3. Select the **TOP** datum plane in the drawing area; you are now prompted to select the deformation tool.

4. Choose the **Twist** tool from the **Warp** dashboard.

5. Enter **120** in the **Angle** edit box in the **Warp** dashboard; the preview of the warp transformation is displayed in the drawing area.

6. Choose the **Build feature** button from the **Warp** dashboard to complete the creation of the feature.

   The model after applying the warp transformation is shown in Figure 10-52.

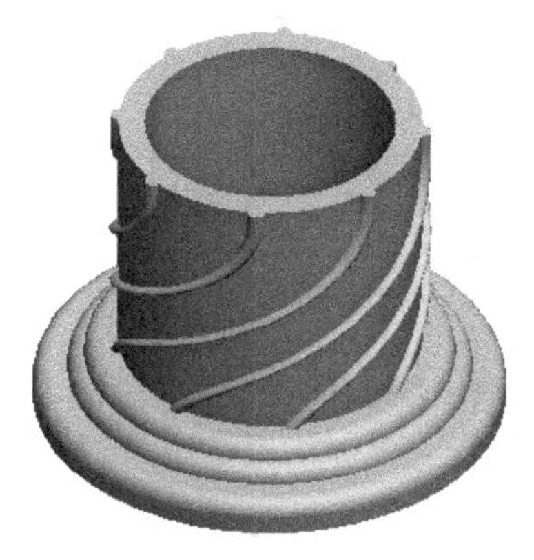

Figure 10-52 *Model after applying the warp transformation*

Saving the Model

1. Choose the **Save** button from the **File** menu to save the model.

Tutorial 3

In this tutorial, you will create the model shown in Figure 10-53. **(Expected time: 30 min)**

Figure 10-53 *Model for Tutorial 3*

The following steps are required to complete this tutorial:

a. Create the extrude feature, refer to Figures 10-54 and 10-55.
b. Create the second extrude feature, refer to Figures 10-56 and 10-57.
c. Create the third extrude feature and pattern it, refer to Figures 10-58 and 10-59.
d. Create the mirror feature, refer to Figure 10-60.
e. Create the toroidal bend feature, refer to Figures 10-61 and 10-62.
f. Create the round features, refer to Figure 10-63.
g. Change the color and appearance of the model, refer to Figure 10-64.

Starting a New Object File

Start a new part file and name it as *c10tut3*.

Creating the First Extrude Feature

1. Draw the sketch for the first extrude feature in the **FRONT** plane, as shown in Figure 10-54.

2. Extrude it to a length of **1200**, refer to Figure 10-55.

Creating the Second Extrude Feature

1. Choose the **Plane** tool from the **Datum** group; the **DATUM PLANE** dialog box is displayed and you are prompted to select the references to place the plane.

2. Select the **TOP** plane from the drawing area. Next, select the **Offset** option from the drop-down list in the **References** selection area and enter **230** in the **Translation** edit box. Choose the **OK** button; the new datum plane is created with the default name **DTM1**.

3. Choose the **Extrude** tool from the **Shapes** group; the **Extrude** dashboard is displayed.

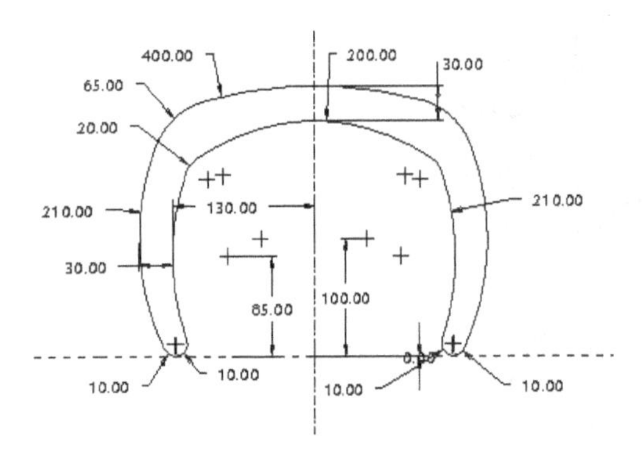

Figure 10-54 Sketch of the extrude feature

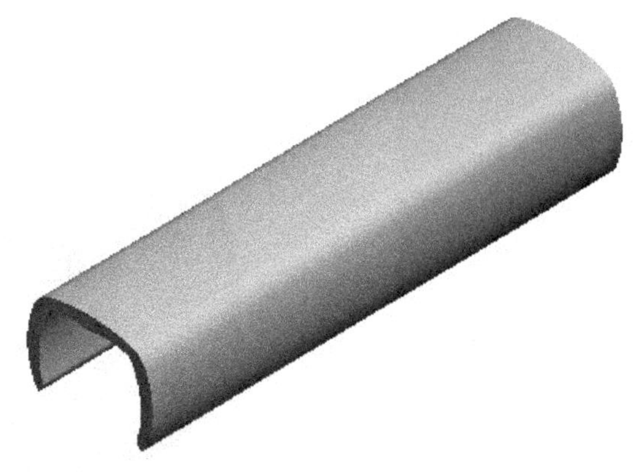

Figure 10-55 Model showing the first extrude feature

4. Choose **DTM1** as the sketching plane and create the sketch of the extrude feature, as shown in Figure 10-56.

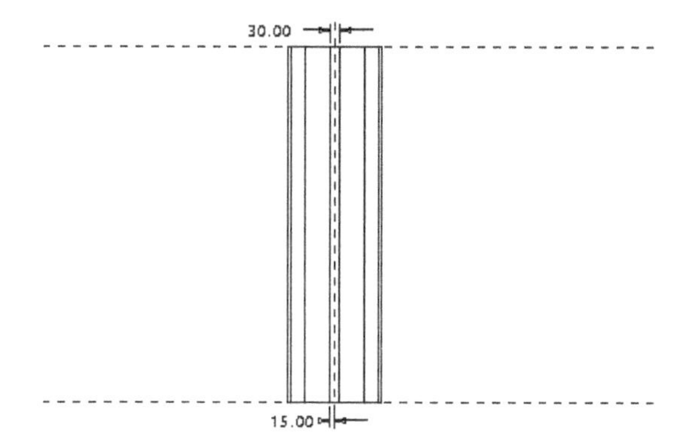

Figure 10-56 *Sketch of the extrude feature*

5. Choose the **Remove Material** button from the **Extrude** dashboard. 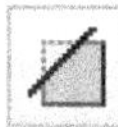

6. Enter **20** in the **Enter the depth value for side 1** edit box.

7. Specify the correct direction of cut and choose the **Build feature** button to complete the creation of the feature.

   The model after creating the second extrude feature is shown in Figure 10-57.

Creating the Next Extrude Feature

1. With the **DTM1** plane as the sketching plane and the **RIGHT** plane oriented to the **Right**, draw the sketch of the tread, as shown in Figure 10-58.

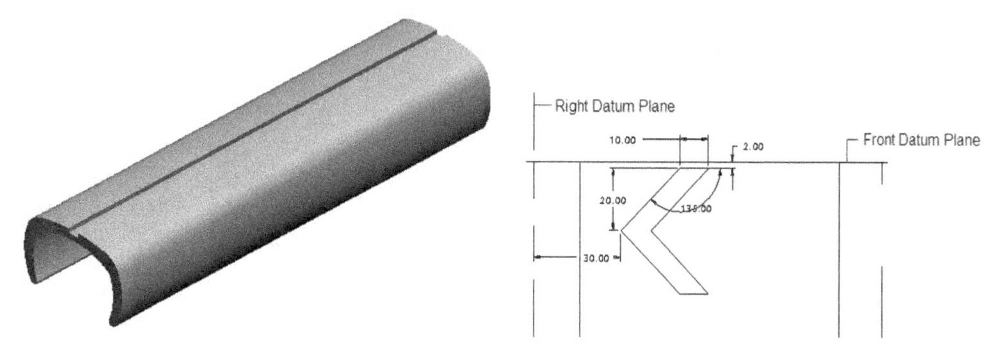

Figure 10-57 *Figure showing the second extrude feature*

Figure 10-58 *Sketch for the third extrude feature*

2. Extrude and remove material from the tire to a depth value of **20**. Note that the feature must be extruded into the model.

Creating the Pattern of the Tread

1. Select the extruded feature created in the previous step from the **Model Tree** or by clicking on it in the drawing area.

2. Choose the **Pattern** tool from the **Pattern** drop-down in the **Editing** group; the **Pattern** dashboard is displayed and you are prompted to select the dimension to vary in the first direction.

3. Choose the **Dimensions** tab in the **Pattern** dashboard and select the dimension value **30** from the feature. Enter **30** in the **Increment** edit box in the **Direction 1** selection area of the **Dimensions** slide-down panel.

4. Click in the **Direction 2** selection area of the **Dimensions** tab; you are prompted to select the dimension to vary in the second direction.

5. Select the dimension from the top edge of the base feature. Enter **40** in the **Direction 2** edit box.

6. Enter **30** in the **Enter the number of pattern members in the second direction** edit box in the dashboard.

7. Choose the **Build feature** button from the dashboard to exit from the **Pattern** tool.

   The model after creating the pattern is shown in Figure 10-59.

Creating the Mirror Feature of the Patterned Feature

1. Mirror the tread pattern on the other side using the **RIGHT** datum plane, as shown in Figure 10-60.

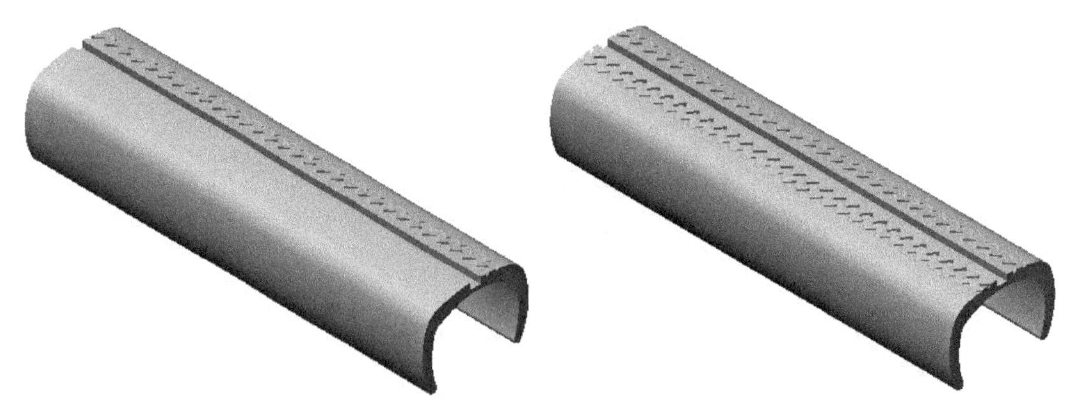

Figure 10-59 *Model after creating the pattern of the tread*

Figure 10-60 *Model after creating the mirror feature of the tread*

Creating the Toroidal Bend Feature

1. Choose the **Toroidal Bend** tool from the **Engineering** group; the **Toroidal Bend** dashboard is displayed.

2. Choose the **References** tab; a slide-down panel is displayed. Select the **Solid Geometry** check box. Also, choose **Define** from the slide-down panel; the **Sketch** dialog box is displayed.

3. Select the front face of the base feature as the sketching plane. Next, choose the **Sketch** button from the **Sketch** dialog box to accept the default orientation and enter the sketcher environment.

4. Draw the section for the toroidal bend feature, as shown in Figure 10-61.

5. Choose the **Geometry Coordinate System** button from the **Datum** group and place the reference coordinate system at the origin, as shown in Figure 10-61.

6. Choose the **OK** button from the **Sketch** dashboard to exit the sketcher mode; the **Toroidal Bend** dashboard is displayed again. Choose the **Options** tab and select the **Standard** radio button, if it is not already selected from the **Curve Bend** area of the slide-down panel.

7. Choose the **Bend Radius** tool; a flyout is displayed. Choose **360 degrees Bend** from the flyout; you are prompted to select two parallel planes to define the length of the bend.

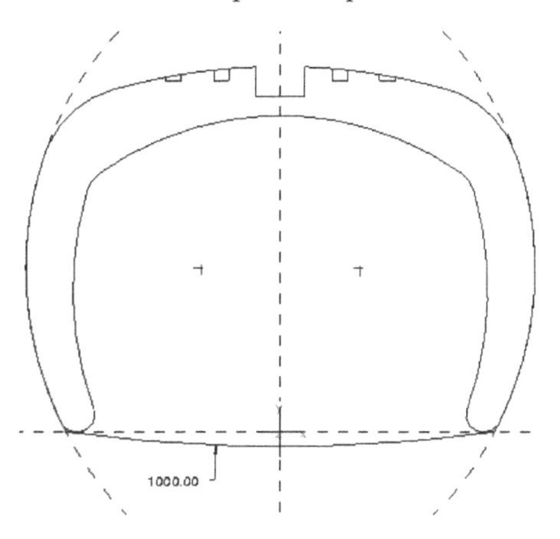

Figure 10-61 *Section created for the toroidal bend feature*

8. Select the front face and then the back face of the base feature; the preview of the toroidal bend feature is displayed.

9. Choose the **Build Feature** button; the model similar to that shown in Figure 10-62 is created in the drawing area.

Creating the Round Feature

1. Create a round feature of radius **5** units by selecting the edges, as shown in Figure 10-63.

Figure 10-62 *Model after creating the toroidal bend feature*

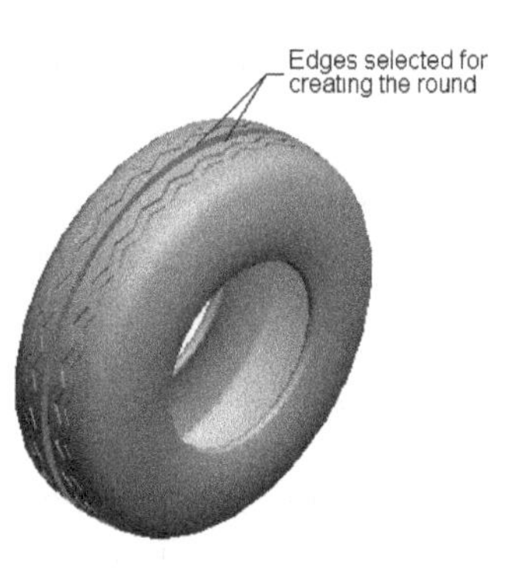

Figure 10-63 *Model showing the edges selected for creating the round feature*

Changing the Color and Appearance of the Model

1. Choose the down-arrow on the right of the **Appearance Gallery** tool from the **Model Display** group available in the **View** tab; a flyout is displayed.

2. Click on the **std-metals.dmt** located on the right of the **Library** head; a flyout is displayed.

3. Click on the **Misc.** node under the **Photolux Library**; a list of categories is displayed.

4. Choose **adv-rubber.dmt** from the categories displayed; the materials available in this category are displayed.

5. Choose the **adv-rubber-black** material; the cursor changes into a paint brush.

6. Select **Part** from the selection filter and then select the part created. Press the middle mouse button; the selected appearance is applied to the entire model.

   Figure 10-64 shows the model after changing the color and appearance.

Figure 10-64 Model after changing the color and appearance

Saving the Model

1. Choose the **Save** button from the **File** menu to save the model.

Self-Evaluation Test

Answer the following questions and then compare them to those given at the end of this chapter:

1. While using the **Sculpt** tool, the _________ drop-down list is used to specify the direction in which the group of points selected move when sculpted.

2. To specify the length of the bend for a 360 degrees toroidal bend feature, you need to select two _________ planes.

3. The **Sculpt** tool deforms the geometry through a _________.

4. The **Switch to next axis** button is used to switch the axis of bend from one _________ to another.

5. The **Spinal Bend** tool is used to bend solids or quilts by continuously _________ the cross-sections along a sketched spine.

6. The **Transform** tool can be used to perform _________, _________, and _________ operations.

7. The **Spine** tool deforms the geometry in either the _________ or the _________ direction.

8. The **Spinal Bend** tool bends solids or quilts along a sketched spine. (T/F)

9. The warp transformations can be performed only in the part mode. (T/F)

10. The toroidal bend feature can be created only on one side of the sketching plane. (T/F)

Review Questions

Answer the following questions:

1. What is the maximum bend angle through which a geometry can be bent using the **Toroidal Bend** tool?

   (a) 90 degrees (b) 180 degrees
   (c) 270 degrees (d) 360 degrees

2. The trajectory for creating a spinal bend on geometry should be sketched __________ to the solid geometry.

3. The __________ tool is used to stretch the geometry along the required direction.

4. The __________ check box, if selected, keeps the original entity selected for the creation of warp feature in the drawing area along with the preview of the transformed feature.

5. The direction reference chosen for the creation of warp feature is any one of the __________ planes.

6. The **Spine** tool reshapes the geometry by manipulating the defining points of a __________.

7. A solid geometry can be rotated by clicking and rotating the active handles of the __________.

8. While using the **Toroidal Bend** tool, the user-defined coordinate system defines the __________ of the bend profile.

9. While creating a model using the **Toroidal Bend** tool, if the sketch defining the bend is a straight line, the profile of the model is a __________.

10. The **Wrap** tool modify a model along its default __________.

Exercises

Exercise 1

In this exercise, you will create the model shown in Figure 10-65. The dimensions for the model can be assumed. The dimensions given in the hint are suggestive.

(Expected time: 20 min)

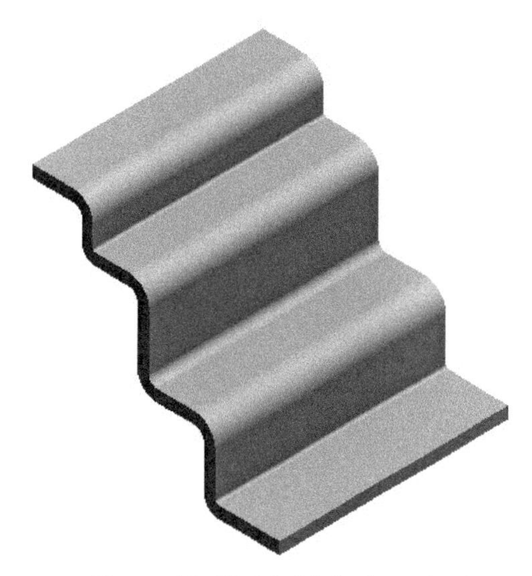

Figure 10-65 *Model for Exercise 1*

Hint

1. Create the sketch of the base feature which is a rectangle of dimensions **500 X 300** and extrude it to a depth of **15**.
2. Create a new datum plane at a distance of **250** from the **RIGHT** datum plane.
3. Choose the **Spinal Bend** tool from the **Engineering** group and select the **FRONT** datum plane as the sketching plane.
4. Create the trajectory for the **Spinal Bend** which should be tangent to the bottom face of the base feature. The shape of the trajectory needs to be in the form of steps with the corners rounded. The dimensions of the trajectory can be assumed.
5. Exit from the sketcher environment; you will be prompted to select a plane to define the volume of the bend.
6. Select the **DTM1** datum plane. The model similar to the one shown in Figure 10-65 is displayed in the drawing area.

Exercise 2

In this exercise, you will create the model shown in Figure 10-66. The dimensions for the base feature can be assumed. **(Expected time: 20 min)**

Figure 10-66 *Model for Exercise 2*

Answers to Self-Evaluation Test

1. Symmetry, **2.** parallel, **3.** mesh, **4.** plane, **5.** repositioning, **6.** translate, rotate, scale, **7.** linear, radial, **8.** T, **9.**T, **10.** F.

Chapter 11

Assembly Modeling

Learning Objectives

After completing this chapter, you will be able to:

- *Understand the top-down assembly approach.*
- *Understand the bottom-up assembly approach.*
- *Assemble components of the assembly using assembly constraints.*
- *Understand the packaging of components.*
- *Create the simplified representations.*
- *Use the View Manager.*
- *Edit assembly constraints after assembling.*
- *Modify the components of an assembly.*
- *Create the exploded state of an assembly.*
- *Add offset lines to exploded components.*
- *Understand the Bill of Material in the assemblies.*

ASSEMBLY MODELING

An assembly is defined as a design consisting of two or more components bonded together at their respective working positions using the assembly constraints. These assembly designs are created in the **Assembly** mode of PTC Creo Parametric. To invoke the **Assembly** mode, choose the **New** button from the **Quick Access** toolbar; the **New** dialog box will be displayed. In this dialog box, select the **Assembly** radio button from the **Type** area and then select the **Design** radio button from the **Sub-type** area, as shown in Figure 11-1. Specify the name of the assembly in the **Name** edit box and choose **OK**; the **Assembly** mode will be activated. Also, the initial screen appearance of the **Assembly** mode in PTC Creo Parametric will be displayed, as shown in Figure 11-2.

*Figure 11-1 Invoking the **Assembly** mode using the **New** dialog box*

IMPORTANT TERMS RELATED TO THE ASSEMBLY MODE

Before proceeding further in this chapter, it is very important for you to understand the following terms.

Top-down Approach

This is the method of assembling the components in which the components of the assembly are created in the same assembly file. In this type of assembly modeling approach, the components are created in the assembly file and then assembled using the assembly constraints. Note that the parts you create in the **Assembly** mode are saved as separate *.prt* files.

Bottom-up Approach

This is the method of assembling the components that are created as separate parts in the **Part** mode and are saved as *.prt* files. Once all parts of an assembly have been created, you will create a new assembly file (*.asm*) and then assemble the parts using the assembly constraints available in the **Assembly** mode. Since the assembly file has information related only to the assembling of components, its size is small and therefore requires less hard disk space.

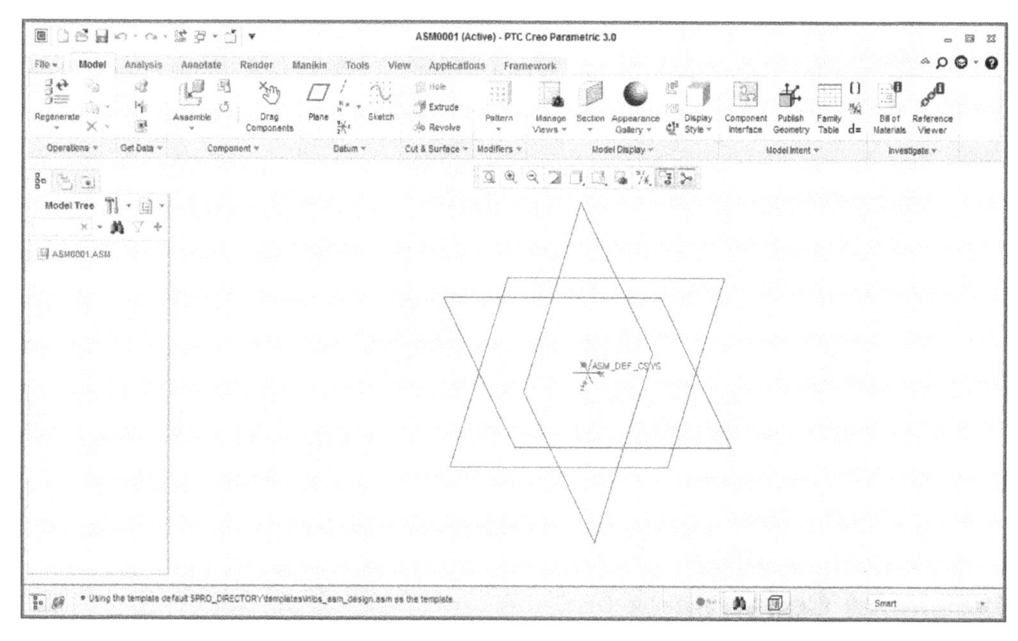

*Figure 11-2 Initial screen appearance of the **Assembly** mode*

Remember that if any of these assembly components is moved from its original location on the hard disk, the assembly will not open next time. This is because PTC Creo Parametric searches for the assembly component in the folder in which it was originally saved. If it does not find the name of the component in that folder, it will show an error while opening the assembly file. The missing component will be displayed in red color and a cross mark will be displayed against the component name in the **Model Tree**. To complete the assembly, either move the required files to the original folder or retrieve the files. To retrieve a file, right-click on the missing component in the **Model Tree**; a flyout will be displayed. Choose the **Retrieve Missing Component** option from the flyout; the **File Open** dialog box will be displayed and you will be prompted to browse to the missing file, refer to Figure 11-3. Select the required file and then choose the **Open** button; the missing component will be assembled on the basis of previously defined relations.

Tip: 1. *It is recommended that you create separate folders for all assemblies that you create in PTC Creo Parametric.*

*2. If you are opening an assembly with the part name same as that in the previous session then the part of the previous session will open instead of the part in the current session. To avoid this, it is recommended that you use the **Erase Not Displayed** button before opening an existing assembly. For example, if Nut was opened in the previous session and you open a component with same name in the current session then PTC Creo Parametric will open the previously opened Nut rather than the new one.*

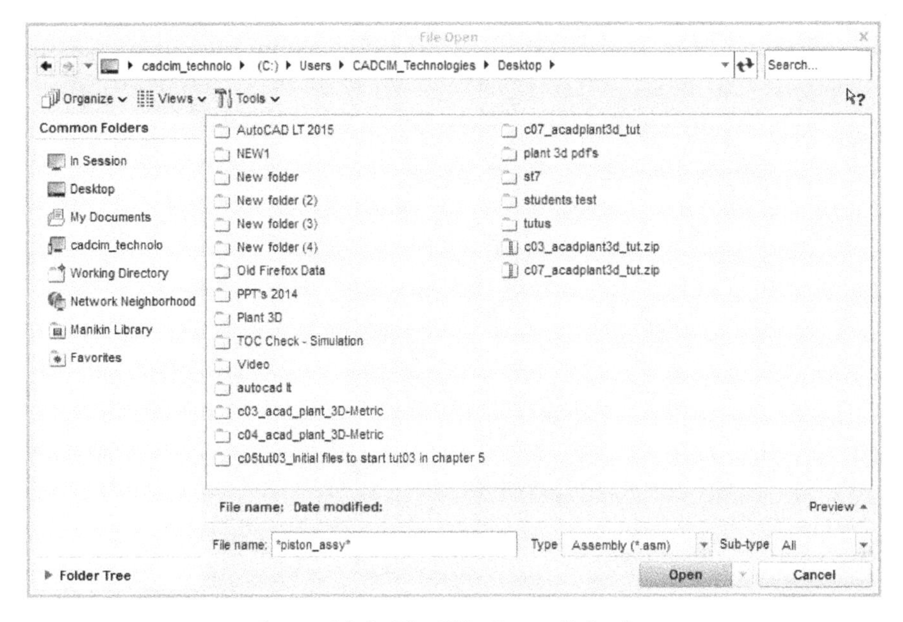

*Figure 11-3 The **File Open** dialog box*

Placement Constraints

The placement constraints are the constraints that are used to rigidly bind the components of the assembly to their respective positions in the assembly. These constraints are also called as assembly constraints. Generally, these constraints are used in combinations and you can constrain upto six degrees of freedom of a component.

Package

It is the state in which the component being assembled is not fully constrained and therefore, is not rigidly placed at its actual location.

CREATING TOP-DOWN ASSEMBLIES

As mentioned earlier, in the top-down assemblies, all components are created in the **Assembly** mode and in the same assembly file. The components created in the **Assembly** mode can be saved as separate *.prt* files. As you already know that to create components, you need the sketcher environment and to invoke the sketcher environment, you need the **Part** mode. You can invoke the **Part** mode from within the **Assembly** mode and create components using the tools in the **Part** mode. The method to create the component from within the **Assembly** mode is discussed next.

Creating Components in the Assembly Mode

Ribbon: Model > Component > Create

To create a component in the **Assembly** mode, choose the **Create** tool from the **Component** group; the **Component Create** dialog box will be displayed, as shown in Figure 11-4. Select the **Part** radio button from the **Type** area and the **Solid** radio button from the **Sub-type** area to create a *.prt* file. Enter the name of the component in the **Name** edit box and then choose the **OK** button; the **Creation Options** dialog box will be displayed, as shown in Figure 11-5. From this dialog box, select the method for creating the component and then choose the **OK** button.

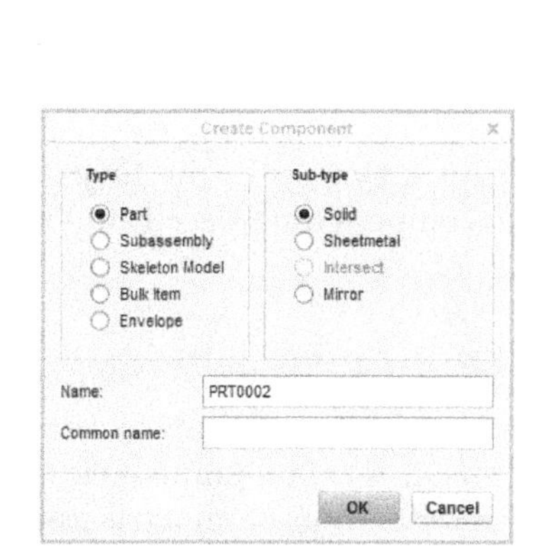

Figure 11-4 The ***Component Create*** *dialog box*

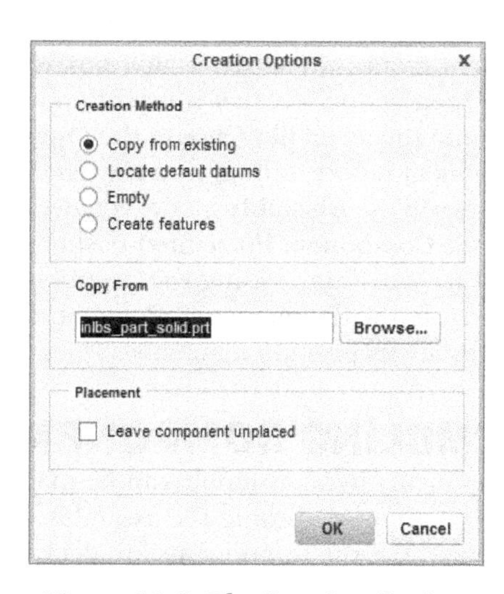

Figure 11-5 The ***Creation Options*** *dialog box*

CREATING BOTTOM-UP ASSEMBLIES

As discussed earlier, the bottom-up assemblies are created by inserting the part files one by one in the assembly file and then assembling them using the assembly constraints. When the first component is inserted in the **Assembly** mode, its three default datum planes are placed in the same orientation as that of the default datum planes of the **Assembly** mode. The method of inserting a component in the assembly is discussed next.

Inserting Components in an Assembly

Ribbon: Model > Component > Assemble > Assemble

When you enter the assembly environment, the three default assembly datum planes will be displayed in the drawing area. If you do not want the default assembly datum planes in the assembly file, then while opening a new file, clear the **Use default template** check box from the **New** dialog box and then choose the **OK** button; the **New File Options** dialog box will be displayed. Choose **Empty** from the **Template** list in this dialog box. It is recommended that you use the assembly datum planes as the first feature of the assembly and assemble the components of the assembly taking the reference of these datum planes. Using the datum planes as the first feature will help you during the modification of the assembly.

Some of the advantages of using the assembly datum planes are:

1. The components of the assembly can be redefined in such a way that the component placed later can be made the first component.

2. The first component can be replaced by some other component.

3. The placement constraint of the first component can be modified.

To insert a component, choose the **Assemble** tool from the **Assemble** drop-down in the **Component** group; the **Open** dialog box will be displayed. This dialog box is the same as the **File Open** dialog box that is used to open an existing file. Select the part that you need to insert into the assembly. Choose the **Open** button to open the part. You can also preview the component before adding it to the assembly. After the components are displayed in the drawing area in the **Assembly** mode, you need to specify constraints in order to assemble them by using the **Component Placement** dashboard. Even if you are placing the first component, you need to constrain it using the assembly constraints. Generally, the first component is assembled with the datum planes. Remember that no component is placed automatically; you need to specify its position manually.

ASSEMBLING COMPONENTS

The components in an assembly can be placed parametrically or non-parametrically. If the components are placed using the assembly constraints, it is called parametric assembly. On the other hand, if the components are packaged, it is called non-parametric assembly.

To assemble the components parametrically, choose the **Assemble** tool; the **Open** dialog box will be displayed. Select the required component and choose the **Open** button from the **Open** dialog box; the **Component Placement** dashboard will be displayed, as shown in Figure 11-6.

Figure 11-6 *Partial view of the **Component Placement** dashboard*

Using the **Component Placement** dashboard, you can apply constraints to assemble the components and also control the display of the component to be assembled in the same or separate windows. The options in the **Component Placement** dashboard are discussed next.

Displaying Components in a Separate Window

The **Show component in a separate window while specifying constraints** button is used to display the component to be assembled in a separate window. When you choose this button, a new window will open, displaying the component to be assembled. Note that viewing the component to be assembled, and the assembly, in separate windows, prevents cluttering of the components in the assembly window.

However, if the components are in the same window, they change their orientation as you apply the constraints, thus giving you an idea of the next constraint to be applied. In other words, you can easily find out which degree of freedom of the component is constrained and which needs to be constrained.

Displaying Components in the Same Window

The **Show component in the assembly window while specifying constraints** button is used to display the component to be assembled in the assembly window itself. This button is chosen by default when you invoke the **Component Placement** dashboard. As you apply constraints, the component in the assembly window changes its position accordingly.

3D Dragger

The 3D Dragger is a tool which is used to move or rotate the component to be assembled. The 3D Dragger can be made visible or hidden by using the **3D Dragger** button available in the dashboard. By using the 3D Dragger, you can move or rotate a component to set its orientation. It has three arrow handles and 3 circular handles, refer to Figure 11-7. The arrow handles are used to move the component and the circular handles are used to rotate the component. If the movement or rotation of a component is restricted in a particular direction then the respective handle of that direction in the 3D Dragger will not be enabled for the movement. It also helps in determining the restricted direction of a component.

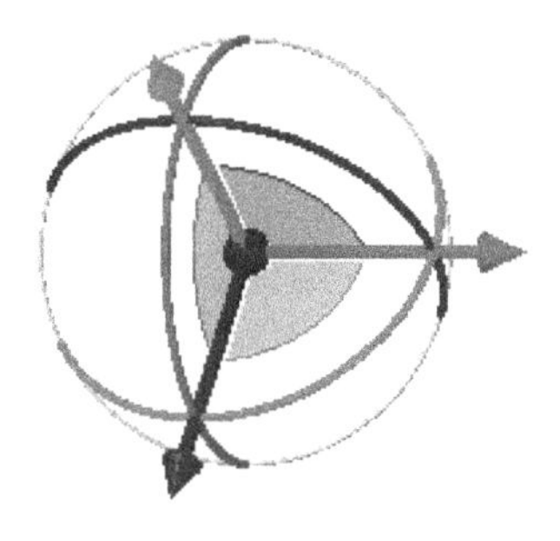

Figure 11-7 The 3D Dragger

Applying Constraints

As discussed earlier, the components are assembled using the assembly constraints. The assembly constraints are also called placement constraints. These constraints help in placing and positioning a component precisely with respect to the other components in the assembly. The **Constraint Type** drop-down list in the **Component Placement** dashboard contains different types of placement constraints, as shown in Figure 11-8. You can select the required constraints from this drop-down list to assemble components in the assembly. These constraints are also available when you choose the **Placement** tab of the **Component Placement** dashboard.

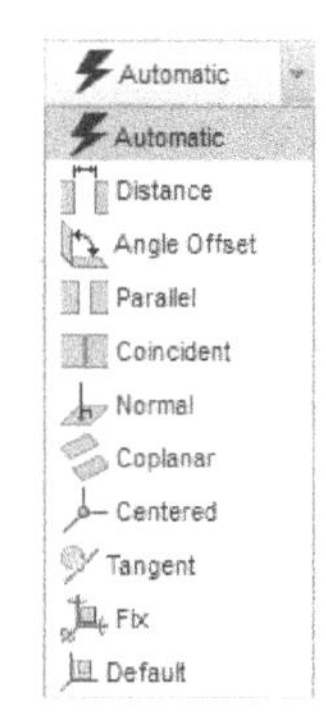

*Figure 11-8 The **Constraint Type** drop-down list*

Automatic

When you choose this constraint, PTC Creo Parametric assumes the constraint and applies it according to the type of entity selected. For example, if you select faces of two components to assemble, PTC Creo Parametric will assume that you want to apply the **Coincident** constraint and the **Coincident** constraint will be applied to the two components.

Distance

The **Distance** placement constraint is used to constrain the distance between the two selected axes, planes, datum planes, faces, or a combination of a datum plane and a face. The selected faces or datums may or may not be in contact with each other. When you select the **Distance** constraint, you are prompted to select one component reference and one assembly reference. Select the references as discussed earlier. When you select the references, the distance between them is displayed in the screen and a handle is attached to the component, refer to Figure 11-9. Using this handle, you can dynamically change the distance. You can also use the **Offset** edit box available in the dashboard or the **Placement** tab to enter the distance value. If you want to flip the direction of offset then you need to choose the **Flip** button next to the **Offset** edit box.

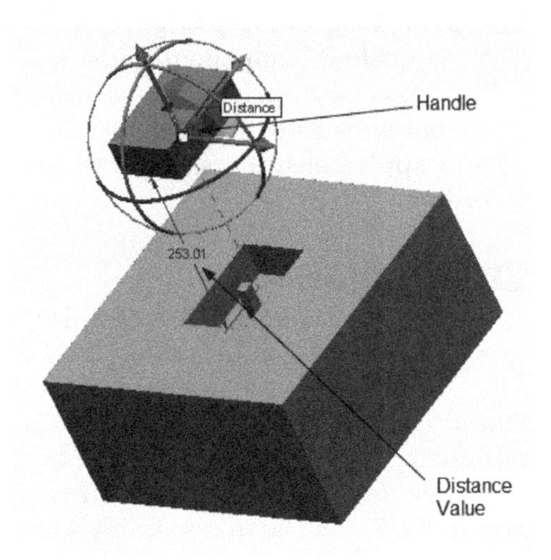

*Figure 11-9 Assembling components by using the **Distance** constraint*

Angle Offset

The **Angle Offset** placement constraint allows you to constrain the angular distance between two selected axes, planes, datum planes, faces, or a combination of a datum plane and a face. The selected faces or datums may or may not be in contact with each other. On selecting this constraint, when you select the references of the component and assembly, the component gets constrained by a default value. You can change the angle value by using the **Offset** edit box.

Parallel

The **Parallel** placement constraint is used to make two selected axes, planes, datum planes, faces, or a combination of a datum plane and a face parallel to each other. The selected faces or datums may or may not be in contact with each other. Figure 11-10 shows the faces selected for applying the **Parallel** constraint. Figure 11-11 shows the components after assembling.

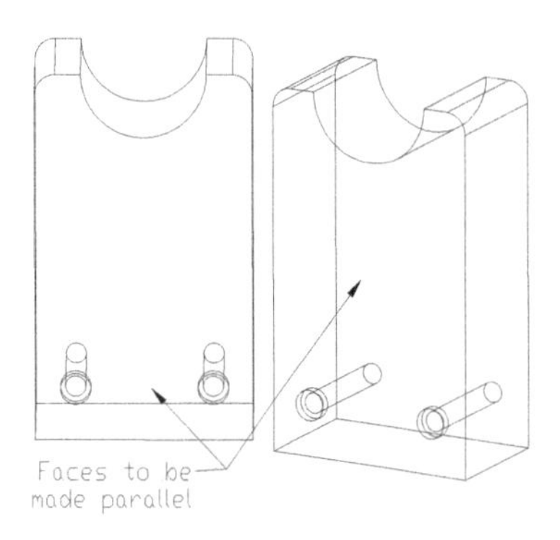

*Figure 11-10 Components before applying the **Parallel** constraint*

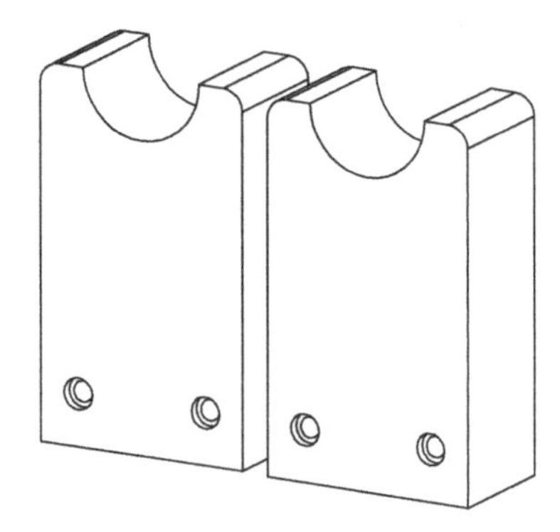

Figure 11-11 Components after assembling

Coincident

The **Coincident** constraint is used to make two selected planes or faces, or two axes coincident to each other. Figure 11-12 shows the axes to be made coincident. Figure 11-13 shows the components after assembling.

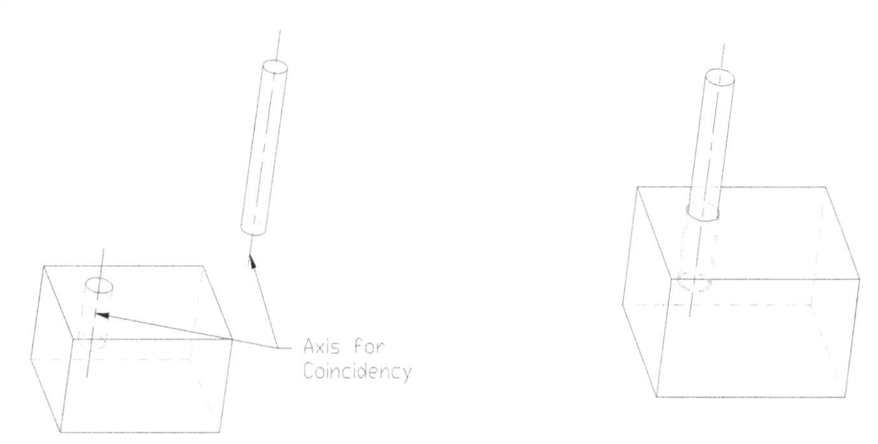

*Figure 11-12 Faces selected for the **Coincident** constraint*

Figure 11-13 Components after assembling

Normal

The **Normal** constraint is used to make two selected planes or faces, or two axes perpendicular to each other. Figure 11-14 shows the faces to be constrained. Figure 11-15 shows the components after assembling.

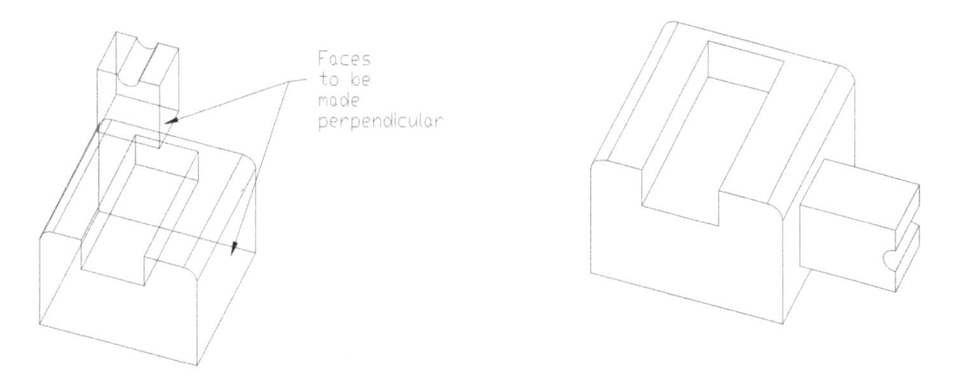

*Figure 11-14 Faces selected for the **Normal** constraint*

Figure 11-15 Components after assembling

Coplanar

The **Coplanar** constraint allows you to make two selected planes, faces, or axes coplanar. Figure 11-16 shows the faces to be constrained. Figure 11-17 shows the components after assembling.

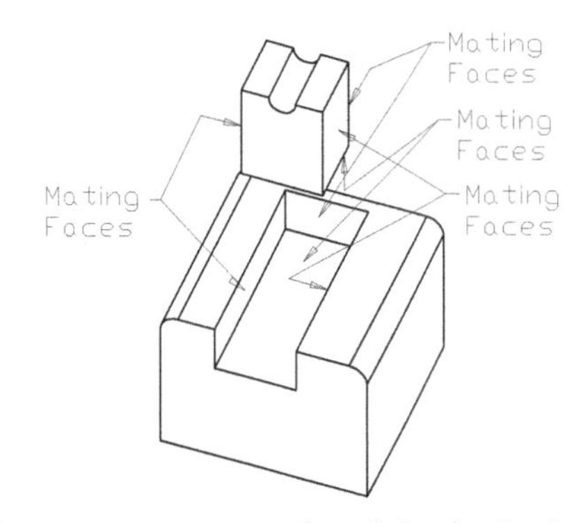

*Figure 11-16 Faces selected for the **Coplanar** constraint*

Figure 11-17 Components after assembling

Centered

The **Centered** constraint is used to assemble the revolved components. On applying this constraint, the revolved components, holes, or a combination of both share the same orientation with respect to the central axis. Figure 11-18 shows the faces selected to be inserted and Figure 11-19 shows the components after assembling. This constraint can also be used to align the coordinate system of the first component with the coordinate system of the second component but the rotation of component will not be constrained.

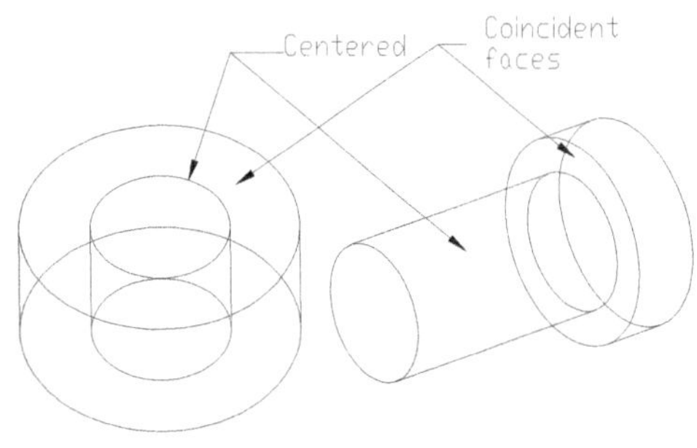

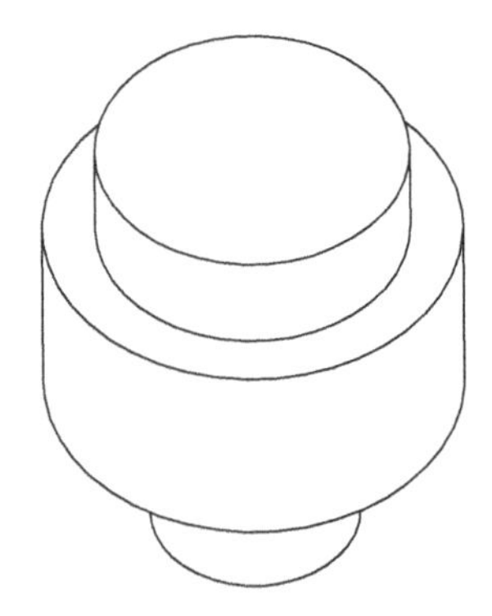

*Figure 11-18 Components to be assembled using the **Centered** constraint*

Figure 11-19 Components after assembling

Tangent

The **Tangent** constraint is used to make the selected circular face tangent to the other selected face or plane.

Fix

The **Fix** constraint is used to fix the component.

Default

The **Default** constraint is used to assemble the component in the assembly by aligning the default coordinate system of the component with the default coordinate system of the assembly. Also, the part datum planes are aligned with the assembly datum planes.

Status Area

The **Status** area in the **Component Placement** dashboard displays the placement status of the component in the assembly. If you choose the **Build feature** button from the **Component Placement** dashboard when the placement status is displayed as **Partially Constrained**, then the component assembled will be displayed in the **Model Tree** with a small square on it. Also, the component is packaged.

The dashboard consists of five tabs; **Placement**, **Move**, **Options**, **Flexibility**, and **Properties**. In this chapter, you will learn about the **Placement** and **Move** tabs.

Placement Tab

The options in this tab are used to apply as well as edit constraints to the components to be assembled. When you choose the **Placement** tab, a panel will be displayed, as shown in Figure 11-20.

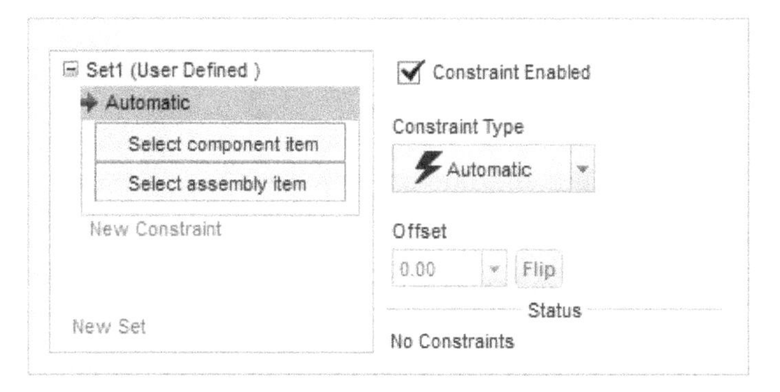

Figure 11-20 *The* ***Placement*** *panel*

The **Constraint Type** and **Offset** drop-down lists have already been discussed. Rest of the options in the panel are discussed next.

New Constraint

This option is used to specify a new set of constraints to the assembly components.

Select component Item Collector

This collector displays the entities of a component that have been chosen to assemble it with the parent component.

Select assembly Item Collector

This collector displays the entities of the parent component that have been chosen to create an assembly.

Note

Most of the options displayed on the dashboard are also available in the shortcut menu, which is displayed on right-clicking when a command is active.

Status Area

This area displays the placement status of the component in the assembly. Sometimes, after adding constraints, a few placement constraints are required to be assumed. This is the reason sometimes the **Allow Assumptions** check box will be displayed in the **Status** area.

Move Tab

The options in the **Move** tab can be used when the component to be assembled is displayed in the assembly itself. This tab is used to move or rotate a component along a degree of freedom that is not constrained. The **Move** slide-down panel is shown in Figure 11-21. The options in this tab are discussed next.

Motion Type

The options in the **Motion Type** drop-down list are discussed next.

Orient Mode

This option, if selected, orients the part in the assembly about its spin center.

Translate

This option, if selected, moves the component from its current location in the assembly. However, remember that the component can be moved only along the degrees of freedom that are not constrained.

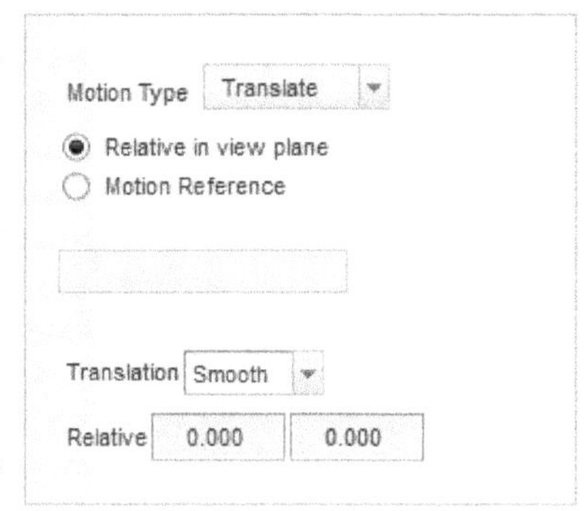

*Figure 11-21 The **Move** slide-down panel*

Rotate

This option, if selected, causes the component to rotate in the assembly around its available degrees of freedom.

Adjust

This option is used to pack a component with reference to the assembly. When you select this option, you will be prompted to select a surface on the packaged component to adjust the reference plane. The component is assembled according to the surface specified.

Tip: *It is recommended that you use the CTRL+ALT+left mouse button to translate a component and CTRL+ALT+middle mouse button to rotate a component. It is always easier to use the above combinations to move the component rather than switching to a separate tab in the **Component Placement** dashboard and using its options to move a component. Moving the components using the keyboard shortcuts can speed up the process of locating the component correctly in the assembly and establishing constraints.*

Relative in view plane

This radio button is selected by default in the **Move** slide-down panel. As a result, the component moves relative to the viewing plane.

Motion Reference

When this radio button is selected, you need to select a reference plane through which the component will be moved. After selecting the reference plane, the **Normal** and **Parallel** radio buttons will be activated.

Normal

On selecting this radio button, the component moves in a plane normal to the selected reference plane.

Parallel

On selecting this radio button, the component moves in a plane that is parallel to the selected reference plane.

Relative

These boxes display the coordinates of the current position of the component.

PACKAGING COMPONENTS

Ribbon: Model > Component > Assemble > Package

Consider a situation when you do not know the exact location of a component in the assembly or you want to specify the exact location later. In such cases, you can non-parametrically place the component in the assembly. This method of non-parametrically placing the components in the assembly is called packaging. You can package a component using the **Package** option. To package a component, choose the **Package** option from the **Assemble** drop-down in the **Component** group in the **Ribbon**; the **PACKAGE** menu will be displayed, as shown in Figure 11-22. The options in this menu are discussed next.

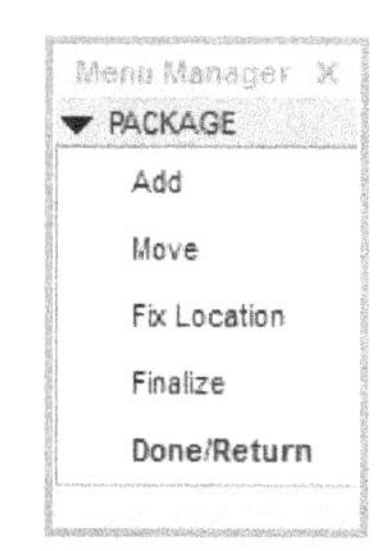

*Figure 11-22 The **PACKAGE** menu*

Add

The **Add** option is used to add the component to be packaged. On choosing this option, the **GET MODEL** sub-menu will be displayed. Using the options of this sub-menu, you can open a new component, select the component from the assembly, or select the last component from the assembly to assemble again.

When you select the component from the **Open** dialog box, the **Move** dialog box will be displayed, as shown in Figure 11-23. The options in this dialog box are similar to those in the **Move** tab of the **Component Placement** dashboard. After you have specified the location of the component in the assembly file, choose **OK** to place the component and close the **Move** dialog box. Remember that you can move only the packaged component using this option.

Move

The **Move** option is used to move the packaged component to a new location. When you choose this option, the **Move** dialog box will be displayed, as shown in Figure 11-23.

Fix Location

The **Fix Location** option is used to fix the location of an existing packaged component. When you choose this option, you are prompted to select the component to be fixed. Select the component to be fixed.

Finalize

As mentioned earlier, the packaged components are not assembled parametrically. Hence, packaging does not relate the component with its neighbouring components in the assembly. Once the location of the component is decided, then it can be finalized using the **Finalize** option. When you choose this option, you will be prompted to select the packaged component. Selected the packaged component; the **Component Placement** dashboard will be displayed. You can parametrically assemble the component by adding the placement constraints to the component by using this dashboard.

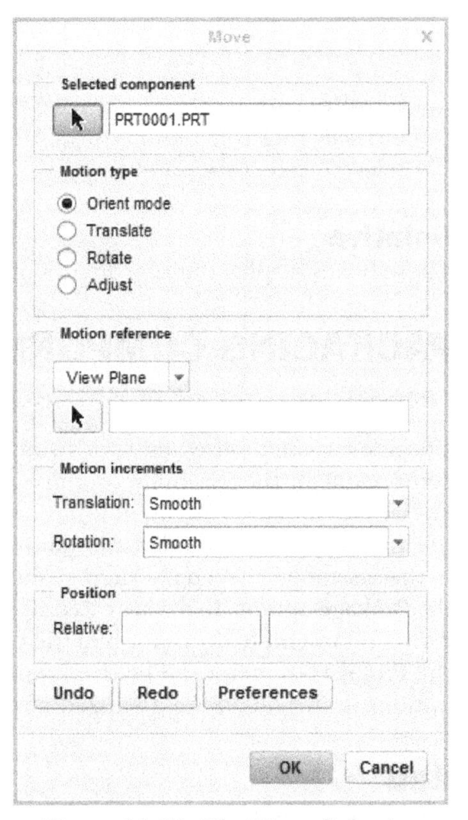

*Figure 11-23 The **Move** dialog box*

Note

*When you exit the **Component Placement** dashboard before fully constraining the component, then the component is assembled as package.*

CREATING SIMPLIFIED REPRESENTATIONS

As you know the assembly designs consist of a number of parts and subassemblies. In case of some complicated assemblies, such as an engine, once the cover is assembled, the components inside it are not visible. So there have to be some means to temporarily remove certain components that are not desired at some point of time from the current display or force them to be displayed in the wireframe so that you can see through them. This process of removing some components or changing the type of display from the current display is called **Simplified Representation**.

Simplified representations can be created using the cascading menu or by using the **View Manager**. The only difference between them is that by using the **View Manager**, you can give a name to the representation. To create a simplified representation by using the cascading menu, select a component from the **Model Tree** and then choose the **Set Representation to** option from the **Manage Views** drop-down in the **Model Display** group of the **Model** tab in the **Ribbon**; a cascading menu will be displayed, as shown in Figure 11-24.

***Figure 11-24** The cascading menu*

The Cascading Menu

The options available in the cascading menu are used to create a simplified representation without saving it. The simplified representation reduces the regeneration time of the assembly, especially the complex assemblies. The options in this menu are discussed next.

Exclude

When you choose this option, the selected component is removed from the display. You can even select more than one component from the **Model Tree** or from the assembly to remove them from the display. However, the component remains in the memory of PTC Creo Parametric and is even displayed in the **Model Tree**. This component can be forced to be displayed again at any time. To display the component again, select the excluded component from the **Model Tree** and then choose the **Master** option from the cascading menu. Figure 11-25 shows a butterfly valve assembly in which the body is removed from the current display. This kind of representation helps in assembling the components inside the body, like the plate and the screw in this case.

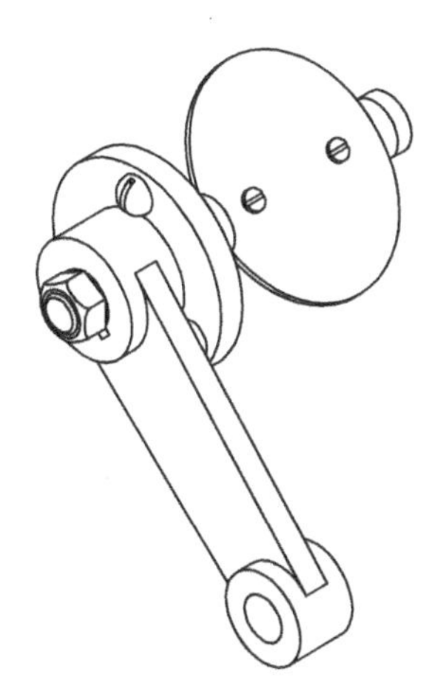

***Figure 11-25** Components removed from the assembly*

Master

This option is used to restore the original representation of the assembly, thereby displaying all components in it. When you use this option, the components that were removed from the current display, after using the **Exclude** option, get redisplayed, see Figure 11-26.

Assembly Only

The **Assembly Only** option will be used to hide all components of an assembly except the components of the subassembly.

Geometry

In this type of simplified representation, the complete geometry of the components will be displayed in the current model display style. This type of simplified representation takes a long time for regeneration. The components that are not selected in this type of representation cannot be modified. However, these components can be redefined.

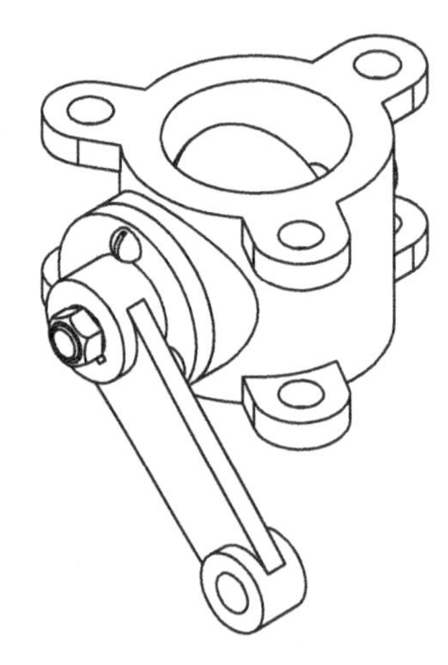

Figure 11-26 *Master representation of the assembly*

Graphics

This type of representation is generally used for assembling the components inside some other components. This type of simplified representation also reduces the regeneration time for large assemblies. The graphics representation of an assembly is shown in Figure 11-27.

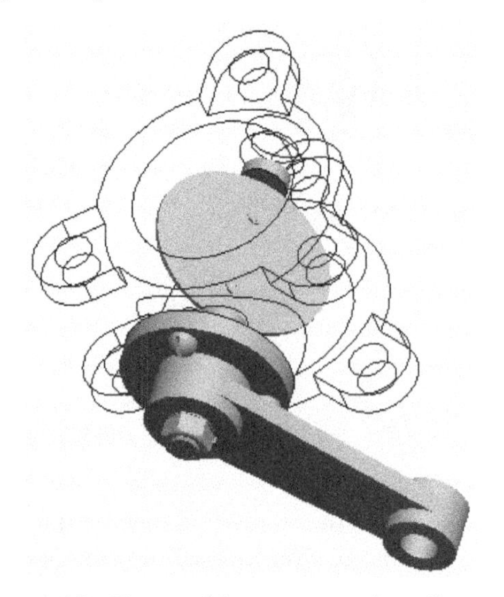

Figure 11-27 *The graphics representation of an assembly*

The View Manager

Ribbon: Model > Model Display > Manage Views > View Manager

The **View Manager** is used to create, modify, and switch between the simplified representations, exploded views, and orientation of the assembly. To invoke the **View Manager**, choose the **View Manager** button from the **Model Display** group; the **View Manager** dialog box will be displayed, as shown in Figure 11-28. The various uses of this dialog box and operations that can be performed on an assembly using this dialog box are discussed next.

Creating a Simplified Representation

When the **View Manager** dialog box is displayed, the **Simp Rep** tab will be chosen by default. The options under this tab are similar to those that were available in the cascading menu. The green arrow on the left of **Master Rep** indicates that this representation is set currently. To set any of the listed representation types, select the representation and right-click to invoke the shortcut menu. Choose the **Activate** option from the shortcut menu to set the display of the assembly to that type of representation or double-click on any one of the representation types.

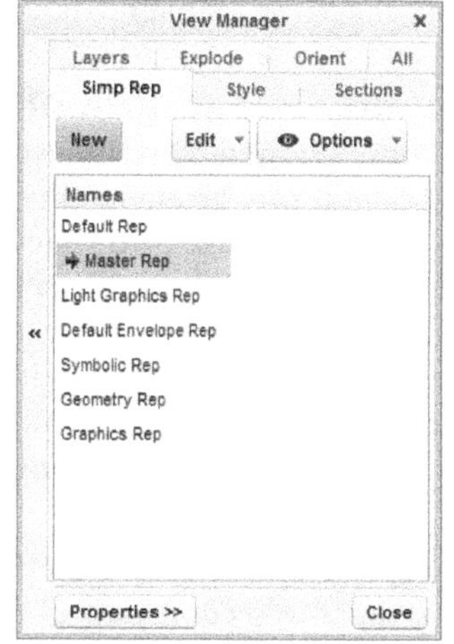

*Figure 11-28 The **View Manager** dialog box*

To create a user-defined representation, choose the **New** button from the **View Manager** dialog box. Specify a name for the representation and press ENTER; the **EDIT** dialog box will be displayed, as shown in Figure 11-29.

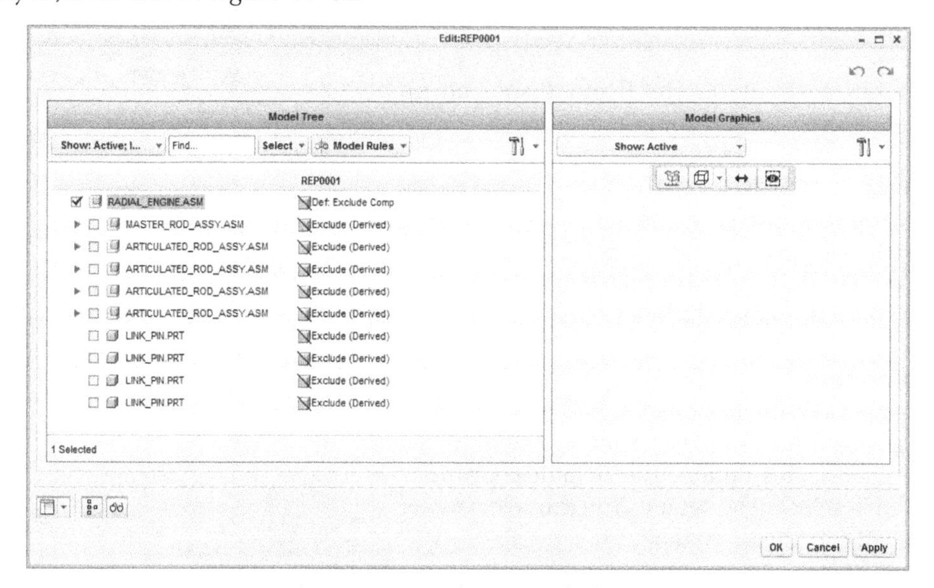

*Figure 11-29 The **EDIT** dialog box*

Using this dialog box, you can select the components that you want to include, exclude, and substitute from the display. You can also redefine the representation type. To do so, select a representation from the list and choose the **Edit** button from the **View Manager** dialog box; a shortcut menu will be displayed. From this shortcut menu, choose the **Redefine** option; the **EDIT** dialog box will be displayed. Using this dialog box, you can redefine the representation.

Creating a Display Style

To create a new display style from the **View Manager** dialog box, choose the **Style** tab. Two styles are available in the **Names** list of the dialog box. You can create a new style or redefine an existing style. You can create new display styles and redefine them using the **New** and **Edit** buttons in the **View Manager** dialog box.

Setting the Orientation of an Assembly

If you choose **Named Views** button from the **Graphics** toolbar, a flyout will be displayed. Various orientations given in this flyout are used to display the assembly as viewed from different directions. Using the **View Manager**, you can invoke the **Orientation** dialog box to set a user-defined orientation. To invoke the **Orientation** dialog box, choose the **Orient** tab and create a new orientation by choosing the **New** button from the **View Manager** dialog box. Name the view and then redefine it to display the **Orientation** dialog box. Use this dialog box to set the orientation of the assembly and then save the orientation by giving it a name listed in the flyout displayed on choosing the **Named Views** button.

Creating Sections of an Assembly

You can create the section view of the assembly using the **View Manager** dialog box. Choose the **Sections** tab from the **View Manager** dialog box and then choose the **New** button; a flyout will be displayed, as shown in Figure 11-30. Select any option from flyout and enter the name of the section in the text box displayed. Next, press the middle mouse button; the **Section** dashboard will be displayed. By using the options in this dashboard, you can create the required section.

The options available in the flyout displayed on choosing the **New** button are discussed next.

Planar

This option is used to create a section using a plane or surface of the model. You can also create the offset section from the selected plane or surface by entering a value.

X Direction

This option is used to create a section along the X direction of the selected coordinate system. When you choose this option, the default coordinate system is selected by default. You can select a user defined coordinate system also.

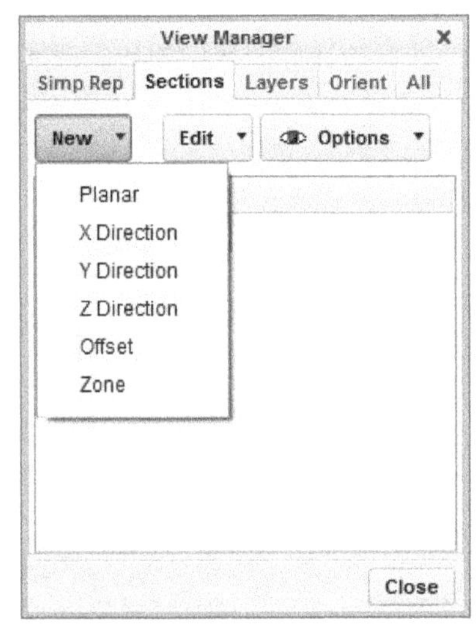

Figure 11-30 *The **New** flyout options*

Y Direction

This option is used to create a section along the Y direction of selected coordinate system. The method of creating a section using this option is similar to that of the **X Direction** option.

Z Direction

This option is used to create a section along the Z direction of selected coordinate system. The method of creating a section using this option is similar to that of the **X Direction** option.

Offset

The **Offset** option is used to create a section by using a user defined sketch. When you select this option, the **Sketch** tab is displayed on the **Section** dashboard. Choose the **Define** button available in the **Sketch** tab to define the sketching plane. In the sketcher environment, draw a line to define the sectioning element.

Zone

This option is used to create a 3D cross-section. You will learn about this option in Chapter 12 of the book.

Tip: *While creating a section, use the **Display a 2D view of the cross section in separate window** button to view the required section in a separate **2D Section Viewer** window.*

REDEFINING THE COMPONENTS OF AN ASSEMBLY

After you have assembled the components of an assembly, you may need to modify or redefine the location constraint or the placement constraint of the components. To do so, select the component from the **Model Tree** and right-click to invoke the shortcut menu. Choose the **Edit the definition of the selected object** option from the **Edit Actions** area of the shortcut menu. You can also right-click on the component to be redefined in the drawing area and choose the **Edit the definition of the selected object** option from the **Edit Actions** area of the shortcut menu. All components assembled after the selected component will become invisible and the **Component Placement** dashboard will be displayed. You can modify the existing placement constraints or add new constraints using this dialog box.

REORDERING COMPONENTS

Ribbon: Model > Component > Component Operations

To change the order of assembling the components, choose the **Component Operations** option from the **Component** group; the **COMPONENT** menu will be displayed, as shown in Figure 11-31. Choose the **Reorder** option from this menu; you will be prompted to select the component to be reordered. When you select a component to reorder from the **Model Tree** or from the drawing area and choose the **OK** button followed by **Done**, a message will be displayed in the **Message Area** mentioning those components before which you can insert the selected component.

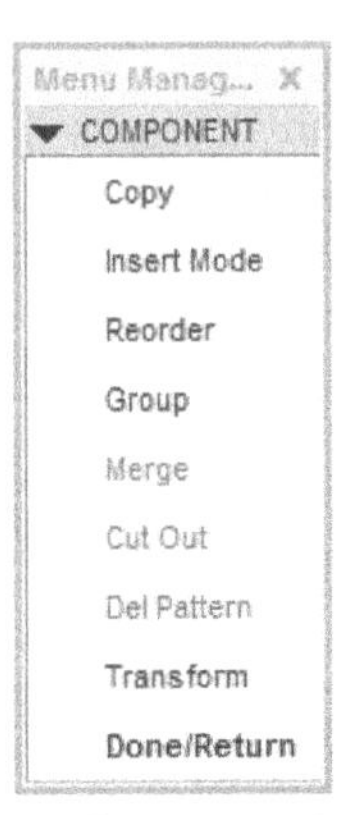

*Figure 11-31 The **COMPONENT** menu*

Tip: *You can also reorder the components in an assembly by using the **Model Tree**. This can be done by selecting the component in the **Model Tree** and then dragging it to the position where it can be placed. Note that this method of reordering does not inform you where the component can or cannot be placed. However, the parent component of the selected child component gets highlighted when you start dragging the child component. This means if a component has a parent, then the parent is also reordered with the child. Remember that in the **Model Tree**, the parent component will always be placed before its child component.*

SUPPRESSING/RESUMING COMPONENTS

If you do not want certain components of the assembly to appear in the current display or in the drawing views, you can suppress them. To suppress the components, select the component from the **Model Tree** or the drawing area and then choose **Suppress > Suppress** from the **Operations** group.

Similarly, the suppressed components can be resumed by choosing the **Resume > Resume** from the same group. The components can also be suppressed/resumed using the shortcut menu that will be displayed when you right-click on the component in the **Model Tree** or in the drawing area. In the shortcut menu, choose **Suppress** to suppress and **Resume** to resume the components.

REPLACING COMPONENTS

Ribbon: Model > Operations > Replace

The existing components of an assembly that are assembled using the assembly constraints can be replaced with some other components, if necessary. To replace a component, choose the **Replace** tool from the **Operations** group; the **Replace** dialog box will be displayed, as shown in Figure 11-32 and you will be prompted to select the components to be replaced. Select the component to be replaced from the drawing area or from the **Model Tree**. The new component can be assembled using the **Family Table**, **Interchange**, **Module or Module Variant**, **Reference Model**, **Notebook**, **By Copy**, or **Unrelated Component** option available in the **Replace By** area. After you have selected one of these options, for example, the **Unrelated**

Component radio button, click on the **Select New Component** collector; you will be prompted to select a model to replace the component. Choose the **Open** button available on the right of the **Select New Component** collector; the **Open** dialog box will be displayed. Select the required model from it and choose the **Open** button; the name of the selected model will be displayed in the **Select New Component** collector in the **Replace** dialog box. Now, choose the **Apply** button and then the **OK** button from the **Replace** dialog box; the **Component Placement** dashboard will be displayed. Now, add the required placement constraints to the model using this dashboard.

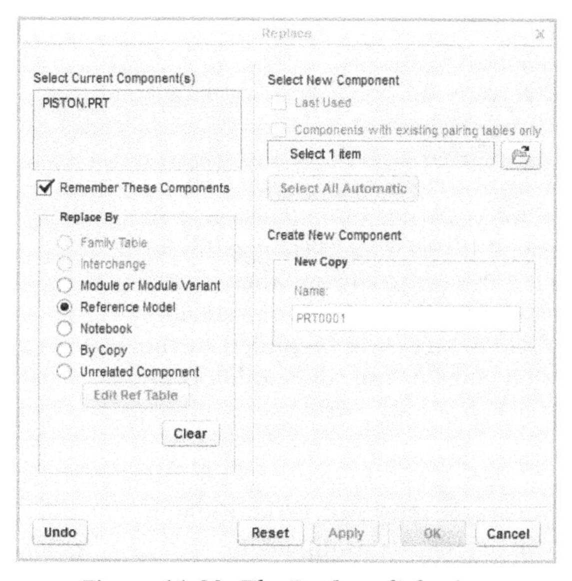

*Figure 11-32 The **Replace** dialog box*

ASSEMBLING REPEATED COPIES OF A COMPONENT

Sometimes, the assembly design demands the assembling of a particular component more than once. One option is that you should assemble the component every time and add the placement constraints all over again. However, this is a very tedious and time-consuming process, especially in the assemblies that have a large number of similar components. To solve this problem, PTC Creo Parametric provides you with an option of assembling multiple copies of the components. To repeat a component in an assembly, select the component to repeat from the **Model Tree** or from the drawing area and then choose the **Repeat** tool from the **Component** group; the **Repeat Component** dialog box will be displayed, as shown in Figure 11-33. The options in the **Repeat Component** dialog box are discussed next.

Component Area

The button in this area is used to select the component whose multiple copies are to be assembled. The name of the selected component will be displayed in the display box provided in this area.

Variable assembly references Area

The **Variable assembly references** area displays the placement constraints that are applied on the selected component. You can keep the placement constraints that need to be present for the new component and modify the remaining constraints by selecting them from this area. The field in this area consists of the following columns.

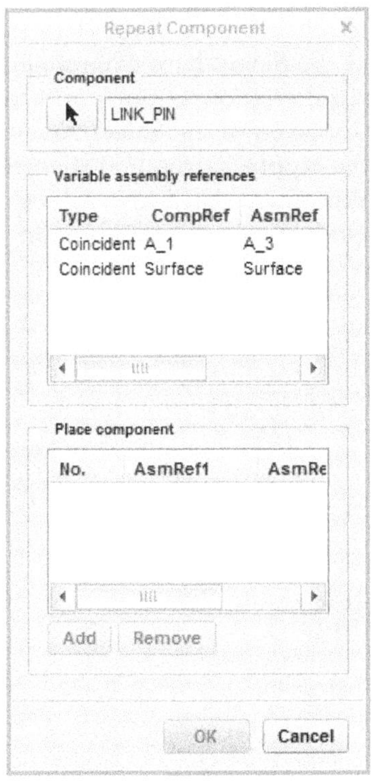

*Figure 11-33 The **Repeat Component** dialog box*

Type

The **Type** column displays the type of constraints that are applied to the component.

CompRef

The **CompRef** column displays the references that were used to assemble the component. If you want to change this for the new component, select this constraint and then choose the **Add** button. Now, select a reference on the assembly. A copy of the selected component will be added to the assembly at a new location. Similarly, to add another copy, repeat this procedure.

AsmRef

The **AsmRef** column displays the reference on the assembly that was used to assemble the selected component.

Offset

The **Offset** column displays the deviation from location of the new copy to the main part which is repeated.

Place component Area

The **Place component** area displays the new references added for the newly placed copies of the selected component.

Add

The **Add** button is chosen to add the required references for the new copy of the selected component. This button will be available only when you select one of the constraints from the **Variable assembly references** area.

Remove

The **Remove** button is chosen to remove the selected reference from the new component. This button is available only when you select one of the references from the **Place component** area.

To create a repeated component in an assembly, select the required component and then invoke the **Repeat Component** dialog box. Next, select the constraint to be copied from the **Variable**

Assembly Refs area in the dialog box and then choose the **Add** button. Now, select the new reference for the new copied component from the drawing area; the selected component will be copied to the new reference. Choose the **OK** button from the **Repeat Component** dialog box to accept the repeated copy of the component.

MODIFYING THE COMPONENTS OF AN ASSEMBLY

Sometimes, during or after the assembly of a component, you may need to modify the dimensions of the component. You can even redefine the selected feature in the **Assembly** mode itself. The methods used to modify and redefine the features are discussed next.

Modifying Dimensions of a Feature of a Component

The dimensions of a selected feature of the component can be modified by selecting the component from the **Model Tree** or from the drawing area. Select the model when the outline of the model turns green. Then right-click to invoke the shortcut menu. Select the **Open** option to open the component as a part file in a separate window and edit the features as required. After modifying the dimensions, regenerate the component, and save the part file. Next, close the part file and activate the assembly window and regenerate the assembly to incorporate the changes made in the component.

Redefining a Feature of a Component

You can redefine the features of a component in the **Assembly** mode. This saves time in opening the part in the **Part** mode and then redefining the features. To redefine the feature of a component, select the component from the **Model Tree** and right-click to invoke the shortcut menu. Choose the **Activate** option from the shortcut menu; a small green button appears on the activated component in the **Model Tree**. After the component is activated, select the feature that you need to redefine from the drawing area. Hold down the right mouse button to invoke the shortcut menu. Choose the **Edit the definition of the selected object** option from the **Edit Actions** area in the shortcut menu displayed; the feature creation tool is activated and the feature can be redefined now.

CREATING THE EXPLODED STATE

Ribbon: Model > Model Display > Edit Position

In assembly design, the components assembled inside other components may not be visible. This could be misleading and may confuse the viewer as the component cannot be viewed. To avoid this confusion, generally an exploded view is provided along with the assembled view. An exploded view is a state in which all the components move from their original position so that they are visible, as shown in Figure 11-34.

The exploded state can be generated by invoking the **Edit Position** tool from the **Model Display** group. On choosing the **Edit Position** option, the default exploded state of an assembly will be displayed. Also, the **Explode Tool** dashboard will be displayed, as shown in Figure 11-35.

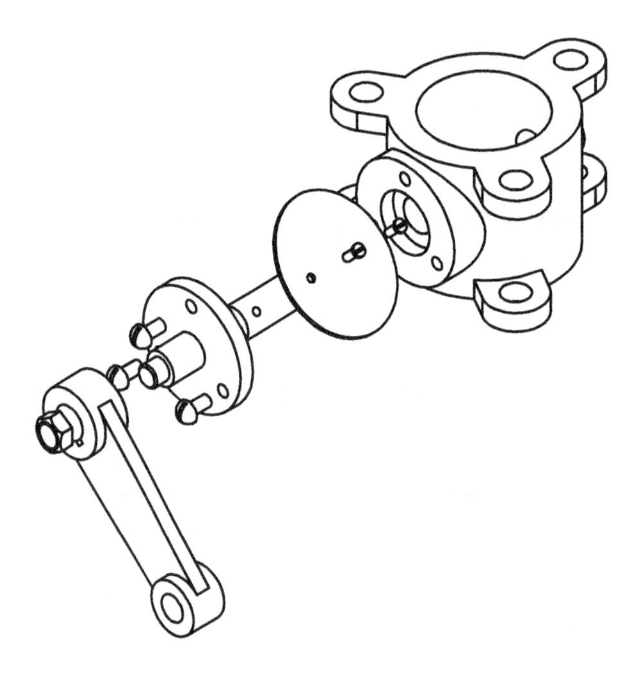

Figure 11-34 The exploded state of an assembly

Figure 11-35 Partial view of the ***Explode Tool*** *dashboard*

Various options in this dashboard are discussed next.

Translate

This button is used to move a component from its current location to a new location in the assembly. To move a component, choose this button; three axes will be displayed on it. Next, move the cursor on the desired axis. When the axis gets highlighted, click and drag the component.

Rotate

This button is used to rotate a component about a selected axis. To rotate a component, choose this button; you will be prompted to select a reference for rotation. Select a reference (edge, axis, planar face/datum plane, any 2 points, or coordinate system); a curved arrow will be displayed. Using this arrow, you can rotate the component.

View Plane

The **View Plane** option is used to select the current view plane (that is, the computer screen) as the reference for exploding a selected component. Choose this button; a small ball will be displayed. Click on the ball and drag to move the component to the desired location.

References Tab

The options in this tab are used to select the components as well as to define a reference for movement. Choose this tab; a slide-down panel will be displayed, as shown in Figure 11-36. The **Components to Move** collector allows you to select and display the name of the selected component. The **Movement Reference** collector allows you to select a reference to define the movement of the selected component. Note that you need to click in the collector so that it gets activated, thereby enabling you to select the required entities.

Options Tab

Choose this tab to display a slide-down panel, as shown in Figure 11-37. The options in this slide-down panel are discussed next.

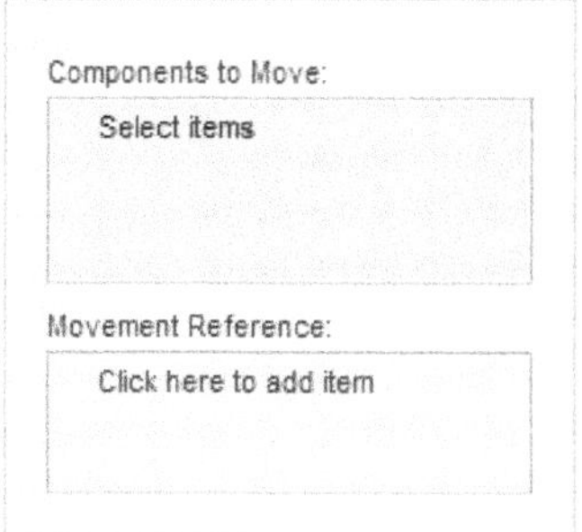

*Figure 11-36 The **References** slide-down panel*

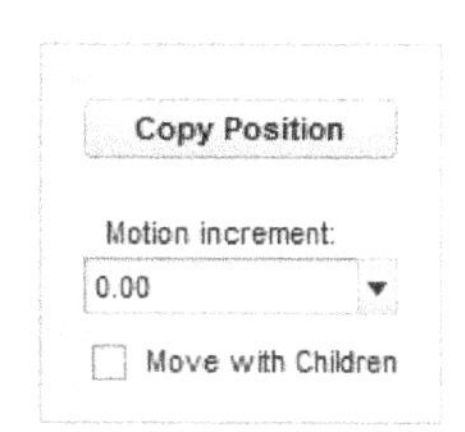

*Figure 11-37 The **Options** slide-down panel*

Copy Position

This button is used to restore the exploded component to its original position. For example, when an assembly explodes, its components move away. To restore components to their original position, choose the **Copy Position** button from the **Options** slide-down panel; the **Copy Position** dialog box will be displayed. Click in the **Components to Move** collector to activate it and select the exploded component moved away from the assembly. Next, click in the **Copy Position From** collector to activate it and then select a component from the assembly that is at a similar location in the assembly. Choose the **Apply** button from this dialog box; the selected component will be assembled at its original position in the assembly.

Motion Increment

You can specify an increment value for movement as well as define a smooth motion using this drop-down list.

Move with Children

On selecting this check box, the selected component moves along with its child components.

Note

*If you have not generated the exploded view of an assembly, a default exploded state will be available on choosing **Explode View** button from the **Model Display** group. To display the unexploded state of the assembly, again choose **Explode View** from the **Model Display** group.*

*Alternatively, you can invoke the **View Manager** and then choose the **Explode** tab. Now, double-click on the required state to activate it. Also, you can create new states, edit existing states, and switch between various states created.*

Explode Lines Tab

Explode lines are used to display the actual path and direction of the mating components. Generally, these lines are used for an easy visualization of the exploded components of the assemblies. The exploded state of an assembly with the offset lines is shown in Figure 11-38.

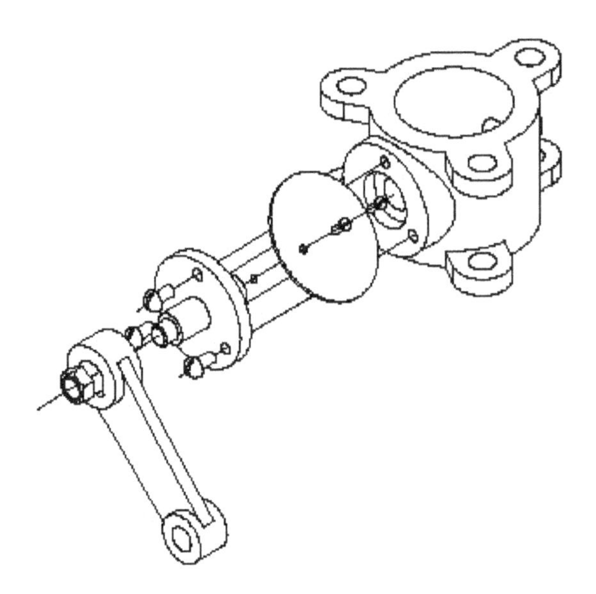

Figure 11-38 *The exploded state of an assembly with the offset lines*

Create cosmetic offset lines to illustrate movement of exploded component Button

To create explode lines, an exploded state should exist in which the explode lines will be created. Choose the **Create cosmetic offset lines to illustrate movement of exploded component** button from the panel that is displayed on choosing the **Explode Lines** tab of the **Explode Tool** dashboard; the **Cosmetic Offset Line** dialog box will be displayed, as shown in Figure 11-39. Using this dialog box, you can select the objects for creating the explode lines. The objects that can be selected are axis, surfaces, edge, or curve.

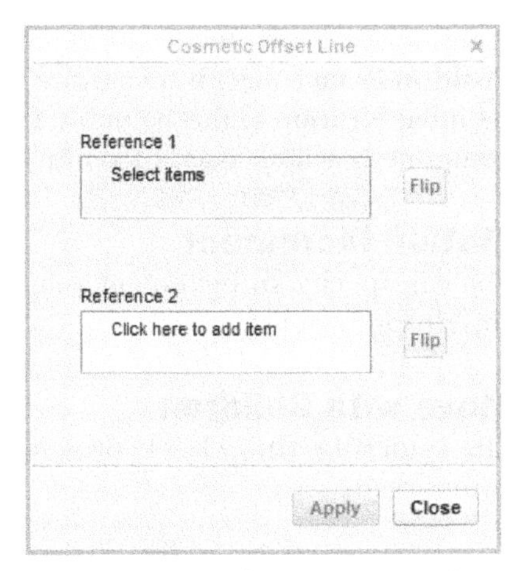

Figure 11-39 *The **Cosmetic Offset Line** dialog box*

Reference 1

Click in this collector to activate it and then select the first reference.

Reference 2
Click in this collector to activate it and then select the second reference.

Next, choose the **Apply** button to accept the creation of the explode lines and continue for the next collection. Choose the **Close** button to close this dialog box.

Note
The display of explode lines is automatically turned off once you display the unexploded state of the assembly.

Edit the selected explode line

This button will be available in the **Explode Lines** tab only when you select the explode line created earlier. To modify the explode line created, select it and then choose this button; two handles will be displayed at both ends of the line. Next, click on the handle and drag it to extend or shorten the explode line segment. To exit the edit mode, right-click; a shortcut menu will be displayed. Choose **Clear** from this shortcut menu.

Delete the selected explode line

To delete the selected explode line, select the exploded line created and then choose the **Delete the selected explode lines** button; the selected explode line will be deleted.

Edit Line Styles

You can edit the attributes of the selected exploded line by choosing the **Edit Line Styles** button. You can modify the attributes such as line style type, color of the line style, or copy the line style from an existing line type.

Default Line Style

Choose this button to set the default attributes of the explode line.

Note
*To delete an explode line, select the explode line and right-click; a shortcut menu will be displayed. Choose **Remove Explode Line** from it. You can also set the line style for the explode lines that you create in an exploded view by using the shortcut menu.*

THE BILL OF MATERIALS

Ribbon: Model > Investigate > Bill of Materials

The Bill of Materials or BOM is the tabular representation of all components of the assembly, along with the information associated with them. The information can be the material of the components, the additional note with the components, and so on. PTC Creo Parametric provides you with a ready-made BOM that can be directly utilized. This BOM will be automatically updated during the assembly of the components. The BOM can be viewed by using the **Bill of Materials (BOM)** dialog box shown in Figure 11-40, which will be displayed on invoking the **Bill of Materials** tool from the **Investigate** group. The options in this dialog box are discussed next.

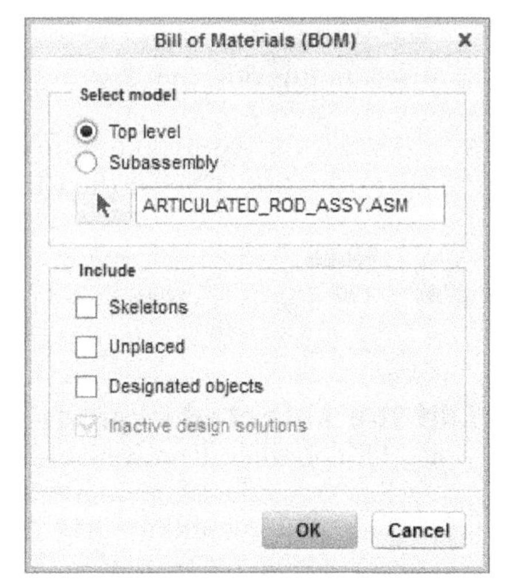

*Figure 11-40 The **Bill of Materials (BOM)** dialog box*

Select Model Area

The options in the **Select Model** area are used to select the type of assembly whose BOM has to be displayed. These options are discussed next.

Top Level

Using the **Top Level** radio button, you can display the top level assembly in the BOM. In this case, the subassemblies inserted in the current assembly are considered as a component.

Subassembly

The **Subassembly** radio button is used to list the components of the selected subassemblies, as individual components in the **BOM**. When you select this radio button, you will be prompted to select the subassembly.

Include Area

The options in this area are discussed next.

Skeletons

In the Skeleton model geometry, the assembly level features such as cuts and holes do not affect the geometry of the skeleton model. The **Skeletons** check box is selected so that the skeletons, if any, in the assembly could also be included in the BOM.

Unplaced

The **Unplaced** check box is selected to include the unplaced components in the BOM.

Designated Objects

The **Designated Objects** check box is selected to include the bulk items, if any, that are used in the assembly. Bulk items are items whose solid models are not created but they are used in the assembly. In PTC Creo Parametric, you can include the weight and other properties of these bulk items to calculate certain parameters related to the assembly like total weight of assembly. Examples of bulk items are paint, glue, cotton lace, and so on.

After you have selected the options in the **BOM** dialog box, choose **OK** to open the browser that displays the BOM for the selected assembly. The file of the Bill of Material is automatically saved with the name of the assembly. The extension of this file is *.bom*. You can later retrieve it in the **Drawing** mode to display the BOM in the drawing views of the assembly.

GLOBAL INTERFERENCE

Ribbon: Analysis > Inspect Geometry > Global Interference

Global Interference is a type of analysis that can be conducted in the **Assembly** and **Drawing** modes. This analysis can be performed on the assembly involving interference fit. Typical examples of interference fit or press fit is press fitting of shaft into bearing and bearing into its housing. Using this analysis, you can check the total volume overlap of the two assembled parts. This volume can be used to calculate the compressive force working on the two assembled parts. The procedure to check the interference between the assembled parts is given next.

1. Invoke the **Global Interference** tool from the **Global Interference** drop-down available in the **Inspect Geometry** group under the **Analysis** tab; the **Global Interference** dialog box will be displayed.

2. Now, choose the **Preview** button from **Global Interference** dialog box; the interfered volume will be displayed in the dialog box, refer to Figure 11-41.

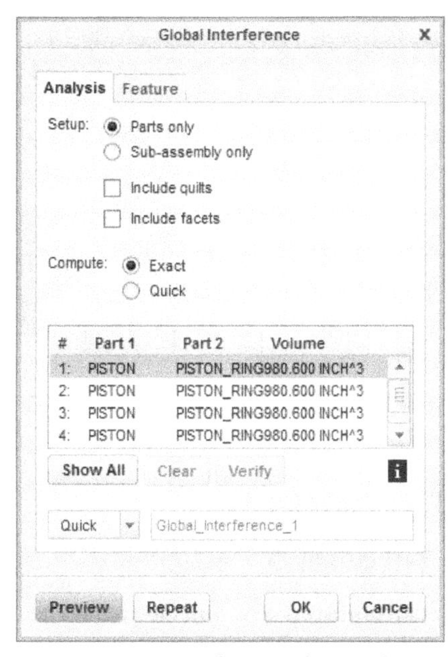

***Figure 11-41** Volume of interference displayed in the dialog box*

In this dialog box, all interfering parts will be displayed with their names and respective interfering volumes. Also, the interfering parts will be highlighted in red color in the model area.

PAIRS CLEARANCE

Ribbon: Analysis > Inspect Geometry > Global Interference > Pairs Clearance

The **Pairs Clearance** analysis is used to check the clearance between two assembled parts. This analysis is available in the **Part**, **Assembly**, **Piping** and **Drawing** modes. In some assemblies, a gap needs to be provided for free motion of the parts called clearance. Using the **Pairs Clearance** analysis, you can find out the clearance between the two assembled parts. Generally two edges or surfaces are required for measuring the clearance. The procedure to check clearance between the assembled parts is given next.

1. Invoke the **Pairs Clearance** tool from the **Global Interference** drop-down available in the **Inspect Geometry** group of the **Analysis** tab; the **Pairs Clearance** dialog box will be displayed.

2. Select the two surfaces or edges of the model in the drawing area.

3. As you choose the surfaces or edges, the clearance value will be displayed in the dashboard as shown in Figure 11-42.

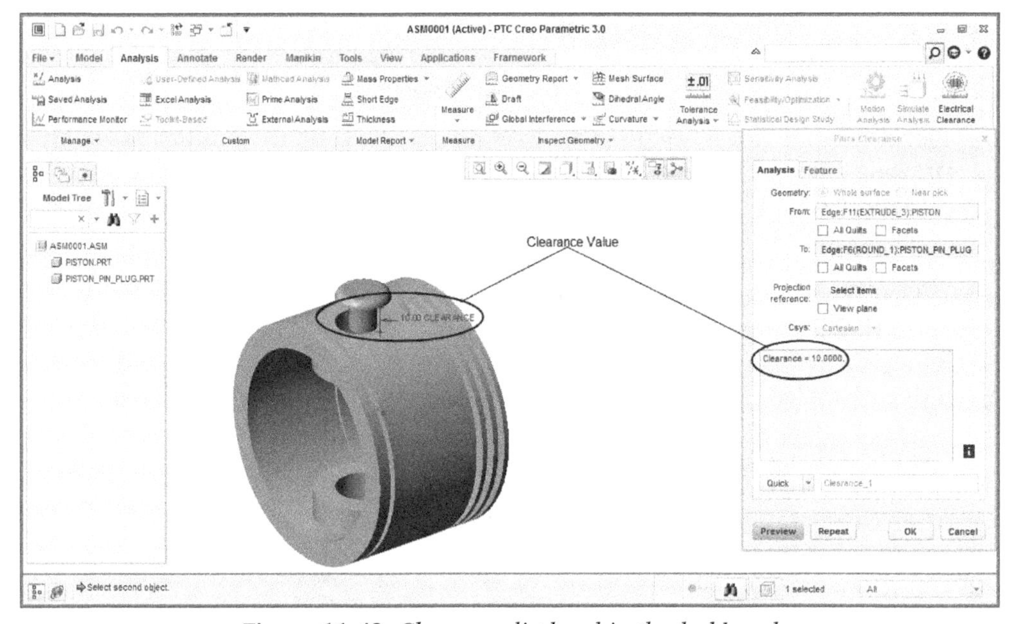

Figure 11-42 Clearance displayed in the dashboard

TUTORIALS

Tutorial 1

In this tutorial, you will create all components of the Shock assembly and then assemble them, as shown in Figure 11-43. Also, you will create an exploded state of the assembly, as shown in Figure 11-44. The BOM is shown in Figure 11-45. The dimensions of the components are shown in Figures 11-46 through 11-53. **(Expected time: 2 hrs)**

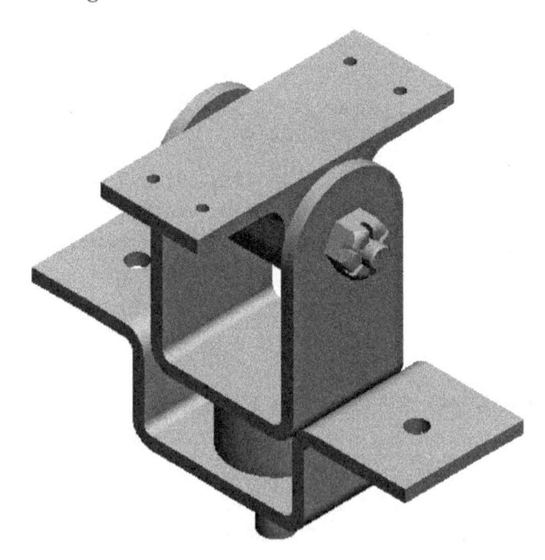

Figure 11-43 The Shock assembly

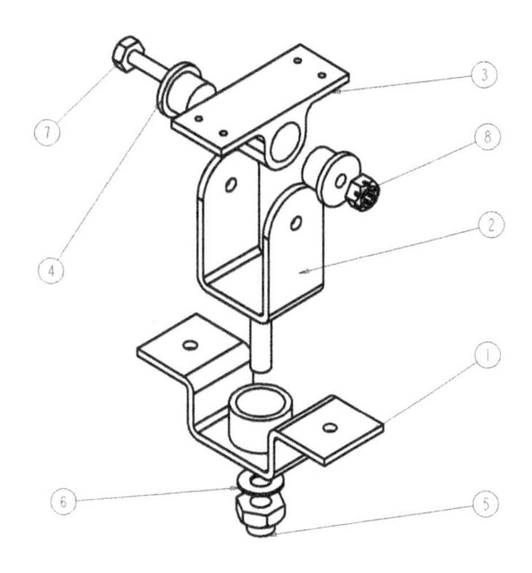

Figure 11-44 *The exploded state of the Shock Assembly*

ITEM	QTY.	NAME	MATERIAL
1	1	BRACKET	STEEL
2	1	U-SUPPORT	STEEL
3	1	PIVOT	STEEL
4	2	BUSHING	BRONZE
5	1	SELF LOCKING NUT	0.625-11UNC
6	1	WASHER	USER DEFINED
7	1	HEXAGONAL BOLT	STEEL
8	1	CASTLE NUT	STEEL

Figure 11-45 *The Bill of Material for the Shock assembly*

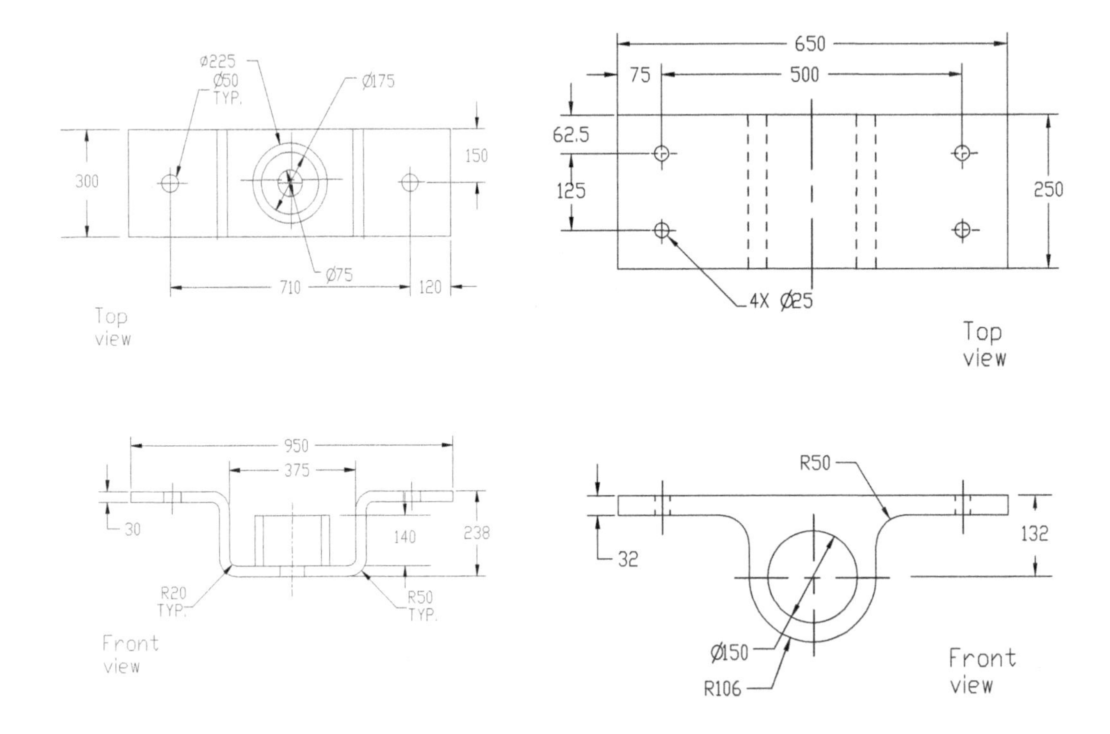

Figure 11-46 *Dimensions for the Bracket*

Figure 11-47 *Dimensions for the Pivot*

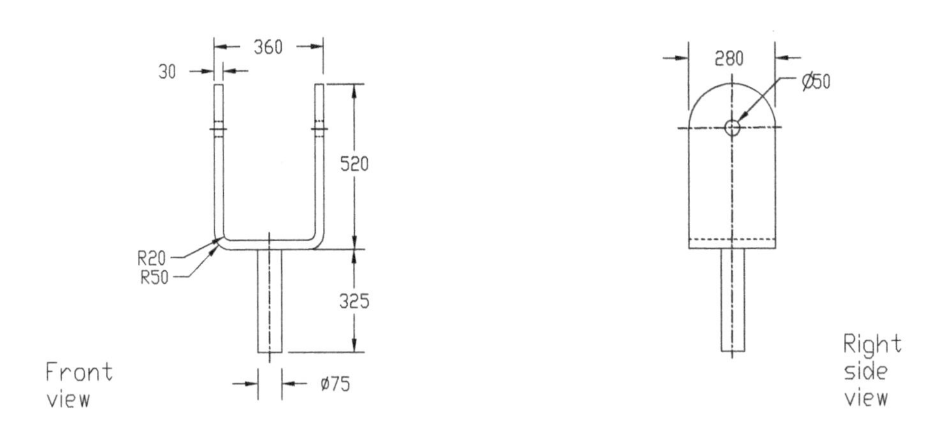

Figure 11-48 *Dimensions for the U-Support*

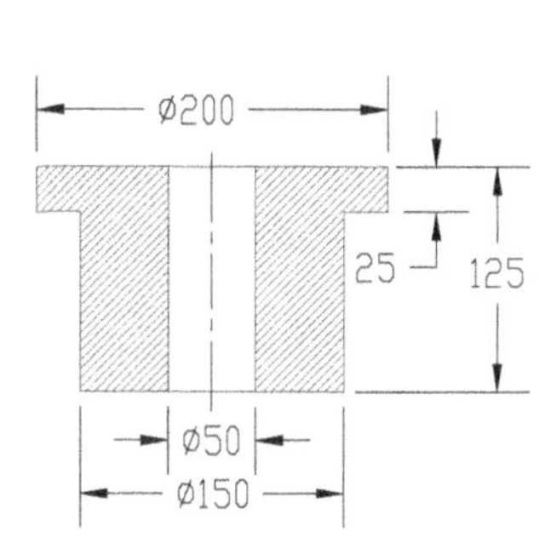

Figure 11-49 Dimensions for the Bushing

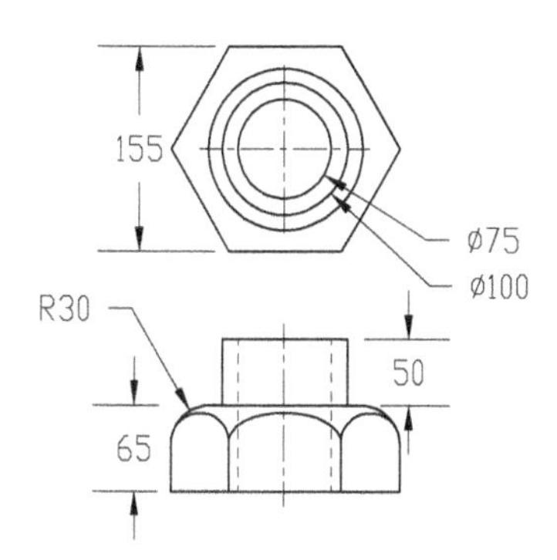

Figure 11-50 Dimensions for the Self locking nut

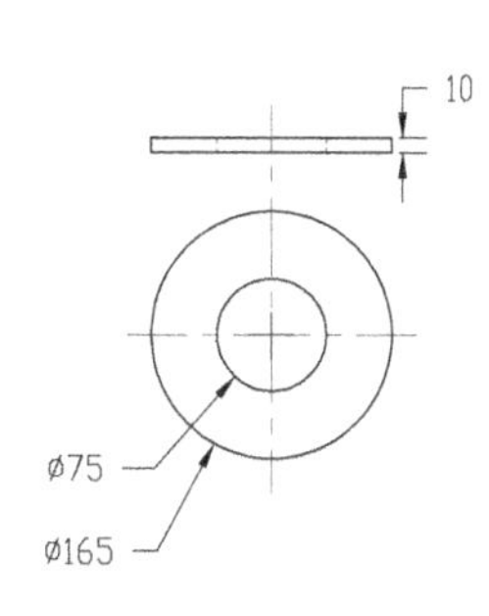

Figure 11-51 Dimensions for the Washer

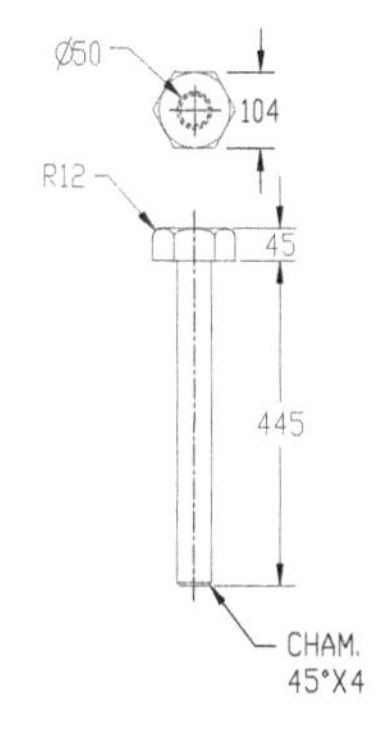

Figure 11-52 Dimensions for the Hexagonal bolt

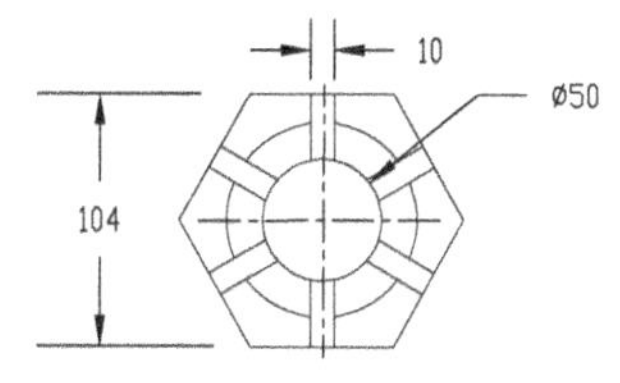

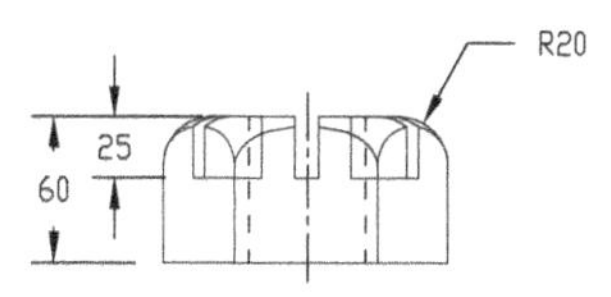

Figure 11-53 Dimensions for the Castle nut

Note

You can download the part files for the assembly from www.cadcim.com. The complete path for downloading the files is as follow: Textbooks > CAD/CAM > Creo Parametric > Creo Parametric 3.0 for Designers

a. Create all components of the assembly as separate part files in the **Part** mode.
b. Create a new file in the **Assembly** mode and then assemble the **U-Support** with the default datum planes.
c. Assemble the **Bushing** with the **U-Support** and then repeat the **Bushing** in the assembly, refer to Figures 11-54 through 11-56.
d. Assemble the **Pivot** with the **Bushing**, refer to Figures 11-57 and 11-58.
e. Suppress the two Bushings and the **Pivot**.
f. Assemble the **Bracket** with the assembly, refer to Figures 11-59 and 11-60.
g. Assemble the **Washer**, **Hexagonal Bolt**, **Castle nut**, and **Self locking nut** in the assembly, refer to Figures 11-61 through 11-63.
h. Unsuppress the suppressed components.
i. After assembling all components, create the exploded state of the assembly, refer to Figure 11-64.
j. Generate offset lines in the exploded state, refer to Figure 11-65.
k. Save the assembly and then exit the **Assembly** mode.

Before you start creating components, set the working directory to *C:\Creo-3.0\c11\Shockassembly*.

Starting the Components of the Assembly

The given assembly is created by using the bottom-up approach. As mentioned earlier, in the assemblies created using the bottom-up approach, the components are created as separate files and are placed in the assembly file. Therefore, first you need to create the components of the assembly.

1. Create all components of the assembly as separate part files and save them in the current working directory.

2. Close the part files, if opened.

Creating a New Assembly File

You need to open a new assembly file to assemble the components.

1. Choose the **New** button from the **Data** group; the **New** dialog box is displayed.

2. Select the **Assembly** radio button in the **Type** area of the **New** dialog box. In the **Sub-type** area of this dialog box, the **Design** radio button is selected by default. Enter the name of the assembly in the **Name** edit box as **SHOCKASSEMBLY**.

Note

*Make sure that the **Use default template** check box is selected in the **New** dialog box. If this check box is not selected, you cannot use the template provided by PTC Creo Parametric. Therefore, in this case, you will need to create the default datum planes by choosing the **Plane** tool from the **Datum** group in the **Ribbon**.*

3. Choose **OK** from the **New** dialog box; the assembly environment is invoked.

Assembling the U-Support with the Default Datum Planes

1. Choose the **Assemble** tool from the **Component** group; the **Open** dialog box is displayed.

2. Select the **U-Support** component from the **Open** dialog box and choose the **Open** button; the **Component Placement** dashboard is displayed and you are prompted to select a reference for auto type constraining.

   In the **Component Placement** dashboard, the **Automatic** constraint type is selected by default in the **Constraint Type** drop-down list.

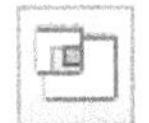

3. Choose the **Show component in a separate window while specifying constraints** button from the **Component Placement** dashboard to display the **U-Support** component in a separate window.

4. Now, select the **FRONT** plane from the drawing area or from the **Model Tree** in the **COMPONENT WINDOW** and select the **ASM_FRONT** plane from the drawing area or from the **Model Tree** in the assembly window. Now, these two datum planes are aligned.

5. Similarly, select the other two datum planes and align them. After all the three planes are aligned, the message **Fully Constrained** is displayed in the **Status** area of the dashboard.

6. Choose the **Build feature** button to complete the assembly of the first part.

Note

*1. You can also assemble the first component in an assembly by selecting the **Default** option from the **Constraint Type** drop-down list in the **Component Placement** dashboard.*

*2. In an assembly, features are not displayed by default in the **Model Tree**, so you may need to display them in the **Model Tree**. To do so, choose on the **Settings** button of the **Model Tree**; a flyout will be displayed. Choose the **Tree Filters** option from the flyout; the **Model Tree Items** dialog box will be displayed. Now, select the **Features** check box from the **Display** area of the dialog box and choose the **OK** button.*

Assembling the Bushing with the U-Support

The next component you need to assemble is the **Bushing**.

1. Choose the **Assemble** tool to display the **Open** dialog box.

2. Select the **Bushing** component and then choose the **Open** button from the **Open** dialog box; the **Component Placement** dashboard is displayed. Choose the **Show component in separate window while specifying constraints** button from the **Component Placement** dashboard to display the **Bushing** component in a separate window, if it is not chosen by default.

3. Choose the **Coincident** option from the **Constraint Type** drop-down list. Select the axis of the **Bushing** component and the axis of the **U-Support** component, as shown in Figure 11-54.

4. Next, select the planar face of the **Bushing** component and the inner planar face of the **U-Support** component, as shown in Figure 11-54. Select the **Coincident** option from the **Constraint Type** drop-down list. Choose the **Build Feature** button; the model similar to the one shown in the Figure 11-55 is displayed in the drawing area.

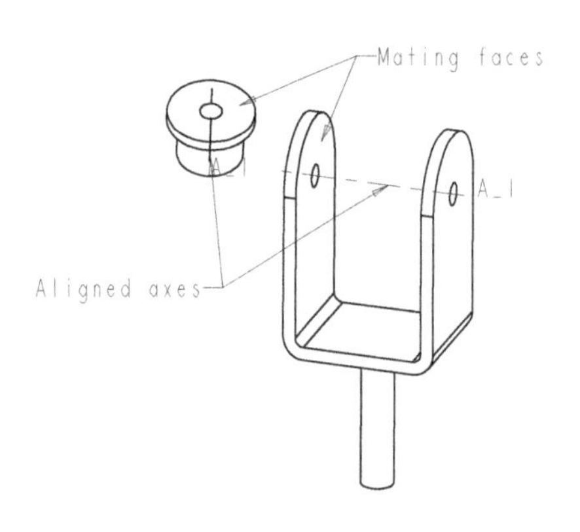

Figure 11-54 *Constraints and the location to apply them*

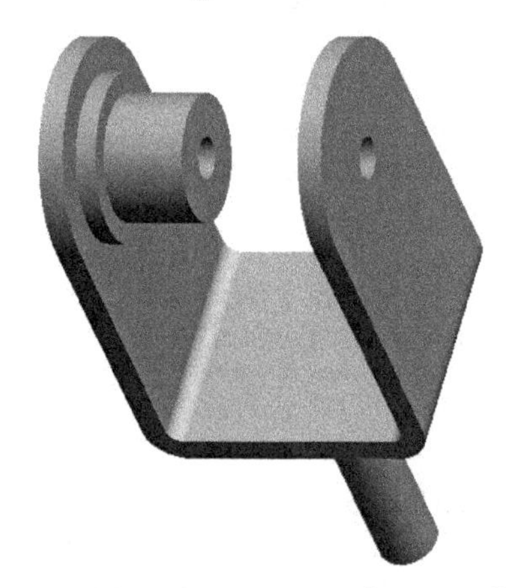

Figure 11-55 *Components after assembling*

Note

*You may need to choose the **Change orientation of constraint** button from the **Component Placement** dashboard to assemble the **Bushing** components in the required orientation.*

Tip: *Viewing the components in separate windows helps the user locate the references for the assembly constraints easily and also gives the user an independent viewing control over the two windows. You can spin, pan, and zoom the components in their respective windows.*

5. The next instance of the **Bushing** can be placed directly by using the **Repeat** tool. To repeat the **Bushing** component in the assembly, select it from the **Model Tree** or from the drawing area and then choose the **Repeat** tool from the **Component** group. On doing so, the **Repeat Component** dialog box is displayed and two **Coincident** constraints are displayed in the **Variable assembly references** area.

6. Select the second **Coincident** constraint from this area and then choose the **Add** button.

7. Select the inner right face of the **U-Support** component as the mating face. The copy of the **Bushing** component is assembled, as shown in Figure 11-56. Choose **OK** from the **Repeat Component** dialog box.

Figure 11-56 *Assembled Bushings and the U-Support*

Assembling the Pivot with the Bushing

The next component that you need to assemble is the **Pivot**.

1. Choose the **Assemble** tool to display the **Open** dialog box.

2. In this dialog box, select the **Pivot** component and then choose the **Open** button to display the **Component Placement** dashboard.

3. Choose the **Placement** tab to display the panel. Select the axis of the **Pivot** and then the axis of the hole of the **Bushing**, as shown in Figure 11-57, and apply the **Coincident** constraint on them.

4. Next, click on the **New Constraint** option in the **Placement** slide-down panel. Select the **Parallel** option from the **Constraint Type** drop-down list.

5. Select the top planar face of the **Pivot** and the planar face of the **U-Support**, as shown in Figure 11-57.

6. Now, select the planar face of the **Pivot** and the planar face of the **Bushing**, as shown in Figure 11-57; the **Distance** constraint is automatically applied between them and is displayed by default in the **Constraint Type** drop-down list in the **Placement** tab. Select the **Coincident** option from the drop-down list so that the selected faces are made coincident.

7. Choose the **Build feature** button to complete the assembly. The model similar to the one shown in Figure 11-58 is displayed in the drawing area.

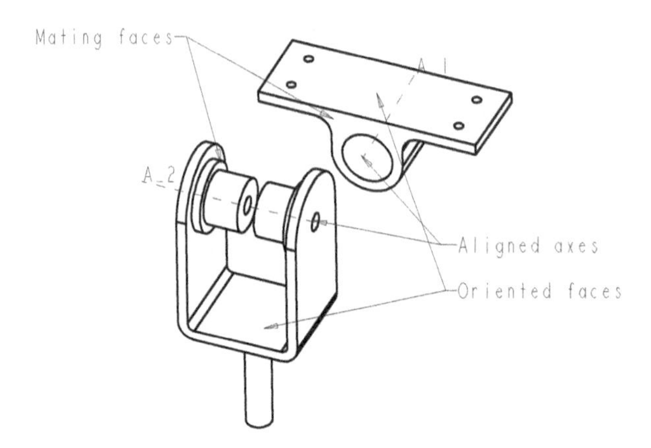

Figure 11-57 Constraints and the location to apply them

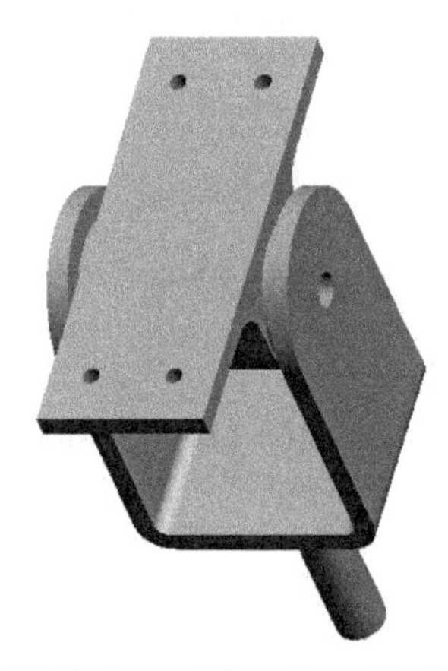

*Figure 11-58 Assembled view of the **Pivot**, **U-Support**, and **Bushing***

Suppressing Components

Next, you need to suppress both the **Bushing** and the **Pivot**. This is because when the components in the assembly increase, some components may be hidden behind the others. As a result, it gets difficult to select them. Also, more the number of components, more are the datum planes. Therefore, it becomes difficult to make selections on the components. However, when you suppress a component, its datum planes are also suppressed and the complications in the drawing area are reduced.

1. Spin the model and select the first **Bushing** from the drawing area or from the **Model Tree** and then right-click and hold; a shortcut menu is displayed. Choose the **Suppress** option from the shortcut menu to suppress the component; the **Suppress** message box is displayed. Choose the **OK** button from the message box; the **Pivot** gets suppressed because it was dependent on the first bushing.

2. Similarly, suppress the other **Bushing** to make the assembly of the other components easy. Choose **OK** from the **Suppress** window to exit.

Note

*Very often you need to use the datum planes and datum axes to assemble the components. Therefore, you need to turn their display on or off from the **Graphics** toolbar, whenever required.*

Assembling the Bracket with the Assembly

The next component you need to assemble is the **Bracket**.

1. Choose the **Assemble** tool from the **Component** group to display the **Open** dialog box.

2. Select the **Bracket** component to open; the **Component Placement** dashboard is displayed.

3. Choose the **Placement** tab to display the panel and select the **Coincident** option from the **Constraint Type** drop-down list.

4. Select the axis of the **Bracket** and the axis of the **U-Support**, as shown in Figure 11-59.

5. Click on the **New constraint** option in the **Placement** panel and select the **Coincident** option from the **Constraint Type** drop-down list. You can also invoke the shortcut menu and choose the **New Constraint** option from it.

6. Now, select the faces to mate, refer to Figure 11-59; the preview of the assembly is displayed.

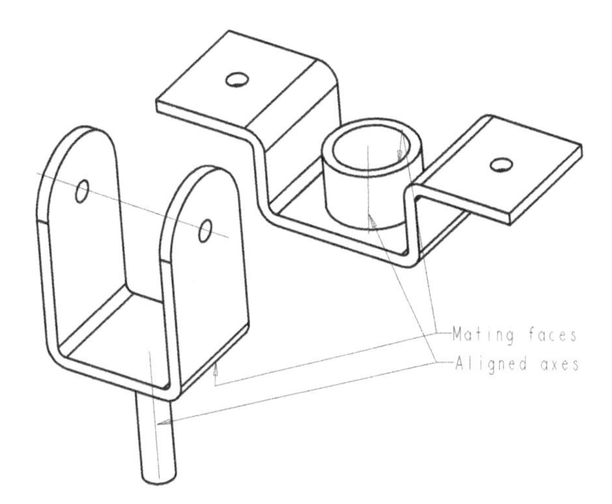

***Figure 11-59** Constraints and the location to apply them*

7. Choose the **Build feature** button from the **Component Placement** dashboard to complete the assembly. The model similar to the one shown in Figure 11-60 is displayed in the drawing area.

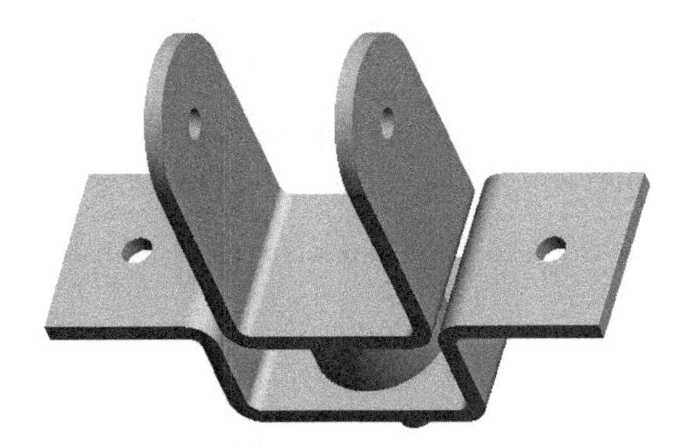

Figure 11-60 *Assembled view of the* ***Bracket*** *with the* ***U-Support***

Assembling the Washer with the Assembly

The next component that you need to assemble is the **Washer**.

1. Choose the **Assemble** tool to display the **Open** dialog box.

2. Open the **Washer** component to display the **Component Placement** dashboard.

3. Choose the **Placement** tab from the dashboard to display the panel and select the **Coincident** option from the **Constraint Type** drop-down list to apply the align constraint.

4. Select the axis of the **Washer** and then select the axis of the **U-Support** from the drawing area to align the axes of the two components.

5. Now, click on the **New Constraint** option in the **Placement** panel and select the **Coincident** option from the **Constraint Type** drop-down list.

6. Select the top planar face of the **Washer** and then the bottom planar face of the **Bracket** to assemble the **Washer**; the message **Fully constrained** is displayed in the **Status** area of the dashboard.

7. Choose the **Build feature** button from the **Component Placement** dashboard to complete the assembly. The model similar to the one shown in Figure 11-61 is displayed in the drawing area.

Figure 11-61 *Assembled view of the **Washer** with the **U-Support***

Inserting the Hexagonal Bolt in the Assembly

The next component that you need to assemble is the **Hexagonal bolt**.

1. Choose the **Assemble** tool to display the **Open** dialog box.

2. Open the **Hexagonal bolt** component to display the **Component Placement** dashboard.

3. Choose the **Placement** tab to invoke a panel and select the **Coincident** option from the **Constraint Type** drop-down list to add the align constraint.

4. Select the axis of the **Hexagonal bolt** and then the axis of the hole on the **U-Support**. Now, the two axes are aligned.

5. Click on the **New Constraint** option in the **Placement** panel and then select the **Coincident** option from the **Constraint Type** drop-down list.

6. Select the bottom planar face of the **Hexagonal bolt** and then select the left outer planar face of the **U-Support**; the preview of the assembly is displayed.

7. Choose the **Build feature** button from the **Component Placement** dashboard to complete the assembly. The model similar to the one shown in Figure 11-62 is displayed in the drawing area.

Assembling other Components

Assemble the **Castle nut** and the **Self locking nut** to the assembly.

Unsuppressing the Components

All components are assembled at their required positions. Now, you need to unsuppress the suppressed components.

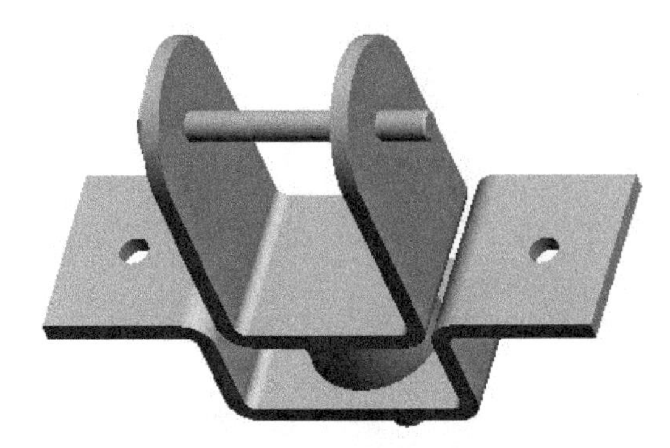

*Figure 11-62 Assembly after assembling the **Hexagonal bolt***

1. Choose the **Resume All** option from the **Resume** flyout of the **Operations** group; the suppressed components appear in the assembly where they were assembled. The completed assembly is shown in Figure 11-63.

Figure 11-63 Completed assembly with all components

Creating the Exploded State of the Assembly

To view all the components in an assembly clearly, you need to create the exploded state of the assembly.

1. Invoke the **Edit position** tool from the **Model Display** group; the **Explode Tool** dashboard is displayed. Also, the default exploded state of the assembly is displayed.

   Now, you need to edit the position of the component in the exploded view.

2. Choose the **References** tab in the **Explode Tool** dashboard and click in the **Movement Reference** collector on the dashboard; you are prompted to select a reference to define the direction of movement.

3. Select the vertical axis of the **U-Support** from the drawing area; you are prompted to select the component to be moved.

4. Select the **Self locking** nut from the drawing area; a direction handle is displayed. Move the cursor to the required direction handle and drag to place it at the required position. Similarly, select and move the **Washer** and the **Bracket** to the required positions.

   Now, you need to move the components in the top half of the assembly. These components are moved with reference to the axis of the hole on the **U-Support**. Therefore, you need to select this axis as the motion reference.

5. Click in the **Movement Reference** collector on the dashboard; you are prompted to select the reference. Select the axis of the hole on the **U-Support** from the drawing area as the reference; you are now prompted to select the components to move.

6. Select and move the **Hexagonal bolt**, **Pivot**, **Bushings** and **Castle nut** to the required locations in the drawing area.

   The model similar to the one shown in Figure 11-64 is displayed in the drawing area.

7. Choose the **Build feature** button from the **Explode Tool** dashboard.

   Now, you need to save this exploded view.

8. To save this view, invoke the **View Manager** dialog box and then choose the **Explode** tab.

9. Choose the **New** button; the **Modified State Save** message box is displayed confirming to save the default exploded view. Choose the **Yes** button from the message box; the **Save Display Elements** dialog box is displayed.

10. Next, enter the **EXP1** in the **Explode** edit box and choose **OK** button from the dialog box and close the **View Manager** dialog box.

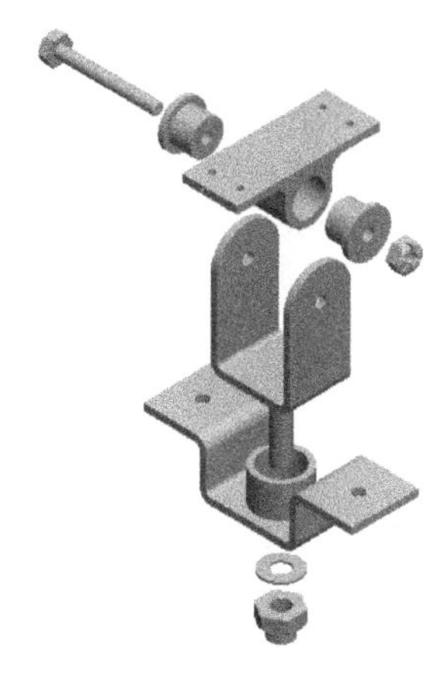

Figure 11-64 *Exploded state of the assembly*

Tip: *You can also view the default exploded state of an assembly by choosing the **Explode View** button from the **Model Display** group in the **View** tab.*

Creating Offset Lines

1. Invoke the **Edit position** tool from the **Model Display** group and choose the **Create cosmetic offset lines to illustrate movement of exploded component** button from the **Explode Tool** dashboard; the **Cosmetic Offset Line** dialog box is displayed.

2. Click in the **Reference 1** collector and then select the axis of the **U-Support**.

3. Next, click in the **Reference 2** collector and then select the axis of the **Self locking nut**. Next, choose the **Apply** button from the **Cosmetic Offset Line** dialog box; the blue colored explode line is created.

4. Similarly, create the offset lines by selecting the axis of the hole on the **U-Support,** followed by the **Bushing** and then the **Castle nut**. Similarly, create offset lines by selecting the axis of the hole on the **U-Support,** followed by the **Bushing** and then the **Hexagonal bolt**.

   The model similar to the one shown in Figure 11-65 is displayed in the drawing area.

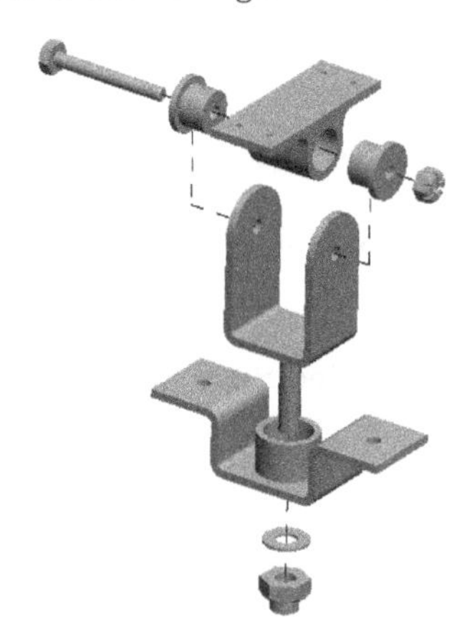

Figure 11-65 *Offset lines displayed in the exploded state*

5. Next, exit the **Cosmetic Offset Line** dialog box by choosing the **Close** button.

6. Choose the **Build feature** button from the **Explode Tool** dashboard.

Saving the Assembly File

1. Choose the **Save** button from the **File** menu and save the model.

Closing the Window

1. Choose the **Close** button from the **Quick Access** toolbar to close the assembly file.

Tutorial 2

In this tutorial, you will create all components of the Pedestal Bearing assembly and then assemble them, as shown in Figure 11-66. You will also create the exploded state, shown in

Figure 11-67, displaying the offset lines. The dimensions of the components are shown in Figures 11-68 through 11-71. The BOM is shown in Figure 11-72. **(Expected time: 2 hrs)**

Figure 11-66 *Assembly of the Pedestal Bearing*

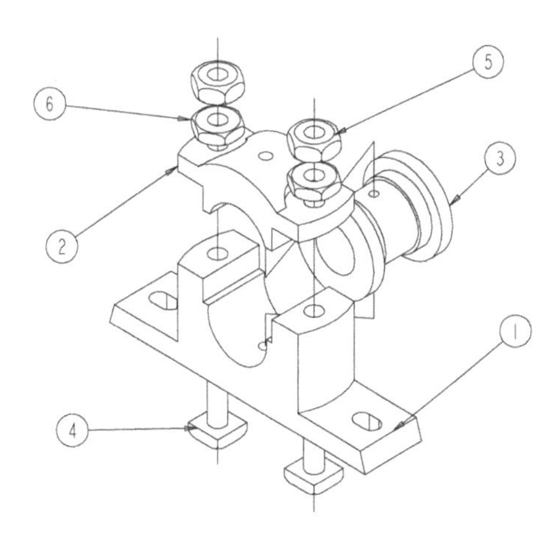

Figure 11-67 *The exploded state of the Pedestal Bearing*

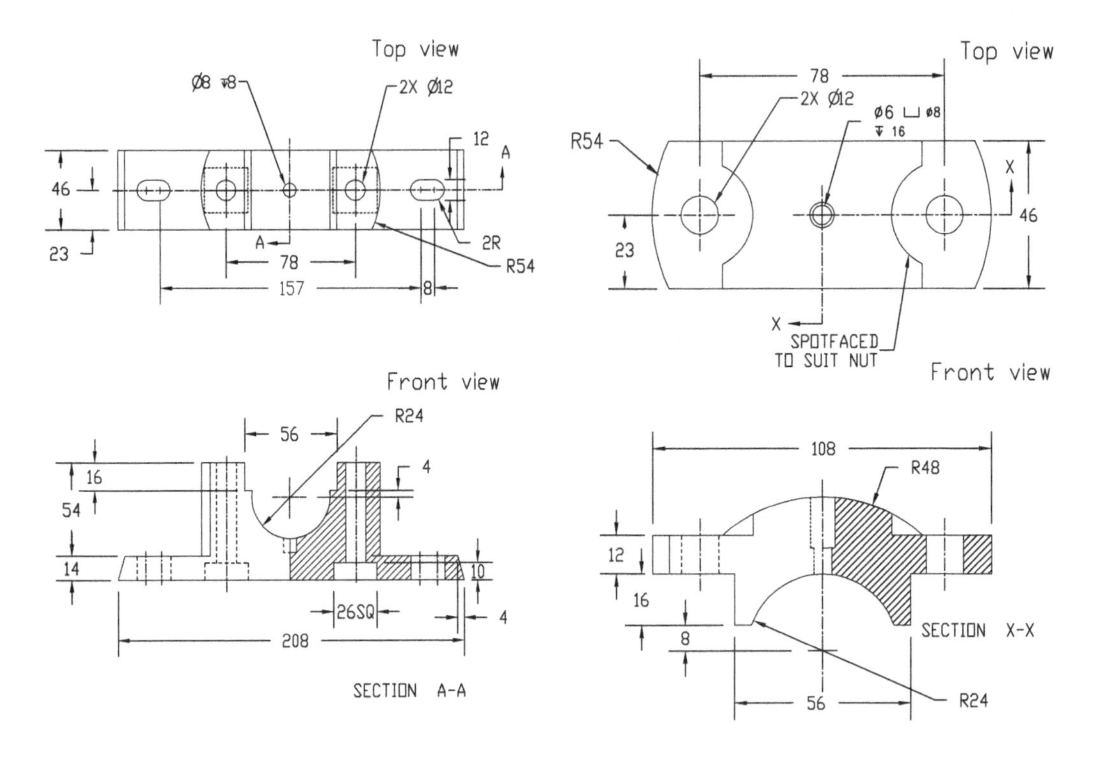

Figure 11-68 Dimensions for the Casting

Figure 11-69 Dimensions for the Cap

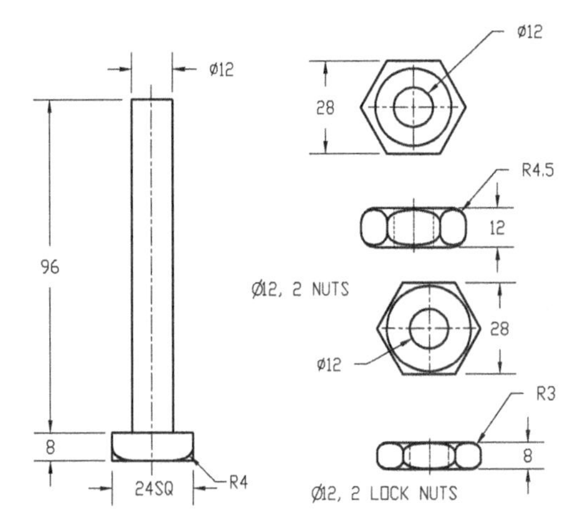

Figure 11-70 Dimensions for the components of the assembly

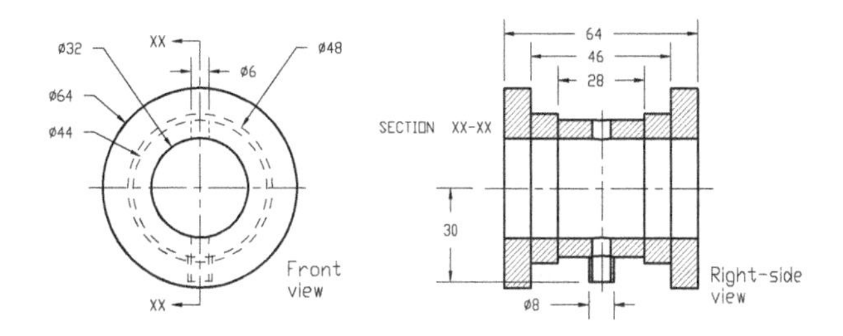

Figure 11-71 *Dimensions for the Brasses*

S. NO.	NO. OFF	PART'S NAME	MATERIAL
1	1	BODY	CAST IRON
2	1	CAP	CAST IRON
3	1	BRASSES	GUN METAL
4	2	SQUARE HEADED BOLT	MILD STEEL
5	2	NUT	MILD STEEL
6	2	LOCK NUT	MILD STEEL

Figure 11-72 *The Bill of Material for the Pedestal Bearing*

The following steps are required to complete this tutorial:

a. Create all components of the assembly as separate part files in the **Part** mode.
b. Create a new file in the **Assembly** mode and then assemble the **Casting** with the default assembly datum planes.
c. Assemble the **Cap** with the **Casting**, refer to Figures 11-73 and 11-74.
d. Suppress the **Cap** from the assembly.
e. Assemble the **Brasses** with the **Casting**, refer to Figures 11-75 and 11-76.
f. Insert the **Square headed bolt**, **Nut**, and **Lock nut** in the assembly and assemble them, refer to Figure 11-77.
g. Resume the suppressed components.
h. Create the exploded state of the assembly, refer to Figure 11-78.
i. Save the assembly and close the file.

Set the working directory to *C:\Creo-3.0\c11\Pedestalbearing*.

Creating Components for the Assembly

To create the assembly, all components must be created first in the **Part** mode. Also, you need to use the bottom-up approach to create the assembly.

1. Create all components of the assembly as separate part files and then save them in the current working directory.

2. Close the part files, if opened.

Creating a New Assembly File

As mentioned earlier, all components created are assembled in an assembly file that has an extension *.asm*. Therefore, you need to open a new *.asm* file.

1. Choose the **New** button from the **File** menu to display the **New** dialog box.

2. Select the **Assembly** radio button in the **Type** area of the **New** dialog box. In the **Sub-type** area of the **New** dialog box, the **Design** radio button is selected by default. Enter the name of the assembly in the **Name** edit box as **PEDESTALBEARING**.

3. Choose the **OK** button to enter into the assembly modeling environment.

Assembling the Casting with the Default Datum Planes

In the new assembly file, the three default assembly datum planes are displayed in the drawing area and the **Model Tree** is displayed at the left of the drawing area. If the display of the **Model Tree** was turned off in the previous tutorial, it would not appear. Now, you can start assembling components.

1. Choose the **Assemble** tool to display the **Open** dialog box.

2. Select **Casting** from the **Open** dialog box and choose the **Open** button; the **Component Placement** dashboard is displayed.

Note

It is recommended to choose the ***Show component in a separate window while specifying constraints*** *button to display the component to be assembled in a new window.*

3. Select the **FRONT** datum plane of the model and then select the **ASM_FRONT** plane from the assembly to align them. Similarly, align the other two default planes with the respective assembly datum planes.

Note

If the datum planes are not displayed in the drawing area, you need to turn on their display by choosing the ***Plane Display*** *tool from the* ***Datum Display Filters*** *drop-down in the* ***Graphics*** *toolbar. After assembling the first component, you may need to turn off the display of planes. This is because the datum planes clutter the drawing area and then the selection of the references for applying assembly constraints becomes difficult. You can turn their display on when the datum planes are required.*

Assembling the Cap with the Casting

At first, you need to assemble the **Cap** with the **Casting**.

1. Choose the **Assemble** tool to display the **Open** dialog box.
2. Select the **Cap** file and then choose the **Open** button from the **Open** dialog box to display the **Component Placement** dashboard.
3. Choose the **Placement** tab to display the panel and select the **Coincident** option from the **Constraint Type** drop-down list.
4. Select the axis of the **Cap** and then the axis of the **Casting**, as shown in Figure 11-73.

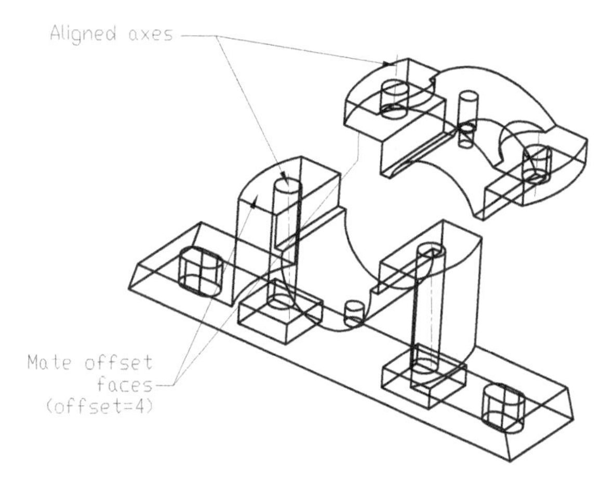

Figure 11-73 *Constraints references used for assembling the components*

5. Choose the **New Constraint** option from the **Placement** tab and then select the **Distance** option from the **Constraint Type** drop-down list in the **Placement** panel.
6. Select the two mating faces and enter **4** in the edit box in the **Component Placement** dashboard.
7. Choose the **Build feature** button from the **Component Placement** dashboard to complete the creation of the assembly. The model similar to the one shown in Figure 11-74 is displayed in the drawing area.

*Figure 11-74 Model after assembling the **Cap** with the **Casting***

Suppressing the Cap from the Assembly

You need to suppress the **Cap** because the next component has to be assembled with the **Casting**. When you suppress the **Cap** from the assembly, it becomes easier to assemble the new component. The **Cap** will be unsuppressed later.

1. Select the **Cap** from the drawing area and right-click to display the shortcut menu.

2. Choose the **Suppress** option from the shortcut menu; you are prompted to confirm the suppression. Choose the **OK** button.

Assembling Brasses with the Casting

Next, you need to assemble the **Brasses** with the **Casting**.

1. Choose the **Assemble** tool to display the **Open** dialog box.

2. Select **Brasses** and choose the **Open** button from the dialog box; the **Component Placement** dashboard is displayed.

3. Choose the **Placement** tab to display the panel and select the **Coincident** option from the **Constraint Type** drop-down list.

4. Select the axis of the **Brasses** and then the axis of the **Casting**, as shown in Figure 11-75.

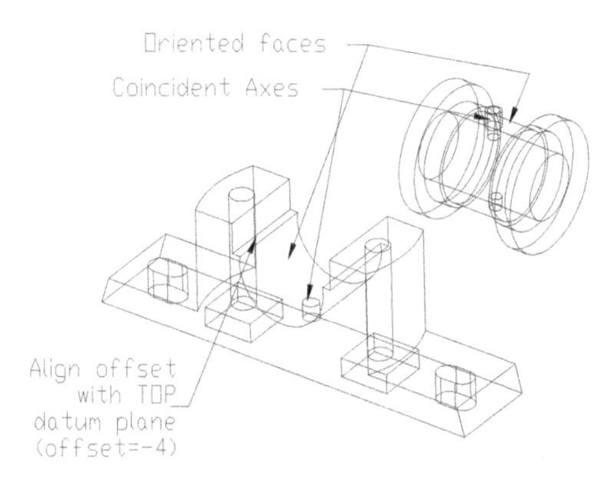

Figure 11-75 Constraints references used for assembling the components

Note

*You can choose the **Show component in separate window while specifying constraints** button from the dashboard to display the component in a separate window, which makes it easier to select the faces and then apply the constraints.*

5. Click on the **New Constraint** option in the **Placement** panel and select the faces of the **Brasses** and the **Casting**, refer to Figure 11-75.

6. As you select the faces, the **Oriented** option from the **Constraint Type** drop-down list will get selected.

7. Choose the **Change orientation of the constraint** button to change the orientation. Choose the **Build feature** button from the **Component Placement** dashboard; the assembly model similar to the one shown in Figure 11-76 is displayed in the drawing area.

*Figure 11-76 Assembling the **Brasses** with the assembly*

8. Now, unsuppress the **Cap** by choosing **Resume > Resume All** from the **Operations** group. Similarly, assemble the remaining components. The final assembly is shown in Figure 11-77. You can choose the **Repeat** option to copy the repeated items.

Figure 11-77 *The final Pedestal Bearing assembly*

Creating the Exploded State of the Assembly

To view all components in an assembly clearly, you need to create its exploded state.

1. Choose the **Edit Position** button from the **Model Display** tab; the **Explode Tool** dashboard and the default exploded state of the assembly are displayed.

   Now, you need to edit the position of the component in the exploded view.

2. Choose the **References** tab from the dashboard and then click in the **Movement Reference** collector; you are prompted to select a reference to define the direction of movement.

3. Select the vertical axis of the **Cap** from the drawing area; you are prompted to select the component to be moved.

4. Select the **Casting** from the drawing area; a direction handle is displayed. Move the cursor to the required direction handle and drag to place it at the required position. Similarly, place the **Cap**, two **Square headed bolts**, two **Lock nuts** and then two **Nuts** at the required location.

   Now, you need to move the **Brasses** to the required location in the drawing area.

5. Click in the **Movement Reference** collector again; you are prompted to select the reference. Select the axis of the **Brasses** from the drawing area as the reference; you are now prompted to select the component to be moved.

6. Select and move the **Brasses** to the required location in the drawing area.

   The model similar to the one shown in Figure 11-78 is displayed in the drawing area.

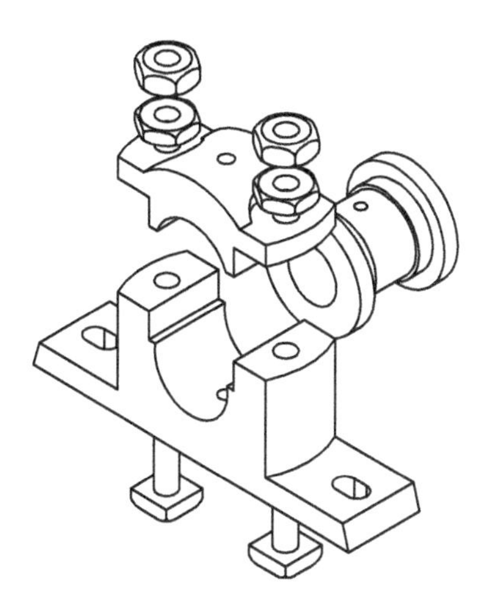

Figure 11-78 *The Exploded Pedestal Bearing assembly*

7. Now, you need to save this exploded view. To save this view, invoke the **View Manager** dialog box and choose the **Explode** tab.

8. Choose the **New** button; the **Modified State Save** message box is displayed confirming to save the default exploded view. Choose the **Yes** button from the message box; the **Save Display Elements** dialog box is displayed.

9. Next, enter the **EXP1** in the **Explode** edit box and choose **OK** button from the dialog box and close the **View Manager** dialog box.

Creating the Offset Lines

1. Choose the **Edit Position** button from the **Model Display** tab and choose the **Create cosmetic offset lines to illustrate movement of exploded component** button from the **Explode Tool** dashboard; the **Cosmetic Offset Line** dialog box is displayed.

2. Click in the **Reference 1** collector and select the axis of the right hole on the **Casting**.

3. Next, click in the **Reference 2** collector and then select the axis of the **Square Headed bolt** on the right half of the **Casting** to display the explode line. Similarly, select the axes from the other components to create the offset lines between them.

   The model similar to the one shown in Figure 11-79 is displayed in the drawing area.

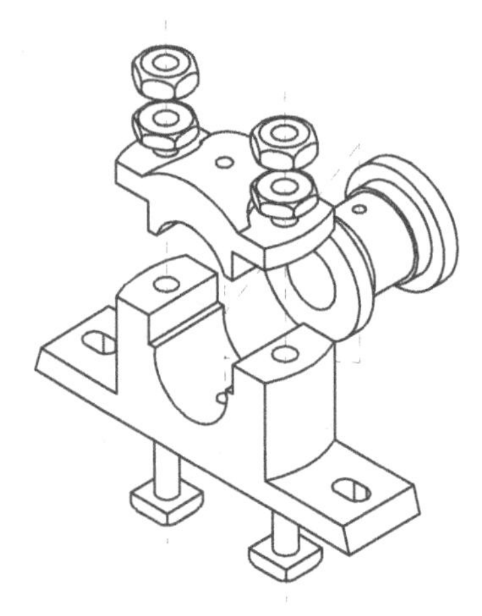

Figure 11-79 Exploded state of the assembly displaying the offset lines

4. Close the **Cosmetic Offset Line** dialog box and choose the **Build feature** button.

Saving the Assembly

1. Choose the **Save** button from the **File** menu and then save the assembly.

Closing the Window

Now, you have saved the assembly so the window can be closed.

1. Choose the **Close** button from the **Quick Access** toolbar to close the file.

Self-Evaluation Test

Answer the following questions and then compare them to those given at the end of this chapter:

1. On applying the **Coincident** constraint to two selected faces, the selected faces get connected to each other. (T/F)

2. An exploded view is a state in which all components are moved from their actual locations so that they are visible. (T/F)

3. Dimensions of a feature cannot be modified in the **Assembly** mode. (T/F)

4. You can display the unexploded state of the assembly by choosing the **Explode View** button from the **Modal Display** group. (T/F)

5. Offset lines are invisible by default in the unexploded state of the assembly. (T/F)

6. In the ____________ type of simplified representation, the components removed from the current display are displayed again.

7. The ______________ type of simplified representation reduces the regeneration time in case of large assemblies.

8. The ___________ constraint is similar to the **Coincident** constraint with the only difference that this constraint allows you to specify some offset distance between the two coplanar faces.

9. The _________ constraint is used to assemble circular components.

10. The __________ is a tabular representation of all components of the assembly along with the information associated with them.

Review Questions

Answer the following questions:

1. Which of the following constraints is used to align the selected datum point or vertex on the first part with the selected surface or datum plane on the second part?

   (a) **Orient** (b) **Pnt on Srf**
   (c) **Edge on Srf** (d) **Tangent**

2. Which of the following messages is displayed in the **Placement Status** area of the **Component Placement** dashboard when the assembly is fully constrained?

   (a) **Partially constrained** (b) **Fully constrained**
   (c) **Completely constrained** (d) None of these

3. In which of the following types of simplified representation, the complete geometry of the components is displayed in the current model display style?

   (a) Geometry representation (b) Graphics representation
   (c) Master representation (d) None of these

4. Which of the following constraints is used to assemble two components by making the selected faces or planes coplanar such that the mating faces or planes face in the same direction?

   (a) **Insert** (b) **Orient**
   (c) **Tangent** (d) **Align**

5. Which of the following groups of the **Model** tab contains the **Explode View** button?

   (a) **Model Display** (b) **Component**
   (c) **Datum** (d) **Modifiers**

6. A component of an assembly assembled by using the assembly constraints cannot be replaced with some other components. (T/F)

7. You can hide a part of an assembly from the assembly. (T/F)

8. Offset lines can only be created in the exploded state of an assembly. (T/F)

9. The order of assembling components in an assembly can be changed. (T/F)

10. PTC Creo Parametric provides you with a ready made BOM that you can directly use. (T/F)

Exercises

Exercise 1

In this exercise, you will create all components of the Crosshead assembly and then assemble them, as shown in Figure 11-80. Also, you will create an exploded state of the assembly, shown in Figure 11-81, which displays the offset lines. The dimensions of the components are shown in Figures 11-82 through 11-86. **(Expected time: 2 hrs)**

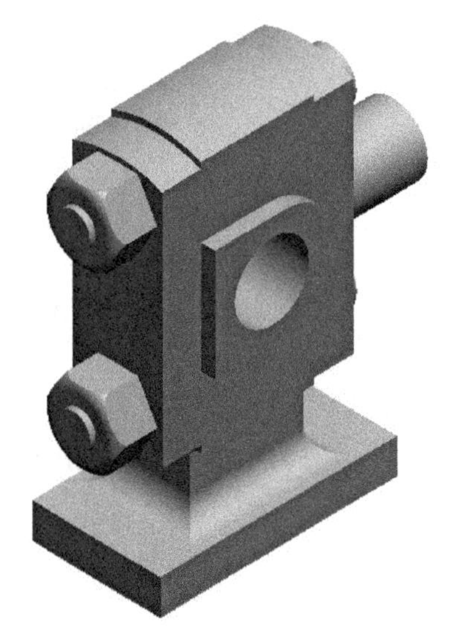

Figure 11-80 *The Crosshead assembly*

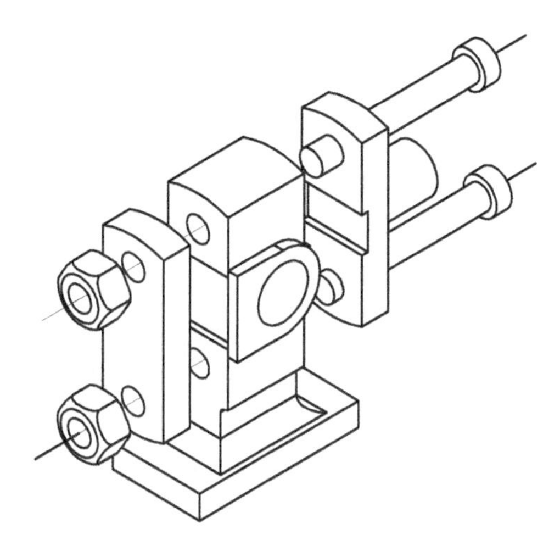

Figure 11-81 *The exploded state of the Crosshead assembly*

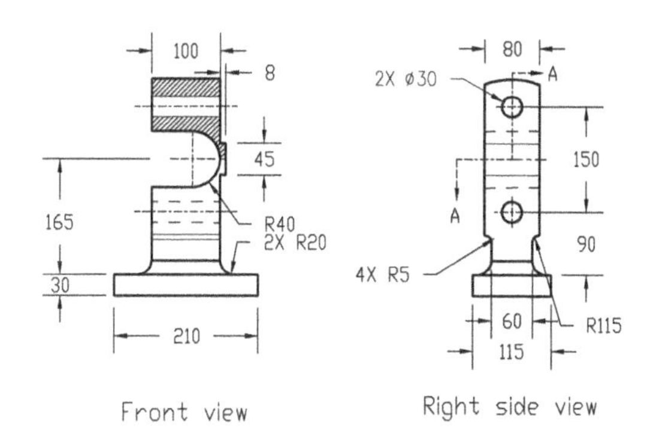

Figure 11-82 *Front view and right-side view of the Body*

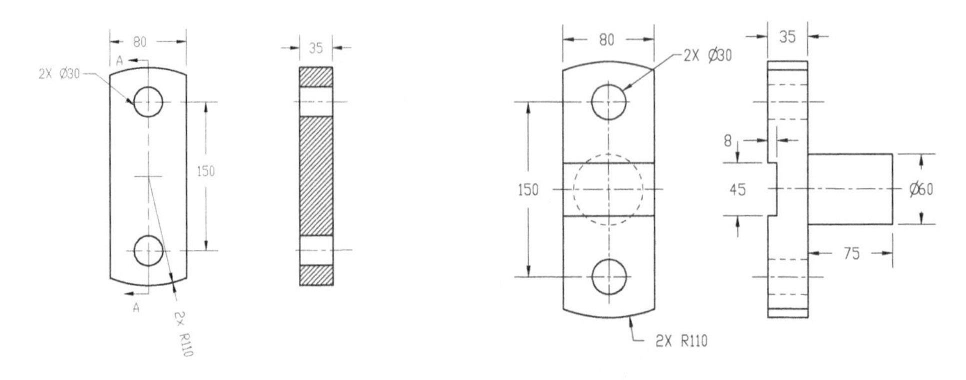

Figure 11-83 Dimensions of the Keep Plate

Figure 11-84 Dimensions of the Piston Rod

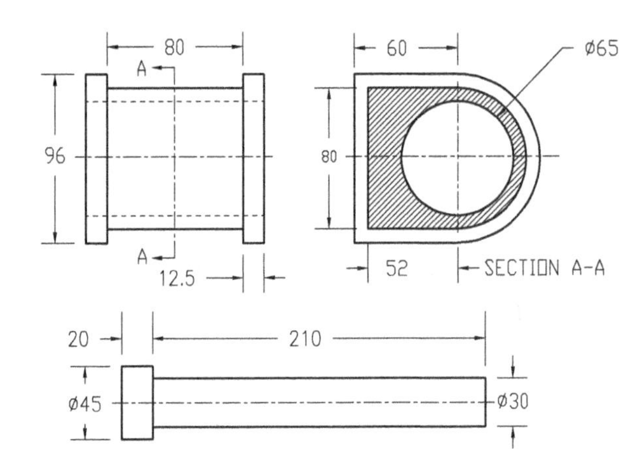

Figure 11-85 Dimensions of the Brasses and Bolt

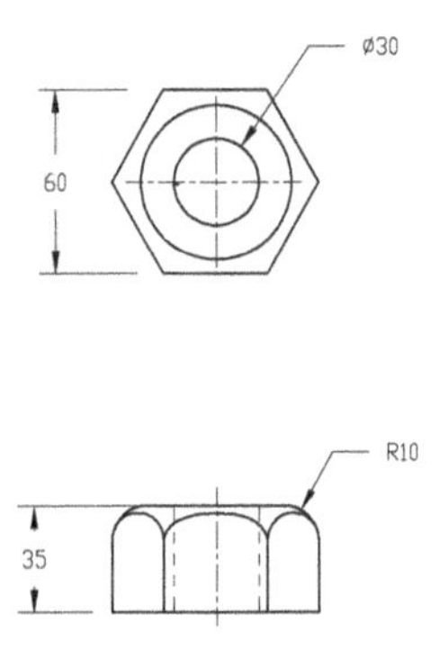

Figure 11-86 Dimensions of the Nut

Exercise 2

Create the assembly of the Radial Engine shown in Figure 11-87. The assembly in the exploded state is shown in Figure 11-88. The dimensions of various parts of this assembly model are given in Figures 11-89 through 11-93. **(Expected time: 4 hrs)**

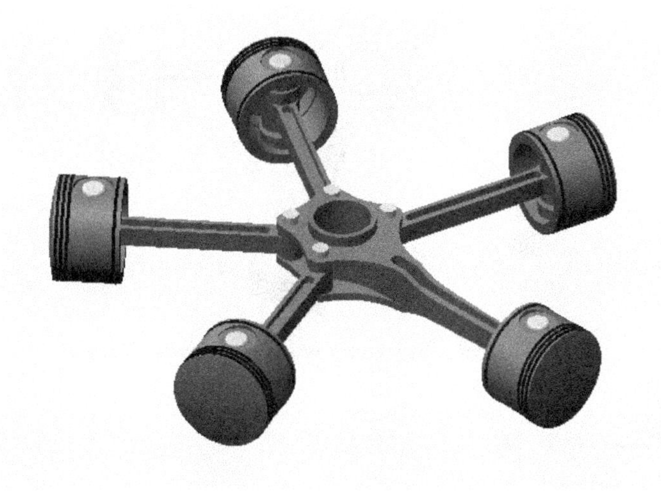

Figure 11-87 The Radial Engine assembly

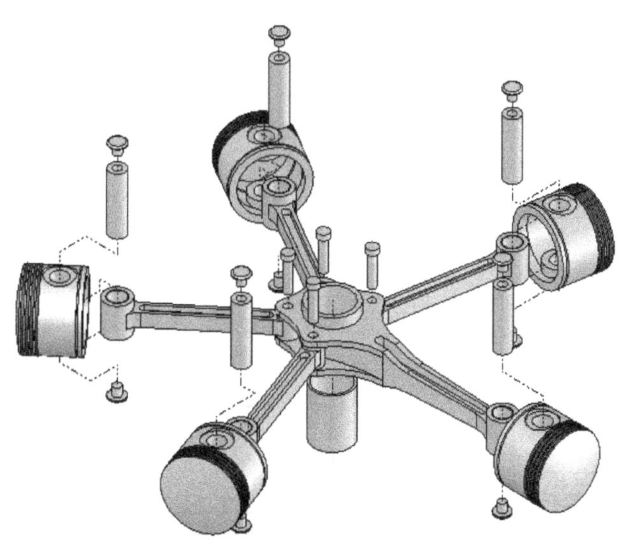

Figure 11-88 Exploded state of the Radial Engine assembly

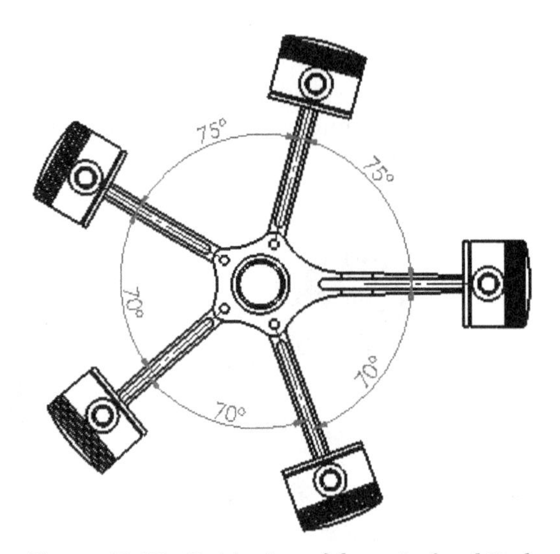

Figure 11-89 *Positioning of the articulated Rods*

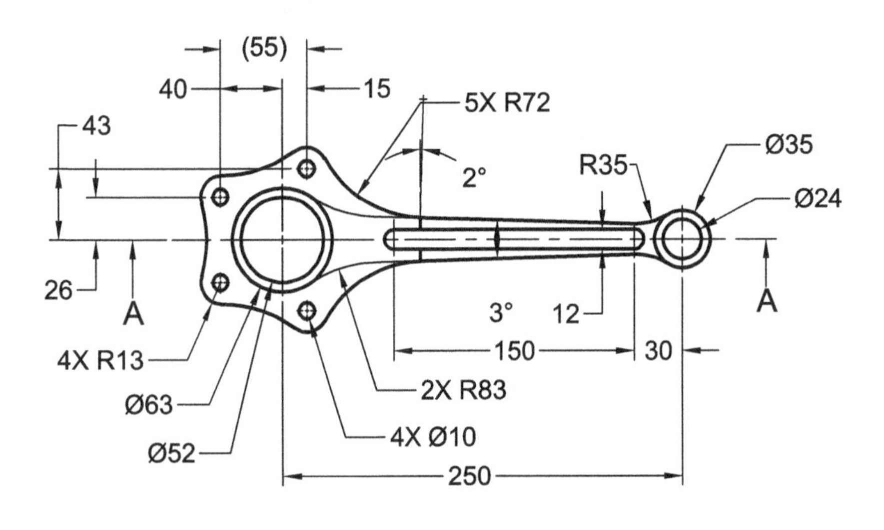

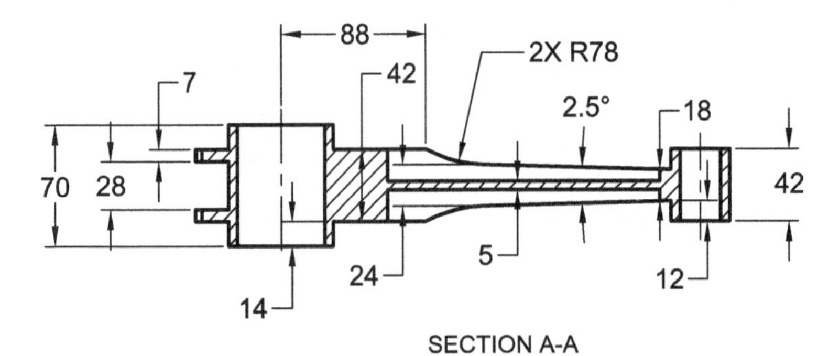

SECTION A-A

Figure 11-90 *Orthographic views and dimensions of the Master Rod*

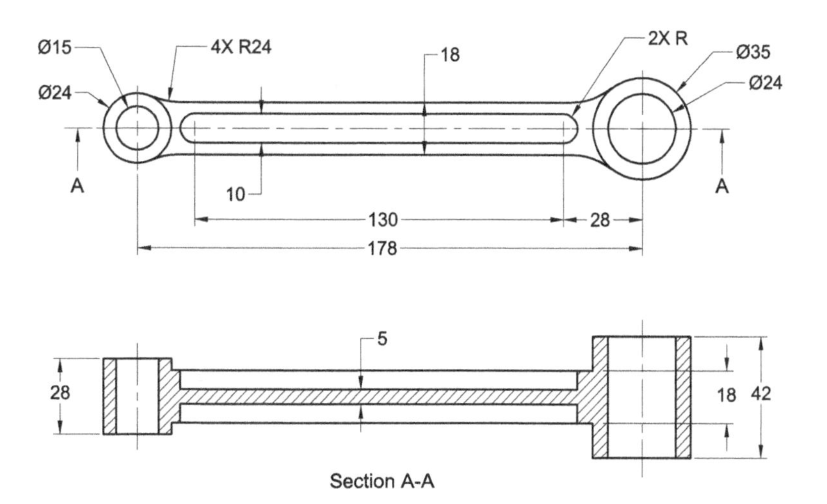

Figure 11-91 *Orthographic views and dimensions of the Articulated Rod*

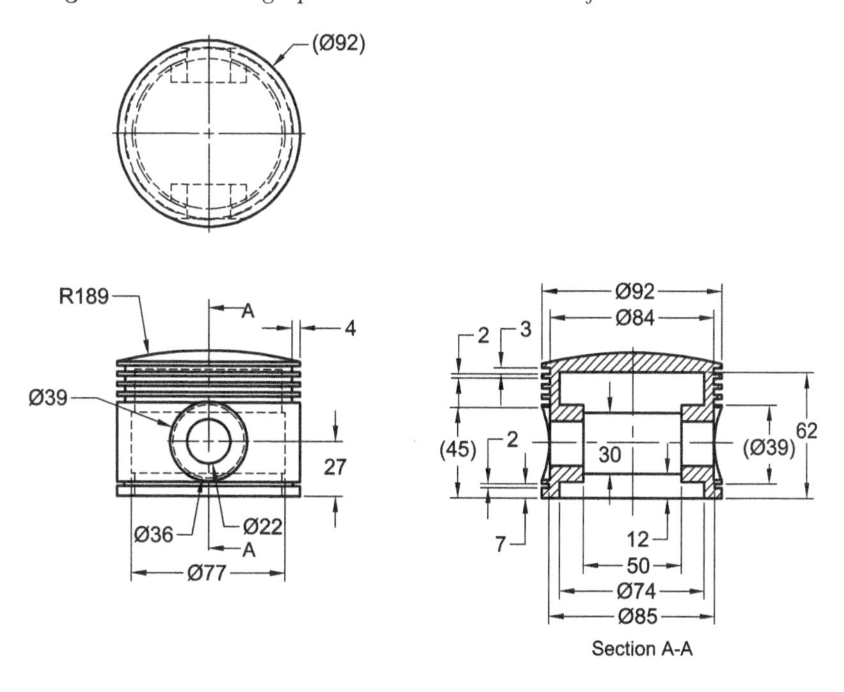

Figure 11-92 *Orthographic views and dimensions of the Piston*

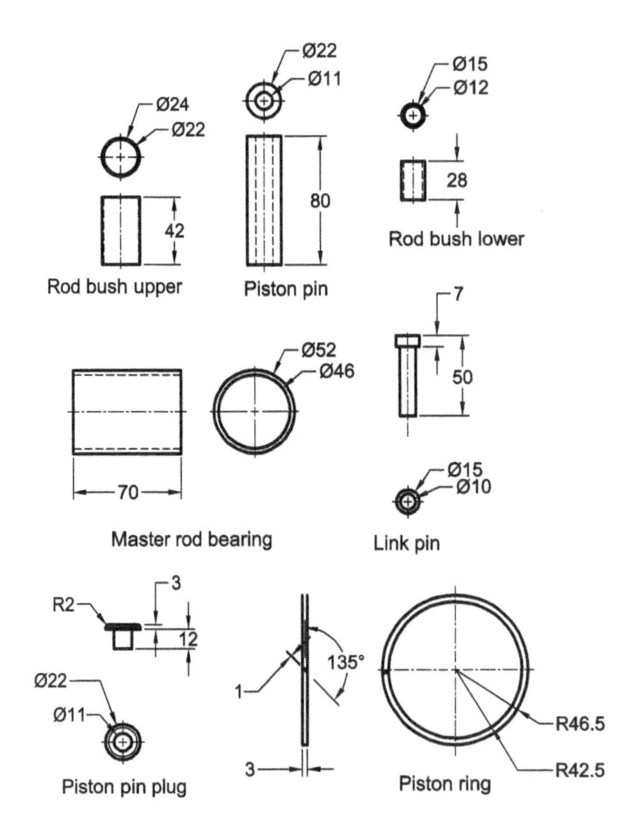

Figure 11-93 Views and dimensions of the other components

Answers to Self-Evaluation Test

1. T, **2.** T, **3.** F, **4.** T, **5.** T, **6.** Master representation, **7.** Graphics representation, **8. Distance**, **9. Insert**, **10.** BOM

Chapter 12

Generating, Editing, and Modifying the Drawing Views

Learning Objectives

After completing this chapter, you will be able to:

- *Create and retrieve the drawing sheet formats.*
- *Generate different drawing views of an existing part.*
- *Edit the existing drawing views and parameters associated with them.*
- *Modify the existing drawing views.*

THE DRAWING MODE

In the earlier chapters, you learned about creating the parts in the **Part** mode and assembling different parts in the **Assembly** mode. In this chapter, you will learn to generate the drawing views of the parts and assemblies created earlier. Drawing views are generated in the **Drawing** mode. One of the major advantages of working with this software package is its bidirectional associative nature. This property ensures that any modifications made in the model in the **Part** mode are reflected in the drawing views of the model and vice-versa. In PTC Creo Parametric, there are two types of drafting methods: Interactive drafting and Generative drafting. In this chapter, you will learn Generative drafting.

To generate the drawing views, first you need to start a new file in the **Drawing** mode of PTC Creo Parametric. Choose the **New** button from the **Quick Access** toolbar to display the **New** dialog box. Select the **Drawing** radio button in the **New** dialog box, as shown in Figure 12-1.

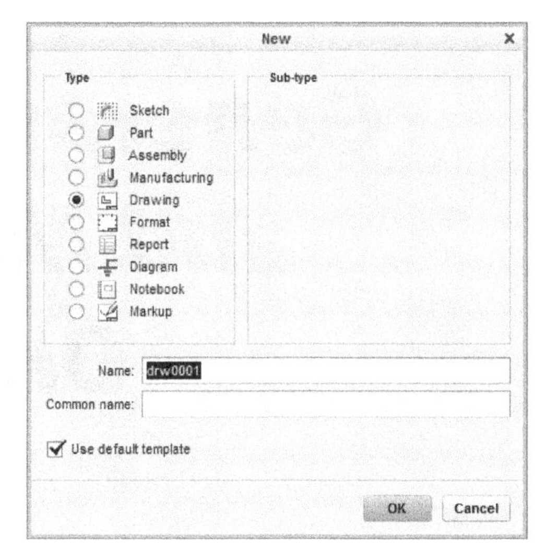

Figure 12-1 The ***New*** *dialog box*

Specify the name of the drawing in the **Name** edit box and then choose **OK** to display the **New Drawing** dialog box, as shown in Figure 12-2.

New Drawing Dialog Box

The **New Drawing** dialog box is used to specify the template that will be used while starting a new file in the **Drawing** mode. The options available in this dialog box are discussed next.

Default Model Area

The **Default Model** area is used to specify the name of the model whose drawing views you want to generate. You can specify the name of the model in the **Name** edit box or select the model using the **Open** dialog box that will be displayed when you choose the **Browse** button. If a part or assembly file is already opened in the current session then the name of that model will be displayed by default in the edit box of the **Default Model** area.

Specify Template Area

The **Specify Template** area is used to specify whether you want to use an empty sheet, predefined formats, or the default template available in PTC Creo Parametric. There are three radio buttons in this area. The **Use template** radio button is selected by default. When the **Empty with format** radio button is selected, the dialog box is modified, as shown in Figure 12-3. When you select the **Empty** radio button, the **New Drawing** dialog box is modified, as shown in Figure 12-4.

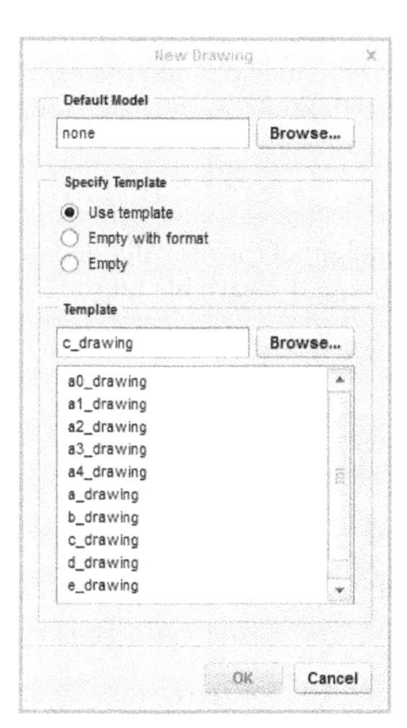

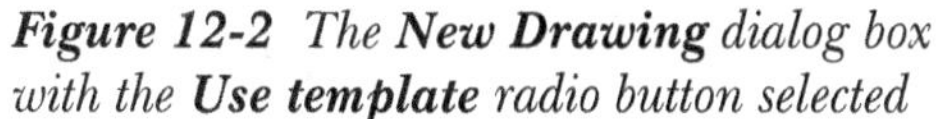
*Figure 12-2 The **New Drawing** dialog box with the **Use template** radio button selected*

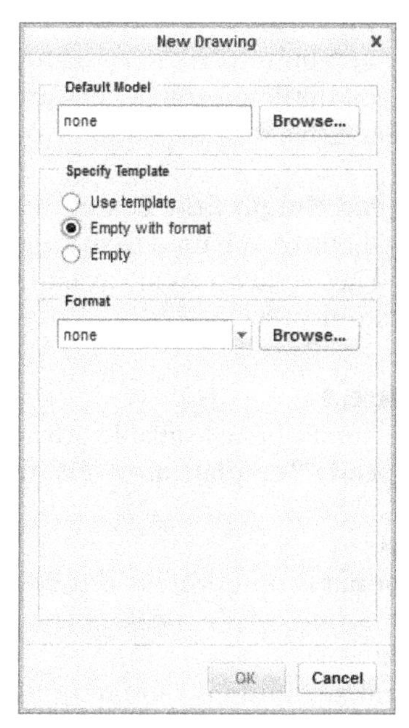

*Figure 12-3 The **New Drawing** dialog box with the **Empty with format** radio button selected*

Orientation Area

The **Orientation** area is available only when you select the **Empty** radio button from the **Specify Template** area. The buttons in this area are used to specify the orientation of the sheet. You can select a standard size sheet with a portrait or a landscape orientation using the **Portrait** or **Landscape** button. You can also specify a sheet with the user-defined size by choosing the **Variable** button.

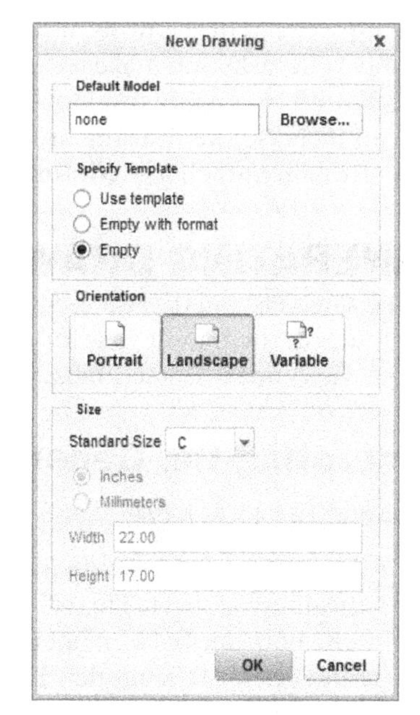

*Figure 12-4 The **New Drawing** dialog box with the **Empty** radio button selected*

Size Area

The options in the **Size** area are used to set the size and the units of the sheet. The **Size** area is available only when you select the **Empty** radio button from the **Specify Template** area. The options in this area are discussed next.

Standard Size

The **Standard Size** drop-down list is used to select a drawing sheet of standard size. This drop-down list is available only when you choose the **Portrait** or **Landscape** button from the **Orientation** area.

Inches and Millimeters Radio Buttons

These radio buttons are selected to set the standards for the user-defined sheets. You can set the size of the sheet in inches or in millimeters. These buttons are available only when you choose the **Variable** button from the **Orientation** area.

Width and Height Edit Boxes

These edit boxes are used to specify the width and the height of the user-defined drawing sheets. These edit boxes are available only when you choose the **Variable** button from the **Orientation** area.

Format Area

The **Format** area will be available only when you select the **Empty with format** radio button from the **Specify Template** area. The options in this area are discussed next.

Format

The **Format** drop-down list is used to select the available formats.

Browse

The **Browse** button is chosen to display the **Open** dialog box for retrieving the drawing formats. By default, there are only eight standard system formats that can be retrieved. However, you can create your own user-defined formats that can be retrieved later.

Choose **OK** from the **New Drawing** dialog box to proceed to the **Drawing** mode; a drawing sheet of the specified size and orientation will be placed on which you can generate the drawing views. The Drawing user interface of PTC Creo Parametric has been enhanced to simplify your tasks and make it easy to perform. Additionally, the **Ribbon** has been introduced to display all frequently used tools collected logically in various groups and tabs. If required, you can increase the drawing area by minimizing the **Ribbon**. To do so, right-click on a tab in the **Ribbon** and choose **Minimize the Ribbon**.

GENERATING DRAWING VIEWS

In PTC Creo Parametric, the first view that you need to generate is the general view. This view mostly acts as the parent view for the remaining views. Different methods used for generating various views are discussed next.

Generating the General View

Ribbon: Layout > Model Views > General

The **General** view is the first view that is generated on the drawing sheet. This view can be the top, front, right, left, bottom, back, trimetric, isometric view, or any user-defined view of the model. To generate the general view, choose the **General** tool from the **Model Views** group in the **Layout** tab of the **Ribbon**; the **Select Combined State** dialog box will be displayed, as shown in Figure 12-5. In this dialog box, if you select the **No Combined State** option from the

***Figure 12-5** The **Select Combined State** dialog box*

Combined state names collector and then choose the **OK** button, you will be prompted to select a center point to place the drawing view in the drawing sheet. Select the center point; the preview of the drawing with unexploded view in the default orientation will be displayed in the drawing sheet and also the **Drawing View** dialog box will be displayed, as shown in Figure 12-6.

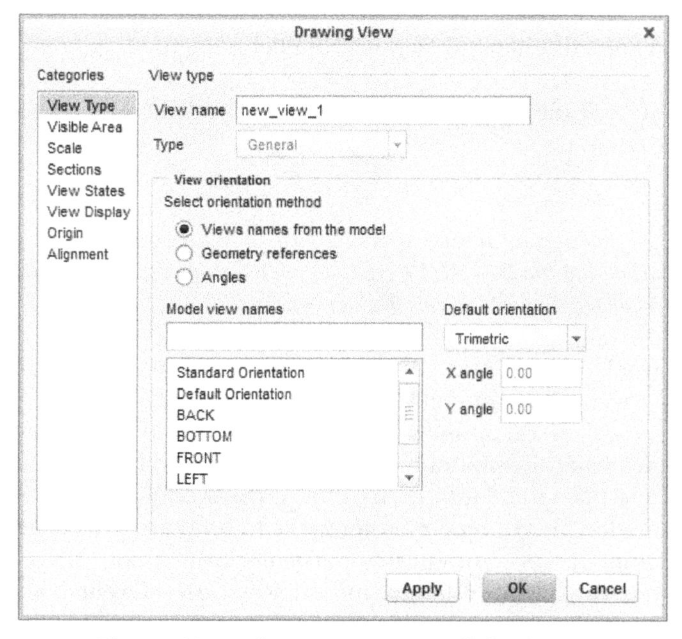

Figure 12-6 *The **Drawing View** dialog box*

If you select the **DEFAULT ALL** option from the **Combined state names** collector of the **Select Combined State** dialog box and choose the **OK** button, you will be prompted to select a center point to place the drawing view in the drawing sheet. Select the center point; the preview of the drawing, with the exploded view in the default orientation, will be displayed in the drawing sheet and also the **Drawing View** dialog box will be displayed, refer to Figure 12-6.

Note

*If you do not want to be prompted for selecting a combination state, select the **Do not prompt for Combined State** check box from the **Select Combined State** dialog box.*

Tip: *1. If you have not specified any model in the **Default Model** area of the **New Drawing** dialog box then choose **Layout > Model Views > General** from the **Ribbon**; the **Open** dialog box will be displayed. You can select the model from the **Open** dialog box.*

*2. PTC Creo Parametric automatically selects the model in the **Default Model** area of the **New Drawing** dialog box. This model is selected from the present session of PTC Creo Parametric. If the model selected automatically is not the required model, then change the model by choosing the **Browse** button in the **New Drawing** dialog box.*

The **Drawing View** dialog box contains the options that are required to generate a general view. The **Categories** list box in the dialog box lists all parameters that are required to generate a drawing view in PTC Creo Parametric. The **View Type** option is selected by default in the **Categories** list box and its related options are displayed on the right side of the **Categories** list box. These parameters/options are discussed later in this chapter.

The procedure to generate a general view is given below:

1. Choose **Layout > Model Views > General** from the **Ribbon**; you will be prompted to select a center point.

Note

If the selected model is an assembly model, then you can create default exploded view of the assembly by choosing the ***DEFAULT ALL*** *option from the* ***Select Combined State*** *dialog box which will be displayed on choosing the* ***General*** *button.*

2. As soon as you select the center point, the **Drawing View** dialog box will be displayed and the default view of the model will be displayed on the drawing sheet. In the **Type** drop-down list, the **General** option is selected by default and in the **View orientation** area, the **Views names from the model** radio button is selected by default. This radio button allows you to select the standard orientations of the model from the **Model view names** list box. If you select the **Geometry references** radio button, you can select the planar faces or the datum planes to orient in a particular orientation. If you select the **Angles** radio button, the **Angle value** edit box and the **Rotation reference** drop-down list will be displayed below the **Reference Angle** area. You can change the orientation of the model by specifying the required angle in the **Angle value** edit box.

3. Select the view that you want to generate as the general view from the list provided below the **Model view names** list box.

Tip: *To set the model orientation using the* ***Geometry references*** *radio button, select the view type from the* ***Reference 1*** *drop-down list and then from the model, select the face or the datum plane that you need to see from the selected view. Similarly, select an option from the* ***Reference 2*** *drop-down list and then select a datum plane or a face from the model. For example, if you select* ***Top*** *from the* ***Reference 1*** *drop-down list, then from the model, you need to select the face or the datum plane that you need to keep in the top orientation in the drawing view.*

4. After selecting the required options, choose the **Apply** button and then choose the **Cancel** button from the **Drawing View** dialog box.

Generating the Projection View

Ribbon: Layout > Model Views > Projection

The projection views are the orthographic views generated by projecting lines normal to the existing view, refer to Figure 12-7. Before generating a projected view, you need to make sure that at least one parent view is already present on the drawing sheet. To generate a projected view, choose **Layout > Model Views > Projection** from the **Ribbon**. Now, move the cursor to a location where you need to place the view and specify a point on this location. Depending on the point selected to place the view, the resulting view will be a top, front, right-side, or left-side view.

The scale factor of these views will be the same as that of the parent view from which they are projected. If more than one view exists that can be the parent view of the projection view, then you will be prompted to select a projection parent view for the new view.

Generating the Detailed View

Ribbon: Layout > Model Views > Detailed

Detailed views are used to provide the enlarged view of a particular portion of an existing view. To generate a detailed view, choose the **Detailed** tool from the **Model Views** group of the **Layout** tab in the **Ribbon**; you will be prompted to specify the center point for detail on an existing view. Define a center point on the view whose detail needs to be generated; you will be prompted to sketch a spline without intersecting other splines to define an outline. Draw a closed spline to define the outline of the detail view and exit the drawing mode by clicking the middle mouse button. Next, you need to specify the placement point for placing the detail view. You can change the name, scale, or the reference point of the parent view and can also select the boundary type of the detailed view using the options from the **Drawing View** dialog box, which will be displayed when you double-click on the detailed view.

Figure 12-7 shows a drawing sheet with various drawing views.

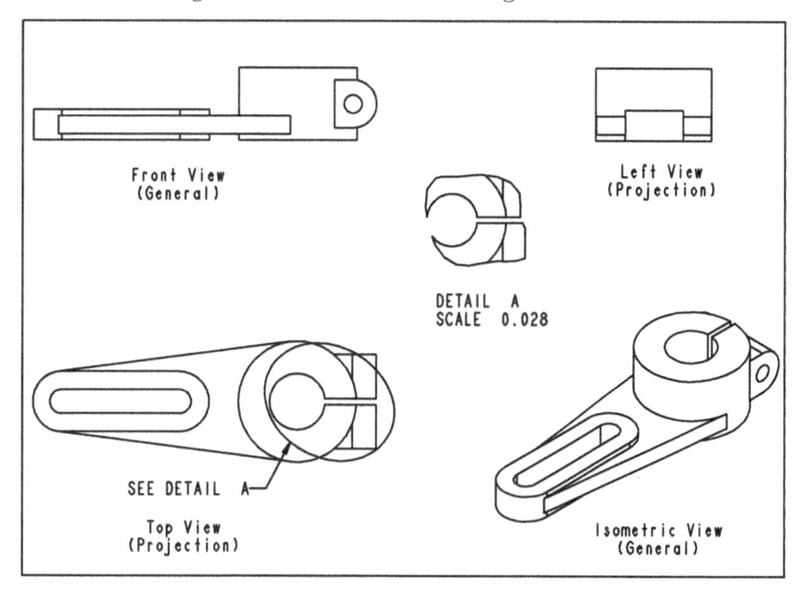

***Figure 12-7** Drawing sheet with various drawing views*

Generating the Auxiliary View

Ribbon: Layout > Model Views > Auxiliary

The auxiliary views are generated by projecting normal lines from a specified edge, axis, or datum plane of an existing view. The view scale will be the same as that of the parent view. To generate an auxiliary view, choose the **Auxiliary** tool from the **Model Views** group of **Layout** tab in the **Ribbon**; you will be prompted to select an edge, axis, or datum plane from the front surface of the main view. Select the edge or surface normal to which you need to place the generated view; a rectangular box will be attached to the cursor. Select a point on the drawing sheet to place the view on the drawing sheet. Figure 12-8 shows the edge to be selected as reference to generate the drawing view and the resulting auxiliary view.

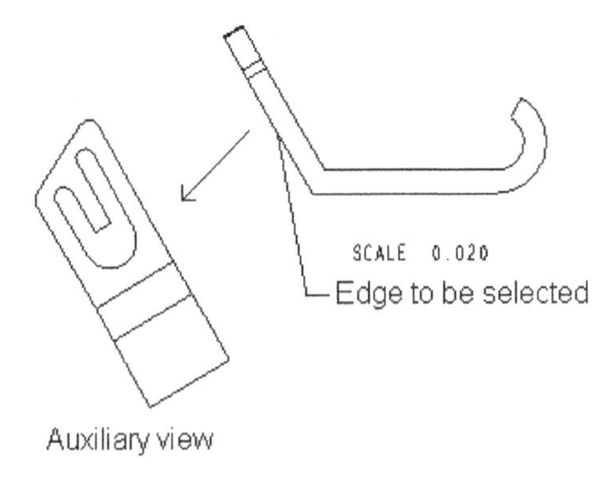

Figure 12-8 *Resulting auxiliary view*

Generating the Revolved Section View

Ribbon: Layout > Model Views > Revolved

A revolved section view is the view that is generated from an existing view by revolving the section through an angle of 90 degree about the cutting plane and then projecting it along the length. Remember that the cutting plane of the revolved section view is normal to the parent view. The procedure to generate a revolved section view is discussed next.

1. Choose the **Revolved** tool from the **Model Views** group in the **Layout** tab of the **Ribbon**; you will be prompted to select a parent view for the revolved section.

2. Select the view from the drawing sheet that will be defined as the parent view for generating the revolved view.

3. Next, you need to select a center point on the drawing sheet to place the view. Select a point anywhere on the drawing sheet; the **Drawing View** dialog box will be displayed. The **Create New** cross-section option is selected by default from the **Cross-section** drop-down list in the **Revolved view properties** area of the **Drawing View** dialog box. The **XSEC CREATE** menu is also displayed on the lower right corner of the screen.

Note
*The **Create New** option is selected by default in the **Revolved view properties** area of the **Drawing View** dialog box, only if there is no existing revolved section present.*

4. Choose **Planar > Single > Done** from the **XSEC CREATE** menu; the message input window will be displayed prompting you to specify the name of the cross-section. Enter the name of the cross-section and accept it; the **SETUP PLANE** menu will be displayed and you will be prompted to select a planar surface or datum plane.

5. Select the plane along which the view will be sectioned, as shown in Figure 12-9. The resulting section view will be placed in-line with the section plane.

6. Choose the **Apply** button and then choose the **Close** button to exit the **Drawing View** dialog box. The resulting sectioned view is shown in Figure 12-10.

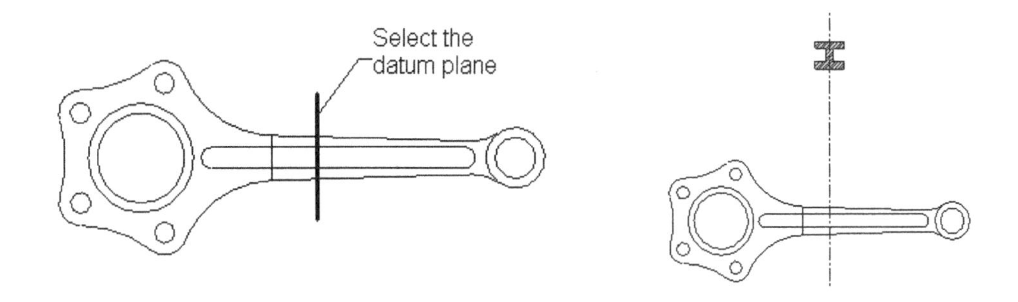

Figure 12-9 *Datum plane for the revolved section view*

Figure 12-10 *Resulting revolved section view*

Note
Copying and aligning of a view will be discussed later in this chapter.

Drawing View Dialog Box Options

The remaining types of drawing views that you can generate in PTC Creo Parametric require the options available in the **Drawing View** dialog box. Therefore, it is important for you to understand these options before proceeding further.

View Type Option

When you invoke the **Drawing View** dialog box, the **View Type** option is automatically selected in the **Categories** list box. The **View name** edit box is used to enter the name of the drawing view. The **View orientation** area lists the options to orient the drawing view. These options are the standard options of orienting the model.

Visible Area Option

The **Visible Area** option is used to set the display of the view. You can display a drawing view as a full view, a partial view, a half view, and a broken view. On choosing the **Visible Area** option, the **Drawing View** dialog box will be displayed, as shown in Figure 12-11. The options in the **View visibility** drop-down list of this dialog box are discussed next.

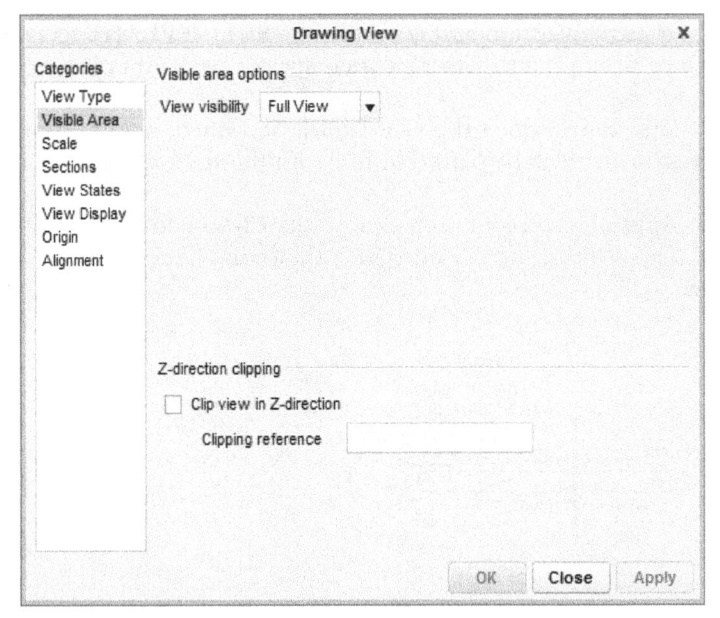

*Figure 12-11 The **Drawing View** dialog box with the **Visible Area** option chosen*

Full View

The **Full View** option is selected by default in the **View visibility** drop-down list. This option can be combined with any view type to generate a drawing view displaying the complete view.

Half View

The **Half View** option can be used on the projection, auxiliary, or general view to generate a drawing view that displays only the half view of the part. Generally, an existing view is selected to display the half view. For generating a half view, double click on an existing view to display the **Drawing View** dialog box. In the **Categories** list box, choose the **Visible Area** option to invoke the **Visible area options** area and then select **Half View** from the **View visibility** drop-down list. On doing so, you will be prompted to select a reference plane that will be used to remove half of the drawing view. The reference plane can be a datum plane or a planar surface, and must be normal to the screen in the selected view. Select the reference plane; an arrow will be displayed attached to the selected plane. This arrow indicates the portion of the view to be removed. You can also flip the direction of the arrow by using the **Flips the side to keep** button available below the **Half view reference plane** collector in the dialog box. Generally, this type of view is generated for symmetric parts. Therefore, you can specify the type of symmetry line using the **Symmetry line standard** drop-down list. Figure 12-12 shows the reference plane to be selected and Figure 12-13 shows the resulting half view.

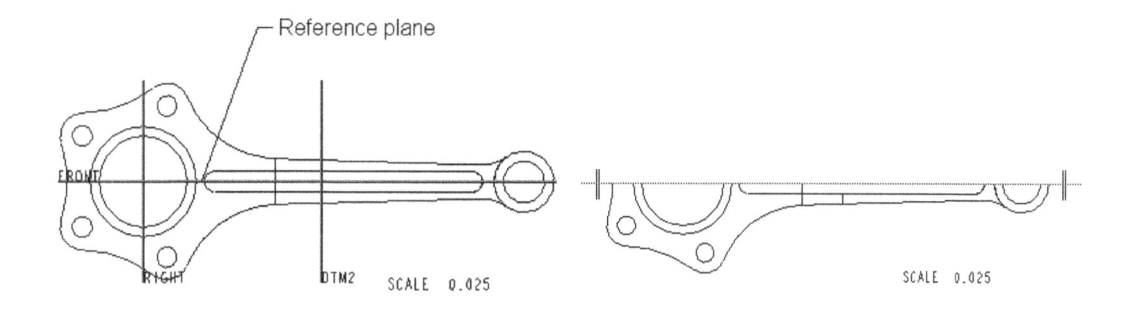

Figure 12-12 *Reference plane to be selected to generate a half view*

Figure 12-13 *Resulting half view*

Partial View

The **Partial View** option can be used on the projection, auxiliary, revolved, or general views to generate a view that displays a specified portion of the view. To convert an existing view to a partial view, double-click on that view. Choose the **Visible Area** option from the **Categories** list box and select the **Partial View** option from the **View visibility** drop-down list; you will be prompted to draw a spline that will be the boundary of the portion of the view you want to display, see Figure 12-14. Select the center point on the view and then draw the spline. Press the middle mouse button and exit the spline creation. Now, choose the **Apply** button from the **Drawing View** dialog box. You will notice that only the area of the view inside the spline is retained and the rest of the portion of the drawing view is cropped, see Figure 12-15.

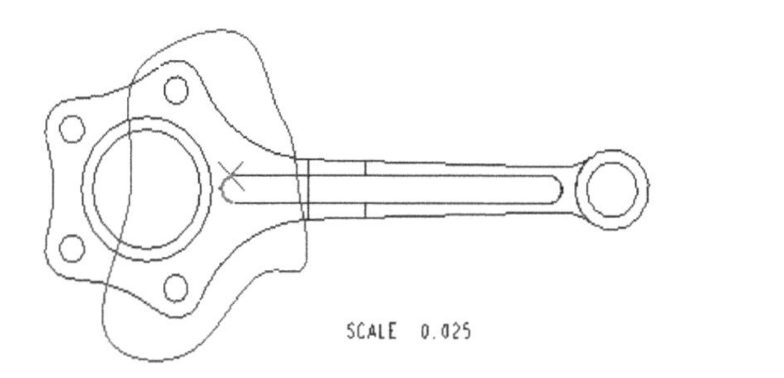

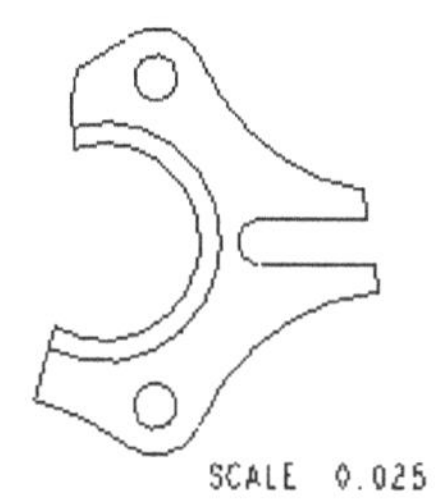

Figure 12-14 *Spline drawn on the view*

Figure 12-15 *Partial view*

Broken View

This type of view is used for the parts having a high length to width ratio. The **Broken View** option can be used on the projection or general view to generate a view that is broken along the horizontal or vertical direction using the horizontal or vertical line. To generate a broken view, select the **Broken View** option from the **View visibility** drop-down list. Choose the **Add break** button in the **Drawing View** dialog box to add break lines. Select an edge from the view and move the cursor vertically and then select a point on the screen to draw the first vertical break line. Move the cursor and select a point up to which you want to break the view, see Figure 12-16. You can also specify the style of the break line using the **Break Line Style** drop-down list available on the extreme right of the list box. Now, choose the **Apply** button from the **Drawing View** dialog box to create the resultant broken view and then choose the **Close** button to exit from it. Figure 12-17 shows the resultant broken view.

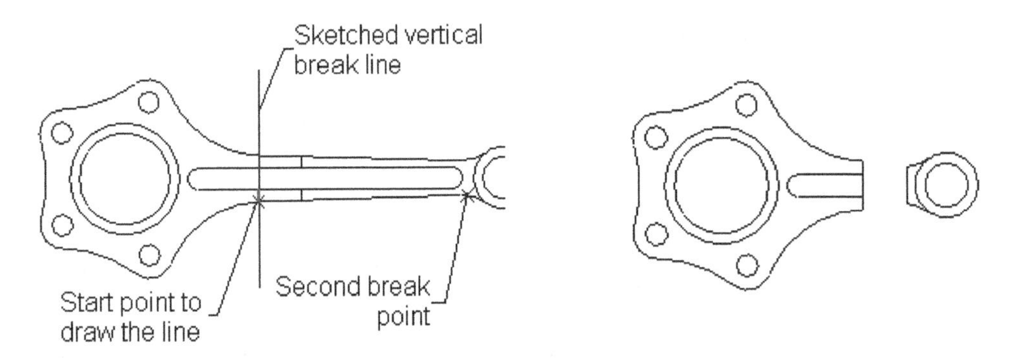

Figure 12-16 *References to break the view*

Figure 12-17 *Resultant broken view*

Figure 12-18 shows various views that have already been discussed in this section.

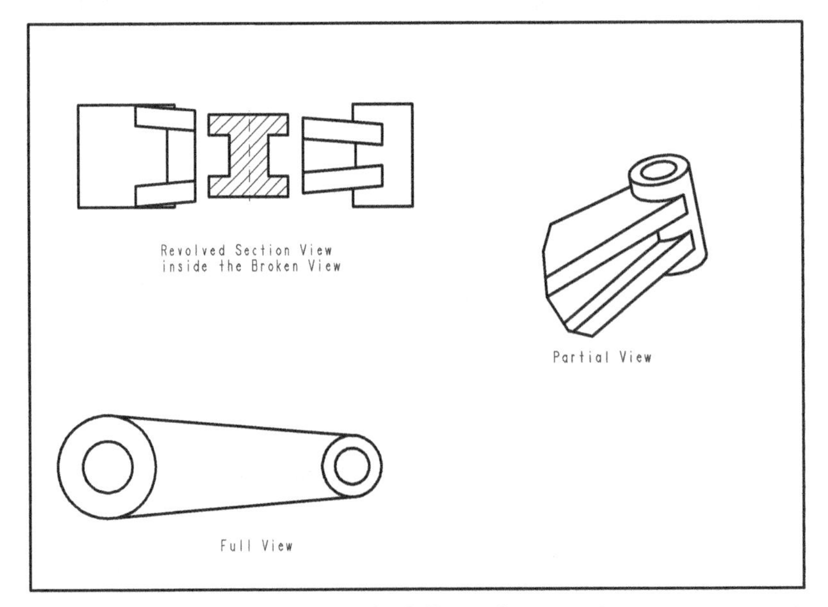

Figure 12-18 *The different drawing views*

Scale Option

When you select the **Scale** option, the **Scale and perspective options** area will be displayed on the right of the **Categories** list box. The **Default scale for sheet** radio button is selected by default. Therefore, all generated views are scaled with the default sheet scale. The second radio button is **Custom scale**. This radio button, when selected, allows you to enter a scale for the drawing view. When you modify the scale factor of a view, the view associated with this parent view is also scaled.

The third radio button is **Perspective**. This radio button is selected to generate a perspective view. After selecting the **Perspective** radio button, you can specify the distance from the eye and the diameter of the view circle. Figure 12-19 shows a perspective view.

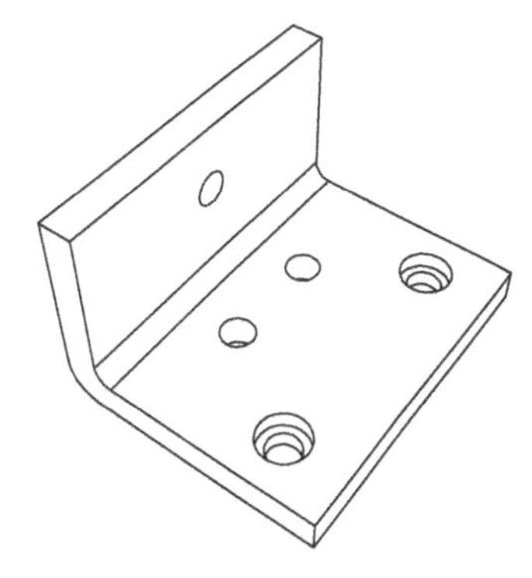

Figure 12-19 *A general perspective view*

Sections Option

The section views are generally used to display drawings of a part that is complicated from inside. As it is not possible to display the inside of the part using the conventional views, therefore these views are cut (sectioned) using a datum plane or a planar surface and the resulting section view is displayed. When you choose the **Sections** option from the **Categories** list box in the **Drawing View** dialog box, the **Section options** area will be displayed. In this area, the **No section** radio button is selected by default. Select the **2D cross-section** radio button and then choose the **Add cross-section to view** button to activate the options to create a section. The options in the **Section options** area and the methods used for creating the section plane are discussed next.

Full Section View

Consider a part that is cut throughout its length, width, or height and the cut portion is removed from the display. The remaining portion, when projected normal to the cutting plane, displays the full section view.

To generate a full section view, you first need to generate a projected view from one of the views existing on the drawing sheet. After generating the projected view, double-click on it to invoke the **Drawing View** dialog box. Choose the **Sections** option from the **Categories** list box. The options available for generating the section view are displayed on the right of the **Drawing View** dialog box. Select the **2D cross-section** radio button to define a section plane. Now, choose the **Add cross-section to view** button to add a section plane. The **Full** option in the **Sectioned area** drop-down list allows you to create a full section view.

When you choose the **Add cross-section to view** button, the **XSEC CREATE** menu will be displayed. Choose **Planar > Single > Done** from the **XSEC CREATE** menu; you will be prompted to specify the name of the cross-section. Specify the name of the cross-section in the message input window. After choosing **OK** from the window, the **SETUP PLANE** menu will be displayed and you will be prompted to select a planar surface or a datum plane. Select the datum plane from the other view; the name of the section view will be displayed in the **Name** drop-down list. Scroll to the right in the dialog box and click once in the **Arrow Display** collector to invoke the selection mode and select the view to display the section view arrows. You can also flip the material side of the section view using the **Flip** button. Choose the **Apply** button from the **Drawing View** dialog box. Figure 12-20 shows the top and section views.

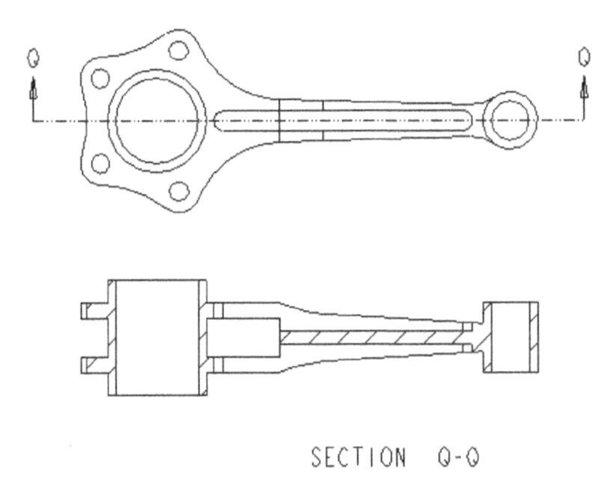

Figure 12-20 *Top and section views*

In the above section, you have learned to generate a full section view by defining a planar section plane. Now, you will learn to generate a full section view by defining an offset section plane. You need to make sure that before generating a section view by defining an offset section plane, the part file of the current model should be closed.

Invoke the **Drawing View** dialog box by double-clicking on an already generated projected view. Now, choose the **Sections** option from the **Categories** list box on the left of this dialog box. Select the **2D cross-section** radio button and choose the **Add cross-section to view** button to define the section plane. If there are some existing section planes, select the **Create New** option from the **Name** drop-down list; the **XSEC CREATE** menu will be displayed. Choose **Offset > Both Sides > Single > Done** from the **XSEC CREATE** menu; the message input window will be displayed. Specify the name of the cross-section.

The model will be displayed in a sub-window on the left of the screen. The **SETUP SK PLN** menu will also be displayed and you need to select a sketching plane for drawing the sketch that represents an offset section plane. Select the sketching plane from the subwindow. Choose **Okay** from the **DIRECTION** submenu; the **SKET VIEW** submenu will be displayed. Choose **Default** from this submenu. Draw the sketch of the offset section using the sketch tools available in the **Sketch** menu of the subwindow, refer to Figure 12-21. After drawing the sketch, choose **Sketch > Done** from the menu bar to exit the sketcher environment.

Note

The sketch created for the section must be a continuous line.

Click once on the **Arrow Display** collector to invoke the selection mode and then select the view in which the section arrows will be displayed. Choose the **Apply** button from the **Drawing View** dialog box. Figure 12-21 shows the top view, offset sectioned front view, and sectioned general view of a part.

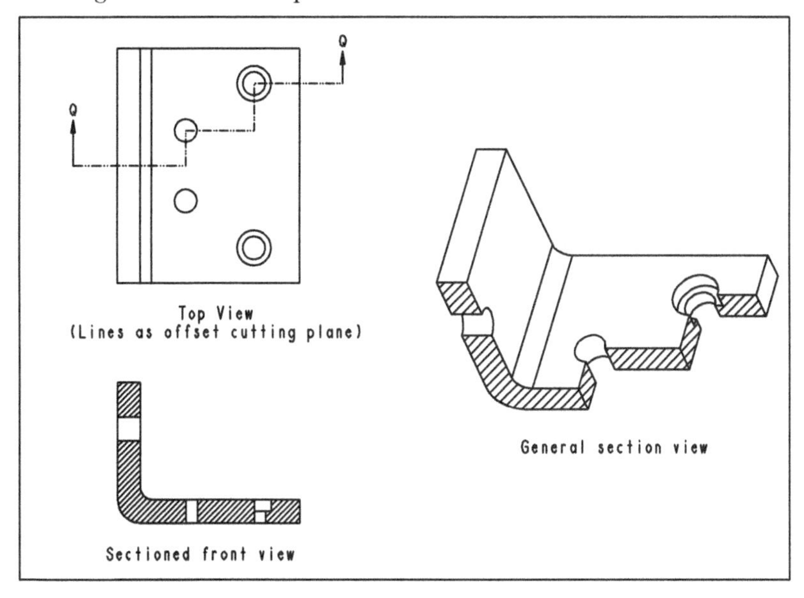

***Figure 12-21** The section views created using the offset section plane*

To generate an isometric section view, generate the general view and switch the orientation of the general view to isometric. Now, select the **Sections** option from the **Categories** list box and sketch a section plane using the subwindow as discussed earlier. Note that you cannot select a datum plane to define the section plane for an isometric section view. Figure 12-22 shows the top, full section front, and full section isometric views.

Half Section View

Consider a part that is cut half way through the length, width, or height and the front cut portion (front quarter) is removed from the display. This part when projected is called the half section view. In this projected view, only half of the part will be displayed sectioned and the other half of the part will be displayed as it is. To generate the half section view, you need to select the **Half** option from the **Sectioned Area** drop-down list while specifying the parameters for sectioning the view. After selecting this option, you are prompted to select the reference plane for the half section creation. Select the plane from the drawing view; you will be prompted to pick the side to keep. Select a point on the side of the section view that you need to keep. Select the parent view where the section arrow will be placed and choose the **Apply** button from the **Drawing View** dialog box. Figure 12-23 shows the resulting half section view.

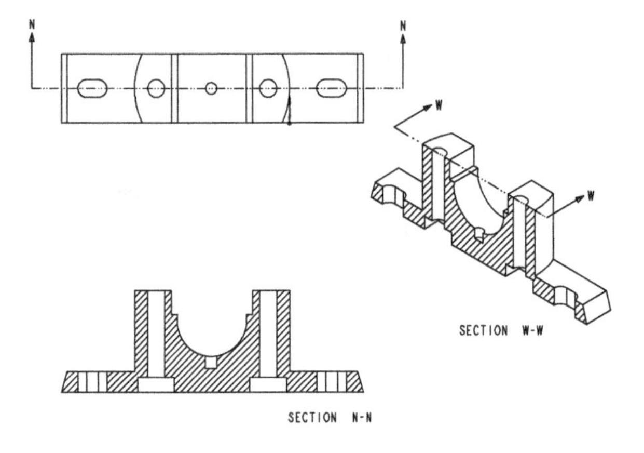

Figure 12-22 *Top view, full section front view, and full section isometric view*

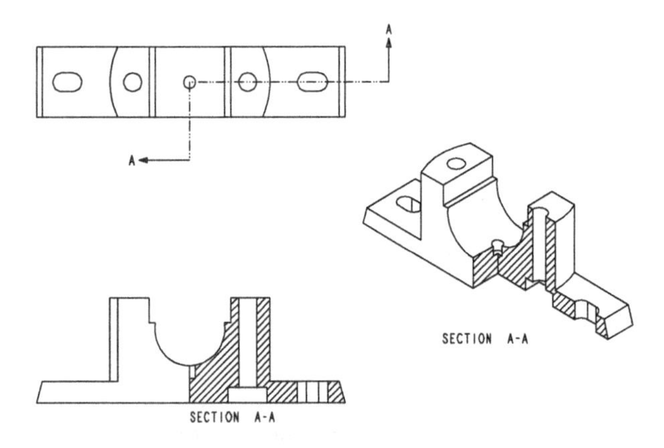

Figure 12-23 *Top view, half section front view, and half section isometric view*

Local Section View

The **Local** section view is used when you want to show a particular portion of the view in the section, and at the same time, not section the remaining view. The local section area is specified by drawing a spline around it. To generate the local section view, you will first generate the full section view, as discussed earlier. After generating the full section view, select the **Local** option from the **Sectioned Area** drop-down list. Next, you need to draw a spline that will define the area of the drawing that needs to be sectioned.

To draw a local section isometric view, you need to draw a section plane and then choose the **Local** option. Define a center point for the local section and sketch the spline. Figure 12-24 shows the local section front view and the local section isometric view.

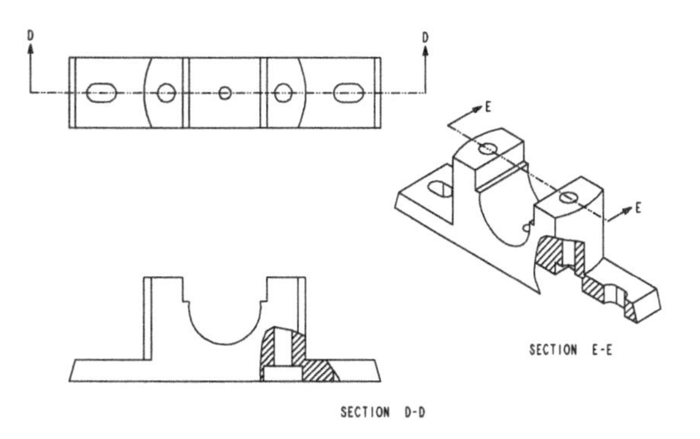

Figure 12-24 *Top view, local section front view, and local section isometric view*

Model edge visibility

PTC Creo Parametric provides you two options for displaying the edges of the section drawing views. If the **Total** radio button is selected in the **Section options** area of the **Drawing View** dialog box, a section view is created that displays all visible edges of the section view, in addition to the section area. This option can be combined with the full, half, local, full (unfold) and full (aligned) views. Figure 12-25 shows the front and total sectioned left view.

If the **Area** radio button is selected, then only the sectioned area of the section view is displayed and no other edges of the view are displayed in the area cross-section view, refer to Figure 12-26. This option can be combined with the full, full (unfold) and full (aligned) views.

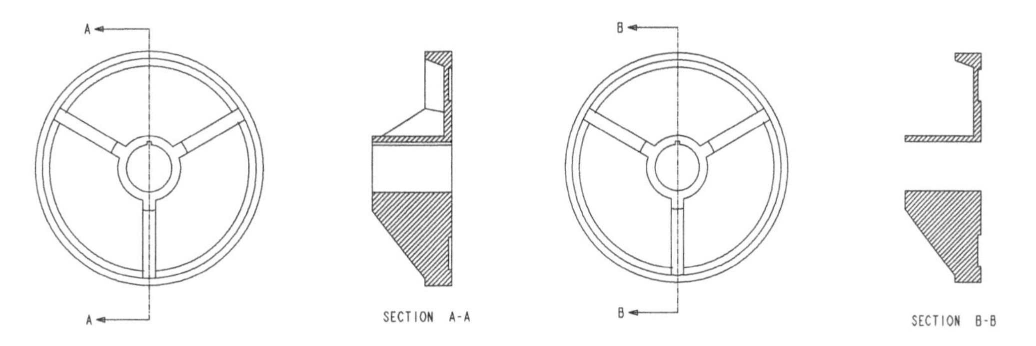

Figure 12-25 *The front view and the total sectioned left view*

Figure 12-26 *The front view and the area cross-section view*

Full(Aligned)

This view is used to section those features that are created at a certain angle to the main section planes. You can align sections to straighten these features by revolving them about an axis that is normal to the parent view. Generally for this section view, the section plane is sketched. Remember that the axis about which the feature is straightened should lie on all cutting planes. This means that the lines that are used to sketch the section plane should pass through the axis about which the feature will be revolved. Figures 12-27 and 12-28 show these views when the **Area** radio button is selected. Figures 12-29 and 12-30 show these views when the **Total** radio button is selected.

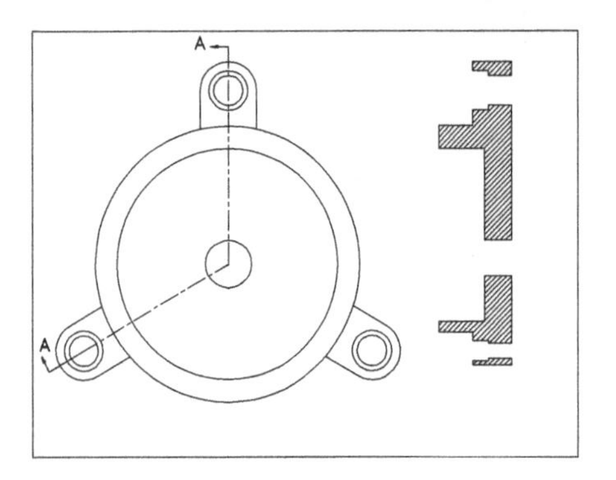

Figure 12-27 *The area cross-section view with normal lines of projection*

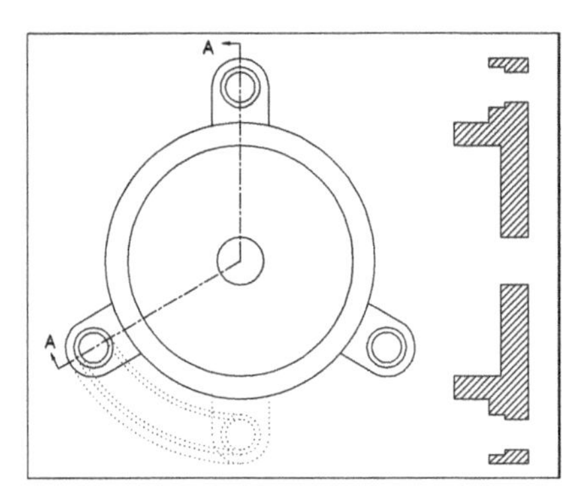

Figure 12-28 *The area cross-section view with aligned lines of projection*

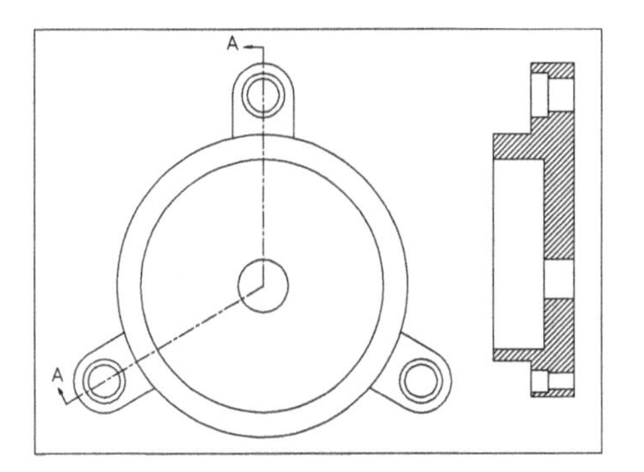

Figure 12-29 *The total cross-section view with normal lines of projection*

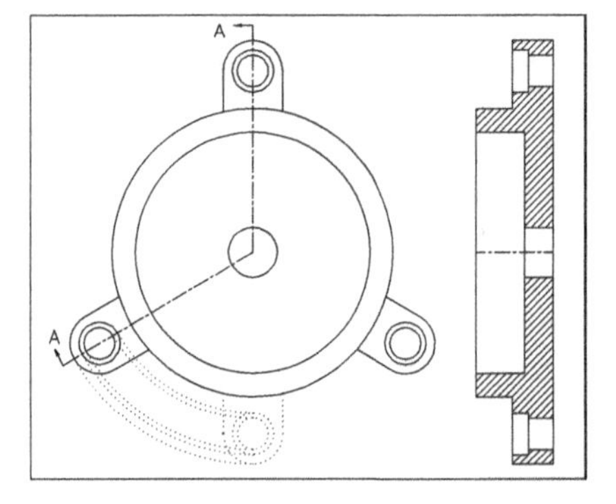

Figure 12-30 *The total cross-section view with aligned lines of projection*

Full(Unfold)

The **Full(Unfold)** option is used to generate the section view by unfolding the section surface of an offset section view. This type of view is only generated using a general view. To generate this type of view, first you need to generate a general view in the required orientation. You can place this general view anywhere on the drawing sheet. Now, choose the **Sections** option from the **Categories** list box on the left of the **Drawing View** dialog box. Now, define the section plane using the offset option and specify the view in which you need to display the arrows. Now, choose the **Full(Unfold)** option from the **Sectioned**

Area drop-down list and complete the creation of the full unfold section view. Figure 12-31 shows the model and the section plane. Figure 12-32 shows an offset section view and a full unfold section view.

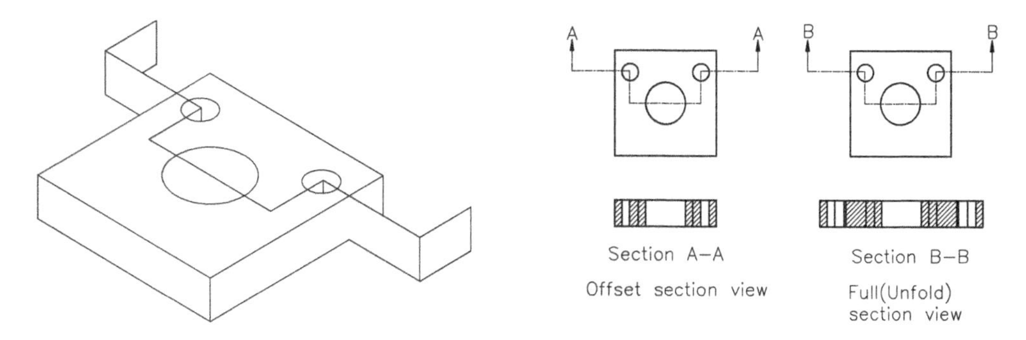

Figure 12-31 *The model and the section plane*

Figure 12-32 *Offset section view and full unfold section view*

Once you have understood the main options of the **Drawing View** dialog box, you can generate the copy and align the view. The method to do so is discussed next.

Generating the Copy and Align View

Ribbon: Layout > Model Views > Copy and Align

The **Copy and Align** tool is used to align views that are generated from an existing partial view. Therefore, to generate this type of drawing view, it is necessary that first a partial view is generated. You will learn more about the partial views later in this chapter.

The procedure to generate a copy and align view is discussed next.

1. To generate the copy and align view, choose the **Copy and Align** tool from the **Model Views** group of the **Layout** tab in the **Ribbon**; you will be prompted to select an existing partial view to be aligned with. Once you have selected a partial view, you will be prompted to select a center point for the drawing view.

2. Select a point anywhere on the drawing sheet to place the view. The view is placed at the selected location. You will be prompted to specify a center point for detail on the current view and to sketch a spline to define the outline of the copy and align view.

3. Specify the center point of the spline and draw the spline. Press the middle mouse button to finish sketching the spline. The drawing view will be cropped along the spline drawn; you will be prompted to select a straight line (axis, segment, datum curve) alignment on the current view.

4. Select an edge from the current view, as shown in Figure 12-33, to align it with the partial view. The copied view is aligned with its parent partial view, as shown in Figure 12-34.

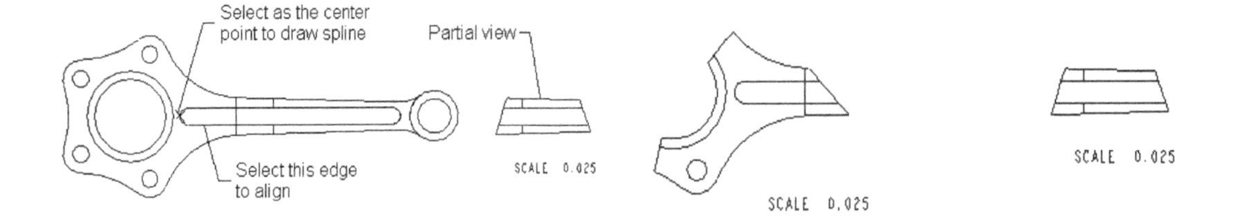

Figure 12-33 Selecting an edge from the current view to align with the partial view

Figure 12-34 Resulting copy and align view

Generating the 3D Cross-Section View

The **3D Cross-section** allows you to display 3D sections in the drawing views. This type of sectioning may be required to show complex cross-sections of a model. Note that the 3D cross-sections are created using zones. Therefore, you first need to create zones and then use them to create 3D cross-sections. The procedure to create zones and 3D cross-sections is discussed next.

1. Open the part file of the model in the **Part** mode and choose **View Manager** from the **Manage Views** drop-down in the **Model Display** group of the **View** tab in the **Ribbon**.

2. Choose the **Sections** tab and then choose the **New** button; a flyout will be displayed, as shown in Figure 12-35. Next, choose the **Zone** option from the flyout to define a new cross-section in the model.

3. Enter the name for the new cross-section and press ENTER; a dialog box with the name that you have specified for the new cross-section will be displayed, refer to Figure 12-36.

Note

*In the example discussed above, the name for the new 3D cross-section is given as **Xsec0001**. Therefore, the dialog box shown in Figure 12-36 is named as **Xsec0001**.*

4. Now, you need to define the planes for the zone creation. Select the datum plane through which you want to create the zone.

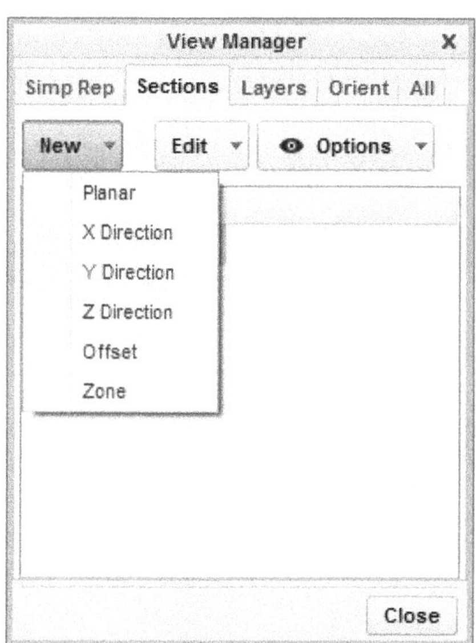

Figure 12-35 The flyout displayed

*Figure 12-36 The **Xsec0001** dialog box*

5. You can add multiple reference entities or datum planes for defining the zone by choosing the **Add a reference to the zone** button from the **Xsec0001** dialog box. You can also remove references specified for the zone by selecting the reference from the **zone scope** area and then choosing the **Remove a reference from the zone** button from the **Xsec0001** dialog box.

6. By default, the direction for creation of the zone is chosen as the negative side of the selected reference entity. You can change the orientation by choosing the **Change orientation** button from the **Xsec0001** dialog box.

7. After specifying all the references for the zone, choose the **OK** button to finish the zone creation and exit the **Xsec0001** dialog box. Next, choose the **Close** button from the **View Manager** dialog box.

Note

*If you want to make the 3D cross-section active in the drawing area, select it from the **View Manager** dialog box and right-click to invoke the shortcut menu. Choose the **Activate** option from the shortcut menu; the 3D cross-section will be activated in the drawing area.*

8. Start a new drawing file and insert a general view in the drawing sheet by choosing the **General** button from the **Model Views** group of the **Ribbon**.

9. Choose the **Sections** option from the **Categories** list of the **Drawing View** dialog box. Next, select the **3D cross-section** radio button from the **Section options** area; the 3D cross-section with the name **Xsec0001** will be displayed in the drop-down list next to this radio button.

10. The **Show X-Hatching** check box is selected by default in the **Section options** area. If you need to display the hatching lines, then select the **Show X-Hatching** check box and choose the **Apply** button from the **Drawing View** dialog box. The 3D cross-section will be displayed on the drawing sheet.

    Figure 12-37 shows the trimetric view and the 3D cross-section of a part model inserted in an A4 drawing sheet.

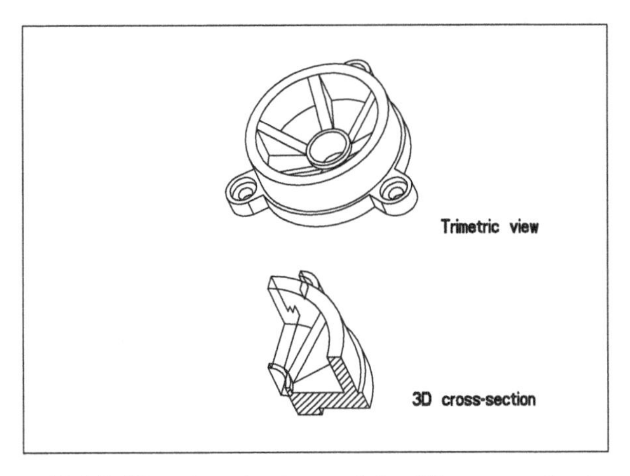

Figure 12-37 The trimetric view and the 3D cross-section in an A4 drawing sheet

Note

*1. The model created in Exercise 4 of Chapter 7 is used to generate a 3D cross-section, refer to Figure 12-37. The **FRONT** and **RIGHT** datum planes will be taken as the references for creating the zone of the model.*

*2. You can also display the 3D cross-section in the **Shading** mode.*

EDITING THE DRAWING VIEWS

In PTC Creo Parametric, you can edit a drawing view as well as the items in the drawing views. All options of modifying the drawing views or items related to it can be chosen from the shortcut menu that will be displayed when you right-click and hold the right mouse button after selecting the view or the item. However, these options can also be chosen from the groups available in the **Ribbon**. PTC Creo Parametric allows you to perform the following types of editing operations on the drawing views.

Moving the Drawing View

When a view is generated, its movement is locked by default. To unlock the drawing view, choose **Lock View Movement** either from the shortcut menu that will be displayed on right-clicking and holdinhg the right mouse button on the drawing view or directly from the **Drawing Tree**. Alternatively, choose the **Lock View Movement** toggle button from the

Document group to unlock the movement of the views. When you select the drawing view, it will be displayed inside a boundary. To move the drawing view, select it and then drag it to the required location on the drawing sheet. Remember that if you select the view that has some child views, the child views will also move along with the parent view in order to maintain their alignment with the parent view. Also, the projected views can be moved only in the direction of projection.

Note

*The **General** and **Detailed** views can be moved to any new location because they are not the projections of any view.*

Erasing the Drawing View

The **Erase View** button available in the **Display** group of the **Layout** tab in the **Ribbon** is used to temporarily remove the selected drawing view from the sheet. However, the view still remains in the memory of the drawing and can be resumed at any point of time. As the view is not completely removed from the memory, you can also erase a view that has some child views associated to it and it will not affect the child view. Once a view is erased, a box will be displayed in place of the view displaying the name of the view. To resume the view, choose the **Resume View** button from the **Display** group of the **Layout** tab in the **Ribbon**. Alternatively, select the erased drawing view from the drawing or directly from the **Drawing Tree** and then right-click on it. Choose **Resume View** from the shortcut menu displayed.

Note

When you erase a view, the leaders and dimensions that were attached with the view are also erased. When you resume an erased view, the leaders and dimension values are displayed again.

Deleting the Drawing View

To delete a drawing view from the drawing sheet, select the view from the drawing or from the **Drawing Tree** and hold down the right mouse button to invoke a shortcut menu. Choose the **Delete** option from the shortcut menu. Once the view is deleted, no information related to the deleted view remains in the memory of the drawing. Remember that if a view has some child views associated with it then it is also deleted. Before deletion, PTC Creo Parametric informs you that views associated with this view will also be deleted. You can choose the **Yes** button to delete the view or choose the **No** button to continue with the view from the displayed dialog box.

Adding New Parts or Assemblies to the Current Drawing

Ribbon: Layout > Model Views > Drawing Models

To generate the drawing views, you can add more parts or assemblies, apart from the default part or assembly, for. To do so, choose **Layout > Model Views > Drawing Models** from the **Ribbon**; the **DWG MODELS** menu will be displayed, as shown in Figure 12-38.

Next, choose the **Add Model** option from the **DWG MODELS** menu to add a model to the current drawing. When you choose this option, the **Open** dialog box will be displayed. You can select the new model to be added using this dialog box. Remember that the latest model added will be the current model and the drawing views generated will be of this model. You can change the current model by choosing the **Set Model** option from the **DWG MODELS** menu. Similarly, you can also delete a model from the current drawing. However, only the model that does not have any view generated from it can be deleted.

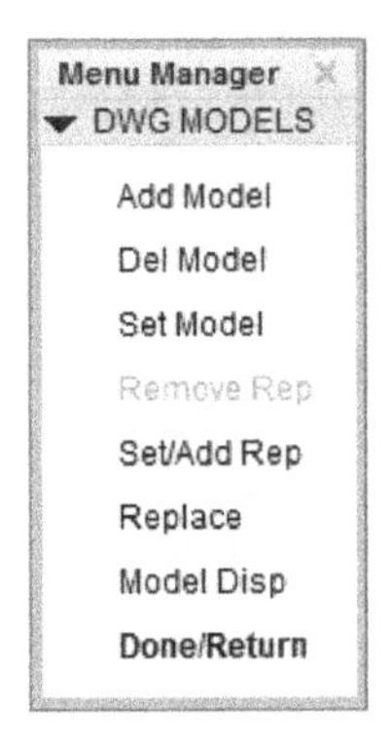

Figure 12-38 The ***DWG MODELS*** *menu*

MODIFYING THE DRAWING VIEWS

You can also make the following modifications in the existing drawing views.

Changing the View Type

Select the drawing view that needs to be modified and hold the right mouse button to invoke the shortcut menu. Choose **Properties** from the shortcut menu; the **Drawing View** dialog box will be displayed, as shown in Figure 12-39. The **Drawing View** dialog box can also be invoked by double-clicking on the view to be modified. The **View Type** option in the **Categories** area is chosen by default. You can modify the view type by selecting any option from the list displayed under **Model view names** collector. Remember that only the general, projection, and auxiliary view types can be modified using this option.

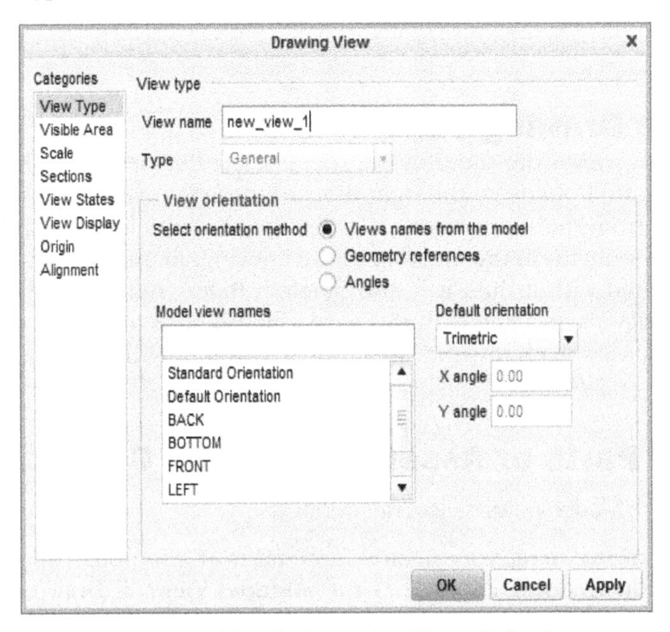

Figure 12-39 The ***Drawing View*** *dialog box*

Changing the View Scale

The scale of a view can be modified using the **Scale** option from the **Categories** list box of the **Drawing View** dialog box. You can modify the scale factor of the views that are generated using the general view and detailed view options.

Note

The scale factor represents the scaling of the drawing model. For example, if the scale is set to 0.25, PTC Creo Parametric scales the drawing views to one-quarter (1/4) of the actual size of the model.

Reorienting the Views

Double click on the general view to be reoriented to display the **Drawing View** dialog box. You can use the options from this dialog box to change the orientation of the selected view. However, if some child view is associated with the general view, then the child view will also be reoriented accordingly.

Modifying the Cross-sections

You can flip the side of sectioning and replace, delete, or rename the sections in the views using the **Sections** option of the **Categories** list in the **Drawing View** dialog box.

Modifying Boundaries of Views

You can modify the boundaries of the detailed or partial views using the option in the **Drawing View** dialog box. To modify the boundary of a detailed view, double-click on it to invoke the **Drawing View** dialog box. The selection mode is invoked by default in the **Reference point on parent view** selection area. Therefore, you can specify the center point for the boundary of the detailed view. To sketch the boundary of the detailed view again, click once in the **Spline boundary on parent view** selection area and draw the boundary on the parent view. To modify the boundary of a partial view, invoke the **Drawing View** dialog box and choose the **Visible Area** option from the **Categories** list box. Using the options available on the right of this dialog box, you can modify the boundary of the detailed view.

Adding or Removing the Cross-section Arrows

If you have not specified the cross-section arrows while generating the section views, you can specify them by selecting the section view and holding down the right mouse button to invoke the shortcut menu. Choose the **Add Arrows** option from the shortcut menu; you will be prompted to select the view where the arrows should be displayed. Select the view in which you need to display the arrows.

To remove arrows, select them and choose the **Delete** button from the right-click shortcut menu.

Modifying the Perspective Views

Double-click on the perspective view to display the **Drawing View** dialog box and choose the **Scale** option from the **Categories** list box. The options related to the perspective view are available. You can modify the eye point distance or the view diameter using these options.

MODIFYING OTHER PARAMETERS

Apart from modifying the drawing views, you can also modify other parameters related to the drawing views. For example, you can modify the size and the style of the text, scale factor of all drawing views, cross-section hatching, and so on. All this is done by selecting the item. When the item is highlighted, hold down the right mouse button to invoke the shortcut menu. You can select the options from the shortcut menu to modify that item.

You can modify any parameter associated with the drawing views. Depending on the item selected to modify, the options related to it vary. The options displayed in the shortcut menu vary from item to item.

Editing the Cross-section Hatching

Select the hatching from a drawing view. When the hatching turns green in color, hold down the right mouse button to invoke the shortcut menu. Choose the **Properties** option from the shortcut menu; the **MOD XHATCH** menu will be displayed, as shown in Figure 12-40.

***Figure 12-40** The **MOD XHATCH** menu*

The parameters related to cross-section hatching that can be modified are: the spacing of the hatching, angle of the hatching lines, offset value, and the line style of the hatching lines. There are also some standard hatch patterns that are available in PTC Creo Parametric. You can retrieve these standard patterns by using the **Open** dialog box displayed upon choosing **Retrieve** from the **MOD XHATCH** menu.

TUTORIALS

Tutorial 1

In this tutorial, you will generate the drawing views of the model created in Exercise 4 of Chapter 7, as shown in Figure 12-41. Select the A4 size drawing sheet and generate all drawing views shown in Figure 12-42. **(Expected time: 45 min)**

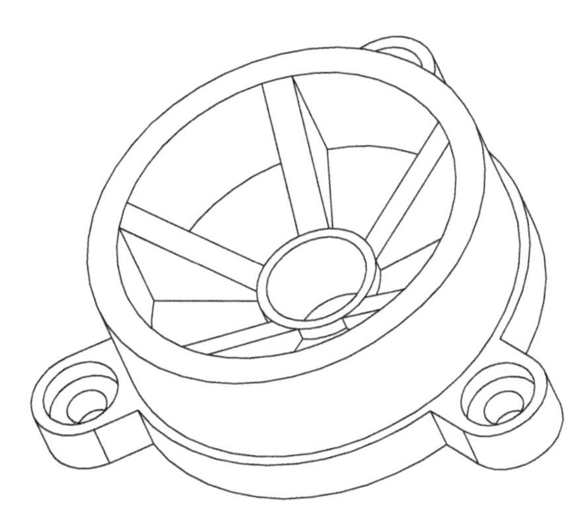

Figure 12-41 *Part for generating drawing views*

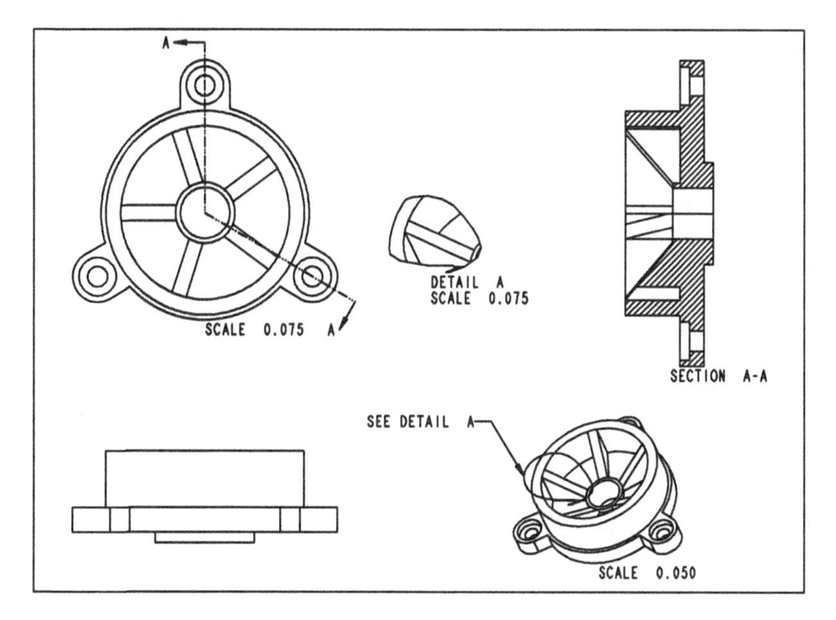

Figure 12-42 *The drawing views to be generated*

The following steps are required to complete this tutorial:

a. Start a new drawing file and select the required drawing sheet.
b. Generate the top view, refer to Figure 12-44.
c. Generate the front view taking the top view as the parent view, refer to Figure 12-44.
d. Generate the sectioned right view by defining a plane on the top view of the model, refer to Figures 12-45 and 12-46.
e. Generate the default 3D view of the model, refer to Figure 12-46.
f. Generate the detail view from the 3D view, refer to Figure 12-47.

Before you start generating drawing views, set *C:\Creo3.0\c12* as working directory. Copy the model file *c07ex4.prt* from the *c07* folder and paste it in the folder named *c12*. The drawing generated in this tutorial will be saved in the *c12* directory with an extension *.drw*. The part file *(.prt)* and the drawing file *(.drw)* should lie in the same directory or folder.

Note

You cannot open a drawing file if the part file, from which the drawing view was generated, has been deleted. Also, if the part file is removed from the folder where its drawing file is placed or if you rename the part file after generating the drawing views, the drawing file will not open.

Starting a New Drawing File

To generate drawing views in PTC Creo Parametric, you first need to start a new file in the **Drawing** mode.

1. Choose the **New** button from the **File** menu to display the **New** dialog box.

2. Select the **Drawing** radio button and then enter *c12tut1* as the name of the file.

3. Choose **OK** from the **New** dialog box to display the **New Drawing** dialog box.

4. Choose the **Browse** button and select *c07exr4.prt* for generating the drawing views.

5. Select the **Empty** radio button from the **Specify Template** area.

6. Choose the **Landscape** button from the **Orientation** area, if it is not chosen by default, as shown in Figure 12-43.

7. Select **A4** from the **Standard Size** drop-down list in the **Size** area. Choose **OK** to proceed to the **Drawing** mode.

***Figure 12-43** The **New Drawing** dialog box with the **Landscape** button chosen*

An A4 size sheet with the landscape orientation is displayed in the drawing area.

Generating the Top View

You can generate any view as the first view on the drawing sheet. However, in this tutorial, you will generate the top view as the first view.

1. Choose **Layout > Model Views > General** from the **Ribbon**; the **Select Combined State** dialog box is displayed. Choose the **OK** button from the dialog box; you are prompted to specify the center point for the drawing view.

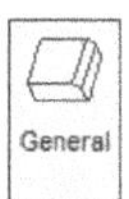

2. Specify the center point for the first view, which is on the top left corner of the sheet, as shown in Figure 12-44. On specifying the center point, the **Drawing View** dialog box is displayed along with the preview of the drawing view in the default orientation. In the **Model view names** list box, select the **TOP** option and choose the **Apply** button. The default view is oriented as the top view.

Note
Although you can move a view anywhere on the sheet, yet you should try to place the drawing view inside the sheet and also leave space for the other views to be generated.

3. Select the **Scale** option from the **Categories** list box in the **Drawing View** dialog box; the **Scale and perspective options** area is displayed. Select the **Custom scale** radio button and enter **0.075** as the scale factor for the view. Choose the **Apply** button; the scale of the view is modified.

4. Choose the **Close** button from the **Drawing View** dialog box to complete the placing of the top view.

5. Turn off the display of datum planes, axes, points, and coordinate system by choosing the respective buttons from the **Datum Display Filters** drop-down in the **Graphics** toolbar.

6. Change the model display to no hidden by choosing the **No Hidden** button from the **Display Style** drop-down. Next, choose the **Repaint** button from the **Graphics** toolbar to repaint the screen.

Tip: *If the view that you have placed on the drawing sheet is not at the proper location, you need to move the drawing view. To do so, select the drawing view; the selected drawing view is enclosed in a green box. Next, right-click and hold the right mouse button to display a shortcut menu. Choose **Lock View Movement** from the shortcut menu displayed and then press the left mouse button inside the box and drag it to place it at the required location.*

Generating the Front View

The front view of the model is generated from the top view, which is already placed on the drawing sheet.

1. Choose **Layout > Model Views > Projection** from the **Ribbon**; you are prompted to specify a center point for the drawing view.

2. Specify the center point for the front view below the top view, as shown in Figure 12-44.

3. Click anywhere on the drawing sheet to clear the current selection set.

If the front view of the model is not similar to the one shown in Figure 12-44, then you need to change the first angle projection to the third angle projection. To do so, choose **Prepare > Drawing Properties** from the **File** menu; the **Drawing Properties** dialog box is displayed. Choose the **change** button available in front of **Detail Options** in this dialog box; the **Options** dialog box is displayed. Enter **projection_type** in the **Option** text box; the options in this dialog box are displayed in the **Value** drop-down list. Select **third_angle** from the drop-down list and then choose the **Add/Change** button. Next, choose the **OK** button to accept the changes and exit the dialog box.

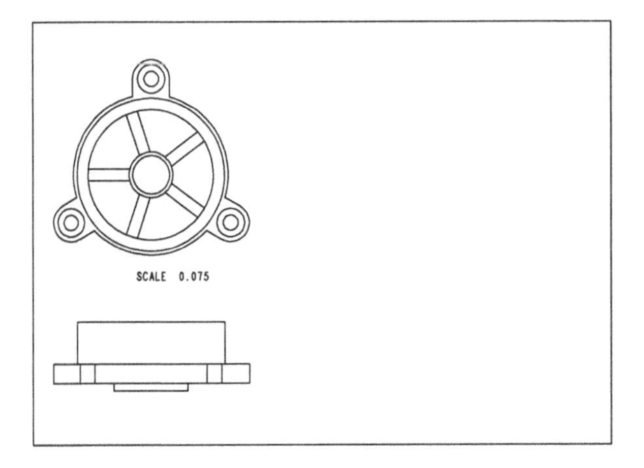

***Figure 12-44** The top and front views*

Tip: *It is recommended that you use the **Repaint** button available in the **Graphics** toolbar to remove any temporary information in the drawing area and to refresh the screen.*

Generating the Section View

To generate a section view, a section must be defined on the model. You will sketch this section to generate the section view for this tutorial.

1. Choose **Layout > Model Views > Projection** from the **Ribbon**; you are prompted to select a projection parent view.

2. Select the top view as the projection parent view; you are prompted to specify a center point for the drawing view.

3. Specify the center point on the right side of the top view to place the view.

4. Once the view is placed, double-click on it to display the **Drawing View** dialog box.

5. Select **Sections** from the **Categories** list box in the **Drawing View** dialog box; the section options are invoked in the **Drawing View** dialog box.

6. Select the **2D cross-section** radio button and choose the **Add cross-section to view** button; the window area below it is activated. Select the **Create New** option from the **Name** drop-down list; the **XSEC CREATE** menu is displayed.

7. Choose **Offset > Both Sides > Single > Done** from the **XSEC CREATE** menu.

8. Specify the name of the cross-section as **A** in the message input window and press ENTER. Once you have specified the name of the cross-section, a separate window displaying the part appears. Also, you are prompted to select a sketching plane.

Note

If the model in the ***Part*** *mode is opened in another window, the subwindow will not appear and you have to manually change the window. Choose* ***View > Window > Windows*** *from the* ***Ribbon*** *and select the part file that is opened. Now, you can continue with the sketcher environment.*

9. Choose the **Plane Display** tool from the **Show** group of the **View** tab in the **Ribbon** from the original window to display datum planes. You may need to repaint the screen in the subwindow by choosing the **Repaint** button from the **View** menu to view the datum planes, if required.

10. Select the **TOP** datum plane from the subwindow. Choose **Okay** from the **DIRECTION** submenu; the **SKET VIEW** submenu is displayed.

11. Choose the **Bottom** option from the **SKET VIEW** submenu and select the **FRONT** datum plane from the subwindow.

12. Choose the **Line** tool from the **Line** flyout in the **Sketch** menu and sketch the lines, as shown in Figure 12-45. These lines define the section plane.

13. Align the lines and modify the angular dimension to **120**, as shown in Figure 12-45.

Figure 12-45 *Sketch for generating the total aligned section view*

14. Choose **Sketch > Done** from the menu bar of the subwindow.

15. Select the **Full(Aligned)** option from the **Sectioned Area** drop-down list; you are prompted to select an axis.

16. Select the central axis of the model from the front drawing view. You may need to turn on the display of the central axis if its display is turned off.

17. Scroll the bar in the **Drawing View** dialog box to the right edge. Click in the field below the **Arrow Display** column; you are prompted to pick a view for arrows where the section is perpendicular. Select the top view to display the arrows.

18. Choose the **Apply** button from the **Drawing View** dialog box and then exit the dialog box; the section view is displayed.

19. Turn off the display of the datum planes and the datum axes.

Tip: *If you do not want to display the section arrows in any view, you can use the middle mouse button (in case of three button mouse) to abort the creation of arrows.*

Modifying the Hatching

You need to modify the spacing between the hatch lines to make the hatching dense. In this section, you will select the type of filter from the **Filter** drop-down list to select the hatch lines.

1. Select the **X-Section** filter from the **Filter** drop-down list available in the **Status Bar** below the drawing area. Select the hatching lines from the drawing sheet; the hatching lines turn green.

2. Press and hold the right mouse button to invoke the shortcut menu. Choose the **Properties** option from the shortcut menu; the **MOD XHATCH** menu is displayed. Alternatively, double-click on the hatching lines to invoke the **MOD XHATCH** menu.

3. In the **MOD XHATCH** menu, choose the **Spacing** option to display the **MODIFY MODE** submenu.

4. The spacing between the hatching lines needs to be reduced. To do so, choose the **Half** option twice and then choose **Done** in the **MOD XHATCH** menu. On doing so, the hatching appears more dense, refer to Figure 12-46.

5. Set the filter back to the **General** option using the **Filter** drop-down list.

Generating the General View

In this section, the **General** view needs to be generated to show the 3D view of the model, which is the trimetric view.

1. Choose the **General** tool from the **Model Views** drop-down in the **Layout** group of the **Ribbon**; the **Select Combined State** dialog box is displayed. Choose the **OK** button from the dialog box; you are prompted to specify a center point for the placement of the view.

2. Specify the center point for placing the general view below the section view, as shown in Figure 12-46; the **Drawing View** dialog box is displayed.

3. Set the value of the scale for the new view to **0.05** using the **Scale** option from the **Drawing View** dialog box.

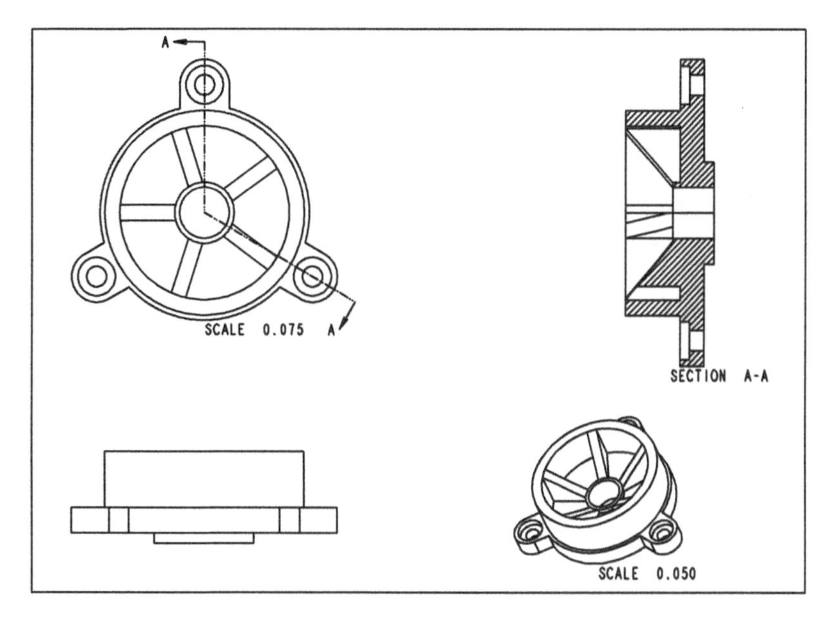

Figure 12-46 *Different drawing views*

4. Choose the **Apply** button and then exit the dialog box.

Generating the Detail View

As mentioned earlier, the detail view is required to provide details of a particular portion of the drawing view. In this section, you need to give the details of one of the ribs of the model.

1. If the general view is selected (highlighted in green box), then you need to click once on the drawing sheet to exit the current selection set. Next, choose the **Detailed** tool from the **Model Views** group of the **Layout** tab in the **Ribbon**; you are prompted to select a center point for detail on an existing view. This center point will be the center of the detailed view.

2. Select the center point for the detail on one of the ribs in the trimetric view, refer to Figure 12-47.

   On doing so, you are prompted to sketch a spline about the center point that you selected.

3. Draw the spline and then press the middle mouse button to exit the **Spline** tool.

4. Select the center point for the placement of the drawing view. The detailed view is generated and scaled by default.

5. Press and hold the right mouse button on the detail view to invoke a shortcut menu and choose the **Properties** option from the shortcut menu; the **Drawing View** dialog box is displayed.

6. Select **Spline** from the **Boundary type on parent view** drop-down list.

7. Select the **Scale** option from the **Categories** list box. Set the value of the custom scale to **0.075**. Choose the **Apply** button and exit the dialog box.

   A note is also displayed with an arrow attached to the trimetric view. You may need to move the note to a suitable position by selecting it and then dragging the cursor. The final sheet after generating all views is shown in Figure 12-47. If you are unable to select the note, then choose the **Annotate** tab from the **Ribbon**. On doing so, you will be able to select it and move it.

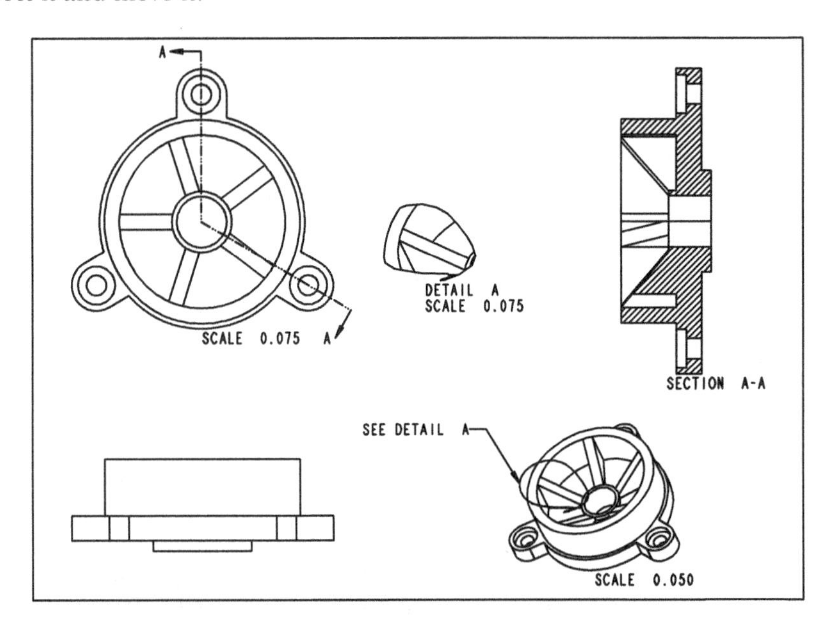

***Figure 12-47** The drawing views for Tutorial 1*

Tip: *If any view or the text on the drawing is overlapping or is not at the desired place on the sheet, select it and then move it by dragging the mouse.*

Saving the Drawing File

In this section, you need to save the drawing file that you have created.

1. Choose the **Save** button from the **Quick Access** toolbar; the **Save Object** dialog box is displayed with the name of the drawing file that you have entered earlier.

2. Choose **OK** to confirm the action.

Closing the Drawing File

After saving the drawing file, you need to close it.

1. Choose **File > Close** from the menu bar to close the drawing window.

Tutorial 2

In this tutorial, you will generate the drawing views of the part created in Tutorial 1 of Chapter 8. The part is shown in Figure 12-48. The drawing views that need to be generated are shown in Figure 12-49. **(Expected time: 45 min)**

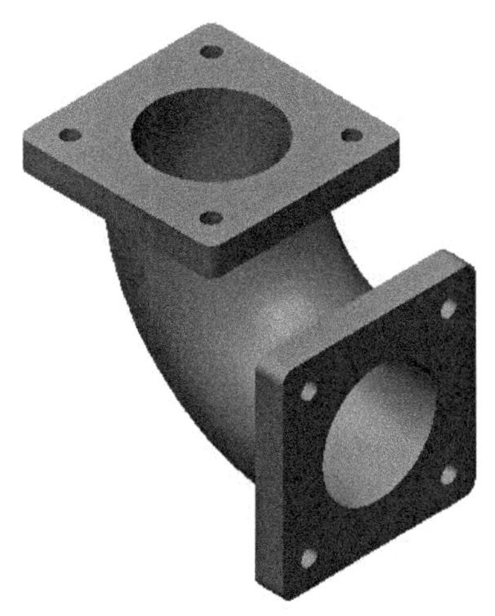

Figure 12-48 *Model for generating drawing views*

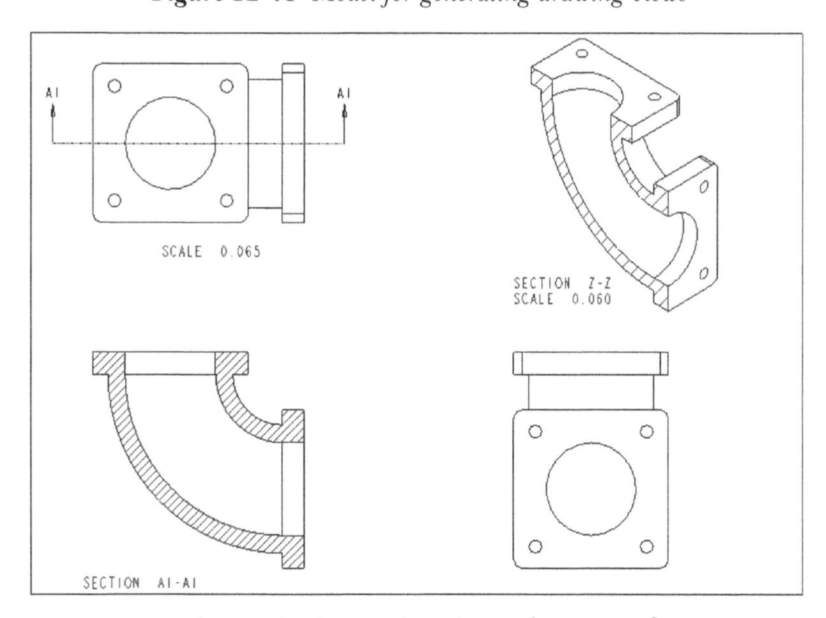

Figure 12-49 *Drawing views to be generated*

Before you start generating drawing views, copy the file *c08tut1.prt* from the *c08* folder in the current directory.

The following steps are required to complete this tutorial:

a. Start a new drawing file and select the size of the drawing sheet.
b. Generate the top view, refer to Figure 12-50.
c. Generate the sectioned front view by defining the **FRONT** datum plane as the section plane, refer to Figure 12-50.
d. Generate the right view of the sectioned front view, refer to Figure 12-51.
e. Generate the isometric sectioned view. The section will be defined by drawing a line on the **TOP** datum plane, refer to Figures 12-52 and 12-53.

Before you start generating the drawing views, set the working directory to *C:\Creo3.0\c12*. The part *(.prt)* file and the drawing *(.drw)* file should lie in the same directory or folder.

Starting a New Drawing File

To generate the drawing views, you first need to start a new drawing file.

1. Choose the **New** button from the **Data** group in the **Ribbon** to display the **New** dialog box.

2. Select the **Drawing** radio button and then enter **c12tut2** as the name of the file.

3. Choose **OK** from the **New** dialog box to display the **New Drawing** dialog box.

4. Choose the **Browse** button to select *c08tut1.prt* from the *c08* folder for generating drawing views and then choose **Open**.

5. Select the **Empty** radio button from the **Specify Template** area.

6. Choose the **Landscape** button from the **Orientation** area.

7. Select **A4** from the **Standard Size** drop-down list in the **Size** area. Choose **OK** from the **New Drawing** dialog box to proceed to the **Drawing** mode.

Generating the Top View

First, you need to generate the top view. All other views, except the sectioned isometric view, will be the child views of the top view. You need to generate the top view first because the required sectioned front view can be generated only from the top view. The right view can also be generated independently, but then this view will not help generate any other required view.

1. Choose the **General** button from the **Model Views** group in the **Ribbon**; the **Select Combined State** dialog box is displayed. Choose the **OK** button from this dialog box; you are prompted to specify a center point for the placement of the view.

2. Specify the center point for the placement of the top view close to the upper left corner of the drawing sheet; the **Drawing View** dialog is displayed.

3. Select the **TOP** option from the **Model view names** list box and then choose the **Apply** button.

4. Select the **Scale** option from the **Categories** list box. Next, select the **Custom scale** radio button and enter **0.065** in the edit box.

5. Choose **OK** to exit the **Drawing View** dialog box.

   If necessary, move the view, as shown in Figure 12-50, and then choose **No Hidden** from the **Display Style** drop-down in the **Graphics** toolbar. You may also need to repaint the screen.

Generating the Sectioned Front View

The sectioned front view of the model is generated from the top view. Before proceeding further, use the **Plane Display** button from the **Datum Display Filters** drop-down in the **Graphics** toolbar to turn on the display of datum planes and repaint the screen.

1. Choose the **Projection** tool from the **Model Views** group of the **Layout** tab in the **Ribbon**; you are prompted to specify a center point for placement of the projection view.

2. Specify the center point for the placement of the front view below the top view, as shown in Figure 12-50.

3. Select the newly generated view and invoke the shortcut menu. Choose the **Properties** option to display the **Drawing View** dialog box.

4. Select the **Sections** option from the **Categories** list box to display the section related options in the dialog box.

5. Select the **2D cross-section** radio button and choose the **Add cross-section to view** button; the window area below it gets activated. Select the **Create New** option from the **Name** drop-down list, if it is not selected; the **XSEC CREATE** menu is displayed.

6. Choose **Planar > Single > Done** from the **XSEC CREATE** menu.

7. Enter **A1** as the name of the cross-section in the message input window and then press ENTER to display the **SETUP PLANE** submenu; you are prompted to select a planar surface or a datum plane.

8. Select the **FRONT** datum plane (the plane that cuts the part horizontally through the center of the cylindrical feature in the top view) from the drawing area.

9. Scroll the bar in the **Drawing View** dialog box to the right. Click on the field below the **Arrow Display** column; you will be prompted to pick a view for arrows where the section is perpendicular. Select the top view to display the arrows.

10. Choose the **Apply** button and then exit the dialog box.

Modifying the Hatching

The offset distance between the hatching lines in the front sectioned view is large. Therefore you need to reduce the distance between the hatching lines.

1. Select the **X-Section** filter from the **Filter** drop-down list in the **Status Bar**. Select the hatching lines from the sectioned front view in the drawing sheet; the hatching lines turn green. Hold the right mouse button to invoke the shortcut menu. Choose the **Properties** option from the shortcut menu; the **MOD XHATCH** menu is displayed.

2. In the **MOD XHATCH** menu, choose the **Spacing** option to display the **MODIFY MODE** submenu.

3. The spacing between the hatching lines has to be reduced. Choose the **Half** option twice and then choose **Done** in the **MOD XHATCH** menu. Now, the hatching appears more dense.

4. Click once in the drawing area to remove the X-hatch from the current selection set. The sheet after placing these two views should look similar to the one shown in Figure 12-50.

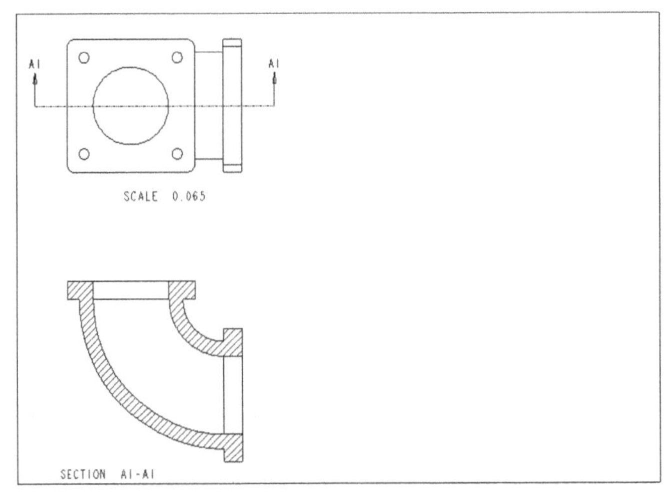

***Figure 12-50** Drawing sheet after generating the top view and the sectioned front view with the display of datum planes turned on*

5. Set the selection filter back to **General**.

Generating the Right View

The right view is the projection of the sectioned front view. Before proceeding further, turn off the display of the datum planes.

1. Choose **Layout > Model Views > Projection** from the **Ribbon**; you are prompted to select the projection parent view.

2. Select the sectioned front view as the parent view.

3. Specify the center point for the placement of the drawing view on the right of the sectioned front view. The right view of the model is placed, as shown in Figure 12-51.

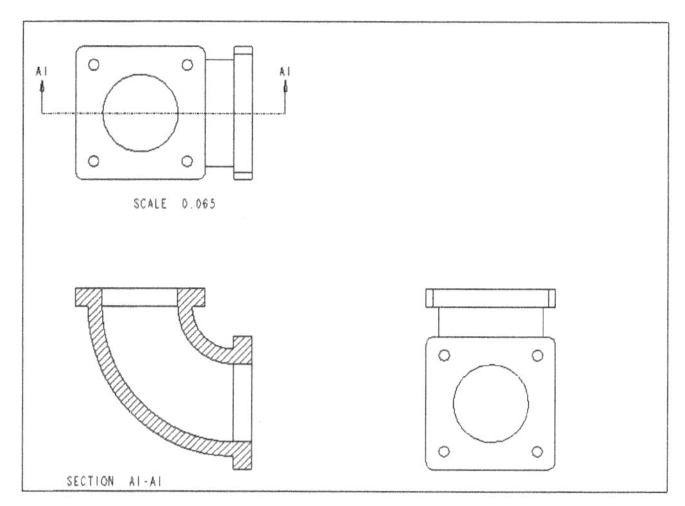

Figure 12-51 *The drawing sheet after generating the top, sectioned front, and right views*

Generating the Isometric Section View

The isometric section view is an independent view and it will be generated by using the **General** option.

1. Choose the **General** button from the **Model Views** group in the **Ribbon**; the **Select Combined State** dialog box is displayed. Choose the **OK** button from this dialog box; you are prompted to specify a center point for the placement of the view.

2. Specify the center point on the upper right corner of the drawing sheet for placing the view; the **Drawing View** dialog is displayed.

   The default view is a trimetric view but you need to display the isometric view. Therefore, you need to change the orientation.

3. Select the **Isometric** option from the **Default orientation** drop-down list; the isometric view of the model is displayed on the drawing sheet.

4. Choose the **Scale** option from the **Categories** list box. Select the **Custom scale** radio button and enter **0.06** as the scale factor in the edit box displayed. Next, choose the **Apply** button.

5. Select the **Sections** option from the **Categories** list box to display the section options in the dialog box.

6. Select the **2D cross-section** radio button and then choose the **Add cross-section to view** button. The collector below it gets activated. Select **Create New** from the name drop-down list; the **XSEC CREATE** menu manager is displayed.

7. Choose **Offset > Both Sides > Single > Done** from the **XSEC CREATE** menu; the message input window is displayed.

8. Enter **Z** as the name of the section in the message input window and press ENTER. Once you have specified the name of the section, a separate window is displayed with the model.

9. Select the **TOP** datum plane from the subwindow. You may need to turn on the display of the datum planes. Choose **Okay** from the **DIRECTION** submenu; the **SKET VIEW** submenu is displayed.

10. Choose the **Right** option from the **SKET VIEW** submenu and select the **RIGHT** datum plane from the subwindow.

11. Choose **Sketch > Line > Line** from the menu bar and draw a line, as shown in Figure 12-52. This line creates a section plane.

12. Choose **Sketch > Constraint > Coincident** from the menu bar and align both ends of the line to the edges.

13. Choose **Sketch > Done** from the menu bar.

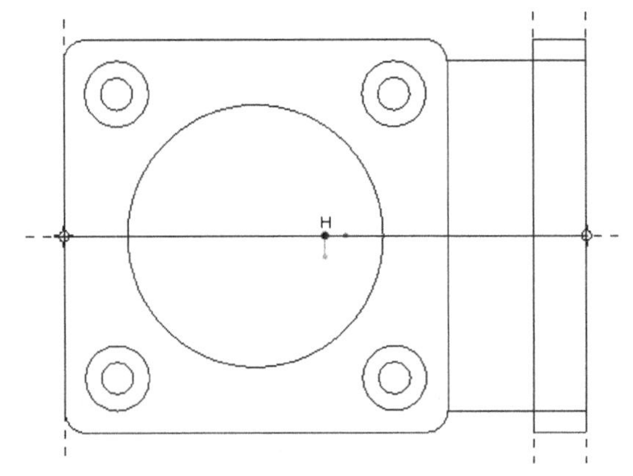

Figure 12-52 *The sketch for the section line with the constraint*

14. Choose the **Apply** button to place the view and then exit the dialog box.

Note

If the drawing view placed on the sheet overlaps the boundary of the drawing sheet, then you can move the drawing view. To do so, select the drawing view; the selected drawing view is enclosed in a green box. Now, press the left mouse button inside the box and drag it to place it at the desired location.

Tip: *You can lock or unlock the drawing view using the* ***Lock View Movement*** *button from the* ***Document*** *group of the* ***Layout*** *tab.*

Modifying the Hatching

The spacing between the hatching lines in the sectioned isometric view is large. Therefore, you need to reduce the distance between them. Use the same procedure that was discussed earlier in this tutorial to modify the spacing between the hatch lines. The sheet after placing all views should look similar to the one shown in Figure 12-53.

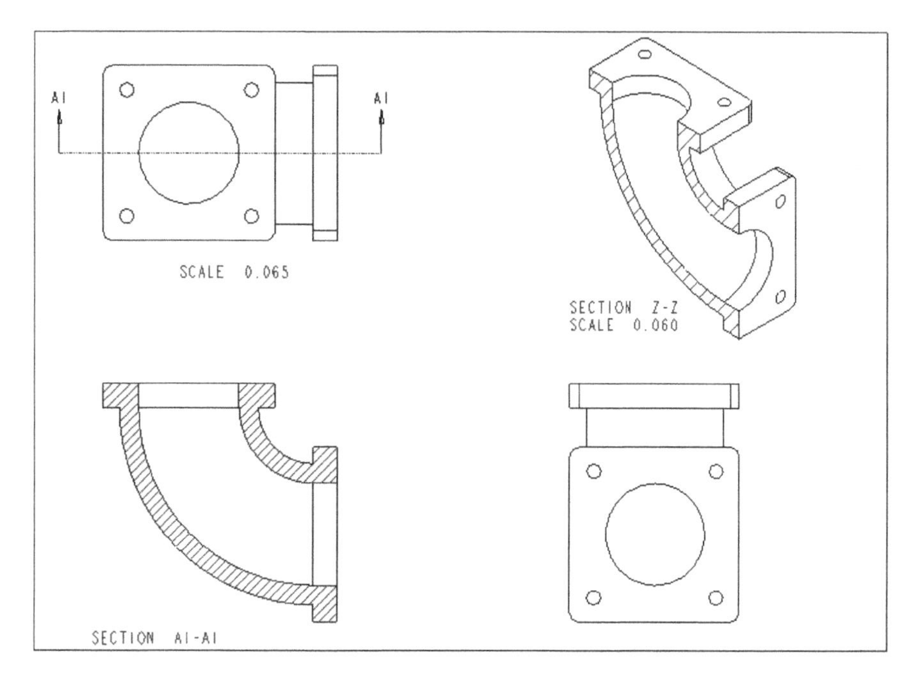

Figure 12-53 *Different drawing views generated*

Saving the Drawing File

In this section, you need to save the drawing file.

1. Choose the **Save** button from the **Quick Access** toolbar; the **Save Object** dialog box is displayed with the name of the drawing file that you have entered earlier.

2. Press ENTER to save the file.

Closing the Drawing File

After you have saved the drawing file, close the drawing file.

1. Choose **File > Close** from the menu bar; the drawing window is closed.

Self-Evaluation Test

Answer the following questions and then compare them to those given at the end of this chapter:

1. The bidirectional associative nature of a software package implies that when one file related to the part model is modified, the modifications are reflected in other related files as well. (T/F)

2. General view is the first view that is generated in the sheet. (T/F)

3. The **Full** section view is the most widely used section view. (T/F)

4. Section views are generally created for the models having features that are not clearly visible in the standard view. (T/F)

5. Broken views are used for the parts that have a high length to width ratio. (T/F)

6. The __________ view is used when you want to show a particular portion of the view in a section and at the same time, do not want to section the remaining view.

7. The __________ option allows you to temporarily remove the selected drawing view from the sheet.

8. The __________ option is used to redisplay the drawing views that are erased using the **Erase View** option.

9. Using the __________ dialog box, you can reorient the general view.

10. The cross-section hatching on the sectioned portion can be modified using the __________ menu.

Review Questions

Answer the following questions:

1. Which of the following options, when chosen, displays all the erased drawing views on the sheet?

   (a) **Delete** (b) **Resume View**
   (c) **Move** (d) **Erase**

2. Which of the following buttons in the **Orientation** group of the **View** tab is used to make all the objects in the drawing area visible on the screen?

   (a) **Zoom Out** (b) **Zoom In**
   (c) **Refit** (d) None of these

3. Which of the following options, when selected, displays only the area of the section view that is sectioned?

   (a) **Total** (b) **Area**
   (c) **Align** (d) None of these

4. Which of the following options is used to permanently remove a drawing view from the sheet?

   (a) **Resume** (b) **Erase**
   (c) **Delete** (d) **Move**

5. Which of the following options is used to modify the numeric value associated with the drawing views?

   (a) **Xhatching** (b) **Any Item**
   (c) **Value** (d) None of these

6. In PTC Creo Parametric, you can change the view type of an existing view. (T/F)

7. In PTC Creo Parametric, you can reorient only the general views. (T/F)

8. In PTC Creo Parametric, you can flip the side of the cross-section views. (T/F)

9. The **Orientation** area in the **New Drawing** dialog box will be available only when you select the **Empty** radio button from the **Specify Template** area. (T/F)

10. The orientation of a model saved in the **Part** mode can be used to orient the drawing view in the **Drawing** mode. (T/F)

Exercise

Exercise 1

Generate the drawing views of the model created in Tutorial 3 of Chapter 9 on an A4 size sheet. The model is shown in Figure 12-54. The drawing views to be generated are shown in Figure 12-55. **(Expected time: 45 min)**

Figure 12-54 *Part for generating the drawing views*

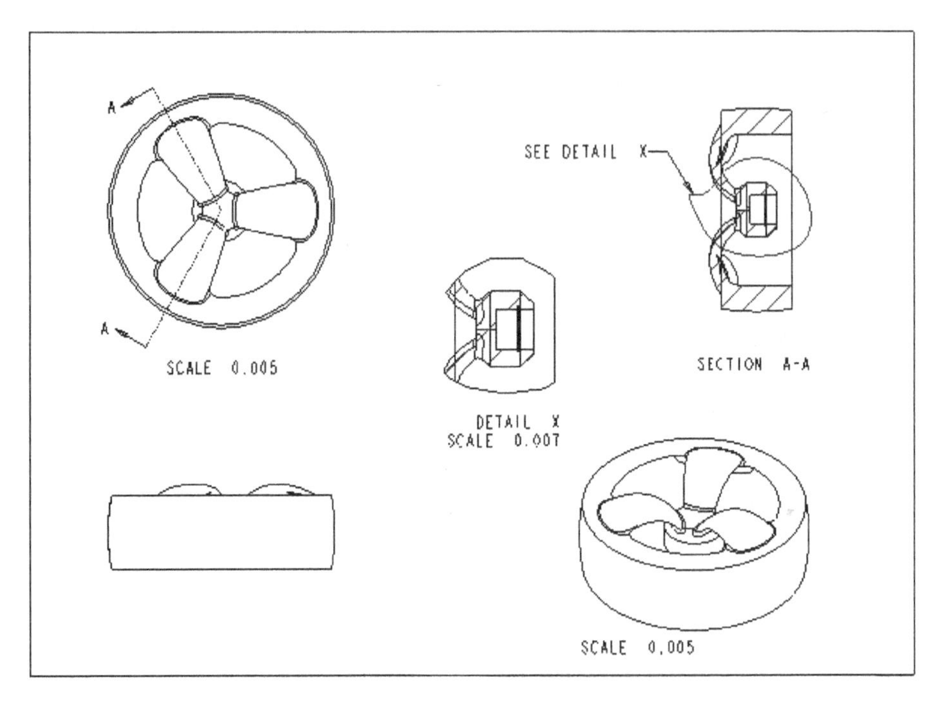

Figure 12-55 *Drawing views to be generated in Exercise 1*

Answers to Self-Evaluation Test

1. T, **2.** T, **3.** T, **4.** T, **5.** T, **6. Local Section**, **7. Erase**, **8. Resume View**, **9. Drawing View**, **10. MOD XHATCH**

Chapter 13

Dimensioning the Drawing Views

Learning Objectives

After completing this chapter, you will be able to:

- *Show or erase dimensions in drawing views.*
- *Add notes to drawings.*
- *Add dimensional and geometric tolerances to drawing views.*
- *Edit the geometric tolerance.*
- *Add balloons to exploded assembly views.*
- *Add reference datums to drawing views.*
- *Modify and edit dimensions.*

DIMENSIONING THE DRAWING VIEWS

Once you have generated the drawing views, you need to generate dimensions, add notes, symbols, balloons, and so on in the drawing views. These dimensions are assigned to each entity of sketches in the model or the features associated with the model. These dimensions are associative in nature and they can be used to modify or drive the dimensions of a part; therefore these dimensions are called driving dimensions. Dimensions can be generated using the **Show Model Annotations** dialog box, refer to Figure 13-1. This dialog box is discussed next.

Note

*You can also create dimensions in the **Drawing** mode, but these dimensions cannot drive the dimensions of the part. Therefore, the dimensions that you create in PTC Creo Parametric are called driven dimensions.*

Show Model Annotations Dialog Box

Ribbon: Annotate > Annotations > Show Model Annotations

When you import a 3D model to a 2D drawing, by default, the dimensions and other information of the model are stored in the drawing but they remain invisible. You can let the required model information be shown or erased on a particular view by using the options in the **Show Model Annotations** dialog box. The visible items are referred to as shown and the items that are not visible are referred to as erased. The erased annotations are not permanently deleted from the model as they remain stored in the database. To do so, choose **Show Model Annotations** from the **Annotations** group of the **Annotate** tab in the **Ribbon**; the **Show Model Annotations** dialog box will be displayed, as shown in Figure 13-1. This dialog box enables you to generate various parameters related to the model in the drawing view. There are six tabs in this dialog box. These are **Show the model dimensions**, **Show the model gtols**, **Show the model notes**, **Show the model surface finishes**, **Show the model symbols**, and **Show the model datums**. These tabs are discussed next.

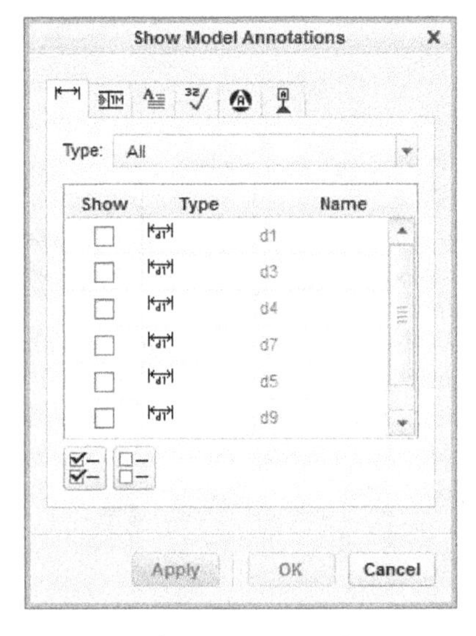

*Figure 13-1 The **Show Model Annotations** dialog box*

Show the model dimensions Tab

The **Show the model dimensions** tab is chosen by default in the **Show Model Annotations** dialog box. The options in this tab enable you to generate parametric dimensions in the drawing views. The bidirectional associative nature of the PTC Creo Parametric enables you to modify the model by modifying the dimensions from the drawing or the 3D environment.

Type Drop-down list

This drop-down list contains options that can be used to specify which dimension you want to be displayed in the drawing. These options act as filters to display the dimensions. The options displayed in this drop-down are **All**, **Driving dimension annotation elements**, **All driving dimensions**, **Strong driving dimensions**, **Driven dimensions**, **Reference dimensions**, and **Ordinate dimensions**. By default, **All** is selected.

List box

The list box consists of the dimensions to be displayed, their names, and a check box on the left of each dimension. You need to select a check box to display the corresponding dimension value in the selected drawing view. By default, all check boxes are cleared. If you want all the dimensions to be displayed in that view, then choose the **Select All** button. If you want to erase all the dimensions displayed in the view, then you need to choose the **Erase All** button. These two buttons are displayed below the list box.

Show the model gtols Tab

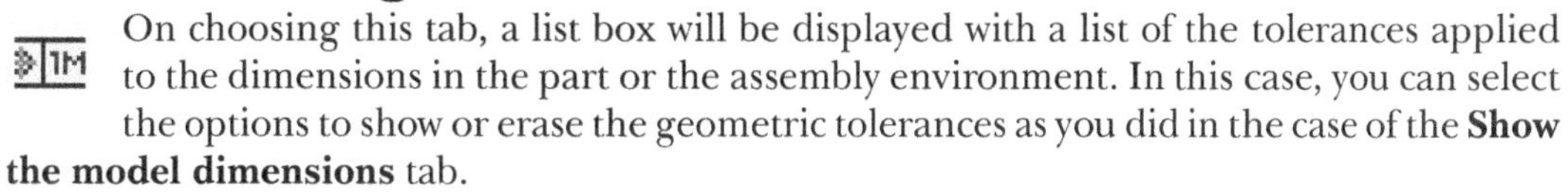

On choosing this tab, a list box will be displayed with a list of the tolerances applied to the dimensions in the part or the assembly environment. In this case, you can select the options to show or erase the geometric tolerances as you did in the case of the **Show the model dimensions** tab.

Show the model notes Tab

Choose this tab to show or erase the notes that were created in the **Part** mode. If any change is made to a note in the **Drawing** mode, the note in the **Part** mode is automatically modified. Similarly, if you modify this note in the **Part** mode, the modifications are reflected automatically in the **Drawing** mode.

Show the model surface finishes Tab

Choose this tab to show or erase the surface finish symbols associated with the part shown in the drawing views.

Show the model symbols Tab

Choose this tab to show or erase any symbol associated with the model.

Show the model datums Tab

Choose this tab to show or erase the critical measurement locations on the drawing view.

After selecting the required parameters, choose the **Apply** button to accept the settings and continue the selection process. Choose the **OK** button to accept the settings and exit the dialog box or choose the **Close** button to close the dialog box and cancel the settings performed.

Figure 13-2 shows some dimensioned drawing views.

The tools in the **Annotations** group of the **Annotate** tab of the **Ribbon** are discussed next.

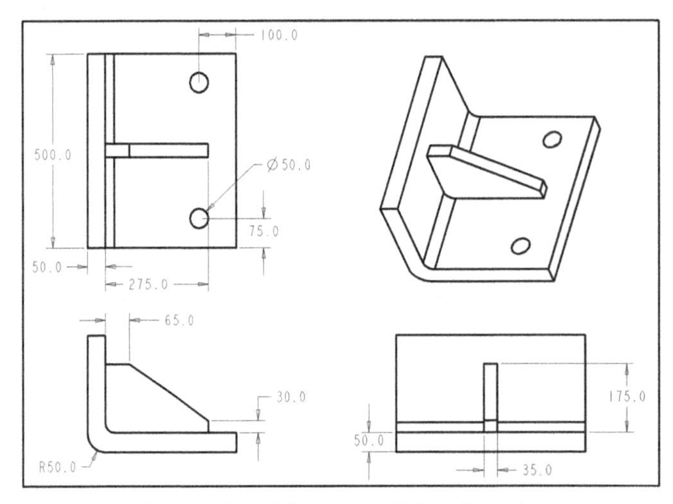

Figure 13-2 Dimensioned drawing views

Dimension

Ribbon: Annotate > Annotations > Dimension

The **Dimension** tool is used to create dimensions with new references in the drawing views. To create dimensions, choose the **Dimension** tool; the **Select Reference** dialog box will be displayed. You can select the required type of reference button from the dialog box to dimension an entity.

Surface Finish

Ribbon: Annotate > Annotations > Surface Finish

The **Surface Finish** tool is used to create the surface finish symbols in the drawing views. To create symbols, choose the **Surface Finish** tool; the **Open** dialog box will be displayed. Select the required symbol name from this dialog box and choose the **Open** button. The **Surface Finish** dialog box will be displayed along with the preview of the symbol and you will be prompted to place the symbol. Select the entity to which the surface finish symbol is to be assigned. You can also specify the roughness height in the **roughness_height** text box of the **Variable Text** tab. After selecting the symbol, if you want to change it then choose the **Browse** button from the **Definition** area of the **General** tab and select the required symbol.

Reference Dimension

Ribbon: Annotate > Annotations > Reference Dimension

The **Reference Dimension** tool is chosen to generate reference dimensions in the drawing views.

Jog

Ribbon: Annotate > Annotations > Jog

The **Jog** tool is used to create a jog. On choosing the **Jog** tool, you will be prompted to select a dimension or leader type to create a jog. Select the required entity to create a jog.

ADDING NOTES TO THE DRAWING

Ribbon: Annotate > Annotations > Note

The notes added to a model in the **Part** mode can be easily displayed in the drawing views by using the **Show Model Annotations** dialog box. However, if you have not added any note to the model in the **Part** mode and want to add it in the drawing views, then you have to create them. To create a note, choose **Anntotate > Annotations > Note** from the **Ribbon**; a flyout will be displayed, as shown in Figure 13-3. In this flyout, different types of options to create note are available. Choose the required type of note from the flyout. For example, if you choose **Unattached Note** option from the flyout, the **Select Point** dialog box will be displayed. Choose the required button from the **Select Point** dialog box. Next, by using the left mouse button, specify a point on the drawing where you want to place the note; the **Format** tab will be added to the **Ribbon** and the text window will be displayed in the drawing area. Now, choose the required style and symbol from the **Style** and **Text** groups in the **Format** tab. Next, click twice outside the text box to finish the note. The note will be placed at the specified point. Now, you can drag it to any required location.

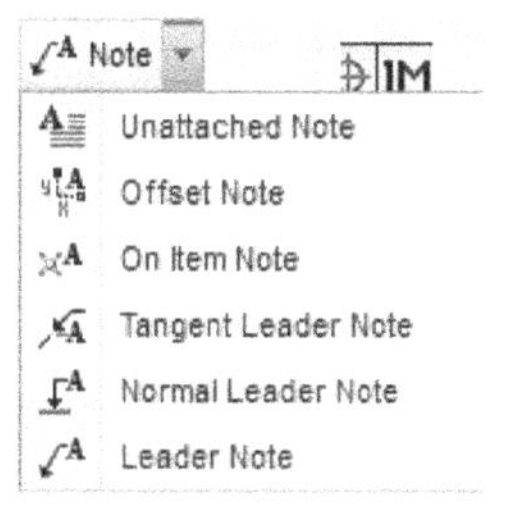

Figure 13-3 The ***NOTE*** *flyout*

ADDING TOLERANCES IN THE DRAWING VIEWS

You can create geometric tolerances in the dimensions. Tolerance is defined as the difference between the maximum and minimum variations in the dimensions of the selected component. It is almost impossible to manufacture a component to the exact dimensions. In such cases, the tolerance value is added to dimensions to make sure that some variation that occurs during manufacturing can be taken care of. However, when you actually send a part for manufacturing, there are other parameters along with the dimension tolerances that may vary and require some tolerances. Depending upon these factors, the tolerances are divided into two types: dimensional tolerances and geometric tolerances.

Dimensional Tolerances

These are the variation in the standard dimensional values. You can select and modify the type of dimensional tolerance from the **Properties** tab of the **Dimension Properties** dialog box. The **Dimension Properties** dialog box is discussed later in this chapter. Figure 13-4 shows the drawing view having dimensions with the tolerance.

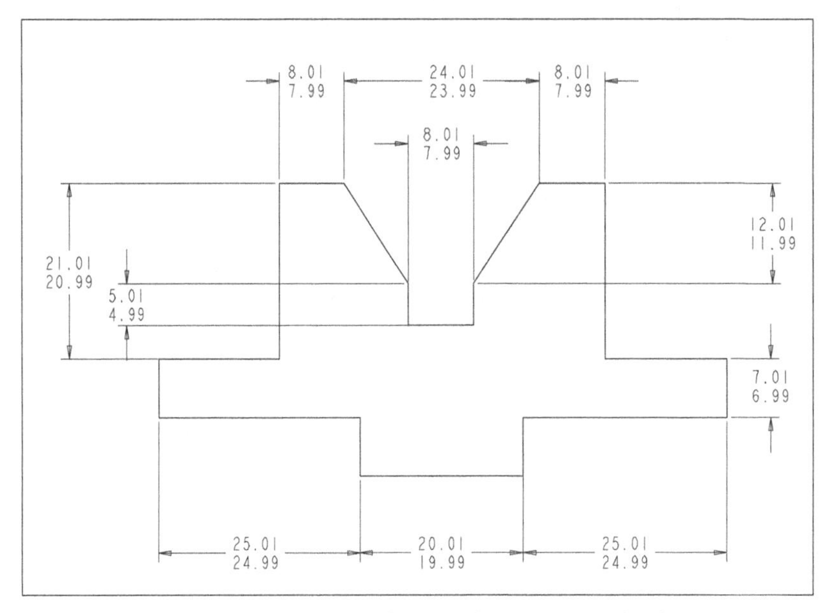

Figure 13-4 Drawing view showing dimensions with tolerances

Geometric Tolerances

Ribbon: Annotate > Annotations > Geometric Tolerance

When you send a component for manufacturing, you have to provide various other parameters along with the dimensions and dimensional tolerances. These parameters can be the geometric condition, surface profile, material condition, and so on. All these parameters are defined using the geometric tolerances. The geometric tolerances (will be called *gtol* henceforth) can be added to a drawing using the **Geometric Tolerance** dialog box, shown in Figure 13-5. This dialog box will be displayed when you choose **Annotate > Annotations > Geometric Tolerance** from the **Ribbon**. The options in the **Geometric Tolerance** dialog box are discussed next.

Model refs Tab

The options in this tab are used to select the model or drawing where the gtol have to be placed and also to specify the model references.

Model

This drop-down list is used to select the model to display the gtol. You can select the model from this drop-down list or by using the **Select Model** button.

Reference

The options in this area are used to select the reference for adding the gtol. The reference can be an edge, axis, surface, feature, datum, or an entity. You can select the required reference type from the **Type** drop-down list in this area and then choose the **Select Entity** button to select the reference from drawing area.

Placement

The options in this area will be available only when you select a reference type for placing the gtol. These options are used to specify the placement type for the gtol. The placement type can be selected from the **Type** drop-down list.

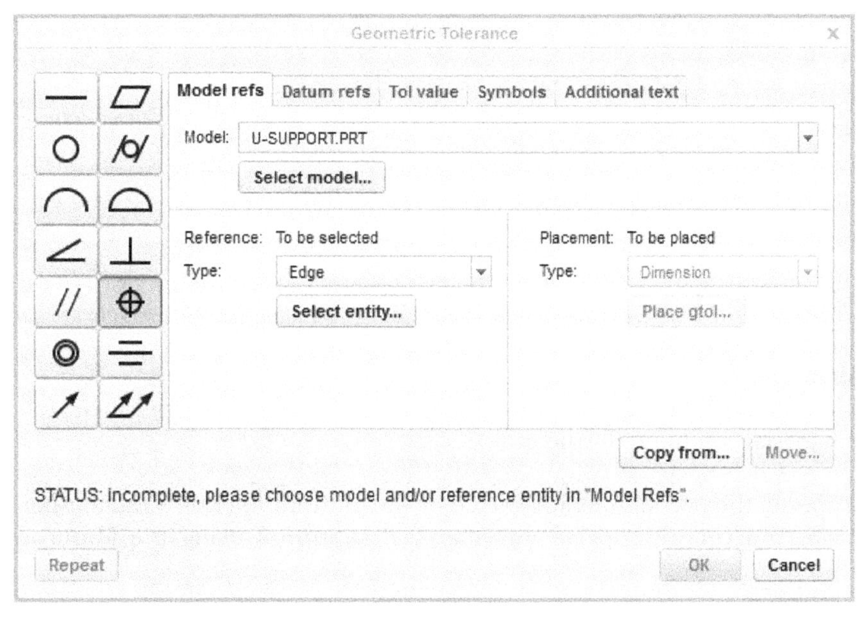

*Figure 13-5 The **Geometric Tolerance** dialog box with the **Model Refs** tab chosen*

Datum refs Tab

The options in the **Datum Refs** tab, shown in Figure 13-6, are used to select the primary, secondary, or tertiary datum references for the gtol.

You can also assign the composite tolerances to the primary, secondary, or tertiary datum references by using the **Composite tolerance** check box in this area. The **Composite tolerance** check box is available only when the **Position** or **Surface Profile** buttons are chosen from the types of gtol buttons on the left of the **Geometric Tolerance** dialog box. The value of composite reference can be specified in the **Value** edit box that will be available only when you select the **Composite tolerance** check box. The datum reference for the composite tolerance can be selected using the **Datum Reference** drop-down list that is available only when you select the **Composite tolerance** check box.

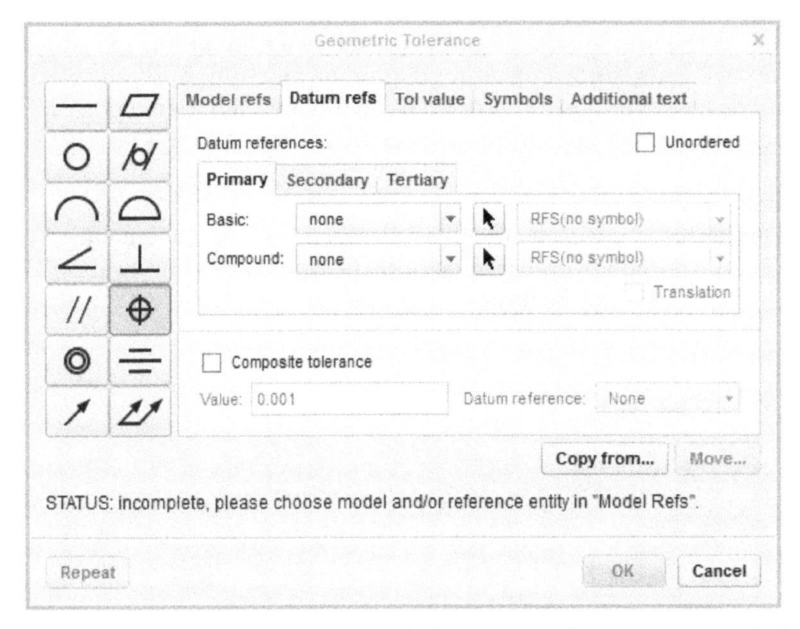

***Figure 13-6** The **Geometric Tolerance** dialog box with the **Datum refs** tab chosen*

Tol value Tab

The options in the **Tol value** tab, shown in Figure 13-7, are used to specify the tolerance value and the material condition for the gtol.

You can specify the overall tolerance value or the per unit tolerance value using the options in this tab. You can also specify the material condition using the **Material condition** drop-down list.

Symbols Tab

The options in the **Symbols** tab, shown in Figure 13-8, are used to specify the symbols for the gtol. The symbols and modifiers for the gtol can be selected from the **Symbols and modifiers** area whereas the projected tolerance zone symbol can be selected from the **Projected tolerance zone** area. The location of the projected tolerance zone symbol and its height can be specified using the options in this area. You can also specify the profile boundary symbol for the profile gtol from the **Profile boundary** area that appears in the place of **Projected tolerance zone** area. The **Profile boundary** area is available only if you choose the **Surface profile** or the **Line profile** tool from the left of the **Geometric Tolerance** dialog box.

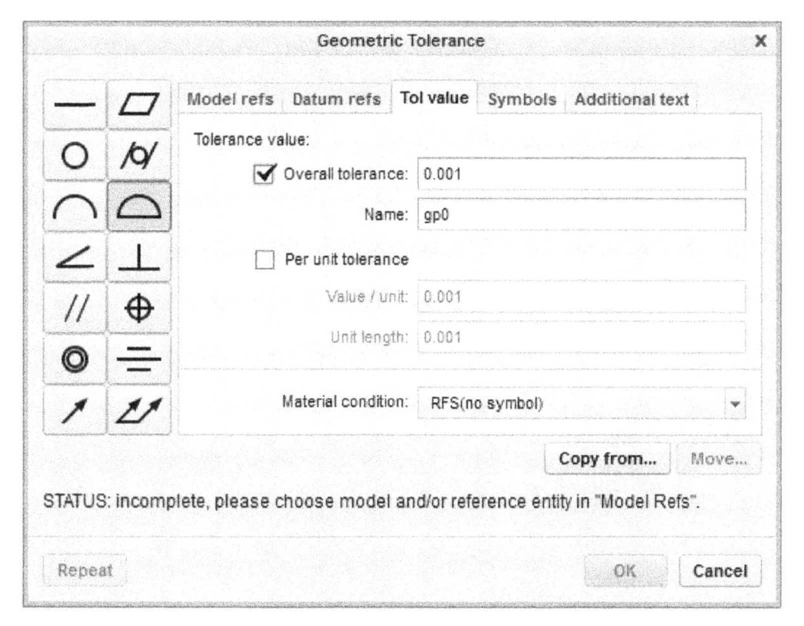

*Figure 13-7 The **Geometric Tolerance** dialog box with the **Tol Value** tab chosen*

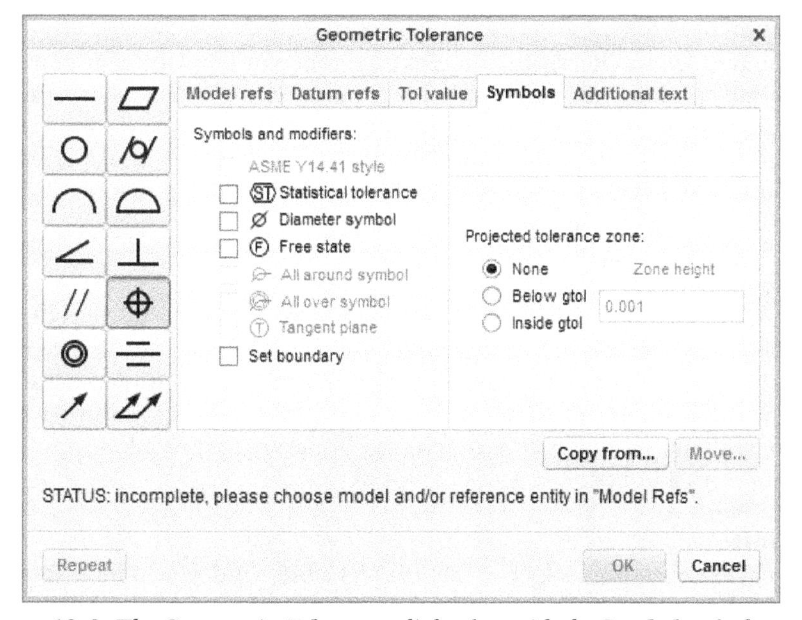

*Figure 13-8 The **Geometric Tolerance** dialog box with the **Symbols** tab chosen*

Additional text Tab

The options in the **Additional text** tab, shown in Figure 13-9, are discussed next.

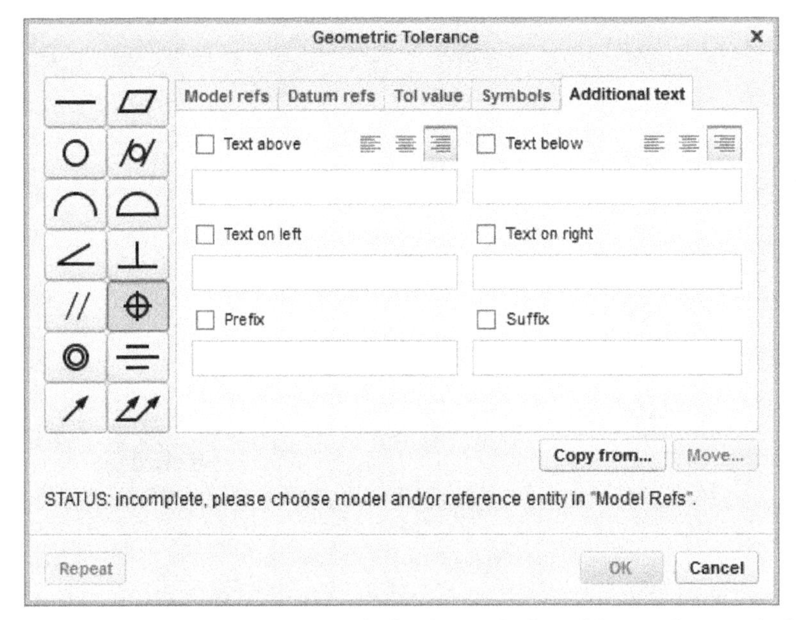

*Figure 13-9 The **Geometric Tolerance** dialog box with the **Additional Text** tab chosen*

Text above

Select the **Text above** check box; the **Text Symbol** window will be displayed. Using the **Text Symbol** window, you can insert the required text symbols above the gtol control frame. You can also insert additional text above the gtol control frame using the **Enter additional text above the gtol control frame** edit box that is available below the **Text above** check box. The text symbols that you have selected from the **Text Symbol** window will also be displayed in this edit box.

Text below

Select the **Text below** check box; the **Text Symbol** window will be displayed. You can insert the required text symbols below the gtol control frame using the **Text Symbol** window. You can also insert additional text below the gtol control frame using the **Enter additional text below the gtol control frame** edit box that is available below the **Text below** check box. The text symbols that you have selected from the **Text Symbol** window will also be displayed in this edit box.

Text on left

Select the **Text on left** check box; the **Text Symbol** window will be displayed. You can insert the required text symbols on the left of the gtol control frame using the **Text Symbol** window. You can also insert additional text on the left of the gtol control frame using the **Enter additional text on the left of the gtol control frame** edit box that is available below the **Text on left** check box. The text symbols that you have selected from the **Text Symbol** window will also be displayed in this edit box.

Text on right
Select the **Text on right** check box; the **Text Symbol** window will be displayed. You can insert the required text symbols on the right of the gtol control frame using the **Text Symbol** window. You can also insert additional text on the right of the gtol control frame using the **Enter additional text on the right of the gtol control frame** edit box that is available below the **Text on right** check box. The text symbols that you have selected from the **Text Symbol** window will also be displayed in this edit box.

Prefix
The **Prefix** check box is used to insert text and text symbols in the gtol control frame. On selecting this check box, the text and text symbols are inserted as prefix to the geometric tolerance text.

Suffix
The **Suffix** check box is used to insert text and text symbols in the gtol control frame. On selecting this check box, the text and text symbols are inserted as suffix to the geometric tolerance text.

Repeat
The **Repeat** button in the **Geometric Tolerance** dialog box is chosen to accept the current gtol and place a new gtol.

Move
The **Move** button is chosen to change the location of the current gtol.

Figure 13-10 shows the drawing view with reference datums and gtol.

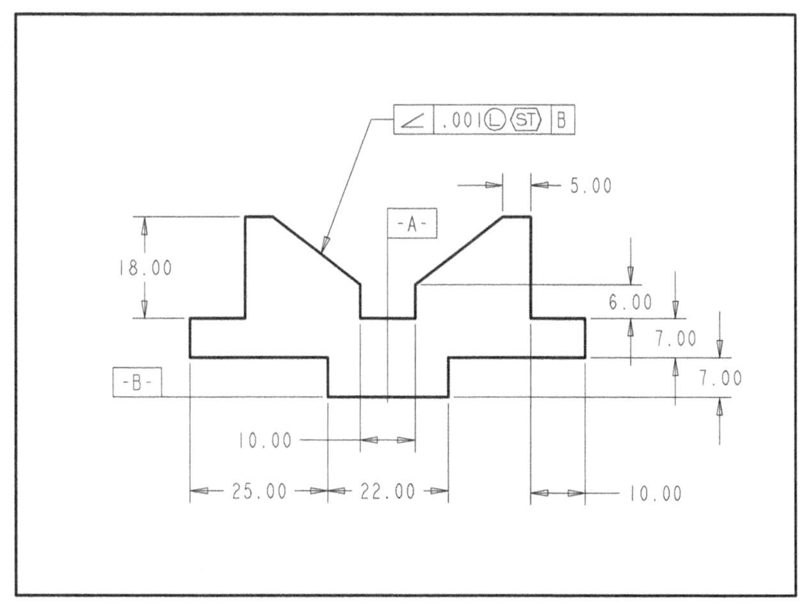

***Figure 13-10** The drawing view with reference datums and gtol*

EDITING THE GEOMETRIC TOLERANCES

The gtol added to the drawing views can be edited using the **Geometric Tolerance** dialog box. This dialog box will be displayed when you select the text of the tolerance from the drawing view and then right-click and hold to display the shortcut menu. Choose the **Properties** option from the shortcut menu to display the **Geometric Tolerance** dialog box. All options in the various tabs of the **Geometric Tolerance** dialog box are similar to those discussed earlier.

ADDING BALLOONS TO THE ASSEMBLY VIEWS

Ribbon: Annotate > Annotations > Balloon Note

The **Balloon Note** tool is used to create balloons associated with the assembly in the drawing views. You can add a balloon to the assembly drawing views by choosing **Annotate > Annotations > Balloon Note** from the **Ribbon**; the **NOTE TYPES** menu will be displayed. In this menu, choose the type of leader to be used, if required. Next, choose the **Make Note** option from this menu; the **Select Point** dialog box will be displayed. Choose the required button from the **Select Point** dialog box. Next, specify a point on the drawing where you want to place the balloon by using the left mouse button. Enter the text in the **Message Input Window** and choose the required symbol from the **Text Symbol** window displayed at the bottom. Choose the **Accept value** button to accept the text. On doing so, you will again be prompted to enter a note in the **Message Input Window**. Now, choose the **Cancel input** button; the balloon will be placed at the specified point, see Figure 13-11.

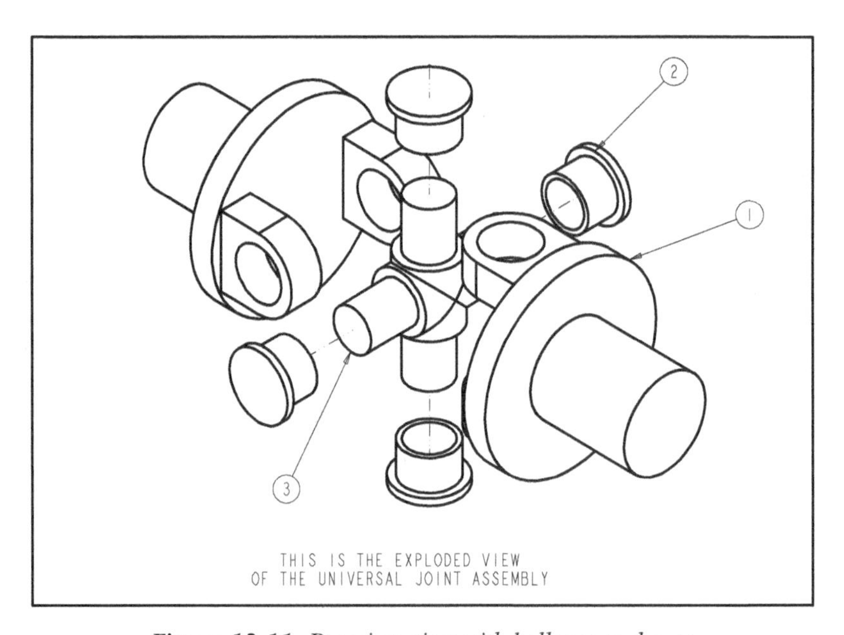

Figure 13-11 *Drawing view with balloons and note*

Tip: *The notes or balloons can be modified and edited using the options in the shortcut menu that will be displayed when you select them.*

ADDING REFERENCE DATUMS TO THE DRAWING VIEWS

Ribbon: Annotate > Annotations > Model Datum > Model Datum Plane

Reference datums are used as references for geometric tolerance. Datums are needed to be set before adding the geometric tolerance.

To create a reference datum, choose **Annotate > Annotations > Model Datum > Model Datum Plane** from the **Ribbon**; the **Datum** dialog box will be displayed, as shown in Figure 13-12. You can create a reference datum from this dialog box. You can also rename the selected datum and then choose the second button from the **Display** area to enclose the datum inside a feature control frame. You can place these reference datums freely in the drawing views or in a selected dimension. Figure 13-13 shows the drawing views with reference datums.

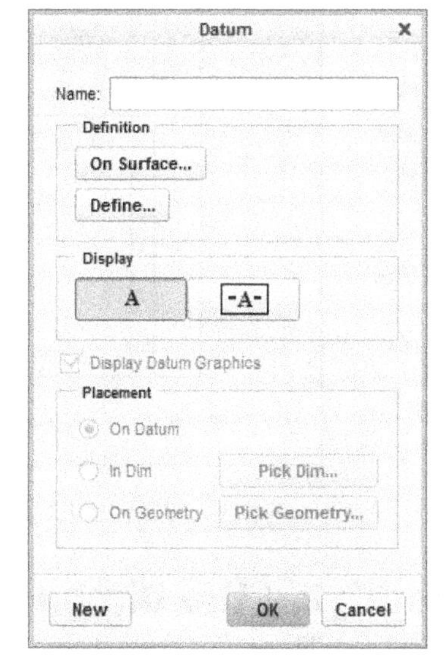

*Figure 13-12 The **Datum** dialog box*

MODIFYING AND EDITING DIMENSIONS

PTC Creo Parametric allows you to edit and modify the dimensions assigned to the drawing views. There are various methods of editing and modifying the driving dimensions. Before these methods are discussed, it is important for you to learn the terminology used in these methods. Figure 13-14 shows the terminology that is used in the **Drawing** mode of PTC Creo Parametric.

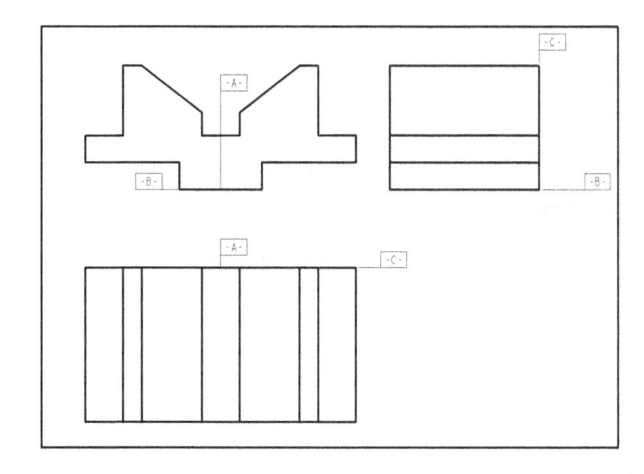

Figure 13-13 Drawing views with reference datums

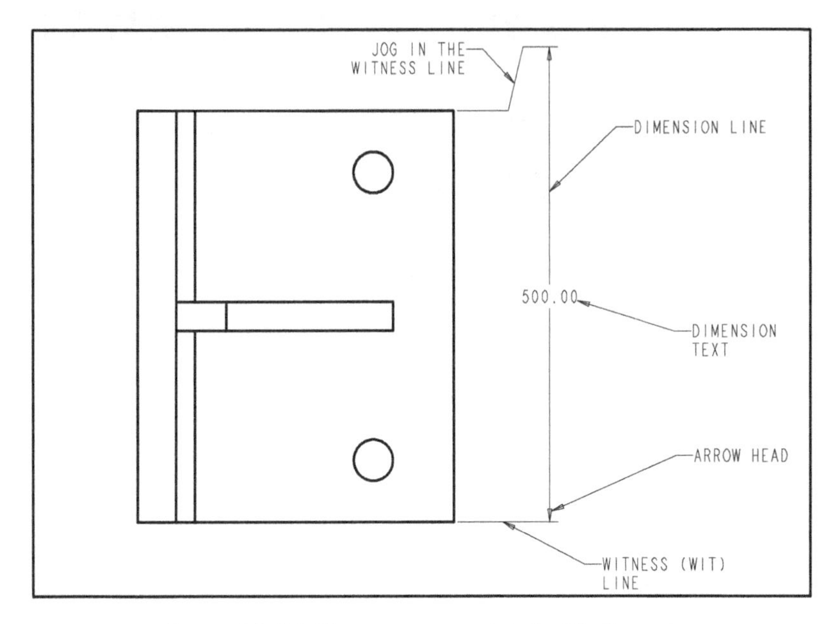

***Figure 13-14** Parameters associated with dimensions*

Modifying the Dimensions Using the Dimension Properties Dialog Box

To modify the dimensions using this option, select the dimension and right-click and hold to invoke a shortcut menu. Choose the **Properties** option from the shortcut menu; the **Dimension Properties** dialog box will be displayed, as shown in Figure 13-15. The options in this dialog box are discussed next.

Properties Tab

The options in this tab are discussed next.

Name

You can specify a name for the dimension in the **Name** edit box.

Value and Display Area

The options in this area are related to the display of dimensions in the drawing. You can set the nominal, override and tolerance values of dimensions, number of decimal places in the dimension, and whether to round the dimension value or not in this area.

Tolerance Area

The **Tolerance mode** option in this area are related to the tolerance in the drawing.

Tolerance mode: This drop-down list will be available only when the value of **tol_display** is set to **yes** in the configuration file. The options in this drop-down list are used to select the type of tolerance mode. Depending on the option selected from this drop-down list, the other options in this area change.

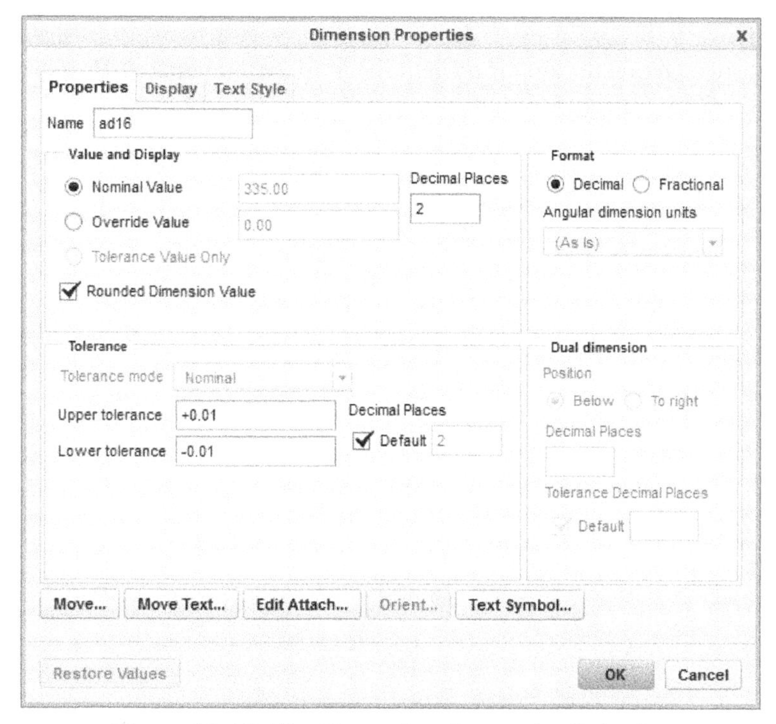

*Figure 13-15 The **Dimension Properties** dialog box*

Format Area

The options in this area are discussed next.

Decimal: Select the **Decimal** radio button to display the dimensions in the decimal format. You can set the number of decimal values in the **Value and Display** area.

Fractional: Select the **Fractional** radio button to display dimensions in the fractional format. The largest denominator can be set in the **Value and Display** area using the **Largest Denominator** edit box that will be displayed when you select the **Fractional** radio button.

Tip: *After you modify the nominal value of a dimension in the **Value and Display** area, you will have to regenerate the model.*

Angular dimension units: This drop-down list is used to specify the angular units for dimensions.

Dual dimension Area

The options under the **Dual dimension** area will be available only when an appropriate value is set for **dual_dimensioning** in the configuration file. The options in this area are used to specify the placement point for the dual dimensioning and the number of decimal places.

Display Tab

The options displayed on choosing the **Display** tab in the **Dimension Properties** dialog box are shown in Figure 13-16. The options in this tab are discussed next.

*Figure 13-16 The **Dimension Properties** dialog box with the **Display** tab chosen*

Display Area

The options in this area are discussed next.

Basic: This radio button, if selected, displays the dimension in a rectangular box.

Inspection: This radio button, if selected, displays the dimension in an elliptical box.

Neither: This radio button is used when you do not require the display of dimensions in the Basic or Inspection type.

Prefix: The **Prefix** edit box is used to add some additional text as prefix to the default dimension text.

Suffix: The **Suffix** edit box is used to add some additional text as suffix to the dimension text.

Text orientation: The options in the **Text orientation** drop-down list are used to orient the text around the leader. The options available in this drop-down list are: **Default**, **Horizontal**, **Parallel to and above leader**, **Parallel to and below leader**, and **(As Is).** These options are used to orient the text or dimension.

Configuration: The options in the **Configuration** drop-down list are used to set the leader type for dimension text. The options available in this drop-down list are **Linear** and **Center Leader.**

Flip Arrows: This button is used to toggle the display of arrows inside and outside the dimension extension lines.

Set dimension text: The **Set dimension text** text box displays the text of the selected dimensions. You can edit the text using this edit box.

Tip: *The modifications made using the **Dimension Properties** dialog box are applied dynamically on the dimensions. You can view these modifications in the background by moving the **Dimension Properties** dialog box on the side of the screen.*

Text Style Tab

The options displayed on choosing the **Text Style** tab in the **Dimension Properties** dialog box are shown in Figure 13-17. The options in this tab are discussed next.

Copy from Area

The **Copy from** area displays the options to modify the text style of the dimension text. You can edit the text style using this area.

Character Area

The **Character** area displays the options to modify the font style of the dimension text. Various parameters like height, font name, width of font, font style, and so on can be modified using the options available in this area.

Note/Dimension Area

The **Note/Dimension** area displays the options to modify the dimension or the text in the form of a note. The parameters that can be modified using this area are text justification, color, line spacing, angle, and so on.

Preview Button

The **Preview** button is used to preview the changes that are made using the **Dimension Properties** dialog box.

Reset Button

The **Reset** button is used to reset the default values.

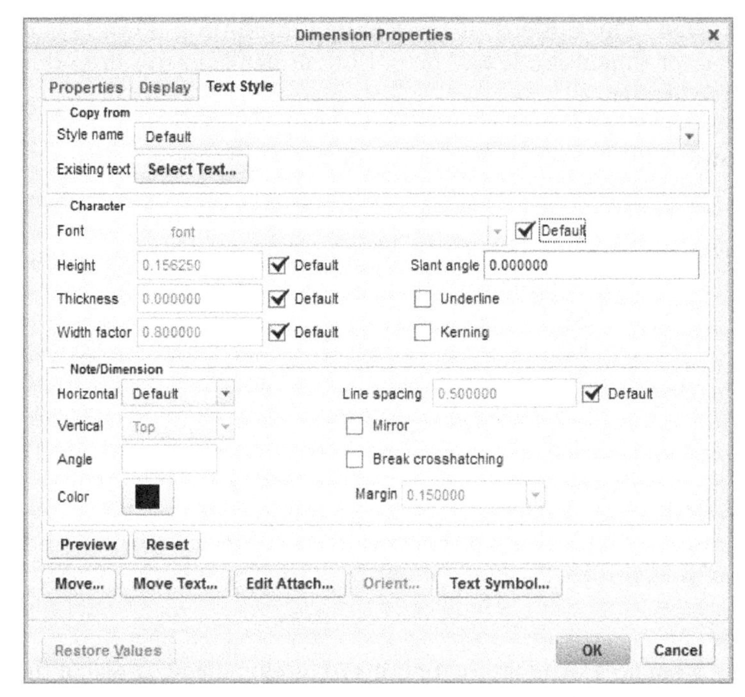

Figure 13-17 The ***Dimension Properties*** *dialog box with the* ***Text Style*** *tab chosen*

In addition to the options in various tabs, the following five buttons are available at the lower portion of this dialog box.

Move

The **Move** button is chosen to move the dimension to a new location on the current sheet. You can also move the dimension by selecting it and then dragging it by using the mouse.

Move Text

The **Move Text** button is chosen to move the dimension text to a new location on the screen. Remember that only the text will be moved. The witness lines and the dimension lines do not move from their original location.

Edit Attach

The **Edit Attach** button is chosen to replace an existing selected dimension with a new dimension in a user-defined location.

Orient

The **Orient** button is chosen to redefine the orientation and the placement of dimension text.

Text Symbol

The **Text Symbol** button is chosen to display the **Text Symbol** window. Using this window, you can add special symbols to the default dimension text.

Modifying the Drawing Items Using the Shortcut Menu

When you select an item from a drawing view and right-click, a shortcut menu with various options will be displayed. These options depend on the item selected to modify. These options vary from item to item. Some of the items related to the drawing, which can be modified using the shortcut menu, are discussed next.

Value

You can modify the dimension value by selecting a dimension value. When the dimension value and the arrows turn green, click on the dimension value. Right-click and hold to invoke the shortcut menu. From the shortcut menu, choose the **Modify Nominal Value** option; the dimension value will be displayed in the edit box. Now, modify the value in the edit box and press ENTER. Alternatively, double-click on the value; an edit box will be displayed. Enter a new value and press ENTER. Note that you can modify nominal value of the dimensions only when the dimensions are generated by using the **Show Model Annotations** dialog box. To view the effect of the modification, you will have to regenerate the drawing view using the **Regenerate Active Model** tool from the **Quick Access** toolbar. This option is also used to modify the scale of the drawing view and the tolerance values.

Text Style

The parameters that can be modified are the selected text line, entire text material, text height, text style, style library for the text, and the current style.

Miscellaneous Modifications

The modifications that can be made using the shortcut menu are: move a dimension, flip the arrows of the dimensions, switch the selected dimension from one view to the other, make a jog in the witness line, modify the values or text of the dimension, modify the arrow style, break the witness line, remove the break from the witness line, erase the witness line, show the erased witness line, or erase the selected dimension from the drawing view.

Number of Decimal Places

To modify the number of decimal values for the dimension text, choose the **Decimal Places** tool from the **Format group** of **Annotate** tab in the **Ribbon**; you will be prompted to specify the number of decimal places in the **Message Input Window**. Specify the number of decimal places; you will be prompted to select the dimensions to be modified. Select the dimension using the left mouse button and then press the middle mouse button to view the change in the decimal places.

Note

For using the ***Decimal Places*** *option, the dimension value should be in decimal format. To change the format of the value, invoke the* ***Dimension Properties*** *dialog box and then choose the* ***Decimal*** *radio button from the* ***Format*** *area.*

Cleaning Up the Dimensions

Ribbon: Annotate > Annotations > Cleanup Dimensons

Generally, dimensions displayed in the drawing views are scattered and improperly placed. This, in turn, makes the drawing views look very confusing. PTC Creo Parametric helps you to clear this confusion by using two methods. The first method is to place all dimensions manually at the appropriate location in the view. The second and more convenient method is to place these dimensions in a proper order by choosing the **Cleanup Dimensions** tool from the **Annotations** group in the **Annotate** tab of the **Ribbon**. On doing so, the **Clean Dimensions** dialog box will be displayed and you will be prompted to select dimensions or individual views to clean. After you select the dimension to clean, press the middle mouse button; the options in this dialog box will be enabled, as shown in Figure 13-18. These options are discussed next.

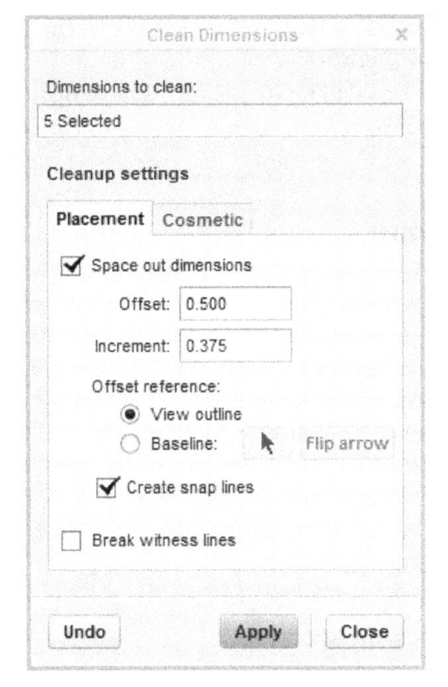

Figure 13-18 *The **Clean Dimensions** dialog box with the **Placement** tab chosen*

Placement Tab

The options in this tab of the **Clean Dimensions** dialog box are used to place the dimensions in the required location. The distance between the two dimensions and the distance between the view boundary and the dimension are set under this tab. The options in this tab are discussed next.

Space out dimensions

This check box is used to maintain the distance between two dimensions. Most of the other options in this tab are available only when you select this check box.

Offset

The **Offset** edit box is used to specify the offset distance between the first dimension and the view boundary.

Increment

The **Increment** edit box is used to specify the offset distance between the two dimension lines.

Offset reference Area

Options in this area are used to define the reference for offset. These options are discussed next.

View outline: The **View outline** radio button is selected to specify the offset distances taking the outline of the view as reference.

Baseline: The **Baseline** radio button is selected to specify the offset distance taking a selected flat edge, datum plane, snap line, detail axis line, or view border as the reference. When you invoke this option, you will be prompted to select either of the above mentioned entities as the reference.

Create Snap Lines

The **Create Snap Lines** check box is selected to create the snap lines, which can be used to clean the dimensions. The snap lines will be created at the distances specified in the **Offset** and the **Increment** edit boxes. If this check box is cleared, the snap lines will not be created.

Break witness lines

The **Break witness lines** check box is selected to break the witness lines if some dimensions overlap each other.

Cosmetic Tab

The options in the **Cosmetic** tab of the **Clean Dimensions** dialog box are used for determining the location of the dimension value with respect to the dimension line, refer to Figure 13-19. The arrows and the orientation of the dimension text can also be controlled using this tab. The options in this tab are discussed next.

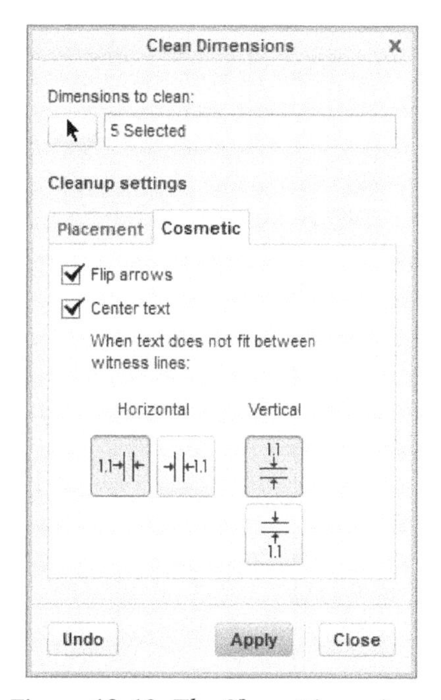

*Figure 13-19 The **Clean Dimensions** dialog box with the **Cosmetic** tab chosen*

Flip Arrows

The **Flip Arrows** check box is selected to flip the arrows of the dimensions that are not easily adjusted inside the dimension lines.

Center Text

The **Center Text** check box is selected to place the dimensions at the center of the dimension lines.

Horizontal and Vertical Buttons

These buttons are used to specify the placement of the horizontal or vertical dimension text outside the dimension line if they do not fit.

Figures 13-20 and 13-21 show the drawing view before and after using the **Clean Dimensions** dialog box.

Note

The system does not show a reference datum plane in a drawing view unless it is perpendicular to the drawing area.

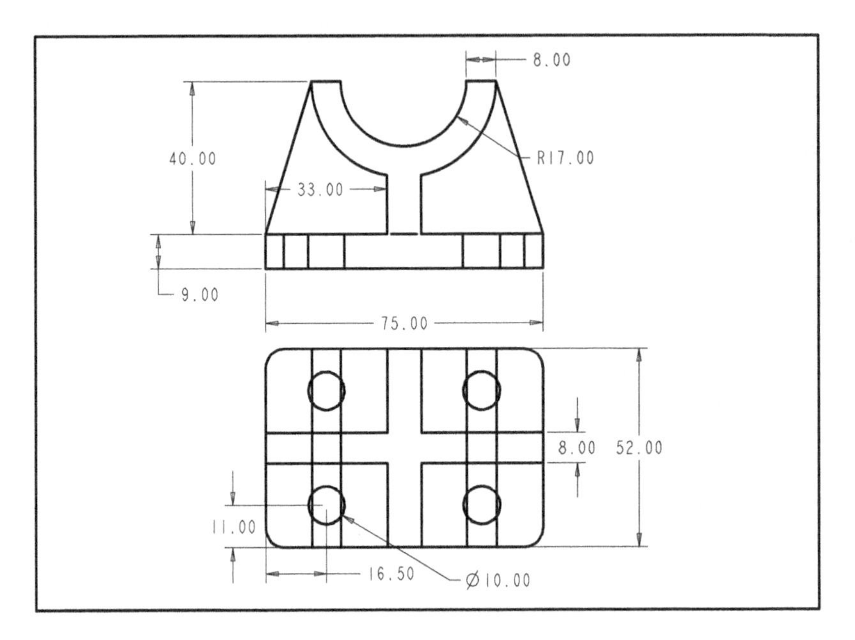

Figure 13-20 Drawing views with dimensions before cleaning

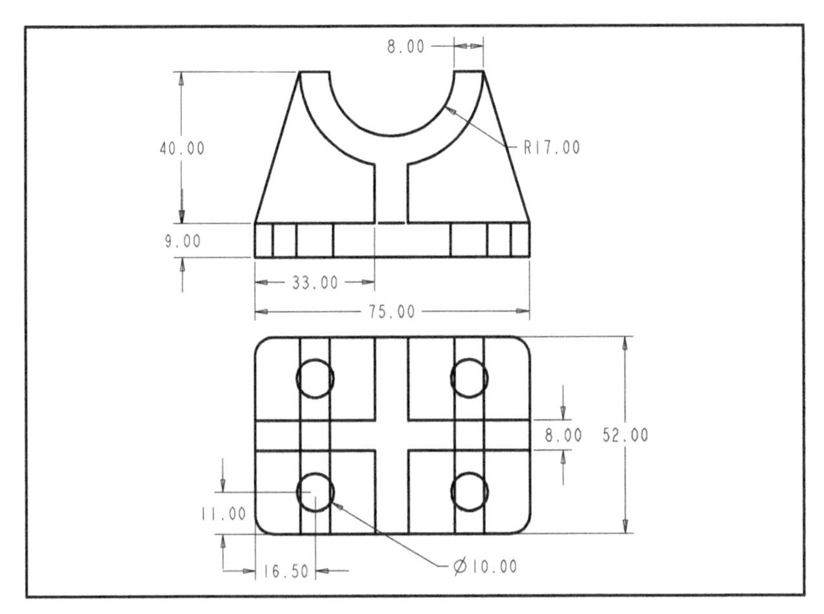

Figure 13-21 Drawing views with dimensions after cleaning

TUTORIALS

Tutorial 1

In this tutorial, you will generate the drawing views of the part shown in Figure 13-22. The views to be generated are shown in Figure 13-23. **(Expected time: 45 min)**

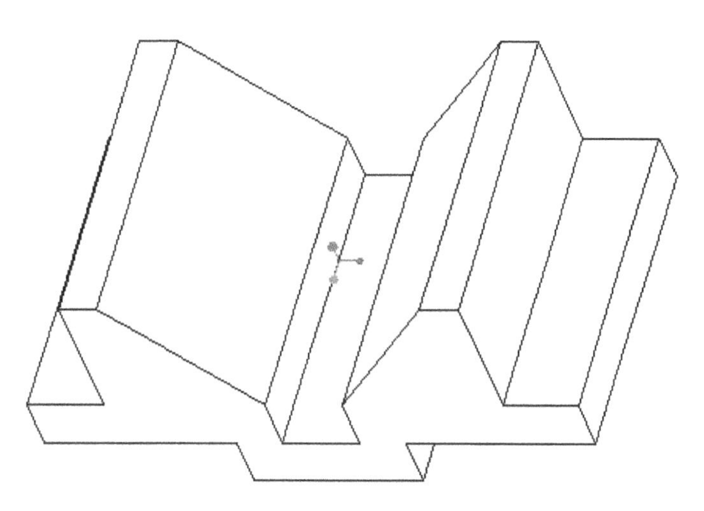

Figure 13-22 *Model for generating drawing views*

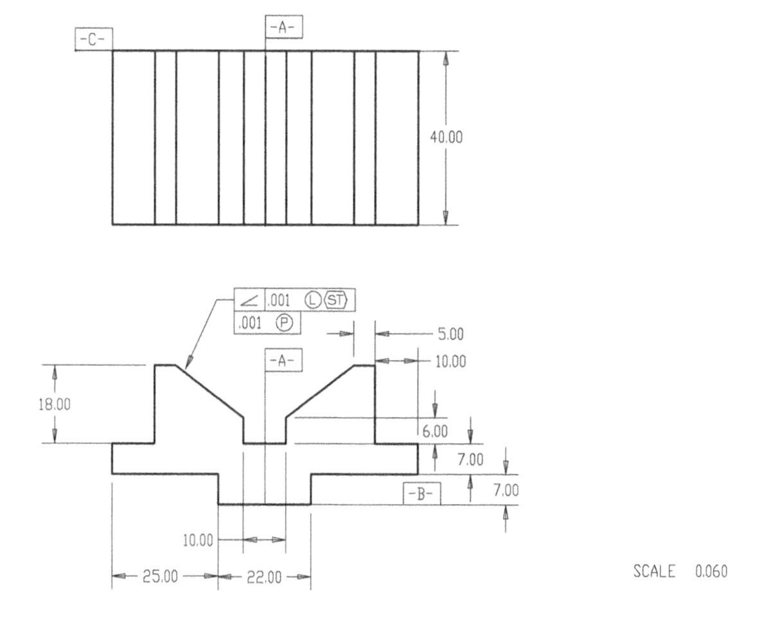

Figure 13-23 *Drawing views to be created alongwith the dimensions*

The following steps are required to complete this tutorial:

a. Create the model in the **Part** mode using the given dimensions and save the file.
b. Open a new drawing file in the **Drawing** mode and generate the top and front views of the model on the drawing sheet, refer to Figure 13-24.
c. Generate the dimensions in the views and clean them using the **Clean Dimensions** dialog box, refer to Figure 13-25.
d. Add geometric tolerance to the required entities, refer to Figure 13-26.

Before you start creating the model and the drawing views, set the working directory to *C:\Creo-3.0\c13*. Make sure that the *c13* folder exists inside the *Creo-3.0* folder.

Creating the Model

First, create the given model in the **Part** mode. Its views will be generated later.

1. Create a new **Part** file and name it as *c13tut1*.

2. Create the model in the **Part** mode and save this model.

Starting a New Drawing File

To generate drawing views, you need to start a new drawing file.

1. Choose the **New** button from the **File** menu; the **New** dialog box is displayed.

2. Select the **Drawing** radio button from the **Type** area and enter the name of the drawing as *c13tut1* in the **Name** edit box.

3. Choose **OK** from the **New** dialog box; the **New Drawing** dialog box is displayed.

4. If the name of the model is not displayed in the **Default Model** area, choose the **Browse** button and select the model. Select the **Empty** radio button from the **Specify Template** area, the **Landscape** button from the **Orientation** area, and the **A4** option from the **Standard Size** drop-down list.

5. Choose **OK** from the **New Drawing** dialog box to enter the **Drawing** mode. The **Drawing** mode is invoked and a new A4 size drawing sheet is displayed.

Generating the Drawing Views

You need to generate the top and front views of the model.

1. Choose the **General** tool from the **Model Views** group of the **Layout** tab in the **Ribbon**; the **Select Combined State** dialog box will be displayed. Choose the **OK** button from the dialog box.

2. Now, specify the center point for the drawing view, refer to Figure 13-24. When you specify the point, the **Drawing View** dialog box is displayed.

3. Select the **TOP** option from the **Model view names** list in the **View orientation** area.

4. Choose the **Scale** option from the **Categories** area and edit the scale factor for the view to **0.06**.

5. Choose the **Apply** button and close the dialog box. The default view is oriented as the top view and scaled to the specified scale factor.

6. Choose **No Hidden** from the **Graphics** toolbar and repaint the screen using the **Repaint** tool from the **Graphics** toolbar. The top view is shown in Figure 13-24.

7. Choose the **Projection** tool from the **Model Views** group of the **Layout** tab in the **Ribbon**. Make sure that the angle of projection is in the third angle projection, else you will not get the desired result.

8. Select the center point for the drawing view below the top view, as shown in Figure 13-24.

   The sheet after generating the two views is shown in Figure 13-24.

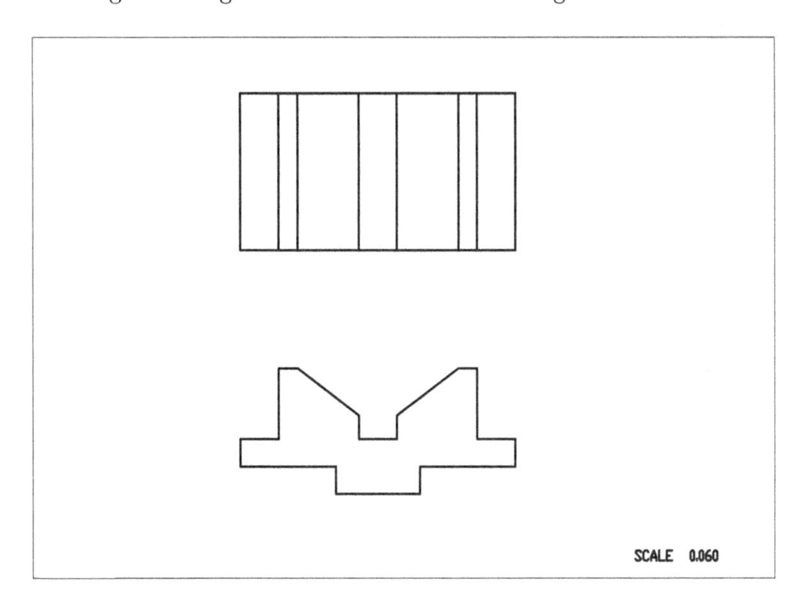

Figure 13-24 *Top and front views of the model*

Renaming the Datum Planes

The datum planes need to be renamed because the gtol values will be applied to the inclined surface with their references. This is evident from Figure 13-23.

1. Choose the **Plane Display** tool from the **Show** group in the **View** tab of the **Ribbon** to display the datum planes and then repaint the screen.

2. Choose the **Annotate** tab from the **Ribbon** and then select the **Datum Plane** from the **Filter** drop-down list in the Status Bar.

3. Select the **RIGHT** datum plane in the top view and when it turns green, press and hold the right mouse button to display a shortcut menu.

4. Choose the **Properties** option from the shortcut menu; the **Datum** dialog box is displayed.

5. Enter **A** in the **Name** edit box and then choose the second button from the **Display** area to enclose the datum in the reference box. Make sure the **On Datum** radio button is selected in the **Placement** area. Choose **OK** to exit the **Datum** dialog box.

6. Repeat steps 3 and 4 to rename the **TOP** and **FRONT** datum planes as **B** and **C**, respectively.

Dimensioning the Drawing Views

1. Choose **Annotate > Annotations > Show Model Annotations** from the **Ribbon**; the **Show Model Annotations** dialog box is displayed with the **Show the model dimensions** tab chosen by default. Also, you are prompted to select a view to display its annotations.

2. Select the front view (the view at the bottom); all dimensions of this view are listed in the dialog box.

3. Choose the **Select All** button available below the list box; all dimensions are displayed as strong dimensions in the selected view.

4. Choose the **Apply** button to accept and continue the settings.

5. Now, select the top view from the drawing area; its dimension is listed in the **Show Model Annotations** dialog box.

6. Select the check box on the left of the dimensions listed; the dimension of the selected item is displayed in the drawing area as a strong dimension.

7. Choose the **OK** button from the dialog box to accept the settings and exit the dialog box.

Placing the Dimensions in Order

The dimensions displayed in the drawing views are scattered and improperly placed. You need to place the dimensions in proper order using the **Clean Dimensions** dialog box.

1. Choose **Annotate > Annotations > Cleanup Dimensions** from the **Ribbon** to display the **Clean Dimensions** dialog box; you are prompted to select the dimensions to be cleaned.

2. Select all dimensions from both the views by using the window selection method; the selected dimensions are highlighted. Press the middle mouse button to activate the options in the **Placement** tab.

3. Enter **0.5** in the **Offset** edit box and **0.5** in the **Increment** edit box.

4. Clear the **Create Snap Lines** check box. Choose **Apply** and then choose **Close** to exit the **Clean Dimensions** dialog box.

5. After cleaning the dimensions, the drawing views should look similar to the ones shown in Figure 13-25. Select and move the dimensions, if necessary.

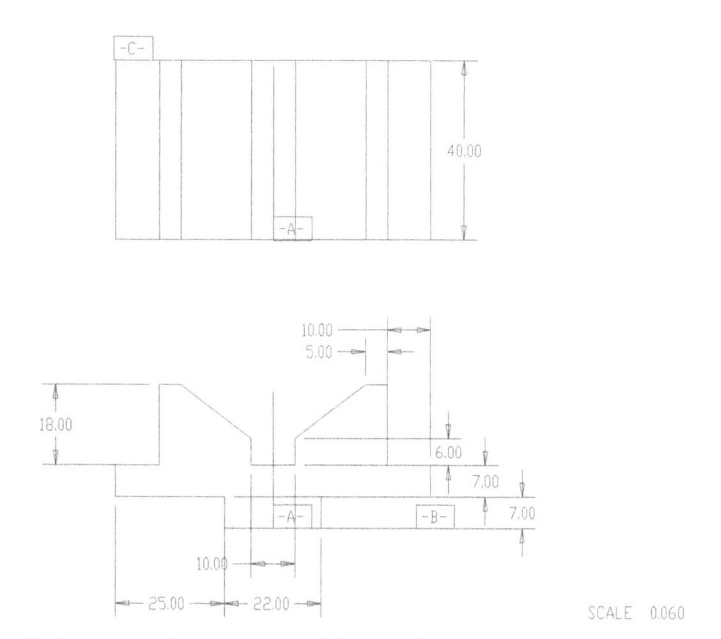

Figure 13-25 *Top and front views of the model with dimensions and datum planes*

Displaying the Geometric Tolerance

1. Choose **Annotate > Annotations > Geometric Tolerance** from the **Ribbon** to display the **Geometric Tolerance** dialog box. You will notice that the name of the model *c13tut1.prt* is displayed in the **Model** drop-down list.

2. Choose the angularity geometric condition by choosing the **Angularity** button from the **Geometric Tolerance** dialog box.

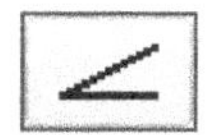

3. Select the **Surface** option from the **Type** drop-down list in the **Reference** area and select the left-inclined edge in the front view.

4. Select **With Leader** from the **Type** drop-down list in the **Placement** area; the **LEADER TYPE** menu is displayed and you are prompted to select an edge for attaching the leader.

5. Select the left-inclined edge from the front view and then click the middle mouse button at the desired place in the drawing area to place the note. Even if the note is at a wrong place in the drawing area, you can move it.

6. Choose the **Tol Value** tab from the **Geometric Tolerance** dialog box.

7. Enter **0.001** in the **Overall tolerance** edit box.

8. Select **LMC** from the **Material condition** drop-down list.

9. Choose the **Symbols** tab and then select the **Statistical tolerance** check box from the **Symbols and modifiers** area in the **Geometric Tolerance** dialog box.

10. Select the **Below gtol** radio button from the **Projected tolerance zone** area and then select the **Zone height** check box. Enter **0.001** in this edit box displayed on selecting the **Zone height** check box. Choose **OK** to exit the **Geometric Tolerance** dialog box.

    The drawing views after adding the geometric tolerances should look similar to the ones shown in Figure 13-26. You need to move the note for the geometric tolerances to a location shown in Figure 13-27 by selecting and dragging it. Similarly, arrange all dimensions and datums on the drawing sheet, as shown in Figure 13-27.

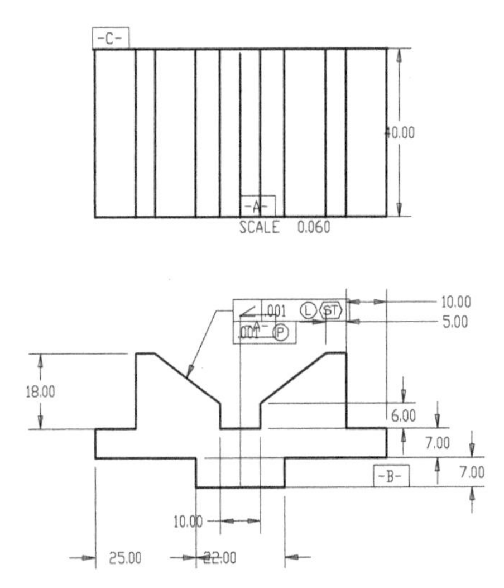

Figure 13-26 *Top and front views of the model with dimensions, datum planes, and geometric tolerances*

Saving the Drawing File

Now, you need to save the drawing file that you have created.

1. Choose the **Save** button from the **Quick Access** toolbar; the **Save Object** dialog box is displayed with the name of the drawing file that you had entered earlier. Press ENTER to confirm the saving of the file.

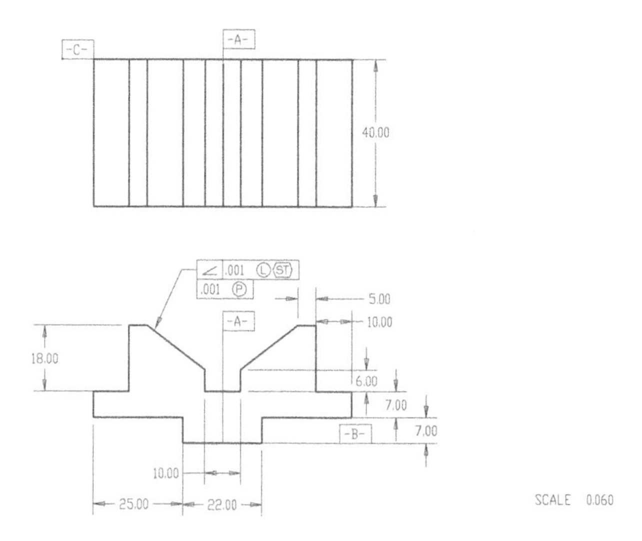

***Figure 13-27** The required drawing views after arranging all dimensions and datums*

Tutorial 2

In this tutorial, you will generate the drawing views of the model shown in Figure 13-28. The dimensioned drawing views are shown in Figure 13-29. **(Expected time: 45 min)**

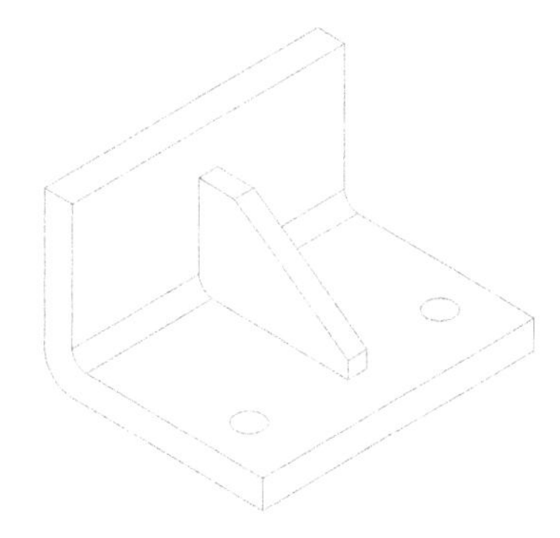

***Figure 13-28** Model for generating drawing views*

The following steps are required to complete this tutorial:

a. Create the model in the **Part** mode using the given dimensions and save the file.
b. Open a new drawing file in the **Drawing** mode and generate the top view, front, right, and detailed views of the model on the drawing sheet, refer to Figure 13-30.
c. Dimension the views and clean the dimensions using the **Clean Dimensions** dialog box, refer to Figure 13-31.

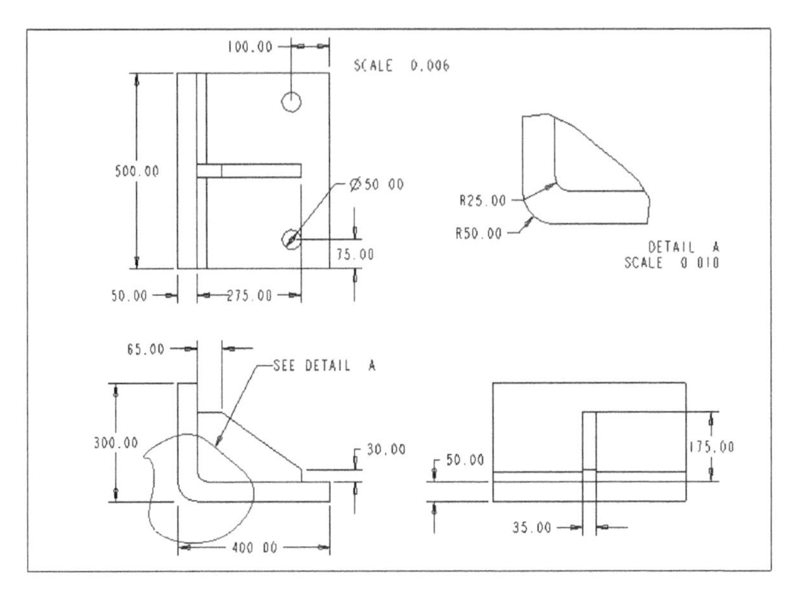

Figure 13-29 Drawing views to be generated in Tutorial 2

Make sure the working directory is set to *C:\Creo 3.0\c13*.

Creating the Model

First, in the **Part** mode you need to create the model whose drawing views are to be generated.

1. Create the model in the **Part** mode using the given dimensions.

2. Save this model created in the **Part** mode with the name *c13tut2*.

Starting a New Drawing File

To generate the drawing views, you first need to start a new drawing file.

1. Choose the **New** button from the **File** menu; the **New** dialog box is displayed.

2. Select the **Drawing** radio button from the **Type** area and enter the name of the drawing as *c13tut2* in the **Name** edit box.

3. Choose **OK** from the **New** dialog box; the **New Drawing** dialog box is displayed. In this dialog box, the name of the model is displayed in the **Default Model** area.

4. If the name of the model is not displayed in the **Default Model** area, choose the **Browse** button and select the model.

5. Select the **Empty** radio button from the **Specify Template** area, choose the **Landscape** button from the **Orientation** area, and select the **A4** option from the **Standard Size** drop-down list.

6. Choose **OK** from the **New Drawing** dialog box; you enter the **Drawing** mode and a new drawing sheet of A4 size is displayed on the screen.

Creating the Drawing Views

First, the top view of the model needs to be generated and then the front view will be generated by projecting it from the top view. The right view will be projected from the front view. The detail view can be placed on the drawing sheet by specifying a scale for it.

1. Choose **Layout > Model Views > General** from the **Ribbon**; the **Select Combined State** dialog box is displayed. Choose the **OK** button from the dialog box.
2. Now, specify the center point for the drawing view on the top-left side of the sheet; the **Drawing View** dialog box is displayed.
3. Select the **TOP** option from the **Model view names** list in the **View orientation** area.
4. Choose the **Apply** button; the default view is oriented as the top view.
5. Choose the **Scale** option in the **Categories** display area and specify the scale factor for the view as **0.006** in the **Custom scale** edit box. Exit the dialog box and then choose **No Hidden** from the **Display Style** drop-down in the **Graphics** toolbar. Repaint the screen using the **Repaint** tool from the **Graphics** toolbar. For the top view, refer to Figure 13-30.

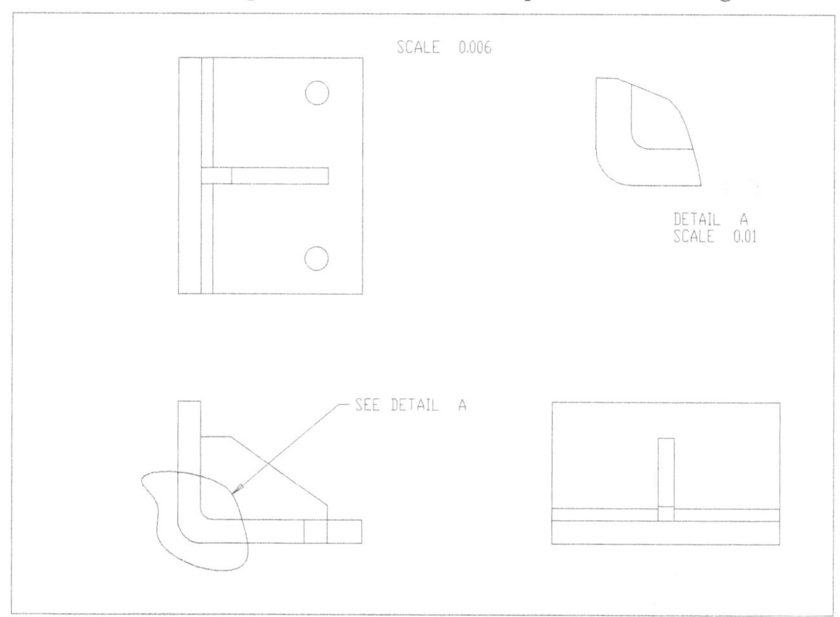

Figure 13-30 *Drawing views before adding dimensions*

The front view of the model needs to be generated from the top view.

6. Choose **Layout > Model Views > Projection** from the **Ribbon**. Next, specify the center point for the drawing view which is below the top view, refer to Figure 13-30. Make sure

that the angle of projection is the third angle projection; else the desired projection will not be displayed.

7. Similarly, place the right and detailed views on the drawing sheet, refer to Figure 13-30.

Dimensioning the Drawing Views

1. Choose the **Show Model Annotations** tool from the **Annotations** group of the **Annotate** tab in the **Ribbon**; the **Show Model Annotations** dialog box is displayed with the **Show the model dimensions** tab chosen by default. Also, you are prompted to select a view to display its annotations.

2. Move the cursor toward the round feature in the detailed view and click when the round feature gets highlighted; the dimensions of the radius of round feature are listed in the dialog box.

3. Choose the dimensions of the round feature listed in the dialog box. All the radial dimensions are displayed in the selected view as strong dimensions.

4. Choose the **Apply** button to accept the settings and continue.

5. Now, select all the three views; all dimensions of the three views are listed in the dialog box.

6. Choose the **Select All** button; all dimensions are displayed on the view as strong dimensions.

7. Choose the **OK** button to accept the settings and exit the dialog box.

Placing the Dimensions in Order

The dimensions displayed in the drawing views are scattered and improperly placed. You need to use the **Clean Dimensions** dialog box to place the dimensions in order.

1. Choose the **Cleanup Dimensions** tool from the **Annotations** group of the **Annotate** tab in the **Ribbon** to display the **Clean Dimensions** dialog box; you are prompted to select the dimensions to clean.

2. Select all dimensions from the top and front views by dragging a rectangle around them. Press the middle mouse button so that the options in the **Clean Dimensions** dialog box are displayed.

   When you select the dimensions, the radius dimensions in the drawing views are not selected. This means that the radius dimensions will remain unaffected by the values that you will enter in the **Clean Dimensions** dialog box.

3. Enter **0.4** in the **Offset** edit box and **0.3** in the **Increment** edit box.

4. Clear the **Create Snap Lines** check box. Choose **Apply** and then choose **Close** to exit the **Clean Dimensions** dialog box.

   After you clean the dimensions, you will notice that the dimensions are not placed in the required order. So, you need to manually place the dimensions in proper order.

5. Select the dimension and drag it to the desired location on the drawing sheet.

   Some dimensions are repeated. For example, the diameter of the hole feature is displayed twice in the drawing view. So, you need to erase the dimensions that are repeated and are not needed.

6. Select the dimension and hold down the right mouse button on the repeated dimensions to display a shortcut menu. Choose the **Erase** option from the shortcut menu.

   In some cases, the dimensions are displayed in the views in which you do not want them to be displayed. In such cases, you can switch those dimensions to the other views. To switch the dimension to other view, select it and hold the right mouse button; a shortcut menu is displayed. Choose the **Move to View** option from it.

   After manually placing the dimensions, the drawing views should look similar to the views shown in Figure 13-31.

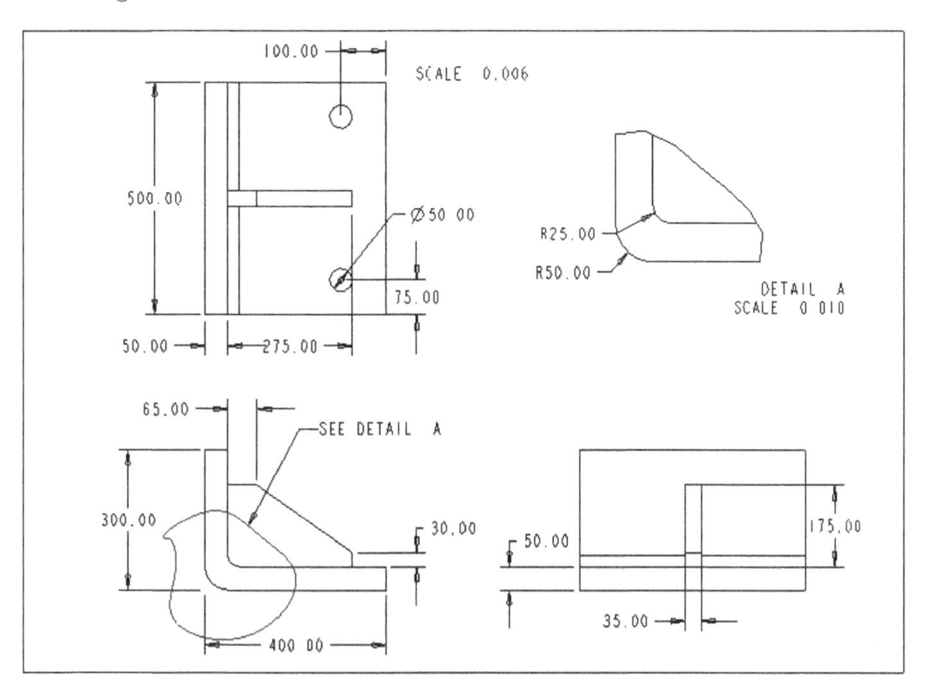

Figure 13-31 *The required drawing views*

Saving the Drawing File

1. Choose the **Save** button from the **Quick Access** toolbar; the **Save Object** dialog box is displayed with the name of the drawing file specified earlier.

2. Press ENTER to confirm the saving of the file.

Self-Evaluation Test

Answer the following questions and then compare them to those given at the end of this chapter:

1. The reference datum planes are used to display geometric tolerances in a drawing view. (T/F)

2. If you modify a dimension value in the **Drawing** mode and then regenerate the drawing view, the modification will also be applied to the model in the **Part** mode. (T/F)

3. The notes created on a model in the **Part** mode can also be displayed in a drawing view in the **Drawing** mode. (T/F)

4. When you select a drawing view to display dimensions, all dimensions are placed in proper order in the drawing view. (T/F)

5. Most of the editing operations in a drawing view can be done using the shortcut menu that is displayed when you hold the right mouse button on the item to be modified. (T/F)

6. The __________ dialog box is used to display dimensions in drawing views.

7. The dimension value can be modified by choosing the __________ option from the shortcut menu displayed on selecting the dimension and then right clicking on it.

8. The __________ dialog box is used to clean dimensions in the drawing views.

9. The configuration file can be opened by choosing __________ from the **File** menu.

10. The **Decimal** and __________ radio buttons are available in the **Format** area of the **Dimension Properties** dialog box.

Review Questions

Answer the following questions:

1. Which of the following dialog boxes is used to display dimensions in a drawing view?

   (a) **Dimension Properties** (b) **Clean Dimensions**
   (c) **Show / Erase** (d) None of these

2. Which of the following dialog boxes is used to modify the properties of the text in a drawing view?

   (a) **Dimension Properties** (b) **Clean Dimensions**
   (c) **Show / Erase** (d) None of these

3. Which of the following dialog boxes is used to display the geometric tolerances in a drawing view?

   (a) **Geometric Tolerance** (b) **Clean Dimensions**
   (c) **Show / Erase** (d) None of these

4. Which of the following options in the shortcut menu is used to switch a dimension from one view to another?

   (a) **Erase** (b) **Flip Arrows**
   (c) **Move Item to View** (d) None of these

5. Which of the following radio buttons in the **Type** area needs to be selected in the **New** dialog box to open a new drawing file?

   (a) **Assembly** (b) **Sketch**
   (c) **Part** (d) None of these

6. The dimensions erased using the **Erase** option of the shortcut menu are also deleted from the model in the **Part** mode. (T/F)

7. Notes or balloons can be modified and edited using the options in the shortcut menu displayed on right-clicking. (T/F)

8. The system does not show a reference datum plane in a drawing view unless it is perpendicular to the drawing area. (T/F)

9. Generally, dimensions displayed in the drawing views are scattered and improperly placed. (T/F)

10. You cannot move text dynamically on a drawing sheet. (T/F)

Exercise

Exercise 1

In this exercise, you will generate the drawing views of the model created in Exercise 2 of Chapter 7, as shown in Figure 13-32. Add dimensions to the views, as shown in the following figure. **(Expected time: 45 min)**

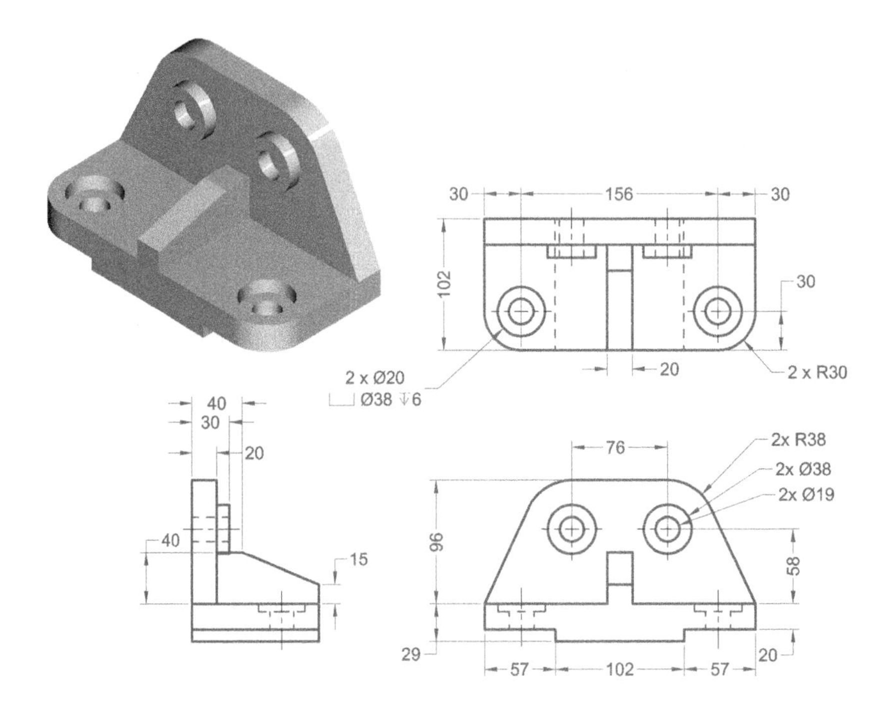

Figure 13-32 *Solid model and its drawing views*

Note

In the given drawing views, the center lines are not generated automatically; instead they are created manually. After creating the center lines, you need to define a line style and then apply it on the center lines.

Answers to Self-Evaluation Test

1. T, **2.** T, **3.** T, **4.** F, **5.** T, **6. Show / Erase**, **7. Modify Nominal Value**, **8. Clean Dimensions**, **9. Options**, **10. Fractional**

Chapter 14

Other Drawing Options

Learning Objectives

After completing this chapter, you will be able to:

- *Sketch in the Drawing mode.*
- *Modify sketched entities and other items in drawing views.*
- *Create a user-defined drawing format for drawing sheets.*
- *Add or remove sheets from the current drawing.*
- *Create tables in the current sheet.*
- *Generate associative Bill of Material in the Drawing mode.*

SKETCHING IN THE DRAWING MODE

Sketching is one of the most important tools available in the **Drawing** mode. Sketching in the **Drawing** mode is called drafting. As discussed earlier, there are two types of drafting techniques in PTC Creo Parametric: Generative drafting and Interactive drafting. Any item on the drawing sheet that is not generated from a model is called a draft entity or a draft item. Drafting is extensively used for creating user-defined formats, drawing tables, and drawing title blocks in the formats. Sketching in the **Drawing** mode is almost similar to that in the other modes of PTC Creo Parametric. Sketching can be done by using the tools in the **Sketch** tab of the **Ribbon**. The tools used to sketch in the **Drawing** mode are discussed next.

Entity Select

The **Entity Select** tool is used to switch to the selection mode. This tool is available in the **Settings** group of the **Sketch** tab in the **Ribbon**. You can select the drawing views, sketched entities, dimensions, notes, and so on from the drawing by simply clicking on them with the left mouse button. This tool is extensively used for moving the drawing views or the items in the drawing view. You can select the entity and drag using the left mouse button to move it.

Tip: *To delete an entity or an item from a drawing view, you can select the item and then press the DELETE key. You can also select the entity and choose the* **Delete** *option from the shortcut menu that is displayed when you hold the right mouse button.*

Line

Ribbon: Sketch > Sketching > Line

The **Line** tool is used to draw line segments. To invoke this tool, choose **Sketch > Sketching > Line** from the **Ribbon**; the **Snapping References** dialog box will be displayed, as shown in Figure 14-1. The options in this dialog box are discussed next.

Figure 14-1 The ***Snapping References*** *dialog box*

Select references

The **Select references** button is used to select an entity to which the line is to be snapped.

Remove

The **Remove** button is used to remove a selected entity from the **Snapping References** dialog box.

Circle

Ribbon: Sketch > Sketching > Circle

The **Circle** tool is used to create circles. To create a circle, you need to specify a center point and a point on the circumference of the circle.

Note

When you choose any sketching tool, the crosshairs of infinite length are displayed attached with the cursor.

Center and Ends Arc / 3-Point / Tangent End

Ribbon: Sketch > Sketching > Arc

The **Center and Ends Arc** and **3-Point / Tangent End** tools are used to draw arcs in the drawing mode.

The other geometric entities that can be drawn by using the tools in the **Ribbon** are splines, construction circles and lines, ellipses, and points. You can also create a fillet and a chamfer between two entities. You can also use the edges of the model and also create offsets.

Chain

Ribbon: Sketch > Settings > Chain

The **Chain** tool is used to draw an object which consists of more than one entity. To invoke this tool, choose **Sketch > Settings > Chain** from the **Ribbon**. For example, when you choose the **Line** tool, you can draw a single line. But if you draw a line after choosing the **Chain** tool, you can draw a continuous chain of lines. This chain can be ended only by pressing the middle mouse button.

Parametric Sketch

Ribbon: Sketch > Settings > Parametric Sketch

The **Parametric Sketch** tool is used to draw sketches by automatically applying constraints.

Sketcher Preferences

Ribbon: Sketch > Settings > Sketcher Preferences

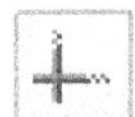

The **Sketcher Preferences** tool is used to apply snapping preferences. When you choose this button, the **Sketch Preferences** dialog box will be displayed, as shown in Figure 14-2. In this dialog box, there are two areas named **Snapping** and **Sketching tools**. These areas are discussed next.

Snapping

The buttons available in this area are used to control the snapping of cursor while sketching. These buttons are discussed next.

Horizontal/Vertical

When this button is chosen, the cursor automatically snaps to horizontal or vertical line while sketching.

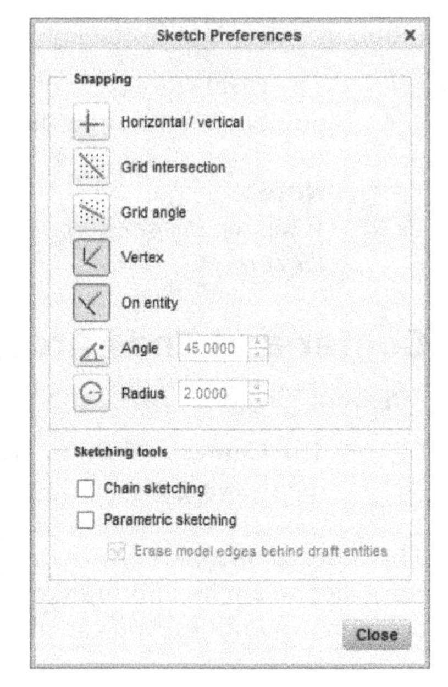

Figure 14-2 *The* ***Sketch Preferences*** *dialog box*

Grid intersection

When this button is chosen, the cursor automatically snaps to the intersection points of the grid.

Grid angle

When this button is chosen, the cursor automatically snaps to the line which is parallel to the grid lines.

Vertex

When this button is chosen, the cursor snaps to the vertices of an entity or an edge automatically.

On entity

When this button is chosen, the cursor automatically snaps to the points on the entity or the edge over which the cursor is moving.

Angle

When this button is chosen, the cursor snaps to the hypothetical line at the angle set in the corresponding spinner in the **Sketch Preferences** dialog box.

Radius

When this button is chosen, the cursor automatically snaps to the curve whose radius is equal to the radius set in the corresponding spinner in the **Sketch Preferences** dialog box.

Sketching tools

In this area, there are three check boxes which are discussed next.

Chain sketching

This check box has the same function as the **Chain** button which is discussed earlier.

Parametric sketching

This check box has the same function as the **Parametric Sketch** tool discussed earlier.

Erase model edges behind draft entities

When this check box is selected, the model edges overlapping the draft entities get deleted. This check box is selected by default.

MODIFYING SKETCHED ENTITIES

There are various options to modify draft entities. The operations or modifications that can be applied on the draft entities are discussed next.

Trimming the Draft Entities

The sketched or draft entities can be trimmed using the tools available in the **Trim** group of the **Sketch** tab in the **Ribbon**. The tools in the **Trim** group are discussed next, refer to Figure 14-3.

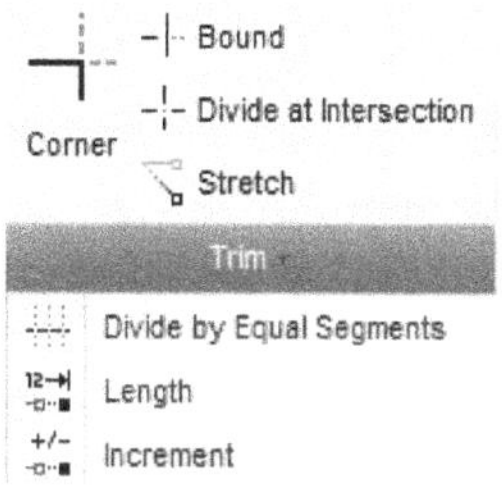

*Figure 14-3 The tools in the **Trim** group of the **Ribbon***

Divide at Intersection

Ribbon: Sketch > Trim > Divide at Intersection

The **Divide at Intersection** tool is used to divide two selected entities at the point of intersection. When you choose this tool, you are prompted to select two entities. Select the two entities to divide them at the intersection point. If a warning message is displayed while selecting the entities, then choose **OK** from it. The message informs you that the command will break all the parametric sketching references.

Divide by Equal Segments

Ribbon: Sketch > Trim > Divide by Equal Segments

This tool is in the extended **Trim** group and is used to divide the draft entities into segments of equal length. To do so, choose the **Divide by Equal Segments** tool and then select the entities to divide. After selecting the entity or entities, press the middle mouse button; the **Message Input Window** will be displayed and you will be prompted to enter the number of equal segments. Enter the number of equal segments into which you need to divide the entity and press ENTER.

Corner

Ribbon: Sketch > Trim > Corner

This tool is used to trim the two selected entities so that a corner is formed after trimming. Make sure you select the entities using the CTRL key.

Bound

Ribbon: Sketch > Trim > Bound

This tool is used for deleting or extending a portion of an entity by defining a bounding entity.

Length

Ribbon: Sketch > Trim > Length

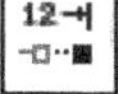

This tool is available in the expanded **Trim** group. Using this tool, you can trim or extend an entity up to a specified length.

Increment

Ribbon: Sketch > Trim > Increment

This tool is available in the expanded **Trim** group. It is used to extend or shorten an entity in specified increments. To extend the entity, enter a positive increment value and to shorten the entity, enter a negative increment value.

Stretch

Ribbon: Sketch > Trim > Stretch

The **Stretch** tool is used to stretch a sketched entity. The vertices of the sketched entity that are included inside the selection window will be moved from their original location and stretched, and the remaining vertices will remain stationary.

Transforming the Draft Entities

The tools in the **Edit** group of the **Sketch** tab of the **Ribbon** are used to translate or move the draft entities. The tools in the **Edit** group are shown in Figure 14-4 and are discussed next.

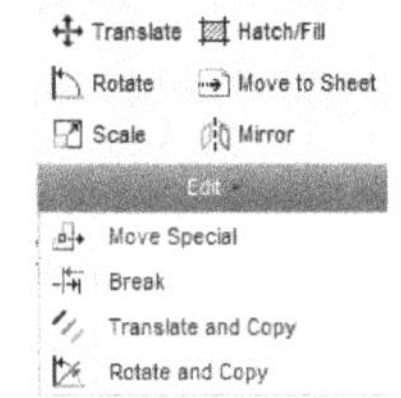

*Figure 14-4 The tools displayed in the **Edit** group*

Translate

Ribbon: Sketch > Edit > Translate

The **Translate** tool is used to move the selected entity or entities on the drawing sheet. To translate the entities, invoke this tool and select the required entities. Next, press the middle mouse button to confirm the selection; the **GET VECTOR** menu along with the **Select Point** dialog box will be displayed, as shown in Figure 14-5 and Figure 14-6. The options in the **GET VECTOR** menu are used to specify the direction of translation.

*Figure 14-5 The **GET VECTOR** menu*

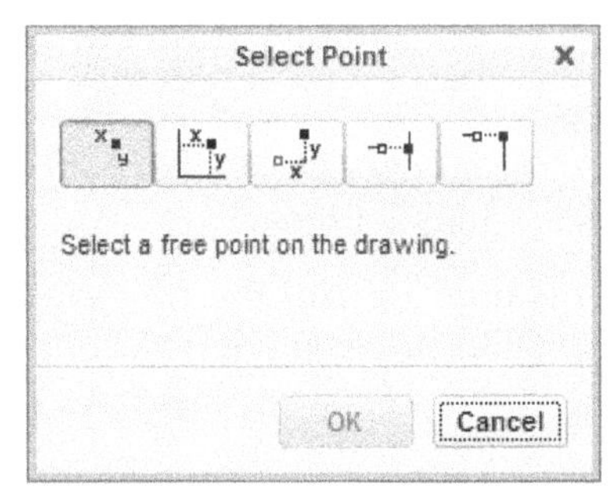

*Figure 14-6 The **Select Point** dialog box*

Translate and Copy

Ribbon: Sketch > Edit > Translate and Copy

The **Translate and Copy** tool is available in the expanded **Edit** group. This tool is used to translate the selected entity and at the same time make a copy of the entity.

After selecting the entities, press the middle mouse button; the **GET VECTOR** menu and the **Select Point** dialog box will be displayed. After defining the translation distance and direction, you need to specify the number of copies you want to create.

Rotate

Ribbon: Sketch > Edit > Rotate

The **Rotate** tool is used to rotate the selected entity. On choosing this tool, you will be prompted to select the entity that you want to rotate. After selecting the entity or entities, press the middle mouse button; you will be prompted to select the center point for rotation. Select the center point; the **Message Input Window** will be displayed and you will be prompted to enter the rotation angle in the counterclockwise direction. The default direction of rotation will be counterclockwise. However, you can rotate the selected entity in the clockwise direction by specifying a negative rotation angle.

Rotate and Copy

Ribbon: Sketch > Edit > Rotate and Copy

The **Rotate and Copy** tool is available in the expanded **Edit** group and is used to create a pattern of the selected draft entity. Choose this tool, select an entity, and press the middle mouse button; the **Select Point** dialog box will be displayed and you will be prompted to select the center point of rotation. Select the center point of rotation in the drawing; the **Message Input Window** will be displayed and you will be prompted to enter the rotation angle in counterclockwise direction. Enter the rotation angle and press the middle mouse button to accept the entered rotation angle value. Now, enter the number of copies in the **Message Input Window**; the rotated copies of the selected entity will be placed on the drawing sheet.

Scale

Ribbon: Sketch > Edit > Scale

The **Scale** tool is used to scale the selected entity. To do so, select the entity to scale and press the middle mouse button; the **Select Point** dialog box will be displayed and you will be prompted to select the origin point for scaling. Select the origin point; the **Message Input Window** will be displayed. The new scale factor can be specified in the **Message Input Window**. A scale factor greater than 1 will increase the size of the selected entity and a scale factor less than 1 will reduce the size of the selected entity.

Mirror

Ribbon: Sketch > Edit > Mirror

The **Mirror** tool is available in the **Edit** group. This tool is used to mirror the selected entities using an existing sketched line. To do so, choose the **Mirror** tool from the **Edit** group of the **Ribbon**. On choosing this tool, you will be prompted to select a 2D entity. Select the 2D entity and choose the **OK** button from the **Select** dialog box; you will be prompted to select a line. Select the line about which you want to create the mirrored copy.

Grouping Entities

Ribbon: Sketch > Group > Draft Group

A draft entity can be grouped with other draft entity, with a note, a geometric tolerance (gtol), or a dimension. In PTC Creo Parametric, the note, gtol, and dimensions are called detail items. After grouping the draft entities, the operation applied on any one of them is also applied to the grouped entity. Choose **Sketch > Group > Draft Group** from the **Ribbon** to form a group among draft entities like lines, arcs, circles, and so on. The operation performed on any one of the grouped entities is performed on all the entities forming that group. When you choose this option, the **DRAFT GROUP** menu will be displayed, as shown in Figure 14-7. Use the required option from this menu to create the draft group.

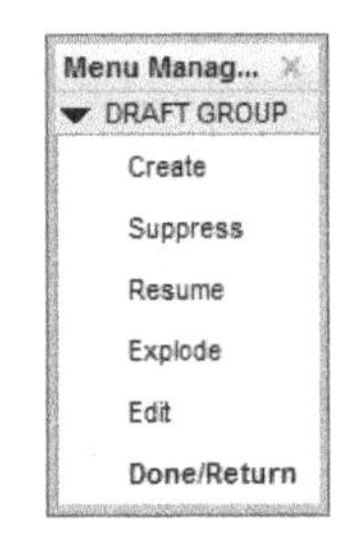

*Figure 14-7 The **DRAFT GROUP** menu*

Relate

Ribbon: Sketch > Group > Relate View

The draft entities can be related to the view so that the draft entities move or re-scale along with the related views, maintaining their position with respect to the related views. The options to relate can be invoked by choosing **Sketch > Group > Relate View** from the **Ribbon**. On doing so, a drop-down menu will be displayed, as shown in Figure 14-8. Note that this drop-down will be activated only when you select the drawing view. The tools in this menu are discussed next.

*Figure 14-8 The **Relate View** drop-down*

Set Default Relate View

Ribbon: Sketch > Group > Relate View drop-down > Set Default Relate View

This tool is available only when the drawing view is selected and is used to set the draft view as the current view. The draft entities that are drawn after setting a default draft view are grouped with the current draft view. Now, if you move the draft view, then the grouped entities will also move with it.

Note

When you move a view that forms a group with draft entities, the draft entities also move. But when you move the entities, the draft view does not move.

Unset Default Relate View

Ribbon: Sketch > Group > Relate View drop-down > Unset Default Relate View

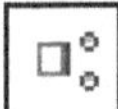

This tool is used to unset the current draft view and is activated only when a current draft view exists.

Relate to View

Ribbon: Sketch > Group > Relate to View

This tool is used to relate a view with a draft entity or entities. This tool is activated in the **Group** group only when a draft entity is selected. On choosing this tool, you will be prompted to select a view to which you want to relate the draft entity. Select the drawing view from the drawing sheet.

Relate to Object

Ribbon: Sketch > Group > Relate to Object

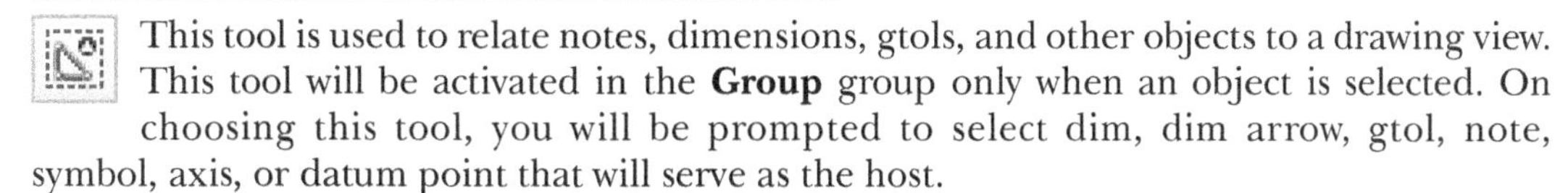

This tool is used to relate notes, dimensions, gtols, and other objects to a drawing view. This tool will be activated in the **Group** group only when an object is selected. On choosing this tool, you will be prompted to select dim, dim arrow, gtol, note, symbol, axis, or datum point that will serve as the host.

Unrelate

Ribbon: Sketch > Group > Unrelate

This tool is used to unrelate the previously related objects. This tool is activated in the **Group** group only when a related object is selected.

Drafting by Using the Drawing Views

You can use the generated drawing views to create an entity or the whole drawing view. To use a generated drawing view, choose the down arrow at right of the **Edge** button in the **Sketching** group of the **Sketch** tab in the **Ribbon**; a drop-down will be displayed with two tools: **Use Edge** and **Offset Edge**. These tools are discussed next.

Offset Edge

Ribbon: Sketch > Sketching > Edge drop-down > Offset Edge

This tool is used to offset the selected edges to the specified offset distance. When you choose this tool, the **OFFSET OPER** menu will be displayed, as shown in Figure 14-9. There are two options in this menu: **Single Ent** and **Ent Chain**. The **Single Ent** option is used to select a single entity from the drawing view, whereas the **Ent Chain** option is used to select a chain of entities from the drawing view. After choosing any one of the options from the **OFFSET OPER** menu, if you move the cursor over the drawing view, each edge of the model in the drawing view will be highlighted in green color. After selecting each edge of the model, you need to define the offset distance.

*Figure 14-9 The **OFFSET OPER** menu*

Use Edge

Ribbon: Sketch > Sketching > Edge drop-down > Use Edge

The **Use Edge** tool is selected to draw a copy of the selected edge in the drawing view. This tool will be available only when you have at least one drawing view in the current drawing sheet and if it is selected.

Converting a Generated View to a Draft View

Ribbon: Layout > Edit > Convert to Draft Group

To convert a generated view into a draft view, select the view to convert and choose the **Convert to Draft Group** tool from the **Edit** group of the **Layout** tab in the **Ribbon**; the **Confirm** dialog box will be displayed with a message that the selected view(s) will be broken up into draft entities. Choose the **Yes** button from this dialog box. After converting a generated view to a draft view, all objects in the view will be converted to individual entities.

Note

*The **Convert to Draft Group** tool is activated in the **Edit** group only when the display style of the view is **Hidden Line, No Hidden,** or **Wireframe**.*

USER-DEFINED DRAWING FORMATS

PTC Creo Parametric provides you with some standard drawing formats for generating drawing views. These standard formats have standard sheet sizes, tables, and title blocks. However, sometimes you may need to create a user-defined drawing format that is specifically designed as per your requirements, including sheet size, tables, and title block. Choose the **New** button from the **File** menu to display the **New** dialog box. In this dialog box, select the **Format** radio button from the **Type** area and then specify the name of the format in the **Name** edit box to create the user-defined format. When you choose **OK**, the **New Format** dialog box will be displayed, as shown in Figure 14-10.

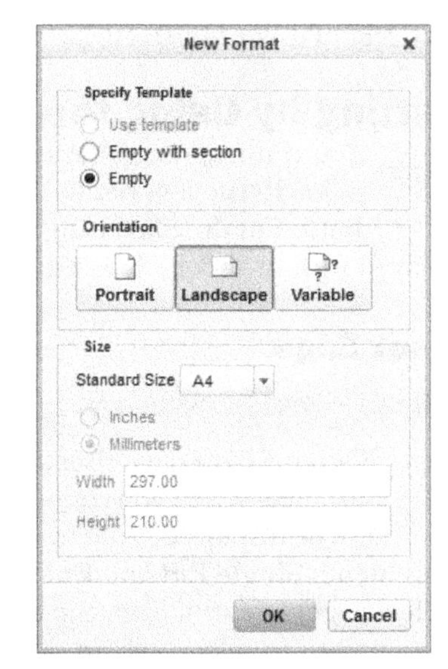

*Figure 14-10 The **New Format** dialog box*

You can set the size and orientation of the format sheet by using this dialog box. After setting the parameters, choose **OK** from this dialog box to proceed to the **Format** mode; the **Format** mode will be invoked.

You can create the desired entities in this format by using the **Sketch** tab of the **Ribbon**. You can also add the text material to the format by using the **Note** button in the **Annotations** group of the **Annotate** tab of the **Ribbon**. Figure 14-11 shows a user-defined format created using the A4 size sheet. This format consists of the user-defined title block. Figure 14-12 shows the drawing views generated on the user-defined format.

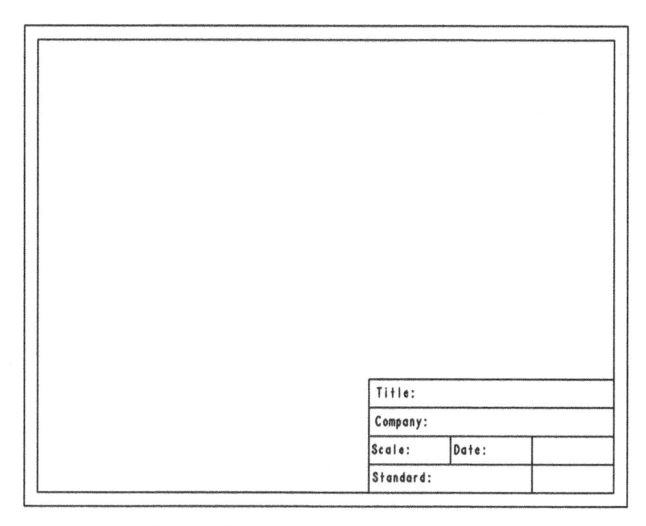

Figure 14-11 *A user-defined format*

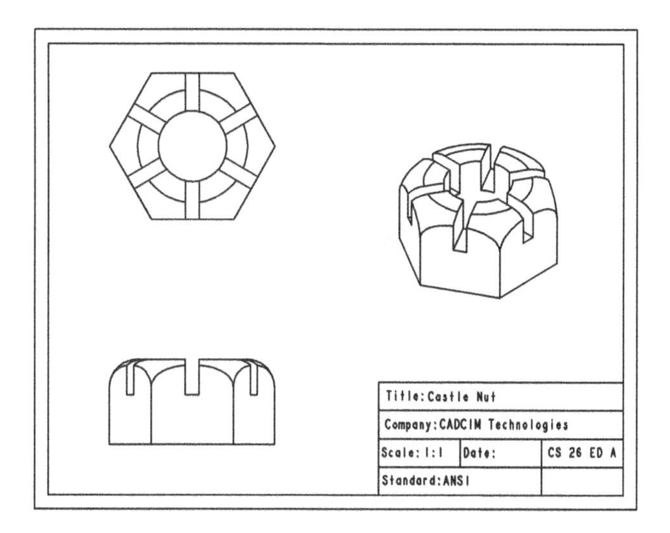

Figure 14-12 *Drawing views generated on the user-defined format*

RETRIEVING THE USER-DEFINED FORMATS IN THE DRAWINGS

Once you have created a user-defined format, you can retrieve and use it in the **Drawing** mode as a sheet for generating the drawing views. To retrieve a user-defined format, select the **Empty with format** radio button and choose the **Browse** button from the **New Drawing** dialog box to invoke the **Open** dialog box. In this dialog box, a folder named **System Formats** will be displayed with some predefined formats. You can retrieve these predefined formats or browse to the location where you have saved a user-defined format created earlier.

Note

*You can add or replace formats in a drawing by using the options in the **Sheet Setup** dialog box that will be displayed when you hold the right mouse button on the drawing sheet. You can add another sheet and then select the format from this dialog box.*

Tip: *The user-defined format that you create is saved in the .frm format. The location of this format will be the working directory that was selected at the time when the format was created.*

ADDING AND REMOVING SHEETS IN THE DRAWING

Ribbon: Layout > Document > New Sheet

To add sheets to the current drawing, choose the **New Sheet** tool from the **Document** group of the **Layout** tab in the **Ribbon**; a new sheet will be displayed and added to the current drawing sheet. Also, the sheet number will be displayed in the **Sheets** bar at the bottom left of the screen. You can also click on the (+) sign displayed at the bottom left of the screen to add new sheet to the current drawing. You can generate various views of a model on multiple sheets. All these drawing views on different sheets are contained in a single drawing file.

To remove a sheet from a drawing, right-click on the required **Sheet** tab available at the bottom of the drawing area to invoke a shortcut menu. Choose the **Delete** option from the shortcut menu to delete that particular sheet; the **Delete Selected Sheets Confirmation** message box will be displayed. Choose **Yes** from the message box to confirm the deletion of the current sheet.

CREATING TABLES IN THE DRAWING MODE

Ribbon: Table > Table > Table

You can create any kind of tabular representation in the **Drawing** mode easily by using the tools in the **Table** tab of the **Ribbon**, as shown in Figure 14-13. The tools in this tab are discussed next.

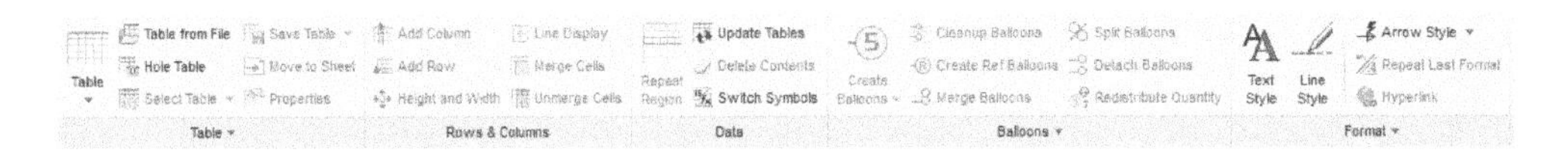

*Figure 14-13 The **Table** tab of the **Ribbon***

Table Tool

To create a table, choose the **Table** tool from the **Table** group in the **Table** tab of the **Ribbon**; a drop-down will be displayed, as shown in Figure 14-14. Move the cursor over required boxes displayed in the drop-down; the number of selected rows and columns of the table will be highlighted on the top-left corner of the drop-down. Click on the position where you get the required number of rows and columns; a rectangle will be attached to the cursor and you will be prompted to define the location. Click on

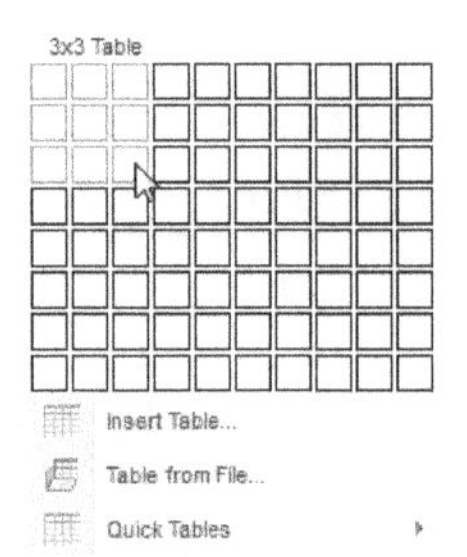

*Figure 14-14 The **TABLE** drop-down*

the desired location to insert the table. You can also create the table by choosing the **Insert Table** tool in the same drop-down. On doing so, the **Insert Table** dialog box will be displayed, as shown in Figure 14-15. In this dialog box, there are four areas which are discussed next.

Direction

In this area, there are four buttons to define the direction of table growth. The direction of table growth is the one in which the new rows and columns will be added later.

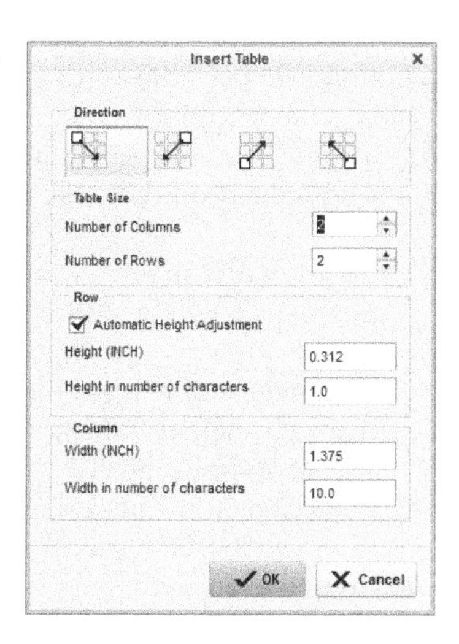

Figure 14-15 *The* ***Insert Table*** *dialog box*

Table Size

In this area, there are two spinners to control the table size. The **Number of Columns** spinner is used to set the number of columns in the table. The **Number of Rows** spinner is used to set the number of rows in the table.

Row

In this area, there are two edit boxes and a check box to control the height of rows. Select the **Automatic Height Adjustment** check box to automatically change the height of row to adjust the text written in the row. If you want to set the height manually, then enter the value of height in inch in the **Height (INCH)** edit box or enter the value of height in number of characters in the **Height in number of characters** edit box. The height of one character is 7 mm by default.

Column

In this area, there are two edit boxes which are used to control the width of the columns. To change the width of columns, you need to enter the value in inch in the **Width (INCH)** edit box. You can also enter the width value in number of character in the **Width in numbers of characters** edit box. The width of one character is 5.95 mm by default.

Table from File Option

To insert a table that already exists in the *.tbl* or *.csv* file format, choose **Table > Table > Table > Table from File** from the **Ribbon**; the **Open** dialog box will be displayed. Browse and select a *.tbl* or *.csv* file from this dialog box to open it.

Add Column

The **Add Column** button is used to add a column to an existing table. To do so, choose **Table > Rows & Columns > Add Column** from the **Ribbon**; you will be prompted to select a point inside the existing table to add a column. Specify a point; a column gets added to the table.

Tip: *To save a table with the .tbl file format, choose* ***Table > Table > Save Table > Save as Table*** *from the* ***Ribbon****. The* ***Save as Table*** *option is activated only when existing table is selected in the drawing sheet.*

Add Row

The **Add Row** button is used to add a row to an existing table. When you choose this button from the **Rows & Columns** group of the **Table** tab of the **Ribbon**, you will be prompted to select a point inside the existing table to add a row.

The following are the editing operations that can be performed on the tables:

1. To move a table, select it and then drag it by holding it from one of the points that are highlighted on the table.

2. To enter text in any of the cells of the table, you need to double-click inside the cell; the **Format** tab will be added to the **Ribbon**.

3. To delete a table from the drawing, click anywhere on the table and choose **Select Table** from the **Select Table** drop-down in the **Table** group of the **Table** tab in the **Ribbon**. Press and hold the right mouse button, and then choose **Delete** from the shortcut menu displayed. You can also use the DELETE key after selecting the required items.

   To delete an individual row/column, move the cursor over the boundary of the required row/column; the entire row/column will be highlighted in green color. Click when the entities are highlighted and then choose the DELETE key.

Tip: *In the **Drawing** mode, most of the modifications and editing can be done by using the shortcut menu that will be displayed when you select the item to modify and then hold the right mouse button. The options in this menu vary and depend on the item selected.*

*It is recommended that you use the options in the **Table** tab of the **Ribbon** to create the title blocks in the formats. The text can easily be added to the title block by double-clicking in the cell.*

GENERATING THE BOM AND BALLOONS IN DRAWINGS

The Bill of Material (BOM) is a representation of the components and their parameters that are used in the assembly. In the **Drawing** mode, the BOM is generated and is associative in nature. Therefore, any modification in the assembly, such as addition or removal of components, is automatically reflected in the BOM.

The procedure to generate the BOM and Balloons in the **Drawing** mode is discussed next.

Consider a case in which you have started a new drawing sheet of standard size A with a user-defined format. Now, follow the steps discussed next to generate the BOM and balloons.

Creating a Table for the BOM

Before proceeding to generate the BOM, first you need to create a table for it. The items to be listed in the BOM will be displayed in this table.

1. Choose **Table > Table > Table > Insert Table** from the **Ribbon** to display the **Insert Table** dialog box.

2. Set the **Number of Columns** spinner to **3** and **Number of Rows** spinner to **2** in the **Table Size** area.

3. Clear the **Automatic Height Adjustment** check box in the **Row** area.

4. Set the height of rows as **0.5** in the **Height (INCH)** spinner in the **Row** area.

5. Set the width of column as **0.5** in the **Width (INCH)** spinner in the **Width** area.

6. Choose the **OK** button to exit; the **Select Point** dialog box will be displayed and you will be prompted to locate the origin of the table.

7. Click at the desired location to place the table; the table gets inserted in the drawing.

8. Select a cell in the second row and choose the **Height and Width** tool from the **Rows & Columns** group of the **Ribbon**; the **Height and Width** dialog box is displayed.

9. Enter the value of width as **3** in the **Width(drawing units)** edit box in the **Columns** area of the dialog box and then choose the **OK** button.

Setting the Repeat Region

Ribbon: Table > Data > Repeat Region

After creating the table for the BOM, you need to define the repeat region in this table. Repeat regions are smart cells that can expand depending on the amount of data inserted in the cell. The second row will be defined as the repeat region.

1. Choose **Table > Data > Repeat Region** from the **Ribbon**; the **TBL REGIONS** menu will be displayed.

2. Choose the **Add** option from the **TBL REGIONS** menu; you will be prompted to locate the corners of the region.

3. Select the first cell of the second row and then the third cell of the second row to define the region. Next, press the middle mouse button twice to exit the menu.

Creating the Column Headers

After creating the table and setting the repeat region, you need to define the column headings in the table.

1. Double-click in the first cell of the first row; the **Format** tab will be added in the **Ribbon**. Enter **Index** and click twice outside the table to exit.

   Similarly, enter **Part Name** and **Qty** in the second and the third cells.

Assigning the Report Symbols in the Repeat Region

The information in the cells that are defined as repeat region is determined by the text that is written in the form of report symbols. These report symbols associate the information in the cells with the assembly file directly.

1. Double-click in the first cell of the second row; the **Report Symbol** dialog box will be displayed. Select **rpt** from the dialog box; the **rpt** options are displayed in the dialog box.

2. Select the **index** option from the dialog box.

3. Double-click in the second cell of the second row; the **Report Symbol** dialog box will be displayed. Select **asm** from the dialog box; the **asm** options are displayed.

4. Select the **mbr** option from the dialog box and then select the **name** option from the dialog boxes that appear. This will ensure that the names of the parts in the assembly are displayed in this column. The number of rows will be automatically modified depending on the number of parts in the assembly.

5. Double-click in the third cell of the second row; the **Report Symbol** dialog box will be displayed. Select **rpt** from the dialog box; the **rpt** options are displayed. Select the **qty** option from the dialog box.

Generating the Exploded Drawing View

Ribbon: Layout > Model Views > General

After setting all the options, you need to generate the isometric view of the exploded state of assembly. Therefore, you should have created an exploded state of assembly. To generate the exploded drawing view, follow the steps listed next.

1. Choose **Layout > Model Views > General** from the **Ribbon**; the **Select Combined State** dialog box will be displayed. Choose the **OK** button from the dialog box and you will be prompted to select the center point for the drawing view.

2. Select the point on the drawing sheet to place the view; the **Drawing View** dialog box will be displayed. Select the **Isometric** option from the **Default orientation** drop-down list in the **Drawing View** dialog box. Choose the **Apply** button. Specify the scale factor, if required.

3. Select the **View States** option from the **Categories** list box. Select the **Explode components in view** check box. Select the name of the exploded state from the **Assembly explode state** drop-down list and then choose the **Apply** button.

4. Close the **Drawing View** dialog box.

Note that the BOM automatically appears on the table by extending the repeat region. Also, note that the **Qty** column is empty. If the BOM is not displayed, choose **Table > Data > Update Tables** from the **Ribbon**.

Setting the No Repeat Option

Ribbon: Table > Data > Repeat Region

If an assembly has more than one quantity of a component, the components are repeated in the BOM. Therefore, you need to set the option to omit the repeated instances in the BOM.

1. Choose **Table > Data > Repeat Region** from the **Ribbon**; the **TBL REGIONS** menu will be displayed.

2. Choose the **Attributes** option from the menu; you will be prompted to select a region.

3. Move the cursor over the table below the first row. All rows, except the first row, are highlighted. The highlighted portion is called region. Select the region; the **REGION ATTR** submenu will be displayed. Choose the **No Duplicates** option and then choose the **Done/Return** button.

4. Press the middle mouse button twice to exit the menu.

5. Choose **Table > Data > Update Tables** from the **Ribbon**, if the information is not updated in the table.

Generating Balloons

Ribbon: Table > Balloons > Create Balloons drop-down > Create Balloons - All

Balloons are generated by using the repeat region. Before generating balloons, you need to set the repeat region for the generation of balloons.

1. Choose the **Create Balloons - All** tool from the **Create Balloons** drop-down available in the **Balloons** group of the **Table** tab in the **Ribbon**; balloons are attached with all the parts of the assembly in the exploded view. The balloons also contain the index number so that the parts in the assembly can be referred to as BOM.

2. Drag the balloons to place them appropriately in the drawing sheet. You can select a balloon and then hold the right mouse button to invoke the shortcut menu. From this menu, you can select the required options to modify the attachment points of balloons.

TUTORIALS

Tutorial 1

In this tutorial, you will create a format of size A and add the title block to the format. Then you will retrieve the format in the **Drawing** mode and create the table for BOM. Next, you will generate the exploded isometric view of the **Pedestal Bearing** assembly created in Tutorial 2 of Chapter 11. Also, you will add balloons to the drawing view, as shown in Figure 14-16.

(Expected time: 45 min)

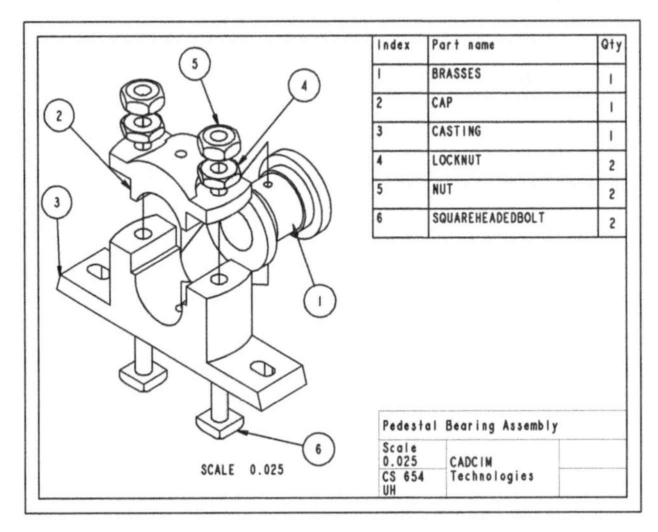

Figure 14-16 *The drawing view of the assembly showing the BOM and balloons*

The following steps are required to complete this tutorial:

a. Start a new file in the **Format** mode. Create the format of the drawing and then add the title block in the format, refer to Figures 14-17 and 14-18.
b. Save the format file and then close it.
c. Start a new drawing file in the **Drawing** mode. Select **Pedestal Bearing** as the model and retrieve the format that you have created.
d. Create the table and enter the headers.
e. Define the repeat region and then assign the report symbols, refer to Figure 14-21.
f. Generate the exploded isometric view of the assembly, refer to Figure 14-22.
g. Add balloons to the drawing, refer to Figure 14-24.

Copy the *Pedestal Bearing* folder from the *c11* folder to the *c14* folder. As this is the first tutorial of the chapter, therefore you need to create the *c14* folder inside the *Creo-3.0* folder. Set the working directory to *C:/Creo-3.0/c14/Pedestal Bearing*.

Starting the Format File

As evident from Figure 14-16, the exploded view of the Pedestal Bearing is generated on a drawing sheet with a customized format. Therefore, you need to create a format before generating the drawing view. This format will be retrieved later.

1. Choose the **New** button from the **Quick Access** toolbar; the **New** dialog box is displayed.

2. Select the **Format** radio button from the **Type** area in the **New** dialog box and enter the name as **Format1** in the **Name** edit box. Choose **OK** from the **New** dialog box; the **New Format** dialog box is displayed.

3. In this dialog box, the **Empty** radio button in the **Specify Template** area is selected by default and the **Landscape** button in the **Orientation** area is also chosen by default. If not so, you need to set them that ways.

4. Select **A** from the **Standard Size** drop-down list in the **Size** area and then choose **OK**; the **Format** mode is invoked. Note that a sheet of size **A** is displayed on the screen. This is evident from the text displayed below the sheet on the screen.

Creating the Format

1. Choose **Sketch > Sketching > Edge > Offset Edge** from the **Ribbon**; the **OFFSET OPER** menu is displayed. Choose the **Ent Chain** option from this menu and then select all the four border lines of the format by drawing a window around them. Press the middle mouse button after the selection has been made; the **Message Input Window** is displayed and also, an arrow pointing outward is displayed on the format boundary. This arrow displays the direction of the offset. To offset the lines in the opposite direction, specify a negative offset value.

2. Enter **-0.25** in the **Message Input Window** and press ENTER. Next, press the middle mouse button twice to exit the **OFFSET OPER** menu.

3. Choose **Table > Table > Table > Insert Table** from the **Ribbon**; the **Insert Table** dialog box is displayed.

4. Choose **Table growth direction: leftward and ascending** button from the **Direction** area to set the direction of table growth.

5. Set the value to **3** in both the **Number of Columns** and **Number of Rows** spinners in the **Table Size** area.

6. Enter **0.5** as height in the **Height (INCH)** edit box and **1.2** as width in the **Width (INCH)** edit box.

7. Choose the **OK** button from the **Insert Table** dialog box; the **Select Point** dialog box is displayed and you are prompted to select a point to place the table. Also, the preview of the table gets attached to the cursor.

8. Click on the desired location in the drawing to place the table, refer to Figure 14-17.

9. Select any cell of the second column and then choose the **Height and Width** button from the **Rows & Columns** group in the **Table** tab; the **Height and Width** dialog box is displayed.

10. Enter the value **2** in the **Width (drawing units)** edit box in the **Columns** area and then choose the **OK** button; the title block created will be similar to the one shown in Figure 14-17. Note that the title block created is not the required one. Therefore, you need to modify it.

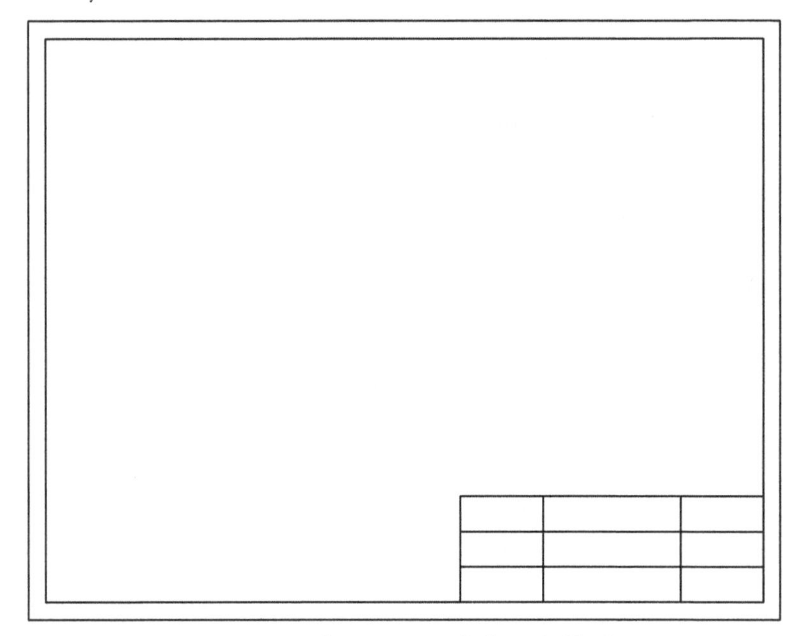

Figure 14-17 *Format with the title block*

11. Select the entire first row and then choose the **Merge Cells** button from the **Rows & Columns** group.

12. Similarly, select the second and third cells in the second column and choose the **Merge Cells** button.

    The table after merging rows and columns should look similar to the one shown in Figure 14-18.

13. You need to set the alignment of the text that will be later entered in the cells of the table. To do so, double-click on any of these cells; the **Format** tab is added.

14. Choose the inclined arrow from the **Style** group of the **Format** tab; the **Text Style** dialog box is displayed.

15. In the **Note/Dimension** area of this dialog box, select the **Middle** option from the **Vertical** drop-down list and select the **Center** option from the **Horizontal** drop-down list. Exit the **Text Style** dialog box by choosing the **OK** button.

16. Now, the text is aligned vertically to the middle of the cell and horizontally to the center of the cell. Similarly, set the alignment in all the cells of the table individually.

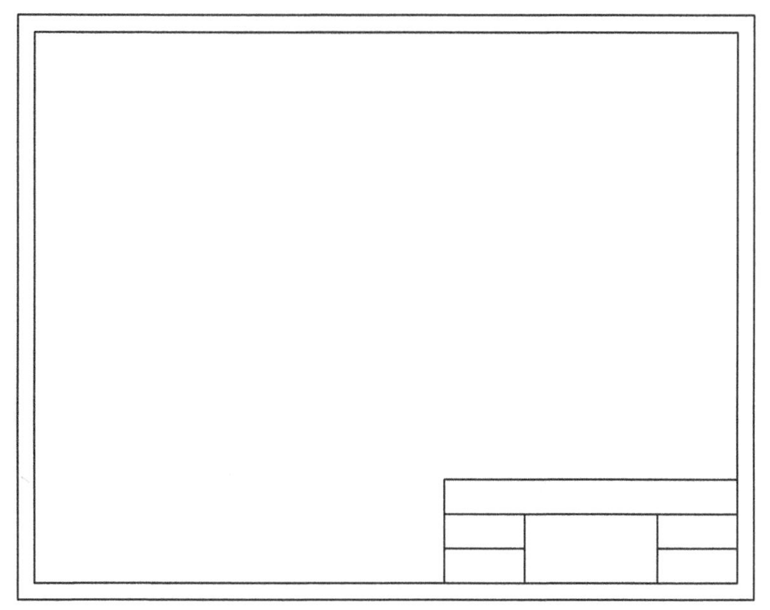

Figure 14-18 *Modified title block*

Saving the Format File

You need to save the format file that you have created so that it can be used as a template in the **Drawing** mode. Using this template, you will generate the exploded drawing view of the Pedestal Bearing. The file will be stored in the *.frm* file format.

1. Choose the **Save** button from the **Quick Access** toolbar; the **Save Object** dialog box is displayed with the name of the file that you entered earlier. Press ENTER.

2. Close the current window by choosing the **Close** button from the **Quick Access** toolbar. 

Starting a New Drawing File

You need to start a new drawing file to generate the exploded drawing view of the Pedestal Bearing.

1. Choose the **New** button from the **File** menu to display the **New** dialog box. In this dialog box, select the **Drawing** radio button and specify the name of the drawing as *c14tut1*. Choose **OK**; the **New Drawing** dialog box is displayed.

2. In the **New Drawing** dialog box, choose the **Browse** button from the **Default Model** area and select the assembly file named *pedestalbearing.asm*.

3. Select the **Empty with format** radio button from the **Specify Template** area in the **New Drawing** dialog box.

4. Select **Format1** from the **Format** drop-down list. If **Format1** is not available in the drop-down list, choose the **Browse** button to locate **Format1**. Then, choose **OK** to start a new drawing file with the selected format.

Creating the Table for BOM

1. Choose **Table > Table > Table > Insert Table** from the **Ribbon**; the **Insert Table** dialog box is displayed.

2. Choose **Table growth direction: leftward and descending** button from the **Direction** area.

3. Set the value to **3** in the **Number of Columns** spinner and **2** in the **Number of Rows** spinner in the **Table Size** area of the dialog box. Now, you need to enter the dimensions of the BOM table.

4. Enter **0.5** as height in the **Height (INCH)** edit box and **1.2** as width in the **Width (INCH)** edit box.

5. Choose the **OK** button from the **Insert Table** dialog box; the **Select Point** dialog box is displayed and you are prompted to select a point to place the table.

6. Click on the desired location to place the table; refer to Figure 14-21.

7. Select any cell of the second column and then choose the **Height and Width** button from the **Rows & Columns** group in the **Table** tab; the **Height and Width** dialog box is displayed.

8. Enter the value **3** in the **Width (drawing units)** edit box in the **Column** area and then choose the **OK** button.

9. Similarly, select any cell of the third column and change its width to **0.8**.

Setting the Repeat Region

Repeat regions are the smart cells in a table that can expand depending on the amount of data inserted in a cell. You need to define the second row as the repeat region.

1. Choose the **Repeat Region** tool from the **Data** group of the **Table** tab in the **Ribbon**; the **TBL REGIONS** menu is displayed.

2. Choose the **Add** option from the **TBL REGIONS** menu; you are prompted to locate the corners of the region.

3. Select the first cell of the second row and then select the third cell of the second row to locate the corners of the region. Press the middle mouse button twice to exit the menu.

Creating the Column Headers

Now, you need to create headings in the table.

1. Double-click in the first cell of the first row; the **Format** tab will be added to the **Ribbon.** Enter **Index** and click outside twice to exit.

2. Similarly, enter **Part Name** and **Qty** in the second and third cells, respectively.

Assigning the Report Symbols in the Repeat Region

The information in the cells that are defined as repeat region is determined by the text written in the form of report symbols. These report symbols associate the information in the cells with the assembly file directly.

1. Double-click in the first cell of the second row; the **Report Symbol** dialog box is displayed, as shown in Figure 14-19. Select **rpt** from this dialog box; the **rpt** options are displayed, as shown in Figure 14-20.

***Figure 14-19** The **Report Symbol** dialog box*

***Figure 14-20** The **Report Symbol** dialog box with the **rpt** options*

2. Select the **index** option from the dialog box; **rpt.index** is added to the cell.

3. Double-click in the second cell of the second row; the **Report Symbol** dialog box is displayed. Select **asm** from this dialog box; the **asm** options are displayed.

4. Select the **mbr** option from the dialog box; the **mbr** options are displayed.

5. Select the **name** option from the dialog box; **asm.mbr.name** is added to the cell.

6. Double-click in the third cell of the second row; the **Report Symbol** dialog box is displayed. Select **rpt** from this dialog box; the **rpt** options are displayed.

7. Select the **qty** option from the dialog box.

   If the text selected to enter in the cells is overlapping the other text or is extending beyond the cell, ignore it. This is because the BOM will be created in the repeat region where this text exists and after the BOM is created, the text of the BOM will fit inside the cells. The drawing sheet after creating the column headers and entering the report symbols is shown in Figure 14-21.

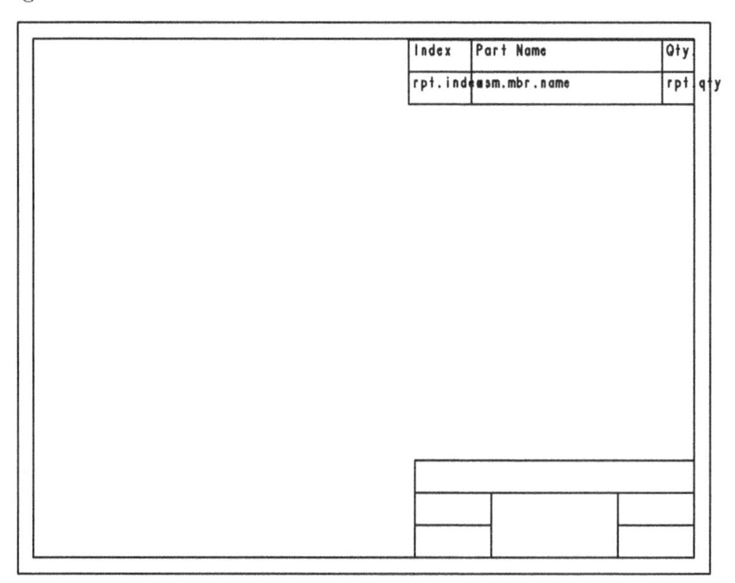

***Figure 14-21** Drawing sheet after adding the report symbols*

Generating the Exploded Drawing View

In Tutorial 2 of Chapter 11, you created the exploded view of the Pedestal Bearing in the **Assembly** mode. The name of the exploded view that you created was **EXP1**. As a result, the exploded view of the Pedestal Bearing is integrated with the assembly file that you copied from the *c11* folder. Now, you will use this exploded state to generate the exploded drawing view.

1. Choose the **General** tool from the **Model Views** group in the **Layout** tab of the **Ribbon**; the **Select Combined State** dialog box is displayed.

2. Select the **Do not prompt for Combined State** check box and then choose **OK** to exit the dialog box; you are prompted to select the center point for the drawing view.

3. Select a point on the left of the drawing sheet; the **Drawing View** dialog box is displayed. Select the **Isometric** option from the **Default Orientation** drop-down list in the **Drawing View** dialog box and then choose the **Apply** button.

4. Select the **Scale** option from the **Categories** list box. Next, select the **Custom scale** radio button from the **Scale and perspective options** area. Enter the value **.025** in the edit box and then choose the **Apply** button.

5. Select the **View States** option from the **Categories** list box. Next, select the **Explode components in view** check box and select the **EXP1** from the **Assembly explode state** drop-down list. Next, choose the **Apply** button.

6. Close the **Drawing View** dialog box to complete the placement of the required view. If the model is not placed properly inside the boundaries of the drawing sheet, unlock it and then drag the drawing view and place it, as shown in Figure 14-22. Note that the BOM automatically appears in the table by extending the repeat region and the Qty column is empty.

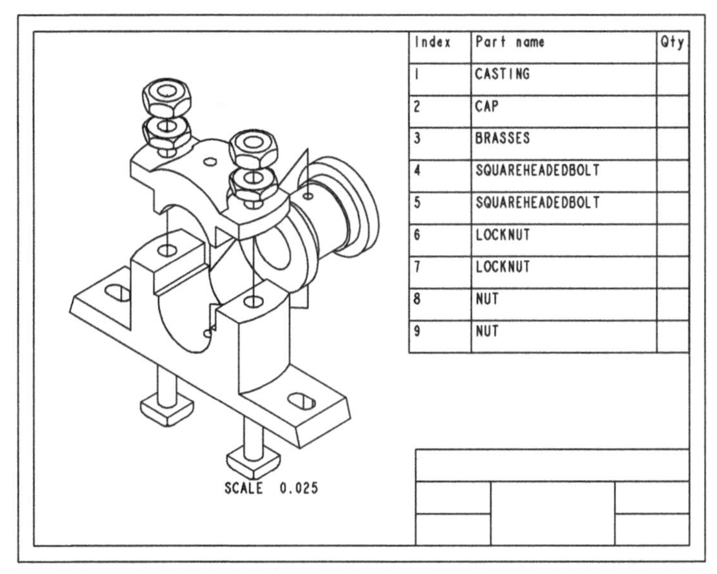

Figure 14-22 *Exploded isometric view of the assembly*

Note

*1. If the BOM is not displayed, then choose **Table > Data > Update Tables** from the **Ribbon**.*

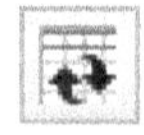

*2. You may need to select the **No Hidden** option from the **Display Style** drop-down list from the **Graphics** toolbar to change the view.*

Setting the No Repeat Option

As some components are repeated in the assembly, they are also repeated in the BOM. The BOM needs to be set such that no part name is repeated.

1. Choose the **Repeat Region** tool from the **Data** group of the **Table** tab in the **Ribbon**; the **TBL REGIONS** menu is displayed.

2. Choose the **Attributes** option from the menu; you are prompted to select a region.

3. Move the cursor over the table below the first row. All rows, except the first row, are highlighted in green. The highlighted portion is called region. Select the region; the **REGION ATTR** submenu is displayed. Choose the **No Duplicates** option and then choose the **Done/Return** button from the submenu.

4. Press the middle mouse button twice and exit the menu; the drawing sheet appears, as shown in Figure 14-23.

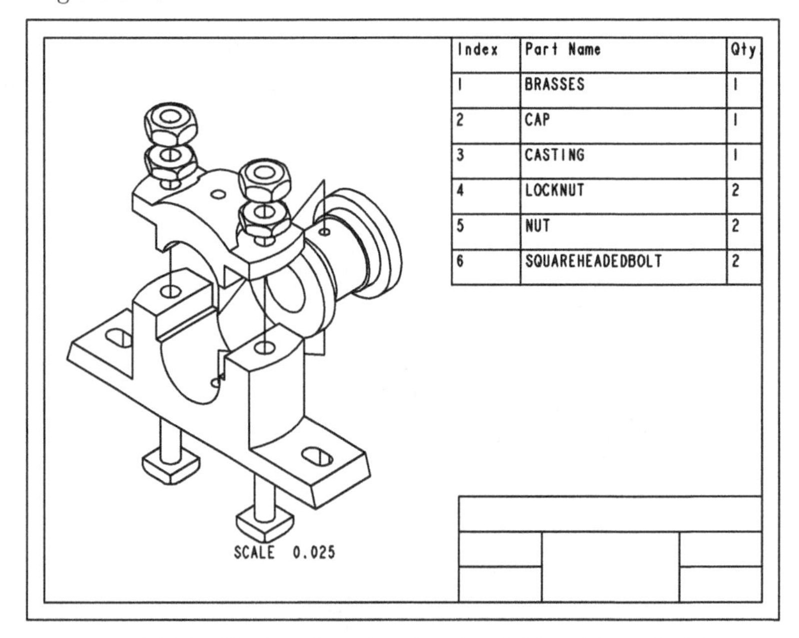

Figure 14-23 *The drawing view showing BOM with no duplicates*

5. If the information is not updated in the table, choose the **Update Tables** tool from the **Data** group of the **Table** tab from the **Ribbon** to update the information.

Generating Balloons

Balloons are generated by using the repeat region. Before generating the balloons, you need to set the repeat region for generating balloon.

1. Choose the **Create Balloons - All** option from the **Create Balloons** drop-down list available in the **Balloons** group of the **Table** tab in the **Ribbon**; balloons are attached with all the parts in the exploded view. The balloons contain the index number so that the parts in the assembly can be referred to as BOM.

2. Drag the balloons to place them appropriately in the drawing sheet, as shown in

Figure 14-24. You can select a balloon and then hold the right mouse button to invoke the shortcut menu. From this menu, you can select the required options to modify the attachment points of balloons.

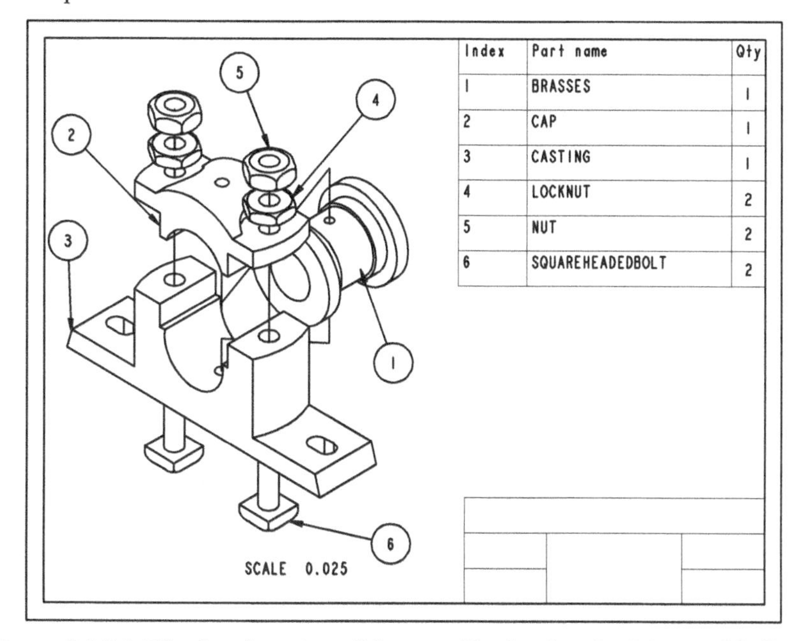

Figure 14-24 The drawing view of the assembly showing the BOM and balloons

Entering Text in the Title Block

The text alignment in various cells of the title block was defined while creating the format. As a result, when you enter the text, it will automatically be aligned to the middle left.

1. Double-click on the top cell in the table; the **Format** tab is added to the **Ribbon**.

2. Now, enter **Pedestal Bearing Assembly** in the cell. Next, choose the inclined arrow from the **Style** group of the **Format** tab; the **Text Style** dialog box is displayed. Now, select the **Default** check box at the right of the **Height** property in the character area and choose **OK**.

3. Double-click on the cell that is in the first column and the second row; the **Format** tab is added to the **Ribbon**.

4. Now, enter **Scale** and then press ENTER. Now, enter **0.025** and click outside twice to exit; the text **Scale: 0.025** is entered in the two lines in the title block.

5. Similarly, enter the text in all the remaining cells one by one, as shown in Figure 14-25.

Saving the Drawing File

1. Choose the **Save** button from the **Quick Access** toolbar to save the drawing file; the **Save Object** dialog box is displayed. Now, save the file by pressing ENTER.

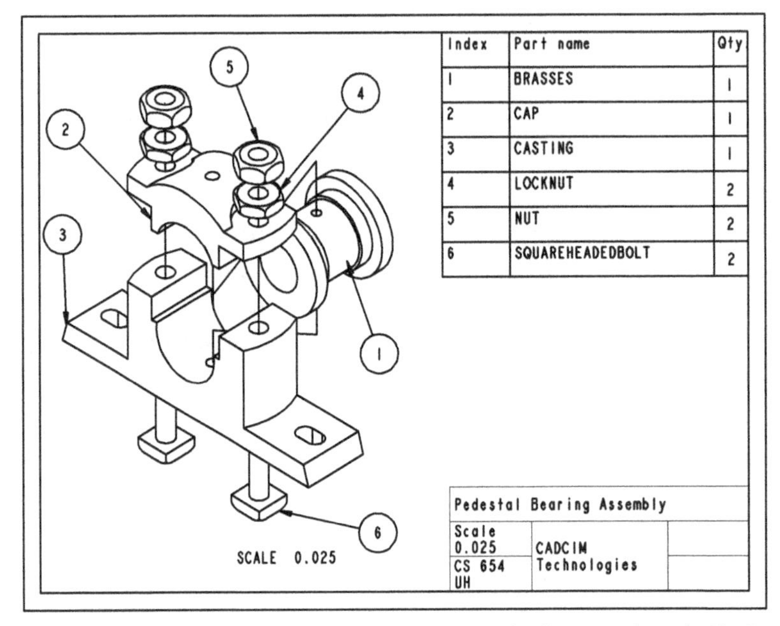

Figure 14-25 The drawing view of the assembly showing the title block, balloons, and the BOM

Closing the Window

The drawing file has been saved and now you can exit the **Drawing** mode.

1. Choose **File > Close** from the menu bar to close the current session.

Note

The occurrence of components in BOM depends on the order in which they were assembled in the assembly. In other words, the component placed first will be placed first in the BOM.

Tutorial 2

In this tutorial, you will create a format of size A and add the title block in the format. You will retrieve the format in the **Drawing** mode and generate the exploded isometric view of the **Shock Assembly** created in Tutorial 1 of Chapter 11. Also, you will add the associative Bill of Material and the Balloons to the drawing view, as shown in Figure 14-26.

(Expected time: 45 min)

The following steps are required to complete this tutorial:

a. Start a new file in the **Format** mode. Create the format of the drawing and add the title block in the format, refer to Figures 14-27 and 14-28.
b. Save the format file and then close it.
c. Start a new drawing file in the **Drawing** mode, select *shockassembly.asm* as the model, and retrieve the format created earlier.
d. Create the table for BOM and enter the headers.

e. Define the repeat region and then assign the report symbols, see Figure 14-29.
f. Generate the exploded isometric view of the assembly and generate balloons in the drawing view, refer to Figures 14-30 and 14-31.
g. Add text in the title block, refer to Figure 14-32.

Copy the *Shock Assembly* folder from the *c11* folder to the *c14* folder and set this folder as the working directory.

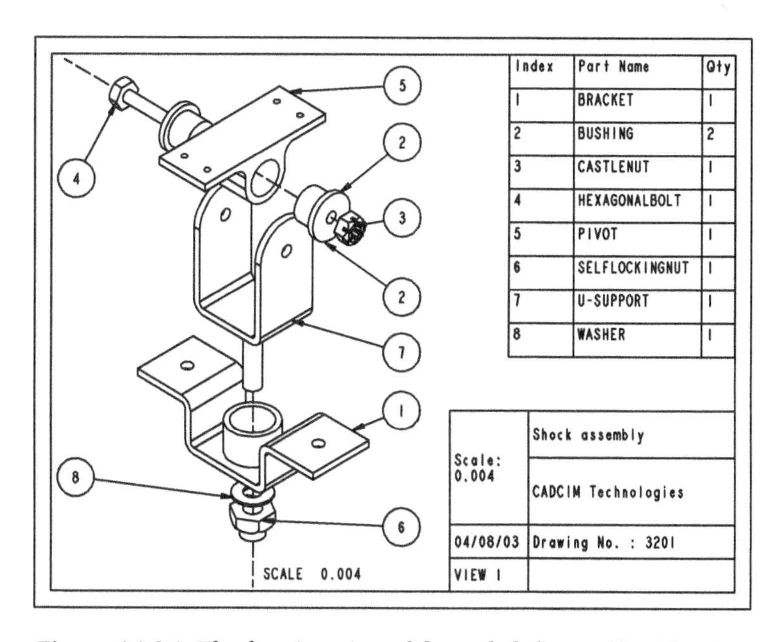

Figure 14-26 *The drawing view of the exploded assembly with BOM*

Starting the Format File

As mentioned in the tutorial description, you need to create a format of size A that will be retrieved later to create the drawing view of the Shock Assembly.

1. Choose the **New** button from the **Quick Access** toolbar to display the **New** dialog box.

2. Select the **Format** radio button from the **Type** area in the **New** dialog box and name the file as *Format2*. Choose **OK** from the **New** dialog box; the **New Format** dialog box is displayed.

3. Select the **Empty** radio button from the **Specify Template** area and the **Landscape** button from the **Orientation** area of the **New Format** dialog box, if they are not already selected.

4. Select **A** from the **Standard Size** drop-down list in the **Size** area and then choose **OK** to invoke the **Format** mode. Note that a sheet of size A is displayed on the screen. This is evident from the text displayed below the sheet on the screen.

Creating the Format

1. Choose **Sketch > Sketching> Edge > Offset Edge** from the **Ribbon**; the **OFFSET OPER** menu is displayed. Choose the **Ent Chain** option and then select all the four border lines of the format by drawing a window around them. Press the middle mouse button after the selection has been made; an arrow is displayed pointing outward. This arrow displays the direction of the offset. Since you need to offset the lines in the opposite direction, you will specify a negative offset distance.

2. Enter **-0.25** in the **Message Input Window** and press ENTER. Press the middle mouse button twice to exit the **OFFSET OPER** menu.

3. Choose **Table > Table > Table > Insert Table** from the **Ribbon**; the **Insert Table** dialog box is displayed.

4. Choose the **Table growth direction: leftward and ascending** button from the **Direction** area.

5. Set the value to **2** in the **Number of Columns** spinner and **4** in the **Number of Rows** spinner in the **Table Size** area.

6. Enter **0.5** as height in the **Height (INCH)** edit box and **1.2** as width in the **Width (INCH)** edit box.

7. Choose the **OK** button from the **Insert Table** dialog box; the **Select Point** dialog box is displayed and you are prompted to select a point to place the table.

8. Click on the required location to place the table, refer to Figure 14-27; the table is placed at the specified place.

9. Select any cell of the second column and then choose the **Height and Width** button from the **Rows & Columns** group in the **Table** tab; the **Height and Width** dialog box is displayed.

10. Enter the value **3.2** in the **Width (drawing units)** edit box in the **Column** area and then choose the **OK** button.

11. Select any cell in the second row and choose the **Height and Width** button from the **Rows & Columns** group in the **Table** tab; the **Height and Width** dialog box is displayed.

12. Enter the value **1** in the **Height (drawing units)** edit box in the **Row** area of the dialog box and then choose the **OK** button. Note that you may need to clear the **Automatic height adjustment** check box to enter the **Height (drawing units)**.

13. Select any cell in the first row and choose the **Height and Width** button from the **Rows & Columns** group in the **Table** tab; the **Height and Width** dialog box is displayed.

14. Now, clear the **Automatic height adjustment** check box in the **Row** area. Enter the value

0.7 in the **Height (drawing units)** edit box in the **Row** area and then choose the **OK** button.

The title block created will be similar to the one shown in Figure 14-27. However, this title block is not the required one. Notice the difference between the title block shown in Figure 14-27 and the one shown in Figure 14-28. So, you need to modify the title block that you have created such that it is similar to the one shown in Figure 14-28.

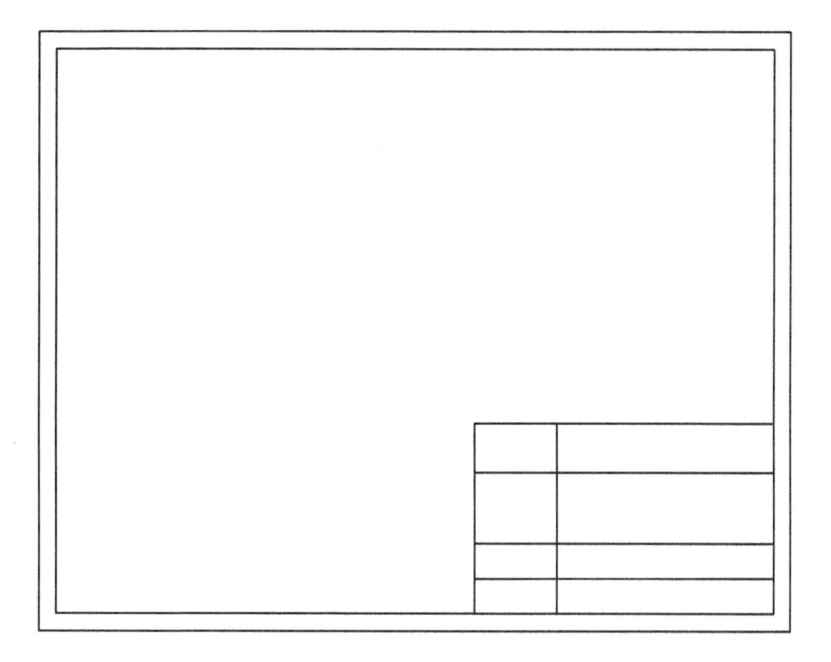

Figure 14-27 *Format with the title block*

15. Select a cell in the first row and a cell in the second row of the first column to merge them. Then, choose the **Merge Cells** tool from the **Rows & Columns** group in the **Table** tab of the **Ribbon**; the cells are merged.

    The table after merging the rows and columns should look similar to the one shown in Figure 14-28. You need to set the alignment of the text that will be entered in the cells of the table later.

16. Double-click in a cell to invoke the **Format** tab and click on the inclined arrow from the **Style** group of the **Format** tab; the **Text Style** dialog box is displayed.

17. In the **Note/Dimension** area of this dialog box, select the **Middle** option from the **Vertical** drop-down list and the **Center** option from the **Horizontal** drop-down list. Next, choose the **OK** button to exit the dialog box.

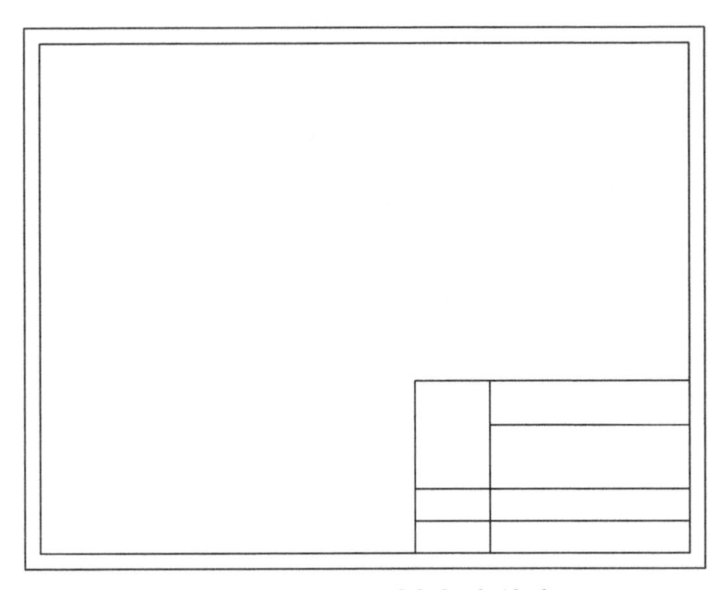

Figure 14-28 Modified title block

Now, the text is aligned vertically to the middle of the cell and horizontally to the center of the cell. Similarly, set the alignment in all the cells of the table individually.

Saving the Format File

You need to save the format file that you have created so that it can be used as a template in the **Drawing** mode. The file will be stored in the *.frm* file format.

1. Choose the **Save** button from the **Quick Access** toolbar; the **Save Object** dialog box is displayed with the name of the file entered earlier. Next, press ENTER to save it.

2. Close the current window by choosing **File > Close** from the menu bar.

Starting a New Drawing File

You need to start a new drawing file for generating the exploded drawing view of the Shock Assembly and for generating the BOM.

1. Choose the **New** button to display the **New** dialog box. Select the **Drawing** radio button from the dialog box, specify the name of the drawing as **c14tut2**, and then choose **OK**; the **New Drawing** dialog box is displayed.

2. Choose the **Browse** button from the **Default Model** area and select the assembly file named *shockassembly.asm*.

3. Select the **Empty with format** radio button in the **Specify Template** area of the **New Drawing** dialog box.

4. Select **Format2** from the **Format** drop-down list. If **Format2** is not available in the drop-down list, choose the **Browse** button to locate **Format2**. Next, choose **OK** to exit the **New Drawing** dialog box.

Creating the Table for BOM

1. Choose **Table > Table > Table > Insert Table** from the **Ribbon**; the **Insert Table** dialog box is displayed.

2. Choose **Table growth direction: leftward and descending** button from the **Direction** area.

3. Set the value to **3** in the **Number of Columns** spinner and **2** in the **Number of Rows** spinner in the **Table Size** area.

4. Enter **0.5** as height in the **Height (INCH)** edit box and **1.2** as width in the **Width (INCH)** edit box.

5. Choose the **OK** button from the **Insert Table** dialog box; the **Select Point** dialog box is displayed and you are prompted to select a point to place the table.

6. Select a point close to the upper right corner of the inner rectangle; the table is placed at the specified point.

7. Select any cell in the second column and then choose the **Height and Width** button from the **Rows & Columns** group in the **Table** tab; the **Height and Width** dialog box is displayed.

8. Enter the value **3** in the **Width (drawing units)** edit box in the **Column** area and then choose the **OK** button. Similarly, change the width of the third column to **0.8**.

Setting the Repeat Region

Repeat regions are the smart cells in a table that can expand depending on the amount of data inserted in the cell. The second row will be defined as the repeat region.

1. Choose **Repeat Region** from the **Data** group of the **Table** tab in the **Ribbon**; the **TBL REGIONS** menu is displayed.

2. Choose the **Add** option from the **TBL REGIONS** menu; you are prompted to locate the corners of the region.

3. Select the first cell of the second row and the third cell of the second row to define the region. Press the middle mouse button twice to exit the menu.

Creating the Column Headers

Next, you need to create headings in the table.

1. Double-click in the first cell of the first row; the **Format** tab is added to the **Ribbon.** Enter **Index** in the cell and click outside twice to exit.

2. Similarly, enter **Part Name** and **Qty** in the second and third cells, respectively.

Entering the Report Symbols in the Repeat Region

As mentioned earlier, the information in the cells that are defined as the repeat region is determined by the text written in the form of report symbols. These report symbols associate the information in the cells with the assembly file directly.

1. Double-click in the first cell of the second row; the **Report Symbol** dialog box is displayed. Select **rpt** from this dialog box; the **rpt** options are displayed.
2. Select the **index** option from this dialog box.
3. Next, double-click in the second cell of the second row; the **Report Symbol** dialog box is displayed. Select **asm** from this dialog box; the **asm** options are displayed.
4. Select the **mbr** option from this dialog box and then select the **name** option from the options that appear.
5. Double-click in the third cell of the second row; the **Report Symbol** dialog box is displayed. Select **rpt** from this dialog box; the **rpt** options are displayed.
6. Select the **qty** option from this dialog box. The drawing sheet after adding the report symbols is shown in Figure 14-29.

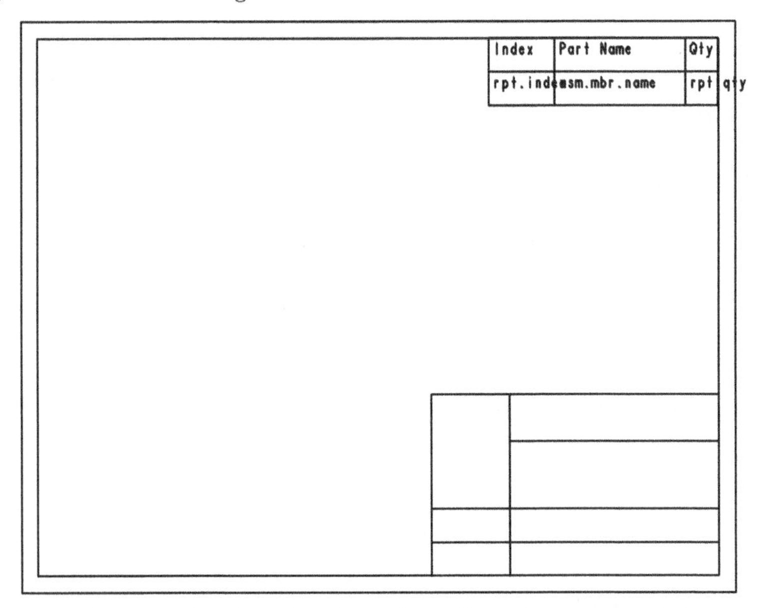

***Figure 14-29** Drawing sheet after adding the report symbols*

Generating the Drawing View

In Tutorial 1 of Chapter 11, you created the exploded view of the Shock Assembly in the **Assembly** mode. The name of the exploded view that you have specified was **EXP1**. As a result, the exploded view of the Shock assembly has been integrated with the assembly file that you copied from the *c11* folder. Now, you will use this exploded state to generate the exploded drawing view in the **Drawing** mode.

1. Choose the **General** tool from the **Model Views** group of the **Layout** tab in the **Ribbon**; the **Select Combined State** dialog box is displayed.

2. Select the **Do not prompt for Combined State** check box and then choose **OK**; the dialog box is closed and you are prompted to select the center point for the drawing view.

3. Select a point on the left of the drawing sheet; the **Drawing View** dialog box is displayed. Select the **Isometric** option from the **Default orientation** drop-down list in the **Drawing View** dialog box and then choose the **Apply** button.

4. Select the **Scale** option from the **Categories** list box. Also, select the **Custom scale** radio button from the **Scale and perspective options** area. Enter **.004** in the edit box and then choose the **Apply** button.

5. Select the **View States** option from the **Categories** list box. Select the **Explode components in view** check box. Select the **EXP1** option from the **Assembly explode state** drop-down list and then choose the **Apply** button.

6. Choose the **Close** button from the **Drawing View** dialog box; the exploded isometric view of the assembly is displayed.

   If the model is not placed properly inside the boundary of the drawing sheet, unlock it and then drag the drawing view and place it, as shown in Figure 14-30. If the BOM is not displayed, choose **Table > Data > Update Tables** from the **Ribbon** to display it.

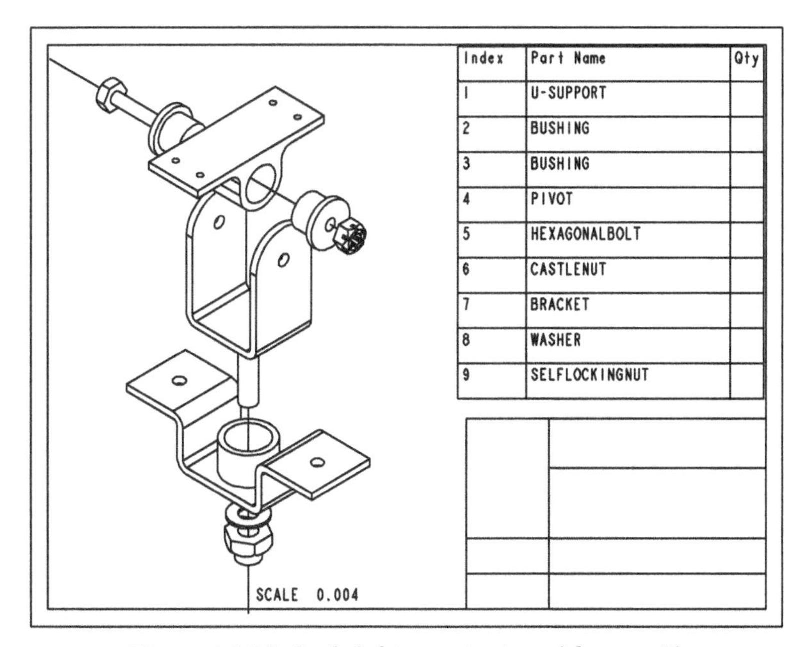

Figure 14-30 *Exploded isometric view of the assembly*

Setting the No Duplicates Option

The Shock assembly consists of two instances of Bushing. Therefore, in BOM, it is also repeated. But, the BOM needs to be set such that no part name is repeated.

1. Choose **Repeat Region** from the **Data** group of the **Table** tab in the **Ribbon**; the **TBL REGIONS** menu is displayed.

2. Choose the **Attributes** option from the menu; you are prompted to select a region.

3. Move the cursor on the table below the first row; all the rows, except the first row, are highlighted in green. Select the region; the **REGION ATTR** submenu is displayed. Choose the **No Duplicates** option and then choose the **Done/Return** option.

4. Press the middle mouse button twice to exit the menu. The drawing sheet appears, as shown in Figure 14-31.

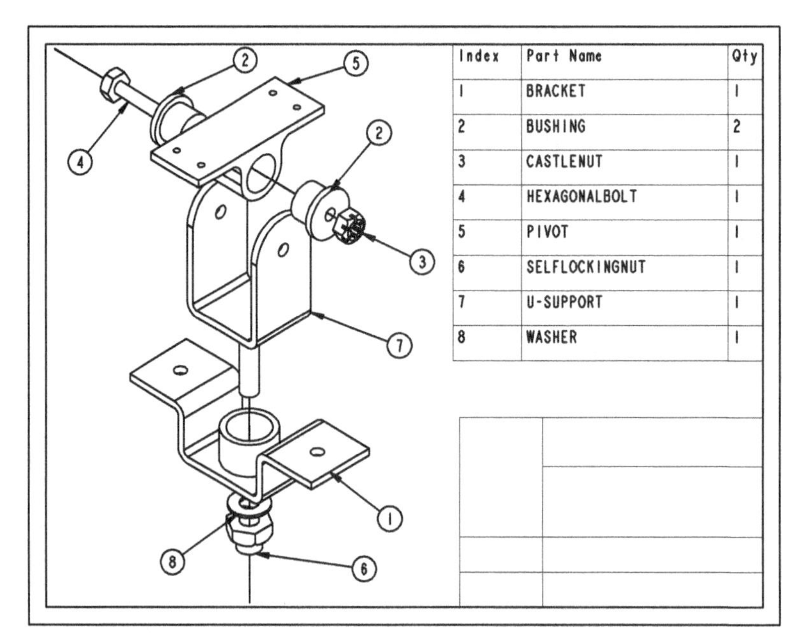

Figure 14-31 *The drawing view of the assembly showing the BOM and balloons*

5. If the information is not updated in the table, choose **Table > Data > Update Tables** from the **Ribbon** to update it.

Generating Balloons

Balloons are generated by using the repeat region. Before generating the balloons, you need to set the repeat region for the balloon generation.

1. Choose the **Create Balloons - All** option from the **Create Balloons** drop-down list available in the **Balloons** group of the **Table** tab of the **Ribbon**; balloons get attached with all the

parts of assembly in the exploded view. The balloons contain the index number so that the parts in the assembly can be referred to as BOM.

2. Drag the balloons to place them appropriately on the drawing sheet, refer to Figure 14-31. You can modify the attachment points of a balloon by selecting the balloon and right-clicking. Next, choose the required option from the shortcut menu displayed to modify the attachment points.

Entering Text in the Title Block

1. Double-click in the cell formed by merging the first row and the second column; the **Format** tab is added to the **Ribbon**.

2. Enter **Shock assembly** in the cell. Next, click on the inclined arrow from the **Style** group of the **Format** tab; the **Text Style** dialog box is displayed and select the **Default** check box at the right of the **Height** property and choose **OK**.

3. Now, select the cell from the second row and the second column. Enter **CADCIM Technologies** in the cell and click outside twice to exit.

4. Similarly, enter the text in all the remaining cells one by one, as shown in Figure 14-32.

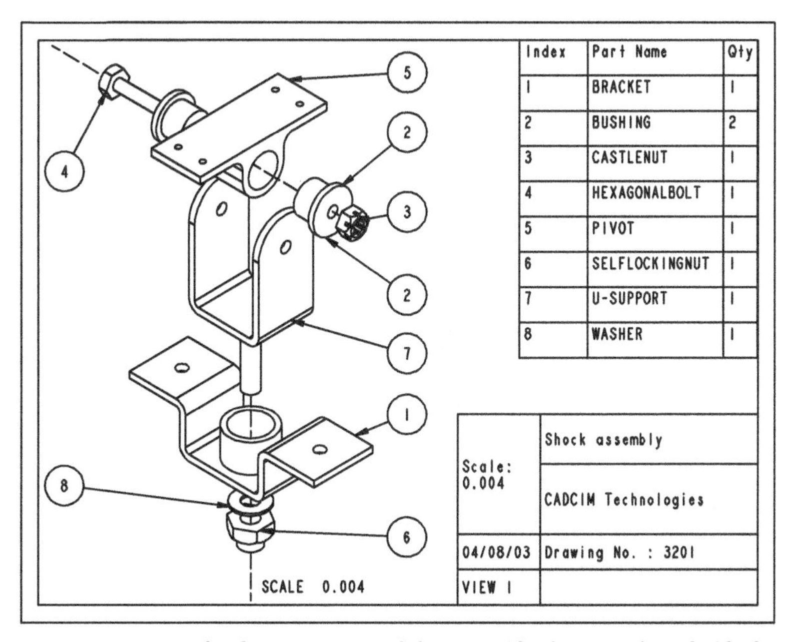

Figure 14-32 The drawing view of the assembly showing the title block, balloons, and the BOM

Saving the Drawing File

1. Choose the **Save** button from the **Quick Access** toolbar to save the drawing file; the **Save Object** dialog box is displayed. Press ENTER to save the file.

Closing the Window

The drawing file has been saved and now you can exit the **Drawing** mode.

1. Choose **File > Close** from the menu bar to close the window.

Self-Evaluation Test

Answer the following questions and then compare them to those given at the end of this chapter:

1. The __________ button of the **Balloons** group in the **Table** tab of the **Ribbon** is used to generate balloons.

2. The __________ tab is added, when you double-click on the cell of a table.

3. The __________ option is chosen to avoid repetition of components in BOM.

4. The __________ key can also be used to delete a draft entity.

5. The __________ button of the **Table** group in the **Table** tab of the **Ribbon** is used to draw tables in the **Drawing** mode.

6. When you move a view that forms a group with draft entities, the draft entities also move. (T/F)

7. PTC Creo Parametric provides you with some standard drawing formats for generating drawing views. (T/F)

8. PTC Creo Parametric allows you to create the user-defined formats. (T/F)

9. The BOM generated by using the repeat region is associative. (T/F)

10. One drawing file has only one drawing sheet. (T/F)

Review Questions

Answer the following questions:

1. Which of the following modes in PTC Creo Parametric is used to create an associative BOM?

   (a) **Part** (b) **Format**
   (c) **Drawing** (d) **Sketch**

2. Which of the following dialog boxes is displayed on double-clicking in the repeat region cell?

   (a) **Drawing View** (b) **Format**
   (c) **Report Symbol** (d) **TBL REGIONS**

3. Which of the following buttons on the **Right Toolchest** enables you to draw geometric entities continuously?

   (a) **Line** (b) **Circle**
   (c) **Chain** (d) **Arc**

4. Which of the following radio buttons is selected in the **New** dialog box to create a user-defined format for the drawing sheet?

   (a) **Part** (b) **Format**
   (c) **Drawing** (d) **Sketch**

5. Which of the following menus in the menu bar has the **Close** option to close the current window?

   (a) **Info** (b) **Window**
   (c) **View** (d) **Model**

6. You can open more than one window in PTC Creo Parametric. (T/F)

7. It is not possible to draw splines in the **Drawing** mode. (T/F)

8. There are two types of drafting in PTC Creo Parametric: Generative drafting and Interactive drafting. (T/F)

9. You can import a text file into the **Drawing** mode. (T/F)

10. The **Draft Group** option in the cascading menu is used to add or remove notes, surface symbols, or gtols from the selected dimension. (T/F)

Exercise

Exercise 1

Create a format of size A and then retrieve it in the **Drawing** mode to place the Right view, and an exploded isometric view of the **Cross Head** assembly created in Exercise 1 of Chapter 11. Add the associative assembly BOM and the balloons to the drawing view, as shown in Figure 14-33. **(Expected time: 45 min)**

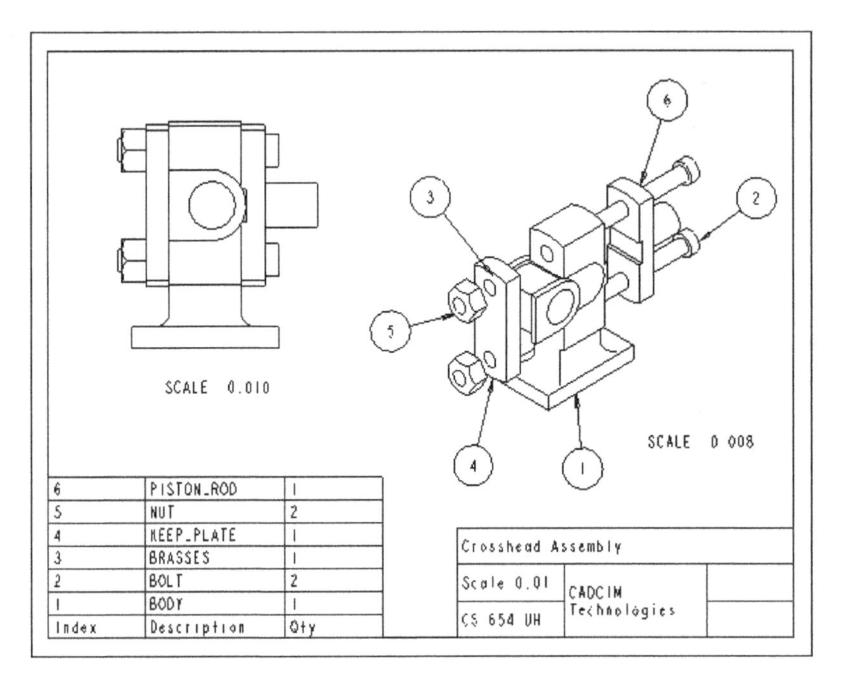

Figure 14-33 *Drawing for Exercise 1*

Answers to Self-Evaluation Test

1. **Create Balloons**, **2.** **Format**, **3.** **No Duplicates**, **4.** DELETE, **5.** **Table**, **6.** T, **7.** T, **8.** T, **9.** T, **10.** F

Chapter 15

Surface Modeling

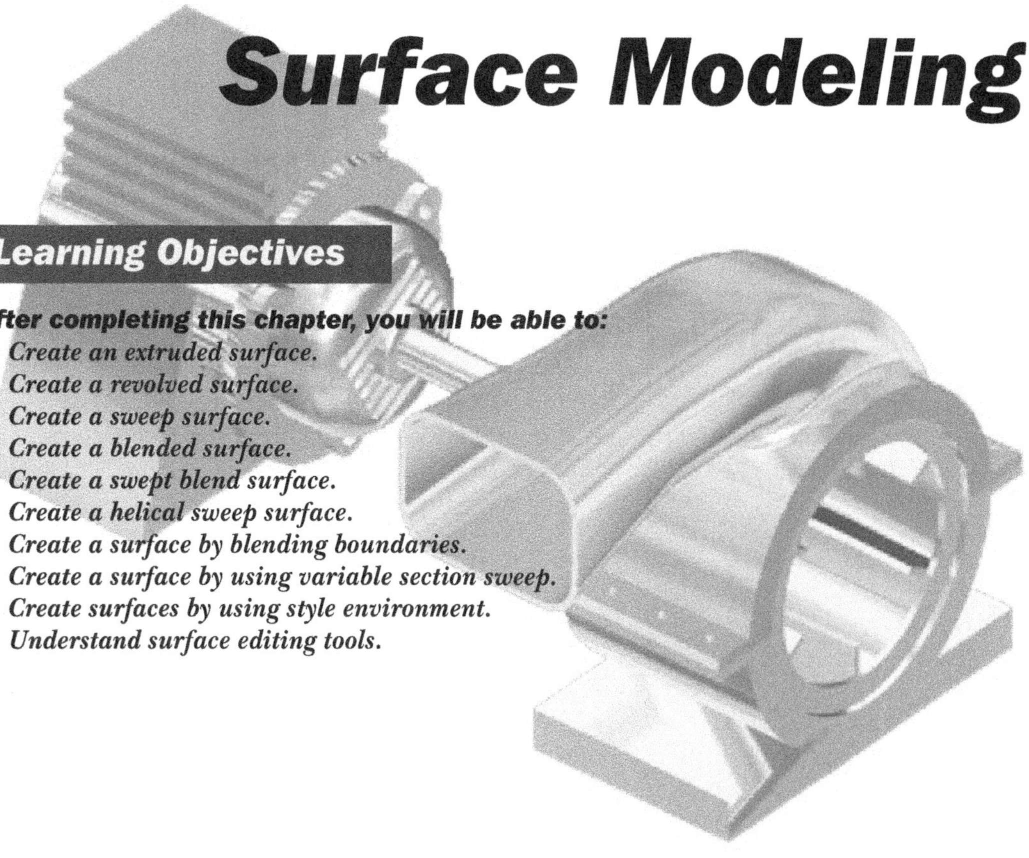

Learning Objectives

After completing this chapter, you will be able to:

- *Create an extruded surface.*
- *Create a revolved surface.*
- *Create a sweep surface.*
- *Create a blended surface.*
- *Create a swept blend surface.*
- *Create a helical sweep surface.*
- *Create a surface by blending boundaries.*
- *Create a surface by using variable section sweep.*
- *Create surfaces by using style environment.*
- *Understand surface editing tools.*

SURFACE MODELING

Surface models are a type of three-dimensional (3D) models with no thickness. These models are widely used in industries like automobile, aerospace, plastic, medical, and so on.

Surface models should not be confused with thick models. The thick models are the models having mass properties. Surface models do not have thickness whereas thick or solid models have a user-defined thickness. In PTC Creo Parametric, the surface modeling techniques and feature creation tools are the same as those used in solid modeling. A solid model created of any shape can also be created using the surface modeling techniques. The only difference between the solid model and the surface model is that the solid model has mass properties but the surface model does not. Sometimes, complex shapes are difficult to create using solid modeling. Such models can be easily created using surface modeling and then the surface model can be converted into the solid model. It becomes easy for a person to learn surface modeling if he is familiar with solid modeling feature creation tools.

CREATING SURFACES IN PTC Creo Parametric

In PTC Creo Parametric, you can toggle between a solid feature and a surface feature while creating a protrusion. The two buttons that are used to toggle between the solid feature and a surface feature are available on dashboards.

Creating an Extruded Surface

Ribbon: Model > Shapes > Extrude

To create an extruded surface, choose the **Extrude** tool from the **Shapes** group; the **Extrude** dashboard will be displayed, as shown in Figure 15-1.

Figure 15-1 *Partial view of the* ***Extrude*** *dashboard*

In this dashboard, the **Extrude as solid** button is chosen by default. Choose the **Extrude as surface** button to extrude the sketch and create a surface model. All attributes of a solid model that were discussed in Chapter 4 are the same for a surface model. Some examples of these attributes are sketching plane, both-side or one-side extrusion, depth of extrusion, and so on.

A surface model can be extruded with capped ends or with open ends. Figure 15-2 shows the open ends surface model and Figure 15-3 shows the capped ends surface model. Remember that to create the capped end surface model, the sketch should be a closed loop. However, a surface can be created with the open sketch.

To create a surface with capped ends, select the **Capped Ends** check box in the **Options** slide-down panel.

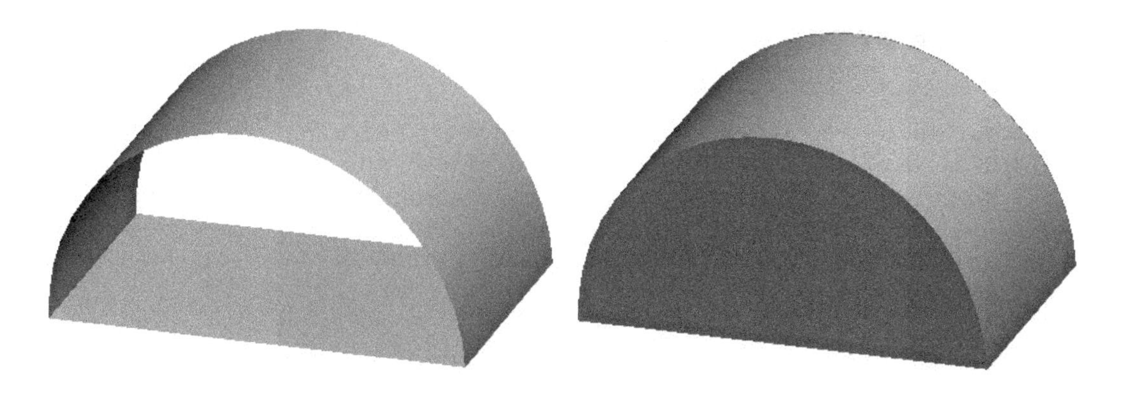

Figure 15-2 *Surface with open ends* *Figure 15-3* *Surface with capped ends*

Creating a Revolved Surface

Ribbon: Model > Shapes > Revolve

To create a revolved surface, choose the **Revolve** tool from the **Shapes** group; the **Revolve** dashboard will be displayed, as shown in Figure 15-4. This feature creation tool works in the same way as in the case of solid modeling.

Figure 15-4 *Partial view of the* ***Revolve*** *dashboard*

In the **Revolve** dashboard, the **Revolve as solid** button is chosen by default. Choose the **Revolve as surface** button to create a revolve surface. You can create a revolved capped end surface or an open end surface. The **Capped Ends** check box in the **Options** slide-down panel is available only when the sketch is closed and the angle of revolution is less than 360 degrees. Figure 15-5 shows the revolved surface with open ends and Figure 15-6 shows the revolved surface with capped ends.

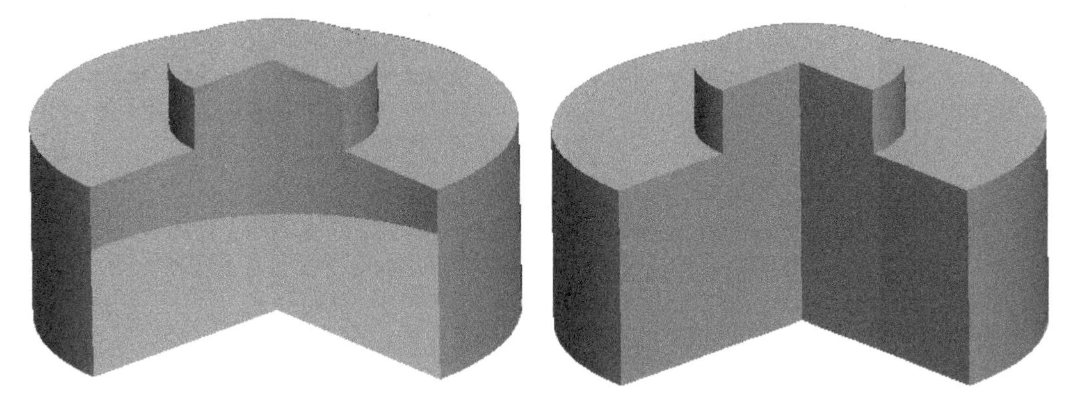

Figure 15-5 *Revolved surface with open ends* *Figure 15-6* *Revolved surface with capped ends*

Creating a Sweep Surface

Ribbon: Model > Shapes > Sweep > Sweep

To create a sweep surface feature, choose the **Sweep** tool from the **Sweep** drop-down in the **Shapes** group of the **Model** tab in the **Ribbon**; the **Sweep** dashboard will be displayed. The method to create a surface sweep feature is the same as that to create a solid sweep feature. To create a solid sweep feature, refer to Chapter 8. The additional option of capping the ends that was available in the **Extrude** and **Revolve** tools is also available in the **Sweep** tool.

Figures 15-7 and 15-8 show the sweep surfaces with the open and capped ends, respectively.

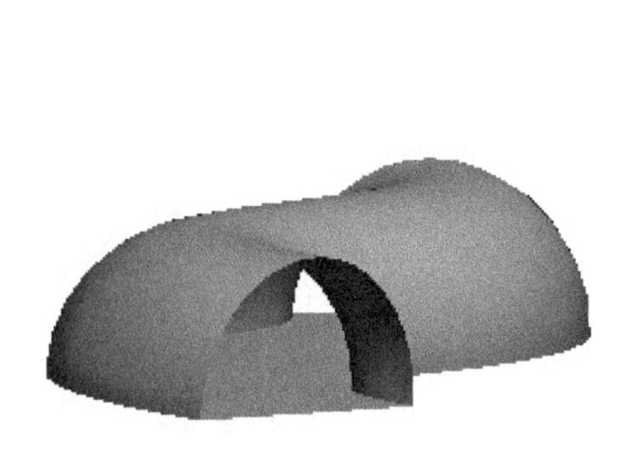

Figure 15-7 Sweep surface with open ends created using a closed sketch

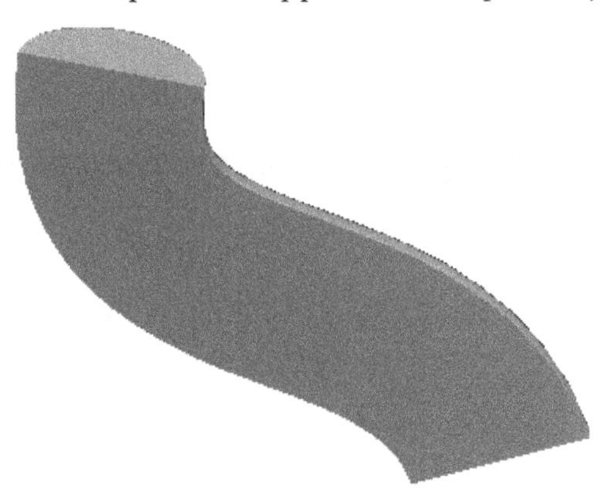

Figure 15-8 Sweep surface with capped ends created using a closed sketch

Creating a Blend Surface

Ribbon: Model > Shapes > Blend

To create a surface blend, choose the **Blend** tool from the expanded **Shapes** group in the **Ribbon**; the **Blend** dashboard will be displayed. In this dashboard, the **Blend as solid** button is chosen by default. Choose the **Blend as surface** button to blend the sketch and create a surface model. The method to create a blend surface is the same as that to create a solid blend. To create a solid blend feature, refer to Chapter 8. Blended surfaces can be with open ends or capped ends. Figure 15-9 shows the blend surface with open ends and Figure 15-10 shows the blend surface with capped ends.

Tip: *If you want to create a surface blend with capped ends, you need to create a closed sketch. PTC Creo Parametric does not accept an open sketch for a capped end blend surface.*

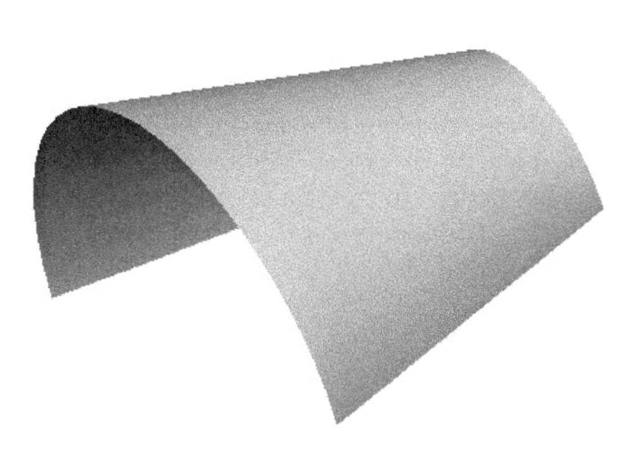

Figure 15-9 Blend surface with open ends

Figure 15-10 Blend surface with capped ends

Creating a Rotational Blend Surface

Ribbon: Model > Shapes > Rotational Blend

To create a rotational blend surface, choose the **Rotational Blend** tool from the expanded **Shapes** group in the **Ribbon**; the **Rotational Blend** dashboard will be displayed. In this dashboard, the **Blend as solid** button is chosen by default. Choose the **Blend as surface** button to create a surface model. The method to create a rotational blend surface is the same as that to create a solid rotational blend. To create a solid rotational blend feature, refer to Chapter 8. Rotational blend surface can be created with open ends or capped ends. Figure 15-11 shows the rotational blend surface with open ends and Figure 15-12 shows the rotational blend surface with capped ends.

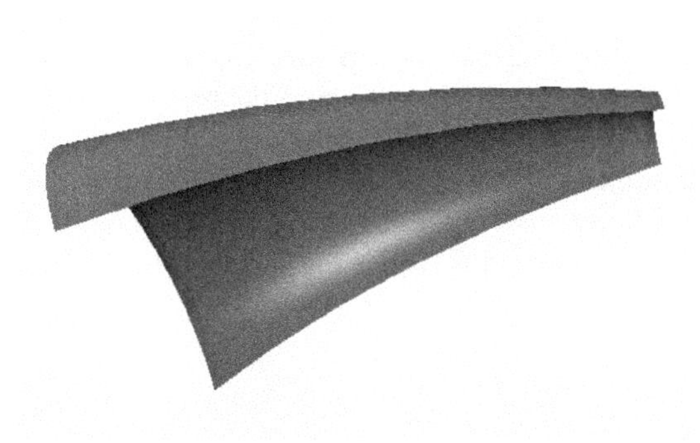

Figure 15-11 Rotational blend surface with open ends

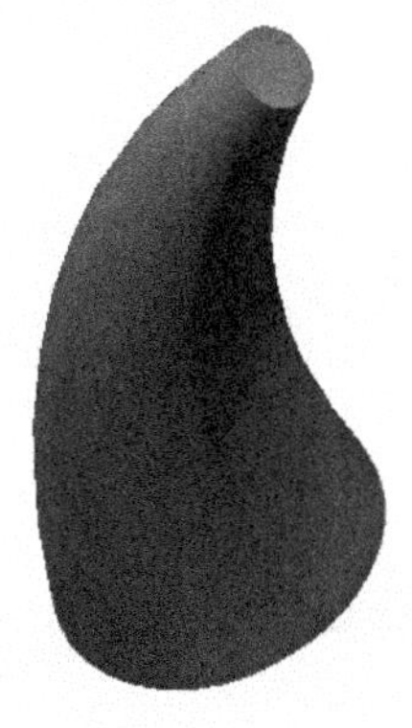

Figure 15-12 Rotational blend surface with capped ends

Creating a Swept Blend Surface

Ribbon: Model > Shapes > Swept Blend

To create a swept blend surface, choose the **Swept Blend** tool from the **Shapes** group in the **Ribbon**; the **Swept Blend** dashboard will be displayed. Choose the **Create a surface** button to create a sweep surface. The method to create a swept blend surface is the same as that to create a solid swept blend feature. To create a solid swept blend feature, refer to Chapter 9. Figure 15-13 shows the swept blend with open ends and Figure 15-14 shows the swept blend with capped ends.

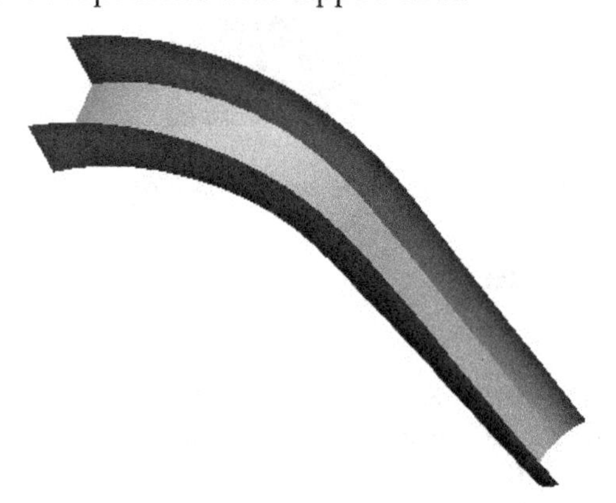

Figure 15-13 Swept blend surface with open ends

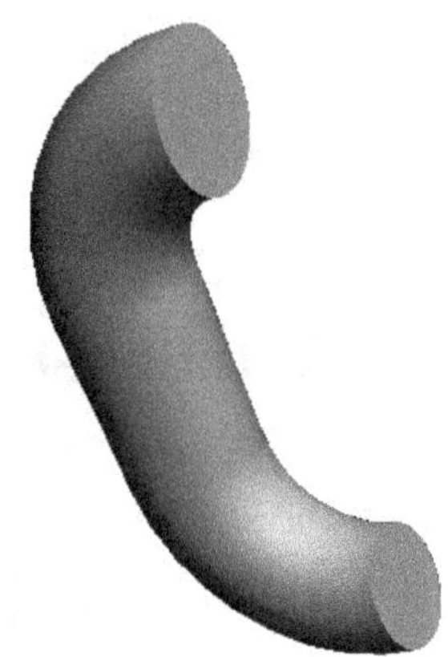

Figure 15-14 Swept blend surface with capped ends

Creating a Helical Sweep Surface

Ribbon: Model > Shapes > Sweep > Helical Sweep

To create a helical sweep surface, choose the **Helical Sweep** tool from the **Sweep** drop-down of the **Shapes** group in the **Ribbon**; the **Helical Sweep** dashboard will be displayed. Choose the **Create a surface** button to create a sweep surface. The method to create a helical sweep surface feature is the same as that to create a solid helical sweep feature. For more information on creating solid helical sweep features, refer to Chapter 9. Figure 15-15 shows the helical sweep surface with open ends and Figure 15-16 shows the helical sweep surface with capped ends.

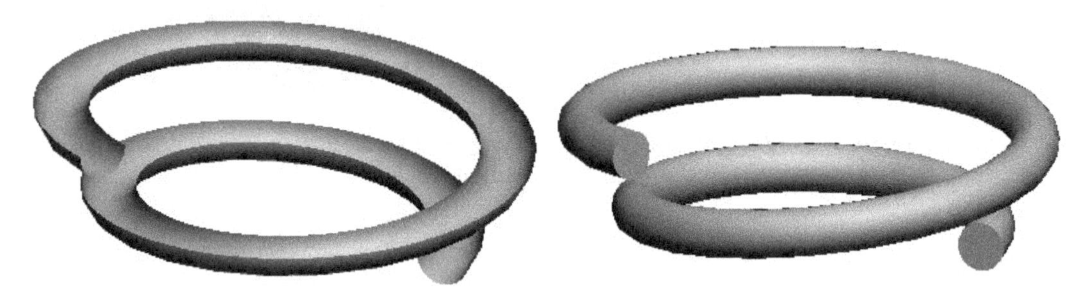

Figure 15-15 Helical sweep surface with open ends created using an open sketch

Figure 15-16 Helical sweep feature with capped ends created using the closed sketch

Creating a Surface by Blending Boundaries

Ribbon: Model > Surfaces > Boundary Blend

To create a surface by blending boundaries, datum curves, or points, choose the **Boundary Blend** tool from the **Surfaces** group in the **Ribbon**; the **Boundary Blend** dashboard will be displayed, as shown in Figure 15-17 and you will be prompted to select two or more curve chains to define a blended surface. The options and buttons in this dashboard are discussed next.

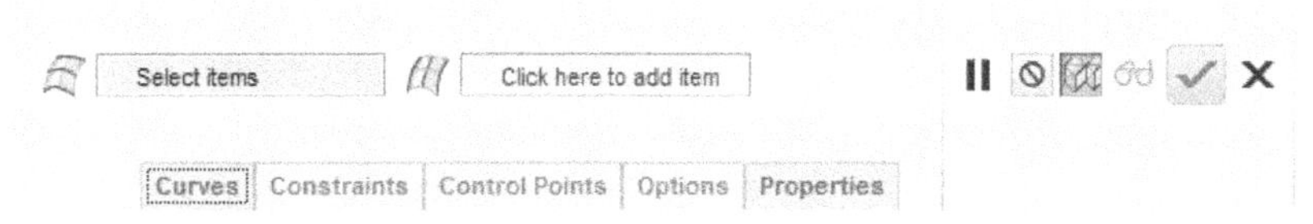

Figure 15-17 *Partial view of the* ***Boundary Blend*** *dashboard*

Curves Tab

When you choose the **Curves** tab, the slide-down panel will be displayed. Choose a curve from the drawing area; the curve will be highlighted in green and the value **0.00** will be displayed on both ends of the curve. When you modify this value, the curve will change accordingly at the corresponding end. Press CTRL+left mouse button to select the second curve; the second curve will also get highlighted, refer to Figure 15-18. These curves will be numbered as per the sequence of selection. The surface created by using these two blending boundaries is shown in Figure 15-19.

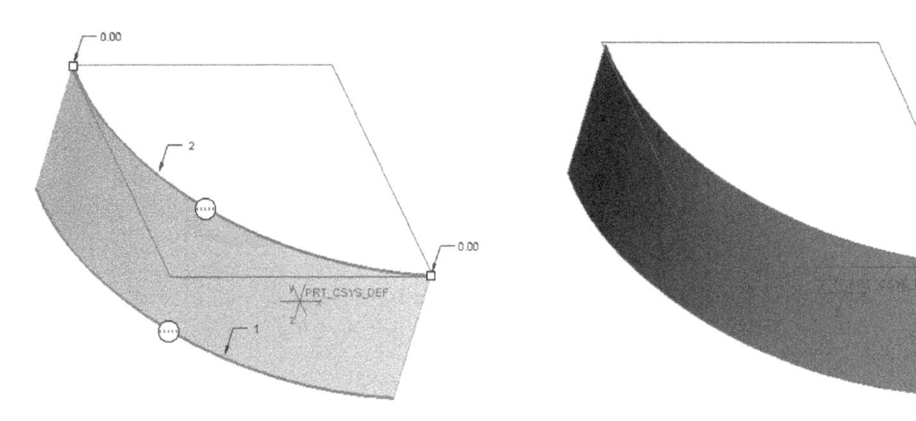

Figure 15-18 *Curves selected as the blending boundary*

Figure 15-19 *Surface created by blending boundaries*

The collector present above the **Curves** tab shows **2 Chains**. The number of curves selected in the first direction will be displayed in this collector.

Now, invoke the **Curves** slide-down panel, refer to Figure 15-20 and select **2 Chain** from the **First direction** collector. The **Move up** and **Move down** buttons available in the slide-down panel are used to change the order of selection of the curves. The **Close blend** check box is used to close the surfaces.

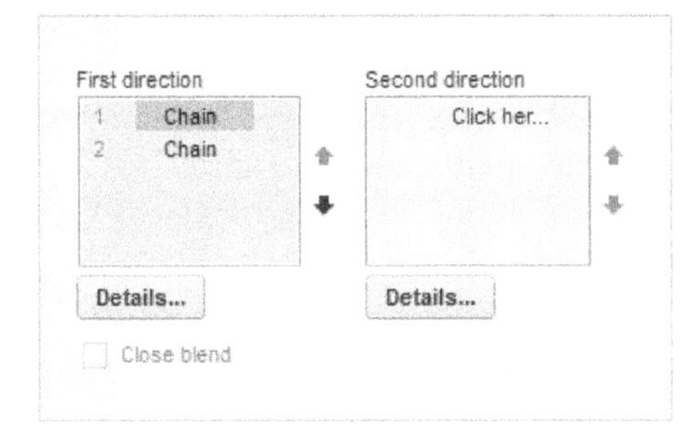

Figure 15-20 *The* ***Curves*** *slide-down panel*

Tip: *To delete the curves from the collector, select the curve in the collector and right-click and choose the* ***Remove all*** *option from the shortcut menu; all the curves available in the collector will be deleted.*

Figure 15-21 shows the surface created by using the four curves and Figure 15-22 shows the surface created by selecting the **Close blend** check box.

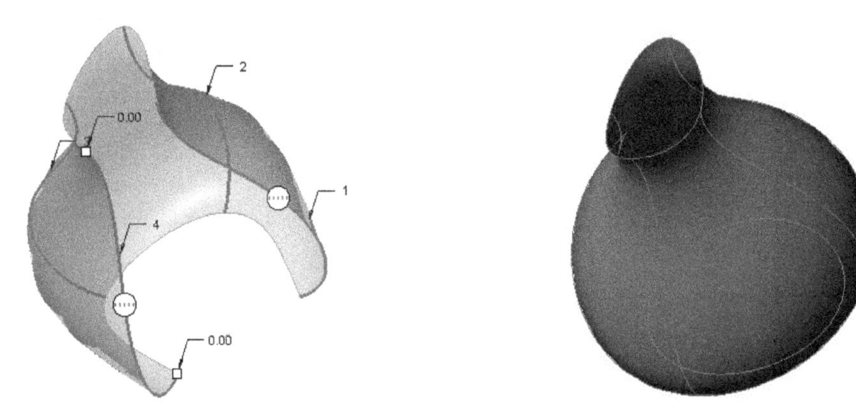

Figure 15-21 *Surface created after selecting the curves*

Figure 15-22 *Surface created by selecting the* ***Close blend*** *check box*

The **Second direction** collector in the **Curves** slide-down panel is used to select curves in the second direction. The second direction curves are usually drawn in a direction other than that of the first direction. Figure 15-23 shows the first and second direction curves and Figure 15-24 shows the surface created after selecting the curves shown in Figure 15-23.

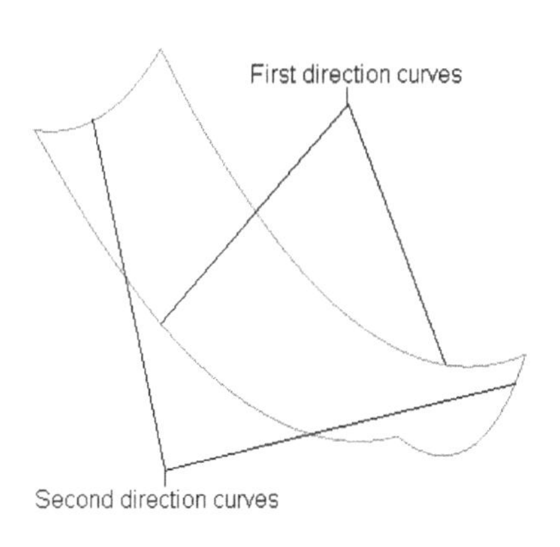

***Figure 15-23** Datum curves*

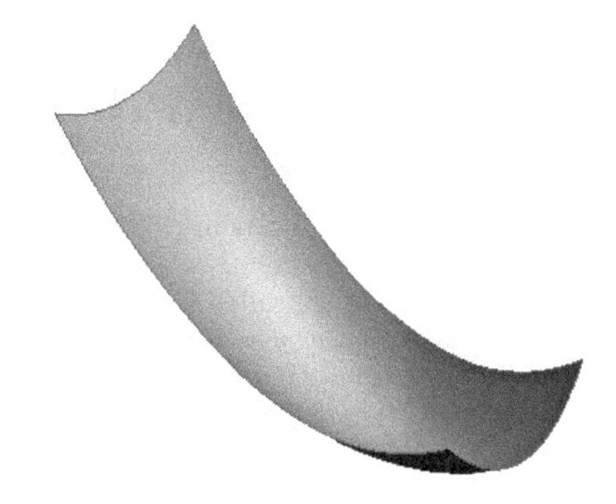

***Figure 15-24** Surface created by using the first and second direction curves*

Creating a Variable Section Sweep Surface by Using the Sweep Tool

Ribbon: Model > Shapes > Sweep

To create the variable section sweep surface, choose the **Sweep** tool from the **Shapes** group in the **Ribbon**; the **Sweep** dashboard will be displayed. Choose the **Sweep as surface** button from the dashboard. To learn more about the variable section sweep, refer to Chapter 9. The procedure of creating a variable section sweep surface is similar to the procedure for creating variable section sweep as was discussed in Chapter 9.

Figure 15-25 shows the section and trajectories used to create the variable section sweep surface. You have an option to keep the ends open or capped. This option is available in the **Options** slide-down panel.

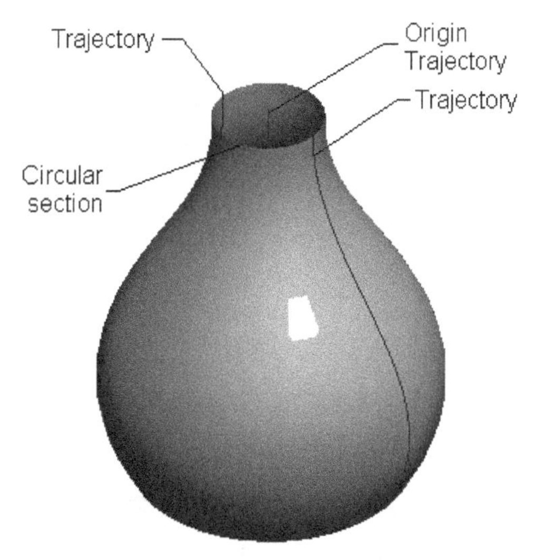

***Figure 15-25** Variable section sweep surface with open ends*

CREATING SURFACES BY USING THE STYLE ENVIRONMENT OF PTC Creo Parametric

Ribbon: Model > Surfaces > Style

Style is an environment available in PTC Creo Parametric and is used to draw free style curves and create surfaces by joining them. The surfaces created using the Style environment are called Super features. This is because these features can contain any number of curves or surfaces. The Style surfaces can be joined with the PTC Creo Parametric surfaces. They can have the parent-child relationship among themselves and also with PTC Creo Parametric features.

To enter the **Style** environment, choose the **Style** tool from the **Surfaces** group. Figure 15-26 shows the appearance of the **Style** environment.

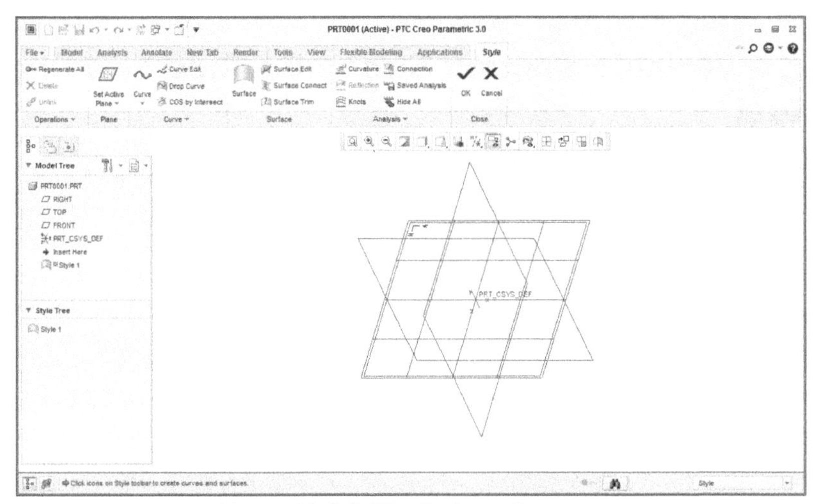

*Figure 15-26 The **Style** environment*

Style Dashboard

Figure 15-27 shows the **Style** dashboard in the **Style** environment. The options in this dashboard are discussed next.

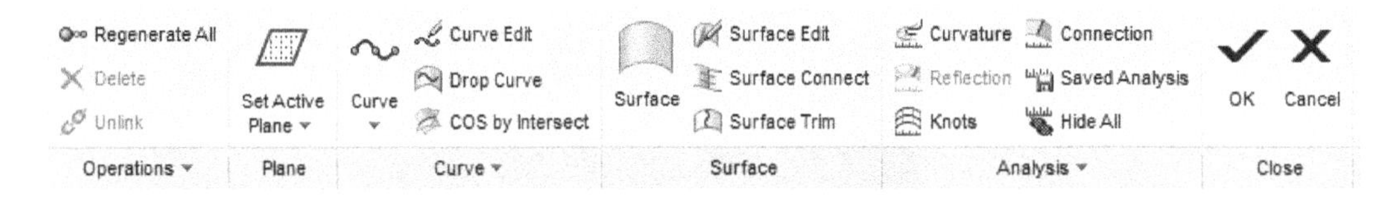

*Figure 15-27 The **Style** dashboard*

Set Active Plane

Ribbon: Style > Plane > Set Active Plane > Set Active Plane

The **Set Active Plane** tool is available in the **Set Active Plane** drop-down in the **Plane** group of the **Style** dashboard in the **Ribbon**. This tool is used to select the datum

plane on which the drawing or the editing operation needs to be performed. The datum plane that you select is highlighted by a mesh. Also, horizontal and vertical directions are shown on the selected plane.

Internal Plane

Ribbon: Style > Plane > Set Active Plane > Internal Plane

The **Internal Plane** tool is available in the **Set Active Plane** drop-down in the **Plane** group of the **Style** dashboard. This tool is used to create an internal datum plane in the **Style** environment. When you choose this tool, the **Datum Plane** dialog box will be displayed. This dialog box is used to create a datum plane in a similar procedure that was discussed in Chapter 5. The datum planes are named as DTM1, DTM2, and so on.

It should be noted that the datum planes created by using this tool will be displayed in the drawing area only when you are in the **Style** environment. Once you exit the **Style** environment, these datum plane become invisible. Any feature created in the **Style** environment will be displayed in the **Style Tree** as a style feature.

Curve

Ribbon: Style > Curve > Curve > Curve

The **Curve** tool is available in the **Curve** drop-down of the **Curve** group in the **Style** dashboard. This tool is used to draw curves. When you choose this tool, the **Style: Curve** dashboard will be displayed, as shown in Figure 15-28. The options in this dashboard are discussed next.

***Figure 15-28** The **Style:Curve** dashboard*

Create a Free Curve

In the **Style:Curve** dashboard, the **Create a Free Curve** button is chosen by default. As a result, you will be prompted to define the points for creating the curve. To create the curve, click on the screen. A green point will be displayed at the location where you have clicked. Now, again click to define the second point of the curve. The two points are joined by a line. When you click to define the location of the third point, you will notice that the curve that you are drawing is defined by a spline. After defining the points, press the middle mouse button to create the curve. While specifying a point, if you press the SHIFT key, then the point is snapped to the entity already present on the screen.

Remember that the curve drawn using the **Create a Free Curve** button will lie on the active datum plane. To draw a 3D curve, you need to snap the point on the existing entity. You can also draw a 3D curve by choosing the **Show All Views** button from the **Graphics** toolbar and then selecting a point each from any of the two different windows. When you choose the **Show All Views** button, the four-window view is displayed. In PTC Creo Parametric, this type of display is called a 4-view display mode. The 4-view display mode shows the top, default, right, and front views. You can select a point in one window and then select the second point in the other window. By specifying points in different windows, the 3D curve can be drawn. To switch back to the single window display mode, choose the **Show All Views** button again.

Tip: *To undo the last operation, choose the **Undo** button from the **Quick Access** toolbar or press CTRL+Z keys.*

Create a Planar Curve

This tool, when selected, allows you to create the curve on the datum plane that is highlighted by the mesh. This datum plane is called the active plane. The active plane can be selected after invoking the **Style: Curve** dashboard by choosing the **Set Active Plane** button from the **Style: Curve** dashboard. The **Set Active Plane** button is available on the right of the dashboard.

Tip: *Using the **Planar** option, you can project a point of an existing entity on the active datum plane. This can be done by selecting the point on the entity using the SHIFT key. The selected point is projected on the active datum plane.*

Create a Curve On Surface

This tool is used to draw curves on surfaces. The points that you define on a surface are constrained to that surface. When you click to define the location of the first point of the curve, the point is placed. Now, this surface will be selected and the points placed hereafter should lie on the same surface. If you click outside this surface, then the point is not placed on the surface. After the curve is drawn, press the middle mouse button. The orange curve is converted to a black curve indicating that the curve is completed. The curve drawn on the surface is the child of the surface.

Edit this curve using control points

If this toggle button is chosen while drawing the curve, then on editing the curve the control points will be displayed.

Proportional Update

If you create a curve, when this check box is selected, then the curve created can be edited proportionally.

Tip: *Using the **Free** option, you can draw a curve on a surface. To draw a curve on a surface, press SHIFT to select a point on the surface. The surface will be highlighted as you select a point on it and then the point will be placed on it. This method of selecting points on a surface can be used to draw curves that join points on two separate surfaces.*

Circle

Ribbon: Style > Curve > Curve > Circle

The **Circle** tool is used to create circles. To create circles, choose **Circle** from the **Curve** drop-down in the **Curve** group of the **Style** dashboard; the **Style: Circle** dashboard will be displayed. Click anywhere in the drawing area; a circle with default radius will be placed at that location. The circle thus created can be free or planar circle. By default, the **Create a Free Curve** button is chosen in the dashboard, and therefore the curve created will be a free curve. You can adjust the radius of the circle by dragging the handle or enter the radius value in the edit box. The circles can also be edited similar to curves. You can also select the **Proportional Update** check box to edit the circle proportionally.

Similarly, you can create arcs by choosing the **Arc** button from the **Curve** drop-down.

Curve Edit

Ribbon: Style > Curve > Curve Edit

The **Curve Edit** button is used to edit the curves that are created as style features. If you choose this button, the **Style: Curve Edit** dashboard will be displayed, as shown in Figure 15-29 and you will be prompted to select a curve. Select the curve; the curve will be displayed in the collector of the **Curve Edit** dashboard.

The options in the **Style: Curve Edit** dashboard are discussed next.

*Figure 15-29 The partial view of **Style: Curve Edit** dashboard*

Curve Collector

When you select a curve to edit, the id of the curve will be displayed in this collector.

Change to a Free Curve

The **Change to a Free Curve** button is used to change the selected curve into a free curve.

Change to a Planar Curve

The **Change to a Planar Curve** button is used to change the selected curve into a planar curve.

Change to a Curve on Surface

If the curve selected for editing is drawn using the **Curve On Surface** option, then the **Change to a Curve on Surface** button will be selected by default.

Proportional Update Check Box

If the curve selected for editing is drawn by selecting the **Proportional Update** check box, then the curve will be edited proportionately with the points.

Edit this curve using control points

If the curve that for editing is drawn using the **Edit this curve using control points** button, then the control points will be displayed on the curve. Using these control points, you can modify the shape of the curve. You can also choose this button to display the control points on a curve drawn without choosing this button.

Tangent

Choose the **Tangent** tab to define the nature of contact between the curve and the adjacent surface. Alternatively, select one of the end control points of the curve, the tangent vector of the curve will be highlighted in green color. Now, right-click on the green vector to display the shortcut menu, as shown in Figure 15-30.

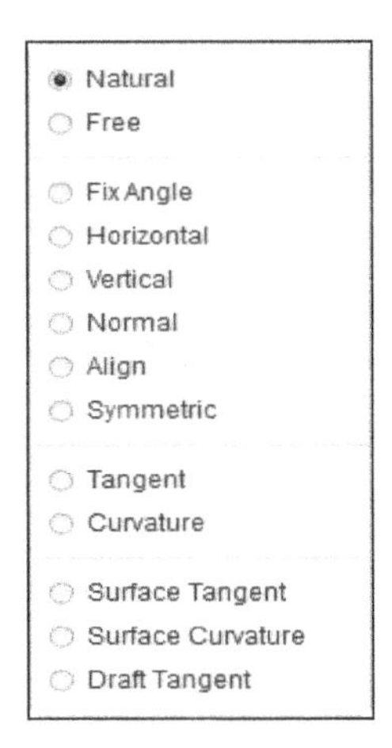

Figure 15-30 *Shortcut menu for the end control point of the curve*

By default, a curve has a natural contact with the adjacent surface. This is evident from the circle mark on the left of the **Natural** option in the shortcut menu. Figure 15-31 shows the curve that is connected to the adjacent surface using the **Natural** option. The curve is drawn using the **Free** option. The point on the cylindrical surface will be selected by using SHIFT+left mouse button and similarly another point will be selected on the surface at the base. Figure 15-32 shows the curve whose contact type is changed to the **Surface Tangent** option by choosing it from the shortcut menu.

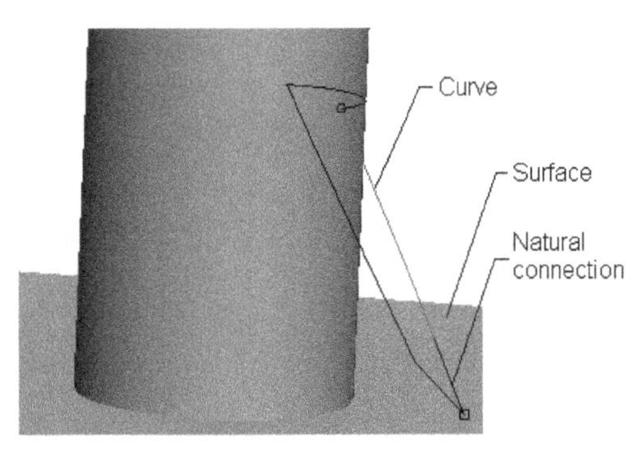

Figure 15-31 *Curve joining the two surfaces*

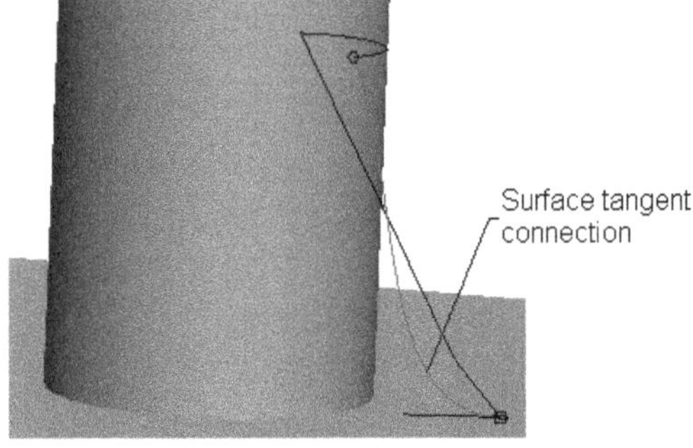

Figure 15-32 *Curve joining the base surface tangentially*

Drop Curve

Ribbon: Style > Curve > Drop Curve

Using the **Drop Curve** tool, a curve created in the **Style** environment can be projected onto the selected surface.

To project a curve on a selected surface, choose the **Drop Curve** tool from the **Curve** group of the **Style** dashboard to create COS (Curve On Surface); you will be prompted to select the curve that you need to drop onto the surface. Select the curve and press the middle mouse button. Now, you will be prompted to select the surface on which you need to drop the curve. Select the surface; you will be prompted to select the plane normal to which the curve will be dropped. Select the plane normal to which the curve will be projected and exit the dashboard.

Curve on Surface by Intersect

Ribbon: Style > Curve > COS by Intersect

The **Curve on Surface by Intersect** tool is used to create a curve at the intersection of two sets of surfaces. To do so, choose **COS By Intersect** from the **Curve** group in the **Style** dashboard; you will be prompted to select the first set of surfaces. Select the required surfaces. Next, click the middle mouse button to accept the selection and then select the second set of surfaces or planes and choose the **Build Feature** button to create the COS at the intersection.

Surface

Ribbon: Style > Surface > Surface

The **Surface** tool is used to create a surface among a closed boundary of curves. Choose the **Surface** tool from the **Surface** group in the **Style** dashboard; the **Style: Surface** dashboard will be displayed and you will be prompted to select boundary curves to define a surface. For example, to create surface using four curves, as shown in Figure 15-33, select these four curves with the CTRL key pressed. After selecting these four curves, press the middle mouse button; the surface will be created, as shown in Figure 15-34.

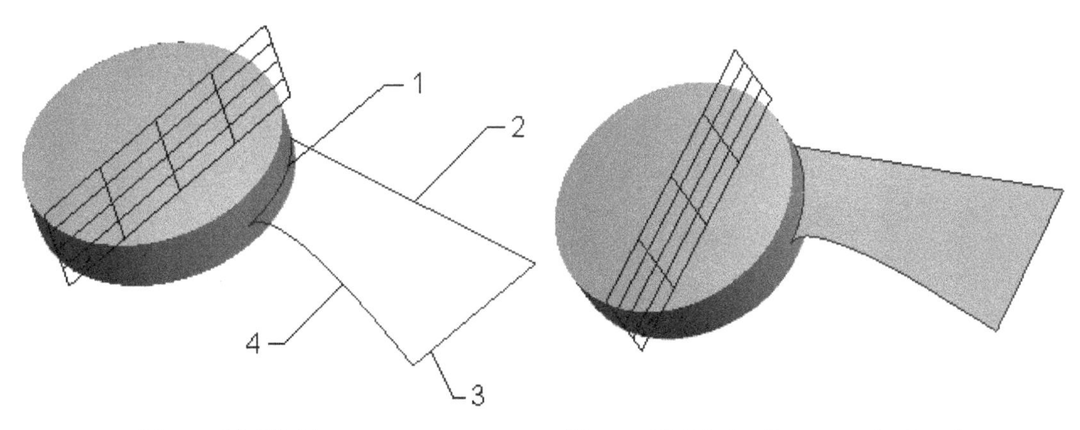

Figure 15-33 *Four curves*

Figure 15-34 *Surface created using the curves*

Surface Connect

Ribbon: Style > Surface > Surface Connect

The **Surface Connect** tool is used to connect two Style surfaces. The Style surface can also be connected to the PTC Creo Parametric surface using this tool. When you choose this tool, the **Style: Surface Connection** dashboard will be displayed and you will be prompted to select two surfaces. By pressing the CTRL key, select the two surfaces, as shown in Figure 15-35, and then press the middle mouse button; the connections will be applied to the two surfaces. These connections may be of two types: curvature and tangent. If the tangent connection is applied, then the arrow will be displayed and if the curvature connection is applied, then a dashed line will be displayed on the surfaces. Figure 15-36 shows the two surfaces where the tangent connection as well as the curvature connection has been applied.

Choose the **Surface Connect** tool from the **Surface** group of the **Style** dashboard; the **Style: Surface Connection** dashboard will be displayed, as shown in Figure 15-37, and you will be prompted to select the two surfaces. Select the surfaces. To apply the connection, click on any one end of the dashed line; the dashed line will be converted to an arrow, indicating that the two surfaces are connected. To remove the connection, use SHIFT+left click on the arrow. Figure 15-38 shows the style surface when it is connected by using curvature connection and Figure 15-39 shows the surface when it is connected tangentially.

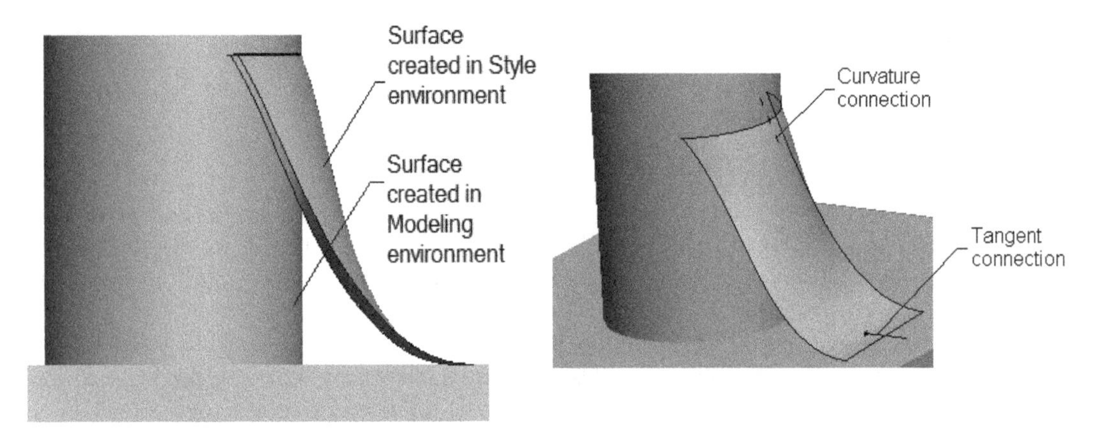

Figure 15-35 *The two surfaces*

Figure 15-36 *Surfaces after applying connections*

Figure 15-37 *The partial view of* ***Style: Surfaces Connection*** *dashboard*

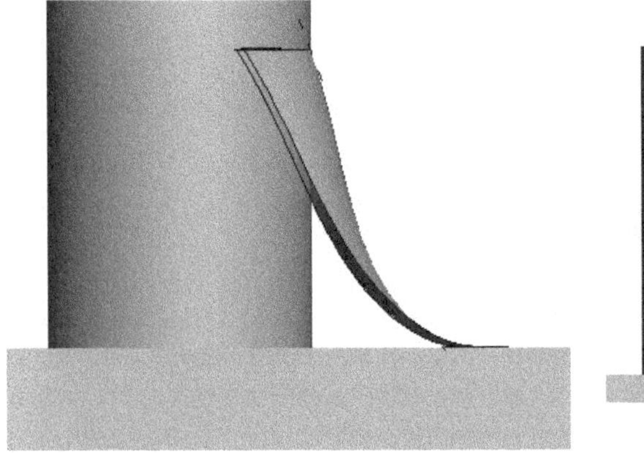

Figure 15-38 *Surface connected at top by using curvature connection*

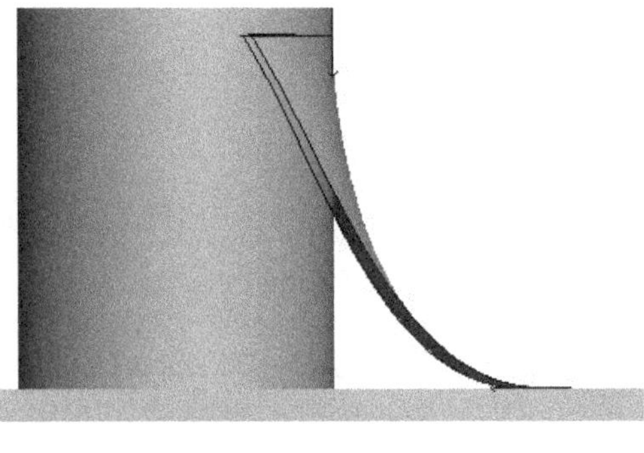

Figure 15-39 *Surface connected at top by using the tangent connection*

Note

To delete a curve, select the curve and when it turns green, press the DELETE key.

Surface Trim

Ribbon: Style > Surface > Surface Trim

The **Surface Trim** tool is used to trim a surface. When you choose this tool, the **Style: Surface Trim** dashboard will be displayed and you will be prompted to select the surface(s) to trim. Select the surface to trim and then press the middle mouse button; you will be prompted to select the curve that can be used to trim the surface. Select the curve and press the middle mouse button. The selected surface will be highlighted in two portions. Select the portion to delete. Choose the **OK** button from the dashboard to exit the trim tool.

Figure 15-40 shows the surface and the curve that are selected for trimming. This figure also show the surface divided into two portions. The portion defined by the curve will be selected to delete. Figure 15-41 shows the surface after trimming.

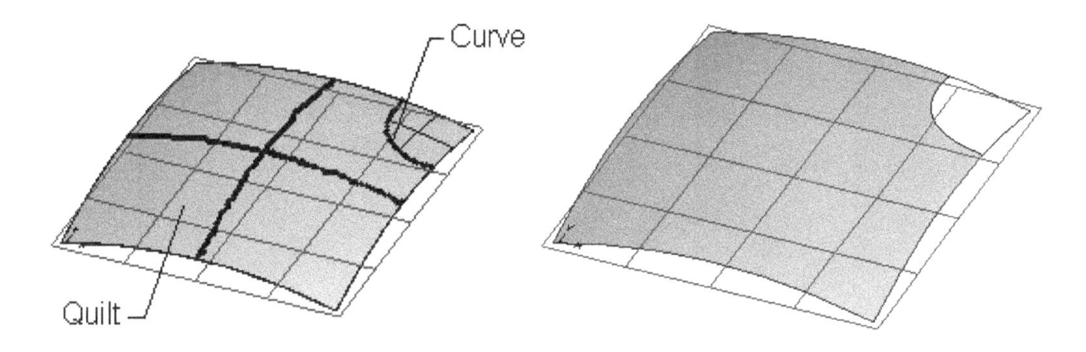

Figure 15-40 *Surface and curve selected for trimming*

Figure 15-41 *Surface after trimming*

Note

*After completing the Style feature creation, choose the **OK** button from the **Close** group in the **Style** dashboard to exit the Style environment.*

SURFACE EDITING TOOLS

The surface editing tools help in decreasing the modeling time. They also help in creating complex surface models. The surface editing tools that you will be learning in the next section are as follows:

1. Mirror
2. Merge
3. Trim
4. Fill
5. Intersect
6. Offset
7. Thicken
8. Solidify
9. Vertex Round

Mirroring the Surfaces

Ribbon: Model > Editing > Mirror

The **Mirror** tool is used to mirror the surface about a plane. This tool is available in the **Editing** group only when a surface is selected. When you choose this tool, the **Mirror** dashboard will be displayed, as shown in Figure 15-42.

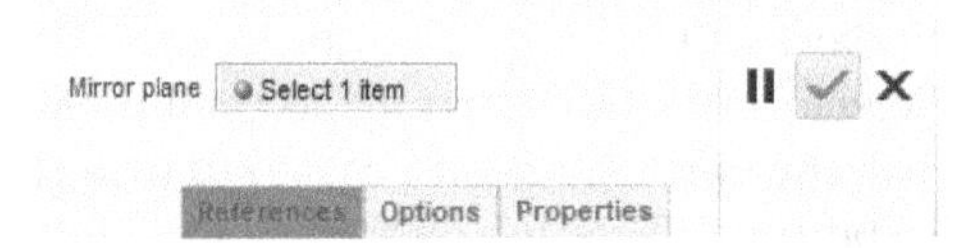

***Figure 15-42** Partial view of the **Mirror** dashboard*

Click in the **Mirror plane** collector, you can choose the mirroring plane from drawing area. The **Dependent Copy** check box in the **Options** tab is selected by default. This makes sure that the parent-child relationship is maintained between the mirrored and original surfaces. Figure 15-43 shows the mirror plane and Figure 15-44 shows the mirrored feature.

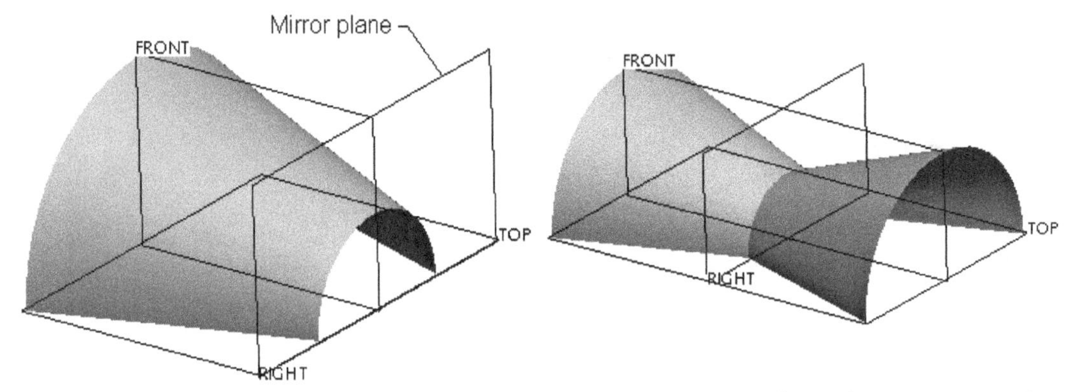

***Figure 15-43** The mirror plane and the surface to be mirrored*

***Figure 15-44** Surfaces after mirroring and keeping the original surface*

Merging the Surfaces

Ribbon: Model > Editing > Merge

The **Merge** tool is used to merge two surfaces and make them a single surface. To convert a surface into a solid, it is necessary that the surfaces are merged. While merging the surfaces, this tool also trims them. This tool will be enabled in the **Editing** group only when the two surfaces to be merged are selected. When you choose the **Merge** tool, the **Merge** dashboard will be displayed, as shown in Figure 15-45.

***Figure 15-45** Partial view of the **Merge** dashboard*

The following steps explain the procedure to merge the surfaces, as shown in Figure 15-46.

1. Select the **Quilts** option from the **Filter** drop-down list. Select the two surfaces and when they are highlighted, choose the **Merge** tool; the **Merge** dashboard will be displayed. While merging the two surfaces, the part of the surfaces that will be retained after the two surfaces are merged is highlighted by dots on it. The arrows point to the direction in which the surfaces are retained. The direction of arrow can be toggled by clicking on them or by using the **Change side of first quilt to keep** and **Change side of second quilt to keep** buttons of the **Merge** dashboard. You can also click on the arrows to flip the direction in which the surfaces are retained.

2. Choose the **Change side of first quilt to keep** button and then choose the **Change side of second quilt to keep** button. Notice that the outer side of the surfaces are highlighted with dots, as shown in Figure 15-47.

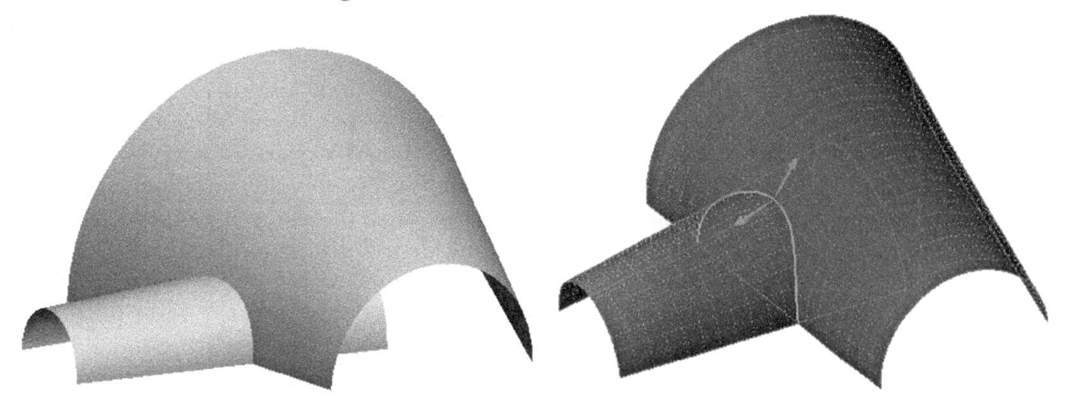

Figure 15-46 *Two surfaces to be merged*

Figure 15-47 *Arrows showing the part of the surface to retain*

3. Choose the **Preview** button to verify the desired trim and then exit the dashboard by choosing the **Build Feature** button. The resulting merged surface is shown in Figure 15-48. This merged surface is a single surface and now can be converted to a solid feature.

The **Reference** tab of the **Merge** dashboard shows the selected quilts. In the **Options** tab, you can select between **Intersect** and **Join** options. The **Join** option can be used when the edge of one quilt lies on the other quilt.

Note

While working with more than two surfaces or quilts, it should be checked that the selection of quilts in the ***References*** *tab is in proper order.*

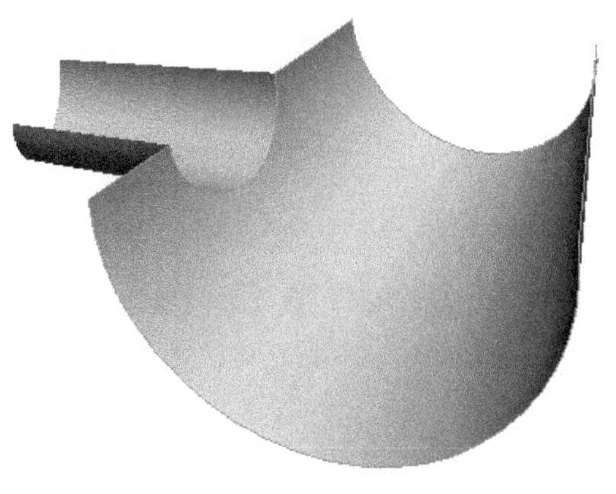

Figure 15-48 *Resulting merged surface*

Trimming the Surfaces

Ribbon: Model > Editing > Trim

As the name suggests, the **Trim** tool is used to trim the selected surfaces by using a trimming object. Select the surface that you need to trim and then choose the **Trim** tool from the **Editing** group; the **Trim** dashboard will be displayed, as shown in Figure 15-49. Also, you will be prompted to select the trimming object. This trimming object can be a curve, plane, edge, or a surface.

***Figure 15-49** Partial view of the **Trim** dashboard*

The part of the surface that is to be retained is highlighted with dots. Also, an arrow is displayed, pointing in the direction of the surface to be retained after trimming. You can choose the **Flip between one side, other side, or both sides of trimmed surface to keep** button from the dashboard or click on the arrow to toggle the direction. By default, the trimming surface is deleted after the surfaces are trimmed. If you need to keep the trimming object, select the **Keep trimming surface** check box from the **Options** slide-down panel. Figure 15-50 shows the surface selected as the trimming object, the trimming surface, and the arrow. It is evident from this figure that the arrow is pointing toward the right; therefore, the right portion of the surface will be retained after trimming. Figure 15-51 shows the surface obtained after trimming.

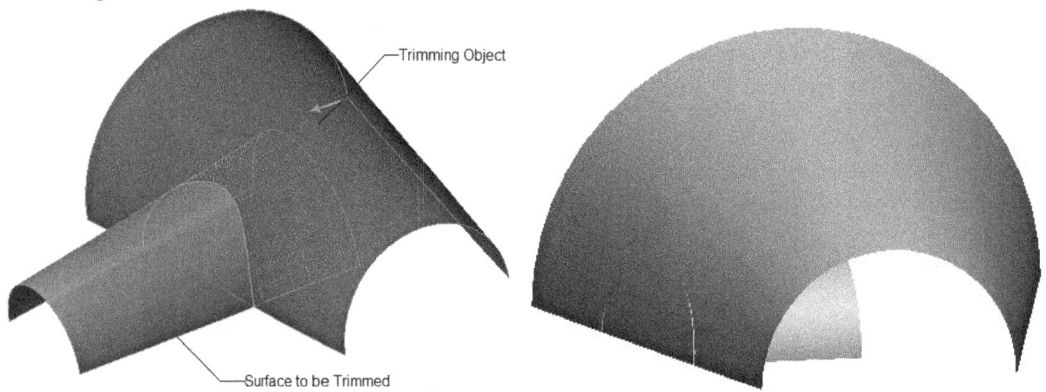

***Figure 15-50** Surface and object to be trimmed* ***Figure 15-51** Surface obtained after trimming*

Creating the Fill Surfaces

Ribbon: Model > Surfaces > Fill

The **Fill** option is used to create a planar surface by sketching its boundaries. When you choose this option from the **Surfaces** group of the **Ribbon**, the **Fill** dashboard will be displayed, as shown in Figure 15-52.

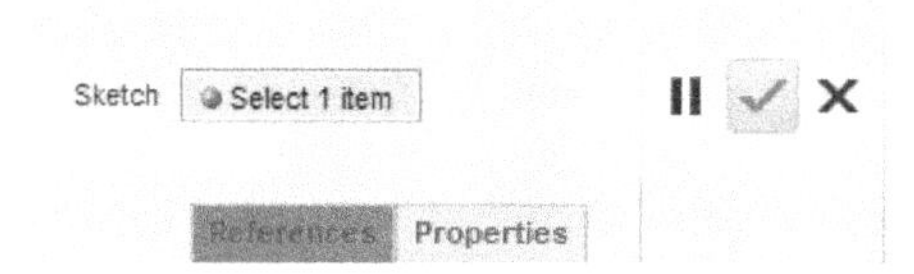

Figure 15-52 *Partial view of the* ***Fill*** *dashboard*

Choose the **References** tab; a slide-down panel will be displayed. Choose the **Define** button from it to select the sketching plane and draw the sketch. Figure 15-53 shows the sketch plane and Figure 15-54 shows the surface that is created by using the **Fill** option.

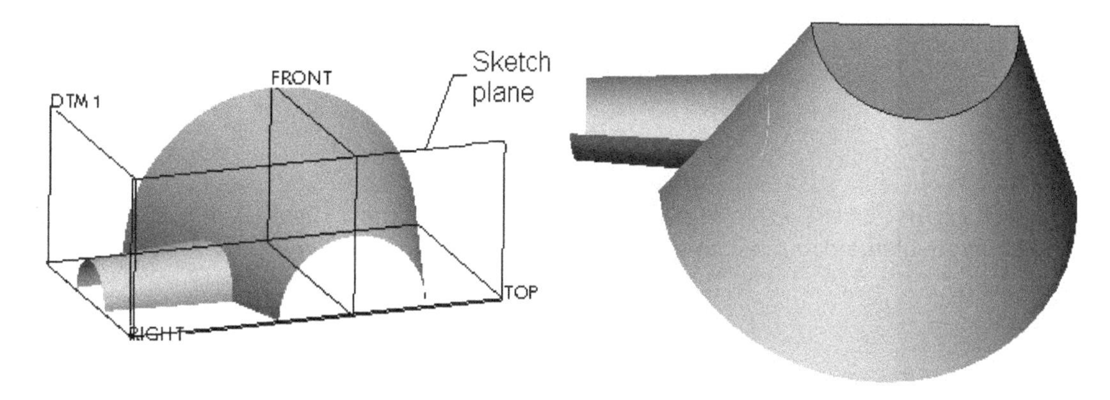

Figure 15-53 *The sketch plane for creating the fill surface*

Figure 15-54 *Fill surface created using the* ***Fill*** *option*

Creating the Intersect Curves

Ribbon: Model > Editing > Intersect

The **Intersect** option is used to create a curve at the intersection of two surfaces. The intersect curve can then be used for various purposes. The **Intersect** option is available in the **Editing** group only when you have selected a surface. When you choose this option from the **Editing** group, the **Surface Intersection** dashboard will be displayed, as shown in Figure 15-55.

Figure 15-55 *Partial view of the* ***Surface Intersection*** *dashboard*

When you select the second surface, the intersecting curve will be created, as shown in Figure 15-56. Make sure to select the second surface by holding the CTRL key. The curve created can be copied, moved, and so on. One of the uses of the intersect curve is shown in Figures 15-57 and 15-58.

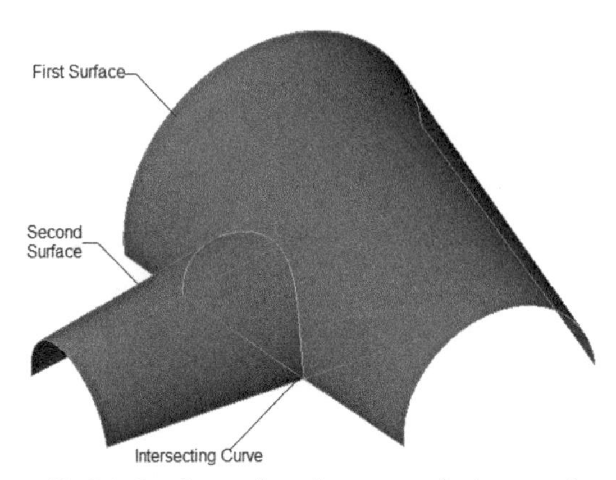

Figure 15-56 *Surfaces selected to create the intersecting curve*

In Figure 15-57, the intersecting curve is copied at a distance of 150. The **Boundary Blend** tool is used to create the surface shown in Figure 15-58. To create the boundary blend, the intersecting curve will be selected and then the curve edge of the surface will be selected. Both the curves are blended and the tangency is increased by dragging the handles.

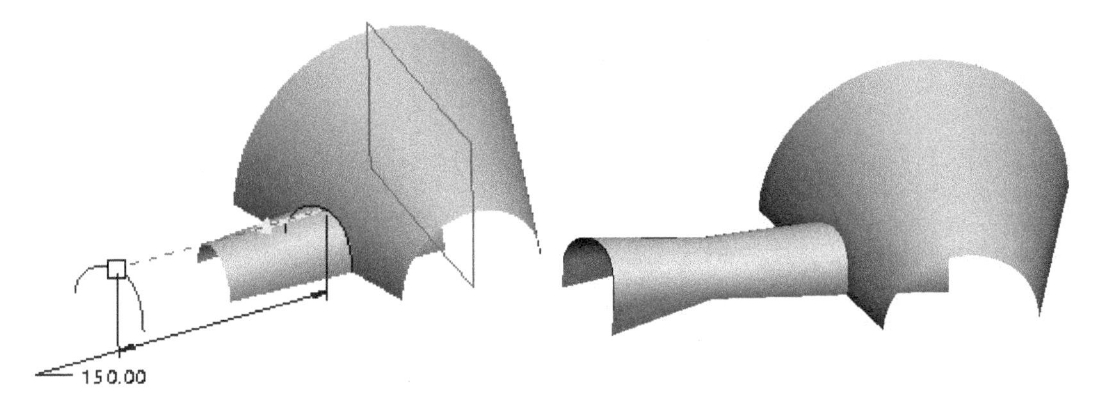

Figure 15-57 *Copied curve*

Figure 15-58 *Boundary blend created using the intersecting curve*

Creating the Offset Surfaces

Ribbon: Model > Editing > Offset

A surface can be copied to an offset distance. To offset a surface, select the surface to offset and then choose the **Offset** tool from the **Editing** group in the **Ribbon**; the **Offset** dashboard will be displayed, as shown in Figure 15-59. The **Offset** tool will be available only when you select a surface to offset.

In PTC Creo Parametric, there are three methods to offset a surface. These methods are as follows:

1. Create the offset of the whole surface using the **Standard Offset Feature** option.
2. Sketch a section and offset the area inside the section with the draft using the **With Draft Feature** option.
3. Sketch a section and offset the area inside or outside the section using the **Expand Feature** option.

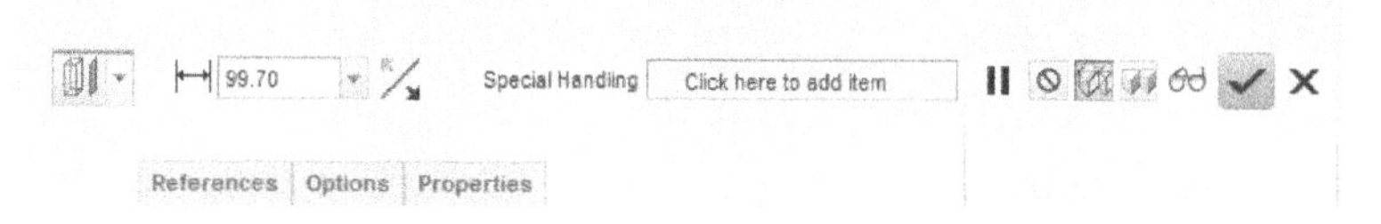

Figure 15-59 *Partial view of the* ***Offset*** *dashboard*

In the **Offset** dashboard, first you need to specify the type of offset surface you need to create. The types of offset that can be created in PTC Creo Parametric are as follows: Standard Offset Feature, With Draft Feature, Expand Feature, and Replace Surface Feature.

Standard Offset Feature

The **Standard Offset Feature** tool is present at the upper-left corner of the **Offset** dashboard and is chosen by default. You can enter the offset value in the dimension edit box.

This option can be used to offset the surface as a whole. From the drop-down list in the **Options** slide-down panel, the **Normal to Surface** option is selected by default. This option allows you to offset a surface normal to the surface. If you select the **Automatic Fit** option from the drop-down list, then PTC Creo Parametric automatically fits the surface, or controls the direction of the offset in the X, Y, and Z axes. If you select the **Controlled Fit** option from the drop-down list in the **Options** slide-down panel, you need to select a coordinate system and specify the direction to offset. You can also join the offset surface with the side surfaces by selecting the **Create side surface** check box. Figure 15-60 shows the original surface and the offset surface.

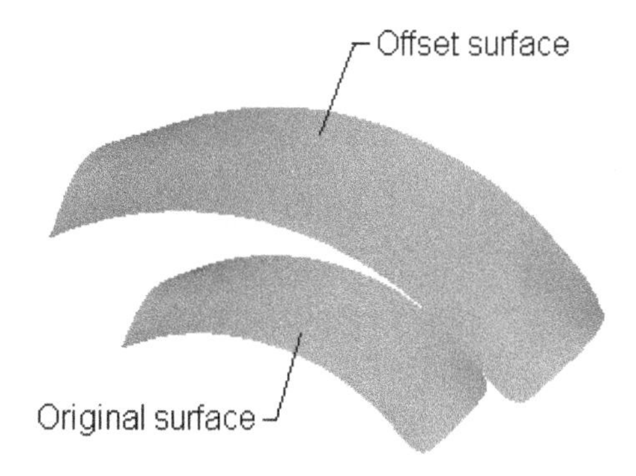

Figure 15-60 *The original and offset surfaces*

With Draft Feature

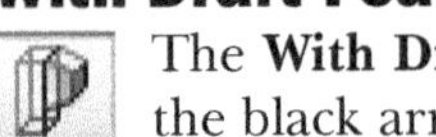

The **With Draft Feature** tool is available in the drop-down that appears by selecting the black arrow present on the right of the **Standard Offset Feature** tool. Using this tool, you can sketch the section and then give a draft angle to the side surfaces. Choose the **Define** button from the **References** slide-down panel to define a sketching plane and to create the sketch of the draft surface. Figure 15-61 shows the draft offset surface with the **Straight** radio button selected from the **Options** slide-down panel. The section drawn on the sketching plane is circular. Similarly, Figure 15-62 shows the draft offset surface with the **Tangent** radio button selected from the **Options** slide-down panel.

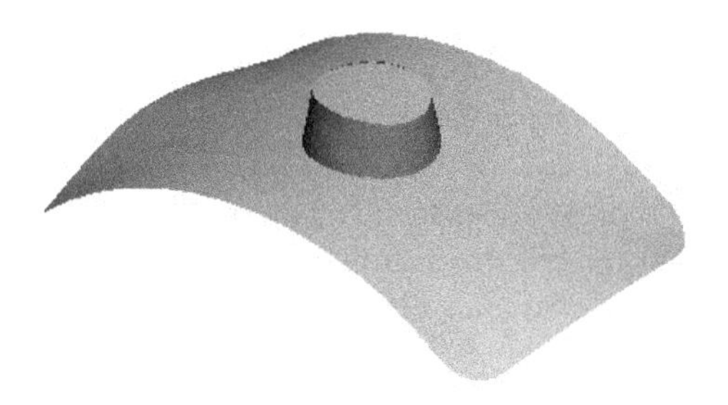

Figure 15-61 *Draft offset surface with straight profile*

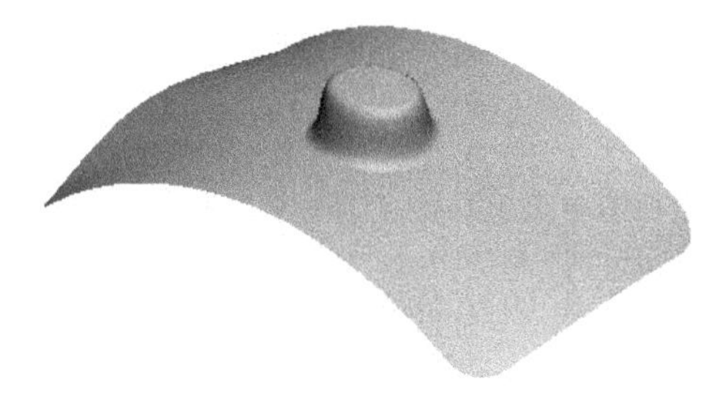

Figure 15-62 *Draft offset surface with tangent profile*

Expand Feature

The **Expand Feature** tool can be chosen from the list of tools that appear by selecting the down arrow present on the right of **Standard Offset Feature** tool. Using this option, you can sketch the section and then choose whether to offset the inside or the outside of the sketch. For this purpose, you need to choose the **Change direction of offset to the other side** button from the dashboard. Alternatively, click on the arrow displayed on the model. Figure 15-63 shows the offset surface when the inside of the sketch is selected to offset. The section that was drawn on the sketching plane was rectangular. Choose the **Define** button from the **Options** slide-down panel to define the sketching plane and create the sketch. Figure 15-64 shows the draft offset surface when the outside of the sketch is selected to offset.

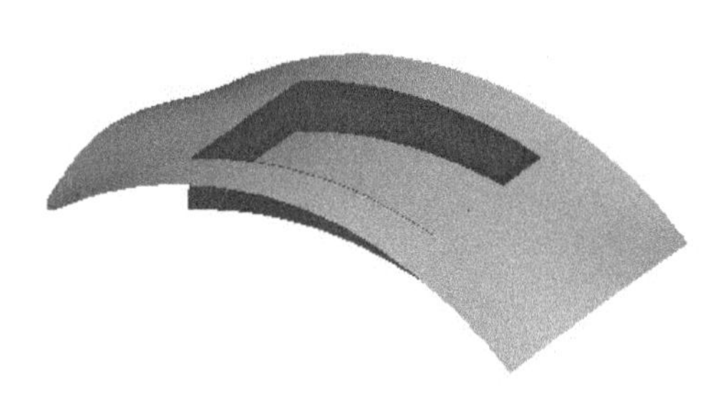

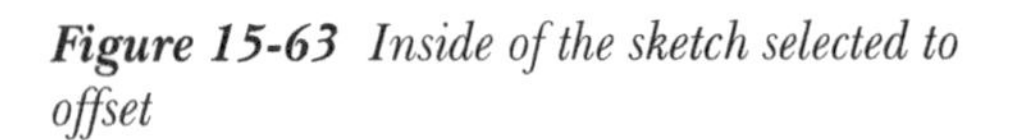

Figure 15-63 *Inside of the sketch selected to offset*

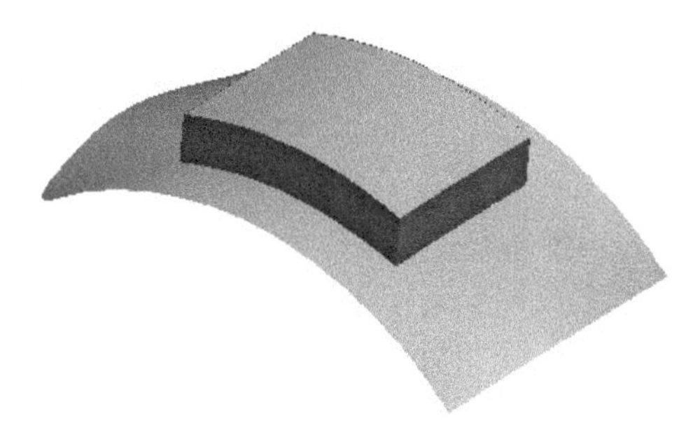

Figure 15-64 *Outside of the sketch selected to offset*

Replace Surface Feature

The **Replace Surface Feature** tool is used to replace a specified face of a solid with a quilt or datum plane. It can add or remove material from the solid. To create a replacement surface, select the face to be replaced and then choose the **Offset** tool from the **Editing** group of the **Ribbon**; the **Offset** dashboard will be displayed. Choose the down arrow present on the right of the **Standard Offset Feature** tool; a list of tools will be displayed. Choose the **Replace Surface Feature** button; you will be prompted to select the surface or datum plane to replace. Select the replacement surface; the preview of the modified solid feature will be displayed. Choose the **Build Feature** button to accept the preview and create the feature. Figure 15-65 shows the face of the solid selected to be offset and the replacement surface. Figure 15-66 shows the preview of the modified solid feature.

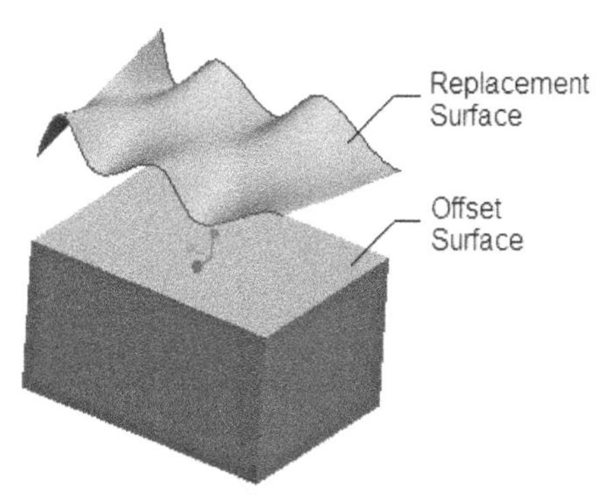

Figure 15-65 *The offset and replacement surfaces*

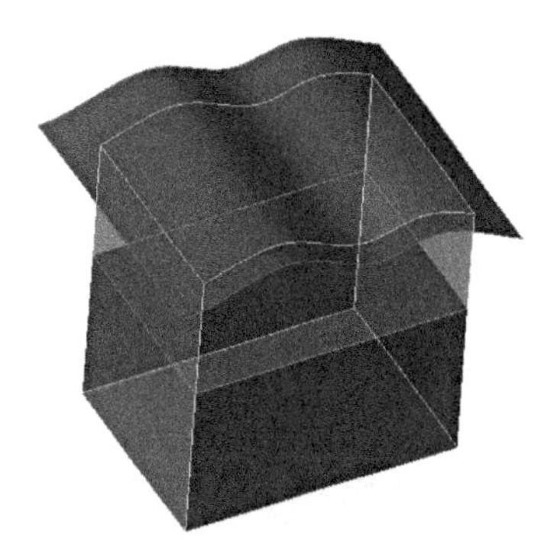

Figure 15-66 *Preview of the modified solid feature*

Adding Thickness to a Surface

Ribbon: Model > Editing > Thicken

To add thickness to a quilt or a surface, select the quilt and choose the **Thicken** option from the **Editing** group; the **Thicken** dashboard will be displayed, as shown in Figure 15-67.

Figure 15-67 *Partial view of the* ***Thicken*** *dashboard*

Drag the handle to set thickness of the quilt or enter the thickness value in the dimension edit box. You can even remove material from the quilt by choosing the **Removes material from inside thickened quilt** button from the dashboard.

Using the drop-down list in the **Options** slide-down panel, you can give thickness to the quilt normal to the surface. PTC Creo Parametric automatically scale the surface along the axes and fit the original surface with respect to the coordinate system. If you select the **Controlled Fit** option from the drop-down list in the **Options** slide-down panel, you need to select a coordinate system and specify the direction to scale.

Figures 15-68 and 15-69 show the surfaces after adding thickness by using the **Normal to surface** and **Automatic fit** options, respectively.

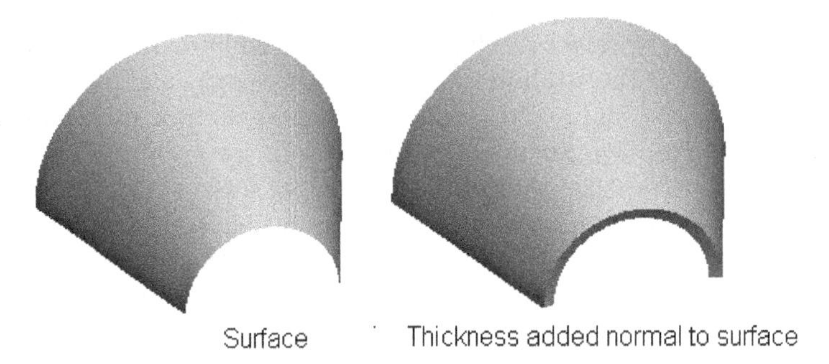

Figure 15-68 *Thickening the surface by using the* ***Normal to surface*** *option*

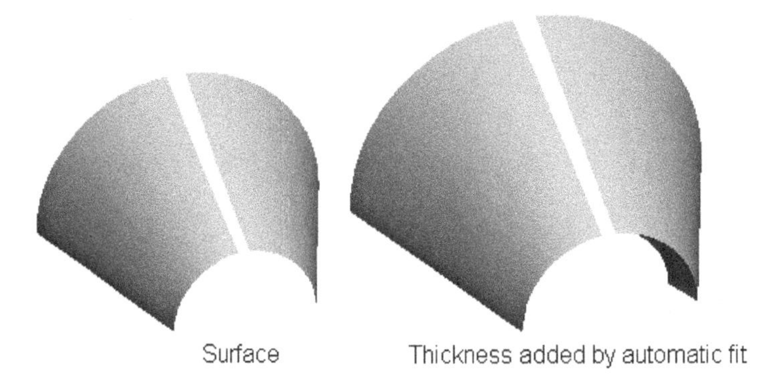

Figure 15-69 *Thickening the surface using the* ***Automatic fit*** *option*

Converting a Surface into a Solid

Ribbon: Model > Editing > Solidify

You can convert a closed surface into a solid by choosing **Solidify** from the **Editing** group of the **Ribbon**. This option is available only when a closed surface with capped ends is selected. This option fills the hollow surface with material.

Creating a Round at the Vertex of a Surface

Ribbon: Model > Surfaces > Vertex Round

The vertices of a surface or quilt can be rounded by using the **Vertex Round** option. To do so, choose **Vertex Round** from the **Surfaces** group of the **Ribbon**; the **Vertex Round** dashboard will be displayed, as shown in Figure 15-70. Also, you will be prompted to select a vertex to add round.

Figure 15-70 *Partial view of the* ***Vertex Round*** *dashboard*

Select the corner vertex(s) to be rounded. Select the first vertex and then press the CTRL key to select the other vertex, as shown in Figure 15-71. After selecting the vertices, enter the radius value in edit box available in the dashboard. Choose the **Build** button from the **Vertex Round** dashboard; the vertices will be rounded, as shown in Figure 15-72.

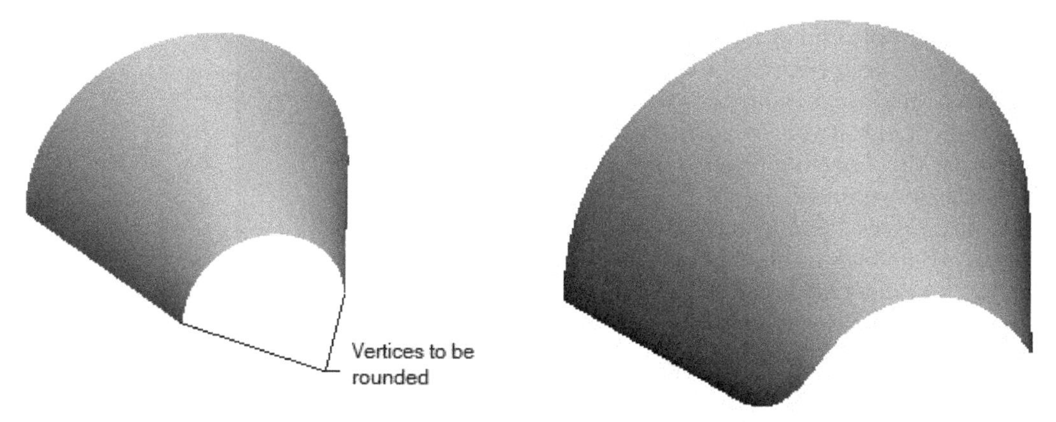

Figure 15-71 *Vertices selected to be rounded*

Figure 15-72 *Vertices after creating round*

FREESTYLE MODELING ENVIRONMENT

Ribbon: Model > Surfaces > Freestyle

Freestyle

The Freestyle Modeling is a surfacing environment which encompasses qualities of both the NURBS and polygon surfaces. NURBS surfaces are smooth surfaces controlled by CV points and the polygon surfaces give the flexibility to extrude a specific area of the surface for detailing. Features created in the Freestyle Modeling environment are called Freestyle features. These features can be created at the beginning of modeling in PTC Creo Parametric. You can invoke the Freestyle Modeling environment by choosing the **Freestyle** tool from the **Surfaces** group of the **Model** tab in the **Ribbon**.

Freestyle Dashboard

When you invoke the **Freestyle** tool from the **Surfaces** group of the **Model** tab in the **Ribbon**, the **Freestyle** dashboard is displayed, as shown in Figure 15-73. The options in this dashboard are discussed next.

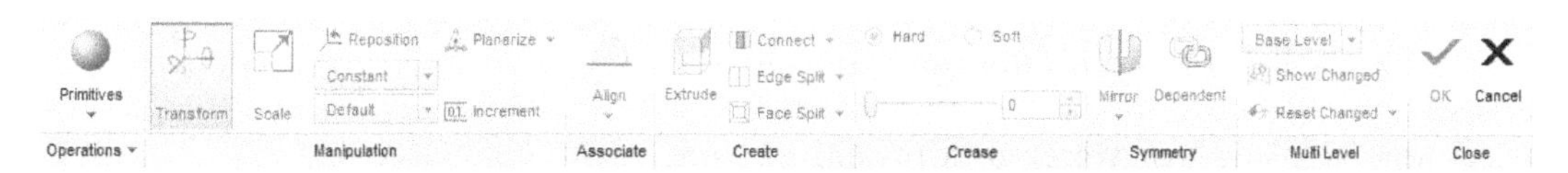

Figure 15-73 *Partial view of the* ***Freestyle*** *dashboard*

Primitives

A primitive is the first feature required to proceed further to create a model in the Freestyle Modeling environment. All operations available in the Freestyle Modeling environment are performed on the primitive. Primitives are available in the **Primitives** drop-down of the **Operations** group in the **Freestyle** dashboard, refer to Figure 15-74. It is important to note that you can create only one primitive at a time in the Freestyle Modeling environment. If you want to create more primitives then you need to invoke the Freestyle environment again. There are two categories of primitives available in the **Primitives** drop-down: **Open Primitives** and **Closed Primitives**. The **Open Primitives** category contains two-dimensional primitives that are created in the XY plane. The **Closed Primitives** category contains three-dimensional primitives which are created with coordinate system at its center. To create a primitive, click on the desired primitive icon; the primitive will be created in the drawing area. You can change the reference coordinate system by using the **Options** dialog box. This dialog box is displayed when you choose the **Options** button from the **Primitives** drop-down, refer to Figure 15-75.

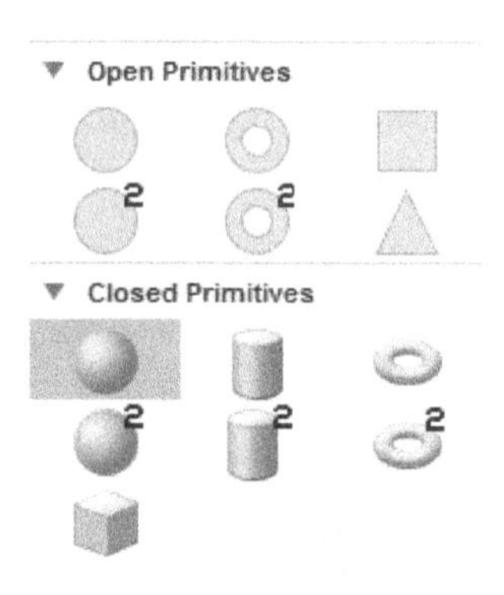

*Figure 15-74 The **Primitives** drop-down*

To change the coordinate system, click in the **Reference CSYS** collector in the **References** area and select the desired reference coordinate system. You can also change the increment value of the 3D Dragger in the linear or angular direction.

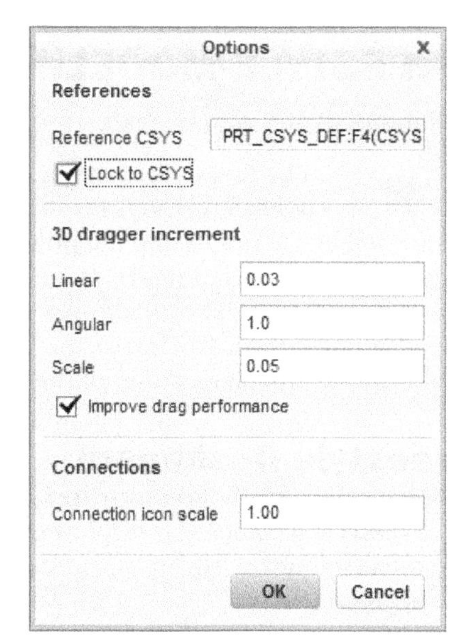

*Figure 15-75 The **Options** dialog box*

Transform

The **Transform** tool in the **Manipulation** group of the **Freestyle** dashboard is used to transform the primitive. This tool will be activated only when you select any face, edge, or vertex of the primitive. Figure 15-76 shows a rectangle primitive with a vertex selected. Figure 15-77 shows the deformed primitive after dragging the 3D Dragger along the Y direction. You can rotate and drag the 3D Dragger along any of the direction available.

Scale

The **Scale** tool in the **Manipulation** group of the **Freestyle** dashboard will be activated only when you select any face, edge, or vertex of a primitive. Note that the scaling operation cannot be performed on a vertex. To apply scaling, you need to select an edge or a face. When you select an edge or a face and choose the **Scale** tool from the dashboard, the 3D Dragger is displayed without the circular handles. Select any of the axes or planes of the 3D Dragger to scale the primitive.

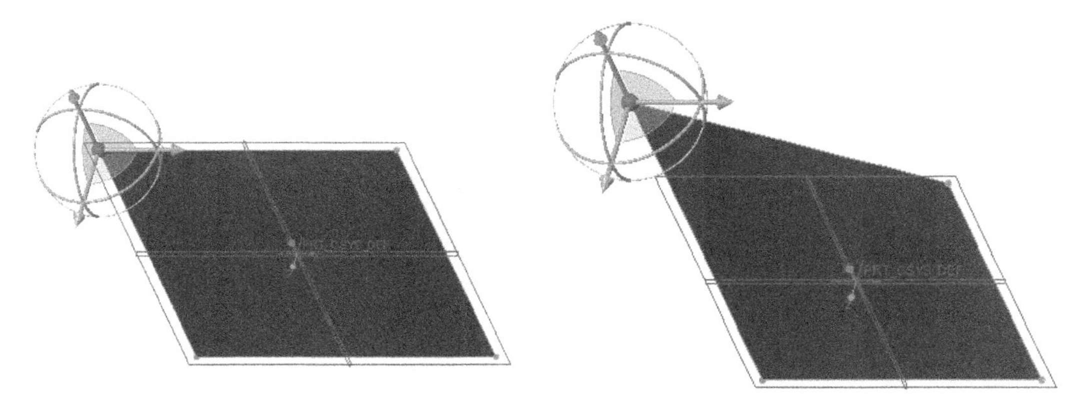

Figure 15-76 A rectangle primitive with a vertex selected

Figure 15-77 The deformed primitive after dragging the 3D Dragger along Y direction

Reposition

The **Reposition** tool is activated in the **Manipulation** group of the **Freestyle** dashboard only when you select any face, edge, or vertex of a primitive. Using the **Reposition** tool, you can reposition the 3D Dragger by rotating or translating it along any of the available axes.

Align

Using the **Align** tool, you can align the primitives along any of the datum planes or planar surfaces. This tool will be activated in the **Manipulation** group of the **Freestyle** dashboard only when you select any face or edge of a primitive. To align a primitive along a datum plane, select the primitive, choose the **Align** tool from the dashboard, and then select the datum plane along which you want the primitive to be aligned.

Planarize

The **Planarize** tool is used to make all the selected vertices lie in a single plane. This tool will be activated in the **Manipulation** group of the **Freestyle** dashboard only when you select any face, edge, or vertex of a primitive. When you invoke this tool after selecting the primitive, a default plane is displayed in which the vertices will lie. You can press the middle mouse button to accept the default plane or you can select a plane along which the vertices will be planarized.

Increment

The **Increment** tool is activated in the **Manipulation** group of the **Freestyle** dashboard only when you select any face, edge, or vertex of a primitive. When the **Increment** toggle button is selected, the increment value will also be displayed during the manipulation of the primitive.

Extrude

The **Extrude** tool is available in the **Create** group of the **Freestyle** dashboard and will be activated only when you select any face or edge of a primitive. The **Extrude** tool available in the Freestyle environment works in the same way as in the Modeling environment. When you select any face and then choose the **Extrude** tool from the dashboard, all the edges of the selected face are converted into walls. Figure 15-78 shows the extruded feature of a rectangle primitive.

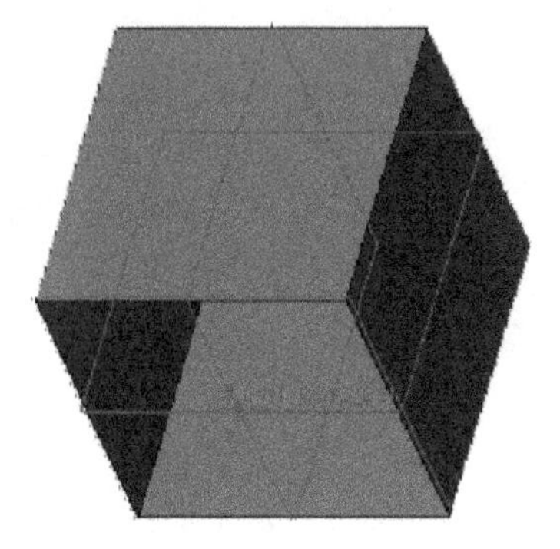

Figure 15-78 The extruded feature of a rectangle primitive

Edge Split

The **Edge Split** drop-down list is available in the **Create** group of the **Freestyle** dashboard and will be activated only when you select an edge of a primitive. After selecting an edge, click on the down arrow at the right of the **Edge Split** button, a drop-down list will be displayed, as shown in Figure 15-79. Now, select the desired option, the selected edge will split into desired number of equal pieces.

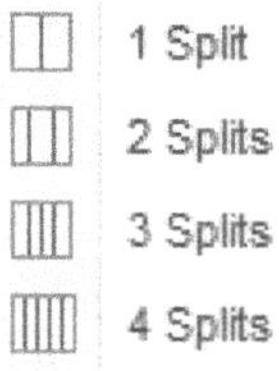

1 Split
2 Splits
3 Splits
4 Splits

***Figure 15-79** The **Edge Split** drop-down list*

Face Split

The **Face Split** drop-down list is available in the **Create** group of the **Freestyle** dashboard and will be activated only when you select a face of a primitive. After selecting the face, click on the down arrow at the right of the **Face Split** button, a drop-down list will be displayed, as shown in Figure 15-80. Select the desired option from the menu; the selected face will split into five pieces of selected offset percentage.

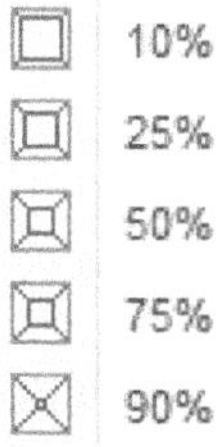

10%
25%
50%
75%
90%

***Figure 15-80** The **Face Split** drop-down list*

Connect

The **Connect** tool is used to join two or more edges. This tool is available in the **Create** group of the **Freestyle** dashboard and will be activated only when you select two or more faces or edges of a primitive. Note that you can select more than one edge or face by holding the CTRL key while selecting the entities. After selecting two or more faces or edges, choose the **Connect** tool from the **Create** group of the dashboard; the faces or edges will be joined together and the resultant surface will be displayed. Figure 15-81 shows a deformed rectangle with two edges to be selected for connecting. Figure 15-82 shows the output after connecting the edges.

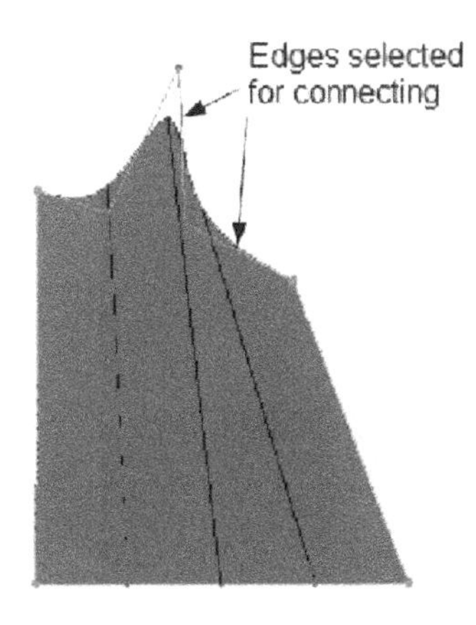

Figure 15-81 A deformed rectangle with two edges to be selected for connecting

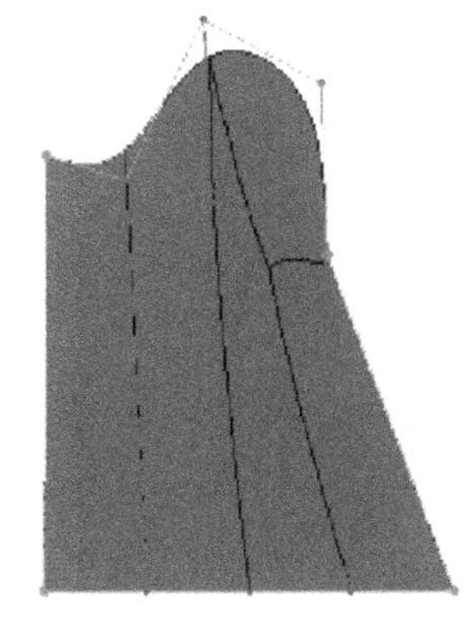

Figure 15-82 The output after connecting the edges

Mirror

The **Mirror** tool is available in the **Mirror** drop-down of the **Symmetry** group in the **Freestyle** dashboard and will be activated only when you select an edge or a face of the primitive. After selecting an edge or a face, choose the **Mirror** tool; you will be prompted to select the datum plane or planar surface as a mirror plane. Select the mirror plane, the edge or face will be mirrored along with the control points.

Connect Mirror

The **Connect Mirror** tool is available in the **Mirror** drop-down of the **Symmetry** group in the **Freestyle** dashboard and will be activated only when you select a face of the surface mirrored earlier. Figure 15-83 shows a mirrored surface and the face to be selected for the mirror connect operation. Figure 15-84 shows the output surface.

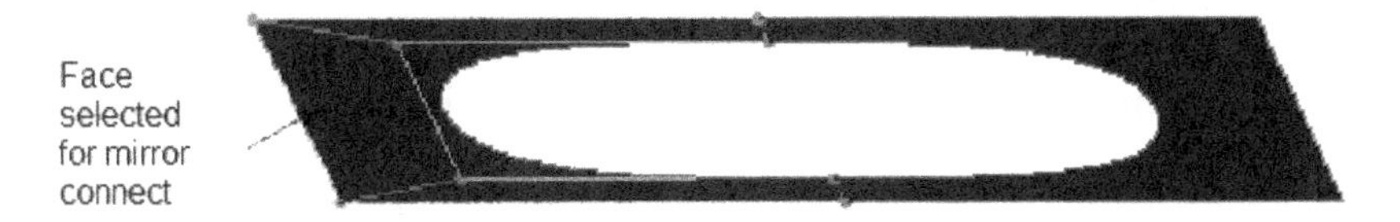

Figure 15-83 A mirrored surface and the face to be selected for the mirror connect operation

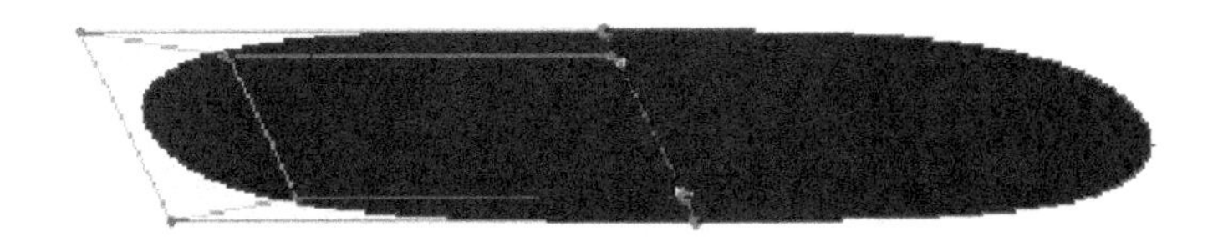

Figure 15-84 The output surface

Dependent

The **Dependent** toggle button is available in the **Symmetry** group of the dashboard. When this toggle button is selected, the control mesh of mirrored feature is dependent on the original feature. But when this button is not selected, the mirrored feature is controlled by its own control mesh. Figure 15-84 shows the default output surface created by using the **Connect Mirror** tool which is a dependent mirrored surface and Figure 15-85 shows the independent mirrored surface.

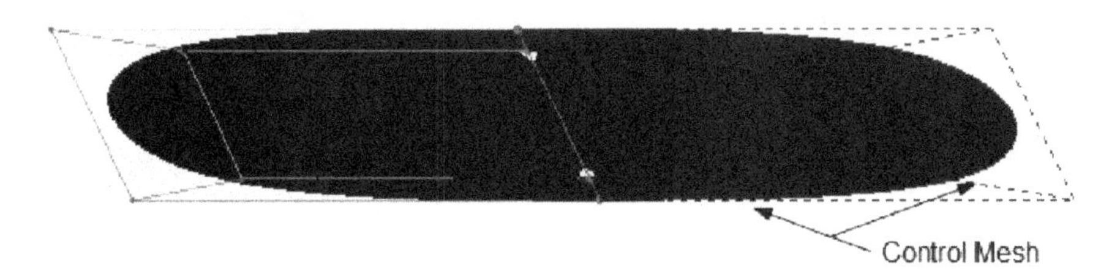

Figure 15-85 *An independent mirrored surface*

Delete Mirror/Change Plane

You can delete the mirrored surface by selecting the **Delete Mirror** option available in the **Mirror** drop-down. You can change the mirror plane by selecting the **Change Plane** option available in the **Mirror** drop-down.

TUTORIALS

Tutorial 1

In this tutorial, you will create the surface model shown in Figure 15-86. The orthographic views of the surface model are shown in Figure 15-87. **(Expected time: 45 min)**

Examine the model and determine the number of features in it, refer to Figure 15-87.

The following steps are required to complete this tutorial:

a. Create the base feature, which is a blend surface, refer to Figures 15-88 through 15-90.
b. Create the second feature, which is a blend feature, refer to Figures 15-91 and 15-92.
c. The third feature is a mirror feature. It will be created by mirroring the second feature about a plane that passes through the center, refer to Figure 15-93.
d. Create the fourth feature, which is also a blend feature, refer to Figures 15-94 and 15-95.
e. Select the surfaces individually and merge them, refer to Figures 15-96 and 15-97.
f. Remaining features are the fill features and they will be used to create surfaces on the blend features, refer to Figures 15-98 through 15-100.
g. Create rounds on the edges and save the model, refer to Figure 15-101.

Starting a New Object File

1. Start a new part file and name it *c15tut1*.

The three default datum planes are displayed in the drawing area. The **Model Tree** is also displayed in the drawing area. Close the **Model Tree** by clicking on the sash present on its right edge.

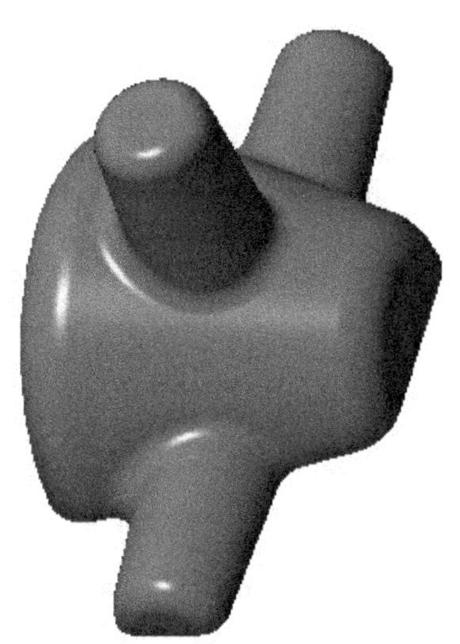

Figure 15-86 *The surface model*

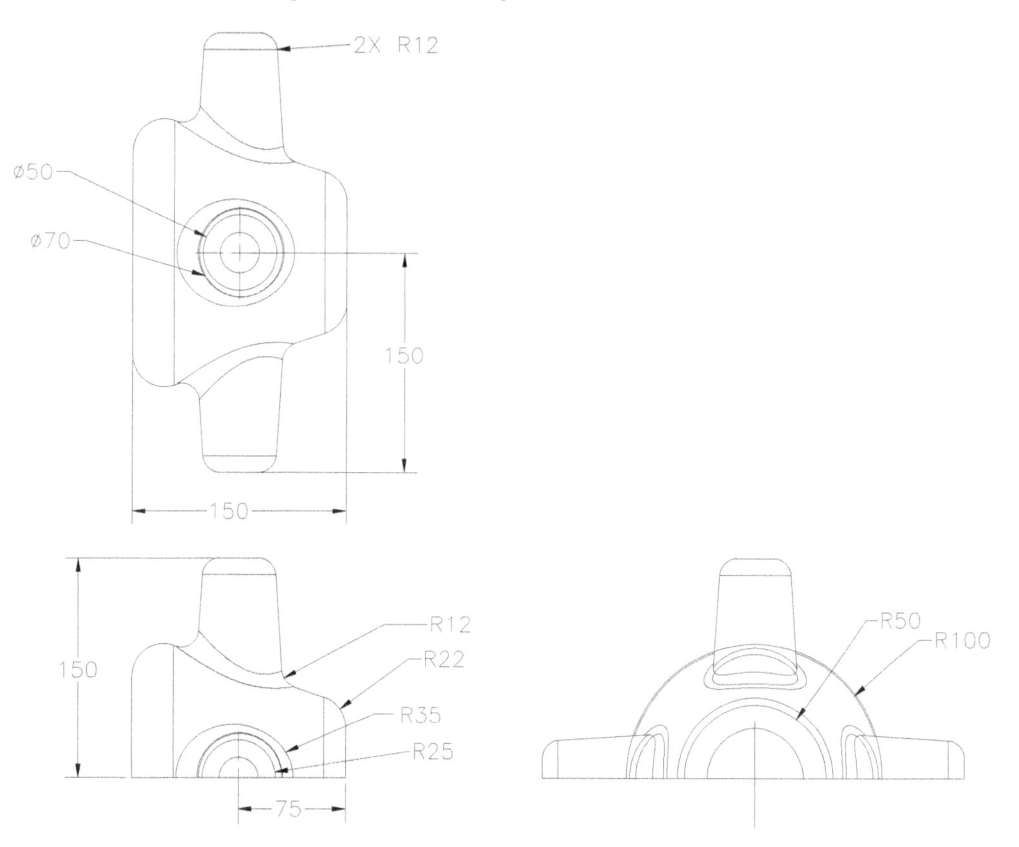

Figure 15-87 *Top, front, and right views of the surface model*

Creating the Base Feature

You need to create the base feature by using the **Blend** option.

1. Choose the **Blend** tool from the expanded **Shapes** group; the **Blend** dashboard is displayed.

2. Choose the **Blend as surface** button from the dashboard. Choose the **Options** tab and select the **Straight** radio button. Choose the **Sections** tab from the dashboard; a slide down panel is displayed. Select the **Sketched sections** radio button and choose the **Define** button from the slide down panel to define the sketching plane.

3. Select the **RIGHT** datum plane as the sketching plane.

4. Select the **TOP** datum plane as reference plane and then select the **Top** option from the **Orientation** drop-down list. Choose the **Sketch** button to enter the sketcher environment.

5. Draw an arc and dimension it, as shown in Figure 15-88.

6. After drawing the first arc, choose **OK** to exit the sketcher environment. Choose the **Insert** button from the **Section** tab; an offset value edit box with a default value will be available in the dashboard and in the **Sections** tab. Enter **-150** in offset value edit box for depth of second section. Choose the **Sketch** button from the **Sections** tab to enter the sketcher environment.

7. Draw the second arc and dimension it, as shown in Figure 15-89.

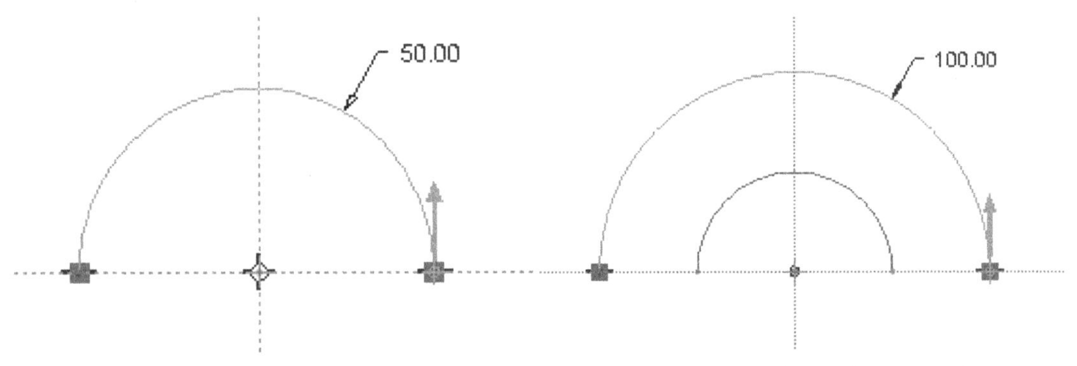

Figure 15-88 *Sketch of the first arc*

Figure 15-89 *Sketch of the second arc*

8. After drawing the second arc, choose the **OK** button to exit the sketcher environment. Choose the **Build feature** button from the dashboard to exit it. The model, similar to the one shown in Figure 15-90, is displayed in the drawing area.

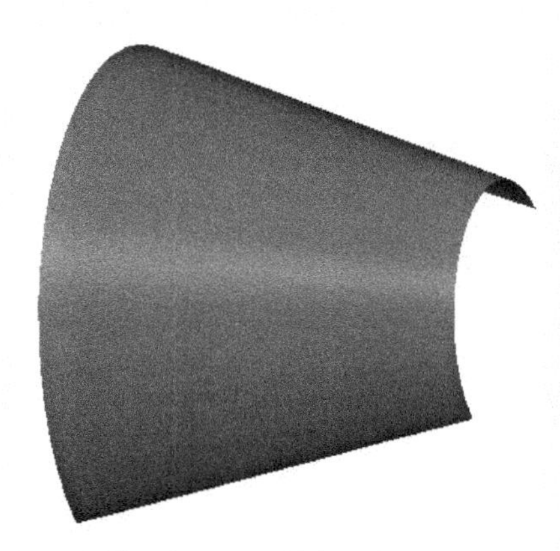

Figure 15-90 *Trimetric view of the base feature*

Creating the Second Feature

The second blend feature will be created on the **FRONT** datum plane and it will pass through the center of the base feature.

1. Choose the **Blend** tool from the expanded **Shapes** group; the **Blend** dashboard is displayed.

2. Choose the **Blend as surface** button from the dashboard. Choose the **Options** tab and select the **Straight** radio button. Choose the **Sections** tab from the dashboard; a slide down panel is displayed. Next, select **Sketched sections** radio button and then choose the **Define** button to define the sketching plane.

3. Select the **FRONT** datum plane as the sketching plane.

4. Select the **TOP** datum plane as the reference plane and then select the **Top** option from the **Orientation** drop-down list. Choose the **Sketch** button to enter the sketcher environment.

5. Sketch an arc of radius 35; for the remaining dimensions refer to Figure 15-87 and then choose **OK** to exit the sketcher environment. Choose the **Insert** button from the **Section** tab and enter **150** in the offset value edit box for depth of second section. Choose the **Sketch** button from the **Sections** tab to enter the sketcher environment.

6. Draw the second arc of radius 25, as shown in Figure 15-91.

7. Choose the **OK** button to exit the sketcher environment. Next, choose the **Build feature** button from the dashboard to exit it. The model, similar to the one shown in Figure 15-92, is displayed in the drawing area.

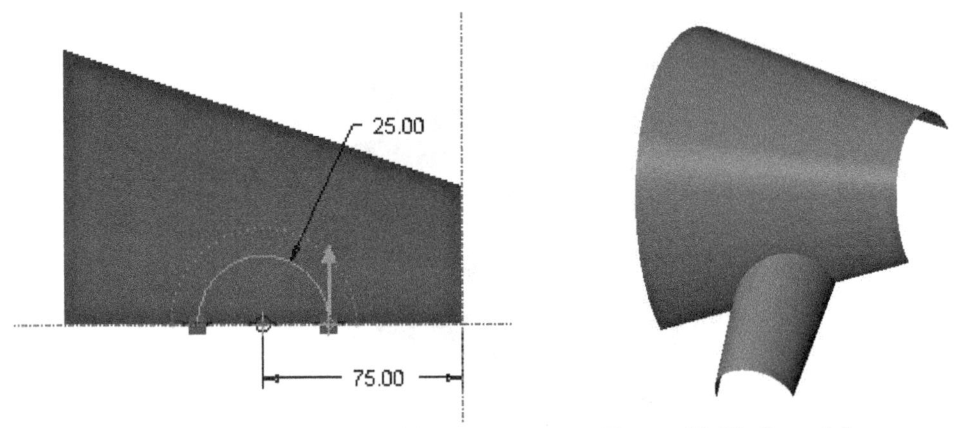

Figure 15-91 Sketch of the second feature

Figure 15-92 Second feature created

Creating the Mirror Copy of the Second Feature

The third blend feature is the same as the second blend feature. Therefore, a mirror copy of the second blend feature will be used to create the third feature.

1. Select the second feature and then choose the **Mirror** tool; the **Mirror** dashboard is displayed and you are prompted to select a plane to mirror about.

2. Select the **FRONT** datum plane and exit the **Mirror** dashboard by choosing the **Build Feature** button. The mirror copy of the second feature is created, as shown in Figure 15-93.

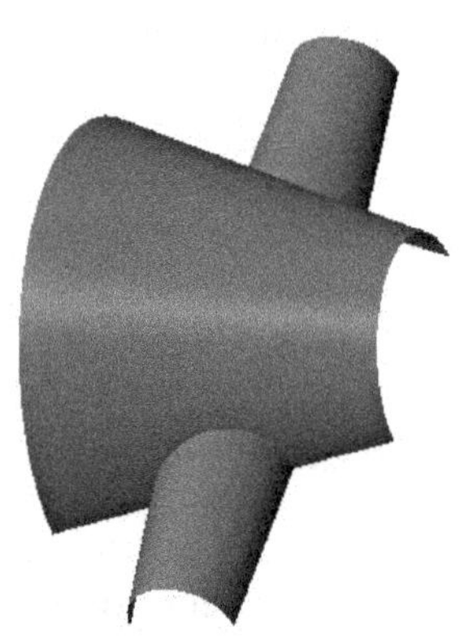

Figure 15-93 Surface model after creating the third feature

Creating the Fourth Blend Feature

The fourth blend feature will be created on the **TOP** datum plane and it will pass through the center of the base feature.

1. Choose the **Blend** tool from the expanded **Shapes** group; the **Blend** dashboard is displayed.

2. Choose the **Blend as surface** button. Choose the **Options** tab and then select the **Straight** radio button. Also, select the **Capped Ends** check box.

3. Choose the **Sections** tab from the dashboard to open the slide-down panel. Choose the **Sketched sections** radio button and then choose the **Define** button to define the sketching plane.

4. Select the **TOP** datum plane as the sketching plane.

5. Select the **RIGHT** datum plane as the reference plane and then select the **Top** option from the **Orientation** drop-down list. Choose the **Sketch** button to enter the sketcher environment.

6. Sketch a circle of diameter 70; for the remaining dimensions refer to Figure 15-87 and then choose **OK** to exit the sketcher environment. Choose the **Insert** button from the **Section** tab and enter **150** in the offset value edit box for depth of second section. Choose the **Sketch** button from the **Sections** tab to enter the sketcher environment.

7. Draw the second circle of diameter 50, as shown in Figure 15-94.

8. Choose the **OK** button to exit the sketcher environment. Choose the **Build feature** button from the dashboard to exit it. The model, similar to the one shown in Figure 15-95, is displayed in the drawing area.

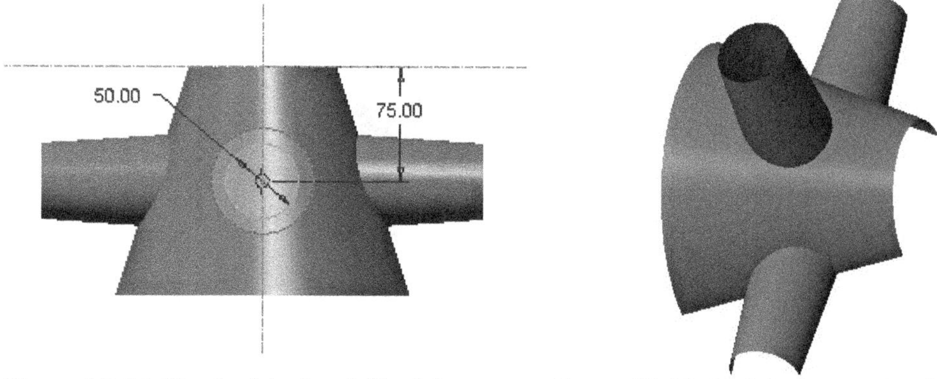

Figure 15-94 *Sketch of the fourth blend feature*

Figure 15-95 *Model after creating the fourth blend surface*

Merging the Surfaces to Create a Quilt

To create a round on the edges, it is necessary to create a common edge where the two surfaces will join. For this purpose, the surfaces need to be merged.

Note

*1. The easier way to select two surfaces for merging is to select them from the **Model Tree**. You should remember that to select more than one surface, you need to press the CTRL key. When you select the surfaces from the **Model Tree**, their boundaries are highlighted, indicating that the surfaces are selected.*

*2. You can also select the **Quilts** option from the **Filter** drop-down list to select the surfaces. The **Filter** drop-down list is available in the **Status Bar** at the right corner below the drawing area.*

1. Select the blend surface on the left, press the CTRL key, and then select the blend in the middle. When the two surfaces are highlighted, choose the **Merge** button; the **Merge** dashboard is displayed with two arrows showing the portion that will be retained after merging.

Note

*The **Merge** button will be available only when two surfaces are selected for merging.*

2. Choose the **Change side of first quilt to keep** button from the dashboard; the direction of the arrow changes.

3. Choose the **Change side of second quilt to keep** button from the dashboard; the direction of the arrow changes. The portion of the surface that is now highlighted will be retained after merging the surfaces.

4. Exit the dashboard by choosing the **Build Feature** button. The model after merging the two surfaces is shown in Figure 15-96.

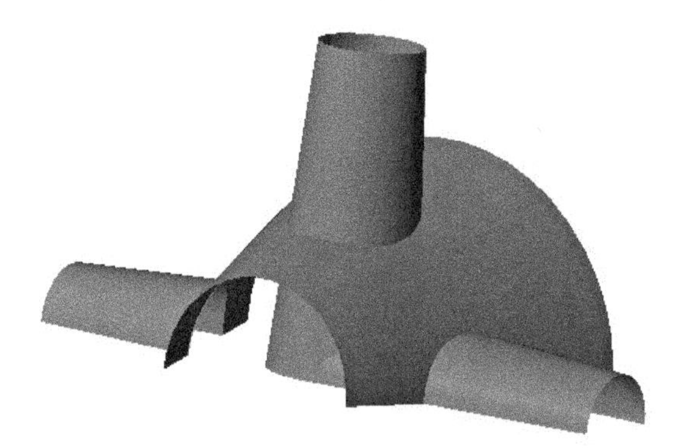

Figure 15-96 *Surface model after merging the two surfaces*

Using the above procedure, merge the blend surface on the right with the merged surface. Next, merge the top blend surface with the merged surface. Figure 15-97 shows the surface model after merging all surfaces and forming a quilt.

Creating the Fill Surfaces

You need to create four surfaces to cap the ends of the blend surfaces. First, you will cap the left blend surface using the **Fill** option.

1. Choose the **Fill** tool from the **Surfaces** group of the **Model** tab in the **Ribbon**; the **Fill** dashboard is displayed.

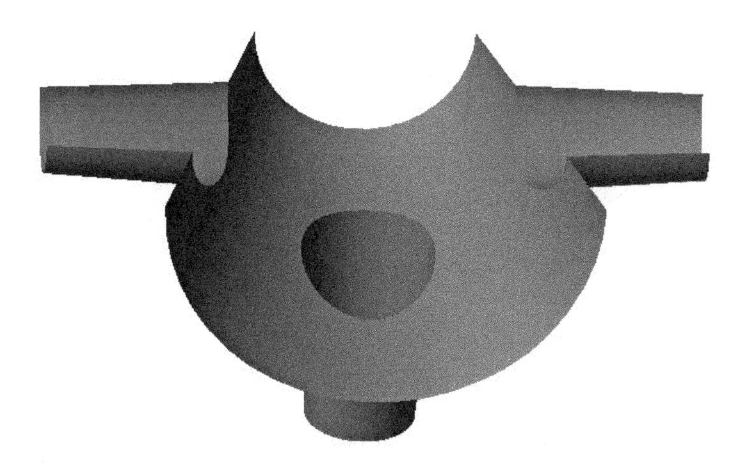

Figure 15-97 *Surface model after merging all the surfaces*

2. From the **References** slide-down panel, choose the **Define** button; the **Sketch** dialog box is displayed and you are prompted to select the sketching plane.

3. Choose the **Plane** button from the **Datum** group of the **Model** tab; the **Datum Plane** dialog box is displayed.

4. Select the two vertices of the left blend surface. To select the second vertex, hold down the CTRL key and then select the **FRONT** datum plane.

5. Select **FRONT** from the **Datum Plane** dialog box; a drop-down list appears in the row where you have clicked. Select the **Parallel** option from the drop-down list and then choose the **OK** button from the **Datum Plane** dialog box.

   The datum plane is created and an arrow pointing in the direction of viewing the sketch is displayed.

6. Choose the **Sketch** button to close the **Sketch** dialog box to enter the **Sketcher** environment.

7. Choose the **Project** button from the **Sketching** group and select the smaller semicircular edge of the blend surface. Next, close the sketch by drawing a horizontal line between the two ends, as shown in Figure 15-98.

8. Exit the **Sketcher** environment and then exit the **Fill** dashboard by choosing the **Build Feature** button; the fill surface is created, as shown in Figure 15-99.

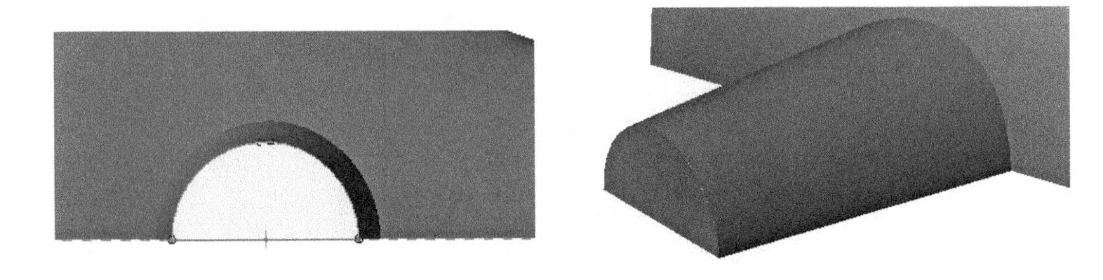

Figure 15-98 Sketch for the fill surface *Figure 15-99 Surface after creating the fill surface*

Similarly, create the fill surfaces to cap the ends of the middle surface blend feature. Mirror the left fill surface to the right. Figure 15-100 shows the surface model after capping all the ends of the blend surfaces.

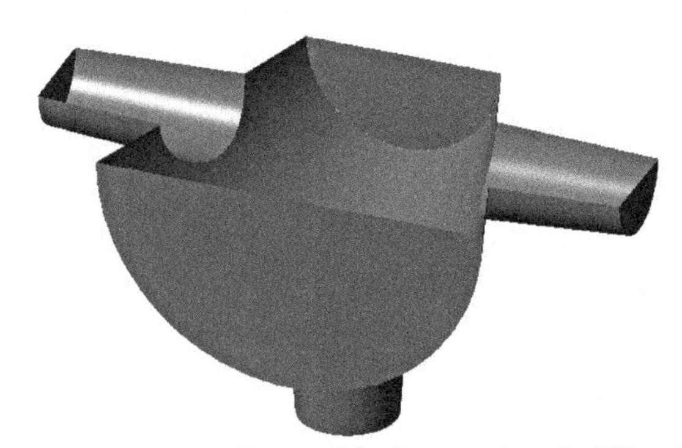

Figure 15-100 Surface model after creating the fill surfaces

Merging the Fill Surfaces

The fill surfaces that you have created should be merged with the other surfaces in order to create a round on their edges.

1. Hold the CTRL key to select the fill surface that is on the left and the blend surface in the middle. When the two surfaces turn green in color, choose the **Merge** button to display the **Merge** dashboard.

2. Exit the dashboard by choosing the **Build Feature** button.

Using the same procedure, merge the remaining fill surfaces individually with the blend surface in the middle. To check whether all the surfaces have been merged, select the surface model. If the whole surface model is highlighted with green border then it indicates that all the surfaces have been merged and they form a quilt.

Creating Rounds

When all the surfaces are merged, edges are formed at the intersection of two surfaces. These edges need to be rounded. In the given surface model, note that there are rounds that have two different radius values. Therefore, you need to create two sets to define two values of rounds.

1. Choose the **Round** button from the **Engineering** group.

2. Select the edges having radius value 12. Remember that to select more than one edge, you need to hold down the CTRL key.

3. After creating the rounds of radii 12, choose the **Sets** tab to display the slide-down panel.

4. Click on **New set**; a set with the name **Set2** is added.

5. Select the two edges that have radius of 22. After creating the rounds of radius 22, exit the **Round** dashboard.

   The surface model after creating the rounds is shown in Figure 15-101.

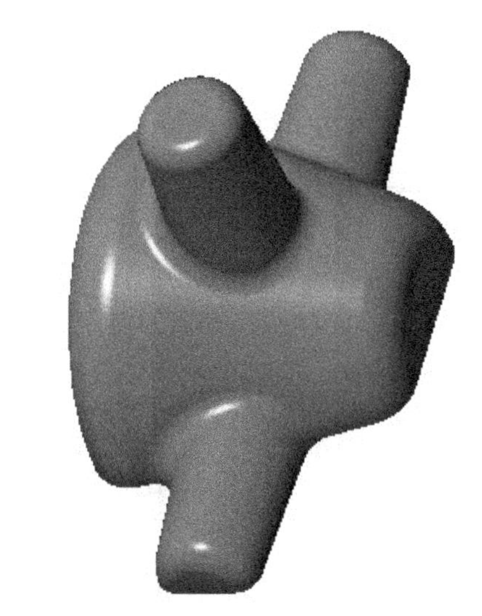

Figure 15-101 *Surface model after creating rounds*

6. Choose the **Save** button from the **File** menu and save the model.

Tutorial 2

In this tutorial, you will create the surface model shown in Figure 15-102. The front and right views of the surface model are shown in Figure 15-103. **(Expected time: 45 min)**

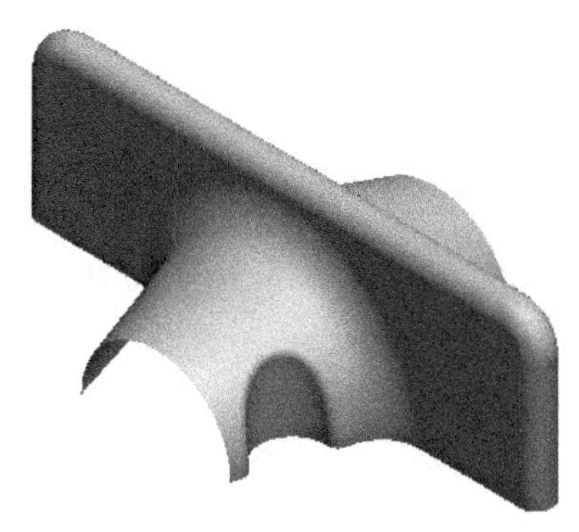

Figure 15-102 *Isometric view of the surface model*

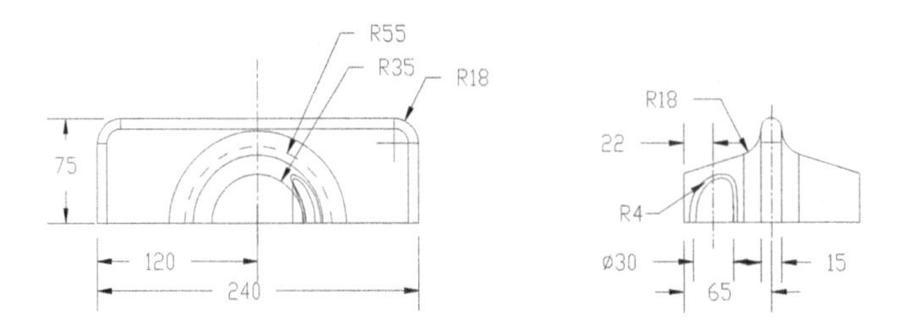

Figure 15-103 Front and right views of the surface model

First, examine the model and determine the number of features in it, refer to Figure 15-103.

The following steps are required to complete this tutorial:

a. Create the base feature, which is an extruded surface with open ends, refer to Figures 15-104 and 15-105.
b. Create the second feature, which is a blend feature created at an offset distance of 65 from the **RIGHT** datum plane, refer to Figures 15-106 and 15-107.
c. The third feature is a mirror copy of the second feature, refer to Figure 15-108.
d. The fourth feature is the cylindrical surface, refer to Figures 15-109 and 15-110.
e. Create the two fill surfaces that will cap the ends of the base surface, refer to Figures 15-111 through 15-113.
f. Merge all surfaces by selecting them individually, refer to Figures 15-114 through 15-116.
g. Create the round features and save the model, refer to Figures 15-117 through 15-119.

Starting a New Object File

1. Start a new part file and name it *c15tut2*.

Creating the Base Feature

The base feature for this model is a surface between two blend surfaces. It will be created on the **RIGHT** datum plane.

1. Choose the **Extrude** tool from the **Shapes** group.

2. Select the **Extrude as surface** button from the **Extrude** dashboard. Select the **RIGHT** datum plane as the sketch plane.

3. Select the **TOP** datum plane as reference plane from the drawing area and then select the **Top** option from the **Orientation** drop-down list.

4. Choose the **Sketch** button to enter the **Sketcher** environment.

5. Once you enter the **Sketcher** environment, create the sketch of the base feature and apply dimensions, as shown in Figure 15-104.

6. After completing the sketch, choose the **OK** button to exit the **Sketcher** environment.

   The **Extrude** dashboard reappears above the drawing area. In this dashboard, the **Extrude from sketch plane by specified depth value** button is chosen by default.

7. Enter **240** in the dimension box present in the **Extrude** dashboard. Choose the **Build feature** button from the **Extrude** dashboard.

   The base feature is created and the default trimetric view of this feature is shown in Figure 15-105.

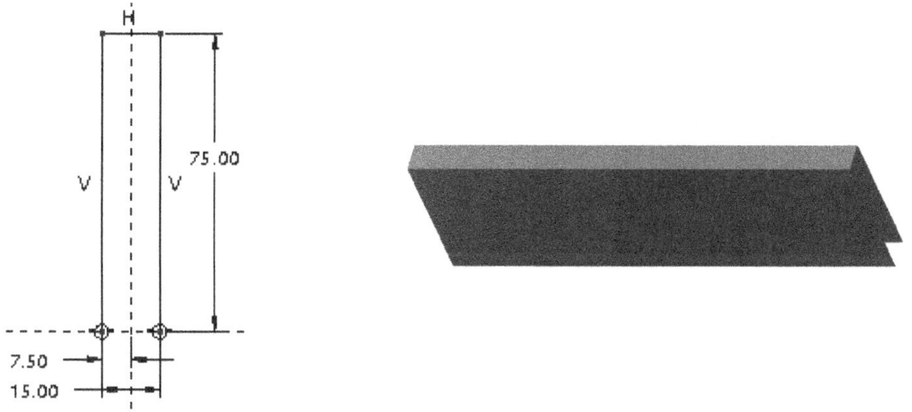

Figure 15-104 Sketch of the base surface

Figure 15-105 Base surface with open ends

Creating the Blend Feature

The second feature is a blend surface. It will be created on the **FRONT** datum plane and passes through the center of the base feature.

1. Choose the **Blend** tool from the expanded **Shapes** group; the **Blend** dashboard is displayed.

2. Choose the **Blend as surface** button from the dashboard. Choose the **Options** tab and select the **Straight** radio button. Choose the **Sections** tab from the dashboard; a slide down panel is displayed. Select the **Sketched sections** radio button and choose the **Define** button to define the sketching plane.

3. Select the **FRONT** datum plane as the sketching plane.

4. Select the **TOP** datum plane as the reference plane and then select the **Top** option from the **Orientation** drop-down list. Choose the **Sketch** button to enter the sketcher environment.

5. Draw an arc of diameter 55; for the remaining dimensions refer to Figure 15-103 and then choose **OK** to exit the sketcher environment. Choose the **Insert** button from the **Section** tab; an offset value edit box with a default value will be available in the dashboard and in the **Sections** tab. Enter **65** in the offset value edit box for specifying the depth of second section. Next, choose the **Sketch** button from the **Sections** tab to enter the sketcher environment.

6. Now, draw another arc of diameter 35, as shown in Figure 15-106.

7. After drawing the second arc, choose the **OK** button to exit the sketcher environment. Choose the **Build feature** button from dashboard to exit it. The model, similar to the one shown in Figure 15-107, is displayed in the drawing area.

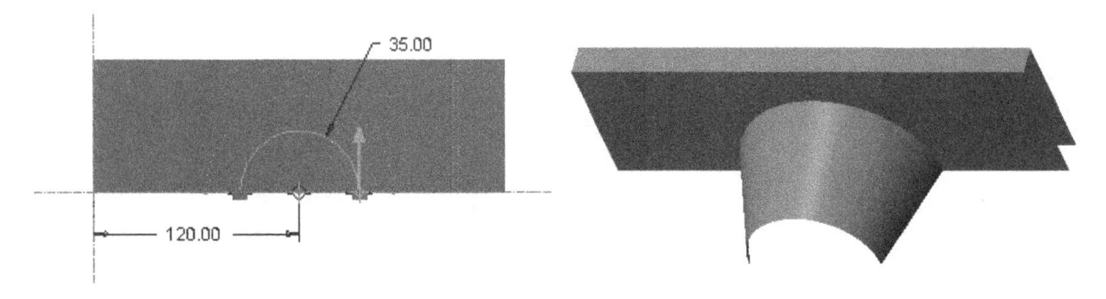

Figure 15-106 Sketch of the blend surface *Figure 15-107 Blend surface*

Mirroring the Blend Surface

Next, you need to mirror the blend surface about the **FRONT** datum plane.

1. Select the blend surface and then choose the **Mirror** tool from the **Editing** group; the **Mirror** dashboard is displayed.

2. Select the **FRONT** datum plane and exit the dashboard; the blend surface is mirrored about the selected datum plane, as shown in Figure 15-108.

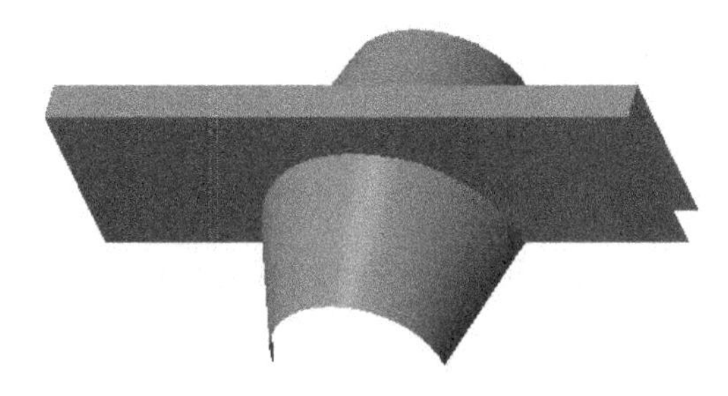

Figure 15-108 Model after creating the mirror copy of the blend surface

Creating the Cylindrical Surface

The cylindrical surface will be created on the **TOP** datum plane.

1. Choose the **Extrude** tool from the **Shapes** group to display the **Extrude** dashboard.

2. Choose the **Extrude as surface** button from the **Extrude** dashboard.

3. Select the **TOP** datum plane as the sketching plane.

4. After entering the **Sketcher** environment, draw a circle and dimension it, as shown in Figure 15-109.

5. Exit the **Sketcher** environment and extrude the sketch to some appropriate depth, refer to Figure 15-110.

   The model after creating the surface extrusion is shown in Figure 15-110.

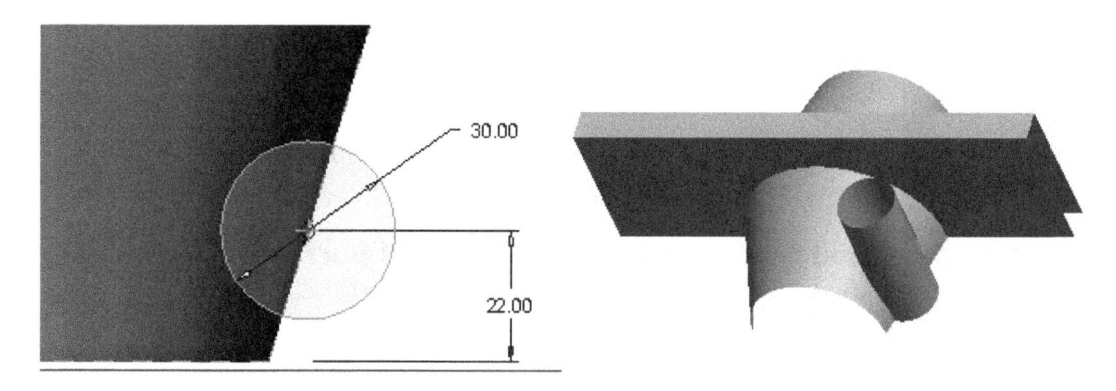

Figure 15-109 Sketch of the cylindrical surface

Figure 15-110 Cylindrical surface

Creating the Fill Surface

The fill surface will be created to cap the ends of the base feature.

1. Choose the **Fill** tool from the **Surfaces** group of the **Model** tab in the **Ribbon**; the **Fill** dashboard is displayed.

2. Choose the **Define** button from the **References** slide-down panel; the **Sketch** dialog box is displayed and you are prompted to select the sketching plane.

3. Select the **RIGHT** datum plane as the sketching plane. Choose the **Flip** button.

4. Select the **TOP** datum plane and select the **Top** option from the **Orientation** drop-down list. Choose the **Sketch** button to enter the **Sketcher** environment.

5. Choose the **Project** button and select the edges of the base feature. Create the line to join the ends of base feature; the sketch is completed, as shown in Figure 15-111.

6. Exit the **Sketcher** environment and then exit the **Fill** dashboard by choosing the **Build Feature** button. On doing so, the fill surface is created, as shown in Figure 15-112.

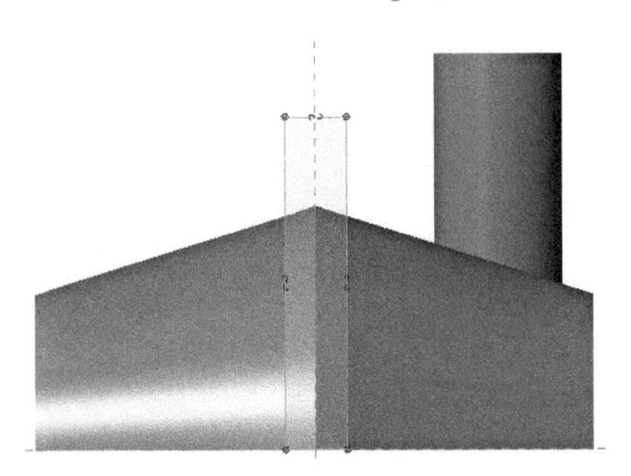

Figure 15-111 Sketch of the fill surface

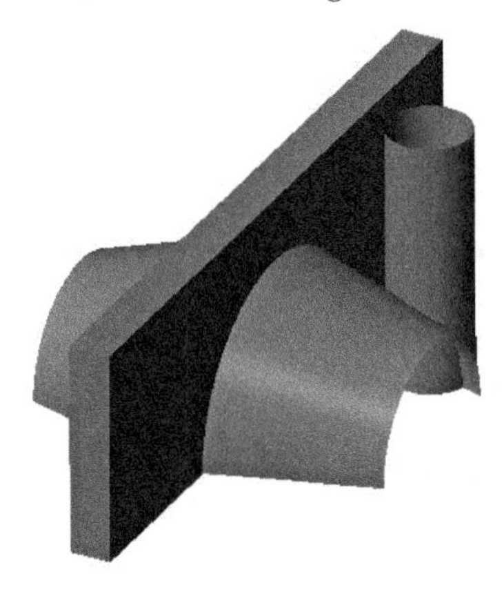

Figure 15-112 Model after creating the fill surface

7. Mirror the fill surface about the datum plane on-the-fly. This datum plane is at an offset distance of 120 from the **RIGHT** datum plane.

   After creating the mirror copy of the fill surface, the other end of the base feature is also capped, as shown in Figure 15-113.

Merging the Blend Surface with the Cylindrical Surface

The blend surface that was the second feature and the cylindrical surface will be merged to get the required circular slot.

1. Select the cylindrical surface and the blend surface by pressing the CTRL key.
2. Choose the **Merge** button from the **Editing** group; the **Merge** dashboard is displayed and the surface that will be retained after merging is highlighted.
3. Choose the **Change side of first quilt to keep** button to change the direction of the pink arrow.
4. Exit the **Merge** dashboard by choosing the **Build Feature** button. The model after merging the two surfaces is shown in Figure 15-114.

Merging the Blend Surface and the Extruded Surface

The blend surface and the extruded surface will be merged to build a single surface.

1. Select the base feature and then select the **Merge 1** feature from the **Model Tree**.

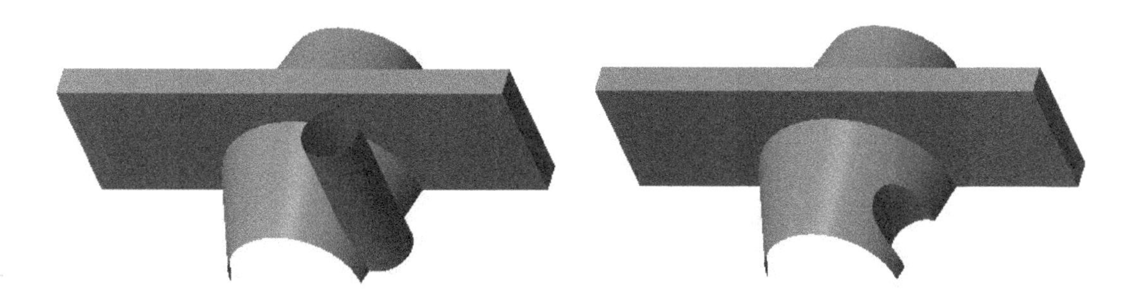

Figure 15-113 Model after creating the mirror copy of the fill surface

Figure 15-114 Model after merging the surfaces

2. Choose the **Merge** button from the **Editing** group; the **Merge** dashboard is displayed and the surface that will be retained after merging is highlighted.

3. Choose the **Change side of first quilt to keep** and **Change side of second quilt to keep** buttons to change the direction of the pink arrow.

4. You can see the preview of the blend surface by choosing the **Preview** button in the dashboard. Exit the **Merge** dashboard by choosing the **Build Feature** button. The model after merging the two surfaces is shown in Figure 15-115.

5. Similarly, merge the mirrored feature and the base feature. The surface model after merging the mirrored surfaces is shown in Figure 15-116.

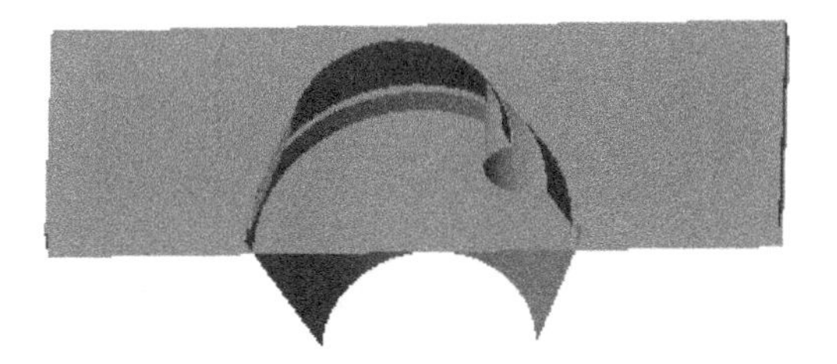

Figure 15-115 Model after merging the blend surface with the base surface

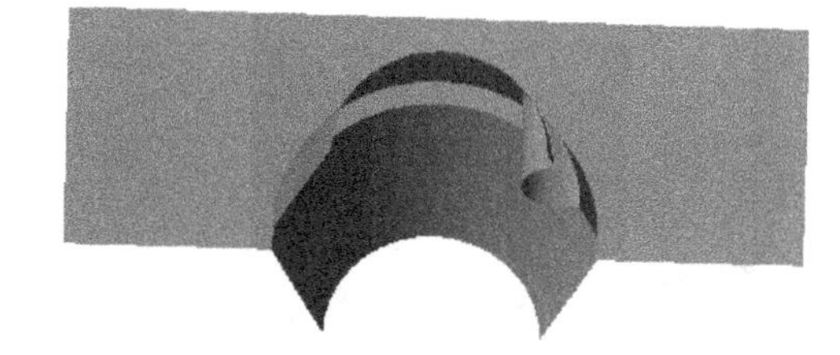

Figure 15-116 Model after merging the mirror copy of the blend surface with the base surface

Merging the Fill Surfaces with the Base Surface

The fill surfaces that you have created should be merged with the base surface in order to create a single quilt or a single surface. After merging the surfaces, you need to round the edges.

1. Select the fill surface and then select the base surface.

Note

*1. It is better to select surfaces to be merged from the **Model Tree**.*

2. To merge two surfaces, it is necessary that they intersect each other.

2. Choose the **Merge** button from the **Editing** group and then choose the **Build feature** button to merge both the surfaces.

3. Similarly, merge the mirror copy of the first fill surface with the base surface.

Creating Rounds

You need to create rounds on the cylindrical slot or edges where the two blend surfaces merge, and on the edges of the base surface.

1. Choose the **Round** tool from the **Engineering** group. Select the edge of the cylindrical slot, refer to Figure 15-117; the preview of the round is highlighted on the selected edge.

2. Enter **4** in the dimension edit box for the radius of the round.

3. Click on **New set** in the **Sets** slide-down panel to add a second set named **Set2**.

4. Select the four edges each of which has radius 18. Among these edges, the two edges are those that are formed by merging the two blend surfaces with the base surface and the other two edges are the top corners of the base surface, refer to Figure 15-117.

5. Choose the **Build feature** button from the **Round** dashboard to create rounds. The surface model after creating the rounds is shown in Figure 15-118.

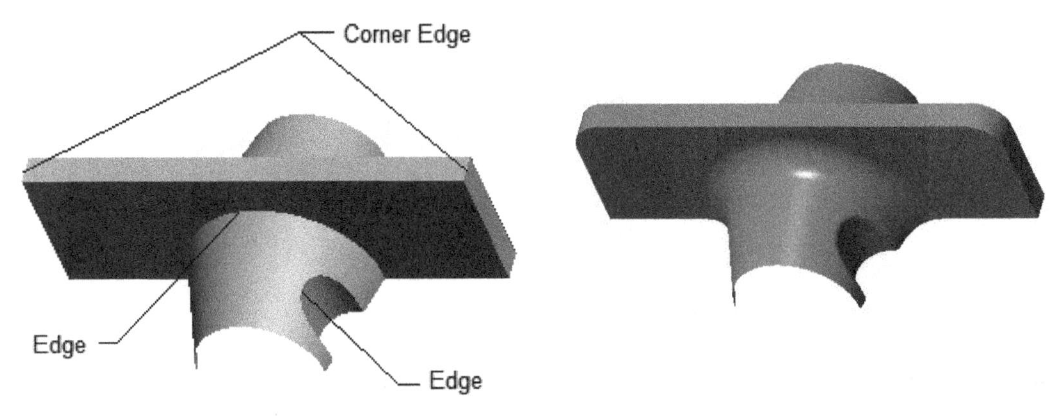

Figure 15-117 *Edges selected to create rounds*

Figure 15-118 *Resulting model after creating rounds on the selected edges*

Creating a Full Round

A full round will be created by selecting the two surfaces. These surfaces are the front and back faces of the base surface.

1. Choose the **Round** tool from the **Engineering** group.

2. Select the two faces, front and back, of the base surface.

3. Invoke the slide-down panel by choosing the **Sets** tab. The selected surfaces are displayed in the **References** collector available in the slide-down panel. The **Full Round** button is chosen by default in the slide-down panel. Now, you need to select the surface to be removed.

4. Select the top face of the base surface; the preview of the round is highlighted on the selected surfaces. Exit the **Round** dashboard by choosing the **Build Feature** button. The round is created, as shown in Figure 15-119.

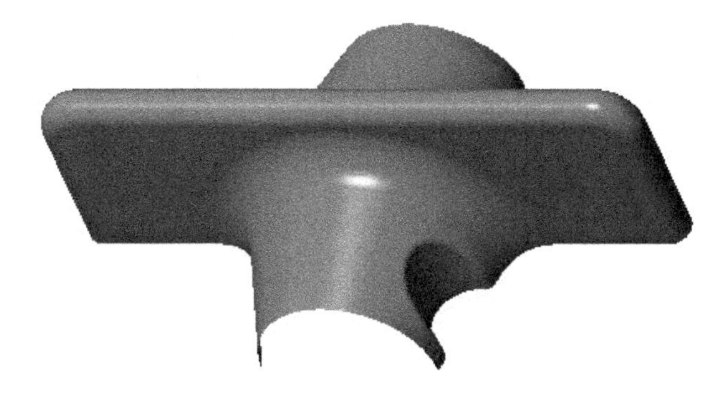

Figure 15-119 *Completed surface model*

5. Choose the **Save** button from the **File** menu and save the model.

Tutorial 3

In this tutorial, you will create the surface model of a table, as shown in Figure 15-120, using the Freestyle Modeling environment. **(Expected time: 45 min)**

First, examine the model and determine the primitive to be used.

The following steps are required to complete this tutorial:

a. Create a new coordinate system.
b. Create the primitive.
c. Create the edge split feature, refer to Figures 15-123 through 15-126.
d. Create the third feature which is transformation of faces and save the model, refer to Figure 15-127 and 15-128.

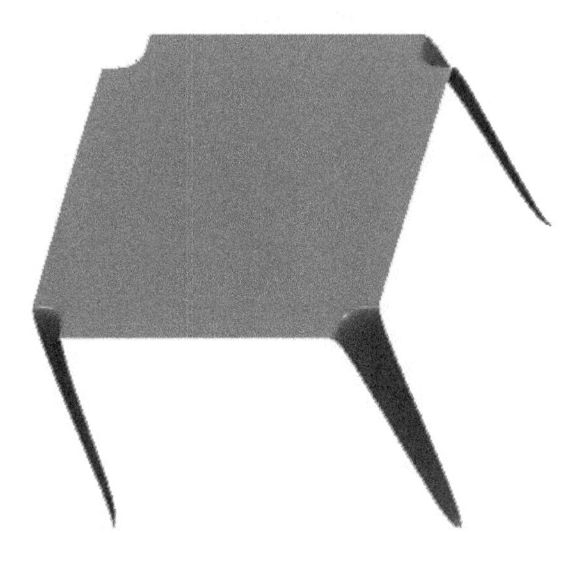

Figure 15-120 *The surface model of a table*

Starting a New Object File

1. Start a new part file and name it *c15tut3*.

Creating a New Coordinate System

It is evident from the Figure 15-120, the model is created by using the square shape primitive on the **Top** plane. Therefore, you need to change the default sketching plane for the placement of primitive.

1. Choose the **Coordinate System** button from the **Datum** group of the **Model** tab in the **Ribbon**; the **Coordinate System** dialog box is displayed.

2. Now, select the **TOP** plane from the **Model Tree** as a reference and the **FRONT** and **RIGHT** planes as offset references and then choose the **OK** button from the dialog box.

Starting the Freestyle Modeling Environment

To start with the Freestyle Modeling environment, you need to follow the step given next.

1. Choose the **Freestyle** tool from the **Surfaces** group in the **Model** tab of the **Ribbon** while working in the Modeling environment; the **Freestyle** dashboard is displayed and Freestyle Modeling environment is activated.

Changing the Sketching Plane

You need to change the sketching plane. To do so, a coordinate system is required.

1. Click on the down arrow below the **Primitives** button from the **Operations** group in the **Freestyle** dashboard; a drop down list is displayed.

2. Choose the **Options** button from this drop-down list; the **Options** dialog box is displayed, as shown in Figure 15-121.

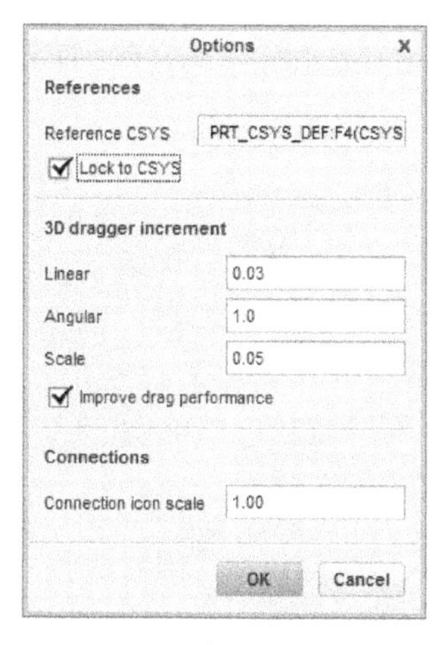

Figure 15-121 *The **Options** dialog box*

3. Select the new coordinate system created earlier and choose the **OK** button from the dialog box.

Creating the Primitive

After examining the model, you can guess that the basic primitive required for this model is a square surface.

1. Click on the down arrow below the **Primitives** button from the **Shapes** group in the **Freestyle** dashboard; a drop down list is displayed.

2. Choose the **Choose a square initial shape** option from the **Open Primitives** area of this drop-down list; a square surface is created in the XY plane, as shown in Figure 15-122.

Figure 15-122 *Square surface created*

Creating the Edge Split Feature

To create the legs of the table, you need to split the edges of the surface.

1. Select any of the four edges and click on the down arrow next to the **Edge Split** button in the **Create** group of the dashboard; a drop-down list is displayed.

2. Select the **4 Splits** option from the drop-down list; the surface is displayed, as shown in the Figure 15-123.

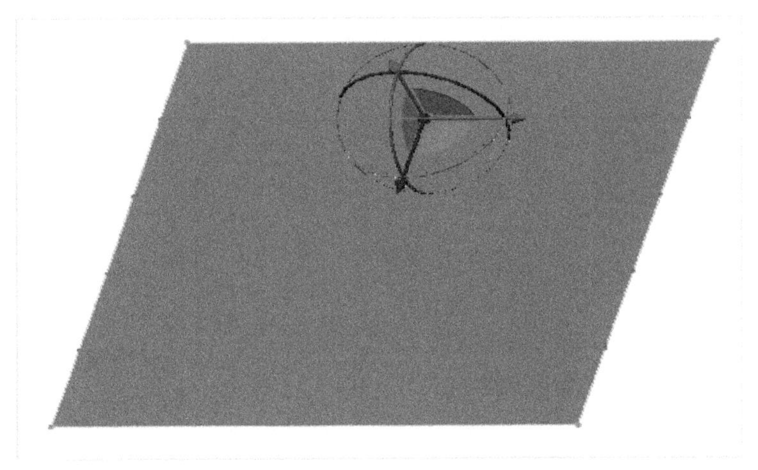

Figure 15-123 *Surface created after using the **4 Splits** option*

3. Select the two corner edges, as shown in Figure 15-124, and then choose the **4 Splits** option again; the surface is split, as shown in Figure 15-125.

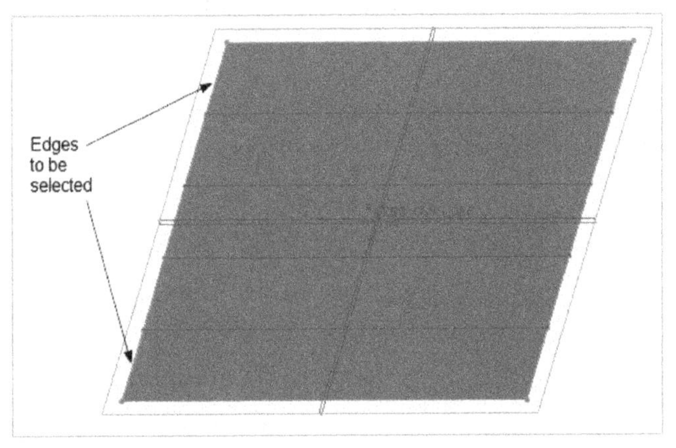

Figure 15-124 *Surface with corners edges to be selected*

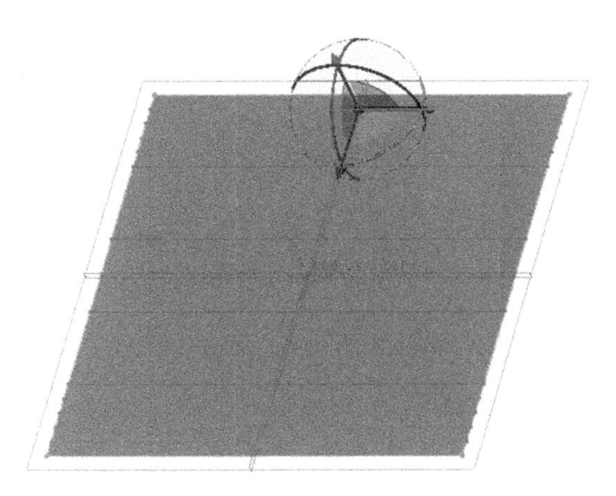

Figure 15-125 *Surface with corner edges split by using the* ***4 Split*** *option*

Similarly, split one edge perpendicular to the earlier selected edge so that the surface is displayed, as shown in Figure 15-126.

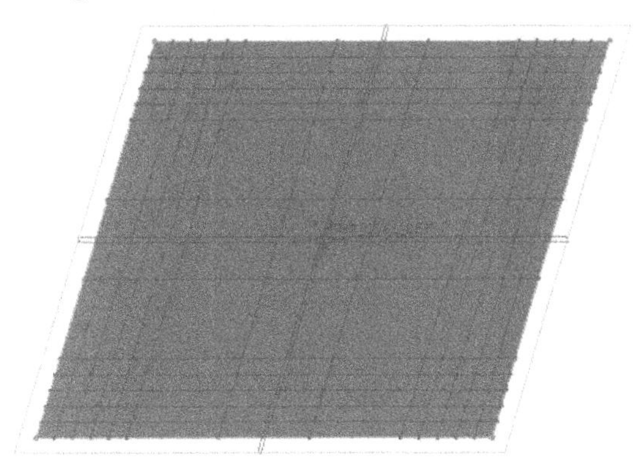

Figure 15-126 *Surface model after splitting is completed*

Transforming the Faces

Now, you need to transform the corner squares to create legs of the table.

1. Select the four small squares, as shown in Figure 15-127, and then drag the orange handle in z-direction such that the surface model looks like the one shown in Figure 15-128.

2. Choose the **Build Feature** button and choose the **Save** button from the **File** menu to save the model.

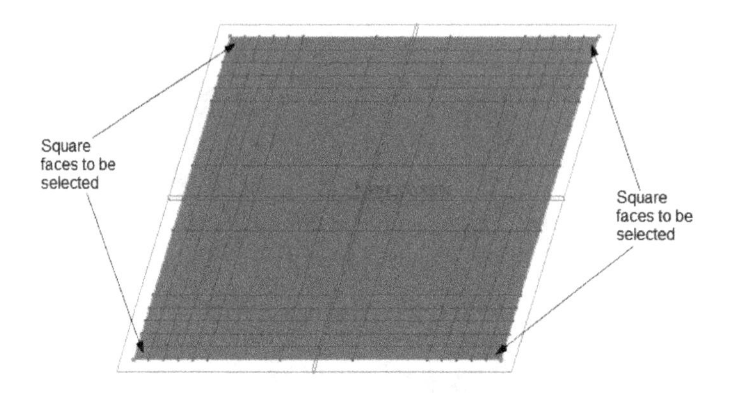

Figure 15-127 *Surface with the squares to be transformed*

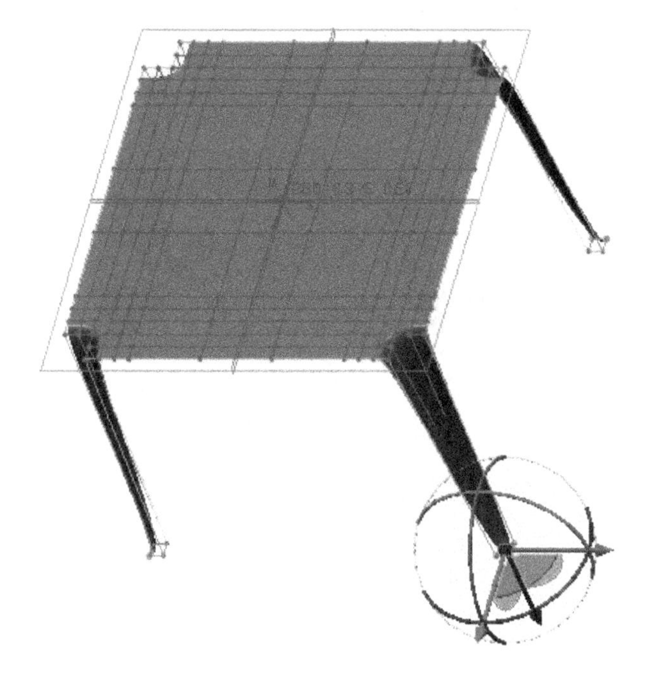

Figure 15-128 *Completed surface model*

Self-Evaluation Test

Answer the following questions and then compare them to those given at the end of this chapter:

1. You can create a surface with capped ends by drawing an open sketch. (T/F)

2. Surface models have no thickness. (T/F)

3. Style features have the parent-child relationship among themselves as well as with the corresponding solid features created in the modeling environment. (T/F)

4. In the **Style** environment, when you press the SHIFT key and select a point on a surface by using the **Free** option, the point is snapped on that surface. (T/F)

5. The procedure to create a helical sweep surface is the same as that of a solid helical sweep feature. (T/F)

6. A feature created in the **Style** environment is displayed in the **Model Tree** as a __________ feature.

7. To enter the **Style** environment, choose the __________ button from the **Surface** group in the **Model** tab.

8. The __________ button is used to merge two surfaces and form an edge.

9. In the **Style** environment, the __________ button is used to draw curves.

10. A quilt is a __________ feature.

Review Questions

Answer the following questions:

1. Which of the following tool is used to create straight or smooth feature profile?

   (a) **Sweep** (b) **Blend**
   (c) **Extrude** (d) None of these

2. Which of the following editing tools is used to create a flat surface by drawing a sketch?

   (a) **Trim** (b) **Copy**
   (c) **Fill** (d) None of these

3. What is the minimum number of sections required for creating a blend feature?

   (a) one (b) two
   (c) three (d) None of these

4. Which of the following editing tools forms an edge between two intersecting surfaces?

   (a) **Merge** (b) **Intersect**
   (c) **Trim** (d) None of these

5. In which one of the following types of blend, sections are translated and rotated about the x, y, and z axes?

   (a) **Parallel** (b) **Rotational**
   (c) **General** (d) None of these

6. The **Intersect** option is used to create an intersect curve. (T/F)

7. In the **Style** environment, the **Curve Edit** button is used to project curves on surfaces. (T/F)

8. Surface models are the 3D models with no thickness. (T/F)

9. In the **Style** environment, the **Surface** button is used to select at least three or four curves and create a surface. (T/F)

10. To undo the last operation performed in the Style environment, choose the **Undo** button from the **Quick Access** toolbar. (T/F)

Exercises

Exercise 1

In this exercise, you will create the surface model shown in Figure 15-129. The orthographic views of the surface model are shown in Figure 15-130. **(Expected time: 40 min)**

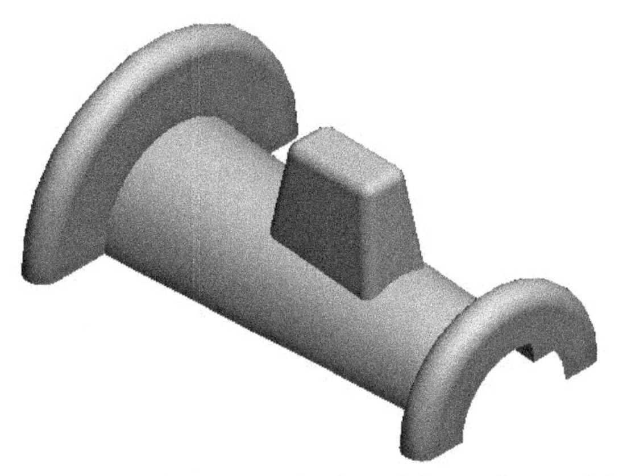

Figure 15-129 *Isometric view of the surface model*

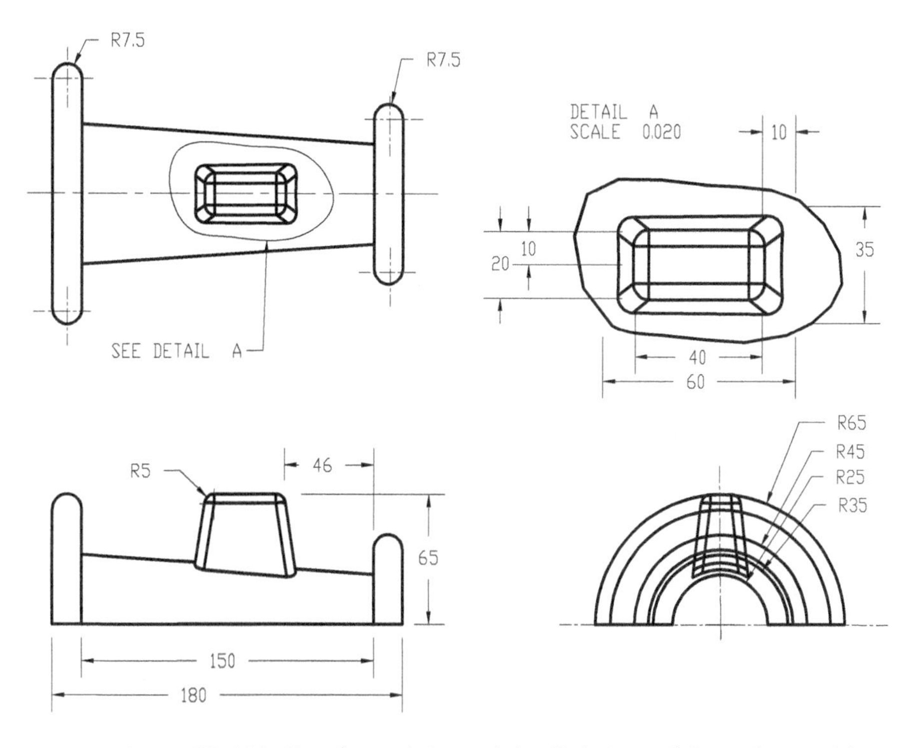

Figure 15-130 Top, front, right, and detailed views of the surface model

Exercise 2

In this exercise, you will create the surface model shown in Figure 15-131. The orthographic views and the detailed view of the surface model are shown in Figure 15-132.

(Expected time: 55 min)

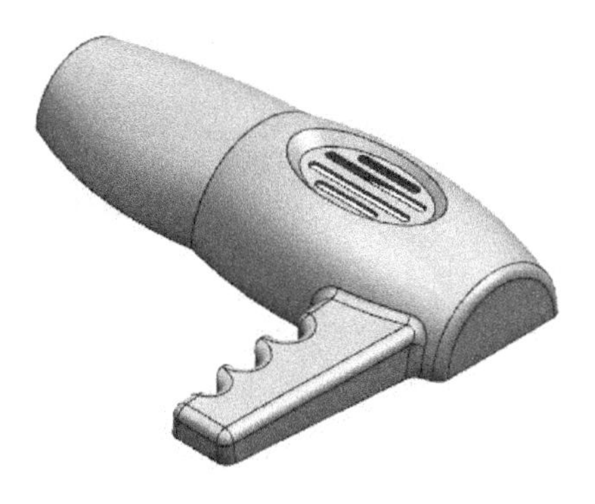

Figure 15-131 Surface model

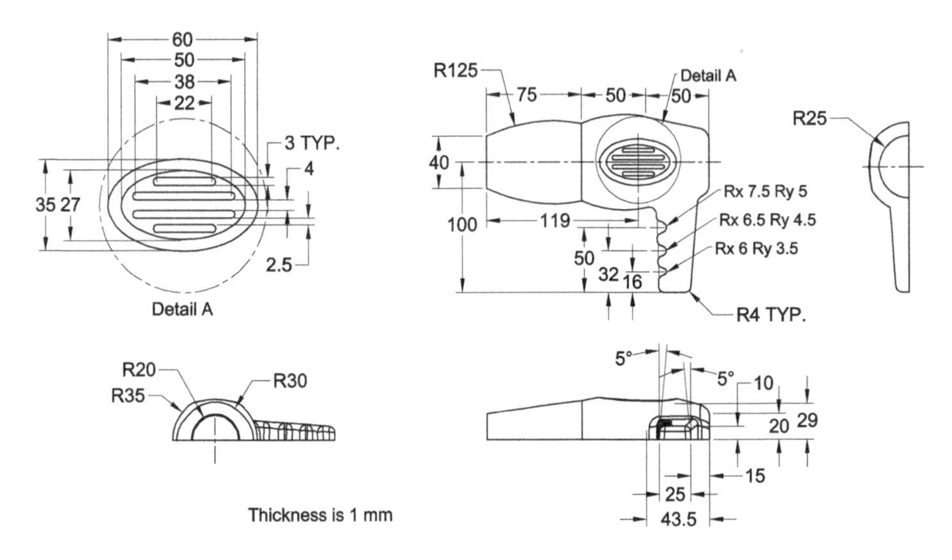

***Figure 15-132** Top, front, right, and detailed views of the surface model*

Exercise 3

In this exercise, you will create the surface model by using the sphere primitive shown in Figure 15-133 in the Freestyle modeling environment. **(Expected time: 55 min)**

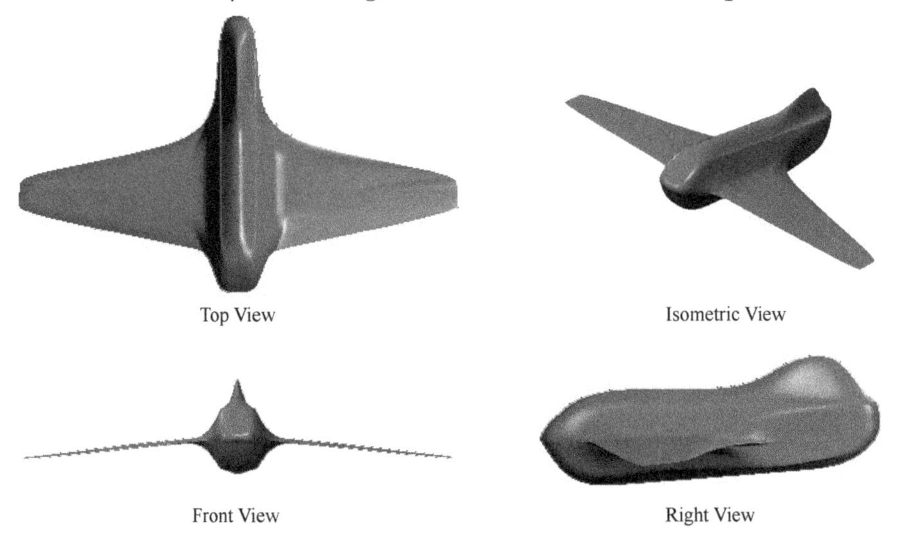

***Figure 15-133** Top, front, right, and isometric views of the surface model*

Answers to Self-Evaluation Test

1. F, **2.** T, **3.** T, **4.** T, **5.** T, **6.** Style, **7. Style**, **8. Merge**, **9. Curve**, **10.** surface

Chapter 16

Working with Sheetmetal Components

Learning Objectives

After completing this chapter, you will be able to:

- *Create unattached flat walls.*
- *Create reliefs in the sheetmetal component.*
- *Create a flat wall.*
- *Create the twist and extend walls.*
- *Add flange walls to sheetmetal components.*
- *Bend a sheetmetal component.*
- *Unbend and bend back the sheetmetal component .*
- *Convert solid models to sheetmetal components.*
- *Create a flat pattern of the sheetmetal component.*

INTRODUCTION TO SHEETMETAL

A sheetmetal component is created by bending, cutting, or deforming a thin sheet of uniform thickness. The sheetmetal designs are created in the **Sheetmetal** mode of PTC Creo Parametric. The **Sheetmetal** mode provides you the options to create primary and secondary walls and convert a solid model into a sheetmetal component.

INVOKING THE SHEETMETAL MODE

To invoke the **Sheetmetal** mode, choose the **New** button from the **Quick Access** toolbar; the **New** dialog box will be displayed. In this dialog box, the **Part** radio button in the **Type** area and the **Solid** radio button in the **Sub-type** area are selected by default. Select the **Sheetmetal** radio button from the **Sub-type** area, as shown in Figure 16-1. Enter the name of the file in the **Name** edit box and choose the **OK** button from the **New** dialog box; the **Sheetmetal** mode will be invoked.

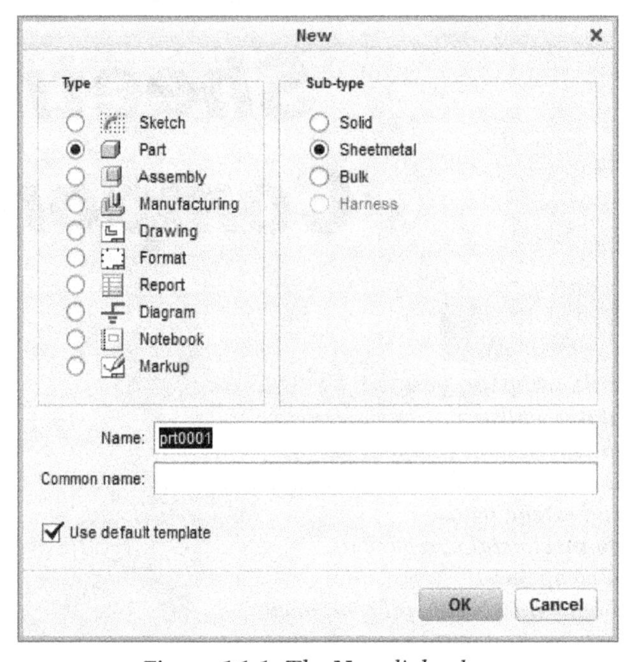

Figure 16-1 *The* ***New*** *dialog box*

Figure 16-2 shows you the initial screen appearance in the **Sheetmetal** mode. The initial screen appearance in the **Sheetmetal** mode is similar to the initial screen appearance in the **Part** mode. The three default datum planes, **FRONT**, **TOP**, and **RIGHT** will be displayed in the drawing area. These datum planes are mutually perpendicular to each other. The **Model Tree** will be displayed to the left of the drawing area.

The process of creating sheetmetal components is similar to that of creating solid or surface protrusions. You first need to create the sketch of the base feature and then use various tools such as Flat, Flange, and so on to create sheetmetal components.

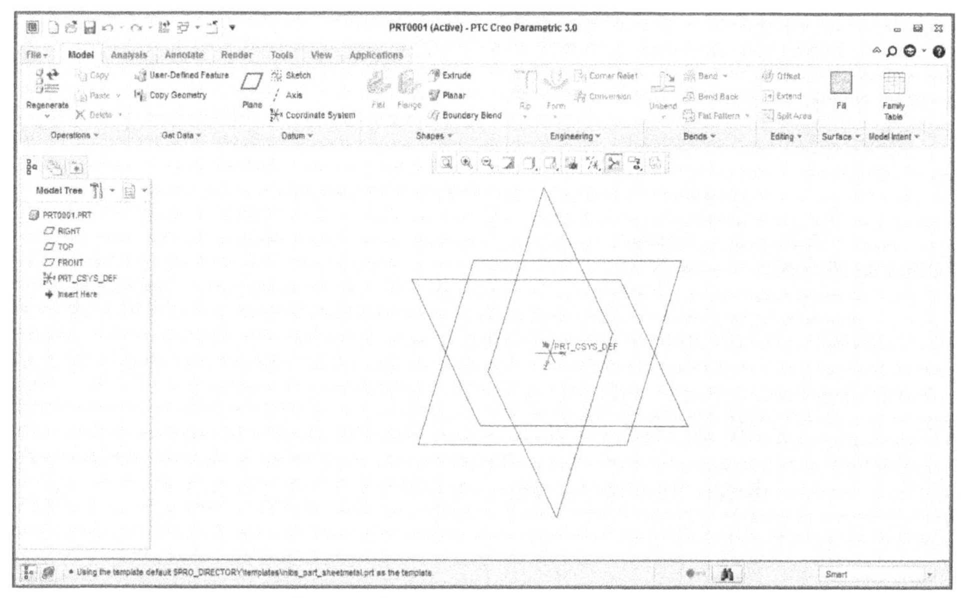

*Figure 16-2 The initial screen appearance in the **Sheetmetal** mode*

INTRODUCTION TO SHEETMETAL WALLS

A wall is any section of the sheetmetal material in your design. There are two types of walls in **Sheetmetal** design of PTC Creo Parametric. They are discussed next.

Primary Walls

Primary walls are created as base features and are independent entities. The different type of primary walls are unattached flat wall, unattached extruded wall, revolved wall, blended wall, offset wall, variable section sweep, swept blend, helical sweep, and so on.

Secondary Walls

Secondary walls are the walls that are dependent on at least one primary wall. They share a parent-child relation with the primary walls. Secondary walls include all the primary walls and also walls such as flat wall, flange wall, extend wall, twist wall, and so on.

To design a part from scratch, you need to create the primary wall as the base feature. The other features such as flange wall, extend wall, and so on can then be created with reference to the primary walls.

Creating the Planar Wall

Ribbon: Model > Shapes > Planar

The first feature created in a sheetmetal design is the base feature. This base feature is known as the unattached flat wall. To create a planar wall, choose the **Planar** tool from the **Shapes** group of the **Ribbon**; the **Planar** dashboard will be displayed, as shown in Figure 16-3. Options in this dashboard are discussed next.

*Figure 16-3 Partial view of the **Planar** dashboard*

References Tab

When you choose the **References** tab, a slide-down panel will be displayed, as shown in Figure 16-4. This slide-down panel helps you to select the sketching plane to create the sketch. Choose the **Define** button to display the **Sketch** dialog box. Select the required sketching plane and orient it. Once you enter the sketcher environment, you need to create the sketch for the wall and then choose the **OK** button to exit the sketcher environment. Next, enter the thickness of the sheet in the edit box in the **Planar** dashboard. Finally, choose the **Build feature** button to complete the creation of the unattached flat wall. Figure 16-5 shows a sheetmetal model created using this method.

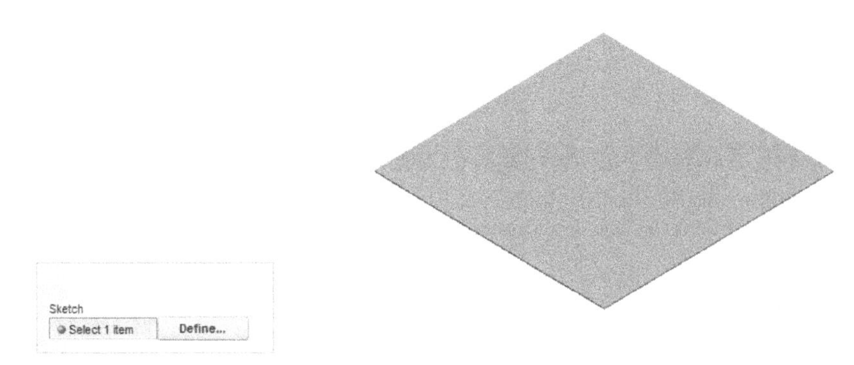

*Figure 16-4 The **References** slide-down panel*

Figure 16-5 Model of the unattached flat wall

Properties Tab

When you choose the **Properties** tab, a slide-down panel will be displayed. This slide-down panel displays the feature identity in the **Name** collector. The **i** button in the slide-down panel, when chosen, opens the browser that displays all information about the feature you are creating.

Note

*The **Options** tab will not be enabled when you are creating the unattached planar wall or any other primary wall. It will be enabled only when the secondary walls are being created.*

Creating the Unattached Revolve Wall

Ribbon: Model > Shapes > Revolve

The unattached revolve wall is a feature created by revolving a sketched section about a centerline. The revolve wall can be a primary or secondary wall. When you choose the **Revolve** button from the expanded **Shapes** group, the **Revolve** dashboard will be displayed, as shown in Figures 16-6.

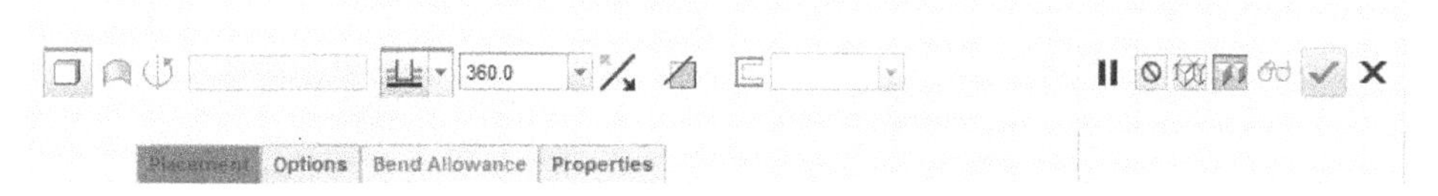

*Figure 16-6 Partial view of the **Revolve** dashboard*

The method of creating a revolve wall is same as of creating a revolve feature, refer to Chapter 4. The section for creating the revolve wall can be closed or open. When you choose the **Options** tab in the dashboard, a slide-down panel will be displayed, as shown in Figure 16-7. Some of the options in this tab are the same as for the **Revolve** feature. The sheetmetal options available in this slide-down panel are discussed next.

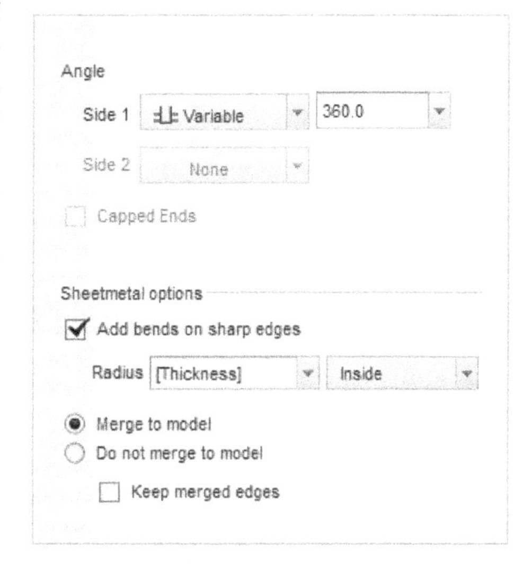

*Figure 16-7 The **Options** slide-down panel*

Add bends on sharp edges

The **Add bends on sharp edges** check box is used to add rounds on the sharp edges. On selecting this check box, the **Radius** drop-down list gets activated. Using the left drop-down list, you can specify the radius value of rounds, and using the right drop-down list, you can specify the location of rounds.

Set driving surface opposite sketch plane

The **Set driving surface opposite sketch plane** check box is used to flip the driving surface. This option will be available only when you select the **Do not merge to model** radio button.

Keep merged edges

The **Keep merged edges** check box is used to merge the revolve surface with the existing walls. This option will be available when you select the **Merge to model** radio button. If you want to keep the edges merged with the existing wall, then you need to select the **Keep merged edges** check box.

Figure 16-8 shows a closed sketch for creating the revolve wall and Figure 16-9 shows the resulting unattached revolved wall.

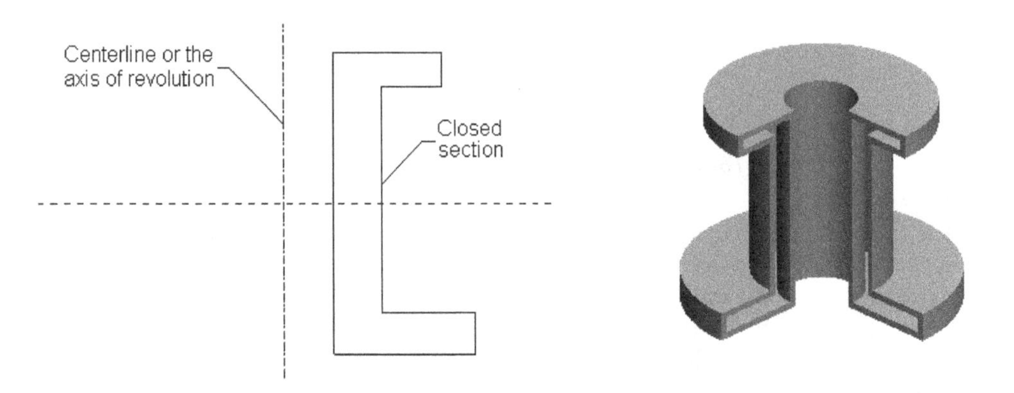

Figure 16-8 Closed sketch for creating the revolved wall

Figure 16-9 Resulting unattached revolved wall

Creating the Unattached Blend Wall

Ribbon: Model > Shapes > Blend

The **Blend** option connects two or more sections by combining the characteristics of each section. The number of entities in each section of the unattached blend wall should be the same. For example, you cannot blend a circle with a rectangle. This is because a rectangle is made up of four entities, whereas a circle is made up of only one entity. Therefore, these two sections can be blended only if the circle is divided into four entities.

The **Blend** option is used where the component to be created has varying cross-sections. To invoke this option, choose the **Blend** tool from the expanded **Shapes** group; the **Parallel Blend** dashboard is displayed, as shown in Figure 16-10. The options in this dashboard are discussed next.

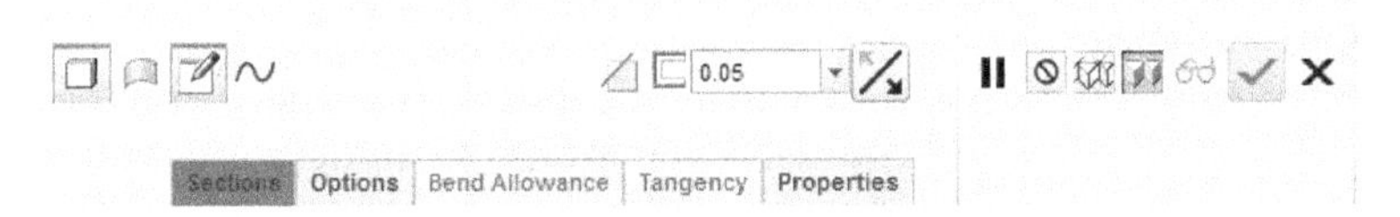

*Figure 16-10 Partial view of the **Parallel Blend** dashboard*

Blend as wall

The **Blend as wall** button is chosen by default and is used to blend two or more sketches in an unattached wall form. On choosing this button, the **Create a thin feature** button will be activated in the dashboard which can be used to define the thickness of the wall.

Blend as surface

The **Blend as surface** button is used to blend two or more sketches in a surface form. The surface form wall has zero mass properties.

Blend with sketched sections

This button is used to create sketches in the drawing area to create the blend. The options for inserting sketches are available in the **Section** tab that is discussed later in this chapter.

Blend with selected sections

This button is used to select sketches available in the drawing area to form the blend. The options for inserting sketches are available in the **Section** tab.

Remove Material

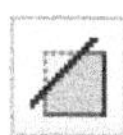

The **Remove Material** button will be available in the **Blend** dashboard only when a wall is created. This is because this button is used to remove the material from an existing wall.

Sections Tab

When you choose the **Sections** tab from the dashboard, the **Sections** slide-down panel will be displayed, as shown in Figure 16-11. This panel has two radio buttons that are discussed next.

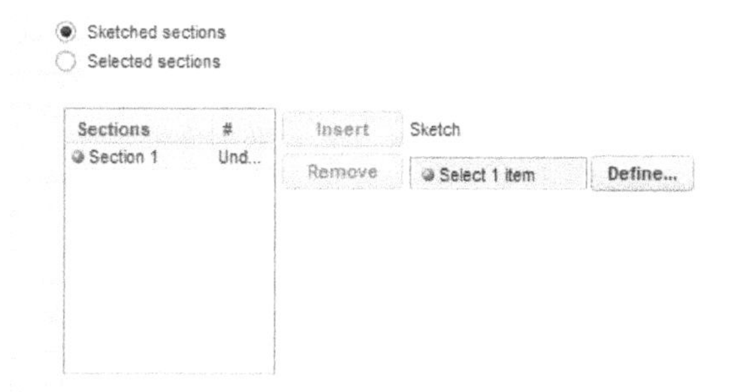

***Figure 16-11** The **Sections** slide-down panel*

Sketched sections

This radio button is used to create sketches in the drawing area to form the blend. When you select this radio button, the **Define** button will be displayed which is used to enter the sketcher environment. You can use the **Insert** button displayed in the panel to add more sketches in blend.

Selected sections

This radio button is used to select sketches from the drawing area to form the blend. You can use the **Insert** button displayed in the panel to add more sketches in blend.

Options Tab

The **Options** tab contains the options related to the shape to be created of the blend feature. The options available in this tab are discussed next.

Straight

The **Straight** radio button is available in the **Blended surfaces** area. This option is used to connect the vertices of all sections in a blend feature with straight lines.

Smooth

The **Smooth** radio button is available in the **Blended surfaces** area. This option is used to connect the vertices of all sections in a blend feature with smooth curves.

Capped Ends

The **Capped Ends** check box is available in the **Start and End Sections** area. This check box will be activated only when you choose the **Blend as surface** option from the dashboard. This option is used to close the sketch ends.

Add bends on sharp edges

The **Add bends on sharp edges** check box is available in the **Sheetmetal options** area. This option is used to provide rounds on the sharp edges of a wall. On selecting this option, you can specify the value of radius in the edit box activated below this option.

Tangency Tab

The **Tangency** tab will only be available when you select the **Smooth** radio button from the **Options** tab. This tab is used to set the boundary of selected or created sketches in the **Free**, **Tangent**, or **Normal** condition.

Figure 16-12 shows the sketch of three sections that will be used to create the parallel blend feature. Figures 16-13 and 16-14 show the parallel blend features with straight edges and smooth edges, respectively. These features are created using the sections shown in Figure 16-12.

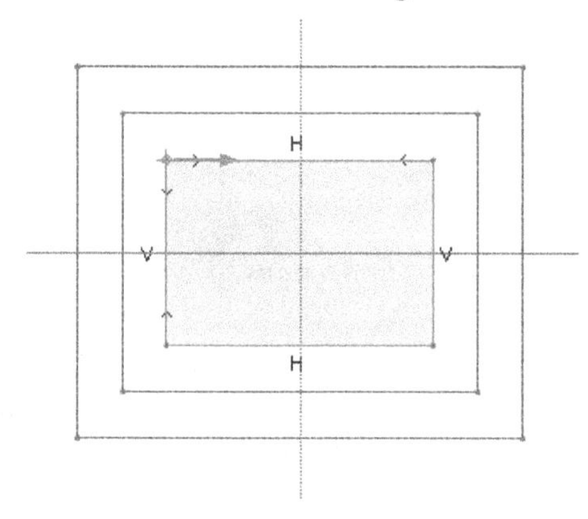

Figure 16-12 *The sketch showing three sections for creating the parallel blend feature*

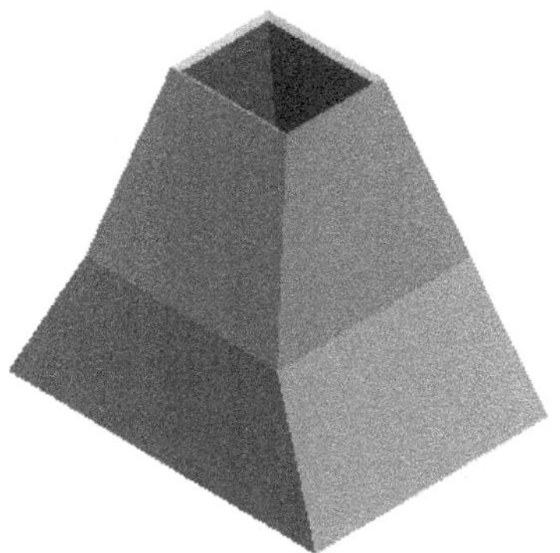

Figure 16-13 *Parallel blend feature with straight edges*

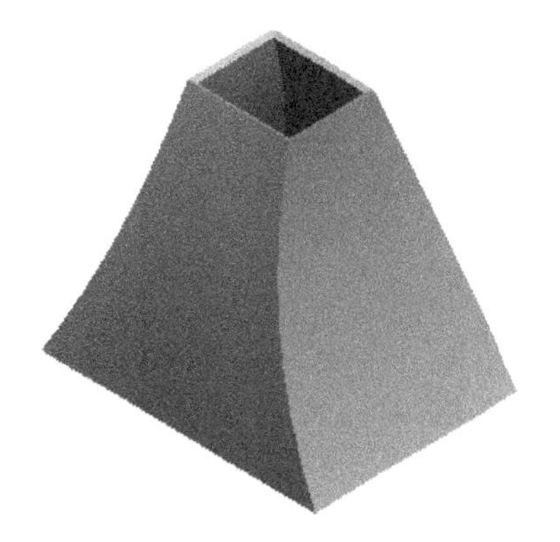

Figure 16-14 *Parallel blend feature with smooth edges*

Note

While drawing sections for creating an unattached blend wall, the start point of all sections should be in the same direction in order to avoid twisted blend features.

Creating the Unattached Rotational Blend Wall

Ribbon: Model > Shapes > Rotational Blend

The rotational blends are created by using the sections that are rotated about the Y-axis up to a maximum of 120-degree. To do so, an angle called rotational blend angle has to be defined between each section. Each section in the rotational blend feature is rotated about a geometric centerline drawn in the first sketch.

To invoke the **Rotational Blend** tool, choose **Rotational Blend** from the expanded **Shapes** group; the **Rotational Blend** dashboard will be displayed, as shown in Figure 16-15.

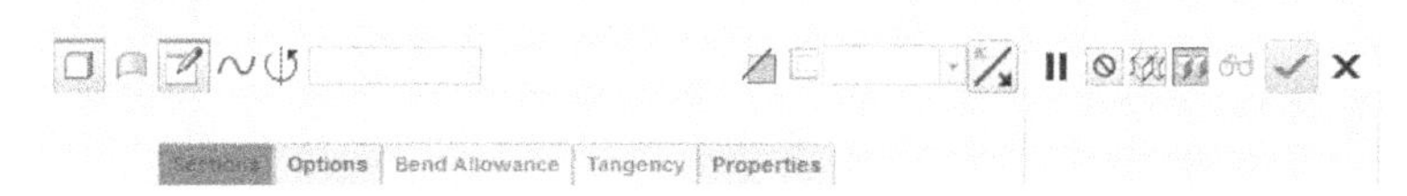

Figure 16-15 *Partial view of the **Rotational Blend** dashboard*

Figure 16-16 shows the sections used to create the rotational blend feature. As evident, in Figure 16-16, two sections are used to create the rotational blend feature and these sections are at an angle of 90-degree to each other. It is also evident from the figure that the second section is dimensioned from the geometric centerline that was defined in the first section. Figure 16-17 shows the shaded model of the rotational blend feature.

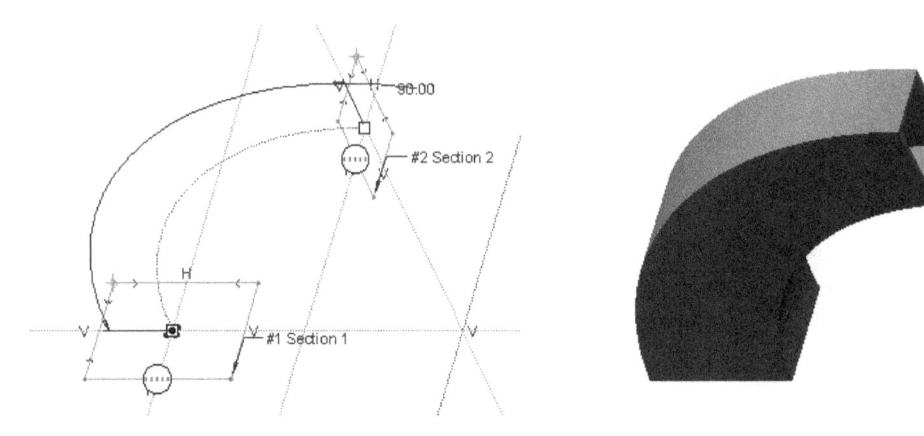

Figure 16-16 *The two sections with dimensions used to create the rotational blend feature*

Figure 16-17 *Shaded model of the rotational blend feature*

Note

*You can select the **Connect end and start sections** check box from the **Options** tab to create a closed blend feature. But for creating a closed blend feature, there should be at least three sections.*

Creating the Unattached Offset Wall

Ribbon: Model > Editing > Offset

The unattached offset wall is a reflection of a quilt or a surface set at a specified distance from the original wall. To create an offset wall, choose the **Offset** button from the **Editing** group; the **Offset** dashboard will be displayed, as shown in Figure 16-18, and you will be prompted to select a quilt or a solid surface to offset.

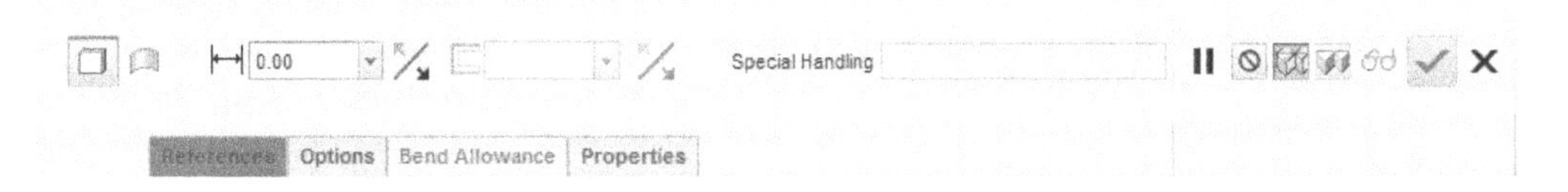

Figure 16-18 *The **Offset** dashboard*

Select the surface that you want to offset; you will be prompted to enter the offset distance. Enter the required distance value in the edit box available in the dashboard. You can also define the type of offset wall using the options in the **Options** tab. When you click on the **Normal to Surface** button from this slide-down panel, a drop-down list will be displayed, as shown in Figure 16-19. The options in the list are discussed next.

Normal to Surface

The **Normal to Surface** option is chosen by default. As a result, the offset wall is created normal to the quilt or selected surface.

Controlled Fit
This option, when selected, creates the offset wall at a controlled distance.

Automatic Fit
This option automatically fits the offset wall at a distance from the quilt or surface. Figure 16-20 shows the offset wall created at a distance of 500 from the top surface of the primary flat wall.

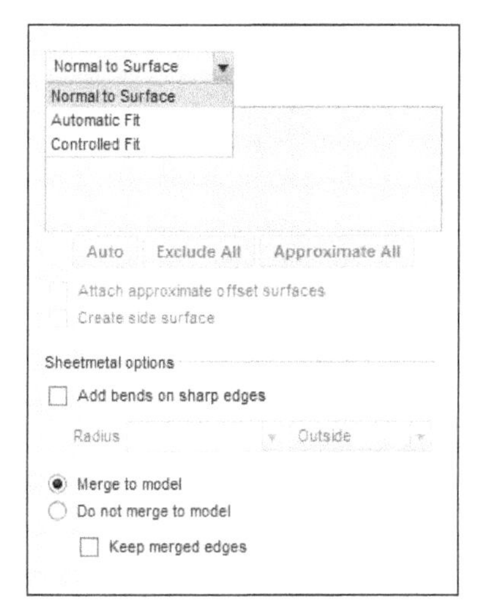

Figure 16-19 *The drop-down list displayed on choosing the* ***Normal to Surface*** *option*

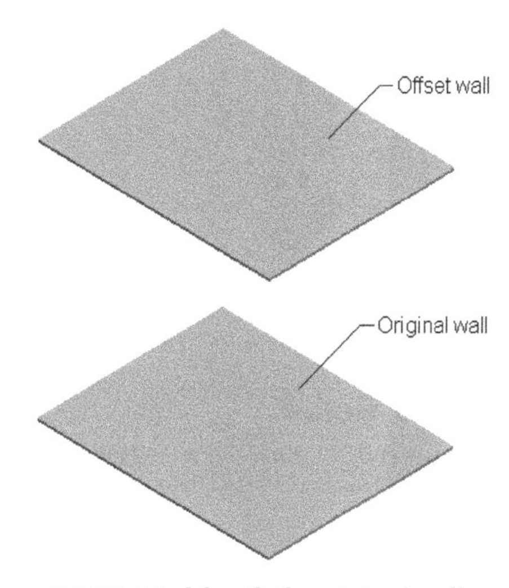

Figure 16-20 *Model with the original wall and the offset wall*

Note
An offset wall cannot be the first feature in your design.

Creating Reliefs in Sheetmetal Components
The reliefs control the sheetmetal material behavior and prevent unwanted deformation. For example, consider a bend without a relief. It might not represent an accurate real life model due to the stretching of material. By adding the appropriate bend relief, the sheetmetal bend will meet your design intent and enable you to create an accurate flat model. Reliefs may be added to the model at various stages of its creation.

Note
You will learn about the options in the ***Reliefs*** *tab later in this chapter.*

The different types of reliefs are discussed next.

Stretch Relief
This type of relief stretches the material, in which the bent area intersects the fixed area. The stretch relief is shown in Figure 16-21.

Rip Relief

This type of relief cuts the material at each bend endpoint. The cuts are made normal to the bend line. The rip relief is shown in Figure 16-22.

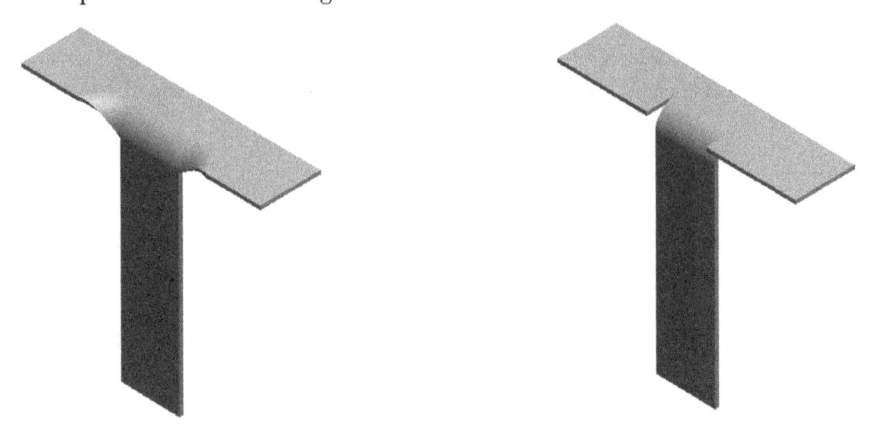

Figure 16-21 Model with the stretch relief *Figure 16-22 Model with the rip relief*

Rectangular Relief

This option adds a rectangular relief at each bend endpoint. The rectangular relief is shown in Figure 16-23.

Obround Relief

This option adds an obround relief at each bend endpoint. The obround relief is shown in Figure 16-24.

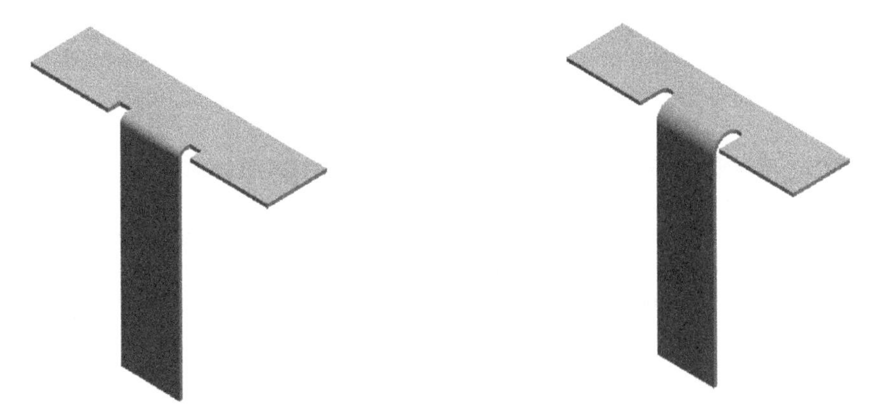

Figure 16-23 Model with the rectangular relief *Figure 16-24 Model with the obround relief*

Creating a Flat Wall

Ribbon: Model > Shapes > Flat

A flat wall is a planar or an unbent section of the sheetmetal. To create a flat wall, first you need to create the primary wall. The creation of the primary wall (unattached flat wall) has been discussed earlier in this chapter. The flat wall can take any shape, but the surface adjacent to the attached edge must be planar. To create a flat wall, choose the **Flat** tool from the **Shapes** group in the **Model** tab; the **Flat** dashboard will be displayed, as shown in Figure 16-25. The options in this dashboard are discussed next.

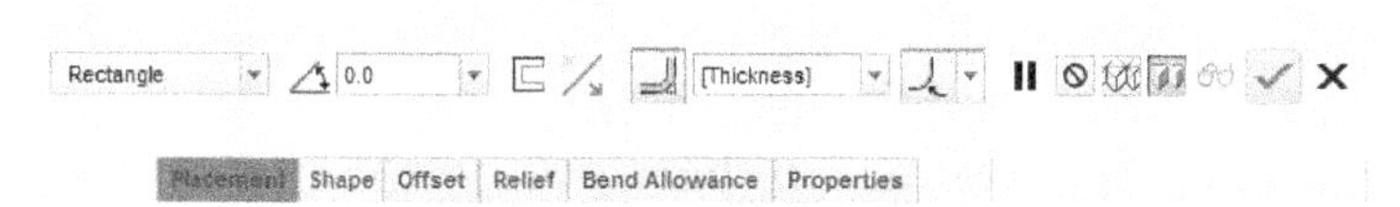

***Figure 16-25** The partial view of the **Flat** dashboard*

Placement Tab

When you choose the **Placement** tab, a slide-down panel will be displayed, as shown in Figure 16-26. The collector in this slide-down panel displays **Select 1 item** because the attachment edge has not yet been defined. When you select the attachment edge, the collector in the slide-down panel will display the edge selected. Also, a preview of the rectangular flat wall will be created perpendicular to the edge selected.

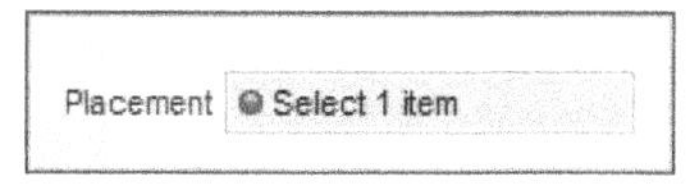

***Figure 16-26** The **Placement** slide-down panel*

Shape Tab

When you choose the **Shape** tab, a slide-down panel will be displayed, as shown in Figure 16-27. This panel displays the predefined wall profiles and their default dimensions.

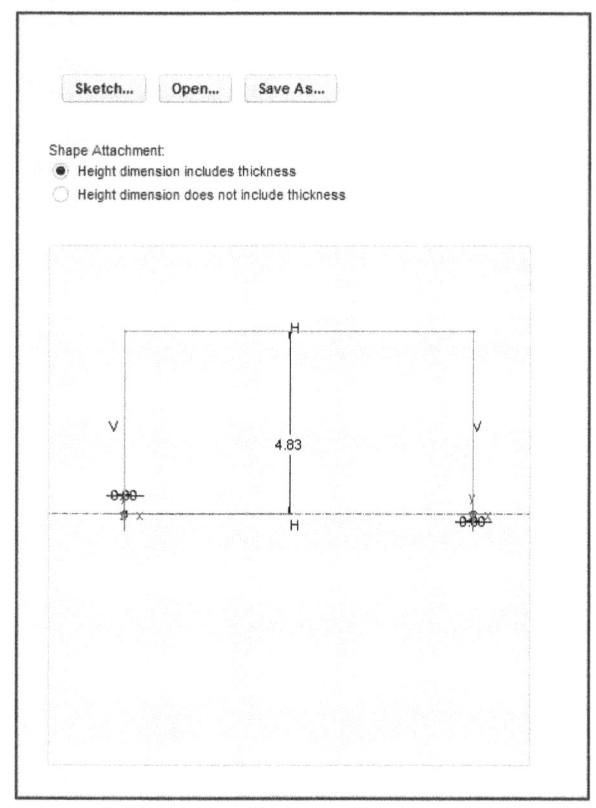

***Figure 16-27** The **Shape** slide-down panel*

By default, **Rectangle** is selected in the Profiles drop-down list in the **Flat** dashboard. The different profiles of walls are **Rectangle, Trapezoid, L, T**, and **User Defined**. These profiles can be selected from the profiles list in the **Flat** wall dashboard.

Offset Tab

The **Offset** tab, when chosen, creates the flat wall at a specified distance from the selected attachment edge. When you choose the **Offset** tab, a slide-down panel will be displayed, as shown in Figure 16-28. When you select the **Offset wall with respect to attachment edge** check box, the **Type** drop-down list will be enabled. The options in this drop-down list are discussed next.

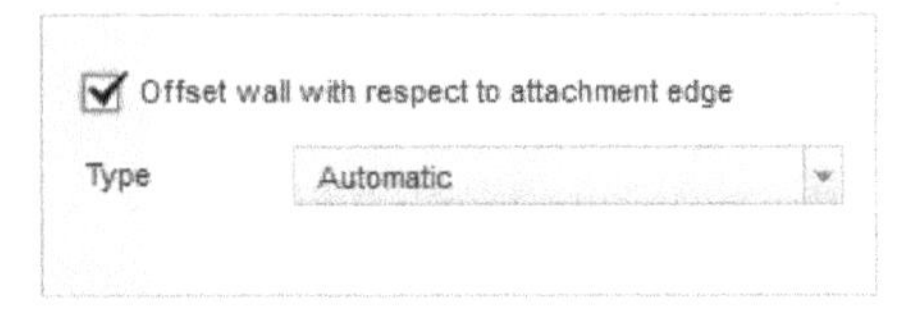

*Figure 16-28 The **Offset** slide-down panel*

Add to Part edge

If you select the **Add to Part edge** option, the wall gets appended to the attachment edge without trimming the height of the attachment edge.

Automatic

The **Automatic** option, when selected, offsets the wall and maintains its original height.

By value

The **By value** option, when selected, offsets the wall at a specified distance from the attachment edge. You can also drag the graphic handle to adjust the offset value.

Relief Tab

The **Relief** tab allows you to choose the type of relief to be applied on the wall. When you choose the **Relief** tab, a slide-down panel will be displayed, as shown in Figure 16-29. When the **Define each side separately** check box is selected, the **Side 1** and **Side 2** radio buttons will be enabled. These options are discussed next.

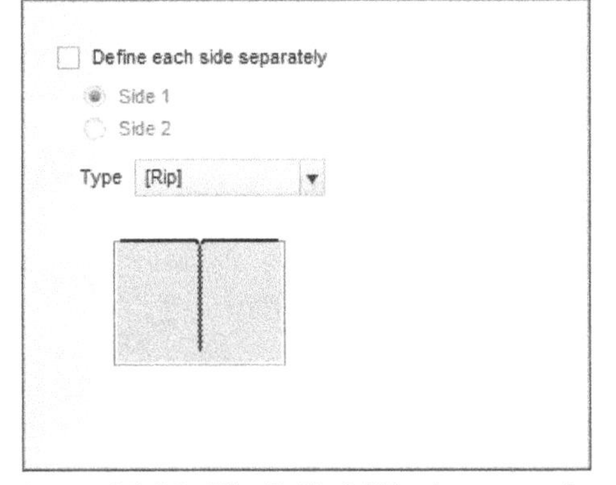

*Figure 16-29 The **Relief** slide-down panel*

Side 1

The **Side 1** radio button is used to apply a relief on one side of the attachment edge.

Side 2

The **Side 2** radio button is used to apply a relief on the other side of the attachment edge.

Type Drop-down List

The **Type** drop-down list specifies different types of reliefs such as **No Relief**, **Rip**, **Stretch**, **Rectangular**, and **Obround**. By default, the **Rip** is selected in the **Type** drop-down list. The types of relief have been discussed earlier.

Bend Allowance Tab

The **Bend Allowance** tab is used to calculate the undeveloped length of the sheetmetal required to create a bend of specified radius and angle. This tab will be available only when the **Adds bend on the attachment edge** button is chosen from the dashboard. When you choose the **Bend Allowance** tab, the slide-down panel will be displayed, as shown in Figure 16-30. When the **A Feature Specific Set Up** check box is selected, the **By K factor**, **By Y factor**, and **By Bend table** options will be enabled. These options are discussed next.

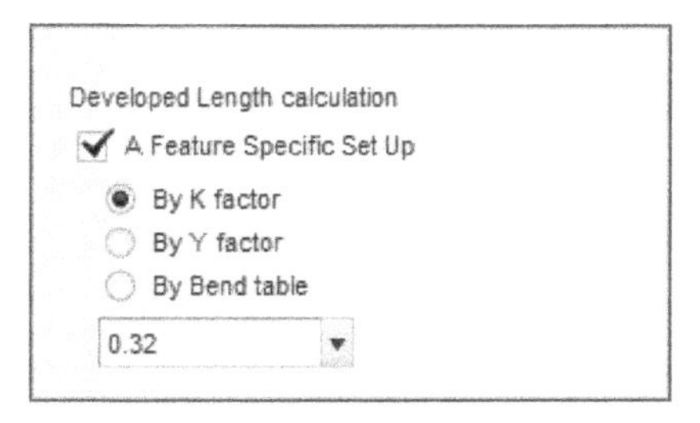

*Figure 16-30 The **Bend Allowance** slide-down panel*

By K factor

K Factor is the ratio of distance between the neutral bend line and the inside bend radius to the thickness of the material.

By Y factor

Y Factor is the ratio between the neutral bend line and thickness of the material.

By Bend table

It controls the calculations required for the length of flat material (developed length) needed to create a bend.

Add Bend on the attachment edge

The **Add Bend on the attachment edge** button creates a bend between the existing sheetmetal wall and the newly created wall. This button is available on the dashboard. There are three methods of specifying the bend parameters of the flat wall and these are discussed next.

Dimension the inner surface of the bend

On choosing this button, you can specify the dimension for the inner surface of the bend.

Dimension the outer surface of the bend

On choosing this button, you can specify the dimension for the outer surface of the bend.

By Parameter

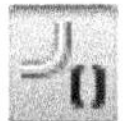

When you choose this button, the radius value and the radius side are automatically taken from the **SMT_DFLT_BEND_RADIUS** and **SMT_DFLT_RADIUS_SIDE** parameters.

Figure 16-31 shows the base wall and Figure 16-32 shows the flat wall created on the rear edge of the base wall.

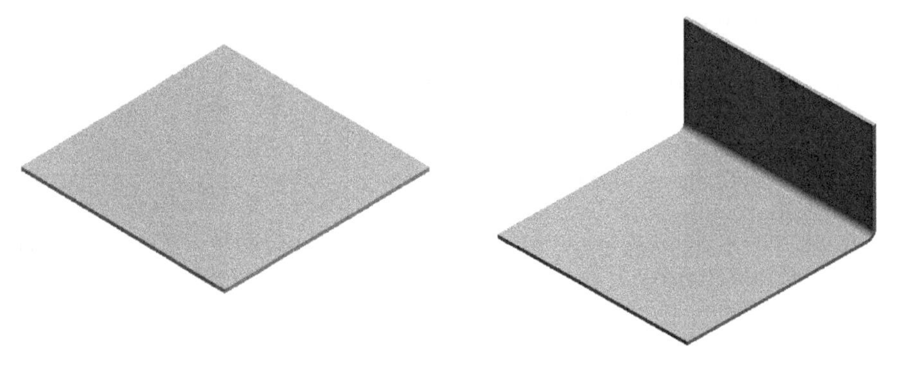

Figure 16-31 Model of the base wall *Figure 16-32 Model with the flat wall*

Creating a Twist Wall

Ribbon: Model > Shapes > Twist

A twist wall is a spiraling or coiling section of sheetmetal. The twist wall is formed around an axis running through the center of the wall, turning the ends of the wall in opposite directions by a relatively small specified angle. You can attach the twist wall to a straight edge or an existing planar wall. The twist wall typically serves as a transition between two areas of sheetmetal.

Choose the **Twist** tool from the expanded **Shapes** group in the **Model** tab to invoke the **TWIST** dialog box and the **FEATURE REFS** menu. The **TWIST** dialog box and the **FEATURE REFS** menu are shown in Figures 16-33 and 16-34, respectively.

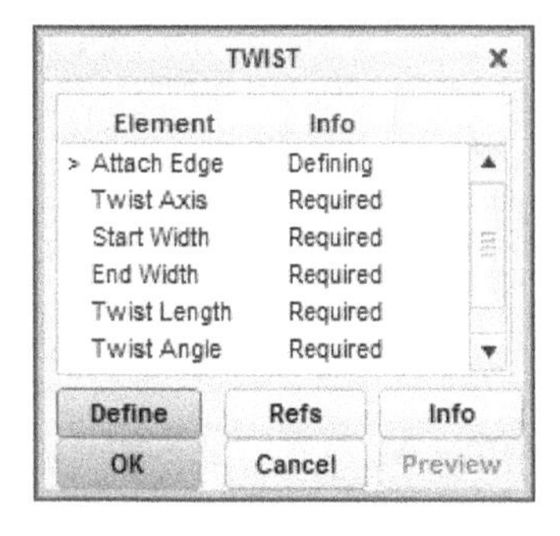

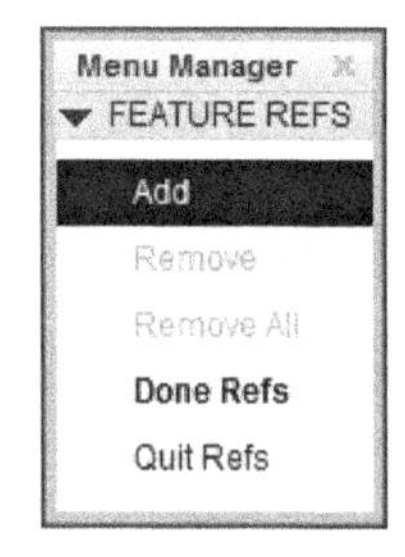

*Figure 16-33 The **TWIST** dialog box* *Figure 16-34 The **FEATURE REFS** menu*

On invoking the **Twist** dialog box, you will be prompted to select the attachment edge on which the twist wall is to be created. Select the attachment edge from the drawing area; the **TWIST AXS PNT** menu will be displayed, as shown in Figure 16-35.

The **TWIST AXS PNT** menu allows you to define a point on the attachment edge that will form the centerline of the twist wall. The two options of defining the start point for the twist wall are discussed next.

Select Point

This option is used to choose an existing datum point on the attachment edge.

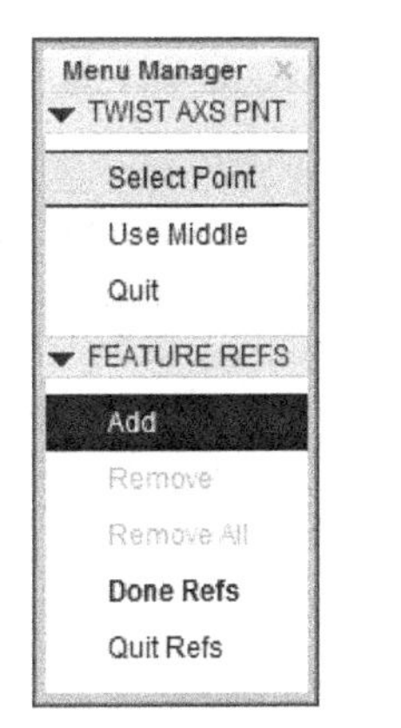

Figure 16-35 The ***TWIST AXS PNT*** *menu*

Use Middle

This option is used to select the middle point of the attachment edge as the centerline for the twist wall.

After specifying the point that defines the twist axis, you will be prompted to specify the dimensions of the wall. These options are discussed next.

Start Width

This option is used to specify the width at the start point of the twist wall.

End Width

This option is used to specify the width at the end point of the twist wall.

Twist Length

This option indicates the length of the twist wall measured from the attachment edge to the end point of the twist axis.

Twist Angle

The **Twist Angle** option is used to specify the angle of twist for the twist wall.

Developed Length

This option indicates the length of the twist wall, when untwisted.

Figure 16-36 shows the twist wall attached to the base wall.

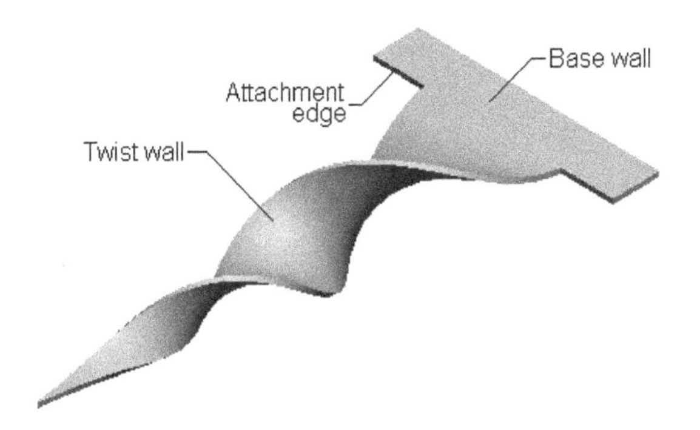

Figure 16-36 *Model with the twist wall attached to the base wall*

Creating an Extend Wall

Ribbon: Model > Editing > Extend

The extend wall lengthens an existing wall. You can extend the wall from a straight edge on an existing wall to a planar surface or to a specified distance. The **Extend** button is located in the **Editing** group. This button will be activated only when you select any straight edge. On choosing this button, the **Extend** dashboard will be displayed, as shown in Figure 16-37. Now, you can enter the value of extension in the **Distance to extend** edit box. The options in the dashboard are discussed next.

*Figure 16-37 The **Extend** dashboard*

Extend the wall by value

This option allows you to extend the wall to a specified distance. You can select a default value, enter a required value, or select two parallel edges to extend.

Extend the wall to intersect the reference plane

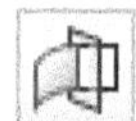

This option allows you to extend the wall to intersect a plane or face. The plane or face to be intersected can be an existing default datum plane, the flat face of another independent flat wall, or a new datum plane.

Extend the wall up to the reference plane

This option allows you to extend the wall up to a plane. The plane or face upto which a wall is to be extended can be an existing default datum plane, the flat face of another independent flat wall, or a new datum plane.

Figure 16-38 shows two independent unattached flat walls, Figure 16-39 shows the extended wall created using the **Extend the wall to intersect the reference plane** button, and Figure 16-40 shows the extended wall created using the **Extend the wall upto the reference plane** button.

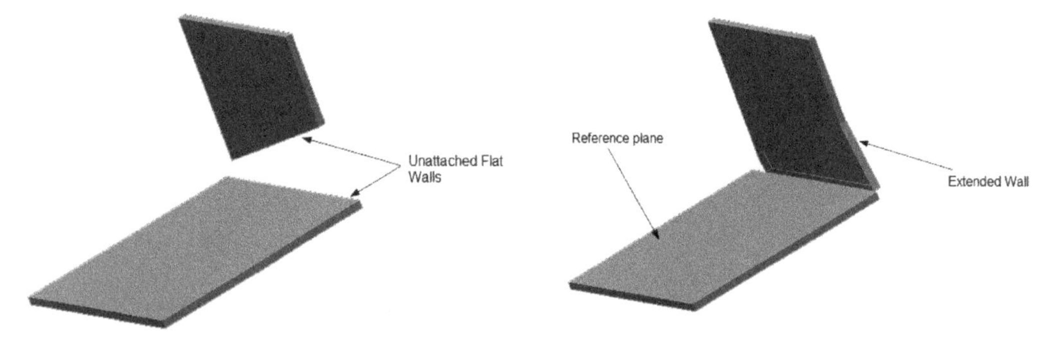

Figure 16-38 Model with two independent unattached flat walls

*Figure 16-39 Extended wall created using the **Extend the wall to intersect the reference plane** button*

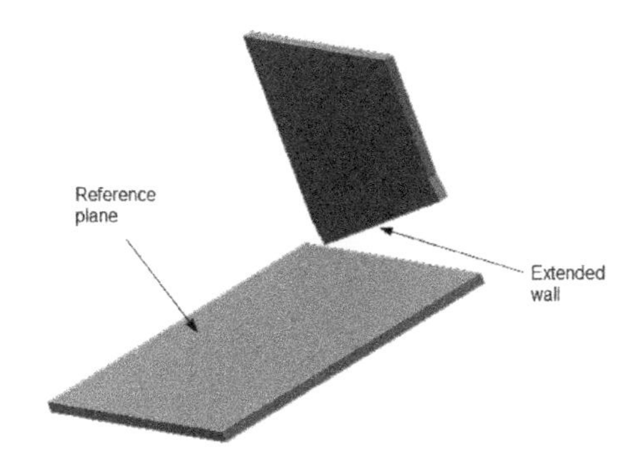

Figure 16-40 Extended wall created using the ***Extend the wall upto the reference plane*** *button*

Creating a Flange Wall

Ribbon: Model > Shapes > Flange

A flange wall is the bent section of a sheetmetal. To create a flange wall, choose **Model > Shapes > Flange** from the **Ribbon**; the **Flange** dashboard will be displayed as shown in Figure 16-41. The options in this dashboard are discussed next.

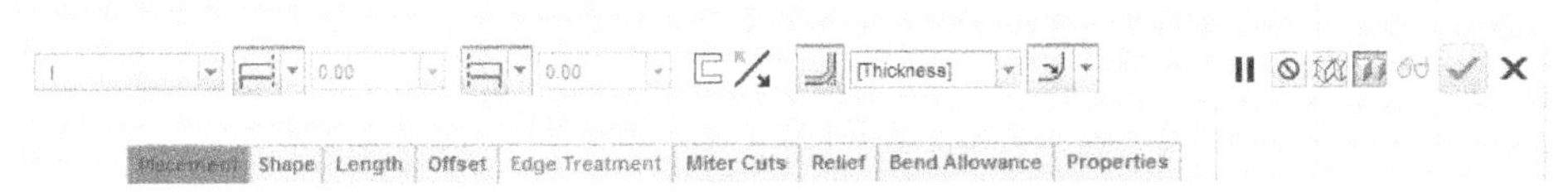

Figure 16-41 *Partial view of the* ***Flange*** *dashboard*

Placement Tab

The **Placement** tab indicates the reference edge chosen to create the flange wall. When you choose the **Placement** tab, a slide-down panel will be displayed, as shown in Figure 16-42.

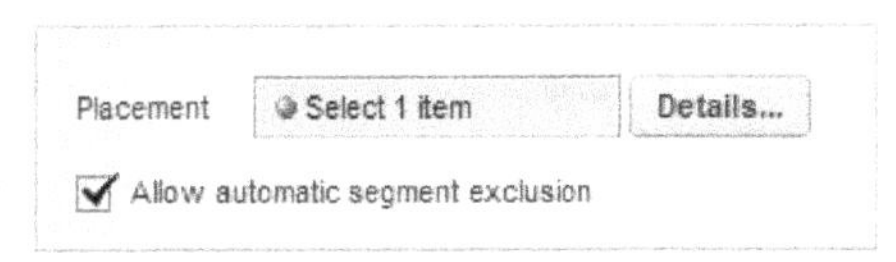

Figure 16-42 *The* ***Placement*** *slide-down panel*

Shape Tab

The **Shape** tab indicates the profiles and dimensions of the predefined flange wall shapes such as **I**, **Arc**, **S**, **Open**, **Flushed**, **Duck**, **C**, **Z**, or **User Defined**. The predefined wall profile can be selected from the drop-down list in the **Flange** dashboard. When you choose the **Shape** tab after selecting the predefined wall profile, a slide-down panel will be displayed with three buttons, as shown in Figure 16-43. The **Sketch** button in the slide-down panel can be used to draw the sketch for the user-defined profile. The **Open** button in the slide-down panel can be used to import sketch. Choose the **Open** button from this slide-down panel; the **Open** dialog box will be displayed. Using this dialog box, you can import the required sketch. The **Shape Attachment** area determines the height of the wall. There are two radio buttons in this area: **Height dimension includes thickness** and **Height dimension does not include thickness**. On selecting the first radio button, the thickness of the sheetmetal is included while calculating the wall height, whereas on selecting the other radio button, the thickness is not included while calculating the wall height. The **Save As** button in the slide-down panel can be used to save the sketch so that you can import it later on.

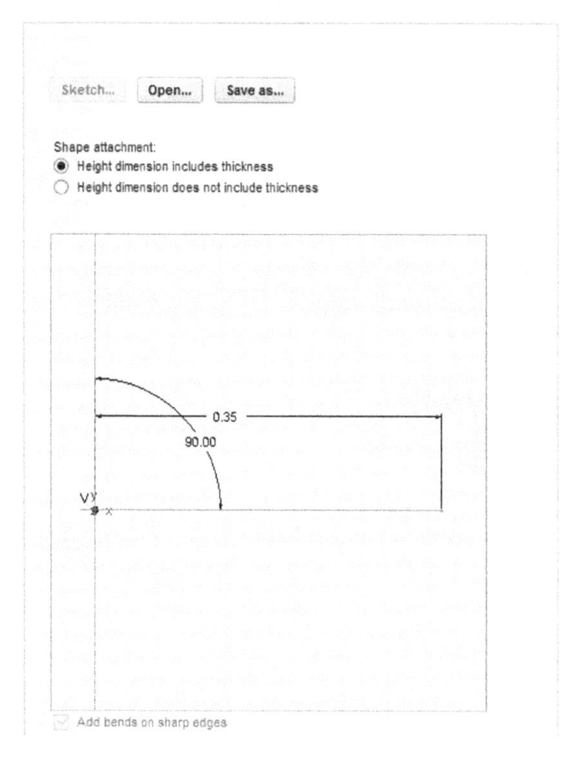

*Figure 16-43 The **Shape** slide-down panel*

Length Tab

The **Length** tab allows you to extend the end faces of the **Flange** independently. The edit boxes in the slide-down panel can be used to specify the distance through which the flange wall has to be extended in either direction. When you choose the **Length** tab, the slide-down panel will be displayed, as shown in Figure 16-44.

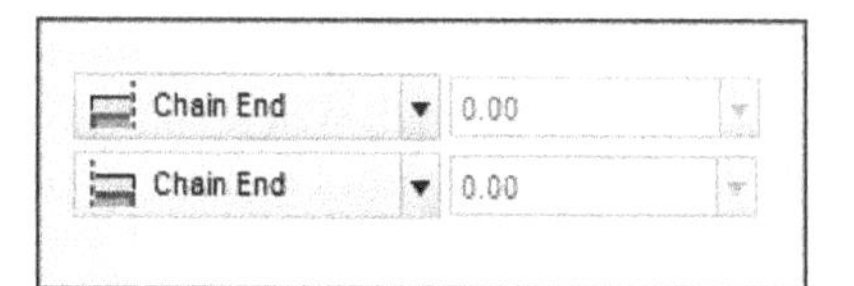

*Figure 16-44 The **Length** slide-down panel*

Offset Tab

The **Offset** tab allows you to offset the flange wall to a specified distance from the selected attachment edge. When you choose the **Offset** tab, the slide-down panel will be displayed, as shown in Figure 16-45.

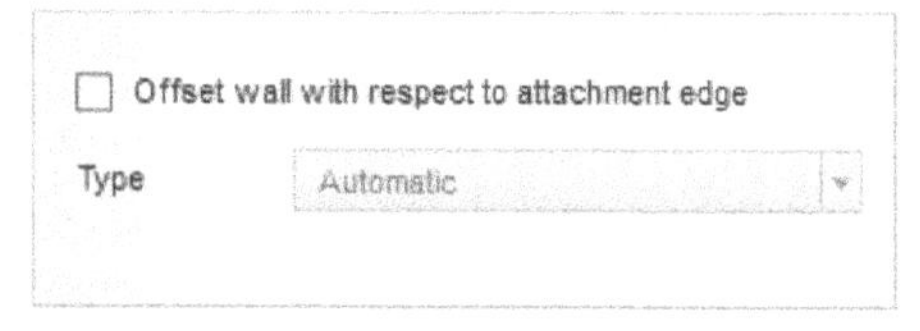

*Figure 16-45 The **Offset** slide-down panel*

When you select the **Offset wall with respect to attachment edge** check box, the **Type** drop-down list will get activated. The options in this drop-down list are: **Add to Part edge**, **Automatic**, and **By value**. These options are discussed next.

Add to Part edge

If you select the **Add to Part edge** option, the wall will be appended to the attachment edge without trimming the height of the attachment edge.

Automatic

Select this option to offset the flange wall and maintains its original height.

By value

Select this option to offset the flange wall at a specified distance.

Miter Cuts Tab

When you choose the **Miter Cuts** tab, a slide-down panel will be displayed, as shown in Figure 16-46. Miter cuts are created to allow bending of two sheetmetal plates created in the same plane.

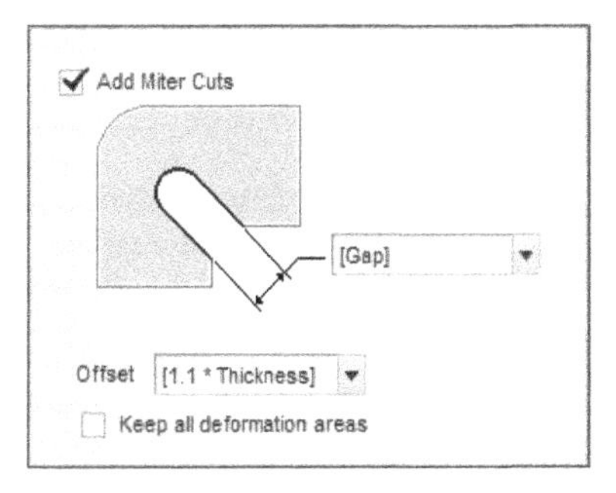

*Figure 16-46 The **Miter Cuts** slide-down panel*

Relief Tab

The **Relief** tab allows you to choose the type of relief to be provided to the sheetmetal wall to prevent any unwanted deformation. This tab can also be used to create bend reliefs and corner reliefs. When you choose the **Relief** tab, a slide-down panel will be displayed, as shown in Figure 16-47.

When the **Define each side separately** check box is selected in the **Relief** slide-down panel, the **Side 1** and **Side 2** radio buttons will be enabled. The bend reliefs available in the **Relief Category** area are **Rip**, **Stretch**, **Rectangular**, and **Obround**.

When you choose the **Corner Relief** option from the **Relief Category**, the **Relief** slide-down panel will be displayed, as shown in Figure 16-48. The corner reliefs available in this area are **V Notch**, **Circular**, **Rectangular**, and **Obround**.

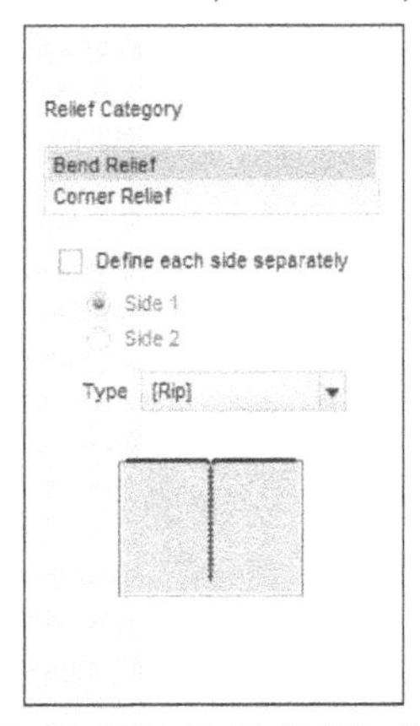

*Figure 16-47 The **Relief** slide-down panel with the **Bend Relief** option selected*

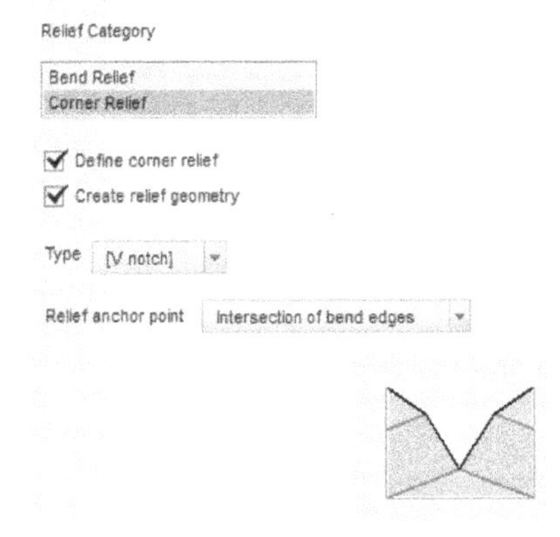

*Figure 16-48 The **Relief** slide-down panel with the **Corner Relief** option selected*

Bend Allowance Tab

When you choose the **Bend Allowance** tab, a slide-down panel will be displayed, as shown in Figure 16-49. When the **A Feature Specific Set Up** check box is selected, the **By K factor** radio button, the **By Y factor** radio button, and the **Use Bend table** check box will be enabled. These options have already been discussed earlier in this chapter.

Properties Tab

When you choose the **Properties** tab, a slide-down panel will be displayed, as shown in Figure 16-50. This slide-down panel displays the feature identity in the **Name** collector. The **i** button in the slide-down panel, when chosen, opens the browser that displays the information about the feature you have created. This slide-down panel also indicates the thickness of the flange wall being created.

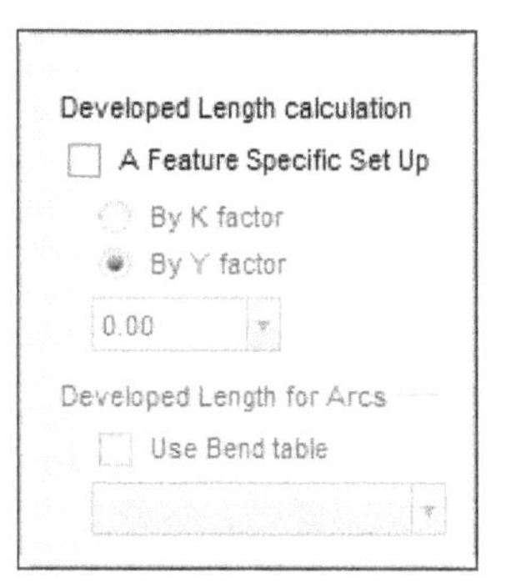

*Figure 16-49 The **Bend Allowance** slide-down panel*

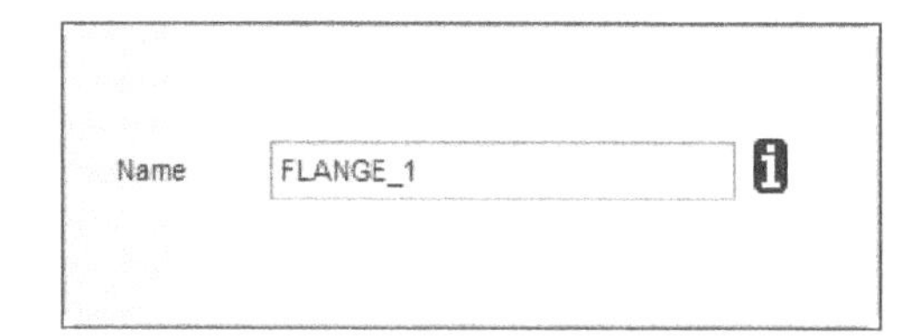

*Figure 16-50 The **Properties** slide-down panel*

Figure 16-51 shows a flange wall with a rectangular relief created on the primary unattached flat wall.

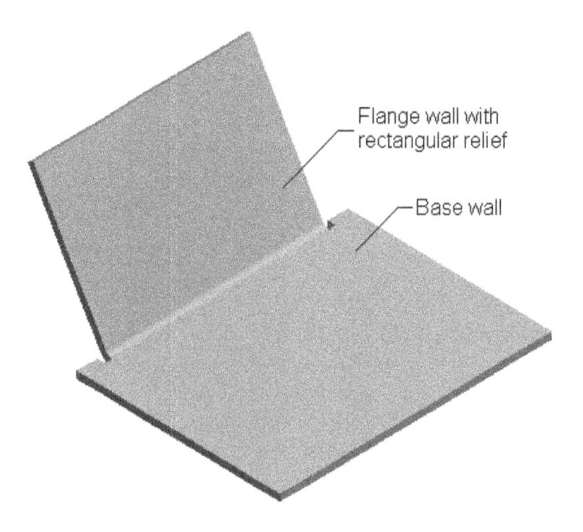

Figure 16-51 *Model with the flange wall attached to the base wall*

CREATING THE BEND FEATURE

Ribbon: Model > Bends > Bend > Bend

The **Bend** tool transforms the sheetmetal wall into an angular or roll shape. You need to sketch a bend line and determine the direction of the bend using the direction arrows or by orienting the sketching view. The bend line is used as the reference for calculating the developed length and creating the bend geometry. Bends can be added at any time during the design process as long as a wall feature exists. You can add bends across form features, but you cannot add them where they cross another bend. Depending on where you place the bend in your sheetmetal design, you may need to add reliefs to the bend.

To create a bend, choose the **Bend** tool from the **Bend** drop-down of the **Bends** group in the **Model** tab of the **Ribbon**; the **Bend** dashboard will be displayed, as shown in Figure 16-52. The options in this dashboard are discussed next.

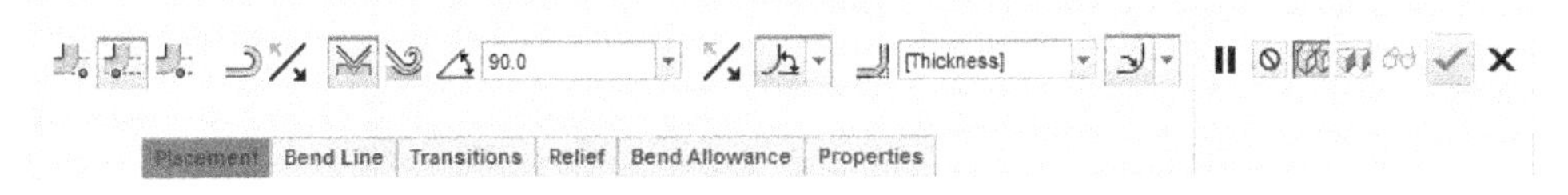

*Figure 16-52 Partial view of the **Bend** dashboard*

Bend material upto bend line

When you choose this button, the outer edge of the bend will become collinear to the bend line. Figure 16-53 shows the bend line and its drag handle, and Figure 16-54 shows preview of the bend created using the **Bend material upto bend line** button.

Bend material on other side of bend line

On choosing this button, the sheetmetal component will bend from the bend line, refer to Figure 16-55.

Bend material on both sides of bend line

When you choose this button, the bend material will be added on both sides of the bend line, refer to Figure 16-56.

Change location of fixed side

Using the **Change location of fixed side** button, you can change the fixed side of the bend. Alternatively, you can click on the horizontal arrow displayed in the preview of bend to flip the fixed side.

Use value to define the bend angle

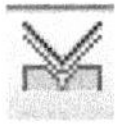

This button is used to create a bend through an angle. The vertical arrow displayed in the preview of bend determines the direction of the bend. When you click on the vertical arrow, the direction of bend gets reversed. Alternatively, you can choose the **Change bending direction** button next to the angle input edit box in the dashboard.

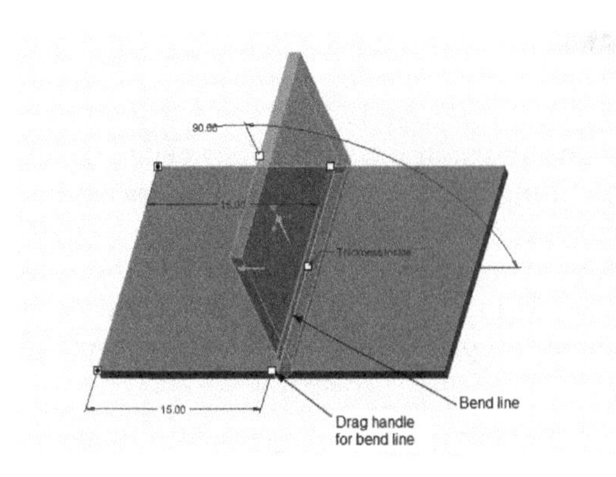

Figure 16-53 Bend line and drag handle of the bend line

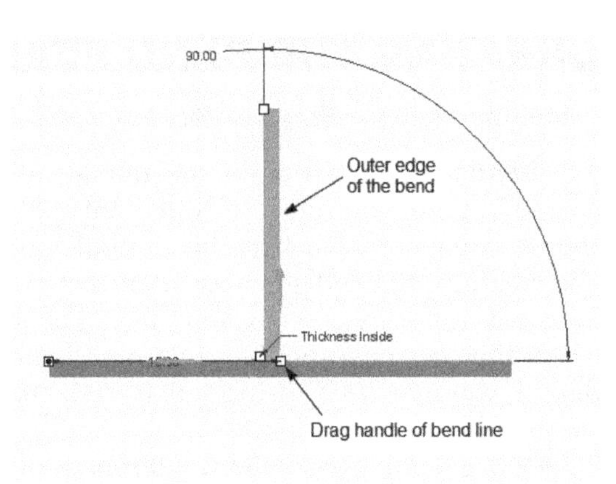

*Figure 16-54 Preview of the bend created using the **Bend material upto bend line** button*

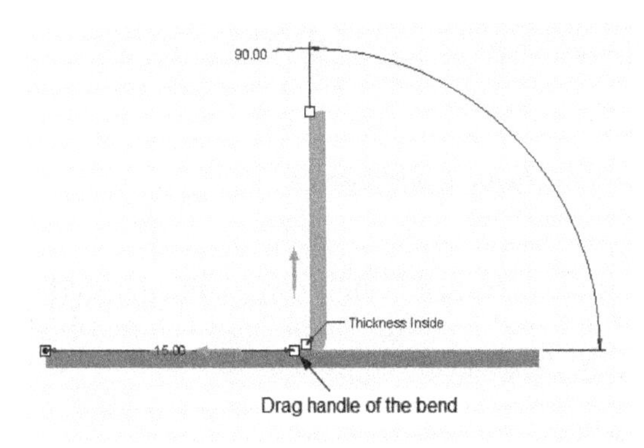

*Figure 16-55 Preview of the bend created using the **Bend material on other side of bend line** button*

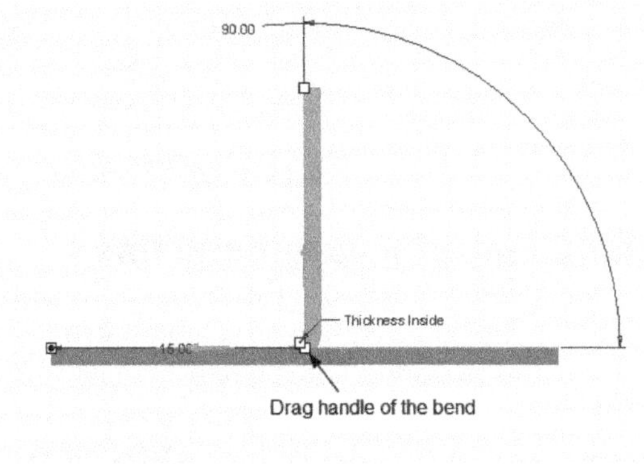

*Figure 16-56 Preview of the bend created using the **Bend material on both sides of bend line** button*

In the dashboard, there is a drop-down list next to the **Change bending direction** button. There are two options available in this drop-down list to measure the bend angle. These options are discussed next.

Measure the resulting internal bend angle

When this option is selected, the bend angle is calculated from the fixed side.

Measure the bend angle deflection from straight

When this option is selected, the bend angle is calculated from the side to be bent.

Bend to end of surface

Using this button, you can create a bend upto the end of the surface selected for bending. When you choose this button, the angle edit box will become inactive. Now, you need to enter the radius value in the edit box next to the **Change bend direction** button. You can use the inner radius or outer radius by using the options in the drop-down list next to the radius value edit box. Figure 16-57 shows the preview of the model after choosing the **Bend to end of surface** button and the **Bend material on other side** button from the dashboard. Therefore, using this tool, you can create a roll of the sheetmetal.

Placement Tab

When you click on the **Placement** tab, a slide-down panel will be displayed, as shown in Figure 16-58. Also, you will be prompted to select a surface or an edge to define the bend surface or bend line. On selecting the edge, the preview of the bend will be displayed and the **Offset bend line** check box will become active. After selecting this check box, you can enter the distance value of the bend line from the selected edge.

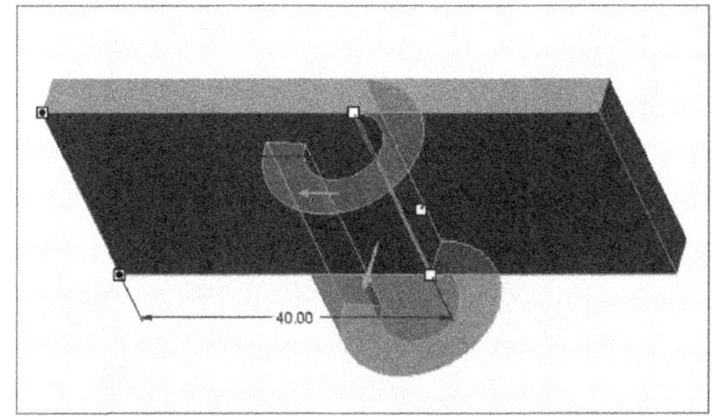

Figure 16-57 *Preview of the model created using the* ***Bend to end of surface*** *and* ***Bend material on other side*** *buttons*

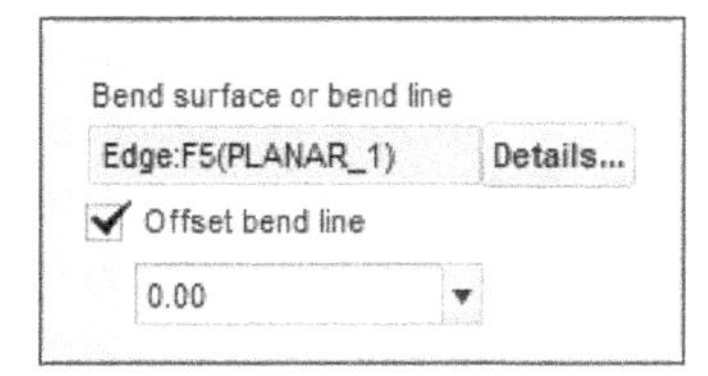

Figure 16-58 *The* ***Placement*** *slide-down panel*

Bend Line Tab

This tab is active only after selecting a surface. If you click on the **Bend Line** tab, a slide-down panel will be displayed, as shown in Figure 16-59. To create a bend line, click on the **Reference** collector in the **Bend line end 1** area and select the desired edge. Now, you need to select an offset reference. Click on the **Offset reference** collector and select the desired edge as offset reference. You can enter the distance value of the offset reference in the edit box adjacent to the offset collector, selected earlier. In this way, the first end point is defined. Use the same procedure to define the second end point of the bend line.

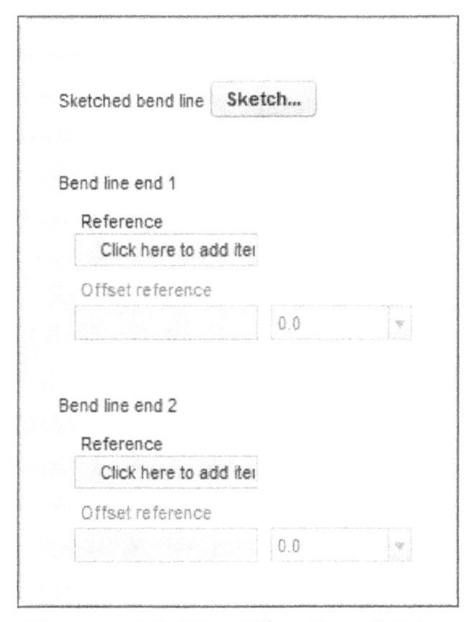

Figure 16-59 *The* ***Bend Line*** *slide-down panel*

Alternatively, select any of the green handles displayed on the selected surface and drag it to the desired edge. Note that when you drag the handle to an edge, the shape of the handle changes to square with circle in it. When you drop the handle on the edge, a new handle connected to the earlier selected handle will be displayed in the screen. Drag this new handle to an edge to define the offset reference. Same procedure will be required for the other handle.

Transitions Tab

When you click on the **Transitions** tab, a slide-down panel is displayed, as shown in Figure 16-60. This slide-down panel is used to add a transition area on the sheetmetal. To define the transition area, click on the **Add Transition** option displayed in the slide-down panel; the **Transition 1** option will be added to the transition list and the **Sketch** button will be activated. You need to choose this button and draw two straight lines to define the transition area. Figure 16-61 and Figure 16-62 show the preview of the bend before and after applying transition, respectively. Rest of the tabs are discussed earlier.

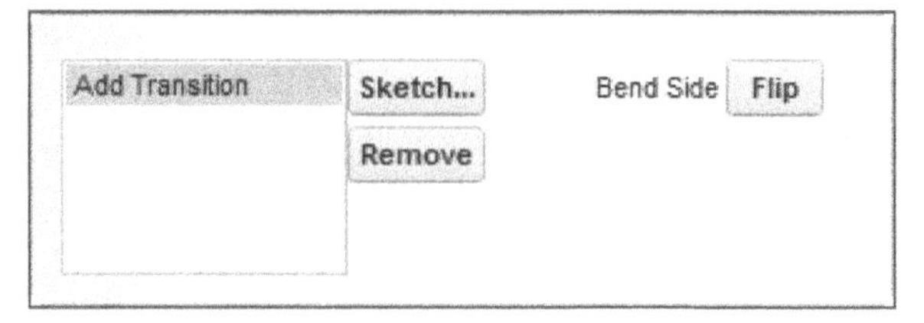

***Figure 16-60** The **Transitions** slide-down panel*

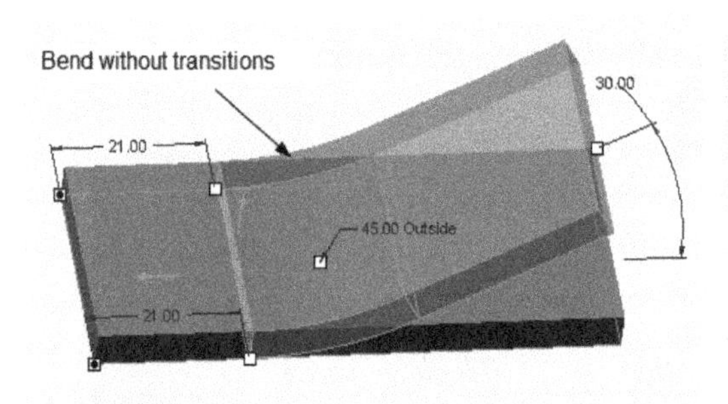

***Figure 16-61** Preview of the bend before applying transition*

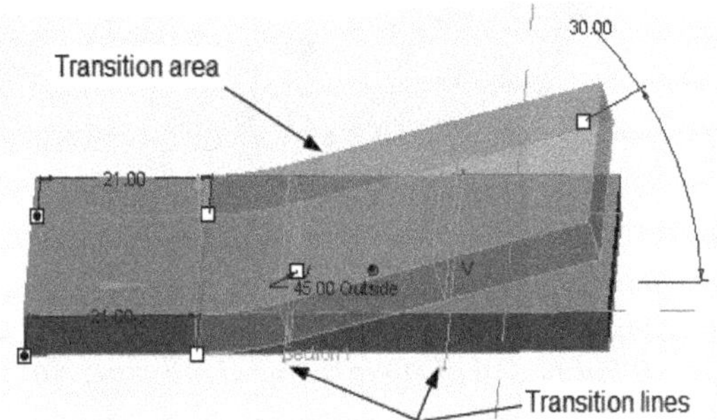

***Figure 16-62** Preview of the bend after applying transition*

Edge Bend

Ribbon: Model > Bends > Bend > Edge Bend

The **Edge Bend** tool is available in the **Bend** drop-down of the **Bends** group in the **Model** tab. You can create rounds at the sharp edges by using this tool. When you choose this tool, the **Edge Bend** dashboard will be displayed, as shown in Figure 16-63. Also, you will be prompted to select edges. Select the edge on which you want to create a round; the preview of the bend will be displayed on the screen.

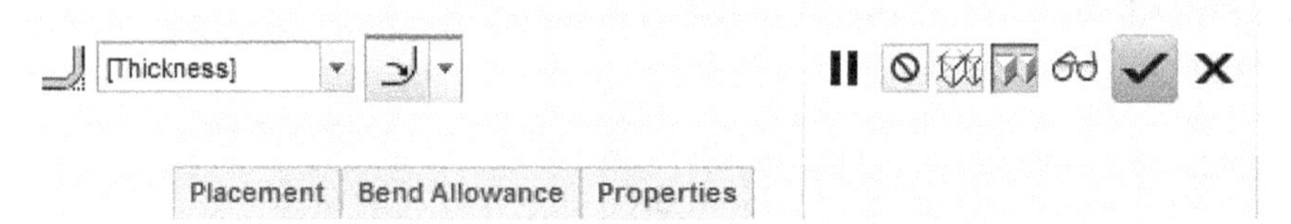

***Figure 16-63** Partial view of the **Edge Bend** dashboard*

Planar Bend

Ribbon: Model > Bends > Bend > Planar Bend

A Planar bend forces the sheetmetal wall around an axis that is normal to the surface and the sketching plane. You need to sketch a bend line and form the planar bend around an axis using the direction arrows which will be displayed in the drawing area. Choose the **Planar Bend** option from the **Bend** drop-down in the **Bends** group of the **Model** tab; the **OPTIONS** menu will be displayed. Choose **Angle > Done** from the **OPTIONS** menu; the **BEND Options:Angle, Planar** dialog box and the **USE TABLE** menu will be displayed. Choose the required option and then choose **Done/Return** from the **USE TABLE** menu. You will next be prompted to select the sketching plane. Select the sketching plane and define its orientation. Sketch the bend line and specify the direction of the bend through the direction arrows, which appear in the drawing area. Figure 16-64 shows the bend line sketched on the base wall and Figure 16-65 shows a bend created on the base wall using the **Planar** option. There are two kinds of **Planar** bends. They are discussed next.

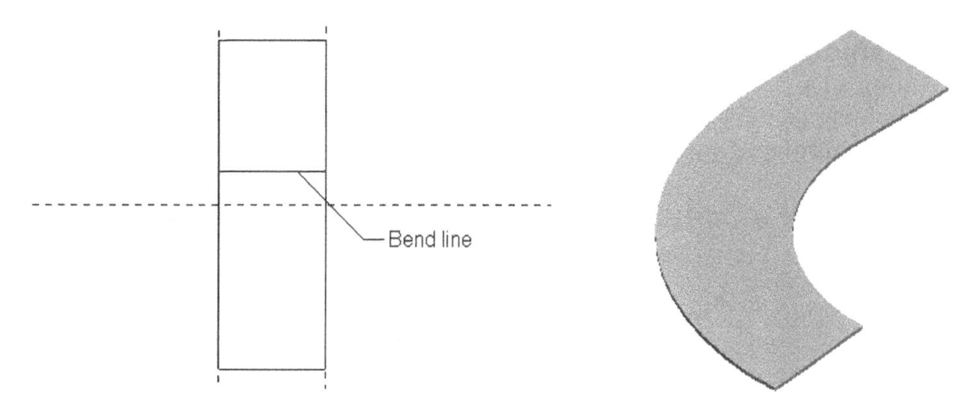

Figure 16-64 *Bend line sketched on the base wall* ***Figure 16-65*** *Bend created on the base wall*

Angle

This option, when selected, bends the sheetmetal wall through a specified radius and angle.

Roll

This option, when selected, bends the sheetmetal wall through a specified radius, but the angle is determined by both the radius and the amount of flat material to be bent.

Note

You cannot copy a bend using the ***Mirror*** *option. While you can generally unbend zero-radius bends, you cannot unbend the bends with cuts across them. If you do modify the developed length, remember that revising the developed length only affects the unbent geometry and not the bend back features.*

CREATING THE UNBEND FEATURE

Ribbon: Model > Bends > Unbend > Unbend

The **Unbend** option flattens any curved surface on the sheetmetal part. The curved surface may be a bend feature or a primary or secondary wall created previously. When you choose the **Unbend** button from the **Unbend** drop-down in the **Bends** group of the **Model** tab, the **Unbend** dashboard will be displayed, as shown in Figure 16-66. The different options in the **Unbend** dashboard are discussed next.

References selected automatically

The **References selected automatically** button is chosen by default in the **Unbend** dashboard. So, on applying the **Unbend** tool, all the bent edges are unbent automatically. Note that the section chosen to create the unbend feature must be developable.

*Figure 16-66 Partial view of the **Unbend** dashboard*

References selected manually

This button is used to select the bend edges manually. You can select more than one edge by holding the CTRL key. When you select the bend edges, the preview of unbent model will be displayed on the screen.

Select a surface or an edge to remain fixed during the unbending

This collector is used to collect a surface or an edge to be kept fixed during unbending.

References Tab

When you click on this tab, the **References** slide-down panel is displayed. In this slide-down panel, two collectors are available, **Bent Geometry** and **Fixed Geometry**. The **Bent Geometry** collector is used to collect the edges or surfaces to be unbent and the **Fixed Geometry** collector is used to collect the surfaces or edges to be fixed.

Deformations Tab

When you click on this tab, the **Deformations** slide-down panel is displayed. Now, you can add a surface to the **Deformation surfaces** collector. The surfaces that deform in more than one direction are added to this collector.

Deformation Control Tab

When you click on this tab, a slide-down panel is displayed. The options in this panel are used to control the type of deformation for the area selected for deformation. There are three radio buttons in this panel. These radio buttons are discussed next.

Blend Boundaries

When this radio button is selected, the boundaries of the deformation area are blended with the adjacent part edges.

Rip area

When this radio button is selected, the deformation area is removed and a flat pattern is created.

Sketch area

When this radio button is selected, the deformation area is created based on the flat impression of the sketch created.

Figure 16-67 shows a model with a flat wall attached to the primary wall and a bend created at either ends. Figure 16-68 shows the unbend model created using the **Unbend** tool.

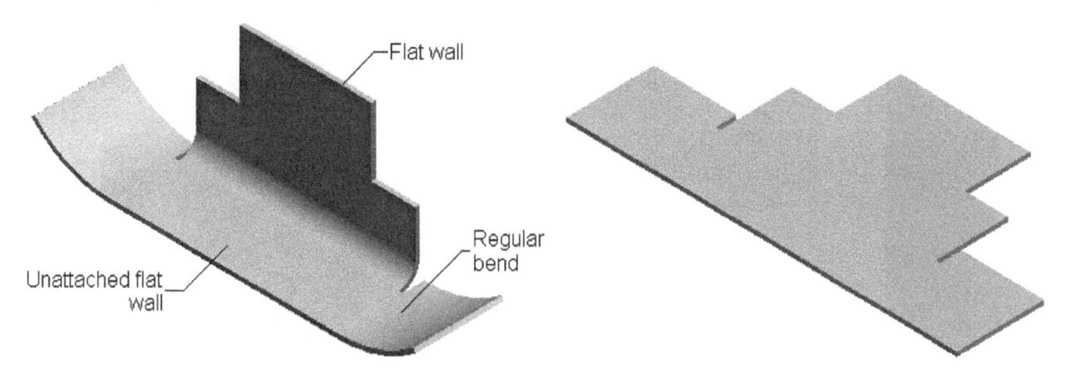

Figure 16-67 *Model with the bend and a flat wall*

Figure 16-68 *Model after unbending*

Creating the Bend Back

Ribbon: Model > Bends > Bend Back

The **Bend Back** tool is used to revert to the unbent position. Remember that ideally, you can bend back only the unbent features. When you choose the **Bend Back** button from the **Bends** group, the **Bend Back** dashboard will be displayed, as shown in Figure 16-69. Next, choose the **Manually select unbent geometry to bent back** button from the dashboard to select the edges or faces to be bent; you will be prompted to select the edge or plane which will remain fixed while creating the feature. When you select the fixed geometry, the preview of bend back feature will be displayed. The options in the **Bend Back** dashboard are discussed next.

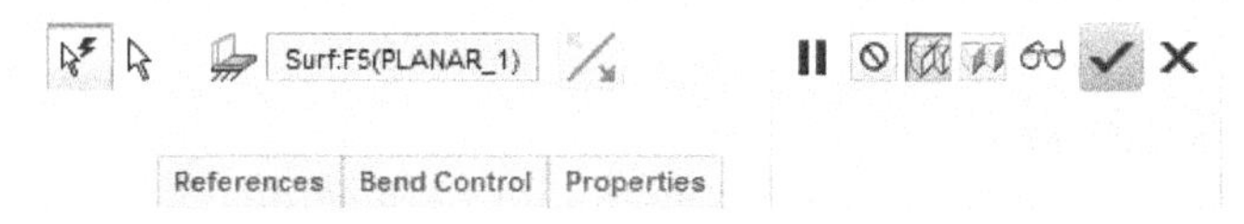

Figure 16-69 *Partial view of the* ***Bend Back*** *dashboard*

References Tab

There are two collectors available under this tab: **Unbent geometry** and **Fixed geometry**. The **Unbent geometry** collector is used to collect the geometries that are to be bent, whereas the **Fixed geometry** collector is used to collect the geometries that are to be fixed.

Bend Control Tab

This tab is used to control the behavior of contours that intersect with the bend line. There are two radio buttons to control the behavior of contours. You can choose the **Bend contour** radio button to bend the contour or you can make them flat by choosing the **Keep flat** radio button.

Figure 16-70 shows the model with the surfaces selected for creating the bend back feature and Figure 16-71 shows the model created by using the **Bend Back** tool.

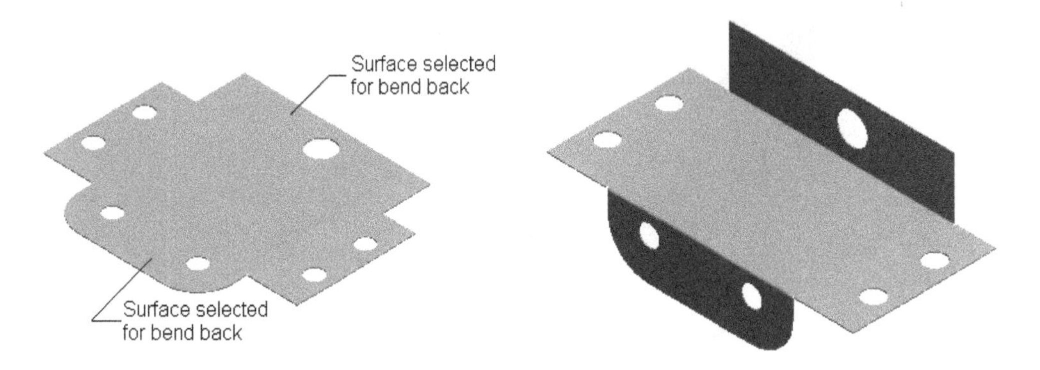

Figure 16-70 *Model with the surfaces selected for creating the bend back feature*

Figure 16-71 *Model after creating the bend back feature*

Note

*The model shown in Figure 16-70 is a base wall, with flange walls attached to its front and rear edges. Both the flange walls are unbent, while the base wall is selected as the section to be kept fixed. The **Bend Back** tool is then applied on the model.*

CONVERSION TO SHEETMETAL PART

Converting solid parts to sheetmetal parts enables you to modify your existing solid design with sheetmetal industry features. The conversion can serve as a shortcut in your design process because you can use existing solid designs to reach your sheetmetal design intent. Also, you can include multiple features within a single conversion feature. After you convert a part to sheetmetal, the part behaves like any other sheetmetal component.

Basic Conversion

The process of converting a solid model into a sheetmetal component by removing one of the surfaces or by making one of the surfaces of the solid model as the driving surface is known as the basic conversion. To make a basic conversion, choose **Convert to Sheetmetal** from the **Operations** group of the **Model** tab in the **Ribbon** in model environment; the **First Wall** dashboard will be displayed with two options. The options are **Shell** and **Driving Surface**. Block-like parts mostly use the **Shell** option for conversion to sheetmetal, while thin protrusions with constant thickness mostly use the **Driving Surface** option.

The basic conversion defines how you want to use the existing solid part in your sheetmetal design. You can create the shell feature on the solid model by selecting the surfaces to be removed, or you can make any one of the surface of the solid model as a driving surface (the driving side).

Figure 16-72 shows the solid model and Figure 16-73 shows the model after shelling and removing the bottom face.

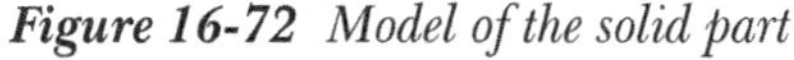

Figure 16-72 Model of the solid part

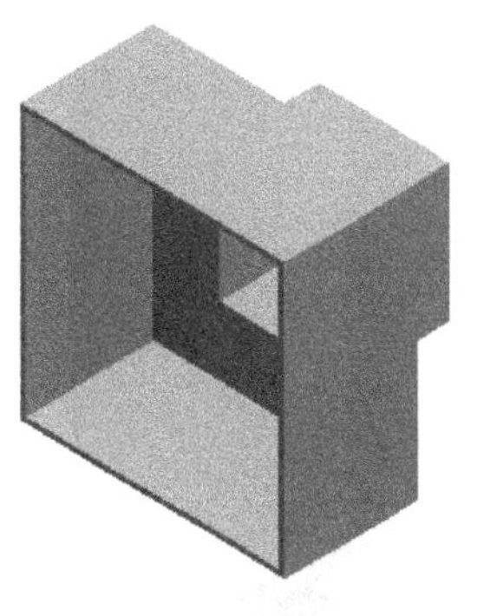

Figure 16-73 Model after converting it into sheetmetal component by removing the bottom face

Sheetmetal Conversion

If the converted part is not manufacturable, then you need to make a sheetmetal conversion by adding features such as rips, bends, and corner reliefs. To do so, choose **Model > Engineering > Conversion** from the **Ribbon** in the Sheetmetal environment; the **Conversion** dashboard will be displayed, as shown in Figure 16-74.

While converting a solid model into a sheetmetal component, you might need to add certain features such as corner reliefs, edge rips, bends, and so on to make the component manufacturable.

Figure 16-74 The partial view of the ***Conversion*** *dashboard*

The options available in the dashboard are discussed next.

Edge Rip

A rip feature can be defined as a zero volume cut that tears or shears the sheetmetal walls. If the sheetmetal component you are designing is a continuous piece of material, you cannot unbend the component without creating the rips. This option creates rips along the selected edge, which enables you to unbend the sheetmetal part.

Rip Connect

The **Rip Connect** option connects existing planar rips with straight line rips. The rip connects are sketched as point-to-point connections, which require you to define rip endpoints. The endpoints can be datum points or vertices and must be the endpoints of a rip.

Edge Bends

This option converts sharp edges to bends. By default, the inner radius of the bend is set to the thickness of the sheetmetal wall.

Corner Reliefs

This option creates reliefs in selected corners.

Figure 16-75 shows a solid model that has been converted into a sheetmetal component and Figure 16-76 shows the model with the edge rip, rip connect, and bend features.

Figure 16-75 *The solid model after being converted into a sheetmetal component*

Figure 16-76 *Model with the edge rip, rip connect, and bend features*

CREATING CUTS IN THE SHEETMETAL COMPONENT

The process of creating cuts in a sheetmetal component is similar to the one followed to create cuts in a solid model. The **Extrude** button in the **Shapes** group can be used to create cuts in a sheetmetal component. Choose the **Extrude** button from the **Shapes** group to invoke the **Extrude** dashboard; the **Extrude** dashboard will be displayed, as shown in Figure 16-77.

Figure 16-77 *Partial view of the* ***Extrude*** *dashboard*

Notice that the **Remove Material** button in chosen by default in the **Extrude** dashboard. This happens only when you have created a primary wall in your design, else it will not be available for function. The other options in the **Extrude** dashboard have the same functionality as discussed in earlier chapters.

The cut features in sheetmetal components are created with reference to the driving surface or the offset surface. Driving surface is the surface on which you create the base feature. When you turn the model display to the **No hidden** mode, the surface highlighted in green color is the driving surface. The offset surface is created by adding material to the driving surface equal to the thickness of the sheetmetal component. When you turn the model display to the **No hidden** mode, the surface highlighted in white color is the offset surface. When you invoke the **Extrude** dashboard, the **Remove material normal to surface** button is chosen by default. There are three options through which you can create cuts in a sheetmetal component.

Remove material normal to both Driving and Offset surfaces

This button, if chosen, removes material normal to both the driving and offset surfaces.

Remove material normal to Driving surface

This button, if chosen, removes material normal to the driving surface. This button is chosen by default.

Remove material normal to Offset surface

This button, if chosen, removes material normal to the offset surface.

CREATING THE FLAT PATTERN

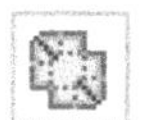

The **Flat Pattern** tool is equivalent to the **Unbend** feature. This feature flattens any curved surface, whether it is a bend feature or a wall. But the most distinctive characteristic of the **Flat Pattern** tool is that it automatically places itself at the end of the **Model Tree** to maintain the flat model view. If new features are added to your design, the flat pattern is suppressed but is automatically resumed after the features are added. If you do not want to flip between the flat pattern and the solid views for each new feature, manually suppress and resume the flat pattern as required. You can create a flat pattern at any point in your design because of the property that the feature jumps to the end of the **Model Tree** to maintain the flat model view.

Note
You can create only one flat pattern per part. Once you create a flat pattern, this option is disabled.

To create a flat pattern, choose **Model > Bends > Flat Pattern** from the **Ribbon**; you are prompted to select the edge or plane that will remain fixed during the process of creating the feature. Select the surface you want as fixed and the flat pattern of the model is created in the drawing area.

Figure 16-78 shows a sheetmetal component chosen to create the flat pattern. Select the primary wall surface that needs to remain fixed. The model after using the **Flat Pattern** tool is shown in Figure 16-79.

Tip: *You can choose the* ***Flat Pattern Preview*** *button from* ***Graphics*** *toolbar to preview the flat sheet. On choosing this button, the* ***Flat Pattern Preview*** *window will be displayed. You can also preview the dimensions of the flat pattern by choosing the* ***Bounding Box*** *button displayed in the window.*

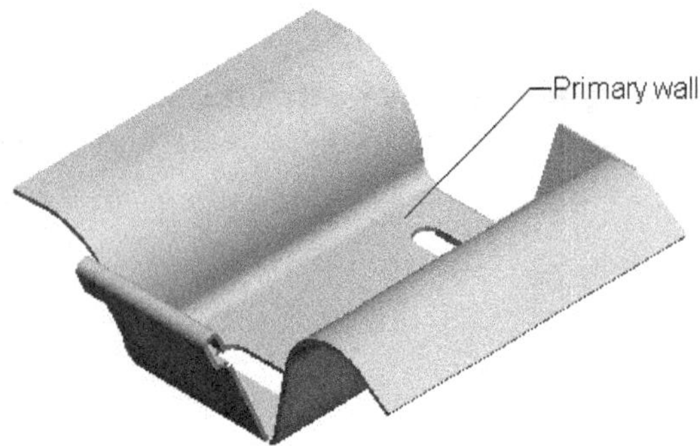

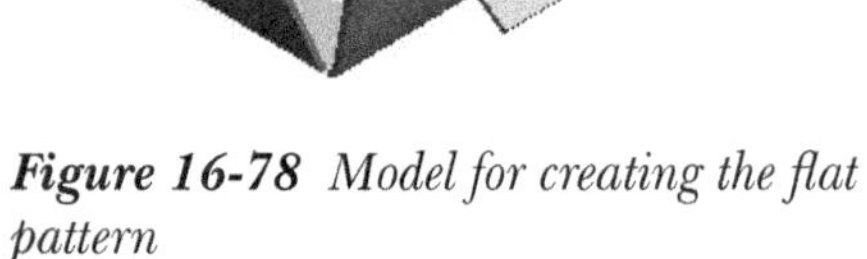

Figure 16-78 Model for creating the flat pattern

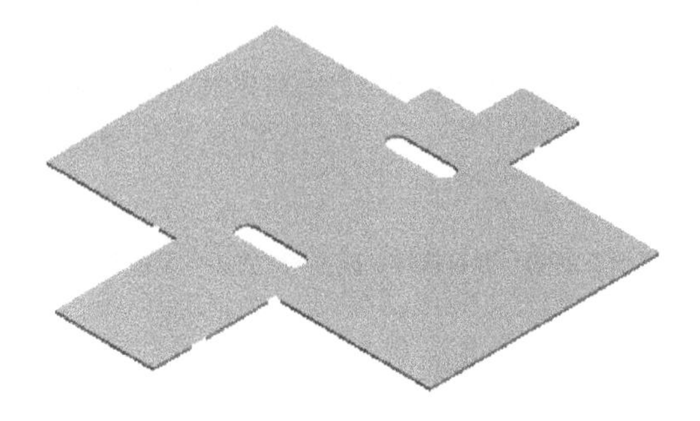

Figure 16-79 Model after creating the flat pattern

TUTORIALS

Tutorial 1

In this tutorial, you will create the sheetmetal component of the Holder Clip shown in Figure 16-80. The flat pattern of the component is shown in Figure 16-81. The dimensions are shown in Figures 16-82 and 16-83. The thickness of the sheet is 1 mm. After creating the sheetmetal component, create its flat pattern. **(Expected time: 45 min)**

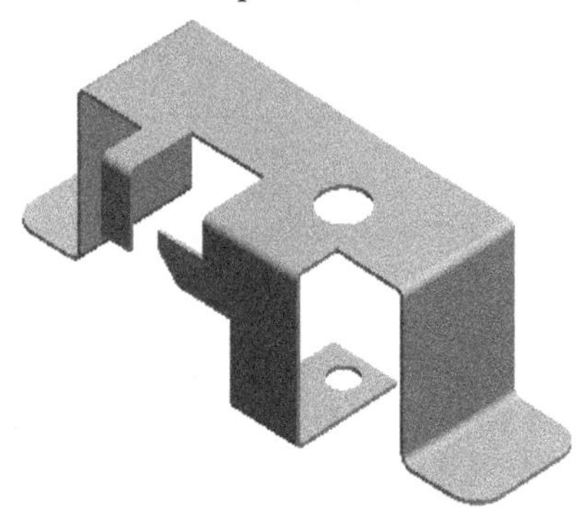

Figure 16-80 Model for Tutorial 1

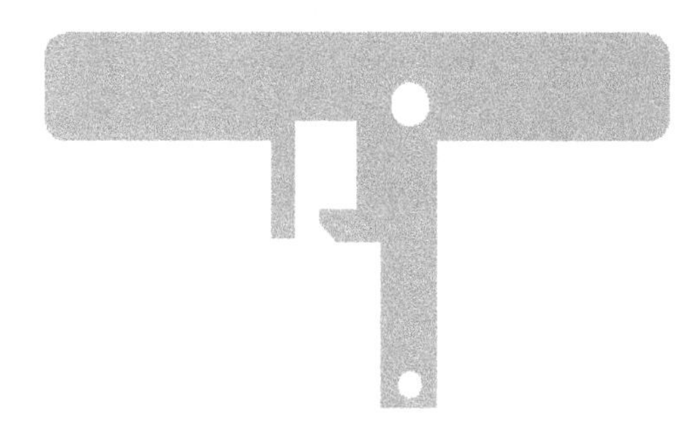

Figure 16-81 The flat pattern of the component

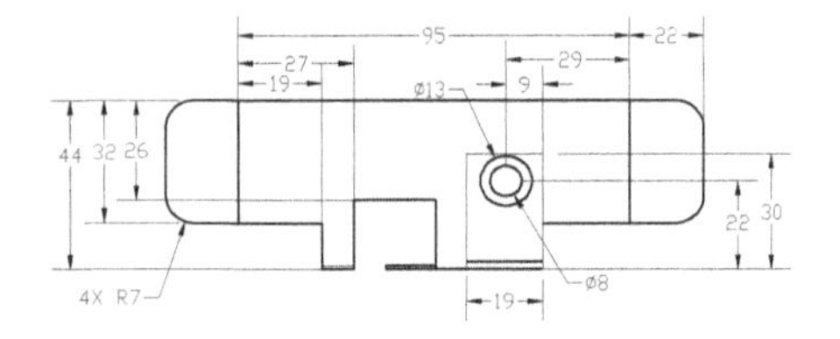

Figure 16-82 Top view of the Holder Clip

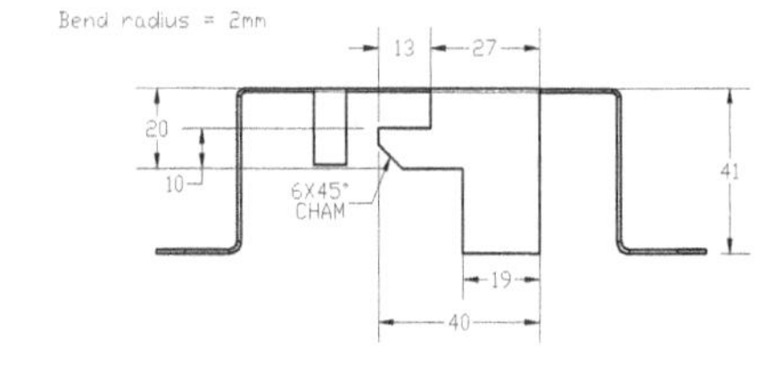

Figure 16-83 Front view of the Holder Clip

The following steps are required to complete this tutorial:

a. Create the base feature, refer to Figures 16-84 and 16-85.
b. Create the flange walls on the right and left edges of the base wall, refer to Figures 16-86 through 16-90.
c. Create the next flange wall on the front edge of the base wall, refer to Figure 16-92. Next, create a cut feature on it, refer to Figures 16-93 and 16-94.
d. Create a flat wall attached to the flange wall created in the previous step, refer to Figures 16-95 and 16-96.
e. Create the next flange wall, refer to Figures 16-97 and 16-98.
f. Create the next flange on the other front edge of the base wall, refer to Figure 16-100.
g. Create the round feature and the chamfer feature, refer to Figures 16-101 and 16-102.
h. Create the hole features on the top and bottom walls, refer to Figures 16-103 and 16-104.
i. Create the flat pattern of the model, refer to Figure 16-105.

Starting a New Object File

1. Start a new file in the **Part** mode and select the **Sheetmetal** from the **Sub-type** area. Next, enter *c16tut1* in the **Name** edit box.

2. Set the template to **mmns_part_sheetmetal**.

Creating the Base Wall

1. Choose the **Planar** button from the **Shapes** group; the **Planar** dashboard is displayed and you are prompted to select a closed sketch.

2. Choose the **References** tab from the **Planar** dashboard to display a slide-down panel and then choose the **Define** button from the slide-down panel; the **Sketch** dialog box is displayed and you are prompted to select a sketching plane.

3. Select the **TOP** datum plane from the drawing area and then choose the **Sketch** button from the **Sketch** dialog box.

4. Draw the sketch for the base wall, as shown in Figure 16-84, and then choose the **OK** button to exit the sketcher environment.

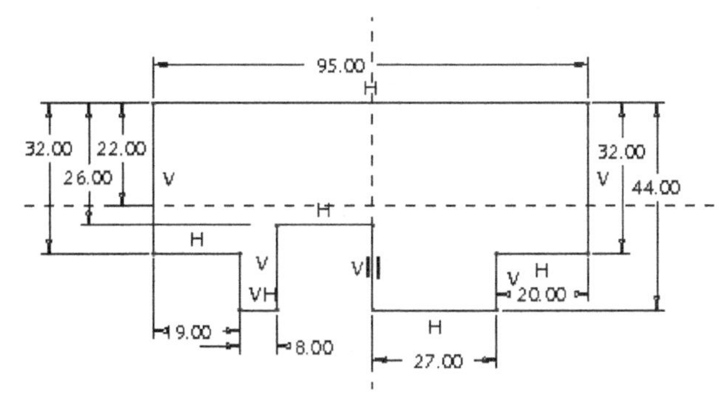

Figure 16-84 *The sketch for the base wall*

5. Enter **1** in the edit box of the **Planar** dashboard and choose the **Build feature** button from the dashboard to complete the creation of the feature. The model similar to the one shown in Figure 16-85 is displayed in the drawing area.

Creating the First Flange Wall

1. Choose the **Flange** button from the **Shapes** group; the **Flange** dashboard is displayed and you are prompted to select the edge to attach the wall.

2. Select the right edge of the base wall, as shown in Figure 16-86; the preview of the flange wall is displayed in the drawing area. The **I** profile is selected by default in the selection box above the **Placement** tab in the **Flange** dashboard. The dimension, indicating the bend radius at the attachment edge, is also displayed in the drawing area. The default value of radius is selected as **Thickness**, which is dimensioned from the inner surface of the bend.

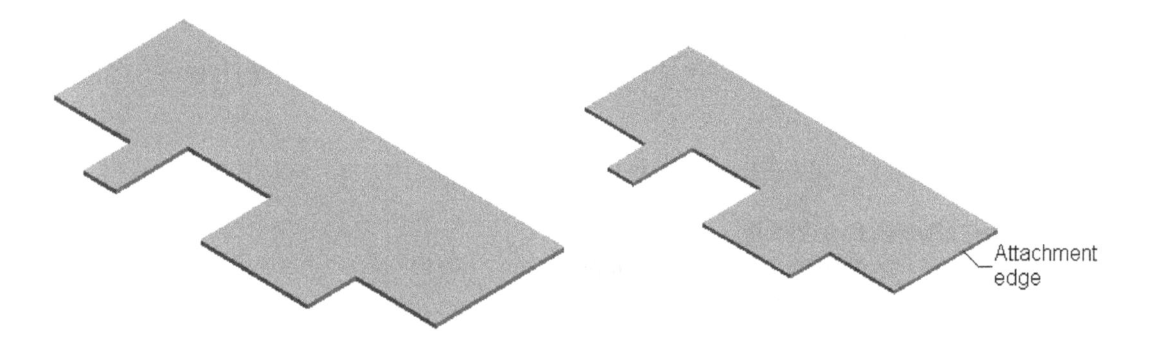

Figure 16-85 *Isometric view of the Model*

Figure 16-86 *The attachment edge for the flange wall*

3. Choose the **Shape** tab from the **Flange** dashboard to display the slide-down panel and then double-click on the default height value in the slide-down panel. Next, enter **41**. The angle at which the wall is created with respect to the base wall is selected as 90-degree by default. You can also drag the handles and adjust the values dynamically. Alternatively, double-click on the values displayed on the model; the edit boxes will be displayed where you can enter new values as per your requirement.

4. Choose the **Dimension the outer surface of the bend** button from the **Flange** dashboard and enter **2** in the edit box; the preview of the wall is displayed in the drawing area.

5. Choose the **Build feature** button from the **Flange** dashboard to complete the creation of the wall; the model similar to the one shown in Figure 16-87 is displayed in the drawing area.

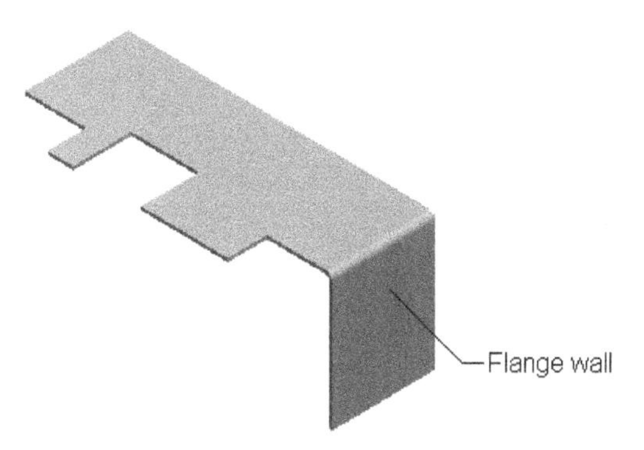

Figure 16-87 Model with the flange wall

Creating the Next Flange Wall

1. Select the attachment edge, which is the inner edge of the previously created flange wall, as shown in Figure 16-88, and create a flange wall as discussed earlier, with **21** as its height value and **2** as its radius value. The model similar to the one shown in Figure 16-89 is displayed in the drawing area.

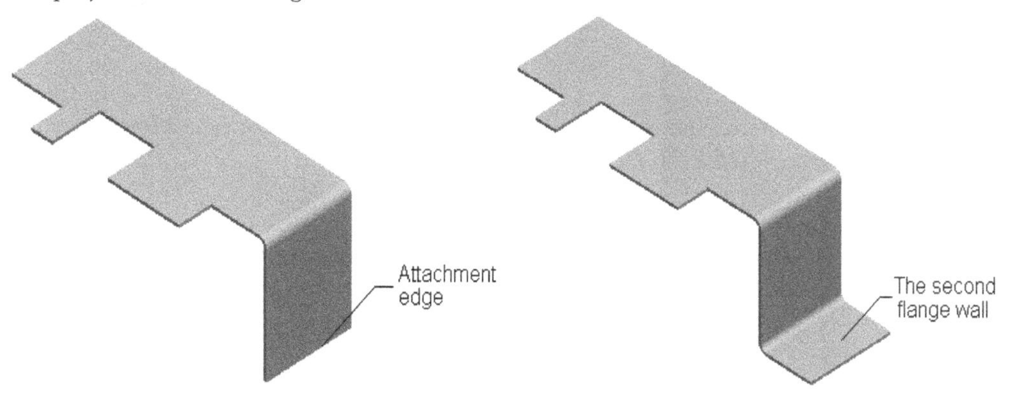

Figure 16-88 The attachment edge chosen for the flange wall

Figure 16-89 Model with the flange wall

Creating the Flange Walls on the Left Edge of the Base Wall

1. Create the flange walls on the left edge of the base wall by following the similar procedure that you used for creating the flange walls on the right edge. The model after creating the flange walls on the left edge of the base wall is shown in Figure 16-90. You can also mirror the left side flange walls about the **RIGHT** plane.

Creating the Next Flange Wall on the Front Edge of the Base Wall

1. Choose the **Flange** button from the **Shapes** group; the **Flange** dashboard is displayed and you are prompted to select the edge to attach the wall.

Figure 16-90 *Model after creating the flange wall on the left edge*

2. Select the front edge of the base wall, as shown in Figure 16-91. The preview of the flange wall is displayed in the drawing area.

3. Choose the **Shape** tab to display the slide-down panel. Specify **41** as the value of height of the wall in the edit box in the slide-down panel. You can also double-click on the height value displayed on the model and enter the new value. The angle at which the wall is created with respect to the base wall is selected as 90-degree by default.

4. Choose the **Dimension the outer surface of the bend** button from the **Flange** dashboard and enter the value **2** in the edit box. The preview of the flange wall is displayed in the drawing area.

5. Choose the **Build feature** button from the **Flange** dashboard; the model similar to the one shown in Figure 16-92 is displayed in the drawing area.

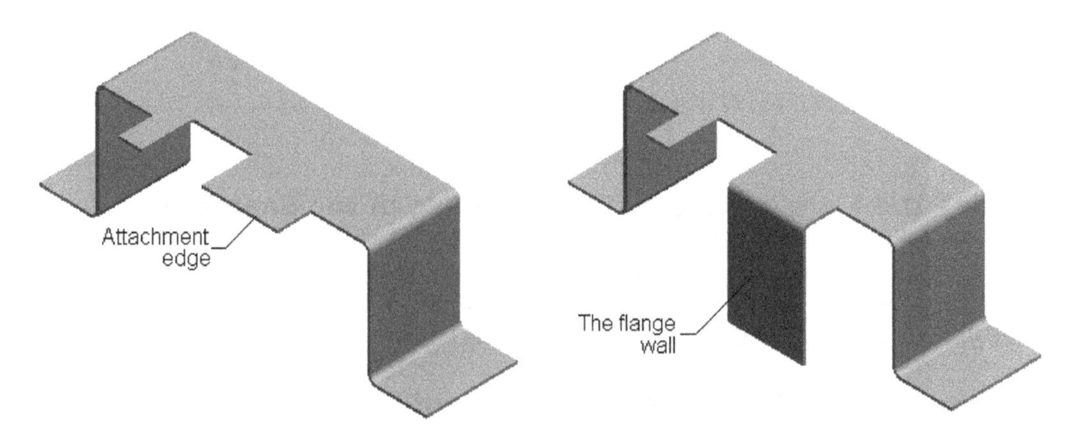

Figure 16-91 *The attachment edge for the flange wall*

Figure 16-92 *Model with the flange wall*

Creating the Cut Feature on the Front Flange Wall

1. Choose the **Extrude** button from the **Shapes** group and select the front face of the front flange wall as the sketching plane. Make sure that the **Remove Material** button is chosen.

2. Draw the sketch for the extruded cut feature, as shown in Figure 16-93.

3. Exit the sketcher environment and extrude the sketch through all faces in the model. The sheetmetal component after creating the cut feature is shown in Figure 16-94.

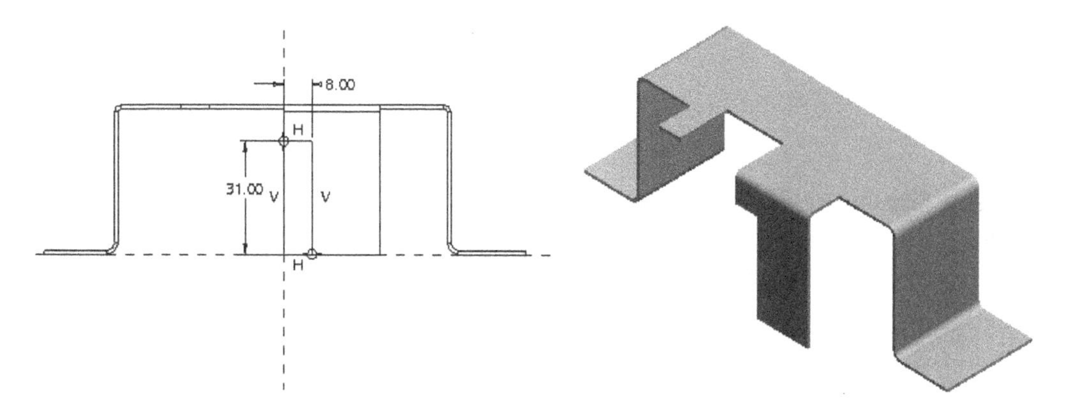

Figure 16-93 The sketch for the extruded cut feature

Figure 16-94 Model after creating the cut feature

Creating the Flat Wall

1. Choose the **Flat** button from the **Shapes** group; the **Flat** dashboard is displayed and you are prompted to select the attachment edge to create the wall.

2. Select the attachment edge, as shown in Figure 16-95; the **Rectangle** profile of the wall is selected by default and the preview of the wall is displayed in the drawing area.

3. Change the angle value to **Flat** in the **Angle** edit box of **Flat** dashboard.

4. Choose the **Shape** tab from the **Flat** dashboard; the slide-down panel is displayed with the default dimensions of the flat wall. Specify **21** as the value for the height of the wall and specify **-21** as the offset distance from the coordinate system in the slide-down panel. The preview of the flat wall is displayed in the drawing area.

5. Choose the **Build feature** button from the **Flat** dashboard to complete the feature creation. The model similar to the one shown in Figure 16-96 is displayed in the drawing area.

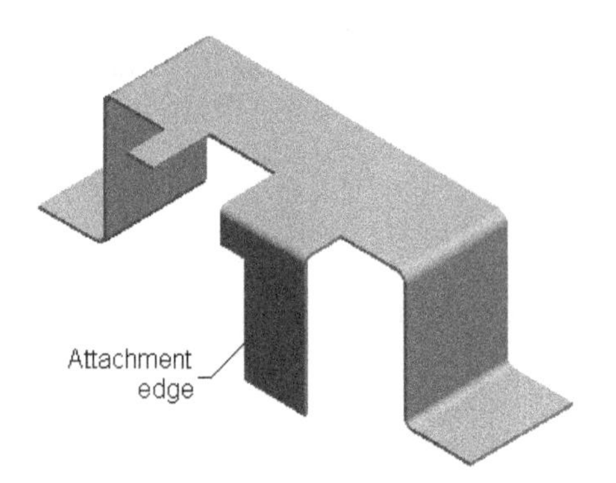

Figure 16-95 The attachment edge for the flat wall

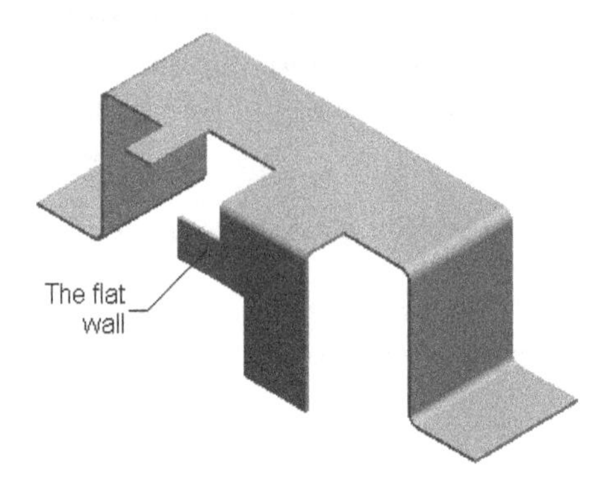

Figure 16-96 Model after creating the flat wall

Creating the Next Flange Wall

1. Choose the **Flange** button from the **Shapes** group; the **Flange** dashboard is displayed and you are prompted to select the attachment edge to create the wall.

2. Select the attachment edge, as shown in the Figure 16-97; the preview of the flange wall is displayed in the drawing area.

3. Choose the **Shape** tab from the **Flange Wall** dashboard to display the slide-down panel. Specify **30** as the value for the height of the wall in the edit box in the slide-down panel. The angle at which the wall is created with respect to the front flange wall is selected as 90-degree by default.

4. Enter **2** as the dimension for creating the bend at the attachment edge which is dimensioned from the outer surface of the bend.

5. Choose the **Build feature** button from the **Flange** dashboard to complete the creation of the feature. The model similar to the one shown in Figure 16-98 is displayed in the drawing area.

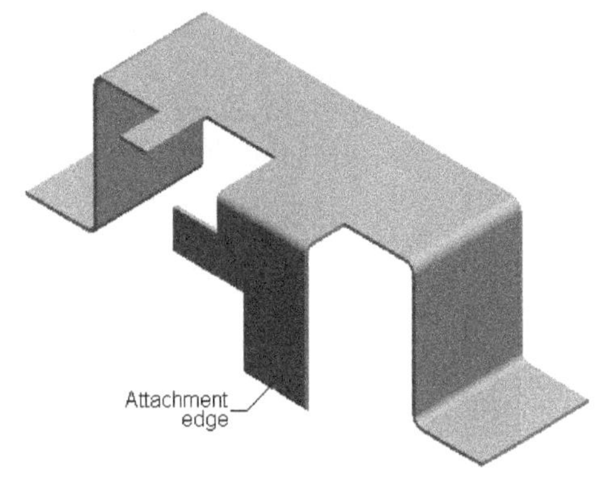

Figure 16-97 The attachment edge for the flange wall

Figure 16-98 Model showing the flange wall

Creating the Next Flange Wall on the Front Edge of the Base Wall

1. Create the next flange wall on the front edge of the base wall by selecting the attachment edge, as shown in Figure 16-99, and specifying the height of the wall as **19**. The model similar to the one shown in Figure 16-100 is displayed in the drawing area.

Figure 16-99 *The attachment edge for the flange wall*

Figure 16-100 *Model showing the flange wall on the front edge of the base wall*

Creating the Round and Chamfer Features

1. Create the rounds with radius 7, as shown in Figure 16-101, using the **Round** tool available in the **Engineering** group.

2. Create the edge chamfer, as shown in Figure 16-102, using the **Edge Chamfer** tool available in the **Engineering** group. For dimensions of the edge chamfer, refer to Figure 16-83.

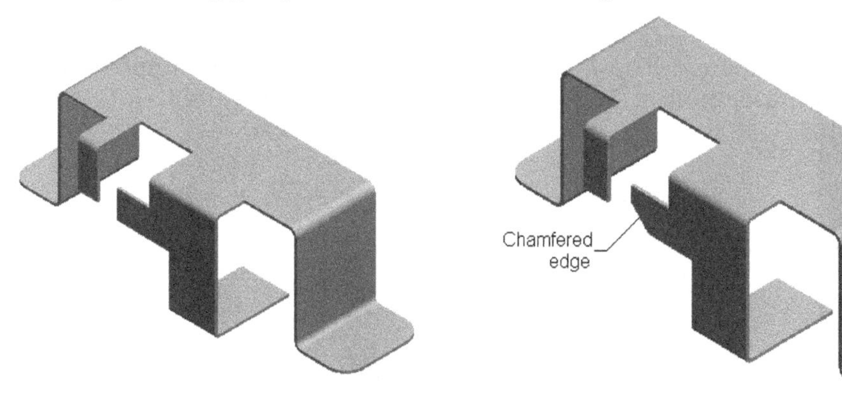

Figure 16-101 *Model after creating the round feature*

Figure 16-102 *Model after creating the chamfer feature*

Creating the Hole Features on the Top and Bottom Walls

1. Choose the **Hole** tool from the **Engineering** group of the **Ribbon**; the **Hole** dashboard is displayed.

2. Select the top surface of the base wall and specify the placement parameters for the hole having diameter 13, refer to Figure 16-82.

3. Choose the **Build feature** button from the **Hole** dashboard to exit it. The model after creating the hole feature on the top surface of the base wall is shown in Figure 16-103.

4. Similarly, create the hole feature on the bottom flange wall by specifying the placement parameters, refer to Figure 16-82. The model after creating the hole feature on the bottom flange wall is shown in Figure 16-104.

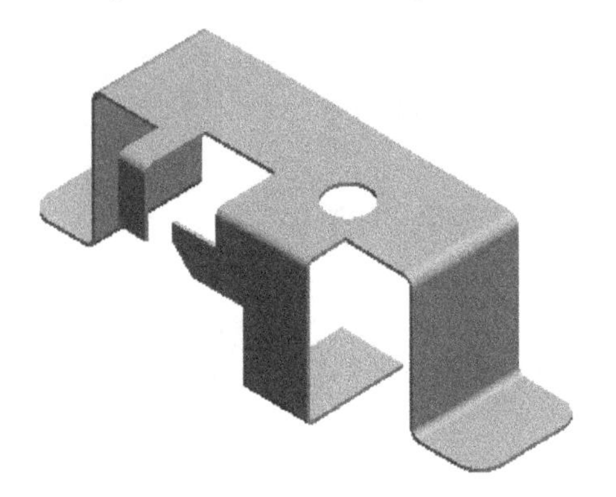

Figure 16-103 *Model with the hole feature on the base wall*

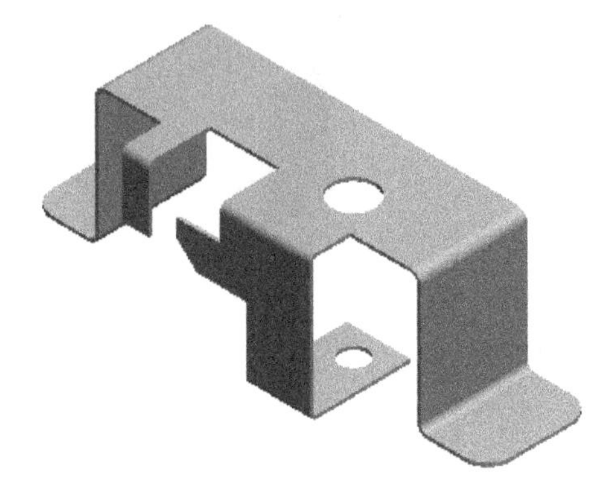

Figure 16-104 *Model with the hole feature on the bottom flange wall*

Creating the Flat Pattern of the Model

1. Choose the **Flat Pattern** button from the **Bends** group; the flat pattern of the model is created with the 3D notes.

2. To clear the 3D notes, choose **Annotation Display Filter > Annotation Display** from the **Graphics** toolbar. The flat pattern of the model is displayed, as shown in Figure 16-105.

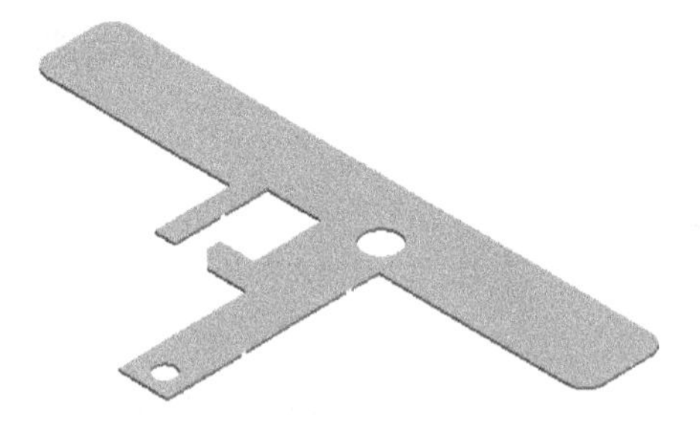

Figure 16-105 *Model after creating the flat pattern*

Saving the Model

1. Choose the **Save** button from the **Quick Access** toolbar to save the model.

Tutorial 2

In this tutorial, you will create the sheetmetal component shown in Figure 16-106. The dimensions are shown in Figures 16-107, 16-108, and 16-109. The flat pattern of the component is shown in Figure 16-110. The thickness of the sheet is 1 mm, Bend Radius is 1 mm, Relief Type is Rectangular, and width value is equal to thickness. After creating the sheetmetal component, create its flat pattern. **(Expected time: 45 min)**

The following steps are required to complete this tutorial:

a. Create the base feature, refer to Figures 16-111 and 16-112.
b. Create the hole feature on the top surface of the base wall, refer to Figure 16-113 and pattern the hole feature, refer to Figure 16-114.
c. Create a flange wall on the right edge of the base wall, refer to Figure 16-115.
d. Create the cut feature, refer to Figure 16-116 and Figure 16-117.
e. Create the next flange wall attached to the wall created previously, refer to Figure 16-119.
f. Create the two flange walls along with the cut feature on the left side of the base wall as created previously, refer to Figure 16-120.
g. Create the flat wall on the front edge of the base wall, refer to Figure 16-122.
h. Create the flat pattern of the component, refer to Figure 16-123.

If required, set the working directory to *C:\Creo-3.0\c16*.

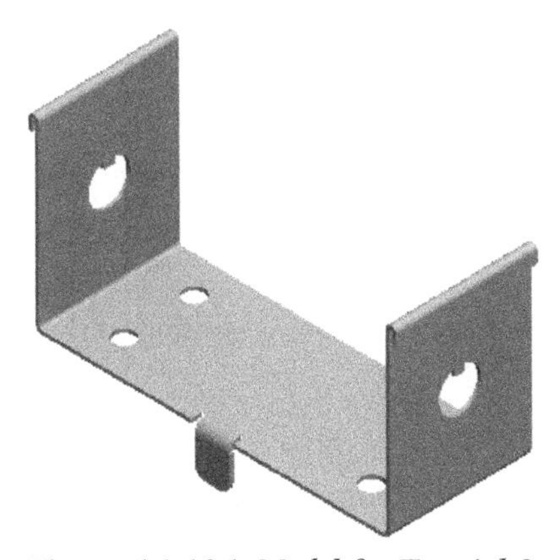

Figure 16-106 *Model for Tutorial 2*

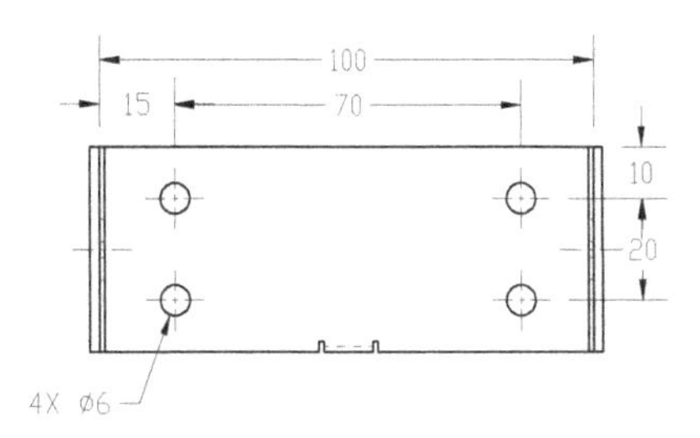

Figure 16-107 *Top view of the model*

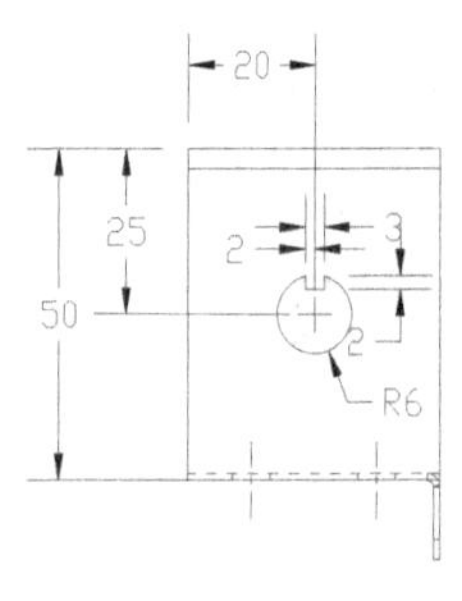

Figure 16-108 *Left-side view of the model*

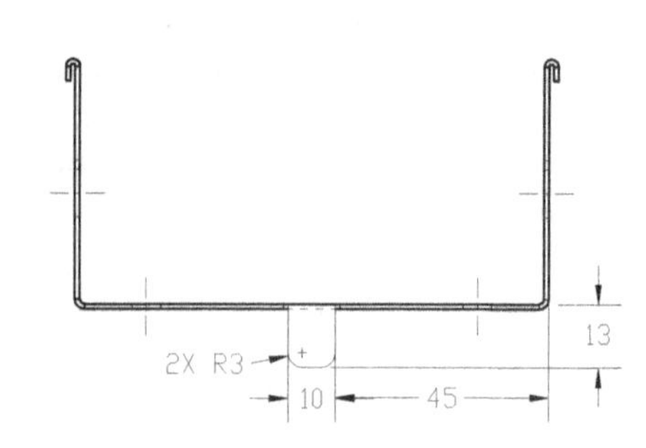

Figure 16-109 *Front view of the model*

Figure 16-110 *Flat pattern of the component*

Starting a New Object File

1. Start a new file in the **Sheetmetal** mode and name it as *c16tut2*.
2. Set the template to **mmns_part_sheetmetal**.

Creating the Base Wall

1. Choose the **Planar** button from the **Shapes** group; the **Planar** dashboard is displayed and you are prompted to select a closed sketch.

2. Choose the **References** tab from the **Planar** dashboard to display the slide-down panel. Choose the **Define** button from the slide-down panel; the **Sketch** dialog box is displayed and you are prompted to select a sketching plane.
3. Select the **TOP** datum plane from the drawing area and choose the **Sketch** button from the **Sketch** dialog box.
4. Draw the sketch for the base wall, as shown in Figure 16-111, and choose the **OK** button to exit the sketcher environment.
5. Enter the value **1** in the edit box in the **Planar** dashboard and choose the **Build feature** button from the dashboard to complete the creation of the feature. The model similar to the one shown in Figure 16-112 is displayed in the drawing area.

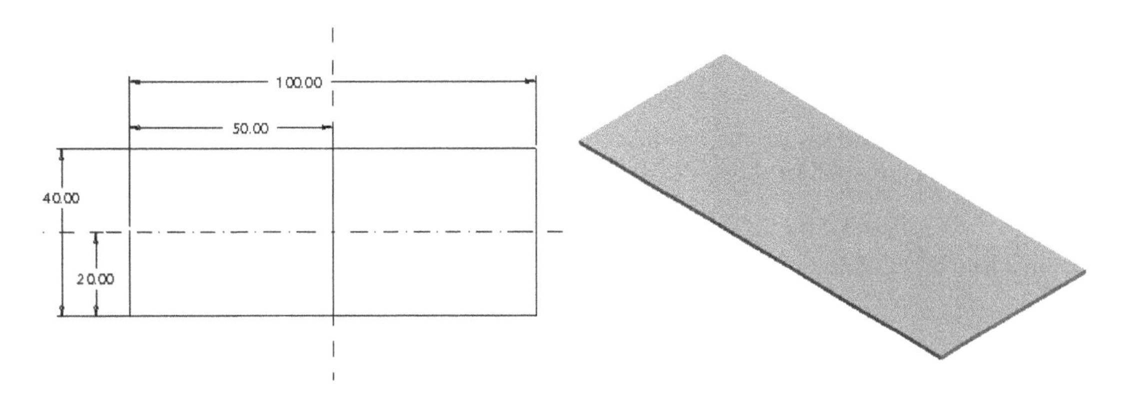

Figure 16-111 *The sketch for the base wall*

Figure 16-112 *Model of the base wall*

Creating the Hole Feature

1. Create a hole on the lower right corner of the base wall, as shown in Figure 16-113, by specifying the dimensions shown in Figure 16-107.

Creating the Pattern of the Hole Feature

1. Create the rectangular pattern of the hole, as shown in Figure 16-114.

Creating the First Flange Wall

1. Choose the **Flange** button from the **Shapes** group; the **Flange** dashboard is displayed and you are prompted to select the edge to attach the wall.

2. Select the right edge of the base wall; the preview of the flange wall is displayed in the drawing area.

3. Choose the **Shape** tab from the **Flange Wall** dashboard to display the slide-down panel and specify **50** as the value for the height of the wall in the edit box.

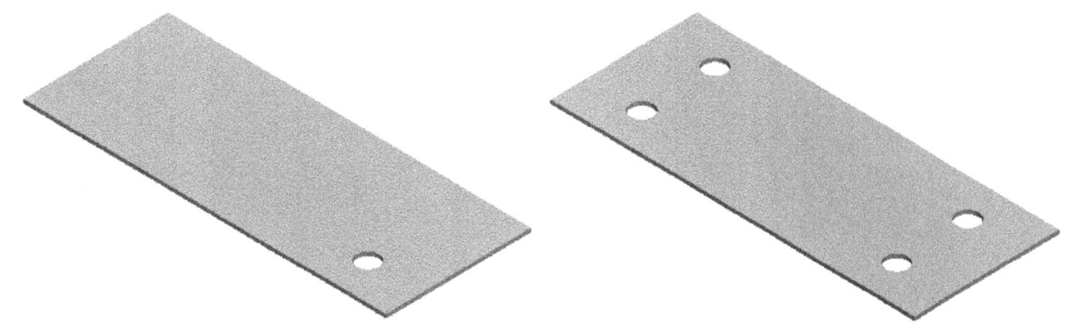

Figure 16-113 *Model after creating the hole feature*

Figure 16-114 *Model after creating the pattern of the hole feature*

4. Choose the **Dimension the outer surface of the bend** button from the **Flange** dashboard and enter **2** as the value in the edit box; the preview of the wall is displayed in the drawing area.

5. Choose the **Build feature** button from the **Flange** dashboard to complete the creation of the wall. Figure 16-115 shows the model after creating the flange wall.

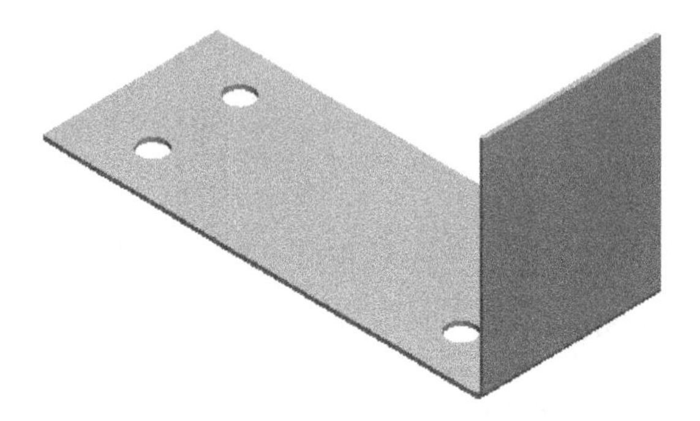

Figure 16-115 *Model with the flange wall*

Creating the Cut Feature

1. Choose the **Extrude** button from the **Shapes** group and select the front face of the flange wall as the sketching plane. Also, make sure that the **Remove Material** button is chosen.

2. Draw the sketch for the cut feature on the front face of the previous flange wall, as shown in Figure 16-116.

3. Extrude the sketch through all faces of the model to create the cut feature, as shown in Figure 16-117.

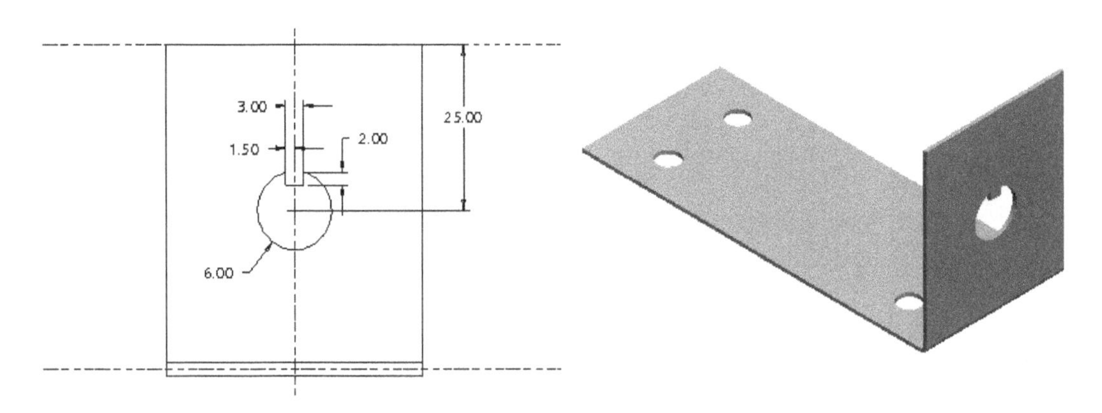

Figure 16-116 *The sketch for the cut feature*

Figure 16-117 *Model with the cut feature*

Creating the Next Flange Wall

1. Choose the **Flange** button from the **Shapes** group; the **Flange** dashboard is displayed and you are prompted to select the edge to attach the wall.

2. Select the attachment edge, as shown in Figure 16-118; the preview of the flange wall is displayed in the drawing area.

3. Select the **Open** option from the Wall profile drop-down list in the dashboard; the preview of the wall is displayed in the drawing area.

4. Accept the default parameters in the **Shape** slide-down panel and choose the **Build feature** button from the **Flange** dashboard to complete the creation of the wall.

   Figure 16-119 shows the model with the flange wall.

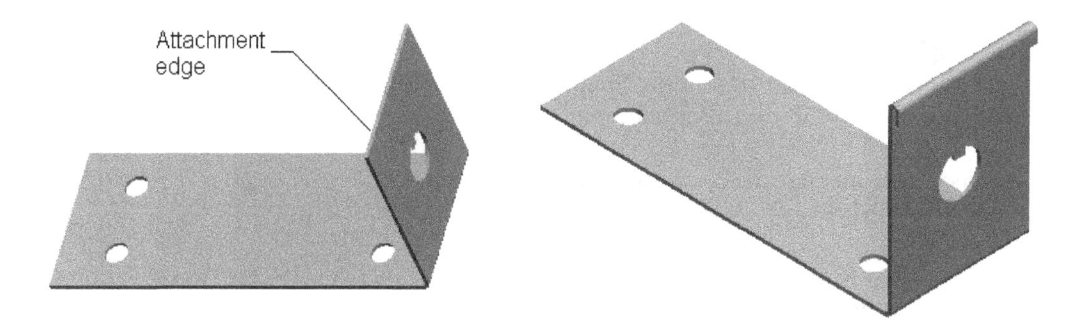

Figure 16-118 *The attachment edge for the flange wall*

Figure 16-119 *Model with the flange wall*

Creating the Flange Walls on the Left Edge of the Base Wall

1. Create the flange walls on the left edge of the base wall by following the same procedure that was used for creating the first two flange walls on the right edge of the base wall, refer to Figure 16-120. You can also mirror the flange walls with respect to the RIGHT plane.

Creating the Cut Feature on the Left Flange Wall

1. Create the cut feature on the left flange wall by following the procedure used for creating the cut on the right flange wall. The model after creating the cut feature is shown in Figure 16-120.

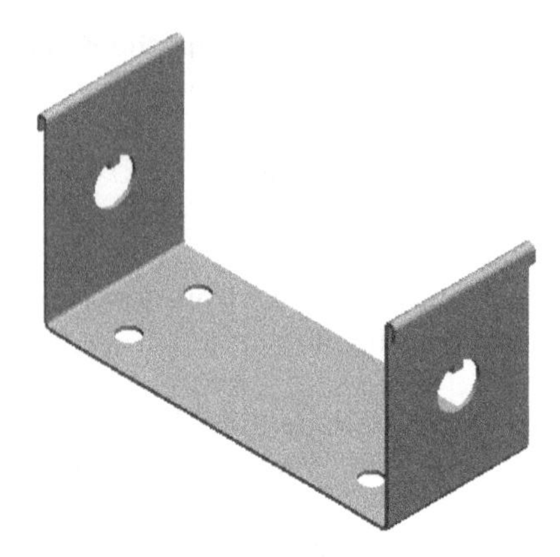

Figure 16-120 *Model after creating the flange walls on the left edge of the base wall and the cut feature on the left flange wall*

Creating the Flat Wall

1. Choose the **Flat** button from the **Shapes** group; the **Flat** dashboard is displayed and you are prompted to select the attachment edge to create the wall.

2. Select the front edge of the base wall; the **Rectangle** profile of the wall is chosen by default and the preview of the wall is displayed in the drawing area.

3. Choose the **Rectangle** option from the dashboard; a flyout is displayed. Choose the **User-Defined** option from the flyout.

4. Choose the **Shape** tab from the **Flat** dashboard to display the slide-down panel. Choose the **Sketch** button from the slide-down panel; the **Sketch** dialog box is displayed.

5. Orient the model such that the top face of the base wall is at the top and draw the sketch for the user-defined flat wall, as shown in Figure 16-121. Note that the sketch drawn should be open at the top. Choose the **OK** button to exit the sketcher environment.

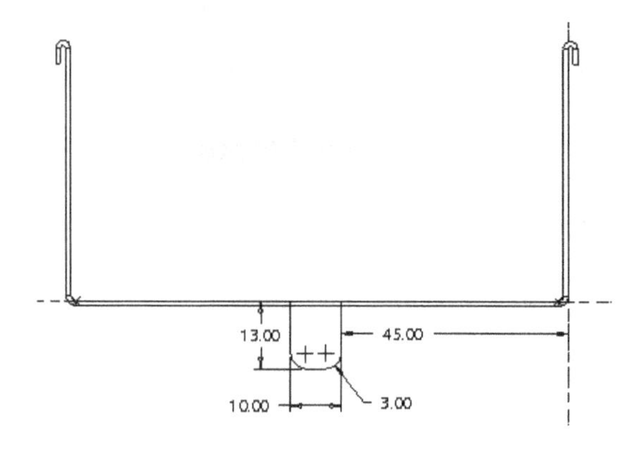

Figure 16-121 *The sketch for the flat wall*

6. Choose the **Relief** tab from the **Flat** dashboard to display the slide-down panel and then select the **Rectangular** option from the **Type** drop-down list. Accept the default parameters for the relief. You need to choose the **Flip** button to flip the direction.

7. Choose the **Build feature** button from the **Flat** dashboard to complete the creation of the flat wall. The model similar to the one shown in Figure 16-122 is displayed in the drawing area.

Creating the Flat Pattern of the Model

1. Choose the **Flat Pattern** button from the **Bends** group; the flat pattern of the model is created and the model similar to the one shown in Figure 16-123 is displayed in the drawing area.

Figure 16-122 Model after creating the flat wall

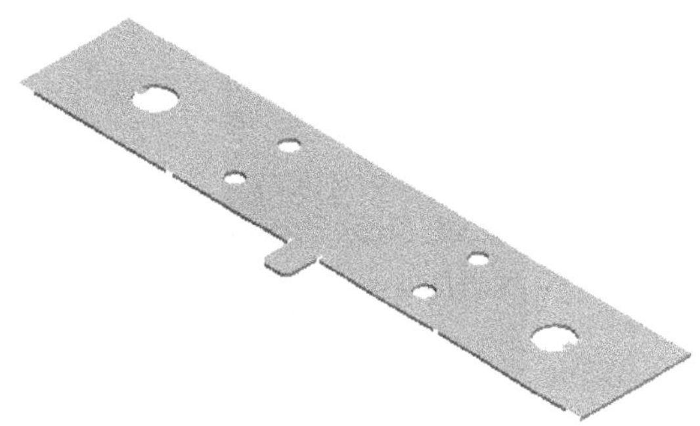

Figure 16-123 Model after creating the flat pattern

Saving the Model

1. Choose the **Save** button from the **Quick Access** toolbar to save the model.

Self-Evaluation Test

Answer the following questions and then compare them to those given at the end of this chapter:

1. In the **Sheetmetal** design of PTC Creo Parametric, the walls can be categorized into two types. (T/F)

2. There are four types of unattached walls in the **Sheetmetal** environment of PTC Creo Parametric. (T/F)

3. The rotational blend walls have sections that can be rotated about the Y-axis upto a maximum angle of 90 degrees. (T/F)

4. __________ , __________ , and __________ are three different types of unattached blend walls.

5. The __________ in a sheetmetal component control the material behavior and prevent the unwanted deformation.

6. In the **Sheetmetal** environment of PTC Creo Parametric, the __________ wall can be a primary or a secondary wall.

7. The __________ wall button can be used to lengthen an existing wall.

8. The __________ is a reference point for calculating the developed length and creating the bend geometry.

9. The __________ option flattens any curved surface, whether it is a bend feature or a curved wall.

10. A __________ feature can be defined as a zero-volume cut.

Review Questions

Answer the following questions:

1. While creating an unattached blend wall, the __________ option is used to connect vertices of all sections in the blend feature with straight lines.

2. Parallel blends have sections that are drawn__________ to each other.

3. While creating a flange wall, the __________ tab is used to specify the dimensions of the predefined wall profiles.

4. The __________ feature bends back only unbent features.

5. The __________feature is defined as a zero volume cut that tears or shears the sheetmetal walls.

6. To convert a solid model into a sheetmetal component, you first need to __________ the solid model.

7. If the rotational blend angle between two sections is equal to 0-degree, then the rotational blend option functions in the same way as the parallel blend option. (T/F)

8. The bend feature can be created on the area that was previously bent. (T/F)

9. The flat pattern can be created only once on a sheetmetal component. (T/F)

10. When you choose the **Extrude** button to create a cut on a sheetmetal component, the **Remove Material** button is chosen by default. (T/F)

Exercises

Exercise 1

Create the sheetmetal component shown in Figure 16-124. The dimensions of the model are shown in Figures 16-125 and 16-126. The flat pattern of the component is shown in Figure 16-127. Assume the missing dimensions. **(Expected time: 30 min)**

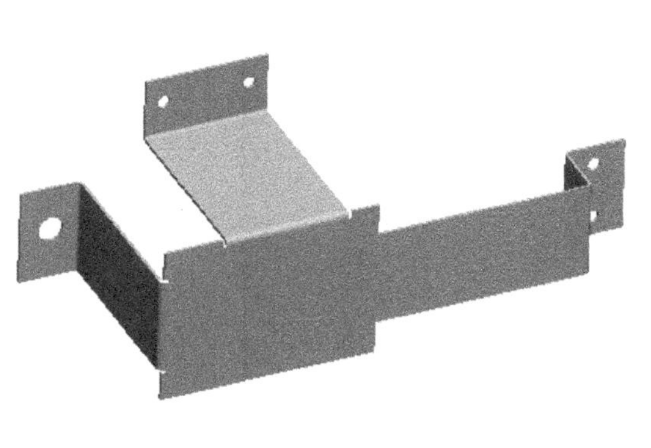

Figure 16-124 Model for Exercise 1

Figure 16-125 Top view of the component

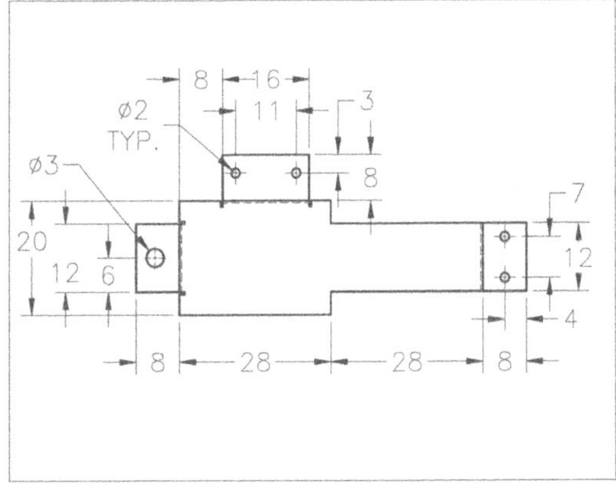

Figure 16-126 Front view of the component

Figure 16-127 Flat pattern of the component

Exercise 2

In this tutorial, you will create the sheetmetal component shown in Figure 16-128. The flat pattern of the component is shown in Figure 16-129. The dimensions of the model are shown in Figure 16-130. The material thickness, bend radius, relief depth, and relief width is 1mm. Assume the missing dimensions. **(Expected time: 30 min)**

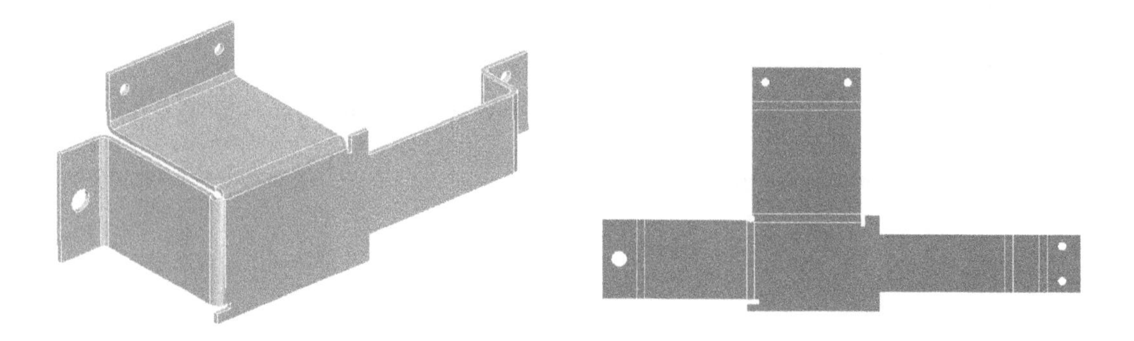

Figure 16-128 Sheetmetal part for Tutorial 1

Figure 16-129 Flat pattern of the part

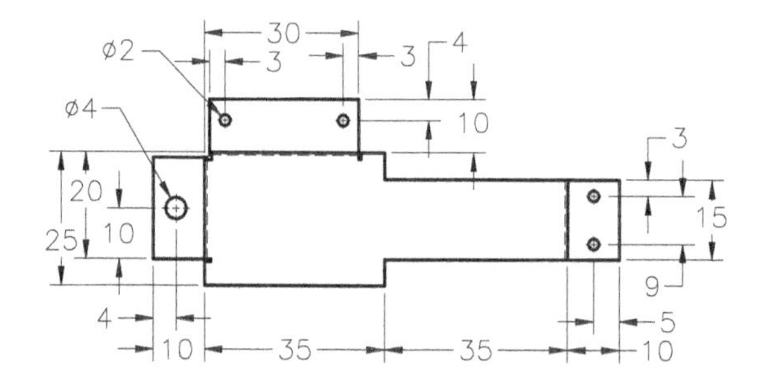

Figure 16-130 Dimensions of the sheetmetal part

Answers to Self-Evaluation Test

1. T, **2.** T, **3.** F, **4.** Parallel, Rotational, General, **5.** reliefs, **6.** flat, **7. Extend**, **8.** bend line, **9. Flat Pattern**, **10. Rip**

Student Projects

Student Project 1

Create different components of the Double Bearing assembly and then assemble them, as shown in Figure 1. Figure 2 shows the exploded view of the assembly. The dimensions of various components are given in Figures 3 through 5.

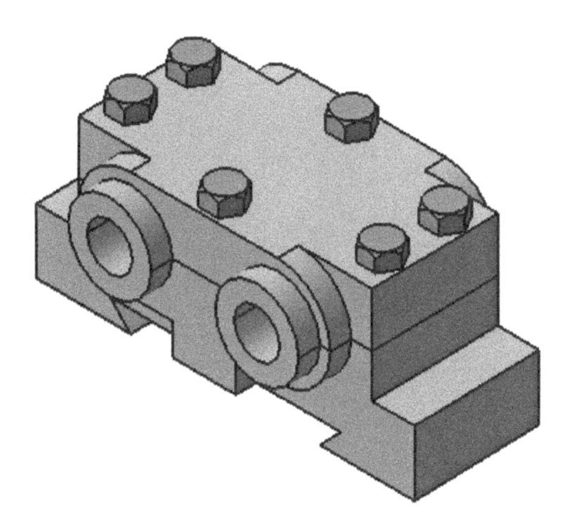

Figure 1 Double Bearing assembly

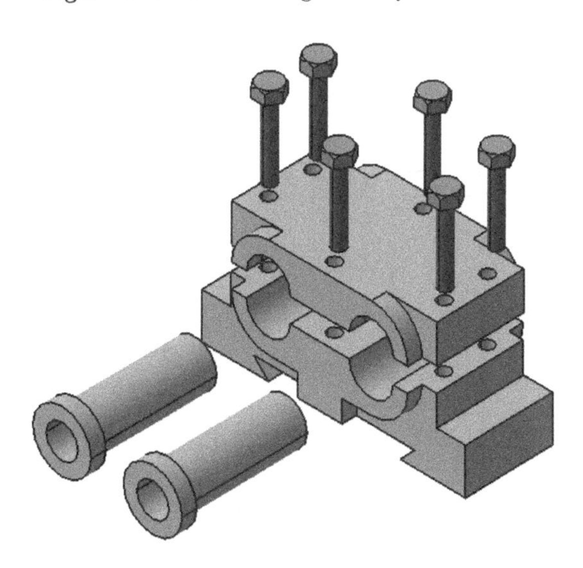

Figure 2 Exploded view of the Double Bearing assembly

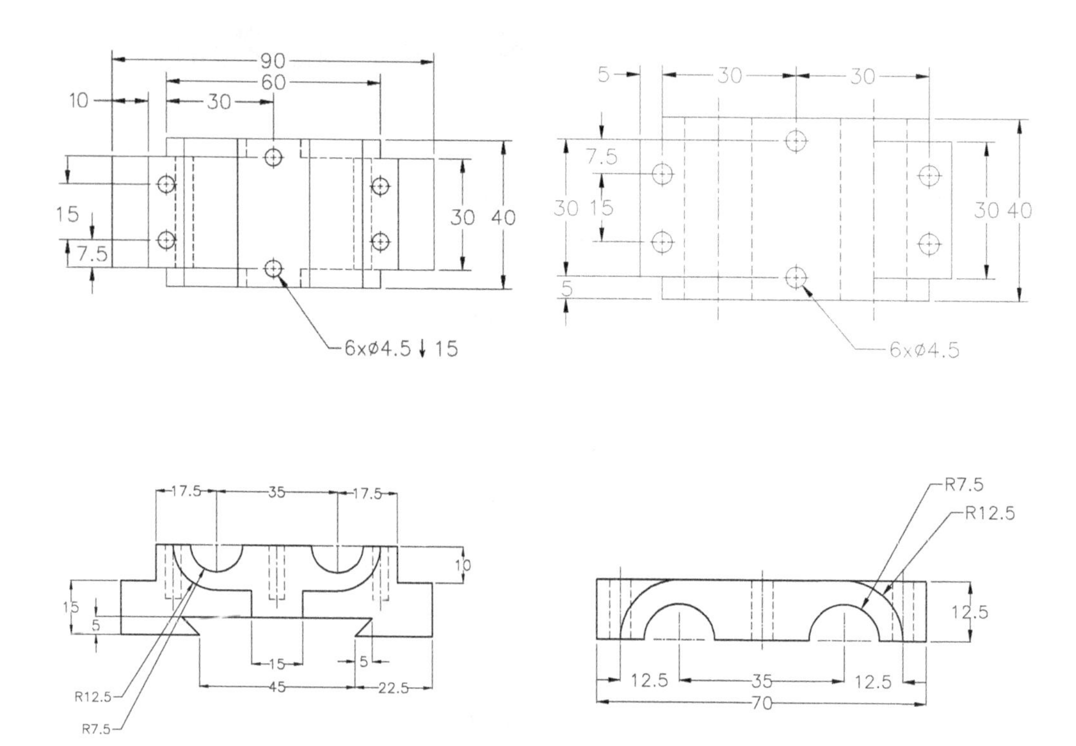

Figure 3 *Top and Front views of the Base*

Figure 4 *Top and Front views of the Cap*

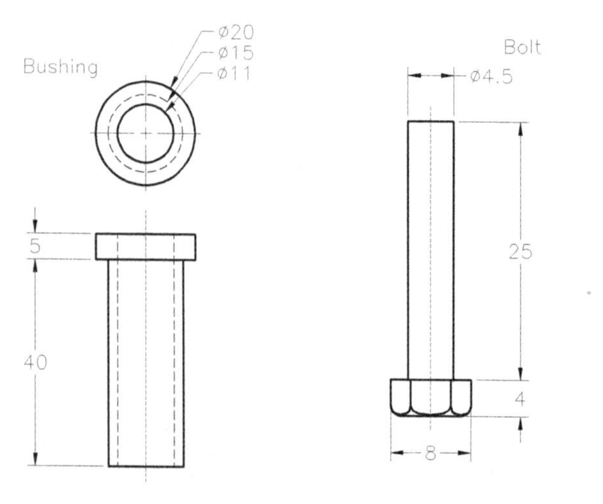

Figure 5 *Top and front views of Bushing, and the front view of Bolt*

Student Project 2

Create all components of the Wheel Support assembly and then assemble them, as shown in Figure 6. The exploded view of the assembly is shown in Figure 7. The dimension of the components are shown in Figures 8 through 11.

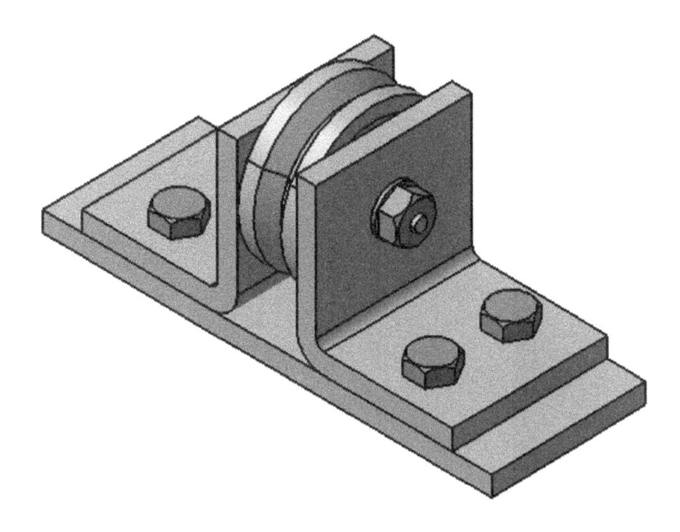

Figure 6 *Wheel Support assembly*

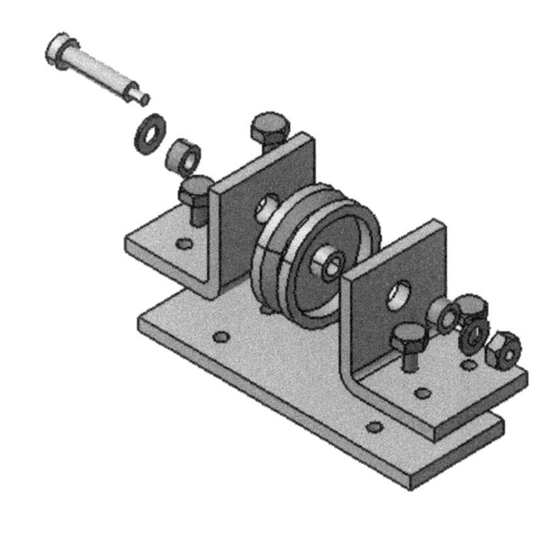

Figure 7 *Exploded view of the Wheel Support assembly*

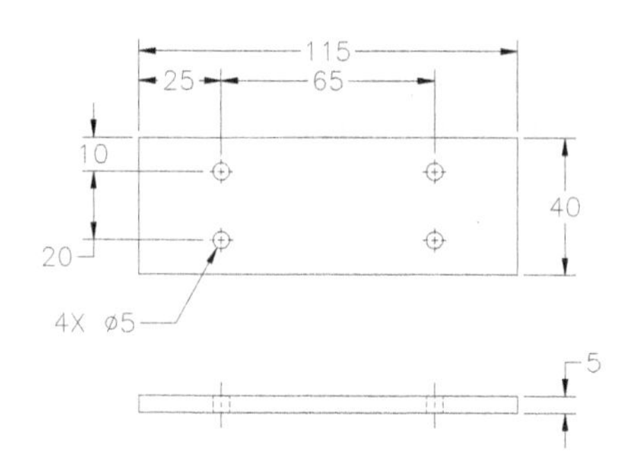

Figure 8 Front and top views of the Base

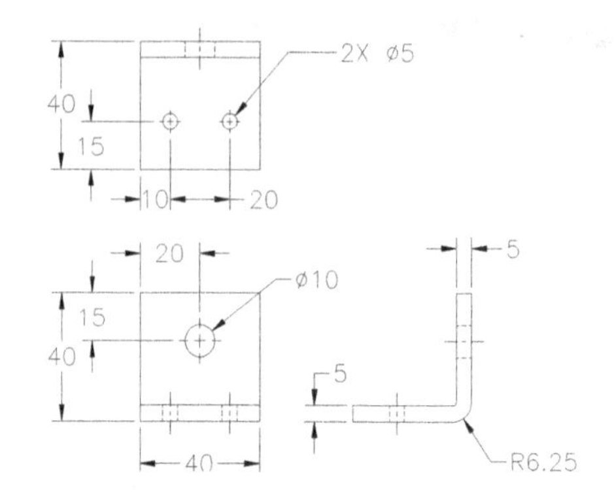

Figure 9 Top, front and right views of the Support

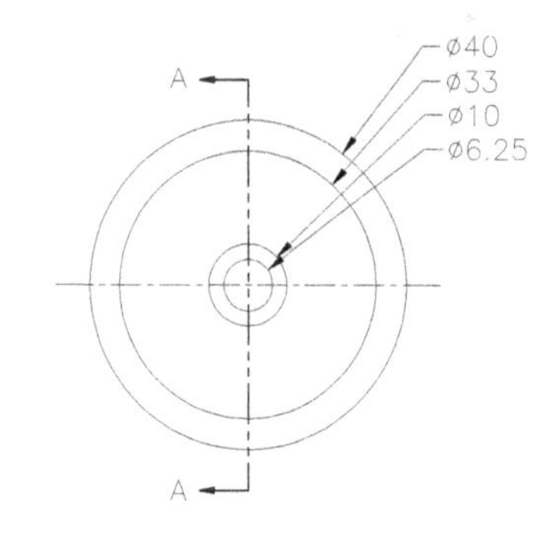

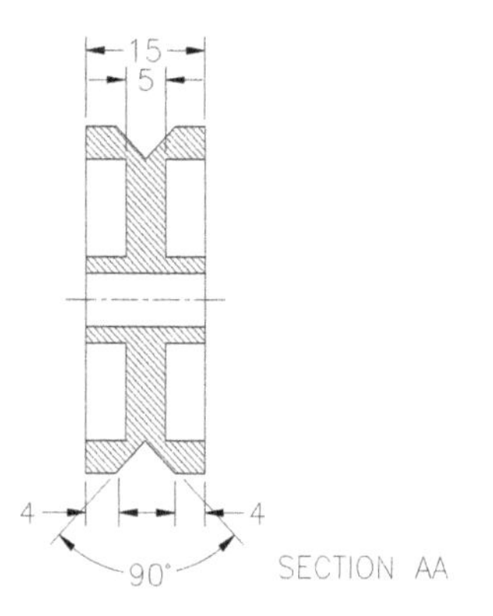

Figure 10 Front view and sectioned side view of the wheel

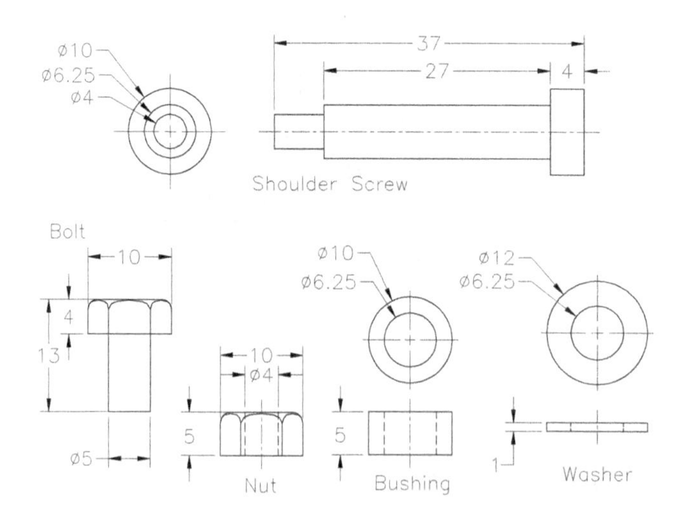

Figure 11 Dimensions of the Shoulder Screw, Bolt, Nut, Bushing, and Washer

Index

D

E

F

G

H

I

J

K

L

M

N

O

P

R

U

V

W

X

Y

Z

Other Publications by CADCIM Technologies

The following is the list of some of the publications by CADCIM Technologies. Please visit www.cadcim.com for the complete listing.

Autodesk Inventor Textbooks

- Autodesk Inventor 2015 for Designers, 15th Edition
- Autodesk Inventor 2014 for Designers

Solid Edge Textbooks

- Solid Edge ST7 for Designers, 12th Edition
- Solid Edge ST6 for Designers, 11th Edition

NX Textbooks

- NX 9.0 for Designers, 13th Edition
- NX 8.5 for Designers

SolidWorks Textbooks

- SolidWorks 2015 for Designers, 13th Edition
- SolidWorks 2014 for Designers, 12th Edition
- SolidWorks 2014: A Tutorial Approach

EdgeCAM Textbooks

- EdgeCAM 11.0 for Manufacturers

CATIA Textbooks

- CATIA V5-6R2014 for Designers, 12th Edition
- CATIA V5-6R2013 for Designers, 11th Edition
- CATIA V5-6R2012 for Designers
- CATIA V5R21 for Designers

Creo Parametric and Pro/ENGINEER Textbooks

- Creo Parametric 2.0 for Designers
- Creo Parametric 1.0 for Designers
- Pro/ENGINEER Wildfire 5.0 for Designers

Autodesk Alias Textbooks

- Learning Autodesk Alias Design 2015, 4th Edition
- Learning Autodesk Alias Design 2012

ANSYS Textbooks

- ANSYS Workbench 14.0: A Tutorial Approach

AutoCAD Textbooks
- AutoCAD 2015: A Problem Solving Approach, Basic and Intermediate, 21st Edition
- AutoCAD 2015: A Problem Solving Approach, 3D and Advanced, 21st Edition

AutoCAD MEP Textbooks
- AutoCAD MEP 2015 for Designers
- AutoCAD MEP 2014 for Designers

Customizing AutoCAD Textbook
- Customizing AutoCAD 2013

AutoCAD LT Textbooks
- AutoCAD LT 2014 for Designers
- AutoCAD LT 2013 for Designers

AutoCAD Electrical Textbooks
- AutoCAD Electrical 2015 for Electrical Control Designers, 6th Edition
- AutoCAD Electrical 2014 for Electrical Control Designers
- AutoCAD Electrical 2013 for Electrical Control Designers

Autodesk Revit Architecture Textbooks
- Autodesk Revit Architecture 2015 for Architects and Designers, 11th Edition
- Autodesk Revit Architecture 2014 for Architects and Designers

AutoCAD Civil 3D Textbooks
- Exploring AutoCAD Civil 3D 2015, 5th Edition
- Exploring AutoCAD Civil 3D 2014

3ds Max Design Textbooks
- Autodesk 3ds Max Design 2015: A Tutorial Approach, 15th Edition
- Autodesk 3ds Max Design 2014: A Tutorial Approach

Coming Soon from CADCIM Technologies
- NX Nastran 9.0 for Designers
- SOLIDWORKS 2015: A Tutorial Approach
- Exploring Primavera P6 V8.1
- Exploring Bentley STAAD Pro V8i